Astronomy

A Beginner's Guide to the Universe

SEVENTH EDITION

Eric Chaisson

Harvard University

Steve McMillan

Drexel University

PEARSON

Boston   Columbus   Indianapolis   New York   San Francisco   Upper Saddle River
Amsterdam   Cape Town   Dubai   London   Madrid   Milan   Munich   Paris   Montreal   Toronto   Delhi
Mexico City   Sao Paulo   Sydney   Hong Kong   Seoul   Singapore   Taipei   Tokyo

Publisher: Jim Smith
Executive Editor: Nancy Whilton
Director of Development: Laura Kenney
Project Editor: Tema Goodwin
Art Development Editor: Barbara Price
Media Producer: Kelly Reed
Director of Marketing: Christy Lesko
Marketing Manager: Will Moore
Managing Editor: Corinne Benson
Production Project Manager: Mary O'Connell
Production Management: Thistle Hill Publishing
    Services, LLC

Compositor: Cenveo Publisher Services/Nesbitt Graphics, Inc.
Design Manager: Mark Ong
Interior Designer: Jerilyn Bockorick, Cenveo Publisher Services/
    Nesbitt Graphics, Inc.
Cover Designer: Mark Ong
Illustrator: Mark Landis
Photo Researcher: Stefanie Ramsay
Image Lead: Maya Melenchuk
Permissions Specialist: Joanna Green
Manufacturing Buyer: Jeff Sargent
Cover Printer: Lehigh-Phoenix Color/Hagerstown
Printer and Binder: Courier/Kendallville

Cover Photo Credit: Chandral/Hubble/Spitzer X-ray/Visible/Infrared Image of M82; NASA, ESA, CXC, and JPL-Caltech.

Library of Congress Cataloging-in-Publication Data
Chaisson, Eric.
  Astronomy : a beginner's guide to the universe / Eric Chaisson. Harvard University, Steve McMillan, Drexel University. — Seventh edition.
      pages cm
  Includes bibliographical references and index.
  ISBN-13: 978-0-321-81535-4 (pbk. : student edition)
  ISBN-10: 0-321-81535-1 (pbk. : student edition)
  ISBN-13: 978-0-321-84061-5 (instructor's review copy)
  ISBN-10: 0-321-84061-5 (instructor's review copy)
  1. Astronomy—Textbooks. I. McMillan, S. (Stephen), 1955– II. Title.
  QB43.3.C43 2013
  520—dc23
                    2012026382

ISBN 10: 0-321-81535-1; ISBN 13: 978-0-321-81535-4 (Student Edition)
ISBN 10: 0-321-84061-5; ISBN 13: 978-0-321-84061-5 (Instructor Review Copy)

3 4 5 6 7 8 9 10 —V011— 17 16 15 14 13

Manufactured in the United States of America.

www.pearsonhighered.com

Brief Contents

PART 1   Foundations   xix

0   Charting the Heavens: The Foundations of Astronomy   2

1   The Copernican Revolution: The Birth of Modern Science   24

2   Light and Matter: The Inner Workings of the Cosmos   42

3   Telescopes: The Tools of Astronomy   68

PART 2   Our Planetary System   98

4   The Solar System: Interplanetary Matter and the Birth of the Planets   100

5   Earth and Its Moon: Our Cosmic Backyard   134

6   The Terrestrial Planets: A Study in Contrasts   162

7   The Jovian Planets: Giants of the Solar System   192

8   Moons, Rings, and Plutoids: Small Worlds Among Giants   216

PART 3   The Stars   242

9   The Sun: Our Parent Star   244

10   Measuring the Stars: Giants, Dwarfs, and the Main Sequence   268

11   The Interstellar Medium: Star Formation in the Milky Way   292

12   Stellar Evolution: The Lives and Deaths of Stars   320

13   Neutron Stars and Black Holes: Strange States of Matter   348

PART 4   Galaxies and the Universe   376

14   The Milky Way Galaxy: A Spiral in Space   378

15   Normal and Active Galaxies: Building Blocks of the Universe   404

16   Galaxies and Dark Matter: The Large-Scale Structure of the Cosmos   432

17   Cosmology: The Big Bang and the Fate of the Universe   458

18   Life in the Universe: Are We Alone?   484

Contents

Preface xi

PART 1 Foundations xix

0 Charting the Heavens
The Foundations of Astronomy 2

0.1 The "Obvious" View 4
0.2 Earth's Orbital Motion 7
   ■ MORE PRECISELY 0-1 Angular Measure 10
0.3 The Motion of the Moon 12
0.4 The Measurement of Distance 17
0.5 Science and the Scientific Method 18
   ■ DISCOVERY 0-1 Sizing up Planet Earth 19

● ANIMATION/VIDEO Summer Solstice 9
● ANIMATION/VIDEO Winter Solstice 9
● ANIMATION/VIDEO The Equinoxes 11
● TUTORIAL Phases of the Moon 12
● ANIMATION/VIDEO Solar Eclipse in Indiana 15
● TUTORIAL Stellar Parallax 18

Chapter Review 21

1 The Copernican Revolution
The Birth of Modern Science 24

1.1 The Motions of the Planets 26
   ■ DISCOVERY 1-1 Ancient Astronomy 28
1.2 The Birth of Modern Astronomy 30
1.3 The Laws of Planetary Motion 32
1.4 Newton's Laws 36
● INTERACTIVE Central of Mass of a Binary Star 38
● ANIMATION/VIDEO Gravity Demonstration on the Moon 39

Chapter Review 40

2 Light and Matter
The Inner Workings of the Cosmos 42

2.1 Information from the Skies 44
2.2 Waves in What? 46
2.3 The Electromagnetic Spectrum 48

2.4 Thermal Radiation 51
   ■ MORE PRECISELY 2-1 The Kelvin Temperature Scale 51
   ■ MORE PRECISELY 2-2 More on the Radiation Laws 54
2.5 Spectroscopy 55
2.6 The Formation of Spectral Lines 58
2.7 The Doppler Effect 63
2.8 Spectral-Line Analysis 64

● TUTORIAL Continuous Spectra and Blackbody Radiation 52
● ANIMATION/VIDEO Solar Eclipse Viewed From Space, in X-rays 54
● ANIMATION/VIDEO Shoemaker-Levy 9 Impact at 2.15 Microns 54
● ANIMATION/VIDEO Multispectral View of Orion Nebula 54
● ANIMATION/VIDEO Earth Aurora in X-rays 54
● TUTORIAL Emission Spectra 55
● TUTORIAL Absorption Spectra 57
● ANIMATION/VIDEO Classical Hydrogen Atom I 59
● ANIMATION/VIDEO Classical Hydrogen Atom II 59
● ANIMATION/VIDEO Photon Emission 60
● TUTORIAL Doppler Effect 64

Chapter Review 65

3 Telescopes
The Tools of Astronomy 68

3.1 Optical Telescopes 70
   ■ DISCOVERY 3-1 The Hubble Space Telescope 74

3.2   Telescope Size 77

   ■ MORE PRECISELY 3-1 Diffraction and Telescope
   Resolution 80

3.3   High-Resolution Astronomy 80

3.4   Radio Astronomy 83

3.5   Space-Based Astronomy 89

● TUTORIAL Chromatic Aberration 71

● TUTORIAL Reflecting Telescopes 72

● TUTORIAL The Optics of a Simple Lens 72

● ANIMATION/VIDEO Hubble Space Telescope in Orbit 74

● ANIMATION/VIDEO Deployment of the James Webb Space Telescope 74

● ANIMATION/VIDEO Gemini Control Room 76

● ANIMATION/VIDEO Speckle Imaging 76

● ANIMATION/VIDEO Adaptive Optics 82

● ANIMATION/VIDEO Chandra Light and Data Paths 92

Chapter Review 95

PART 2 Our Planetary System 98

4 The Solar System
Interplanetary Matter and the Birth of the Planets 100

4.1   An Inventory of the Solar System 102

   ■ MORE PRECISELY 4-1 Measuring Sizes with Geometry 103

4.2   Interplanetary Matter 105

   ■ DISCOVERY 4.1 What Killed the Dinosaurs? 108

4.3   The Formation of the Solar System 117

   ■ MORE PRECISELY 4-2 The Concept of Angular Momentum 122

4.4   Planets Beyond the Solar System 125

● ANIMATION/VIDEO An Astronomical Ruler 102

● ANIMATION/VIDEO Size and Scale of the Terrestrial Planets I 104

● ANIMATION/VIDEO The Gas Giants 104

● ANIMATION/VIDEO Orbiting Eros 109

● ANIMATION/VIDEO NEAR Descent 109

● ANIMATION/VIDEO Comet Hale-Bopp Nucleus Animation 111

● ANIMATION/VIDEO Deep Impact Simulation 112

● ANIMATION/VIDEO Anatomy of a Comet Part I 112

● ANIMATION/VIDEO Anatomy of a Comet Part II 113

● ANIMATION/VIDEO Beta Pictoris Warp 119

● ANIMATION/VIDEO The Formation of the Solar System 120

● ANIMATION/VIDEO Protoplanetary Disk Destruction 122

● ANIMATION/VIDEO Protoplanetary Disks in the Orion Nebula 122

● ANIMATION/VIDEO Survey for Transiting Extrasolar Planets 126

● ANIMATION/VIDEO Hot Jupiter Extrasolar Planet Evaporating 130

Chapter Review 131

5 Earth and Its Moon
Our Cosmic Backyard 134

5.1   Earth and the Moon in Bulk 136

5.2   The Tides 137

   ■ MORE PRECISELY 5-1 Why Air Sticks Around 140

5.3   Atmospheres 140

   ■ DISCOVERY 5-1 Earth's Growing Ozone "Hole" 142

   ■ DISCOVERY 5-2 The Greenhouse Effect and Global
   Warming 144

5.4   Internal Structure of Earth and the
   Moon 145

5.5   Surface Activity on Earth 148

5.6   The Surface of the Moon 151

5.7   Magnetospheres 154

5.8   History of the Earth–Moon System 156

● ANIMATION/VIDEO Earth as Seen by Galileo 136

● TUTORIAL The Greenhouse Effect 142

● ANIMATION/VIDEO Seasonal Changes on the Earth 142

● TUTORIAL Atmospheric Lifetimes 143

● ANIMATION/VIDEO Ozone Hole Over the Antarctic 143

● ANIMATION/VIDEO One Small Step 151

● ANIMATION/VIDEO Full Rotation of the Moon 151

● ANIMATION/VIDEO Ranger Spacecraft Descent to Moon 151

● ANIMATION/VIDEO Northern and Southern Lights 156

Chapter Review 159

6 The Terrestrial Planets
A Study in Contrasts 162

6.1 Orbital and Physical Properties 164

6.2 Rotation Rates 164

6.3 Atmospheres 167

6.4 The Surface of Mercury 169

6.5 The Surface of Venus 170

6.6 The Surface of Mars 173

  ▪ DISCOVERY 6-1 Martian Canals? 174

  ▪ DISCOVERY 6-2 Life on Mars? 184

6.7 Internal Structure and Geological History 185

6.8 Atmospheric Evolution on Earth, Venus, and Mars 187

● ANIMATION/VIDEO Transit of Mercury 166
● ANIMATION/VIDEO Transit of Venus 167
● ANIMATION/VIDEO Topography of Venus 171
● ANIMATION/VIDEO Flight Over Alpha Regio 171
● ANIMATION/VIDEO Flight Over Sif Mons Volcano 172
● TUTORIAL SuperSpaceship–Voyage to Venus 173
● ANIMATION/VIDEO *Hubble* View of Mars 173
● ANIMATION/VIDEO Flight Over Columbia Hills 176
● ANIMATION/VIDEO Flight Over Tharsis 176
● ANIMATION/VIDEO Flight Over Mariner Valley 176
● TUTORIAL Comparative Planetology: Mars 177
● ANIMATION/VIDEO Meteorites Ejected from Mars 183
● ANIMATION/VIDEO Microscopic Martian Fossils? 183
● ANIMATION/VIDEO Mars Exploration Rover Landing 184
● ANIMATION/VIDEO Flight Over Opportunity at Gustav Crater 184
● ANIMATION/VIDEO Terrestrial Planets II 187

Chapter Review 189

7 The Jovian Planets
Giants of the Solar System 192

7.1 Observations of Jupiter and Saturn 194

7.2 The Discoveries of Uranus and Neptune 195

  ▪ DISCOVERY 7-1 Spaceflight in the Solar System 198

7.3 Bulk Properties of the Jovian Planets 197

7.4 Jupiter's Atmosphere 201

7.5 The Atmospheres of the Outer Jovian Worlds 204

7.6 Jovian Interiors 208

  ▪ DISCOVERY 7-2 A Cometary Impact 212

● ANIMATION/VIDEO *Galileo* Mission to Jupiter 195
● TUTORIAL Jupiter—Differential Rotation 199

● ANIMATION/VIDEO Jupiter's Rotation 203
● ANIMATION/VIDEO Saturn Cloud Rotation 205
● ANIMATION/VIDEO The Gas Giants Part II 209

Chapter Review 213

8 Moons, Rings, and Plutoids
Small Worlds Among Giants 216

8.1 The Galilean Moons of Jupiter 218

8.2 The Large Moons of Saturn and Neptune 223

8.3 The Medium-Sized Jovian Moons 227

8.4 Planetary Rings 230

8.5 Beyond Neptune 235

● ANIMATION/VIDEO Galilean Moons Transit Jupiter 218
● ANIMATION/VIDEO Jupiter Icy Moons Orbiter Mission 220
● ANIMATION/VIDEO Io Cutaway 220
● ANIMATION/VIDEO *Galileo's* View of Europa 220
● ANIMATION/VIDEO *Galileo's* View of Ganymede 222
● ANIMATION/VIDEO *Huygens* Landing on Titan 225
● ANIMATION/VIDEO Saturn Satellite Transit 227
● ANIMATION/VIDEO Saturn Ring Plane Crossing 231
● ANIMATION/VIDEO *Voyager* Ring Spokes Animation 233

Chapter Review 239

PART 3 The Stars 242

9 The Sun
Our Parent Star 244

9.1 The Sun in Bulk 246
9.2 The Solar Interior 247
■ DISCOVERY 9-1 Eavesdropping on the Sun 250
9.3 The Solar Atmosphere 251
9.4 The Active Sun 254
9.5 The Heart of the Sun 261
■ MORE PRECISELY 9-1 The Strong and Weak Nuclear Forces 262
■ MORE PRECISELY 9-2 Energy Generation in the Proton–Proton Chain 265

● TUTORIAL SuperSpaceship—Voyage to the Sun 246
● ANIMATION/VIDEO Solar Granulation 252
● ANIMATION/VIDEO Solar Chromosphere 252
● ANIMATION/VIDEO Sunspot 254
● ANIMATION/VIDEO Solar Flare 258
● ANIMATION/VIDEO Coronal Mass Ejections 260

Chapter Review 265

10 Measuring the Stars
Giants, Dwarfs, and the Main Sequence 268

10.1 The Solar Neighborhood 270
10.2 Luminosity and Apparent Brightness 272
■ DISCOVERY 10-1 Naming the Stars 273
■ MORE PRECISELY 10-1 More on the Magnitude Scale 275
10.3 Stellar Temperatures 275
10.4 Stellar Sizes 278
10.5 The Hertzsprung–Russell Diagram 279
■ MORE PRECISELY 10-2 Estimating Stellar Radii 282
10.6 Extending the Cosmic Distance Scale 282
10.7 Stellar Masses 285
■ MORE PRECISELY 10-3 Measuring Stellar Masses in Binary Stars 287

● TUTORIAL Hertzsprung–Russell Diagram 279
● ANIMATION/VIDEO White Dwarfs in Globular Cluster 282
● TUTORIAL Binary Stars—Radial Velocity Curves 285
● TUTORIAL Eclipsing Binary Stars—Light Curves 285
● INTERACTIVE Exploring the Light Curve of an Eclipsing Binary Star System 286

Chapter Review 289

11 The Interstellar Medium
Star Formation in the Milky Way 292

11.1 Interstellar Matter 294
■ DISCOVERY 11-1 Ultraviolet Astronomy and the "Local Forming" 296
11.2 Star-Forming Regions 297
11.3 Dark Dust Clouds 301
11.4 The Formation of Stars Like the Sun 305
11.5 Stars of Other Masses 312
11.6 Star Clusters 313

● INTERACTIVE Multiwavelength Cloud with Embedded Stars 299
● ANIMATION/VIDEO Gaseous Pillars of Star Birth 300
● ANIMATION/VIDEO Pillars Behind the Dust 303
● ANIMATION/VIDEO The Tarantula Nebula 303
● ANIMATION/VIDEO Horsehead Nebula 303
● ANIMATION/VIDEO Orion Nebula Mosaic 308
● ANIMATION/VIDEO Visit to Orion Nebula 308
● ANIMATION/VIDEO Bi-Polar Outflow 311
● ANIMATION/VIDEO Carina Nebula 314

Chapter Review 317

12 Stellar Evolution
The Lives and Deaths of Stars 320

12.1 Leaving the Main Sequence 322
12.2 Evolution of a Sun-like Star 323
12.3 The Death of a Low-Mass Star 326

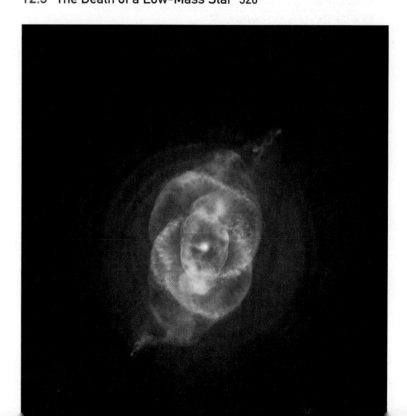

12.4 Evolution of Stars More Massive than the Sun 332

12.5 Supernova Explosions 335

12.6 Observing Stellar Evolution in Star Clusters 339

12.7 The Cycle of Stellar Evolution 342

■ DISCOVERY 12-1 Supernova 1987A 344

● ANIMATION/VIDEO Death of the Sun Part I 326

● ANIMATION/VIDEO Death of the Sun Part II 326

● ANIMATION/VIDEO Helix Nebula 327

● ANIMATION/VIDEO Helix Nebula Animation 327

● ANIMATION/VIDEO Helix Nebula White Dwarf 327

● ANIMATION/VIDEO White Dwarf Cooling Sequence 327

● ANIMATION/VIDEO Bi-Polar Planetary Nebula 327

● ANIMATION/VIDEO Recurrent Nova 331

● ANIMATION/VIDEO Light Echo 333

● INTERACTIVE Life and Death of a High-Mass Star 333

● ANIMATION/VIDEO Supernova Explosions 337

● INTERACTIVE Supernova Remnant Cassiopeia A 337

Chapter Review 345

13 Neutron Stars and Black Holes

Strange States of Matter 348

13.1 Neutron Stars 350

13.2 Pulsars 350

13.3 Neutron Star Binaries 354

13.4 Gamma-Ray Bursts 356

■ DISCOVERY 13-1 Gravity Waves—A New Window on the Universe 359

13.5 Black Holes 360

13.6 Einstein's Theories of Relativity 362

■ MORE PRECISELY 13-1 Special Relativity 364

■ MORE PRECISELY 13-2 Tests of General Relativity 367

13.7 Space Travel Near Black Holes 368

13.8 Observational Evidence for Black Holes 371

● ANIMATION/VIDEO Pulsar in Crab Nebula 352

● ANIMATION/VIDEO Colliding Binary Neutron Stars 358

● TUTORIAL Escape Speed and Black Hole Event Horizons 362

● INTERACTIVE Exploring the Effects of Relative Motion 363

● ANIMATION/VIDEO Energy Released from a Black Hole? 368

● INTERACTIVE Gravitational Time Dilation and Redshift 369

● ANIMATION/VIDEO X-Ray Binary Star 370

● ANIMATION/VIDEO Black Hole and Companion Star 371

● ANIMATION/VIDEO Black Hole in Galaxy M87 372

● ANIMATION/VIDEO Supermassive Black Hole 372

Chapter Review 373

PART 4 Galaxies and the Universe 376

14 The Milky Way Galaxy

A Spiral in Space 378

14.1 Our Parent Galaxy 380

14.2 Measuring the Milky Way 381

14.3 Galactic Structure 384

■ DISCOVERY 14-1 Early Computers 385

14.4 Formation of the Milky Way 388

14.5 Galactic Spiral Arms 390

■ DISCOVERY 14-2 Density Waves 391

14.6 The Mass of the Milky Way Galaxy 393

14.7 The Galactic Center 396

● ANIMATION/VIDEO Cepheid Variable Star in Distant Galaxy 383

● INTERACTIVE Stellar Populations in a Star Cluster 386

● INTERACTIVE Spiral Arms and Star Formations 391

● TUTORIAL Gravitational Lensing 395

● INTERACTIVE Spiral Galaxy M81 397

● ANIMATION/VIDEO Black Hole in the Center of the Milky Way? 400

Chapter Review 401

15 Normal and Active Galaxies

Building Blocks of the Universe 404

15.1 Hubble's Galaxy Classification 406

15.2 The Distribution of Galaxies in Space 412

15.3 Hubble's Law 416

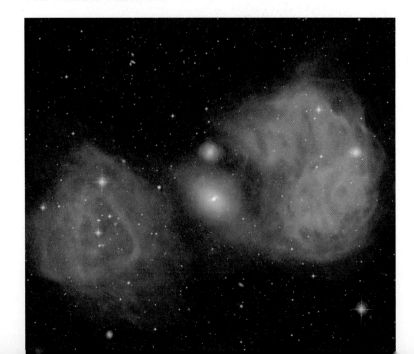

15.4 Active Galactic Nuclei 418
   ◼ MORE PRECISELY 15-1 Relativistic Redshifts and
      Look-Back Time 422
15.5 The Central Engine of an Active Galaxy 425

● ANIMATION/VIDEO The Evolution of Galaxies 411
● ANIMATION/VIDEO Cosmic Jets 426

Chapter Review 429

16 Galaxies and Dark Matter
The Large-Scale Structure of the Cosmos 432

16.1 Dark Matter in the Universe 434
16.2 Galaxy Collisions 437
16.3 Galaxy Formation and Evolution 439
16.4 Black Holes in Galaxies 443
16.5 The Universe on Very Large Scales 448
   ◼ DISCOVERY 16-1 The Sloan Digital Sky Survey 451

● INTERACTIVE Rotation Curve for a Merry-Go-Round 434
● ANIMATION/VIDEO Dark Matter 434
● INTERACTIVE Colliding Galaxies 438
● ANIMATION/VIDEO Galaxy Collision I 439
● INTERACTIVE Starburst Galaxy M82 441
● ANIMATION/VIDEO Hubble Deep Field Zoom I 441
● ANIMATION/VIDEO Hubble Deep Field Zoom II 441
● ANIMATION/VIDEO Galaxy Collision II 443
● ANIMATION/VIDEO Cluster Merger 448
● INTERACTIVE Quasar Spectrum due to Absorption from Distant Clouds 450
● ANIMATION/VIDEO Simulation of Gravitational Lens in Space 453
● ANIMATION/VIDEO The Cosmic Web of Dark Matter 454

Chapter Review 455

17 Cosmology
The Big Bang and the Fate of the Universe 458

17.1 The Universe on the Largest Scales 460
17.2 The Expanding Universe 461
17.3 Cosmic Dynamics and the Geometry of Space 464
17.4 The Fate of the Cosmos 466
   ◼ DISCOVERY 17-1 Einstein and the Cosmological Constant 468
17.5 The Early Universe 471
17.6 The Formation of Nuclei and Atoms 473
17.7 Cosmic Inflation 475
17.8 The Formation of Large-Scale Structure
      in the Universe 478

● INTERACTIVE The Expanding Raisin Cake (Universe) 462
● ANIMATION/VIDEO Gravitational Curvature of Space 469
● INTERACTIVE Creation of the Cosmic Microwave Background 474

● ANIMATION/VIDEO Cosmic Structure 478
● ANIMATION/VIDEO Ripples in the Early Universe 480
● ANIMATION/VIDEO COBE and WMAP View of Cosmic Microwave Background 480

Chapter Review 481

18 Life in the Universe
Are We Alone? 484

18.1 Cosmic Evolution 486
18.2 Life in the Solar System 490
18.3 Intelligent Life in the Galaxy 493
18.4 The Search for Extraterrestrial Intelligence 498

● ANIMATION/VIDEO Icy Organics in Planet-Forming Disc 489

Chapter Review 501

Appendices A-1

APPENDIX 1: Scientific Notation A-1
APPENDIX 2: Astronomical Measurement A-2
APPENDIX 3: Tables A-3

Glossary G-1
Answers to Concept Check Questions AK-1
Answers to Self-Test Questions AK-4
Credits C-1
Index I-1
Star Charts S-1

Preface

We are pleased to have the opportunity to present in this book a representative sample of the known facts, evolving ideas, and frontier discoveries in astronomy today.

Astronomy: A Beginner's Guide to the Universe has been written and designed for students who have taken no previous college science courses and who will likely not major in physics or astronomy. We present a broad view of astronomy, straightforwardly descriptive and without complex mathematics. The absence of sophisticated mathematics, however, in no way prevents discussion of important concepts. Rather, we rely on qualitative reasoning as well as analogies with objects and phenomena familiar to the student to explain the complexities of the subject without oversimplification. We have tried to communicate the excitement that we feel about astronomy and to awaken students to the marvelous universe around us.

We are very gratified that the first six editions of this text have been so well received by many in the astronomy education community. In using those earlier texts, many of you—teachers and students alike—have given us helpful feedback and constructive criticisms. From these, we have learned to communicate better both the fundamentals and the excitement of astronomy. Many improvements inspired by your comments have been incorporated into this edition.

Organization and Approach

As in previous editions, our organization follows the popular and effective "Earth out" progression. We have found that most students, especially those with little scientific background, are much more comfortable studying the relatively familiar solar system before tackling stars and galaxies. With Earth and Moon as our initial planetary models, we move through the solar system. Integral to our coverage of the solar system is a discussion of its formation. This line of investigation leads directly into a study of the Sun.

With the Sun as our model star, we broaden the scope of our discussion to include stars in general—their properties, their evolutionary histories, and their varied fates. This journey naturally leads us to coverage of the Milky Way Galaxy, which in turn serves as an introduction to our treatment of other galaxies. Finally, we reach cosmology and the large-scale structure and dynamics of the universe as a whole. Throughout, we strive to emphasize the dynamic nature of the cosmos—virtually every major topic, from planets to quasars, includes a discussion of how those objects formed and how they evolve.

We place much of the needed physics in the early chapters— an approach derived from years of experience teaching thousands of students. Additional physical principles are developed as needed later, both in the text narrative and in the *Discovery* and *More Precisely* boxes (described on p. xiv). We have made the treatment of physics, as well as the more quantitative discussions, as modular as possible, so that these topics can be deferred if desired. Instructors presenting this material in a one-quarter course, who wish to (or have time to) cover only the essentials of the solar system before proceeding on to the study of stars and the rest of the universe, may want to teach only Chapter 4 and then move directly to Chapter 9 (the Sun).

What's New in This Edition

Astronomy is a rapidly evolving field, and almost every chapter in the seventh edition has been updated with new and late-breaking information. Several chapters have also seen significant internal reorganization in order to streamline the overall presentation, strengthen our focus on the process of science, and reflect new understanding and emphases in contemporary astronomy. Among the many improvements are the following:

- New chapter-opening images reflecting the latest astronomical discoveries.

- Streamlined art program providing more direct and accurate representations of astronomical objects.

- Addition of multiple annotations to about a third of the figures, while simultaneously condensing most figure captions.

- Pedagogical improvements in the placement and usefulness of the visual analogies.

- The Learning Outcomes feature that opens each chapter now connects to an appropriate Review and Discussion question at the end of the chapter.

- Streamlined mathematical level throughout, both within and outside the *More Precisely* boxes.

- A new *Discovery* feature on ancient astronomy in Chapter 1.

- New coverage of the *Fermi* gamma-ray telescope and findings in Chapter 3.

- Updated presentation of asteroid and comet spacecraft missions in Chapter 4.

- Rewritten and updated discussion of extrasolar planets in Chapter 4.

- Updates on the *Lunar Reconnaissance Orbiter (LRO)*, the *Lunar CRater Observation and Sensing Satellite (LCROSS)*, and the *Gravity Recovery and Interior Laboratory (GRAIL)* lunar missions in Chapter 5.

- New material on the Moon's solid inner core in Chapter 5.

- Rewritten discussion of atmospheric evolution and the carbon cycle on terrestrial planets in Chapter 6.

- A new *Discovery* feature on space exploration in Chapter 7.

- Updates on comet collisions with Jupiter and new imagery of storms on Saturn in Chapter 7.

- Updates on the *Cassini Solstice* and *Juno* missions in Chapters 7 and 8.

- New material from *Cassini* on Titan, Enceladus, Iapetus, and Saturn's E ring in Chapter 8, as well as new imagery of Pluto and its moons.

- New coverage in Chapter 9 of the *Solar Dynamics Observatory (SDO)* mission, including several new spectacular images of the Sun.

- Discussion in Chapter 9 of internal circulation currents in the Sun and their effect on the sunspot cycle.

- Updated Milky Way and nebular imagery in Chapter 11.

- Improved cluster color-magnitude diagrams and some imagery in Chapter 12.

- Updated discussion of multiple stellar populations in globular clusters in Chapter 12.

- Updated coverage of SN 1987a in Chapter 12.

- Updated neutron star masses and pulsar numbers in Chapter 13.

- Revised discussion of gamma-ray bursts in Chapter 13.

- Additional coverage in Chapter 14 of stellar tidal streams in the Milky Way halo.

- Updated discussion in Chapter 14 of the Milky Way's Galactic nucleus.

- Revised discussion in Chapter 15 of the currently accepted value of the Hubble constant.

- Updated numbers, redshifts and some of the imagery of the most distant known galaxies and quasars in Chapters 15 and 16.

- Extensive revision of the presentation of cosmology in Chapter 17 for greater clarity.

- Revised discussion of planetary systems and habitable planets in Chapter 18.

The Illustration Program

Visualization plays an important role in both the teaching and the practice of astronomy, and we continue to place strong emphasis on this aspect of our book. We have tried to combine aesthetic beauty with scientific accuracy in the artist's conceptions that enrich the text, and we have sought to present the best and latest imagery of a wide range of cosmic objects. Each illustration has been carefully crafted to enhance student learning; each is pedagogically sound and tightly tied to nearby discussion of important scientific facts and ideas.

Full-Spectrum Coverage and Spectrum Icons

Increasingly, astronomers are exploiting the full range of the electromagnetic spectrum to gather information about the cosmos. Throughout this book, images taken at radio, infrared, ultraviolet, X-ray, or gamma-ray wavelengths are used to supplement visible-light images. As it is sometimes difficult (even for a professional) to tell at a glance which images are visible-light photographs and which are false-color images created with other wavelengths, each photo in the text is accompanied by an icon that identifies the wavelength of electromagnetic radiation used to capture the image.

Other Pedagogical Features

As with many other parts of our text, instructors have helped guide us toward what is most helpful for effective student learning.

Learning Outcomes Studies indicate that beginning students often have trouble prioritizing textual material. For this reason, a few (typically five or six) well-defined learning outcomes are provided at the start of each chapter. These help students determine what mastery they should be able to demonstrate after reading the chapter and then structure their reading accordingly. The outcomes are numbered and keyed to the items in the Chapter Summary and the Review and Discussion section, which in turn refer back to passages in the text. This highlights the most important aspects of the chapter, helping students prioritize information and aiding their review of the material. The outcomes are organized and phrased in such a way as to make them objectively testable, affording students a means of gauging their own progress.

Concept Links The connection between the astronomical material and the physical principles set forth early in the text is crucial. It is important that students, when they encounter, say, Hubble's law in Chapter 15, recall what they learned about spectral lines and the Doppler shift in Chapter 2. Similarly, the discussions of the masses of binary star components (Chapter 10) and of galactic rotation (Chapter 14) both depend on the discussion of Kepler's and Newton's laws in Chapter 1. Throughout, discussions of new astronomical objects and concepts rely heavily on comparison with topics introduced earlier in the text.

It is important to remind students of these links so that they can recall the principles on which later discussions rest and, if necessary, review them. To this end, we have inserted "Concept Links" throughout the text—symbols that mark key intellectual bridges between material in different chapters. The links, denoted by the symbol ∞, signal students that the topic under discussion is related in some significant way to ideas developed earlier and direct them to material that they might wish to review before proceeding.

Interactive Figures and Photos Icons throughout the text direct students to dynamic versions of art and photos on MasteringAstronomy™. Using online applets, students can manipulate factors such as time, wavelength, scale, and perspective to increase their understanding of these figures.

Key Terms Like all subjects, astronomy has its own specialized vocabulary. To aid student learning, the most important astronomical terms are boldfaced at their first appearance in the text. Boldfaced key terms in the Chapter Summary are linked with the page number where the term was defined. In addition, a full alphabetical glossary, defining each key term and locating its first use in the text, appears at the end of the book.

Concept Checks We incorporate into each chapter a number of Concept Checks—key questions that require the reader to reconsider some of the material just presented or attempt to place it into a broader context.

Visual Analogies Revised for clarity, visual analogies link art and analogy, explaining complex astronomical concepts with references to everyday experience.

ANALOGY: A good analogy to a tangled solar magnetic field is a garden hose with loops and kinks.

▼ **Compound Art** It is rare that a single image, be it a photograph or an artist's conception, can capture all aspects of a complex subject. Wherever possible, multiple-part figures are used in an attempt to convey the greatest amount of information in the most vivid way:

- Visible images are often presented along with their counterparts captured at other wavelengths.

- Interpretive line drawings are often superimposed on or juxtaposed with real astronomical photographs, helping students to really "see" what the photographs reveal.

- Breakouts—often multiple ones—are used to zoom in from wide-field shots to close-ups, so that detailed images can be understood in their larger context.

Figure Annotations The seventh edition now incorporates the research-proven technique of strategically placing annotations (which always appear in blue type) within key pieces of art, fostering students' ability to read and interpret complex figures, focus on the most relevant information, and integrate verbal and visual knowledge.

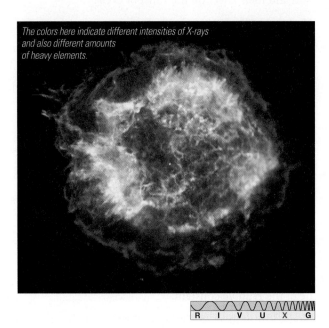

The colors here indicate different intensities of X-rays and also different amounts of heavy elements.

R I V U X G

H–R Diagrams and Acetate Overlays All of the book's H–R diagrams are drawn in a uniform format, using real data. In addition, a unique set of transparent acetate overlays dramatically demonstrates to students how the H–R diagram helps us organize our information about the stars and track their evolutionary histories.

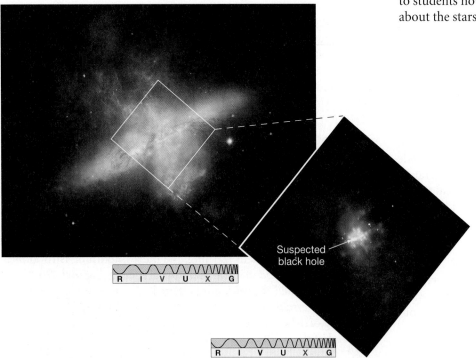

R I V U X G

Suspected black hole

R I V U X G

▲ FIGURE 13.24 **Intermediate-Mass Black Holes?** X-ray observations (inset) of the center of the starburst galaxy M82 reveal a collection of bright sources that may be the result of matter accreting onto intermediate-mass black holes. The black holes are probably young, have masses between 100 and 1000 times the mass of the Sun, and lie relatively far from the center of M82. *(Subaru; NASA)*

▶ **MORE PRECISELY Boxes** These boxes provide more quantitative treatments of subjects discussed qualitatively in the text. Removing these more challenging topics from the main flow of the narrative and placing them within a separate modular element of the chapter design allow instructors greater flexibility in setting the level of their coverage.

9-2 MORE PRECISELY

Energy Generation in the Proton–Proton Chain

Let's look in a little more detail at the energy produced by fusion in the solar core and compare it with the energy needed to account for the Sun's luminosity. As discussed in the text and illustrated below, the net effect of the fusion process is that four protons combine to produce a nucleus of helium-4, in the process creating two neutrinos and two positrons (which are quickly converted into energy by annihilation with electrons).

We can calculate the total amount of energy released by accounting carefully for the total masses of the nuclei involved and applying Einstein's famous formula $E = mc^2$. This allows us to relate the Sun's total luminosity to the consumption of hydrogen fuel in the core. Careful laboratory experiments have determined the masses of all the particles involved in the above reaction: The total mass of the protons is 6.6943×10^{-27} kg, the mass of the helium-4 nucleus is 6.6466×10^{-27} kg, and the neutrinos are virtually massless. We omit the positrons here—their masses will end up being counted as part of the total energy released. The difference between the total mass of the four protons and that of the final helium-4 nucleus, 0.0477×10^{-27} kg, is not great—only about 0.71 percent of the original mass—but it is easily measurable.

Multiplying the vanished mass by the square of the speed of light yields 0.0477×10^{-27} kg $\times (3.00 \times 10^8 \text{ m/s})^2 = 4.28 \times 10^{-12}$ J. This is the energy produced in the form of radiation when 6.69×10^{-27} kg (the rounded-off mass of the four protons) of hydrogen

fuses to helium. It follows that fusion of 1 kg of hydrogen generates $4.28 \times 10^{-12} / 6.69 \times 10^{-27} = 6.40 \times 10^{14}$ J.

Thus we have established a direct connection between the Sun's energy output and the consumption of hydrogen in the core. The Sun's luminosity of 3.86×10^{26} W (see Table 9.1), or 3.86×10^{26} J/s (joules per second), implies a mass consumption rate of 3.86×10^{26} J/s $\div 6.40 \times 10^{14}$ J/kg $= 6.03 \times 10^{11}$ kg/s—roughly 600 million tons of hydrogen every second. That sounds like a lot— the mass of a small mountain and 600 times the mass loss rate in the solar wind—but it represents only a few million million millionths of the total mass of the Sun. Our parent star will be able to sustain this loss for a very long time—this is the subject of Chapter 12.

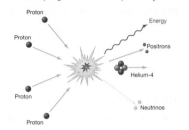

5-1 DISCOVERY

Earth's Growing Ozone "Hole"

During the last two centuries, human technology has begun to produce measurable—and possibly permanent—changes in our planet. You can probably think of many instances, mostly negative, of such changes, from the threat of nuclear war to the reality of air and water pollution worldwide. One example very much in the news and relevant to the discussion in this chapter is the depletion of Earth's ozone layer. (See *Discovery 5–2* for an unrelated, but even more serious, threat to our environment.)

A particularly undesirable by-product of human technology is a group of chemicals known as chlorofluorocarbons (CFCs), relatively simple compounds once widely used for a variety of purposes— propellant in aerosol cans, dry-cleaning products, and in air conditioners and refrigerators. In the 1970s scientists discovered that, instead of quickly breaking down after use as had previously been thought, CFCs accumulate in the atmosphere and are carried high into the stratosphere by convection. There they are broken down by sunlight, releasing chlorine, which quickly reacts with ozone (O_3; see Section 5.3), turning it into oxygen (O_2). In chemical terms the chlorine is said to act as a *catalyst*—it is not consumed in the reaction, and so it survives to react with many more ozone molecules. A single chlorine atom can destroy up to 100,000 ozone molecules before being removed by other, less frequent chemical reactions. In this way,

even a sl
at destroying atmospheric ozone.

As we saw in the text, ozone is part of the protective "blanket" of gases in our atmosphere that protect us from the harsh realities of outer space—in this case, solar ultraviolet radiation. Thus, the result of CFC emission is a substantial increase in ultraviolet radiation levels at Earth's surface, with detrimental effects to most living organisms. The accompanying figure shows the development of an ozone "hole" (colored magenta) over the Antarctic—a region where atmospheric circulation and low temperatures conspire each Antarctic spring to create a vast circumpolar cloud of ice crystals, which act to promote the ozone-destroying reactions, resulting in ozone levels about 50 percent below normal for the region.

Since the hole (really a relative lack of ozone) was discovered in the 1980s, its depth and area have grown each year. Its peak size is now larger than North America. Ozone depletion is not confined to the Antarctic, although the effect is greatest here. Smaller holes have been observed in the Arctic, and occasional ozone depletions of up to 20 percent have been reported at lower northern latitudes.

In the late 1980s, when the effect of CFCs on the atmosphere was realized, the world moved remarkably rapidly to curtail their production and use, with the goal of phasing them out entirely by 2030. Substantial cuts have already been made. Still, scientists think that even if all remaining CFC emissions were to stop today, it would still take several decades for CFCs to leave the atmosphere completely.

◀ **DISCOVERY Boxes** Exploring a wide variety of interesting topics, *Discovery* boxes provide the reader with insight into how scientific knowledge evolves and emphasizes the process of science.

CHAPTER REVIEW

SUMMARY

L01 Jupiter and Saturn were the outermost planets known to ancient astronomers. Uranus was discovered in the 18th century, by chance. Neptune was discovered after mathematical calculations of Uranus's slightly non-Keplerian orbit revealed the presence of an eighth planet.

L02 The four jovian planets are all much more massive than the terrestrial worlds and are of much lower density. They are composed primarily of hydrogen and helium. However, all have large, dense cores, each roughly 10 times Earth's mass and of chemical composition similar to that of the terrestrial planets. The jovian planets all rotate more rapidly than Earth. Having no solid surfaces, they display **differential rotation** (p. 199)—the rotation rate varies with latitude and with depth below the cloud tops. For unknown reasons, Uranus's spin axis lies nearly in the ecliptic plane, leading to extreme seasonal variations in solar heating on the planet as it orbits the Sun.

L03 The cloud layers on all the jovian worlds are arranged into bands of light-colored zones and darker belts crossing the planets parallel to their equators. The bands are the result of convection in the planets' interiors and the planets' rapid rotation. The lighter zones are the tops of upwelling, warm currents, and the darker bands are cooler regions, where gas is sinking. Underlying them is a stable pattern of eastward or westward wind flow called a **zonal flow**. The wind direction alternates as we move north or south away from the equator. Jupiter's atmosphere consists of three main cloud layers. The colors we see are the result of chemical reactions at varying depths below the cloud tops.

L04 A major weather pattern on Jupiter is the **Great Red Spot** (p. 201), a huge hurricane that apparently has been raging for at least three centuries. Other, smaller, atmospheric features—**white ovals** (p. 203) and **brown ovals** (p. 203)—are also observed. Similar systems are found on the other jovian planets, although they are less distinct on Saturn and Uranus. Long-lived storms on Saturn may lie hidden below the planet's clouds. The **Great Dark Spot** (p. 207) on Neptune had many similarities to Jupiter's Red Spot. It disappeared within a few years of its discovery by *Voyager 2*. Storms in the jovian atmospheres are long-lived because there is no solid surface to dissipate their energy.

L05 The atmospheres of the jovian planets become hotter and denser with depth, eventually becoming liquid. In Jupiter and Saturn the interior pressures are so high that the hydrogen near the center exists in a liquid-metallic form rather than as molecular hydrogen. In Uranus and Neptune this change from molecular hydrogen to metallic hydrogen apparently does not occur. Radio waves emitted from the planets' magnetospheres provide a measure of their interior rotation rates. The combination of rapid rotation and conductive interiors means that all four jovian planets have strong magnetic fields and extensive magnetospheres.

L06 Three jovian planets radiate more energy than they receive from the Sun. On Jupiter and Neptune the source of this energy is most likely heat left over from the planet's formation. On Saturn the heating is the result of helium precipitation in the interior, where helium liquefies and forms droplets that then fall toward the center of the planet. This process is also responsible for reducing the amount of helium observed in Saturn's outer layers.

▶ **Chapter Summaries** Key terms introduced in each chapter are listed again, in context and in boldface, along with key figures and page references to the text discussion. Summary items are keyed to the Learning Outcomes presented at the start of the chapter.

End-of-Chapter Questions and Problems Many elements of the end-of-chapter material have seen substantial reorganization:

- Each chapter has 15 Review and Discussion questions, which may be used for in-class review or for assignment. The material needed to answer Review and Discussion questions can be found within the chapter. The Review and Discussion questions explore particular topics more deeply, often asking for opinions, not just facts. As with all discussions, these questions usually have no single "correct" answer. A few (2–4) questions per chapter are marked as directly relevant to the Process of Science theme of the book, and each Learning Outcome is reflected in one of the Review and Discussion questions.

- Each chapter incorporates 15 Self-Test questions, roughly divided between true/false and multiple choice formats, designed to allow students to assess their understanding of the chapter material. As with the Review and Discussion questions, the Self-Test questions can be answered based on material presented in the chapter. Two of the multiple choice questions in each chapter are tied directly to a specific figure or diagram in the text to test students' comprehension of the visual material presented there. Answers to all these questions appear at the end of the book.

- The end-of-chapter material also includes 10 Problems based on the chapter contents and requiring some numerical calculation. In many cases, the Problems are tied directly to quantitative statements made (but not worked out in detail) in the text. The solutions to the Problems are not contained verbatim within the chapter, but the information necessary to solve them has been presented in the text. Answers appear at the end of the book.

- New to this edition, the end-of-chapter material now includes a number of astronomical Activities relevant to the material presented in the text. Activities include both group and individual projects, ranging from basic naked-eye and telescopic observing exercises, to opinion polls, surveys and group discussion, and astronomical research on the Web.

Instructor Resources

MasteringAstronomy™

www.masteringastronomy.com

MasteringAstronomy is the most widely used and most advanced astronomy tutorial and assessment system in the world. By capturing the step-by-step work of students nationally, MasteringAstronomy has established an unparalleled database of learning challenges and patterns. Using this student data, a team of astronomy education researchers has refined every activity and problem. The result is a library of activities of unique educational effectiveness and assessment accuracy. MasteringAstronomy provides students with two learning systems in one: a dynamic self-study area and the ability to participate in online assignments.

MasteringAstronomy, now easier to use than ever, provides instructors with a fast and effective way to assign uncompromising, wide-ranging online homework assignments of just the right difficulty and duration. The tutorials coach 90 percent of students to the correct answer with specific wrong-answer feedback. The powerful post-diagnostics allow instructors to assess the progress of their class as a whole or to quickly identify individual student's areas of difficulty. Tutorials built around text content and all the end-of-chapter problems from the text are also available.

MasteringAstronomy has been revised in this edition with extensive new interactive opportunities in the item library for assignment and in the media-rich study area for open-ended student exploration, which students can use whether the instructor assigns homework or not.

Instructor Resource Manual
Updated for the seventh edition, this manual provides sample syllabi and course schedules, an overview of each chapter, pedagogical tips, useful analogies, suggestions for classroom demonstrations, writing questions, answers to the end-of-chapter Review and Discussion questions, Conceptual Self-Test, and Problems, selected readings, and additional references and resources. The Instructor Resource Manual is available for download on the Pearson Instructor Resource Center (www.pearsonhighered.com/educator) and in the MasteringAstronomy Instructor Resource Area.

Test Bank
This extensive file of approximately 2500 multiple-choice, true/false, fill-in-the-blank, short answer and essay questions is newly updated and revised for the seventh edition. The questions are organized and referenced by chapter section and by question type. The Test Bank is available within TestGen® as well as electronically on the Pearson Instructor Resource Center (www.pearsonhighered.com/educator) and in MasteringAstronomy.

Instructor Resource DVD
This DVD provides virtually every electronic asset you'll need in and out of the classroom. The DVD is organized by chapter and contains all text illustrations, tables, and photos in jpeg and PowerPoint® formats, as well as animations, videos, and self-guided tutorials from the self-study section of MasteringAstronomy.
ISBN 0-321-83491-7

Learner-Centered Astronomy Teaching: Strategies for ASTRO 101

Timothy F. Slater, University of Wyoming; Jeffrey P. Adams, Millersville University

Strategies for ASTRO 101 is a guide for instructors of the introductory astronomy course for nonscience majors. Written by two leaders in astronomy education research, this book details various techniques instructors can use to increase students' understanding and retention of astronomy topics, with an emphasis on making the lecture a forum for active student participation. Drawing from the large body of recent research to discover how students learn, this guide describes the application of multiple classroom-tested techniques to the task of teaching astronomy to predominantly nonscience students.
ISBN 0-13-046630-1

Peer Instruction for Astronomy
Paul Green, Harvard Smithsonian Center for Astrophysics
Peer instruction is a simple yet effective method for teaching science. Techniques of peer instruction for introductory physics were developed primarily at Harvard and have aroused interest and excitement in the physics education community. This approach involves students in the teaching process, making science more accessible to them. Peer instruction is a new trend in astronomy that is finding strong interest and is ideally suited to introductory astronomy classes. This book is an important vehicle for providing a large number of thought-provoking, conceptual short-answer questions aimed at a variety of class levels. While significant numbers of such questions have been published for use in physics, *Peer Instruction for Astronomy* provides the first such compilation for astronomy. ISBN 0-13-026310-9

Student Resources

MasteringAstronomy™
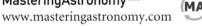
www.masteringastronomy.com
This homework, tutorial, and assessment system is uniquely able to tutor each student individually by providing students with instantaneous feedback specific to their wrong answers, simpler subproblems upon request when they get stuck, and partial credit for their method(s) used. Students also have access to a self-study area that contains practice quizzes, self-guided tutorials, animations, videos, and more.

Pearson eText is available through MasteringAstronomy, either automatically when MasteringAstronomy is packaged with new books, or available as a purchased upgrade online. Allowing the students to access the text wherever they have access to the Internet, Pearson eText comprises the full text, including figures that can be enlarged for better viewing. Within Pearson eText, students are also able to pop up definitions and terms to help with vocabulary and reading. Students also can take notes in Pearson eText using the annotation feature.

Starry Night® College Student Access Code Card (Simulation Curriculum) This access kit provides a one-time download of Starry Night College, the best-selling planetarium software that lets you escape the Milky Way and travel across 700 million light-years of space. Hailed for its breathtaking realism, powerful features, and intuitive interface, Starry Night College is available to be packaged (for a minimal charge) with new copies of introductory astronomy textbooks. This access kit also enables users to download *Starry Night College Activities & Observation and Research Projects for Astronomy Today* by Erin O'Connor and Steve McMillan. ISBN 0-321-71295-1

Starry Night® College Activities & Observation and Research Projects for *Astronomy Today*, Seventh Edition, Chaisson/McMillan This booklet contains over 35 activities and 70 observation and research projects written by Erin O'Connor and Steve McMillan and is based on Starry Night College planetarium software. ISBN 0-321-75307-0

SkyGazer 5.0 Student Access Code Card (Carina Software) This access kit provides a one-time download of SkyGazer 5.0 that combines exceptional planetarium software with informative pre-packaged tutorials. Based on the popular Voyager software, this access kit is available to be packaged at no additional charge with new copies of introductory astronomy textbooks. This access kit also enables users to download the *Astronomy Media Workbook* by Michael LoPresto. ISBN 0-321-76518-4

Astronomy Media Workbook, 7th Edition *Astronomy Media Workbook* by Michael LoPresto includes a wide selection of in-depth activities based on *Voyager*: SkyGazer 5.0 planetarium software and the Interactive Figures™ and RSS Feeds in MasteringAstronomy. These thought-provoking projects are suitable for labs or homework assignments. It is downloadable with a SkyGazer access code. ISBN 0-321-74124-2

Observation Exercises in Astronomy This workbook by Lauren Jones contains a series of astronomy exercises that integrate technology from planetarium software such as Stellarium, Starry Night College, WorldWide Telescope, and SkyGazer. Using these online products adds an interactive dimension to students' learning. ISBN: 0-321-63812-3

Edmund Scientific Star and Planet Locator The famous rotating roadmap of the heavens shows the location of the stars, constellations, and planets relative to the horizon for the exact hour and date you determine. This 8-inch square star chart was plotted by the late astronomer and cartographer George Lovi. The reverse side of the locator is packed with additional data on the planets, meteor showers, and bright stars. Included with each star chart is a 16-page, fully-illustrated, pocket-size instruction booklet.

Norton's Star Atlas and Reference Handbook, 20th edition Now in a superbly redesigned, two-color landmark 20th edition, this combination star atlas and reference work by Ian Ridpath has no match in the field. First published in 1910, *Norton's* owes much of its legendary success to its unique maps, arranged in slices known as gores, each covering approximately one-fifth of the sky. Every star visible to the naked eye under the clearest skies—down to magnitude 6.5—is charted, along with star clusters, nebulae, and galaxies. Extensive tables of data on interesting objects for observation accompany each of the precision-drawn maps. Preceding the maps is the unique and authoritative reference handbook covering timekeeping and positional measurements on the celestial sphere; the Sun, Moon, and other bodies of the solar system; telescopes and other equipment for observing and imaging the sky; and stars, nebulae, and galaxies. Throughout, succinct fundamental principles and practical tips guide the reader into the night sky. The appendices Units and Notation, Astronomical Constants, Symbols and Abbreviations, and Useful Addresses complete what has long been the only essential reference for the stargazer. ISBN 0-13-145164-2

Lecture-Tutorials for Introductory Astronomy, 3rd Edition Edward E. Prather, University of Arizona; Timothy F. Slater, University of Wyoming; Jeffrey P. Adams, Millersville University; Gina

Brissenden, University of Arizona Funded by the National Science Foundation, *Lecture-Tutorials for Introductory Astronomy* is designed to help instructors bring interactive teaching strategies into general education astronomy courses of all sizes. The third edition features new lecture-tutorials entitled the Greenhouse Effect; Dark Matter; Making Sense of the Universe and Expansion; Hubble's Law; Expansion, Lookback Times, and Distances; and Big Bang. Each of the 44 lecture-tutorials is presented in a classroom-ready format, asks students to collaborate in groups of two to three, takes approximately 15 minutes, challenges students with a series of carefully designed questions that spark student discussions, engages students in critical reasoning, and requires no equipment. ISBN 0-321-82046-0

Sky and Telescope This supplement, edited by Evan Skillman, contains nine articles that originally appeared in the popular amateur astronomy magazine, plus a summary and four question sets focusing on the issues professors most want to address: general review, process of science, scale of the universe, and our place in the universe. ISBN 0-321-70620-X

Acknowledgments

Throughout the many drafts that have led to this book, we have relied on the critical analysis of many colleagues. Their suggestions ranged from the macroscopic issue of the book's overall organization to the minutiae of the technical accuracy of each and every sentence. We have also benefited from much good advice and feedback from users of the first six editions of the text and our more comprehensive text, *Astronomy Today*. To these many helpful colleagues, we offer our sincerest thanks.

Reviewers of the Seventh Edition
Ron Armale, *Cypress College*
Jacqueline Dunn, *Midwestern State University*
Michael Frey, *Cypress College*
Alina Gabryzewska-Kukawa, *Delta State University*
Martin Hackworth, *Idaho State University*
Andrei Kolmakov, *Southern Illinois University*
Chris McCarthy, *San Francisco State University*
George Nock, *Northeast Mississippi Community College*
Robert K. Tyson, *University of North Carolina*

Reviewers of Previous Editions
Todd Adams, *Florida State University*
Stephen G. Alexander, *Miami University*
Nadine G. Barlow, *Northern Arizona University*
Michael L. Broyles, *Collin County Community College*
Juan Cabanela, *Haverford College*
Erik Christensen, *South Florida Community College*
Michael L. Cobb, *Southeast Missouri State University*
Anne Cowley, *Arizona State University*
Manfred Cuntz, *University of Texas at Arlington*
James R. Dire, *U.S. Coast Guard Academy*
John Dykla, *Loyola University–Chicago*
Tina Fanetti, *Western Iowa Tech Community College*
Doug Franklin, *Western Illinois University*

Peter Garnavich, *University of Notre Dame*
Richard Gelderman, *Western Kentucky University*
Martin Goodson, *Delta College*
David J. Griffiths, *Oregon State University*
Susan Hartley, *University of Minnesota–Duluth*
David Hudgins, *Rockhurst University*
Doug Ingram, *Texas Christian University*
Marvin Kemple, *Indiana University–Purdue University, Indianapolis*
Linda Khandro, *Shoreline Community College*
Mario Klaric, *Midlands Technical College*
Patrick Koehn, *Eastern Michigan University*
Andrew R. Lazarewicz, *Boston College*
Paul Lee, *Middle Tennessee University*
M.A. K. Keohn Lodhi, *Texas Tech University*
F. Bary Malik, *Southern Illinois University-Carbondale*
Fred Marschak, *Santa Barbara College*
John Mattox, *Francis Marion University*
Chris McCarthy, *San Francisco State University*
George E. McCluskey, Jr., *Leigh University*
Scott Miller, *Pennsylvania State University*
L. Kent Morrison, *University of New Mexico*
Richard Nolthenius, *Cabrillo College*
Edward Oberhofer, *University of North Carolina–Charlotte*
Robert S. Patterson, *Southwest Missouri State University*
Jon Pedicino, *College of the Redwoods*
Cynthia W. Peterson, *University of Connecticut*
Robert Potter, *Bowling Green State University*
Heather L. Preston, *U.S. Air Force Academy*
Andreas Quirrenbach, *University of California–San Diego*
James Regas, *California State University, Chico*
Tim Rich, *Rust College*
Frederick A. Ringwald, *California State University, Fresno*
Gerald Royce, *Mary Washington College*
Louis Rubbo, *Coastal Carolina University*
Rihab Sawah, *Moberly Area Community College*
John C. Schneider, *Catonsville Community College*
C. Ian Short, *Florida Atlantic University*
Philip J. Siemens, *Oregon State University*
Earl F. Skelton, *George Washington University*
Don Sparks, *Los Angeles Pierce College*
Angela Speck, *University of Missouri–Columbia*
Phillip E. Stallworth, *Hunter College of CUNY*
Peter Stine, *Bloomsburg University of Pennsylvania*
Irina Struganova, *Barry University*
Jack W. Sulentic, *University of Alabama*
Jonathan Tan, *University of Florida*
Gregory Taylor, *University of New Mexico*
Gregory R. Taylor, *California State University, Chico*
George Tremberger, Jr., *Queensborough Community College*
Craig Tyler, *Fort Lewis College*
Alex Umantsev, *Fayetteville State University*
Michael Vaughn, *Northeastern University*
Jimmy Westlake, *Colorado Mountain College*
Donald Witt, *Nassau Community College*
J. Wayne Wooten, *University of West Florida*
Garett Yoder, *Eastern Kentucky University*
David C. Ziegler, *Hannibal-LaGrange College*

We would also like to express our gratitude to those that have contributed to our accompanying media products and print supplements, as well as Lola Judith Chaisson for assembling and drawing all the H-R diagrams (including the acetate overlays) for this edition.

The publishing team at Addison-Wesley has assisted us at every step along the way in creating this text. Much of the credit goes to our project editor, Tema Goodwin, for her fine attention to detail and editorial judgment while managing the many variables that go into a multifaceted publication such as this, and our editor, Nancy Whilton, who has successfully navigated us through the twists and turns of the publishing world. Development editor Barbara Price worked on improving the visual analogies in this new edition. Production project manager Mary O'Connell has done an excellent job of tying together the threads of this very complex project, made all the more complex by the necessity of combining text, art, and electronic media into a coherent whole. Special thanks are also in order to the project management team at Thistle Hill Publishing Services, Andrea Archer and Angela Williams Urquhart; interior designer Jerilyn Bockorick; and cover designer Mark Ong for making the seventh edition look spectacular.

We are interested in your feedback on this text. Please email us at aw.astronomy@pearson.com if you find any errors or have comments.

Eric Chaisson
Steve McMillan

About the Authors

ERIC CHAISSON

Eric holds a doctorate in astrophysics from Harvard University, where he spent 10 years on the faculty of Arts and Sciences. For more than two decades thereafter, he served on the senior science staff at the Space Telescope Science Institute and held various professorships at Johns Hopkins and Tufts Universities. He is now back at Harvard, where he teaches natural science and conducts research at the Harvard-Smithsonian Center for Astrophysics. Eric has written 12 books on astronomy, which have received such literary awards as the Phi Beta Kappa Prize, two American Institute of Physics Awards, and Harvard's Smith-Weld Prize for Literary Merit. He has published nearly 200 scientific papers in professional journals, some of which won him Harvard's B. J. Bok Prize for original contributions to astrophysics.

STEVE McMILLAN

Steve holds a bachelor's and master's degree in mathematics from Cambridge University and a doctorate in astronomy from Harvard University. He held post-doctoral positions at the University of Illinois and Northwestern University, where he continued his research in theoretical astrophysics, star clusters, and numerical modeling. Steve is currently Distinguished Professor of Physics at Drexel University and a frequent visiting researcher at Princeton's Institute for Advanced Study and Leiden University in the Netherlands. He has published more than 90 scientific papers in professional journals.

Foundations

Astronomy is the study of the universe—the totality of all space, time, matter, and energy. It is a subject like no other, for it requires us to profoundly change our perspective and to consider sizes, scales, and times unfamiliar to us from everyday experience. To appreciate astronomy, we must broaden our view and expand our minds. We must think big!

Part 1 presents the basic methods used by astronomers to chart the space around us. We describe the progress of scientific knowledge, from stories of chariots and gods to today's well-tested ideas of planetary motion and quantum physics. We also delve into the microscopic realm of atoms and molecules, whose properties hold keys to understanding the universe on macroscopic scales.

The images here illustrate the range of scales encountered in Part 1, from atoms to humans to Earth itself.

Atoms ~ 10^{-10} m

Cells ~ 10^{-5} m

Humans ~ 2 m

Earth ~10^7 m

Mountains ~ 10^4 m

Earth is neither central nor special;
we inhabit no unique place in the universe.

0
Charting the Heavens

The Foundations of Astronomy

Nature offers no greater splendor than the starry sky on a clear, dark night. Silent and jeweled with the constellations of ancient myth and legend, the night sky has inspired wonder throughout the ages—a wonder that leads our imaginations far from the confines of Earth and out into the distant reaches of space and time. Astronomy, born in response to that wonder, is built on two basic traits of human nature: the need to explore and the need to understand. People have sought answers to questions about the universe since the earliest times. Astronomy is the oldest of all the sciences, yet never has it been more exciting than it is today.

MasteringAstronomy® Visit www.masteringastronomy.com for quizzes, animations, videos, interactive figures, and self-guided tutorials.

◄ Stars are the most fundamental visible component of the universe. Roughly as many stars reside in the observable universe as grains of sand in all the beaches of the world. Here, we see high overhead a rich band of stars known as the Milky Way—so-called because it often resembles a milky band of countless stars when seen on a clear, dark night. All these stars and more comprise part of a much larger system of stars called the Milky Way Galaxy, of which our star, the Sun, is one member. This single exposure, entitled *Going to the Stars Road*, was made at night with only the Moon's light illuminating the terrain on the continental divide at Logan Pass in Glacier National Park, near the Montana/Alberta border. (© *Tyler Nordgren*)

Universe

x10⁴

Galaxy

x10⁸

0.1 The "Obvious" View

Our Place in Space

Of all the scientific insights achieved to date, one stands out boldly: Earth is neither central nor special. We inhabit no unique place in the universe. We live on an ordinary rocky **planet** called Earth, one of eight known planets orbiting an average **star** called the Sun, a middle-aged star near the edge of a huge collection of stars called the Milky Way Galaxy, one **galaxy** among countless billions of others spread throughout the observable **universe**. Figure 0.1 illustrates the enormous range of scales of these very different objects (see also the Part 1 opening discussion on page 1).

Today, our scientific understanding of the cosmos extends literally as far as we can see, spanning the vast reaches of intergalactic space. But the modern view of the universe depicted here is in many ways the "punch line" of the story presented in this text. It is the culmination of countless scientific discoveries, large and small, representing the work of generations of astronomers. How have we come to know the universe around us? How have astronomers achieved the perspective sketched in Figure 0.1? Our study of the universe, the science of **astronomy,** begins by examining the sky.

Constellations in the Sky

Between sunset and sunrise on a clear night, we can see some 3000 points of light. Include the view from the opposite side of Earth, and nearly 6000 stars are visible to the unaided eye. A natural human tendency is to see patterns and relationships between objects even when no true connection exists, and people long ago connected the brightest stars into configurations called **constellations.** In the Northern Hemisphere, most constellations were named after mythological heroes and animals. Figure 0.2 shows a constellation especially prominent in the northern night

▼ FIGURE 0.1 **Size and Scale in the Universe**   Bottom right of this figure shows humans on Earth, a view that widens progressively in each of the next four scenes illustrated from bottom to top—Earth, the planetary system, a galaxy, and truly deep space. The numbers within the dashed zooms indicate approximately the increase in scale between successive images: Earth is 10 million times larger than humans, our solar system in turn is some million times larger than Earth, and so on. (See the Preface, p. xi, for an explanation of the icon at the bottom, which here indicates that this image was made in visible light.) *(NASA; J. Lodriguss; NOAA)*

x10⁶

Planetary system

R   I   V   U   X   G

x10⁷

Earth

Humans

INTERACTIVE ▶ FIGURE 0.2 **Constellation Orion**
(a) A photograph of the group of bright stars that make up the constellation Orion. (b) The stars connected to show the pattern visualized by the Greeks: the outline of a hunter. The Greek letters serve to identify some of the brighter stars in the constellation (see Figure 0.3). You can easily find Orion in the northern winter sky by identifying the line of three bright stars in the hunter's "belt." (P. Sanz /Alamy)

(a)

R I V U X G
(b)

sky from October through March: the hunter Orion, named for a mythical Greek hero famed, among other things, for his amorous pursuit of the Pleiades, the seven daughters of the giant Atlas. According to Greek mythology, the gods placed the Pleiades among the stars to protect them from Orion, who still stalks them nightly across the sky. Many other constellations have similarly fabulous connections with ancient cultures.

The stars making up a particular constellation are generally not close together in space. They merely are bright enough to observe with the naked eye and happen to lie in the same direction in the sky as seen from Earth. Figure 0.3 illustrates this point for Orion, showing the true relationships between that constellation's brightest stars. Although constellation patterns have no real significance, the terminology is still used today. Constellations provide a convenient means for astronomers to specify large areas of the sky, much as geologists use continents or politicians use voting precincts to identify certain localities on Earth. In all, there are 88 constellations, most of them visible from North America at some time during the year.

The Celestial Sphere

Over the course of a night, the constellations appear to move across the sky from east to west. However, ancient sky-watchers noted that the *relative* positions of stars (to each other) remained unchanged as this nightly march took place. It was natural for early astronomers to conclude that the stars were attached to a **celestial sphere** surrounding Earth—a canopy of stars like an astronomical painting on a vast heavenly ceiling. Figure 0.4 shows how early astronomers pictured the stars as moving with this celestial sphere as it turned around a fixed Earth. Figure 0.5 shows how stars appear to move in circles around a point in the sky very close to the star Polaris (better known as the Pole Star or the North Star). To early astronomers, this point represented the axis around which the celestial sphere turned.

From our modern standpoint, the apparent motion of the stars is the result of the spin, or **rotation,** not of the celestial sphere, but of

▶ FIGURE 0.3 **Orion in 3D** The true three-dimensional relationships among the most prominent stars in Orion. The distances in light-years were measured by the European *Hipparcos* satellite in the 1990s. (See Section 10.1.)

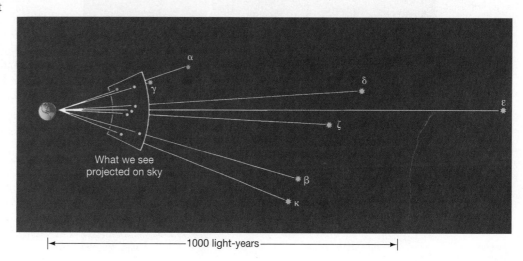

What we see projected on sky

◀————— 1000 light-years —————▶

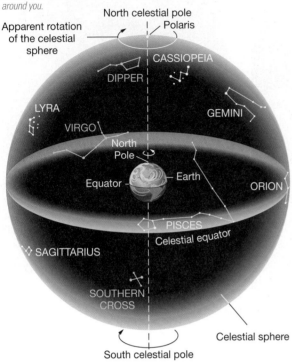

Imagine yourself at the center of this sphere, looking out at the whole sky around you.

Apparent rotation of the celestial sphere

North celestial pole
Polaris
CASSIOPEIA
DIPPER
LYRA
GEMINI
VIRGO
North Pole
Equator — Earth
ORION
PISCES
Celestial equator
SAGITTARIUS
SOUTHERN CROSS
Celestial sphere
South celestial pole

INTERACTIVE ▲ **FIGURE 0.4 The Celestial Sphere** Planet Earth sits fixed at the hub of the celestial sphere. This is one of the simplest possible models of the universe, but it doesn't agree with the facts that astronomers now know **MA** about the universe.

CONCEPT CHECK ▶ Earth isn't really enclosed in a sphere with stars attached. Why then do astronomers find it convenient to retain the fiction of the celestial sphere? What vital piece of information about stars is lost when we talk about their positions "on" the sky?

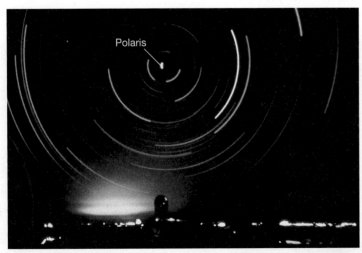

Polaris

The duration of this exposure is about 5 hours,...

...since each star traces out approximately 20 percent of a circle.

R I V U X G

Earth. Even though we now know that a revolving celestial sphere is an incorrect description of the heavens, astronomers still use the idea as a convenient fiction that helps us visualize the positions of stars in the sky. The point where Earth's rotation axis (the line through the center around which the planet rotates) intersects the celestial sphere in the Northern Hemisphere is known as the **north celestial pole;** it is directly above Earth's North Pole. The star Polaris happens to lie close to the north celestial pole, which is why its direction indicates due north. In the Southern Hemisphere, the extension of Earth's axis in the opposite direction defines the **south celestial pole.** (There are no bright stars conveniently located near the south celestial pole and hence no "southern Pole Star.") Midway between the north and south celestial poles lies the **celestial equator,** representing the intersection of Earth's equatorial plane (the plane through Earth's center, perpendicular to the rotation axis) with the celestial sphere.

Celestial Coordinates

The simplest method of locating stars in the sky is to specify their constellation and then rank the stars in that constellation in order of brightness. The brightest star is denoted by the Greek letter α (alpha), the second brightest by β (beta), and so on. For example, Betelgeuse and Rigel, the two brightest stars in the constellation Orion, are also known as α Orionis and β Orionis, respectively (see Figures 0.2 and 0.3). (Precise observations show that Rigel is actually brighter than Betelgeuse, but the names are now permanent.) Because there are many more stars in any given constellation than there are letters in the Greek alphabet, this method is of limited use. However, for naked-eye astronomy, where only bright stars are involved, it is quite satisfactory.

For more precise measurements, astronomers find it helpful to use a system of **celestial coordinates** on the sky. If we think of the stars as being attached to the celestial sphere centered on Earth, then the familiar system of angular measurement on Earth's surface—latitude and longitude (Figure 0.6a)—extends quite naturally to the sky. The celestial analogs of latitude and longitude are called **declination** and **right ascension,** respectively (Figure 0.6b). Just as latitude and longitude are tied to Earth, right ascension and declination are fixed on the celestial sphere. Although the stars appear to move across the sky because of Earth's rotation, their celestial coordinates remain constant over the course of a night.

Declination (dec) is measured in *degrees* (°) north or south of the celestial equator, just as latitude is measured in degrees north or south of Earth's equator (see *More Precisely 0-1*). The celestial equator is at a declination of 0°, the north celestial pole is at +90°, and the south celestial pole is at −90° (the plus sign here just means "north of the celestial equator"; minus means "south"). Right ascension (RA) is measured in angular units called *hours, minutes,* and *seconds,* and it increases in the eastward direction. Like the choice of the Greenwich Meridian as the zero-point of longitude on Earth, the choice of zero right ascension is quite arbitrary—it is conventionally taken to be the position of the Sun in the sky at the instant of the vernal equinox (to be discussed in the next section).

INTERACTIVE ◀ FIGURE 0.5 **The Northern Sky** Time-lapse photograph of the northern sky. Each trail traces the path of a single star across the night sky. The concentric circles are centered near the North Star, Polaris. **MA** (AURA)

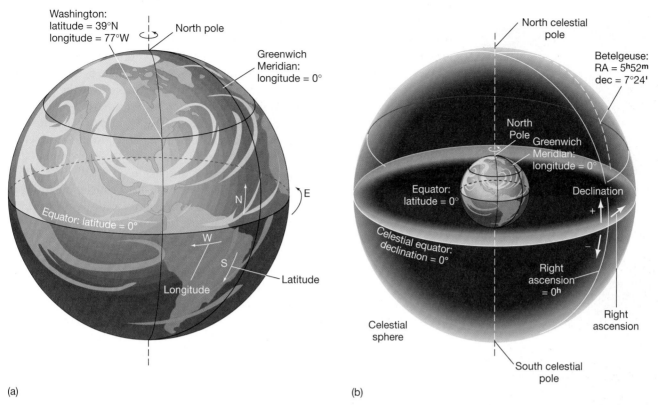

(a)

(b)

▲ **FIGURE 0.6 Right Ascension and Declination** (a) Longitude and latitude allow us to locate a point on the surface of Earth by specifying its distance (as an angle) east or west of the Greenwich Meridian, and north or south of the equator. For example, to find Washington, D.C., on Earth, look 77° west of Greenwich and 39° north of the equator. (b) Similarly, right ascension and declination specify locations on the sky. To locate the star Betelgeuse on the celestial sphere; look 5h52m east of the vernal equinox (the line on the sky with a right ascension of zero) and 7°24' north of the celestial equator.

0.2 Earth's Orbital Motion

Day-to-Day Changes

We measure time by the Sun. The rhythm of day and night is central to our lives, so it is not surprising that the period of time from one sunrise (or noon, or sunset) to the next, the 24-hour **solar day,** is our basic social time unit. As we have just seen, this apparent daily progress of the Sun and other stars across the sky, known as **diurnal motion** is a consequence of Earth's rotation. But the stars' positions in the sky do not repeat themselves exactly from one night to the next. Each night, the whole celestial sphere appears shifted a little compared with the night before—you can confirm this for yourself by noting over the course of a week or two which stars are visible near the horizon just after sunset or just before dawn. Because of this shift, a day measured by the stars—called a **sidereal day** after the Latin word *sidus,* meaning "star"—differs in length from a solar day.

The reason for the difference in length between a solar day and a sidereal day is sketched in Figure 0.7. Earth moves in two ways simultaneously: it rotates on its central axis while at the same time **revolving** around the Sun. Each time Earth rotates once on its axis, it also moves a small distance along its orbit. Therefore, each day Earth has to rotate through slightly more than 360° in order for the Sun to return to the same apparent location in the sky. As a result, the interval of time between noon one day and noon the next (a solar day) is slightly greater than the true rotation period (one sidereal day). Our planet takes 365 days to orbit the Sun, so the additional angle is 360°/365 = 0.986°. Because Earth takes about 3.9 minutes to rotate through this angle, the solar day is 3.9 minutes longer than the sidereal day.

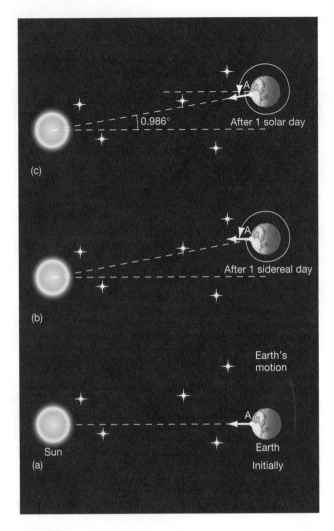

(c)

0.986°

After 1 solar day

(b)

After 1 sidereal day

Earth's
motion

Sun

(a)

A

Earth

Initially

Seasonal Changes

Because Earth revolves around the Sun, our planet's darkened hemisphere faces in a slightly different direction each night. The change is only about 1° per night (Figure 0.7)—too small to be easily discerned with the naked eye from one evening to the next. However, the change is clearly noticeable over the course of weeks and months, as illustrated in Figure 0.8. In 6 months, Earth moves to the opposite side of its orbit, and we face an entirely different group of stars and constellations at night. Because of this motion, the Sun appears, to an observer on Earth, to move slowly (at a rate of 1° per day) relative to the background stars over the course of a year. This apparent motion of the Sun on the sky traces out a path on the celestial sphere known as the **ecliptic.** The 12 constellations through which the Sun passes during the year as it moves along the ecliptic—that is, the constellations we would see looking in the direction of the Sun if they weren't overwhelmed by the Sun's light—had special significance for astrologers of old. They are collectively known as the **zodiac.**

As illustrated in Figure 0.9, the ecliptic forms a great circle on the celestial sphere, inclined at an angle of about 23.5° to the celestial equator. In reality, as shown in Figure 0.10, the plane defined by the ecliptic is *the plane of Earth's orbit around the Sun.* Its tilt is a consequence of the *inclination* of our planet's rotation axis to its orbital plane.

The point on the ecliptic where the Sun is at its northernmost point above the celestial equator (see Figure 0.9) is known as the **summer solstice** (from the Latin words *sol,* meaning "sun," and *stare,* "to stand"). As indicated in Figure 0.10, it represents the point on Earth's orbit where our planet's North Pole is oriented closest to the Sun. This occurs on or near June 21—the exact date varies slightly from year to year because the actual length of a year is not a whole number of days. As Earth rotates on that date, points north of the equator spend the greatest fraction of their time in sunlight, so the summer solstice corresponds to the

▲ **FIGURE 0.7 Solar and Sidereal Days**   A sidereal day is Earth's true rotation period—the time taken for our planet to return to the same orientation in space relative to the distant stars. A solar day is the time from one noon to the next. The difference in duration between the two is easily explained because Earth revolves around the Sun at the same time as it rotates on its axis. Frames (a) and (b) are one sidereal day apart, when Earth rotates exactly once on its axis and also moves a little in its solar orbit—approximately 1°. Consequently, between noon at point A on one day and noon at the same point the next day, Earth actually rotates through about 361° (frame c), and the solar day exceeds the sidereal day by about 4 minutes. Note that the diagrams are not drawn to scale; the 1° angle is actually much smaller than that shown here.

INTERACTIVE ▶ **FIGURE 0.8   The Zodiac** The night side of Earth faces a different set of constellations at different times of the year. The 12 constellations named here make up the astrological zodiac. The arrows indicate the most prominent zodiacal constellations in the night sky at various times of year. For example, in June, when the Sun is "in" Gemini, Sagittarius and Capricornus **(MA)** are visible at night.

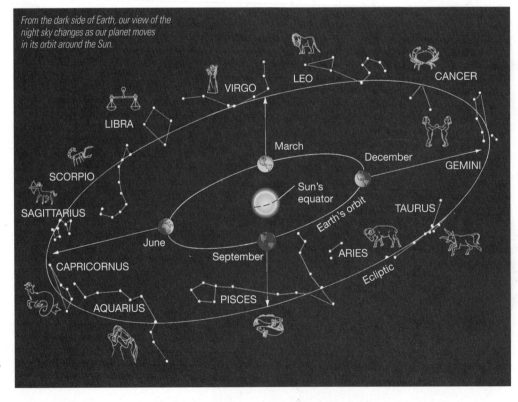

From the dark side of Earth, our view of the night sky changes as our planet moves in its orbit around the Sun.

LEO

VIRGO

CANCER

LIBRA

March

December

GEMINI

SCORPIO

Sun's
equator

Earth's orbit

TAURUS

SAGITTARIUS

June

September

ARIES

CAPRICORNUS

Ecliptic

AQUARIUS

PISCES

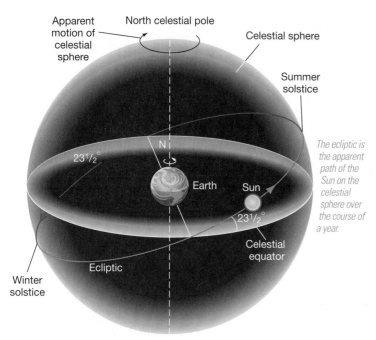

Apparent motion of celestial sphere
North celestial pole
Celestial sphere
Summer solstice
N
23½°
Earth
Sun
23½°
Celestial equator
Winter solstice
Ecliptic

The ecliptic is the apparent path of the Sun on the celestial sphere over the course of a year.

◄ **FIGURE 0.9 Ecliptic** The seasons result from the changing height of the Sun above the celestial equator. At the summer solstice, the Sun is at its northernmost point on its path around the ecliptic. It is therefore highest in the sky, as seen from Earth's Northern Hemisphere, and the days are longest. The reverse is true at the winter solstice. At the vernal and autumnal equinoxes, when the Sun crosses the celestial equator, day and night are of equal length.

longest day of the year (that is, the greatest number of daylight hours—Earth's rotation period doesn't change!) in Earth's Northern Hemisphere and the shortest day in Earth's Southern Hemisphere.

Six months later, the Sun is at its southernmost point below the celestial equator (Figure 0.9)—or, equivalently, Earth's North Pole is oriented farthest from the Sun (Figure 0.10). We have reached the **winter solstice** (December 21), the shortest day in Earth's Northern Hemisphere and the longest in the Southern Hemisphere.

The tilt of Earth's rotation axis relative to the ecliptic is responsible for the **seasons** we experience—the marked difference in temperature between the hot summer and cold winter months. As illustrated in Figure 0.10, two factors combine to cause this variation. First, there are more hours of daylight during the summer than in winter. To see why this is, look at the yellow lines on the surfaces of the Earths in the figure. (For definiteness, they correspond to a latitude of 45 degrees north—roughly that of the Great Lakes or the south of France.) A much larger fraction of the line is sunlit in the summertime, and more daylight means more solar heating. Second, as illustrated in the insets in Figure 0.10, when the Sun is high in the sky in summer, rays of sunlight striking Earth's surface are more concentrated—spread out over a smaller area—than in winter. As a result,

MA ANIMATION/VIDEO Summer Solstice

MA ANIMATION/VIDEO Winter Solstice

INTERACTIVE ▶ **FIGURE 0.10 Seasons** Earth's seasons result from the inclination of our planet's rotation axis with respect to its orbit plane. The summer solstice corresponds to the point on Earth's orbit where our planet's North Pole points most nearly toward the Sun. The opposite is true of the winter solstice. The vernal and autumnal equinoxes correspond to the points in Earth's orbit where our planet's axis is perpendicular to the line joining Earth and the Sun. The insets show how rays of sunlight striking the ground at an angle (during northern winter) are spread over a larger area than rays coming nearly straight down (e.g., during northern summer). As a result, the amount of solar heat delivered to a given area of Earth's surface is greatest when the Sun is high in the sky. **MA**

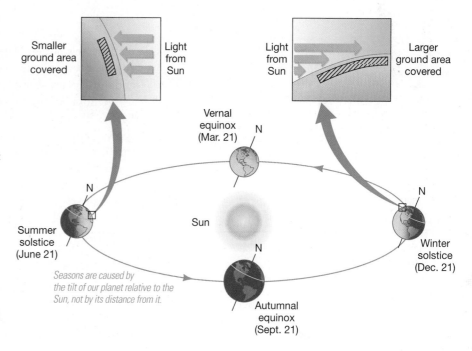

Smaller ground area covered
Light from Sun
Light from Sun
Larger ground area covered

Vernal equinox (Mar. 21) N

N
Summer solstice (June 21)

Sun

N
Winter solstice (Dec. 21)

Seasons are caused by the tilt of our planet relative to the Sun, not its distance from it.

N
Autumnal equinox (Sept. 21)

0-1 MORE PRECISELY

Angular Measure

The size and scale of astronomical objects are often specified by measuring lengths and angles. The concept of length measurement is fairly intuitive. The concept of angular measurement may be less familiar, but it too can become second nature if you remember a few simple facts:

- A full circle contains 360 arc degrees (or 360°). Therefore, the half circle that stretches from horizon to horizon, passing directly overhead and spanning the portion of the sky visible to one person at any one time, contains 180°.
- Each 1° increment can be further subdivided into fractions of an arc degree, called arc minutes; there are 60 arc minutes (60′) in 1 arc degree. Both the Sun and the Moon project an angular size of 30 arc minutes on the sky. Your little finger, held at arm's length, does about the same, covering about a 40-arc minute slice of the 180° horizon-to-horizon arc.
- An arc minute can be divided into 60 arc seconds (60″). Put another way, an arc minute is 1/60 of an arc degree, and an arc second is 1/60 × 1/60 = 1/3600 of an arc degree. An arc second is an extremely small unit of angular measure—it is the angular size of a centimeter-size object (a dime, say) at a distance of about 2 kilometers (a little over a mile).

The accompanying figure illustrates this subdivision of the circle into progressively smaller units.

One final note: Arc degrees have nothing to do with temperature, and arc minutes and arc seconds have nothing to do with time. However, the angular units used to measure right ascension (and only right ascension)—hours (^{h}), minutes (^{m}), and seconds (^{s})—

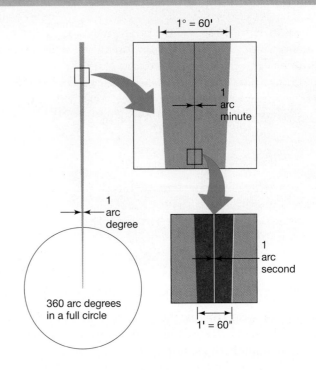

are constructed to parallel the units of time, the connection being provided by Earth's rotation. In 24 hours, Earth rotates once on its axis, or through 360°. In a time interval of 1 hour, Earth rotates through 360°/24 = 15°, or 1^h. In 1 minute of time Earth rotates through 15°/60 = 0.25° = 15′ = 1^m; in 1 second Earth rotates through 15′/60 = 0.25′ = 15″ = 1^s. Just bear in mind that angular hours, minutes, and seconds refer *only* to right ascension and you should avoid undue confusion.

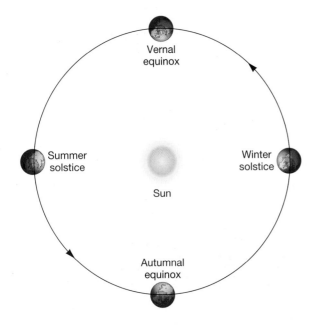

the Sun feels hotter. Therefore summer, when the Sun is highest above the horizon and the days are longest, is generally much warmer than winter, when the Sun is low and the days are short.

A popular misconception is that the seasons have something to do with Earth's distance from the Sun. Figure 0.11 illustrates why this is *not* the case. It shows Earth's orbit face on, instead of almost edge-on, as in Figure 0.10. Notice that the orbit is almost perfectly circular, so the distance from Earth to the Sun varies very little (in fact, by only about 3 percent) over the course of a year—not nearly enough to explain the seasonal changes in temperature. What's more, Earth is actually *closest* to the Sun in early January, the dead of winter in the Northern Hemisphere, so distance from the Sun cannot be the main factor controlling our climate.

◀ **FIGURE 0.11 Earth's Orbit**   Seen face on, Earth's orbit around the Sun is almost indistinguishable from a perfect circle. The distance from Earth to the Sun varies only slightly over the course of a year and is *not* the cause of the seasonal temperature variations we experience on our planet.

The two points where the ecliptic intersects the celestial equator (Figure 0.9)—that is, where Earth's rotation axis is perpendicular to the line joining Earth to the Sun (Figure 0.10)—are known as **equinoxes**. On those dates, day and night are of equal duration. (The word *equinox* derives from the Latin for "equal night.") In the fall (in Earth's Northern Hemisphere), as the Sun crosses from the northern into the southern celestial hemisphere, we have the **autumnal equinox** (on September 21). The **vernal equinox** occurs in spring, on or near March 21, as the Sun crosses the celestial equator moving north. The vernal equinox plays an important role in human timekeeping. The interval of time from one vernal equinox to the next—365.242 solar days—is known as one **tropical year.**

MA ANIMATION/VIDEO The Equinoxes

Long-Term Changes

The time required for Earth to complete exactly one orbit around the Sun, relative to the stars, is called a **sidereal year.** One sidereal year is 365.256 solar days long, about 20 minutes longer than a tropical year. The reason for this slight difference is a phenomenon known as **precession**. Like a spinning top that rotates rapidly on its own axis while that axis slowly revolves about the vertical, Earth's axis changes its direction over the course of time, although the angle between the axis and a line perpendicular to the plane of the ecliptic always remains close to 23.5°. Figure 0.12 illustrates Earth's precession, which is caused by the combined gravitational pulls of the Moon and the Sun (see Chapter 1). During a complete cycle of precession, taking about 26,000 years, Earth's axis traces out a cone. Because of this slow shift in the orientation of Earth's rotation axis, the vernal equinox, which defines the tropical year, drifts slowly around the ecliptic over the course of the precession cycle. Notice in Figure 0.12(b) that most of the time there is *no* bright "Pole Star" marking due north.

The tropical year is the year our calendars measure. If our timekeeping were tied to the sidereal year, the seasons would slowly march around the calendar as Earth precessed—13,000 years from now, summer in the Northern Hemisphere would be at its height in mid-February! By using the tropical year instead, we ensure that July and August will always be (northern) summer months. However, in 13,000 years' time, Orion, now a prominent feature of the northern winter sky, will be a summer constellation.

CONCEPT CHECK ▶ Earth is actually farthest from the Sun during northern summer. Why, then, is it hottest in North America during this season?

INTERACTIVE ▼ FIGURE 0.12 **Precession** (a) Earth's axis currently points nearly toward the star Polaris. Some 12,000 years from now, Earth's axis will point toward a star called Vega, which will then be the "North Star." Five thousand years ago, the North Star was a star named Thuban in the constellation Draco. (b) The yellow circle shows the precessional path of the north celestial pole among some prominent northern stars and depicts the direction toward which Earth's pole points on the sky. Each tick mark represents 1000 years. **MA**

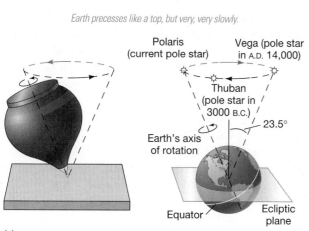

Earth precesses like a top, but very, very slowly.

Polaris (current pole star)
Vega (pole star in A.D. 14,000)
Thuban (pole star in 3000 B.C.)
23.5°
Earth's axis of rotation
Equator
Ecliptic plane

(a)

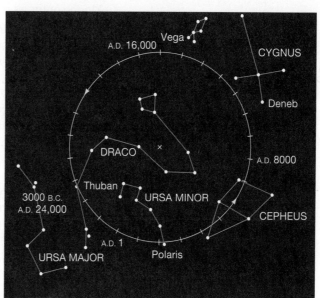

Vega
A.D. 16,000
CYGNUS
Deneb
DRACO
A.D. 8000
Thuban
URSA MINOR
3000 B.C.
A.D. 24,000
CEPHEUS
A.D. 1
URSA MAJOR
Polaris

(b)

0.3 The Motion of the Moon

Early astronomers had very practical reasons for studying the sky. Some stars (such as Polaris) served as navigational guides, while others served as primitive calendars to predict planting and harvesting seasons. By observing repeating patterns in the sky and associating them with events on Earth, astronomers began to establish concrete connections between celestial events and everyday life and took the first steps toward true scientific understanding of the heavens. In a real sense, then, human survival depended on astronomical knowledge. The ability to predict and even explain astronomical events was undoubtedly a highly prized, and perhaps jealously guarded, skill.

The Moon also played an important role in ancient astronomy. Calendars and religious observances were often tied to its phases and cycles, and even today the calendars of most of the world's major religions are still based wholly or partly on the lunar orbit. The Moon's regularly changing appearance (as well as its less regular, but much more spectacular, eclipses) was an integral part of the framework within which ancient astronomers sought to understand the universe. We will study the Moon's physical properties in more detail in Chapter 5. Here we continue our inventory of the sky with a brief description of the motion of our nearest neighbor in space.

Lunar Phases

Apart from the Sun, the Moon is by far the brightest object in the sky. Like the Sun, the Moon appears to move relative to the background stars. Unlike the Sun, however, the explanation for this motion is the obvious one—the Moon really does revolve around Earth.

The Moon's appearance undergoes a regular cycle of changes, or **phases,** taking a little more than 29 days to complete. (The word *month* is derived from the word *Moon.*) Figure 0.13(a) illustrates the appearance of the Moon at different times in this monthly cycle. Starting from the **new Moon,** which is all but invisible in the sky, the Moon appears to *wax* (grow) a little each night and is visible as a growing *crescent* (panel 1 of Figure 0.13a). One week after new Moon, half of the lunar disk (the circular face we would see if the Moon were completely illuminated) can be seen (panel 2). This phase is known as a **quarter Moon.** During the next week, the Moon continues to wax, passing through the *gibbous* phase (more than half of the lunar disk visible, panel 3), until 2 weeks after new Moon the **full Moon** (panel 4) is visible. During the next 2 weeks, the Moon *wanes* (shrinks), passing in turn through the gibbous, quarter, and crescent phases (panels 5–7), eventually becoming new again.

The location of the Moon in the sky, as seen from Earth, depends on its phase. For example, the full Moon rises in the east as the Sun sets in the west, while the first quarter Moon actually rises at noon, but often only becomes visible late in the day as the Sun's light fades. By this time the Moon is already high in the sky. These connections between lunar phase and rising/setting times are indicated on Figure 0.13(a).

Unlike the Sun and the other stars, the Moon emits no light of its own. Instead, it shines by reflected sunlight, giving rise to the phases we see. As indicated in Figure 0.13(a), half of the Moon's surface is illuminated by the Sun at any moment, but not all of the Moon's sunlit face can be seen because of the Moon's position with respect to Earth and the Sun. When the Moon is full, we see the entire "day lit" face because the Sun and the Moon are in opposite directions from Earth in the sky. The Sun's light is not blocked by Earth at the full phase because, as shown in Figure 0.13(b), the Moon's orbit is inclined at a small angle (5.2°) to the plane of the

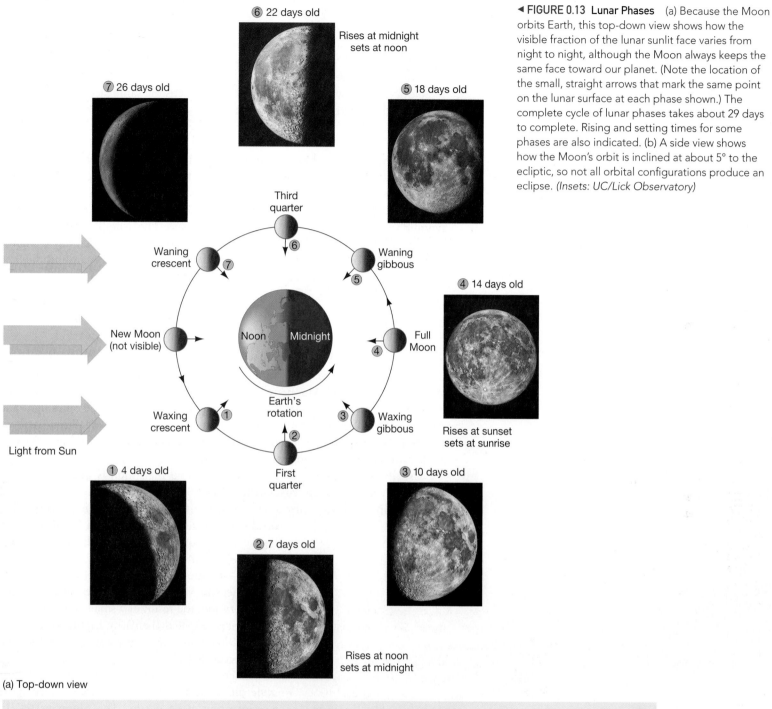

⑥ 22 days old

Rises at midnight
sets at noon

⑦ 26 days old

⑤ 18 days old

Third
quarter

Waning
crescent

Waning
gibbous

New Moon
(not visible)

Noon   Midnight

Full
Moon

④ 14 days old

Earth's
rotation

Light from Sun

Waxing
crescent

Waxing
gibbous

Rises at sunset
sets at sunrise

① 4 days old

First
quarter

③ 10 days old

② 7 days old

Rises at noon
sets at midnight

(a) Top-down view

Moon's shadow

Earth's shadow

New
Moon

5.2°

Earth

Full
Moon

(b) Side view

◀ FIGURE 0.13 Lunar Phases (a) Because the Moon orbits Earth, this top-down view shows how the visible fraction of the lunar sunlit face varies from night to night, although the Moon always keeps the same face toward our planet. (Note the location of the small, straight arrows that mark the same point on the lunar surface at each phase shown.) The complete cycle of lunar phases takes about 29 days to complete. Rising and setting times for some phases are also indicated. (b) A side view shows how the Moon's orbit is inclined at about 5° to the ecliptic, so not all orbital configurations produce an eclipse. (Insets: UC/Lick Observatory)

ecliptic, so the alignment of the three bodies is not perfect. (The sizes of Earth and the Moon are greatly exaggerated in these figures.) In the case of a new Moon, the Moon and the Sun are in nearly the same part of the sky, and the sunlit side of the Moon is oriented away from us—at new Moon the Sun is almost behind the Moon, from our perspective.

Notice, by the way, that the Moon always keeps the same face toward Earth—as indicated on the figure, it rotates on its axis in exactly the same time it takes to orbit Earth. This is called *synchronous rotation*. We will discuss the reason for it in Chapter 5.

As it revolves around Earth, the Moon's position in the sky changes with respect to the stars. In one **sidereal month** (27.3 days), the Moon completes one revolution and returns to its starting point on the celestial sphere, having traced out a great circle in the sky. The time required for the Moon to complete a full cycle of phases, one **synodic month,** is a little longer—about 29.5 days. The synodic month is a little longer than the sidereal month for the same basic reason that a solar day is slightly longer than a sidereal day (Figure 0.7): Because of Earth's motion around the Sun, the Moon must complete slightly more than one full revolution to return to the same phase in its orbit.

Eclipses

From time to time—but only at new or full Moon—the Sun, Earth, and the Moon line up precisely and we observe the spectacular phenomenon known as an **eclipse**. When the Sun and the Moon are in exactly *opposite* directions as seen from Earth, Earth's shadow sweeps across the Moon, temporarily blocking the Sun's light and darkening the Moon in a **lunar eclipse,** as illustrated in Figure 0.14. From Earth we see the curved edge of Earth's shadow begin to cut across the face of the full Moon and slowly "eat" its way into the circular lunar disk.

Usually the alignment of the Sun, Earth, and the Moon is imperfect, so the shadow never completely covers the Moon. Such an occurrence is known as a **partial eclipse.** Occasionally, however, the entire lunar surface is obscured in a **total eclipse** (such as that shown in the inset in Figure 0.14). Total lunar eclipses last only as long as is needed for the Moon to pass through Earth's shadow—no more than about 100 minutes. During that time, the Moon often acquires an eerie, deep red coloration, the result of a small amount of sunlight being refracted (bent) by Earth's atmosphere onto the lunar surface, preventing the shadow from being completely black.

When the Moon and the Sun are in exactly the *same* direction as seen from Earth, an even more awe-inspiring event occurs. The Moon passes directly in front of the Sun, briefly turning day into night in a **solar eclipse** (Figure 0.15). In a *total solar eclipse,* when the alignment is perfect, planets and some stars become visible in the daytime as the Sun's light is blocked. By pure chance the Sun and Moon have almost exactly the same angular size as seen from Earth—the Sun is much bigger than the Moon, but it also lies much farther away (see *More Precisely 0-1*). Consequently, during a total solar eclipse, we can often see the Sun's ghostly outer atmosphere, or *corona*, its faint glow becoming temporarily visible when the rest of the Sun's glare is

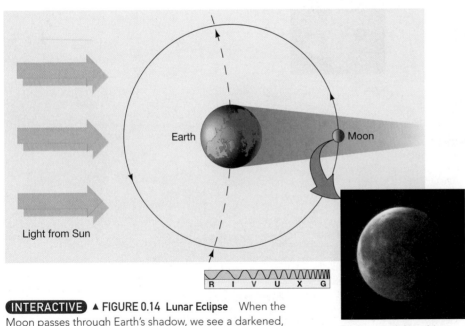

R I V U X G

INTERACTIVE ▲ FIGURE 0.14 Lunar Eclipse   When the Moon passes through Earth's shadow, we see a darkened, copper-colored Moon, as shown by the partial eclipse in the inset photograph. The red coloration is caused by sunlight deflected by Earth's atmosphere onto the Moon's surface. (Inset: G. Schneider)

This is an actual photo of the eclipsed Moon, one of the great light shows visible to the naked eye.

▶ **FIGURE 0.15 Solar Eclipse**   During a total solar eclipse, the Sun's corona becomes visible as an irregularly shaped halo surrounding the blotted-out disk of the Sun. This was the August 1999 eclipse, as seen from the banks of the Danube River near Sofia, Bulgaria. *(Bencho Angelov)*

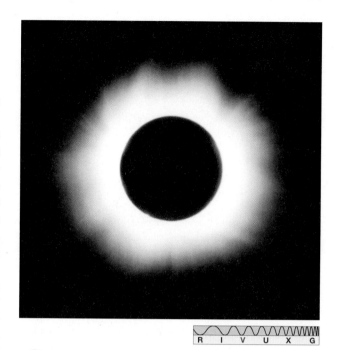

R   I   V   U   X   G

MA **ANIMATION/VIDEO** Solar Eclipse in Indiana

CONCEPT CHECK ▶ If Earth's distance from the Sun were to double, would you still expect to see total solar eclipses? What if the distance became half its present value?

obscured. In a *partial solar eclipse,* the Moon's path is slightly off center, and only a portion of the Sun's face is covered.

Unlike a lunar eclipse, which is simultaneously visible from all locations on Earth's night side, a total solar eclipse can be seen from only a small portion of the daytime side. The Moon's shadow on Earth's surface is about 7000 kilometers wide—roughly twice the diameter of the Moon. Outside that shadow, no eclipse is seen. However, only within the central region of the shadow, in the **umbra,** is the eclipse total. The umbra is the part of a shadow (in this case, the Moon's) where all light from the source is blocked. Within the shadow but outside the umbra, in the **penumbra,** some but not all of the Sun's light is blocked and the eclipse is partial, with less and less of the Sun being obscured the farther one travels from the shadow's center.

The connections between the umbra, the penumbra, and the relative locations of Earth, Sun, and Moon are illustrated in Figure 0.16. The umbra is always very small—even under the most favorable circumstances, its diameter never exceeds 270 kilometers. Because the Moon's shadow sweeps across Earth's surface at a speed of more than 1700 kilometers per hour, the duration of a total solar eclipse at any given point can never exceed 7.5 minutes (270 km divided by 1700 km/hour, times 60 minutes per hour).

The Moon's orbit around Earth is not exactly circular. As a result, the Moon may be far enough from Earth at the moment of an eclipse that its disk fails to cover the disk of the Sun completely, even though their centers coincide. In that case, there is no *region of totality*—the umbra never reaches Earth at all,

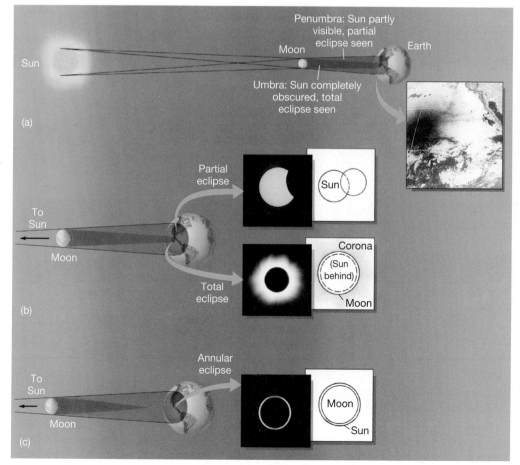

INTERACTIVE ▶ **FIGURE 0.16 Solar Eclipse Types** (a) The Moon's shadow consists of two parts: the umbra, where no sunlight is seen, and the penumbra, where a portion of the Sun is visible. (b) Situated in the umbra, we see a total eclipse; in the penumbra, we see a partial eclipse. (c) If the Moon is too far from Earth at the moment of the eclipse, the umbra does not reach Earth and there is no region of totality; instead, an annular eclipse is seen. (Note that these figures are not drawn to scale.) **MA** *(Insets: NOAA; G. Schneider)*

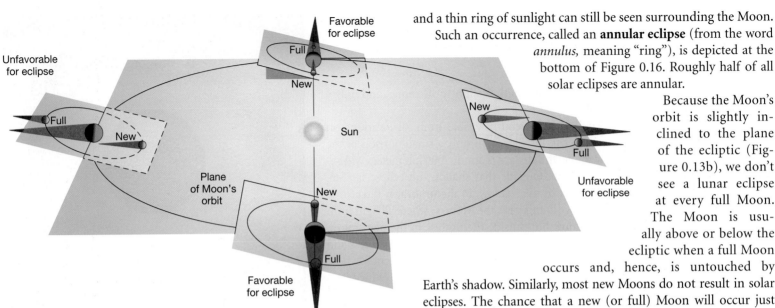

and a thin ring of sunlight can still be seen surrounding the Moon. Such an occurrence, called an **annular eclipse** (from the word *annulus,* meaning "ring"), is depicted at the bottom of Figure 0.16. Roughly half of all solar eclipses are annular.

Because the Moon's orbit is slightly inclined to the plane of the ecliptic (Figure 0.13b), we don't see a lunar eclipse at every full Moon. The Moon is usually above or below the ecliptic when a full Moon occurs and, hence, is untouched by Earth's shadow. Similarly, most new Moons do not result in solar eclipses. The chance that a new (or full) Moon will occur just as the Moon happens to cross the ecliptic plane (so Earth, Moon, and Sun are perfectly aligned, as illustrated in Figure 0.17) is quite low. As a result, eclipses are relatively infrequent events. On average, there are 7 total lunar eclipses and 15 total or annular solar eclipses each decade. Because we know the orbits of Earth and the Moon to great accuracy, we can predict eclipses far into the future. Figure 0.18 shows the location and duration of all total and annular eclipses of the Sun between 2010 and 2030.

▲ **FIGURE 0.17 Eclipse Geometry** An eclipse occurs when Earth, Moon, and Sun are precisely aligned. If the Moon's orbital plane lay in exactly the plane of the ecliptic, this alignment would occur once a month. However, the Moon's orbit is inclined at about 5° to the ecliptic, so not all configurations are actually favorable for producing an eclipse. For an eclipse to occur, the line of intersection of the two planes must lie along the Earth–Sun line. For clarity, only the umbra of each shadow is shown (see Figure 0.16).

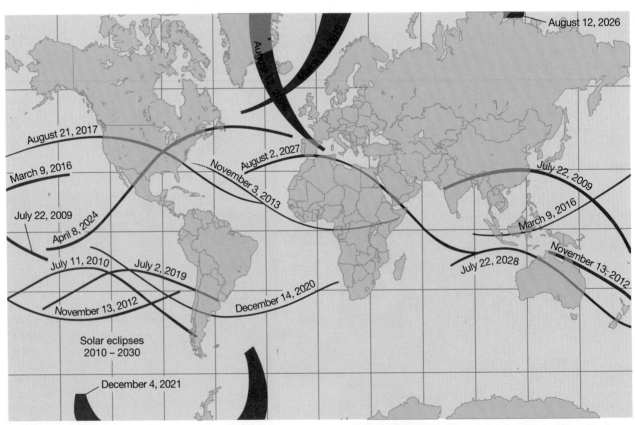

▲ **FIGURE 0.18 Eclipse Tracks** Regions of Earth that saw or will see total solar eclipses between the years 2009 and 2030. Each track represents the path of the Moon's umbra across Earth's surface during an eclipse. High-latitude tracks are broader because sunlight strikes Earth's surface at an oblique angle near the poles and because of the projection of the map.

▶ **FIGURE 0.19 Triangulation** Surveyors often use simple geometry to estimate the distance to a faraway object by triangulation. By measuring the angles at A and B and the length of the baseline, the distance can be calculated without the need for direct measurement (or getting wet!).

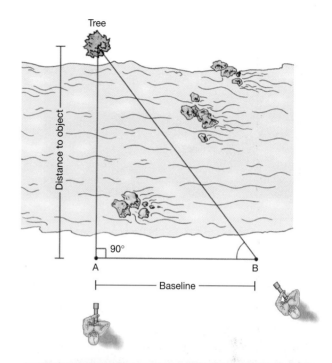

0.4 The Measurement of Distance

So far we have considered only the *directions* to the Sun, Moon, and stars, as seen from Earth. But knowing the direction in which an object lies is only part of the information needed to locate it in space. Before we can make a systematic study of the heavens, we must find a way of measuring *distances*, too. One distance-measurement method, called **triangulation**, is based on the principles of Euclidean geometry and finds widespread application today in both terrestrial and astronomical settings. Modern engineers, especially surveyors, use this age-old geometrical idea to measure indirectly the distance to faraway objects. In astronomy, it forms the foundation of the family of distance-measurement techniques that together make up the **cosmic distance scale.**

Imagine trying to measure the distance to a tree on the other side of a river. The most direct method would be to lay a tape across the river, but that's not always practical. A smart surveyor would make the measurement by visualizing an imaginary triangle (hence, the term *triangulation*), sighting the tree (measuring its direction) on the far side of the river from two positions on the near side, as illustrated in Figure 0.19. The simplest triangle is a right triangle, in which one of the angles is exactly 90°, so it is often convenient to set up one observation position directly opposite the object, as at point A, although this isn't necessary for the method to work. The surveyor then moves to another observation position at point B, noting the distance covered between A and B. This distance is called the **baseline** of the imaginary triangle. Finally, the surveyor, standing at point B, sights toward the tree and notes the angle formed at point B by the intersection of this sight line and the baseline. No further observations are required. Knowing the length of one side (AB) and two angles (the right angle at A and the angle at B) of the triangle, and having some knowledge of elementary trigonometry, the surveyor can construct the remaining sides and angles and so establish the distance from A to the tree.

Obviously, for a fixed baseline, the triangle becomes longer and narrower as the tree's distance from A increases. Narrow triangles cause problems because it is hard to measure the angles at A and B with sufficient accuracy. A surveyor on Earth can "fatten" the triangle by lengthening the baseline, but in astronomy there are limits on how long a baseline we can choose. For example, consider an imaginary triangle extending from Earth to a nearby object in space, perhaps a neighboring planet. The triangle is now extremely long and narrow, even for a relatively nearby object (by cosmic standards). Figure 0.20(a) illustrates a case in which the longest baseline possible on Earth—Earth's diameter, measured from point A to point B—is used. In principle, two observers could sight the planet from opposite sides of Earth, measuring the triangle's angles at A and B. However, in practice it is easier to measure the third angle of the imaginary triangle. Here's how.

The observers each sight toward the planet, taking note of its position *relative to some distant stars* seen on the plane of the sky. The observer at A sees the planet at apparent location A′ (pronounced "A prime") relative to those stars, as indicated in Figure 0.20(a). The observer at B sees the planet at location B′. If each observer takes a photograph of the same region of the sky, the planet will appear at slightly

INTERACTIVE ▶ **FIGURE 0.20 Parallax** (a) A triangle can be imagined to extend from Earth to a nearby object in space. The group of stars at the top represents a background field of very distant stars. (b) Hypothetical photographs of the same star field showing the nearby object's apparent displacement, or shift, relative to the distant undisplaced stars.

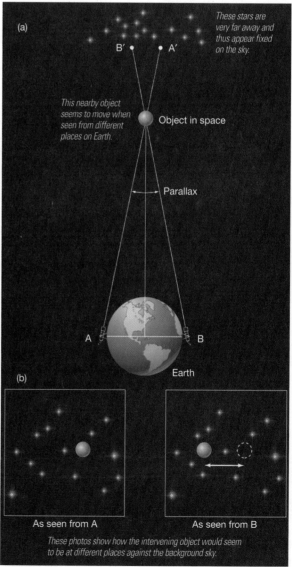

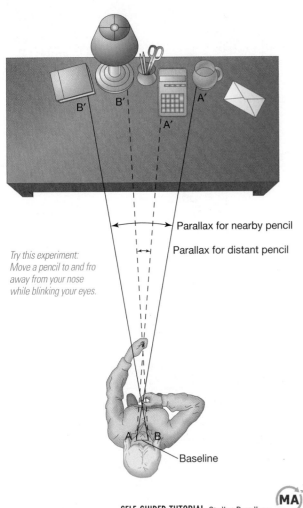

Parallax for nearby pencil

Parallax for distant pencil

Try this experiment: Move a pencil to and fro away from your nose while blinking your eyes.

Baseline

SELF-GUIDED TUTORIAL Stellar Parallax **(MA)**

CONCEPT CHECK ▶ Why is elementary geometry essential for measuring distances in astronomy?

▲ **FIGURE 0.21 Parallax Geometry** Parallax is inversely proportional to an object's distance. An object near your nose has a much larger parallax than an object held at arm's length.

different places in the two images, as shown in Figure 0.20(b). (The positions of the background stars appear unchanged because of their much greater distance from the observer.) This apparent displacement (shift) of a foreground object relative to the background as the observer's location changes is known as **parallax**. The size of the shift in Figure 0.20(b), measured as an angle on the celestial sphere, is equal to the third angle of the imaginary triangle in Figure 0.20(a).

The closer an object is to the observer, the larger the parallax. To see this for yourself, hold a pencil vertically just in front of your nose (see Figure 0.21). Look at some far-off object—a distant wall, say. Close one eye, then open it while closing the other. You should see a large shift in the apparent position of the pencil relative to the wall—a large parallax. In this example, one eye corresponds to point A in Figure 0.20, the other eye to point B. The distance between your eyeballs is the baseline, the pencil represents the planet, and the distant wall the remote field of stars. Now hold the pencil at arm's length, corresponding to a more distant object (but still not as far away as the distant stars). The apparent shift of the pencil will be smaller. By moving the pencil farther away, you are narrowing the triangle and decreasing the parallax. If you were to paste the pencil to the wall, corresponding to the case where the object of interest is as far away as the background star field, blinking would produce no apparent shift of the pencil at all.

The amount of parallax is inversely proportional to an object's distance. Small parallax implies large distance. Conversely, large parallax implies small distance. Knowing the amount of parallax (as an angle) and the length of the baseline, we can easily derive the distance through triangulation.

Surveyors of the land routinely use these simple geometric techniques to map out planet Earth (*Discovery 0-1* presents an early example). As surveyors of the sky, astronomers use the same basic principles to chart the universe.

0.5 Science and the Scientific Method

Science is a step-by-step process for investigating the physical world, based on natural laws and observed phenomena. However, the scientific facts just presented did not come easily or quickly. Progress in science is often slow and intermittent and may require a great deal of patience before significant progress is made. The earliest known descriptions of the universe were based largely on imagination and mythology and made little attempt to explain the workings of the heavens in terms of testable earthly experience. However, history shows that some early scientists did come to realize the importance of careful observation and testing to the formulation of their ideas. The success of their approach changed, slowly but surely, the way science was done and opened the door to a fuller understanding of nature. Experimentation and observation became central parts of the process of inquiry.

To be effective, a **theory**—the framework of ideas and assumptions used to explain some set of observations and make predictions about the real world—must be continually tested. Scientists accomplish this by using a theory to construct a **theoretical model** of a physical object (such as a planet or a star) or phenomenon (such as gravity or light), accounting for its known properties. The model then makes further predictions about the object's properties or perhaps how it might behave or change under new circumstances. If experiments and observations favor those predictions, the theory can be further developed and refined. If they do not, the theory must be reformulated or rejected, no matter how appealing it originally seemed. The process is illustrated schematically in Figure 0.22. This approach to investigation, combining thinking and doing—that is, theory and experiment—is known as the **scientific method.** It lies at the heart of modern science, separating science from pseudoscience, fact from fiction.

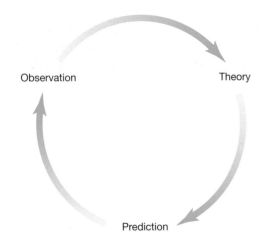

Observation

Theory

Prediction

▲ **FIGURE 0.22 Scientific Method** Scientific theories evolve through a combination of observation, theory, and prediction, which in turn suggests new observations. The process can begin at any point in the cycle, and it continues forever—or until the theory fails to explain an observation or makes a demonstrably false prediction.

0-1 DISCOVERY

Sizing up Planet Earth

In about 200 B.C. a Greek philosopher named Eratosthenes (276–194 B.C.) used simple geometric reasoning to calculate the size of our planet. He knew that at noon on the first day of summer, observers in the city of Syene (now called Aswan), in Egypt, saw the Sun pass directly overhead. This was evident from the fact that vertical objects cast no shadows and sunlight reached to the very bottoms of deep wells, as shown in the insets in the accompanying figure. However, at noon of the same day in Alexandria, a city 5000 *stadia* to the north, the Sun was seen to be displaced slightly from the vertical. The *stadium* was a Greek unit of length, roughly equal to 0.16 km—the modern town of Aswan lies about 780 (5000 × 0.16) km south of Alexandria. Using the simple technique of measuring the length of the shadow of a vertical stick and applying elementary geometry, Eratosthenes determined the angular displacement of the Sun from the vertical at Alexandria to be 7.2°.

What could have caused this discrepancy between the two measurements? As illustrated in the figure, the explanation is simply that Earth's surface is not flat, but *curved*. Our planet is a sphere. Eratosthenes was not the first to realize that Earth is spherical—the philosopher Aristotle had done that over 100 years earlier (Section 0.5)—but he was apparently the first to build on this knowledge, combining geometry with direct measurement to infer our planet's size. Here's how he did it.

Rays of light reaching Earth from a very distant object, such as the Sun, travel almost parallel to one another. Consequently, as shown in the figure, the angle measured at Alexandria between the Sun's rays and the vertical (that is, the line joining Alexandria to the center of Earth) is equal to the angle between Syene and Alexandria, as seen from Earth's center. (For the sake of clarity, this angle has been exaggerated in the drawing.) The size of this angle in turn is proportional to the fraction of Earth's circumference that lies between Syene and Alexandria:

$$\frac{7.2°}{360°} = \frac{5000 \text{ stadia}}{\text{Earth's circumference}}.$$

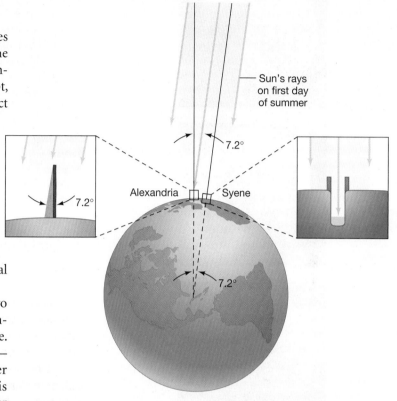

Earth's circumference is therefore 50 × 5000 or 250,000 stadia, or about 40,000 km. Earth's radius is therefore $250,000/2\pi$ stadia, or 6366 km. The correct values for Earth's circumference and radius, now measured accurately by orbiting spacecraft, are 40,070 km and 6378 km, respectively.

Eratosthenes' reasoning was a remarkable accomplishment. More than 20 centuries ago he estimated the circumference of Earth to within 1 percent accuracy, using only simple geometry. A person making measurements on only a small portion of Earth's surface was able to compute the size of the entire planet on the basis of observation and pure logic—an early triumph of scientific reasoning.

Notice that there is no end point to the process depicted in Figure 0.22. A theory can be invalidated by a single wrong prediction, but no amount of observation or experimentation can ever prove it "correct." Theories simply become more and more widely accepted as their predictions are repeatedly confirmed. The process can fail at any point in the cycle. If a theory cannot explain an experimental result or observation, or if its predictions are demonstrated to be untrue, it must be discarded or amended. And if it makes no predictions at all, then it has no scientific value.

Scientific theories share several important defining characteristics:

● They must be *testable;* that is, they must admit the possibility that both their underlying assumptions and their predictions can be exposed to experimental verification. This feature separates science from, for example, religion, since ultimately divine revelations or scriptures cannot be challenged within a religious framework: We can't design an experiment to "understand the mind of God." Testability also distinguishes science from a pseudoscience such as astrology, whose underlying assumptions and predictions have been repeatedly tested and never verified, with no apparent impact on the views of those who continue to believe in it!

- They must continually be *tested,* and their consequences tested, too. This is the basic circle of scientific progress depicted in Figure 0.22.

- They should be *simple.* This is a practical outcome of centuries of scientific experience—the most successful theories tend to be the simplest ones that fit the facts. This view is often encapsulated in a principle known as *Occam's razor* (after the 14th-century English philosopher William of Ockham): If two competing theories both explain the facts and make the same predictions, then the simpler one is better. A good theory should contain no more complexity than is absolutely necessary.

- Finally, most scientists have the additional bias that a theory should in some sense be *elegant.* When a clearly stated simple principle naturally explains several phenomena previously thought to be distinct, this is widely regarded as a strong point in favor of the new theory.

You may find it instructive to apply these criteria to the many physical theories— some old and well established, others much more recent and still developing—we will encounter throughout the text.

The notion that theories must be tested and may be proven wrong sometimes leads people to minimize their importance. We have all heard the expression, "Of course, it's only a theory," used to deride or dismiss an idea that someone finds unacceptable. Don't be fooled! Gravity (see Section 1.4) is "only" a theory, but calculations based on it have guided human spacecraft throughout the solar system. Electromagnetism and quantum mechanics (see Chapter 2) are theories, too, yet they form the foundation for most of 20th- (and 21st-) century technology. Facts about the universe are a dime a dozen. Theories are the intellectual "glue" that combine seemingly unrelated facts into a coherent and interconnected whole.

The birth of modern science is usually associated with the Renaissance, the historical period from the late 14th to the mid-17th century that saw a rebirth (*renaissance* in French) of artistic, literary, and scientific inquiry in European culture following the chaos of the Dark Ages. However, one of the first documented uses of the scientific method in an astronomical context was by Aristotle (384–322 B.C.) some 17 centuries earlier. Aristotle is not normally remembered as a strong proponent of this approach—many of his best-known ideas were based on pure thought, with no attempt at experimental test or verification. Still, his brilliance extended into many areas now thought of as modern science. He noted that, during a lunar eclipse (Section 0.3), Earth casts a curved shadow onto the surface of the Moon. Figure 0.23 shows a series of photographs taken during a recent lunar eclipse. Earth's shadow, projected onto the Moon's surface, is indeed slightly curved. This is what Aristotle must have seen and recorded so long ago.

Because the observed shadow seemed always to be an arc of the same circle, Aristotle theorized that Earth, the cause of the shadow, must be round. Don't underestimate the scope of this apparently simple statement. Aristotle also had to reason that the dark region was indeed a shadow and that Earth was its cause—facts we regard as obvious today, but far from clear 25 centuries ago. On the basis of this *hypothesis*—one possible explanation of the observed facts—he then predicted that any and all future lunar eclipses would show Earth's shadow to be curved, regardless of our planet's orientation. That prediction has been tested every time a lunar eclipse has occurred. It has yet to be proved wrong. Aristotle was not the first person to argue that Earth is round, but he was apparently the first to offer observational proof using the lunar-eclipse method.

▲ FIGURE 0.23 A Lunar Eclipse   These photographs show Earth's shadow sweeping across the Moon during an eclipse. Aristotle reasoned that Earth was the cause of the shadow and concluded that Earth must be round. *(G. Schneider)*

Today, scientists throughout the world use an approach that relies heavily on testing ideas. They gather data, form a working hypothesis that explains the data, and then proceed to test the implications of the hypothesis using experiment and observation. Eventually, one or more "well-tested" hypotheses may be elevated to the stature of a physical law and come to form the basis of a theory of even broader applicability. The new predictions of the theory will in turn be tested, as scientific knowledge continues to grow. Experiment and observation are integral parts of the process of scientific inquiry. Untestable theories, or theories unsupported by experimental evidence, rarely gain any measure of acceptance in scientific circles. *Observation, theory,* and *testing*—these are the cornerstones of the scientific method, a technique whose power will be demonstrated again and again throughout our text.

Don't think that the scientific method is perfect. We will see many examples in this book of how even good scientists make mistakes—following incorrect lines of reasoning, or perhaps placing too much faith in faulty observations. Nevertheless, used properly over a period of time, this rational, methodical approach enables us to arrive at conclusions that are mostly free of the personal bias and human failings of any one scientist. The scientific method is designed to yield—eventually—an objective view of the universe we inhabit.

CONCEPT CHECK ▶ Can a theory ever become a "fact," scientifically speaking?

CHAPTER REVIEW

SUMMARY

LO1 The **universe** (p. 4) is the totality of all space, time, matter, and energy. **Astronomy** (p. 4) is the study of the universe. In order of increasing size, the basic constituents of the cosmos are **planets** (p. 4), **stars** (p. 4), **galaxies** (p. 4), and the universe itself. They differ enormously in scale—a factor of a billion billion from planet Earth to the entire observable universe.

LO2 Early observers grouped the stars visible to the naked eye into patterns called **constellations** (p. 4), which they imagined were attached to a vast **celestial sphere** (p. 5) centered on Earth. Constellations have no physical significance, but are still used to label regions of the sky. **Celestial coordinates** (p. 6) are a more precise way of specifying a star's location on the celestial sphere.

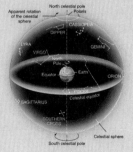

LO3 The nightly motion of the stars across the sky is the result of Earth's **rotation** (p. 5) on its axis. Because of Earth's **revolution** (p. 7) around the Sun, we see different stars at night at different times of the year. The Sun's

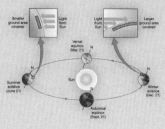

apparent yearly path around the celestial sphere (or the plane of Earth's orbit around the Sun) is called the **ecliptic** (p. 8). We experience **seasons** (p. 9) because Earth's rotation axis is inclined to the ecliptic plane. At the **summer solstice** (p. 8), the Sun is highest in the sky and the length of the day is greatest. At the **winter solstice** (p. 9), the Sun is lowest and the day is shortest. Because of **precession** (p. 11), the orientation of Earth's axis changes slowly over the course of thousands of years.

LO4 As the Moon orbits Earth, it keeps the same face permanently turned toward our planet. We see lunar **phases** (p. 12) as the fraction of the Moon's sunlit face visible to us varies. A **lunar eclipse**

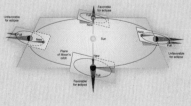

(p. 14) occurs when the Moon enters Earth's shadow. A **solar eclipse** (p. 14) occurs when the Moon passes between Earth and the Sun. An eclipse may be **total** (p. 14) if the body in question (Moon or Sun) is completely obscured, or **partial** (p. 14) if only a portion of the surface is affected. If the Moon happens to be too far from Earth for its disk to completely hide the Sun, an **annular eclipse** (p. 16) occurs. Because the Moon's orbit around Earth is slightly inclined to the ecliptic, solar and lunar eclipses are relatively rare events.

 **L05** Astronomers use **triangulation** (p. 17) to measure the distances to planets and stars, forming the foundation of the **cosmic distance scale** (p. 17), the family of distance-measurement techniques used to chart the universe. **Parallax** (p. 18) is the apparent motion of a foreground object relative to a distant background as the observer's position changes. The larger the **baseline** (p. 17)—the distance between the two observation points—the greater the parallax.

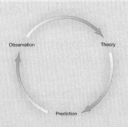 **L06** **Science** (p. 18) is a step-by-step process for investigating the physical world. The **scientific method** (p. 18) is a methodical approach employed by scientists to explore the universe around us in an objective manner. A **theory** (p. 18) is a framework of ideas and assumptions used to explain some set of observations and make predictions about the real world. These predictions in turn are amenable to further observational testing. In this way, the theory expands and science advances.

MasteringAstronomy®   For instructor-assigned homework go to www.masteringastronomy.com

Problems labeled POS explore the process of science. VIS problems focus on reading and interpreting visual information. LO connects to the introduction's numbered Learning Outcomes.

REVIEW AND DISCUSSION

1. **L01** Compare the size of Earth with that of the Sun, the Milky Way Galaxy, and the entire universe.

2. **L02** What is a constellation? Why are constellations useful for mapping the sky?

3. Why does the Sun rise in the east and set in the west each day? Does the Moon also rise in the east and set in the west? Why? Do stars do the same? Why?

4. How and why does a day measured by the Sun differ from a day measured by the stars?

5. How many times in your life have you orbited the Sun?

6. Why do we see different stars at different times of the year?

7. **L03** Why are there seasons on Earth?

8. What is precession, and what is its cause?

9. **L04** If one complete hemisphere of the Moon is always lit by the sun, why do we see different phases of the Moon?

10. What causes a lunar eclipse? A solar eclipse? Why aren't there lunar and solar eclipses every month?

11. **POS** Do you think an observer on another planet might see eclipses? Why or why not?

12. What is parallax? Give an everyday example.

13. Why is it necessary to have a long baseline when using triangulation to measure the distances to objects in space?

14. **L05** What two pieces of information are needed to determine the diameter of a faraway object?

15. **L06 VIS** What is the scientific method? In what ways does science differ from religion?

CONCEPTUAL SELF-TEST: TRUE OR FALSE?/MULTIPLE CHOICE

1. The Milky Way Galaxy is about 1 million times larger than Earth. (T/F)

2. The stars in a constellation are physically close to one another. (T/F)

3. The solar day is longer than the sidereal day. (T/F)

4. The seasons are caused by the precession of Earth's axis. (T/F)

5. A lunar eclipse can occur only during the full phase. (T/F)

6. The angular diameter of an object is inversely proportional to its distance from the observer. (T/F)

7. If we know the distance of an object from Earth, we can determine the object's size by measuring its parallax. (T/F)

8. If Earth rotated twice as fast as it currently does, but its motion around the Sun stayed the same, then (a) the night would be twice as long; (b) the night would be half as long; (c) the year would be half as long; (d) the length of the day would be unchanged.

9. A long, thin cloud that stretched from directly overhead to the western horizon would have an angular size of (a) 45°; (b) 90°; (c) 180°; (d) 360°.

10. When a thin crescent of the Moon is visible just before sunrise, the Moon is in its (a) waxing phase; (b) new phase; (c) waning phase; (d) quarter phase.

11. If the Moon's orbit were a little larger, solar eclipses would be (a) more likely to be annular; (b) more likely to be total; (c) more frequent; (d) unchanged in appearance.

12. If the Moon orbited Earth twice as fast, but in the same orbit, the frequency of solar eclipses would (a) double; (b) be cut in half; (c) stay the same.

13. **VIS** According to Figure 0.8 (The Zodiac), in January the Sun is in the constellation (a) Cancer; (b) Gemini; (c) Leo; (d) Aquarius.

14. **VIS** In Figure 0.19 (Triangulation), using a longer baseline would result in (a) a less accurate distance to the tree; (b) a more accurate distance to the tree; (c) a smaller angle at point B; (d) a greater distance across the river.

15. **VIS** In Figure 0.20 (Parallax), a smaller Earth would result in (a) a smaller parallax angle; (b) a shorter distance measured to the object; (c) a larger apparent displacement; (d) stars appearing closer together.

PROBLEMS

The number of squares preceding each problem indicates its approximate level of difficulty.

1. ■ The vernal equinox is now just entering the constellation Aquarius. In what constellation will it lie in the year A.D. 10,000?

2. ■ Given that the distance from Earth to the Sun is 150,000,000 km, through what distance does Earth move in a second? An hour? A day?

3. ■■■ How, and by roughly how much, would the length of the solar day change if Earth's rotation were suddenly to reverse direction?

4. ■■ Through how many degrees, arc minutes, or arc seconds does the Moon move in (a) 1 hour, (b) 1 minute, (c) 1 second? How long does it take for the Moon to move a distance equal to its own diameter of angular 0.5°?

5. ■■ Given that the distance to the Moon is 384,000 km, and taking the Moon's orbit around Earth to be circular, estimate the speed (in kilometers per second) at which the Moon orbits Earth.

6. ■■■ Use reasoning similar to that illustrated in Figure 0.7 to verify that the length of the synodic month (the time from one full Moon to the next; Section 0.3) is 29.5 days.

7. ■ The baseline in Figure 0.19 is 100 m and the angle at B is 60°. By constructing the triangle on a piece of graph paper, determine the distance from A to the tree.

8. ■■ Use reasoning similar to that in *Discovery 0-1* (but now using a circle centered on the object and containing the baseline) to determine the distance to an object if its parallax, as measured from either end of a 1000-km baseline, is (a) 1°; (b) 1′; (c) 1″?

9. ■ What would the measured angle in *Discovery 0-1* have been if Earth's circumference were 100,000 km instead of 40,000 km?

10. ■ What angle would Eratosthenes have measured (see *Discovery 0-1*) had Earth been flat?

ACTIVITIES

Collaborative

1. Measure the nightly and monthly motion of the Moon. On a clear night, sketch a 10-degree-wide patch of the sky containing the Moon, with the Moon initially toward the west side of the patch. (See Individual Activity 3 below for how to estimate angles on the sky.) Repeat the observation of the same collection of stars every hour over the course of a night (take turns!). You will see that the Moon's position relative to the stars changes noticeably even in a few hours. What is the Moon's angular speed (in degrees per hour)? Now observe the Moon at the *same* time each night over the course of a month. Sketch its appearance and note its position on the sky each night. Can you interpret its changing phase in terms of the relative positions of Earth, the Sun, and the Moon? (See Figure 0.13.)

2. Consider Figure 0.18, which shows solar eclipse paths on a world map. As a group, write a description of which two eclipses you would most like to observe together and where and when you would go to observe them. Explain why you chose the dates and sites you did.

Individual

1. Find the star Polaris, also known as the North Star, in the evening sky. Identify any separate pattern of stars in the same general vicinity of the sky. Wait several hours, at least until after midnight, and then locate Polaris again. Has Polaris moved? What has happened to the nearby pattern of stars? Why?

2. Consider the curved star trails shown in Figure 0.5, a time-lapse photograph of the northern sky. They are arcs of circles centered on Polaris. What was the exposure time used for the photograph? How long would you need to take a similar picture from (a) the planet Mercury? (b) Jupiter's moon Europa? (Jump ahead to Chapters 6 or 8 to find the rotation periods of these bodies.)

3. Hold your little finger out at arm's length. Can you cover the disk of the Moon? The Moon projects an angular size of 30′ (half a degree); your finger should more than cover it. You can use this fact to make some basic sky measurements. As a simple rule, your little finger at arm's length is about 1 degree across, your middle three fingers are about 4 degrees across, and your clenched fist is about 10 degrees across. If the constellation Orion is visible, use this information to estimate the angular size of Orion's belt and the angular distance between Betelgeuse and Rigel. Compare your findings with Figure 0.2(a).

1

The Copernican Revolution

The Birth of Modern Science

Living in the Space Age, we have become accustomed to the modern view of our place in the universe. Images of Earth taken from space leave little doubt that we inhabit just one planet among several, and no one seriously questions the idea that we orbit the Sun. Yet there was a time, not so long ago, when our ancestors maintained that Earth had a special role in the cosmos and lay at the center of all things. Our view of the universe—and of ourselves—has undergone a radical transformation since those early days. Humankind has been torn from its throne at the center of the cosmos and relegated to an unremarkable position on the periphery of the Milky Way Galaxy. But in return we have gained a wealth of scientific knowledge. The story of how this came about is the story of the rise of science and the genesis of modern astronomy.

MasteringAstronomy® Visit www.masteringastronomy.com for quizzes, animations, videos, interactive figures, and self-guided tutorials.

◄ Exploration is at the heart of the modern scientific method used by all scientists around the world. Ideas must be tested against what is observed in nature, and those ideas that fail the test are discarded. In this way, scientists generate progressively better knowledge in the quest to understand the universe. Here, in a colorized piece of historical artwork, a young Nicholas Copernicus is observing a lunar eclipse in Rome in the year 1500. Perhaps more than anyone else, this Polish astronomer began the revolution that overthrew more than a thousand years of philosophical thinking that claimed Earth to be the immovable center of all things. (*S. Terry*; engraving from the 1975 edition of *Vies des Savants Illustres*)

1.1 The Motions of the Planets

Over the course of a night, the stars slide smoothly across the sky. ∞ *(Sec. 0.1)* Over the course of a month, the Moon moves smoothly and steadily along its path in the sky relative to the stars, passing through its familiar cycle of phases. ∞ *(Sec. 0.3)* Over the course of a year, the Sun progresses along the ecliptic at an almost constant rate, varying little in brightness from day to day. ∞ *(Sec. 0.2)* In short, the behavior of the Sun, Moon, and stars seems fairly simple and orderly. This basic predictability of the night sky provided many ancient cultures with a means of tracking the seasons and organizing their activities (see *Discovery 1-1*).

But ancient astronomers were also aware of five other bodies in the sky—the planets Mercury, Venus, Mars, Jupiter, and Saturn—whose behavior was not so easy to grasp. The explanation for their motion would change forever our perception of the universe and our place in it.

Wanderers in the Heavens

Planets do not behave in nearly as orderly a fashion as do the Sun, Moon, and stars. They vary in brightness and don't maintain a fixed position in the sky. Instead, they seem to wander irregularly around the celestial sphere—the word *planet* derives from the Greek word *planetes,* meaning "wanderer." Planets never stray far from the ecliptic and generally traverse the celestial sphere from west to east, as does the Sun. However, they seem to speed up and slow down as they go and at times even appear to loop back and forth relative to the stars, as shown in Figure 1.1. Astronomers refer to the planets' eastward motion as *direct,* or *prograde,* motion. The backward (westward) loops are **retrograde motion.**

Unlike the Sun and stars, but like the Moon, the planets produce no visible light of their own. Instead, they shine by reflected sunlight. Ancient astronomers correctly reasoned that the apparent brightness of a planet in the night sky is related to its distance from Earth—a planet appears brightest when closest to us. However, Mars, Jupiter, and Saturn are always brightest during the retrograde portions of their orbits. The challenge facing astronomers was to explain the observed motions of the planets and to relate those motions to the variations in planetary brightness.

In many societies, people came to believe that there were other benefits to studying the planets. The positions of heavenly bodies at a person's birth were carefully studied by *astrologers,* who used the data to make predictions about that person's destiny. In a sense, astronomy and astrology arose from the same basic impulse—the desire to "see" into the future. Indeed, for a long time the disciplines were indistinguishable from one another. This chapter chronicles the period of human history when astronomy replaced astrology, and modern science was born.

Motions of the planets relative to the stars produce continuous streaks on a planetarium "sky."

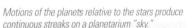

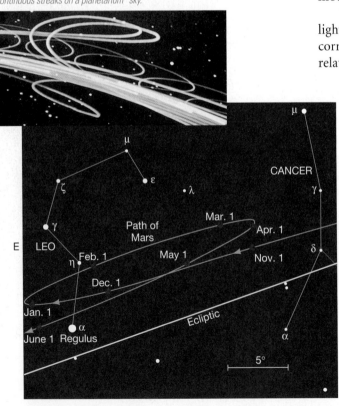

Observed planet motions can be complicated because each planet travels with a different speed around the Sun.

◄ **FIGURE 1.1 Planetary Motions**   Most of the time, planets move from west to east relative to the background stars. Occasionally—roughly once per year—they change direction and temporarily undergo retrograde motion (east to west) before looping back. The main illustration shows an actual retrograde loop in the motion of the planet Mars. The inset depicts the movements of several planets over the course of several years, as reproduced on the inside dome of a planetarium. *(Inset: Museum of Science, Boston)*

The Geocentric Universe

The earliest models of the solar system followed the teachings of the Greek philosopher Aristotle (384–322 B.C.) and were **geocentric,** meaning that Earth lay at the center of the universe and all other bodies moved around it. (Figures 0.4 and 0.9 illustrate the basic geocentric view.) ∞ *(Sec. 0.2)* These models employed what Aristotle had taught was the perfect form: the circle. The simplest possible description—uniform motion around a circle having Earth at its center—provided a fairly good approximation to the orbits of the Sun and the Moon, but it could not account for the observed variations in planetary brightness, nor for their retrograde motion. A better model was needed to describe the planets.

In the first step toward this new model, each planet was taken to move uniformly around a small circle, called an **epicycle,** whose *center* moved uniformly around Earth on a second and larger circle, known as the **deferent** (Figure 1.2). The motion was therefore composed of two separate circular orbits, creating the possibility that at some times the planet's apparent motion could be retrograde. Also, the distance from the planet to Earth would vary, accounting for changes in brightness. By tinkering with the relative sizes of epicycle and deferent, with the planet's speed on the epicycle, and with the epicycle's speed along the deferent, early astronomers were able to bring this "epicyclic" motion into fairly good agreement with the observed paths of the planets in the sky. Moreover, this model predicted fairly well the positions of the known planets, at least to the accuracy of observations at the time.

However, as the number and the quality of observations increased, astronomers had to introduce small corrections into the simple epicyclic model to bring it into line with new observations. The center of the deferents had to be shifted slightly from Earth's center, and the motion of the epicycles had to be imagined uniform with respect not to Earth, but to yet another point in space. Around A.D. 140 a Greek astronomer named Claudius Ptolemaeus (known today as Ptolemy) constructed perhaps the best geocentric model of all time. Illustrated in simplified form in Figure 1.3, it explained remarkably well the observed paths of the five planets then known, as well as the paths of the Sun and the Moon. However, to achieve its explanatory and predictive power, the full **Ptolemaic model** required a series of no fewer than 80 circles. To account for the paths of the Sun, Moon, and all eight planets known today would require a vastly more complex set.

Today our scientific training leads us to seek simplicity, because simplicity in science has so often proved to be an indicator of truth. ∞ *(Sec. 0.5)* The intricacy of a model as complicated as the Ptolemaic system is a clear sign of a fundamentally flawed theory. With the benefit of hindsight, we now recognize that the major error lay in the assumption of a geocentric universe, compounded by the insistence on uniform circular motion, the basis of which was largely philosophical, rather than scientific, in nature.

Actually, history records that some ancient Greek astronomers reasoned differently about the motions of heavenly bodies. Foremost among them was Aristarchus of Samos (310–230 B.C.), who proposed that all the planets, including Earth, revolve around the Sun and, furthermore, that Earth rotates on its axis once each day. This, he argued, would create an apparent motion of the sky—a simple idea that is familiar to anyone who has ridden on a merry-go-round and watched the landscape appear to move past in the opposite direction. However, Aristarchus's description of the heavens, though essentially correct, did not gain widespread acceptance during his lifetime. Aristotle's influence was too strong, his followers too numerous, his writings too comprehensive.

The Aristotelian school did present some simple and (at the time) compelling arguments in favor of their views. First, of course, Earth doesn't *feel* as if it's moving.

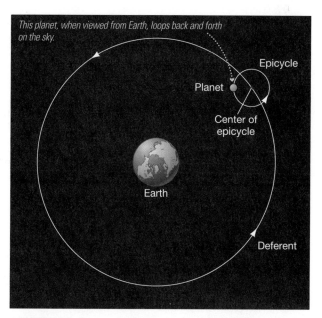

▲ **FIGURE 1.2 Geocentric Model** In the geocentric model of the solar system, the observed motions of the planets made it impossible to assume that they moved on simple circular paths around Earth. Instead, each planet was thought to follow a small circular orbit (the epicycle) about an imaginary point that itself traveled in a large, circular orbit (the deferent) about Earth.

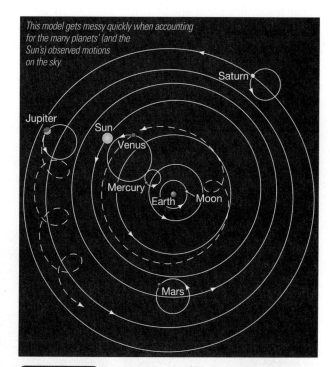

INTERACTIVE ▲ **FIGURE 1.3 Ptolemaic Model** The basic features, drawn roughly to scale, of Ptolemy's geocentric model of the inner solar system, a model that enjoyed widespread popularity prior to the Renaissance. To avoid confusion, partial paths (dashed) of only two planets, Venus and Jupiter, are drawn here.

1-1 DISCOVERY

Ancient Astronomy

Many ancient societies took a keen interest in the changing night-time sky. Seafarers needed to navigate their ships and farmers had to know when to plant their crops. Cultures all over the world built elaborate structures to serve, at least in part, as primitive calendars to predict celestial events. Often, the keepers of the secrets of the sky enshrined their knowledge in myth and ritual, and these astronomical sites were also used for religious ceremonies.

(S. Pitamitz)

Perhaps the best known such site is Stonehenge, in England, shown above. This ancient stone circle dates from the Stone Age and probably served as a kind of three-dimensional almanac. Many of the stones are aligned (within a degree or so) with important astronomical events, such as the rising Sun on the summer solstice and the rising and setting of the Sun and the Moon at other key times of the year, allowing its builders to track the seasons.

The Big Horn Medicine Wheel in Wyoming (below) is similar to Stonehenge in design, and perhaps also in intent. Some researchers have identified alignments between the Medicine Wheel's spokes and the ris-

(G. Gerster)

(H. Lapahie Jr.)

ing and setting Sun at solstices and equinoxes, and with some bright stars, suggesting that its builders—the Plains Indians—were familiar with the changing

(J. Cornell)

night sky. Others dispute the accuracy of the alignments, however, suggesting that the wheel's purpose was more symbolic than practical. A similar controversy swirls around the Caracol temple (bottom right of the same figure) in the Mayan city of Chitzen Itza, built around A.D. 1000 on Mexico's Yucatán peninsula. Were its windows aligned with astronomical events, such as the rising of Venus, or did it have some different purpose? Experts disagree. Researchers do seem to agree that the Sun Dagger (bottom right), in Chaco Canyon, New Mexico, is a genuine astronomical calendar that probably aided agriculture. The rock is sculpted so that a thin streak of light passes precisely through the center of a carved spiral pattern at noon on the summer solstice.

The ancient Chinese also observed the heavens. Their astrology attached particular importance to "omens" such as comets and "guest stars" that appeared suddenly in the sky and then slowly faded away, and they kept careful records of such events. Perhaps the best-known guest star was one that appeared in A.D. 1054 and was visible in the daytime sky for many months. We now know that the event was a *supernova*, the explosive death of a giant star (see Chapter 13). It left behind a remnant that is still detectable today, nine centuries later. The Chinese data are a prime source of historical information for supernova research.

A vital link between the astronomy of ancient Greece and that of modern Europe was provided by astronomers in the Muslim world. This 16th-century illustration shows Persian astronomers at work. From the depths of the Dark Ages to the beginning of the Renaissance, Islamic astronomy flourished, preserving and adding to the knowledge of the Greeks. Its influence on modern astronomy is widespread. Many techniques used in trigonometry were developed by Islamic astronomers in response to practical problems, such as determining the precise dates of holy days or the direction of Mecca at any given location on Earth, and many astronomical terms, such as *azimuth, zenith,* and the names of many stars, are of Muslim origin.

(Bridgeman Art Library)

Second, if it were moving, wouldn't there be a strong wind or force of some sort as we move at high speed around the Sun? For that matter, if the Sun is at the center of the universe, then why do things fall down and not up? And finally, considering that the vantage point from which we view the stars changes over the course of a year, why don't we see stellar parallax? ⟳ *(Sec. 0.4)*

Today we might dismiss the first points as naive, but the last is a valid argument and the reasoning essentially sound. We now know that there *is* stellar parallax as Earth orbits the Sun. However, because the stars are so distant, it is less than 1 arc second, even for the closest stars. Early astronomers simply would not have noticed it (indeed, it was conclusively measured only in the latter half of the 19th century). We will encounter many instances in astronomy where correct reasoning led to the wrong conclusions because it was based on inadequate data.

The Heliocentric Model of the Solar System

The Ptolemaic picture of the universe survived, more or less intact, for almost 13 centuries until a 16th-century Polish cleric, Nicholas Copernicus (Figure 1.4), rediscovered Aristarchus's **heliocentric** (Sun-centered) model. Copernicus asserted that Earth spins on its axis and, like all other planets, orbits the Sun. Not only does this model explain the observed daily and seasonal changes in the heavens, as we have seen, but it also naturally accounts for planetary **retrograde motion** and brightness variations. The critical realization that Earth is not at the center of the universe is now known as the **Copernican revolution.**

Figure 1.5 shows how the Copernican view explains both the changing brightness of a planet (in this case, Mars) and its apparent looping motions. If we suppose that Earth moves faster than Mars, then every so often Earth "overtakes" that planet. Each time this happens, Mars appears to move backward in the sky, in much the same way as a car we overtake on the highway seems to slip backward relative to us. Furthermore, at these times Earth is closest to Mars, so Mars appears brightest, in agreement with observations. Notice that in the Copernican picture the planet's looping motions are only apparent—in the Ptolemaic view, they were real.

Copernicus's major motivation for introducing the heliocentric model was simplicity. Even so, he was still influenced by Greek thinking and clung to the idea of circles to model the planets' motions. To bring his theory into agreement with observations, he was forced to retain the idea of epicyclic motion, though with the deferent centered on the Sun rather than on Earth and with smaller epicycles than in the Ptolemaic picture. Thus, he retained unnecessary complexity and actually gained little in accuracy over the geocentric model. The heliocentric model did rectify some small discrepancies and inconsistencies in the Ptolemaic system, but for Copernicus the primary attraction of heliocentricity was its simplicity, its being "more pleasing to the mind." To this day, scientists still are guided by simplicity, symmetry, and beauty in modeling all aspects of the universe.

Copernicus's ideas were never widely accepted during his lifetime. By relegating Earth to a noncentral and undistinguished place within the solar system, heliocentricity contradicted the conventional wisdom of the time and violated the religious doctrine of the Roman Catholic Church. Copernicus surely discussed and debated his theory with his fellow scholars, but, possibly because he wished to avoid direct conflict with the Church, his book *On the Revolution of the Celestial Spheres* was not published until 1543, the year he died. Only much later, when others extended and popularized the heliocentric model—and as supporting observational evidence began to mount—did the Copernican theory gain widespread recognition.

CONCEPT CHECK ► How do the geocentric and heliocentric models of the solar system differ in their explanations of planetary retrograde motion?

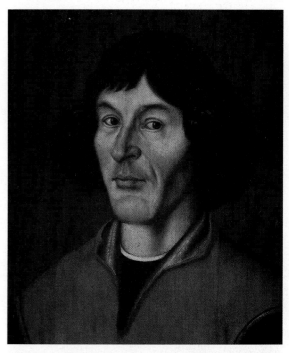

▲ **FIGURE 1.4 Nicholas Copernicus (1473–1543)** *(E. Lessing/Art Resource, NY)*

Retrograde Motion The Copernican model of the solar system explains both the varying brightnesses of the planets and the phenomenon of retrograde motion. Here, for example, when Earth and Mars are relatively close to one another in their respective orbits (as at position 6), Mars seems brighter. When they are farther apart (as at position 1), Mars seems dimmer. Also, because the (light blue) line of sight from Earth to Mars changes as the two planets orbit the Sun, Mars appears to loop back and forth in retrograde motion. Follow the blue lines in numerical order, and note how the line of sight moves backward relative to the stars between locations 5 and 7. That's because Earth, on the inside track, moves faster in its orbit than does Mars. The white curves are actual planetary orbits. The red curve is Mars's motion as seen from Earth.

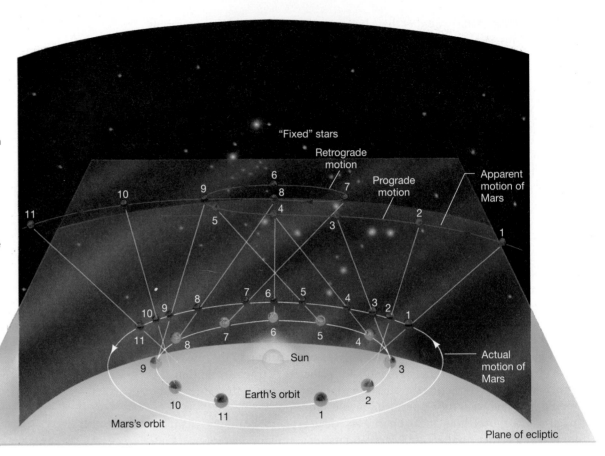

In terms of the scientific method presented in Chapter 0, what were the principal advantages of the heliocentric theory over the geocentric model? ○○ *(Sec. 0.5)*

▲ FIGURE 1.6 Galileo Galilei (1564–1642) *(Scala/Art Resource, NY)*

1.2 The Birth of Modern Astronomy

In the century following the death of Copernicus and the publication of his theory of the solar system, two scientists—Galileo Galilei and Johannes Kepler—made indelible imprints on the study of astronomy. Each achieved fame for his discoveries and made great strides in popularizing the Copernican viewpoint.

Galileo's Historic Observations

Galileo Galilei (Figure 1.6) was an Italian mathematician and philosopher. By his willingness to perform experiments to test his ideas—a radical approach in those days (Section 1.5)—and by embracing the brand-new technology of the telescope, he revolutionized the way science was done, so much so that he is now widely regarded as the father of experimental science. The telescope was invented in Holland in the early 17th century. Having heard of the invention (but without having seen one), Galileo built a telescope for himself in 1609 and aimed it at the sky. What he saw conflicted greatly with the philosophy of Aristotle and provided much new data to support the ideas of Copernicus.

Using his telescope, Galileo discovered the following:

- The Moon has mountains, valleys, and craters—terrain in many ways reminiscent of that on Earth.
- The Sun has imperfections—dark blemishes now known as *sunspots* (see Chapter 9). By noting the changing appearance of these sunspots from day to day, Galileo inferred that the Sun *rotates,* approximately once per month, around an axis roughly perpendicular to the ecliptic plane.

► FIGURE 1.7 **Galilean Moons** The four Galilean moons of Jupiter, as sketched by Galileo in his notebook, on seven nights between January 7 and 15, 1610. (*From Sidereus Nuncius*)

- Four small points of light, invisible to the naked eye, orbit the planet Jupiter. He realized that they were *moons* (Figure 1.7) circling that planet just as our Moon orbits Earth. To Galileo, the fact that another planet had moons provided the strongest support for the Copernican model. Clearly, Earth was not the center of all things.
- Venus shows a complete cycle of *phases* (Figure 1.8), much like the familiar monthly changes exhibited by our own Moon. This finding can be explained only by the planet's motion around the Sun.

All these observations ran directly counter to the accepted scientific beliefs of the day. They strongly supported the view that Earth is not the center of all things and that at least one planet orbited the Sun.

In 1610 Galileo published his observational findings and his controversial conclusions supporting the Copernican theory, challenging both scientific orthodoxy and the religious dogma of the day. In 1616 his ideas were judged heretical, both his works and those of Copernicus were banned by the Church, and Galileo was

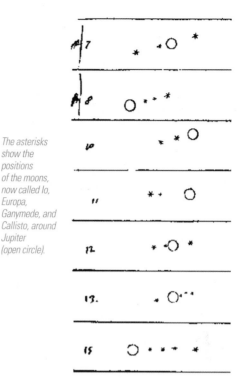

The asterisks show the positions of the moons, now called Io, Europa, Ganymede, and Callisto, around Jupiter (open circle).

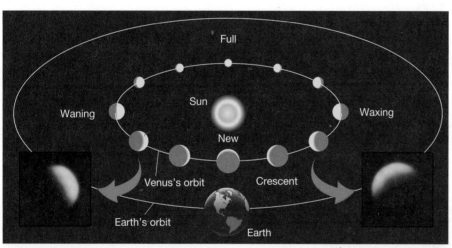

(a) Sun-centered model

R I V U X G

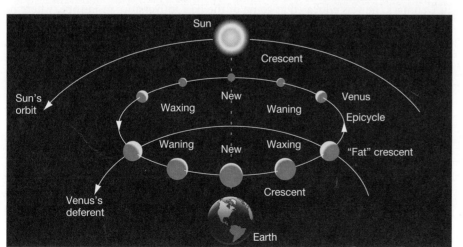

(b) Ptolemy's model

INTERACTIVE ◄ FIGURE 1.8 **Venus Phases** Both the Ptolemaic and the Copernican models of the solar system predict that Venus should show phases as it moves in its orbit. (a) In the Copernican picture, when Venus is directly between Earth and the Sun, its unlit side faces us and the planet is invisible to us. As Venus moves in its orbit, progressively more of its illuminated face is visible from Earth. Note the connection between the orbital phase and the apparent size of the planet: Venus seems much larger in its crescent phase than when it is full because it is much closer to us during its crescent phase. This is the behavior actually observed. The insets at bottom left and right are photographs of Venus taken at two of its crescent phases (*Courtesy of New Mexico State University*). (b) The Ptolemaic model (see also Figure 1.3), is unable to account for these observations. In this model, the full phase of the planet cannot be explained. Seen from Earth, Venus reaches only a "fat crescent," yet never a full phase, then begins to wane as it nears the Sun. (Both these views are from a sideways perspective; from overhead, both orbits are very nearly circular, as shown in Figure 1.14.)

instructed to abandon his astronomical pursuits. This he refused to do and instead continued to collect and publish data supporting the heliocentric view. These actions brought Galileo into direct conflict with the Church. The Inquisition forced him, under threat of torture, to retract his claim that Earth orbits the Sun, and he was placed under house arrest in 1633. He remained imprisoned for the rest of his life. Not until 1992 did the Church publicly forgive Galileo's "crimes." But the damage to the orthodox view of the universe was done, and the Copernican genie was out of the bottle once and for all.

Ascendancy of the Copernican System

The Copernican revolution is a case study of how the scientific method, although affected at any given time by the subjective whims, human biases, and even sheer luck of individual researchers, does ultimately lead to a definite degree of objectivity. ∞ *(Sec. 0.5)* Over time, groups of scientists checking, confirming, and refining experimental tests can neutralize the subjective attitudes of individuals. In the case of heliocentricity, objective confirmation was not obtained until about three centuries after Copernicus published his work and more than 2000 years after Aristarchus had proposed the concept. Nonetheless, objectivity *did in fact* eventually prevail.

CONCEPT CHECK ▶ In what ways did Galileo's observations of Venus and Jupiter conflict with the prevailing view at the time?

1.3 The Laws of Planetary Motion

At about the time Galileo was becoming famous for his telescopic observations, Johannes Kepler (Figure 1.9), a German mathematician and astronomer, announced his discovery of a set of simple empirical (that is, based on observation) laws that accurately described the motions of the planets. While Galileo was the first "modern" observer who used telescopic observations of the skies to confront and refine his theories, Kepler was a pure theorist. He based his work almost entirely on the observations of another scientist (in part because of his own poor eyesight). Those observations, which predated the telescope by several decades, had been made by Kepler's employer, Tycho Brahe (1546–1601), arguably one of the greatest observational astronomers who ever lived.

Brahe's Complex Data

Tycho, as he is often called, was an eccentric aristocrat and a skillful observer. Born in Denmark, he was educated at some of the best universities in Europe, where he studied astrology, alchemy, and medicine. Most of his observations, which predated the invention of the telescope by several decades, were made at his own observatory, named Uraniborg, in Denmark (Figure 1.10). There, using instruments of his own design, Tycho maintained meticulous and accurate records of the stars, the planets, and noteworthy celestial events.

In 1597, having fallen out of favor with the Danish court, Tycho moved to Prague. Kepler joined Tycho in Prague in 1600 and was put to work trying to find a theory that could explain Brahe's planetary data. When Tycho died a year later, Kepler inherited not only Brahe's position as Imperial Mathematician of the Holy Roman Empire, but also his most priceless possession: the accumulated observations of the planets, spanning several decades. Tycho's observations, though made with the naked eye, were nevertheless of very high quality. Kepler set to work seeking a unifying principle to explain the motions of the planets without the need for epicycles. The effort was to occupy much of the remaining 29 years of his life.

▲ **FIGURE 1.9 Johannes Kepler (1571–1630)** *(E. Lessing/Art Resource, NY)*

Kepler's goal was to find a simple description of the solar system, within the basic framework of the Copernican model, that fit Tycho's complex mass of detailed observations. In the end, he had to abandon Copernicus's original simple notion of circular planetary orbits, but even greater simplicity emerged as a result. Kepler determined the shapes and relative sizes of each planet's orbit by triangulation, not from different points on Earth, but from different points on Earth's orbit, using observations made at many different times of the year. ∞ *(Sec. 0.4)* Noting where the planets were on successive nights, he was able to infer the speeds at which they moved. After long years working with Brahe's planetary data and after many false starts and blind alleys, Kepler succeeded in summarizing the motions of all the known planets, including Earth, in the three **laws of planetary motion** that now bear his name.

Kepler's Simple Laws

Kepler's first law addresses the *shapes* of the planetary orbits:

> I. The orbital paths of the planets are elliptical (*not* necessarily circular), with the Sun at one focus.

An **ellipse** is simply a flattened circle. Figure 1.11 illustrates a means of constructing an ellipse using a piece of string and two thumbtacks. Each point at which the string is pinned is called a **focus** (plural: foci) of the ellipse. The long axis of the ellipse, containing the two foci, is known as the *major axis*. Half the length of this long axis is referred to as the **semimajor axis,** a conventional measure of the ellipse's size. The **eccentricity** of the ellipse is equal to the distance between the foci divided by the length of the major axis. The length of the semimajor axis and the eccentricity are all we need to describe the size and shape of a planet's orbital path. (A circle is an ellipse in which the two foci happen to coincide, so the eccentricity is zero. The semimajor axis of a circle is simply its radius.) These two numbers—semimajor axis and eccentricity—are all that are needed to describe the size and shape of a planet's orbital path. Figure 1.12 illustrates how two other useful quantities—the planet's **perihelion** (its point of closest approach to the Sun) and its **aphelion** (greatest distance from the Sun)—can be computed from them.

In fact, no planet's elliptical orbit is nearly as elongated as the one shown in Figure 1.11. With one exception (Mercury), the planets' orbits have such small eccentricities that our eyes would have trouble distinguishing them from true circles. Only because the orbits are so nearly circular were the Ptolemaic and Copernican models able to come as close as they did to describing reality.

Kepler's second law, illustrated in Figure 1.13, addresses the *speed* at which a planet traverses different parts of its orbit:

> II. An imaginary line connecting the Sun to any planet sweeps out equal areas of the ellipse in equal intervals of time.

While orbiting the Sun, a planet traces the arcs labeled A, B, and C in Figure 1.13 in equal times. Notice, however, that the distance traveled along arc C is greater than the distance traveled along arc A or arc B. Because the time is the same and the distance is different, the speed must vary: When a planet is close to the Sun, as in sector C, it moves much faster than when farther away, as in sector A.

These laws are not restricted to planets. They apply to *any* orbiting object. Spy satellites, for example, move very rapidly as they swoop close to Earth's surface, not because they are propelled by powerful onboard rockets, but because their highly eccentric orbits are governed by Kepler's laws.

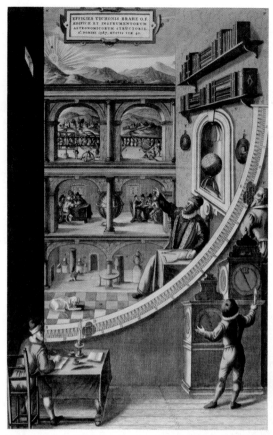

▲ **FIGURE 1.10 Tycho Brahe** The astronomer in his observatory Uraniborg, on the island of Hveen, in Denmark. *(Newberry Library/SuperStock)*

CONCEPT CHECK ▶ In what ways did Galileo and Kepler differ in their approach to science? In what ways did each advance the Copernican view of the universe?

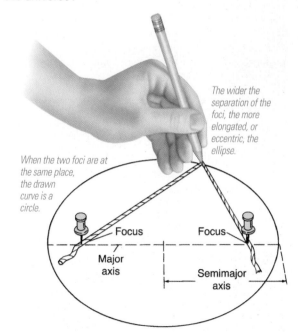

The wider the separation of the foci, the more elongated, or eccentric, the ellipse.

When the two foci are at the same place, the drawn curve is a circle.

Focus   Focus

Major axis   Semimajor axis

INTERACTIVE ▲ **FIGURE 1.11 Ellipse** An ellipse can be drawn using a string, a pencil, and two thumbtacks.

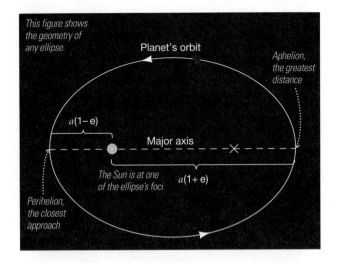

This figure shows the geometry of any ellipse.

Planet's orbit

Aphelion, the greatest distance

$a(1-e)$

Major axis

The Sun is at one of the ellipse's foci.

$a(1+e)$

Perihelion, the closest approach

◄ FIGURE 1.12 **Orbital Properties** A planet's perihelion and aphelion are related in a simple way to the orbital semimajor axis a and eccentricity e. Note that while the Sun resides at one focus, the other focus (depicted by an X) is empty and has no particular significance. No planet in our solar system has an orbital eccentricity as large as shown here (see Table 1.1), but some meteorites and all comets do (see Chapter 4). **MA**

Kepler published his first two laws in 1609, stating that he had proved them only for the orbit of Mars. Ten years later he extended them to all the known planets (Mercury, Venus, Earth, Mars, Jupiter, and Saturn) and added a third law relating the size of a planet's orbit to its sidereal orbital **period,** defined as the time needed for the planet to complete one circuit around the Sun.

Kepler's third law states:

III. The square of a planet's orbital period is proportional to the cube of its semimajor axis.

This law becomes particularly simple when we choose the (Earth) year as our unit of time and the *astronomical unit* as our unit of length. One **astronomical unit** (AU) is the semimajor axis of Earth's orbit around the Sun—the average distance between Earth and the Sun. Using these units for time and distance, we can conveniently write Kepler's third law for any planet in the form

$$P^2 \text{ (in Earth years)} = a^3 \text{ (in astronomical units)},$$

where P is the planet's sidereal orbital period and a is its semimajor axis.

Table 1.1 presents some basic data describing the orbits of the eight major planets of the solar system. The semimajor axis of each planet's orbit is measured in astronomical units—that is, relative to the size of Earth's orbit—and all orbital periods are measured in years. Renaissance astronomers knew these properties for the innermost six planets only. However, *all* the known bodies orbiting the Sun obey Kepler's laws *not just the six planets on which he based his conclusions.* The rightmost column lists the ratio P^2/a^3. As we have just seen, in the units used in the table, Kepler's third law implies that this number should equal 1 in all cases. The small deviations of P^2/a^3 from 1 in the cases of Uranus and Neptune are caused by the gravitational attraction between those two planets (see Chapter 7).

The laws developed by Kepler were far more than mere fits to existing data. They made definite, testable predictions about the future locations of the planets.

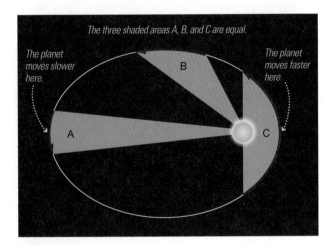

The three shaded areas A, B, and C are equal.

The planet moves slower here.

The planet moves faster here.

B

A

C

▲ FIGURE 1.13 **Kepler's Second Law** A line between a planet and the Sun sweeps out equal areas in equal intervals of time. Any object traveling along the elliptical path would take the same amount of time to cover the distance indicated by the three red arrows. Planets move faster when closer to the Sun. **MA**

TABLE 1.1	Some Planetary Properties			
Planet	Orbital Semimajor Axis, *a* (AU)	Orbital Period, *P* (Earth Years)	Orbital Eccentricity, *e*	P^2/a^3
Mercury	0.387	0.241	0.206	1.002
Venus	0.723	0.615	0.007	1.001
Earth	1.000	1.000	0.017	1.000
Mars	1.524	1.881	0.093	1.000
Jupiter	5.203	11.86	0.048	0.999
Saturn	9.537	29.42	0.054	0.998
Uranus	19.19	83.75	0.047	0.993
Neptune	30.07	163.7	0.009	0.986

Those predictions have been borne out to high accuracy every time they have been tested by observation—the hallmark of any credible scientific theory. ∞ *(Sec. 0.5)*

CONCEPT CHECK ▶ Why is it significant that Kepler's laws apply to Uranus and Neptune?

The Dimensions of the Solar System

Kepler's laws allow us to construct a scale model of the solar system, with the correct shapes and *relative* sizes of all the planetary orbits, but they do not tell us the *actual* size of any orbit. We can express the distance to each planet only in terms of the distance from Earth to the Sun—in other words, only in astronomical units. Why is this? Because Kepler's measurements were based on triangulation using a portion of Earth's orbit as a baseline, so his distances could be expressed only relative to the size of that orbit, which was not itself determined.

Our model of the solar system would be analogous to a road map of the United States showing the *relative* positions of cities and towns but lacking the all-important scale marker indicating distances in kilometers or miles. For example, we would know that Kansas City is about three times farther from New York than it is from Chicago, but we would not know the actual mileage between any two points on the map. If we could somehow determine the value of the astronomical unit— in kilometers, say—we would be able to add the vital scale marker to our map of the solar system and compute the exact distances between the Sun and each of the planets.

The modern method for deriving the absolute scale of the solar system uses a technique called *radar ranging*. The word **radar** is an acronym for **ra**dio **d**etection **a**nd **r**anging. Radio waves are transmitted toward an astronomical body, such as a planet. Their returning echo indicates the body's direction and range, or distance, in absolute terms—in other words, in kilometers rather than in astronomical units. Multiplying the round-trip travel time of the radar signal (the time elapsed between transmission of the signal and reception of the echo) by the speed of light (300,000 km/s, which is also the speed of radio waves), we obtain twice the distance to the target planet.

CONCEPT CHECK   Why don't Kepler's laws tell us the value of the astronomical unit?

We cannot use radar ranging to measure the distance to the Sun directly, because radio signals are absorbed at the solar surface and are not reflected back to Earth. Instead, the planet Venus, whose orbit brings it closest to Earth, is the most common target for this technique. Figure 1.14 is an idealized diagram of the Sun–Earth–Venus orbital geometry. Neglecting for simplicity the small eccentricities of the two planets' orbits, we see from Table 1.1 that the distance from Venus to the Sun is roughly 0.7 AU. Hence (from Figure 1.14), the distance from Earth to Venus at closest approach is approximately 0.3 AU. Radar signals bounced off Venus at that instant return to Earth in about 300 seconds, indicating that Venus lies 300,000 km/s × 300 s ÷ 2 (dividing by 2 for a one-way trip) = 45,000,000 km from Earth. Since 0.3 AU is 45,000,000 km, it follows that 1 AU is 45,000,000 km/0.3, or 150,000,000 km.

Through precise radar ranging, the astronomical unit is now known to be 149,597,870 km. In this text, we will round this value off to 1.5×10^8 km. Having determined the value of the astronomical unit, we can re-express the sizes of the other planetary orbits in more familiar units, such as miles or kilometers. The entire scale of the solar system can then be calibrated to high precision.

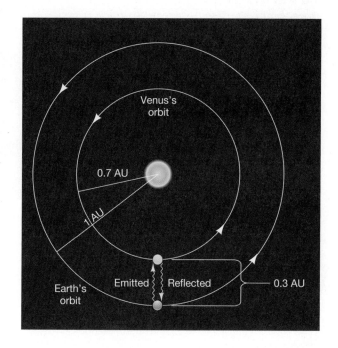

INTERACTIVE ▶ FIGURE 1.14 Astronomical Unit   The wavy blue lines represent the paths along which radar signals are transmitted toward Venus and received back at Earth when Venus is at its minimum distance from Earth. Because the radius of Earth's orbit is 1 AU and that of Venus is about 0.7 AU, the one-way distance covered by the signal is 0.3 AU. Thus, we can calibrate the astronomical unit in kilometers.

▲ FIGURE 1.15 Isaac Newton (1642–1727)   *(S. Terry)*

1.4 Newton's Laws

Kepler's three laws, which so simplified the solar system, were discovered *empirically*. In other words, they resulted solely from the analysis of observational data, rather than being derived from a theory or mathematical model. Indeed, Kepler did not have any appreciation for the physics underlying his laws. Nor did Copernicus understand the basic reasons *why* his heliocentric model of the solar system worked. Even Galileo, the father of modern physics, failed to understand why the planets orbit the Sun.

What causes the planets to revolve about the Sun, apparently endlessly? What prevents them from flying off into space or falling into the Sun? Of course, the motions of the planets obey Kepler's three laws, but only by considering something more fundamental can we truly understand planetary motion. The critical test of any complete scientific theory is its ability to explain physical phenomena, not simply describe them.

The Laws of Motion

In the 17th century, the British physicist and mathematician Isaac Newton (Figure 1.15) developed a deeper understanding of the way *all* objects move and interact with one another. Newton's theories form the basis for what today is known as **Newtonian mechanics.** Three basic laws of motion, the law of universal gravitation, and a little calculus (which Newton also helped invent) are sufficient to explain and quantify virtually all of the complex dynamic behavior we see on Earth and throughout the universe.

Figure 1.16 illustrates *Newton's first law of motion,* which states:

> I. An object at rest remains at rest, and a moving object continues to move forever in a straight line with constant speed, unless some external **force** changes their state of motion.

An example of an external force would be the force exerted by, say, a brick wall when a rolling ball glances off it, or the force exerted on a pitched ball by a baseball bat. In either case, a force changes the original motion of the object. The tendency of an object to keep moving at the same speed and in the same direction unless acted upon by a force is known as **inertia.** A familiar measure of an object's inertia is its **mass**—loosely speaking, the total amount of matter the object contains. The greater an object's mass, the more inertia it has and the greater is the force needed to change its motion.

Newton's first law contrasts sharply with the view of Aristotle, who maintained (incorrectly) that the natural state of an object was to be *at rest*—most probably an opinion based on Aristotle's observations of the effect of friction. To simplify our discussion, we will neglect friction—the force that slows balls rolling along the ground, blocks sliding across tabletops, and baseballs moving through the air. In any case, this is not an issue for the planets, as there is no appreciable friction in outer space. The fallacy in Aristotle's argument was first exposed by Galileo, who conceived of the notion of inertia long before Newton formalized it into a law.

The rate of change of the velocity of an object—speeding up, slowing down, or simply changing direction—is called its **acceleration.** *Newton's second law* states:

> II. The acceleration of an object is directly proportional to the net applied force and inversely proportional to the object's mass.

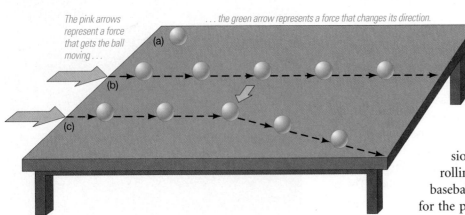

The pink arrows represent a force that gets the ball moving . . .

. . . the green arrow represents a force that changes its direction.

(a)
(b)
(c)

▲ FIGURE 1.16 **Newton's First Law**   An object at rest will remain at rest (a) until some force acts on it. When a force (represented by the red arrow) does act (b), the object will remain in that state of uniform motion until another force acts on it. When a second force (green arrow) acts in a direction different from the first force (c), the object changes its direction of motion.

In other words, the greater the force acting on the object, or the smaller the mass of the object, the greater its acceleration. Thus, if two objects are pulled with the same force, the more massive one accelerates less. If two identical objects are pulled with different forces, the one experiencing the greater force accelerates more.

Finally, *Newton's third law* tells us that forces always occur in pairs:

III. To every action there is an equal and opposite reaction.

If body A exerts a force on body B, then body B necessarily exerts a force on body A that is equal in magnitude but oppositely directed.

Gravity

Forces can act either *instantaneously* or *continuously*. The force from a baseball bat that hits a home run can reasonably be thought of as instantaneous. A good example of a continuous force is the one that prevents the baseball from zooming off into space—**gravity,** the phenomenon that started Newton on the path to the discovery of his laws. Newton hypothesized that any object having mass exerts an attractive **gravitational force** on all other massive objects. The more massive an object, the stronger its gravitational pull.

Consider a baseball thrown upward from Earth's surface, as illustrated in Figure 1.17. In accordance with Newton's first law, the downward force of Earth's gravity steadily modifies the baseball's velocity, slowing its initial upward motion and eventually causing the ball to fall back to the ground. Of course, the baseball, having some mass of its own, also exerts a gravitational pull on Earth. By Newton's third law, this force is equal in magnitude to the weight of the ball (the weight of an object is a measure of the force with which Earth attracts that object), but oppositely directed. By Newton's second law, however, Earth has a much greater effect on the light baseball than the baseball has on the much more massive Earth. The ball and Earth each feel the same gravitational force, but Earth's *acceleration* as a result of this force is much smaller and can be safely ignored.

Now consider the trajectory of a baseball batted from the surface of the Moon. The pull of gravity is about one-sixth as great on the Moon as on Earth, so the baseball's velocity changes more slowly—a typical home run in a ballpark on Earth would travel nearly half a mile on the Moon. The Moon, less massive than Earth, has less gravitational influence on the baseball. The magnitude of the gravitational force, then, depends on the *masses* of the attracting bodies. In fact, it is directly proportional to the product of the two masses.

Studying the motions of the planets reveals a second aspect of the gravitational force. At locations equidistant from the Sun's center, the gravitational force has the same strength and is always directed toward the Sun. Furthermore, detailed calculation of the planets' accelerations as they orbit the Sun reveals that the strength of the Sun's gravitational pull decreases in proportion to the *square* of the distance from the Sun. (Newton is said to have first realized this fact by comparing the accelerations not of the planets, but of the Moon and an apple falling to the ground—the basic reasoning is the same in either case.) The force of gravity is said to obey an **inverse-square law** (Figure 1.18a).

We can combine the preceding statements about mass and distance to form a law of gravity that dictates the way in which *all* massive objects (i.e., objects having some mass) attract one another:

> Every particle of matter in the universe attracts every other particle with a force that is directly proportional to the product of the masses of the particles and inversely proportional to the square of the distance between them.

As shown in ure 1.18(b), inverse-square forces decrease rapidly with distance from their source. For example, tripling the distance makes the force $3^2 = 9$ times weaker, while multiplying the distance by five results in a force that is $5^2 = 25$ times

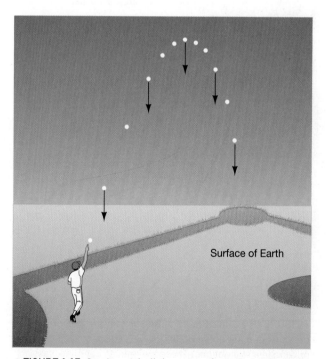

▲ **FIGURE 1.17 Gravity** A ball thrown up from the surface of a massive object, such as a planet, is pulled continuously down (arrows) by the gravity of that planet—and, conversely, the gravity of the ball continuously pulls the planet (although very, very little).

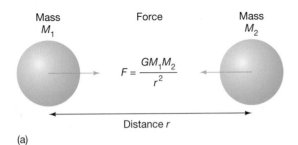

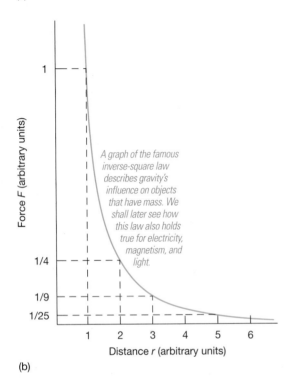

Mass M_1

Force

Mass M_2

$$F = \frac{GM_1M_2}{r^2}$$

Distance r

(a)

Force F (arbitrary units)

A graph of the famous inverse-square law describes gravity's influence on objects that have mass. We shall later see how this law also holds true for electricity, magnetism, and light.

Distance r (arbitrary units)

(b)

◀ FIGURE 1.18 Gravitational Force (a) The gravitational force between two bodies is proportional to the mass of each and is inversely proportional to the square of the distance between them. (b) Inverse-square forces weaken rapidly with distance from their source, never quite reaching zero no matter how far away.

weaker. Despite this rapid decrease, the force never quite reaches zero. The gravitational pull of an object having some mass can never be completely extinguished.

Orbital Motion

The mutual gravitational attraction of the Sun and the planets, as expressed by Newton's law of gravity, is responsible for the observed planetary orbits. As depicted in Figure 1.19, this gravitational force continuously pulls each planet toward the Sun, deflecting its forward motion into a curved orbital path. Because the Sun is much more massive than any of the planets, it dominates the interaction. We might say that the Sun "controls" the planets, not the other way around.

The planet–Sun interaction sketched here is analogous to what occurs when you whirl a rock at the end of a string above your head. The Sun's gravitational pull is your hand and the string, and the planet is the rock at the end of that string. The tension in the string provides the force necessary for the rock to move in a circular path. If you were suddenly to release the string—which would be like eliminating the Sun's gravity—the rock would fly away along a tangent to the circle, in accordance with Newton's first law.

In the solar system, at this very moment, Earth is moving under the combined influence of these two effects: the competition between gravity and inertia. The net result is a stable orbit, despite our continuous rapid motion through space. In fact, Earth orbits the Sun at a speed of about 30 km/s, or some 70,000 mph. (Verify this for yourself by calculating how fast Earth must move to complete a circle of radius 1 AU—and, hence, of circumference 2π AU, or 940 million km—in 1 year, or 3.2×10^7 seconds.)

Kepler's Laws Reconsidered

Newton's laws of motion and his law of universal gravitation provided a theoretical explanation for Kepler's empirical laws of planetary motion. Just as Kepler modified the Copernican model by introducing ellipses in place of circles, so too did Newton make corrections to Kepler's first and third laws. Because the Sun and a planet feel equal and opposite gravitational forces (by Newton's third law), the Sun must also move (by Newton's first law) due to the planet's gravitational pull. As a result, the planet does not orbit the exact center of the Sun but, instead, both the planet and the Sun orbit their common **center of mass**—the "average" position of the matter comprising the two bodies (see Figure 1.20)—and Kepler's first law becomes:

> I. The orbit of a planet around the Sun is an ellipse having the center of mass of the planet–Sun system at one focus.

As illustrated in Figure 1.20, the center of mass of a system consisting of two objects of comparable mass does not lie within either object. Hence, for identical masses orbiting one another (Figure 1.21a), the orbits are identical ellipses with a common focus located midway between the two objects. For unequal masses (Figure 1.21b), the elliptical orbits still share a focus and have the same eccentricity, but the more massive object moves more slowly and on a tighter orbit. (Note that Kepler's second law continues to apply without modification to each orbit separately, but the rates at which the two orbits sweep out area are different.) In the extreme case of a planet orbiting the much more massive Sun (Figure 1.21c), the orbit of the Sun's center lies entirely within the Sun.

The change to Kepler's third law is small in the case of a planet orbiting the Sun but may be very important in other circumstances, such as the orbits of two stars that are gravitationally bound to one another. Following through the mathematics of Newton's theory, we find the true relationship between the semimajor axis

CONCEPT CHECK ▶ Explain, in terms of Newton's laws of motion and gravity, why planets orbit the Sun.

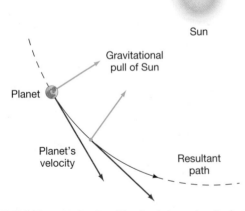

Sun

Gravitational pull of Sun

Planet

Planet's velocity

Resultant path

▲ FIGURE 1.19 Sun's Gravity The Sun's inward pull of gravity on a planet competes with the planet's tendency to continue moving in a straight line. These two effects combine, causing the planet to move smoothly along an intermediate path, which continuously "falls around" the Sun. This unending tug-of-war between the Sun's gravity and the planet's inertia causes the planet to orbit the Sun.

INTERACTIVE Center of Mass of a Binary Star **MA**

a (measured in astronomical units) of the planet's orbit relative to the Sun and its orbital period *P* (in Earth years) is

$$P^2 \text{ (in Earth years)} = \frac{a^3 \text{ (in astronomical units)}}{M_{\text{total}} \text{ (in solar units)}},$$

where M_{total} is the *combined* mass of the two objects, expressed in terms of the mass of the Sun. Notice that this restatement of Kepler's third law preserves the proportionality between P^2 and a^3, but now the proportionality also includes M_{total} so it is *not* quite the same for all the planets. The Sun's mass is so great, however, that the differences in M_{total} among the various combinations of the Sun and the other planets are almost unnoticeable. Therefore, Kepler's third law, as originally stated, is a very good approximation.

This modified form of Kepler's third law is true in all circumstances, inside or outside the solar system. Most importantly, it provides a means of measuring mass *anywhere* in the universe, so long as we can determine the orbital properties—separation and period—of the bodies involved. In fact, this is how *all* masses are measured in astronomy. When we need to know an object's mass, we always look for its gravitational influence on something else. This principle applies to planets, stars, galaxies, even clusters of galaxies—very different objects but all subject to the same fundamental physical laws. Newton's laws extend our intellectual reach far beyond the tiny fraction of the universe we can actually visit or touch.

The Circle of Scientific Progress

The progression from the complex Ptolemaic model of the universe to the elegant simplicity of Newton's laws is a case study in the scientific method. ⬡ *(Sec.0.5)* Copernicus made a radical conceptual leap away from the Ptolemaic view, gaining much in insight but little in predictive power. Kepler made critical changes to the Copernican picture and gained both accuracy and predictive power, but still fell short of a true physical explanation of planetary motion within the solar system, or of orbital motion in general. Eventually, Newton showed how all known planetary motion could be explained in detail by the application of four simple, fundamental laws—the three laws of motion and the law of gravity. The process was slow, with many starts and stops and a few wrong turns, but it worked!

In a sense, the development of Newton's laws and their application to planetary motion represent the end of the first "loop" around the schematic diagram shown in Figure 0.22. The practical and conceptual questions raised by ancient observations of retrograde motion were finally resolved, and new predictions, themselves amenable to observational testing, became possible. But Newton's laws went much further. Unlike the essentially descriptive models of Ptolemy, Copernicus, and Kepler, Newtonian mechanics is not limited to the motions of planets. They apply to moons, comets, spacecraft, stars, and even the most distant galaxies, extending the range of our scientific inquiries across the observable universe.

Newton's laws are still being tested today. Every time a comet appears in the night sky right on schedule or a spacecraft reaches the end of a billion-kilometer journey within meters of its target and seconds of the predicted arrival time, our confidence in the laws is further strengthened. Only in extreme circumstances do Newton's laws break down, and this fact was not realized until the 20th century, when Albert Einstein's theories of relativity once again revolutionized our view of gravity and the universe (see Chapter 13).

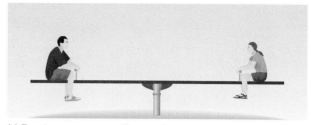

(a) Equal masses   Center of mass

(b) Unequal masses   Center of mass

▲ **FIGURE 1.20 Center of Mass**   (a) The center of mass of two bodies of equal mass lies midway between them. (b) As the mass of one body increases, the center of mass moves toward it. Experienced seesawers know that when both sides are balanced, the center of mass is at the pivot point.

(MA)   **ANIMATION/VIDEO** Gravity Demonstration on the Moon

PROCESS OF ▶
SCIENCE CHECK   Describe some ways in which Newtonian mechanics superseded Kepler's laws as a model of the solar system.

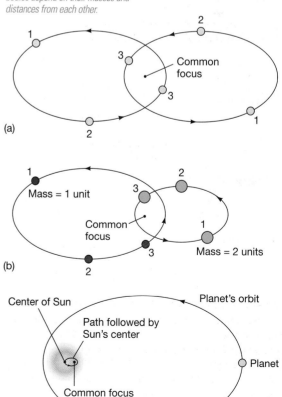

The resulting orbits for mutually gravitating bodies depend on their masses and distances from each other.

▶ **FIGURE 1.21 Orbits**   (a) The orbits of two bodies with equal masses, under the influence of their mutual gravity, are identical ellipses with a common focus. The pairs of numbers (e.g., the two 2s in each orbit) indicate the positions of the two bodies at three different times. (b) The orbits of two bodies, one twice as massive as the other, are again elliptical and with the same eccentricity, but according to Newton's laws, the more massive body moves more slowly and in a smaller orbit. (c) In the case of an extremely small planet orbiting the massive Sun, the common focus of the two orbits could be inside the Sun.

CHAPTER REVIEW

SUMMARY

L01 **Geocentric** (p. 27) models of the universe, such as the **Ptolemaic model** (p. 27), have the Sun, Moon, and planets all orbiting Earth. They explain **retrograde motion** (p. 29) as a real backward motion of a planet as it moves along its epicyclic path.

L02 The **heliocentric** (p. 29) view of the solar system, due to Aristarchus and later Copernicus, holds that Earth, like all the other planets, orbits the Sun. This model explains both retrograde motion as Earth overtakes other planets in its orbit and the observed brightness variations of the planets. The realization that the solar system is Sun centered, not Earth centered, is known as the **Copernican revolution** (p. 29).

L03 Copernicus suggested that if the planets orbited the Sun, rather than Earth, this would provide a simpler explanation of their observed motion. Galileo Galilei was the first experimental scientist. His telescopic observations of the Sun, Moon, and planets provided experimental evidence against the geocentric theory and supporting Copernicus's heliocentric model. Johannes Kepler constructed a set of three simple

laws describing the motions of the planets around the Sun, explaining Tycho Brahe's detailed observational data.

L04 Kepler's three **laws of planetary motion** (p. 33) state that (1) planetary orbits are **ellipses** (p. 33) having the Sun as one **focus** (p. 33), (2) a planet moves faster as its orbit takes it closer to the Sun, and (3) the orbital **semimajor axis** (p. 33) is simply related to the planet's orbital **period** (p. 34).

L05 The average distance from Earth to the Sun is one **astronomical unit** (p. 34), today precisely determined by bouncing **radar** (p. 35) signals off the planet Venus. Once this is known, the distances to all other planets can be inferred from Kepler's laws.

L06 To change a body's velocity, a **force** (p. 36) must be applied. The rate of change of velocity, called the **acceleration** (p. 36), is equal to the applied force divided by the body's mass. **Gravity** (p. 37) attracts the planets to the Sun. Every object having any mass exerts a **gravitational force** (p. 37) on all other objects, and the strength of this force decreases with distance according to an **inverse-square law** (p. 37).

MasteringAstronomy®   For instructor-assigned homework go to www.masteringastronomy.com

Problems labeled **POS** explore the process of science.   **VIS** problems focus on reading and interpreting visual information. **LO** connects to the introduction's numbered Learning Outcomes.

REVIEW AND DISCUSSION

1. **LO1** Describe the strengths and weaknesses of the geocentric model of the universe.

2. **POS** The benefit of our current knowledge lets us see flaws in the Ptolemaic model of the universe. What is its basic flaw?

3. What was the great contribution of Copernicus to our knowledge of the solar system?

4. **LO2** How did the heliocentric model explain planetary motions and brightness variations?

5. **LO3 POS** How did Galileo help confirm the views of Copernicus?

6. Why is Galileo often thought of as the first experimental scientist?

7. **LO4** Briefly describe Kepler's three laws of planetary motion.

8. **LO3** How did Tycho Brahe contribute to Kepler's laws?

9. **POS** What does it mean to say Kepler's laws are empirical?

10. **LO5** If radio waves cannot be reflected from the Sun, how can radar be used to find the distance from Earth to the Sun?

11. List the two modifications made by Newton to Kepler's laws.

12. Why do we say that a baseball falls toward Earth, and not Earth toward the baseball?

13. Why would a baseball thrown from the surface of the Moon go higher than one thrown with the same velocity from Earth's surface?

14. **LO6** According to Newton, why does Earth orbit the Sun?

15. What would happen to Earth if the Sun's gravity were suddenly "turned off"?

CONCEPTUAL SELF-TEST: TRUE OR FALSE?/MULTIPLE CHOICE

1. Aristotle proposed that all planets revolve around the Sun. (T/F)

2. During retrograde motion, planets actually stop and move backward in space. (T/F)

3. The heliocentric model of the universe holds that Earth is at the center and everything else moves around it. (T/F)

4. Copernicus's theories gained widespread scientific acceptance during his lifetime. (T/F)

5. Galileo's observations of the sky were made with the naked eye. (T/F)

6. The speed of a planet orbiting the Sun is independent of the planet's position in its orbit. (T/F)

7. Kepler's laws hold only for the six planets known in his time. (T/F)

8. You throw a baseball to someone; before the ball is caught, it is temporarily in orbit around Earth's center. (T/F)

9. A major flaw in Copernicus's model was that it still had (a) the Sun at the center; (b) Earth at the center; (c) retrograde loops; (d) circular orbits.

10. VIS An accurate sketch of Mars's orbit around the Sun would show (a) the Sun far off center; (b) an oval twice as long as it is wide; (c) a nearly perfect circle; (d) phases.

11. A calculation of how long it takes a planet to orbit the Sun would be most closely related to Kepler's (a) first law of orbital shapes;

(b) second law of orbital speeds; (c) third law of planetary distances; (d) first law of inertia.

12. An asteroid with an orbit lying entirely inside Earth's (a) has an orbital semimajor axis of less than 1 AU; (b) has a longer orbital period than Earth's; (c) moves more slowly than Earth; (d) has a highly eccentric orbit.

13. If Earth's orbit around the Sun were to double in size, the new "year" would be (a) less than 2; (b) 2; (c) more than 2 current Earth years.

14. VIS As shown in Figure 1.8 (Venus Phases), Galileo's observations demonstrated that Venus must be (a) orbiting Earth; (b) orbiting the Sun; (c) larger than Earth; (d) similar to the Moon.

15. VIS Figure 1.17 (Gravity), showing the motion of a ball near Earth's surface, depicts how gravity (a) increases with altitude; (b) causes the ball to accelerate downward; (c) causes the ball to accelerate upward; (d) has no effect on the ball.

PROBLEMS

The number of squares preceding each problem indicates its approximate level of difficulty.

1. ■■ Tycho's observations were accurate to about 1 arc minute (1′). To what distance does this correspond at the distance of (a) the Moon, (b) the Sun, (c) Saturn at closest approach? (⚬ *Problem 0.8*)

2. ■ How long would a radar signal take to complete a round-trip between Earth and Mars when the two planets are 0.7 AU apart?

3. ■■ Seen from Earth, through what angle will Mars appear to move relative to the stars over the course of 24 hours when the two planets are at closest approach? Assume that Earth and Mars move on circular orbits of radii 1.0 AU and 1.5 AU, respectively, in the same plane. Will the apparent motion be prograde or retrograde?

4. ■■ An asteroid has a perihelion distance of 2.0 AU and an aphelion distance of 4.0 AU. Calculate its orbital semimajor axis, eccentricity, and period (see Figure 1.12).

5. ■■ Halley's Comet has a perihelion distance of 0.6 AU and an orbital period of 76 years. What is its aphelion distance from the Sun?

6. ■■ Using the data in Table 1.1, calculate how much farther Mercury is from the Sun at aphelion than at perihelion.

7. ■■■ Jupiter's moon Callisto orbits the planet at a distance of 1.88 million kilometer. Callisto's orbital period about Jupiter is 16.7 days. What is the mass of Jupiter? [Assume that Callisto's mass is negligible compared with that of Jupiter, and use the modified version of Kepler's third law (Section 1.4).]

8. ■■ The acceleration due to gravity at Earth's surface is 9.80 m/s². What is the acceleration at altitudes of (a) 100 km? (b) 1000 km? (c) 10,000 km? Take Earth's radius to be 6400 km.

9. ■ Use Newton's law of gravity to calculate the force of gravity between you and Earth. Convert your answer, which will be in newtons, to pounds using the conversion 4.45 N = 1 pound. What do you normally call this force?

10. ■■ The Moon's mass is 7.4×10^{22} kg, and its radius is 1700 km. What would be the period and the speed of a spacecraft moving in a circular orbit just above the lunar surface?

ACTIVITIES

Collaborative

1. You have been asked to arbitrate a dispute between a tour bus company and a nearby Native American tribe concerning a recently discovered ancient medicine wheel. Using sketches as necessary, compose a document describing what a medicine wheel is designed to do astronomically, and summarize the pros and cons of letting the public have unrestricted access to the site.

2. Select what you believe to be Galileo's single most important astronomical observation, state why you think it was most important, and explain using sketches what he observed.

Individual

1. Look in an almanac for the dates of opposition of Mars, Jupiter, and Saturn. At opposition, these planets are at their closest points

to Earth and are at their largest and brightest in the night sky. Observe these planets. How long before opposition does each planet's retrograde motion begin? How long afterward does it end?

2. Use a small telescope to replicate Galileo's observations of Jupiter's largest moons. Note the brightness and location of the moons relative to Jupiter. If you watch over a period of several nights, draw what you see; the moons' positions will change as they orbit the planet.

3. Draw an ellipse (see Figure 1.11). You'll need two pins, a piece of string, and a pencil. Tie the string in a loop and place around the pins. Place the pencil inside the loop and run it around the inside of the string, holding the loop taut. The two pins will be at the foci of the ellipse. What is the eccentricity of the ellipse you have drawn? How does its shape change as you vary the distance between the pins?

2

Light and Matter

The Inner Workings of the Cosmos

Astronomical objects are more than just things of beauty in the night sky. Planets, stars, and galaxies are of vital significance if we are to understand fully the big picture—the grand design of the universe. Every object is a source of information about the universe—its temperature, its chemical composition, its state of motion, its past history. The starlight we see tonight began its journey to Earth decades, centuries, even millennia ago. The faint rays from the most distant galaxies have taken billions of years to reach us. The stars and galaxies in the night sky show us not just the far away, but also the long ago. In this chapter, we begin our study of how astronomers extract information from the light emitted by astronomical objects. The observational and theoretical techniques that enable researchers to determine the nature of distant atoms by the way they emit and absorb light are the indispensable foundation of modern astronomy.

MasteringAstronomy® Visit www.masteringastronomy.com for quizzes, animations, videos, interactive figures, and self-guided tutorials.

◄ Stars change from birth to maturity to death, much like living things, but on vastly longer timescales. Our own star, the Sun, is midway through its evolutionary cycle. About 5 billion years ago, it emerged from a stellar nursery much like the one shown here. This is a composite image of invisible radiation captured by two telescopes now orbiting Earth. The *Spitzer Space Telescope* senses heat in the infrared (shown here mostly in red), and the *Chandra X-ray Observatory* captures X-rays (purple). By acquiring radiation outside the visible part of the spectrum, astronomers can peer inside this young nebula, named NGC281, with its towering pillars of cool gas and dust illuminated with radiation from warm embryonic stars. This region is about 9000 light-years away and spans roughly 70 light-years across. *(NASA)*

2.1 Information from the Skies

Figure 2.1 shows our nearest large galactic neighbor, which lies in the constellation Andromeda. On a dark, clear night, far from cities or other sources of light, the Andromeda galaxy, as it is generally called, can be seen with the naked eye as a faint, fuzzy patch on the sky, comparable in diameter to the full Moon. Yet the fact that it is visible from Earth belies this galaxy's enormous distance from us. It lies roughly 2.5 million light-years away. An object at such a distance is truly inaccessible in any realistic human sense. Even if a space probe could miraculously travel at the speed of light, it would need two and a half million years to reach this galaxy and two and a half million more to return with its findings. Considering that civilization has existed on Earth for fewer than 10,000 years (and its prospects for the next 10,000 are far from certain), even this unattainable technological feat would not provide us with a practical means of exploring other galaxies—or even the farthest reaches of our own galaxy, several tens of thousands of light-years away.

Light and Radiation

How do astronomers know anything about objects far from Earth? How can we obtain detailed information about any planet, star, or galaxy too distant for a personal visit or any kind of controlled experiment? The answer is that we use the laws of physics, as we know them here on Earth, to interpret the light, or **electromagnetic radiation** emitted by these objects. *Radiation* is any way in which energy is transmitted through space from one point to another without the need for any physical connection between those two locations. The term *electromagnetic* refers to the fact that the energy in a beam of light is actually carried in the form of rapidly changing *electric* and *magnetic* fields, as discussed further in Section 2.2. Virtually all we know about the universe beyond Earth's atmosphere has been learned from analysis of electromagnetic radiation received from afar.

Visible light is the particular type of electromagnetic radiation to which the human eye happens to be sensitive. But, just as there are sound waves that humans can't hear, there is also *invisible* electromagnetic radiation, which goes completely undetected by our eyes. **Radio, infrared**, and **ultraviolet** waves, as well as **X-rays** and **gamma rays**, all fall into this category. You should recognize that, despite the different names, the words *light, rays, electromagnetic radiation*, and *waves* really all refer to the same thing. The names are just historical accidents, reflecting the fact that it took many years for scientists to realize that these apparently very different types of radiation are in reality one and the same physical phenomenon. Throughout this text, we will use the general terms *light* and *electromagnetic radiation* more or less interchangeably.

Wave Motion

All types of electromagnetic radiation travel through space in the form of **waves**. To understand the behavior of light, then, we must know a little about this kind of motion. Simply stated, a wave is a way in which energy is transferred from place to

◀ **FIGURE 2.1 Andromeda Galaxy**   The pancake-shaped Andromeda galaxy is about 2.5 million light-years away and contains a few hundred billion stars. *(R. Gendler)*

place without physical movement of material from one location to another. In wave motion, the energy is carried by a *disturbance* of some sort that occurs in a distinctive, repeating pattern. Ripples on the surface of a pond, sound waves in air, and electromagnetic waves in space, despite their many obvious differences, all share this basic defining property.

As a familiar example, imagine a twig floating in a pond (Figure 2.2). A pebble thrown into the pond at some distance from the twig disturbs the surface of the water, setting it into up-and-down motion. This disturbance propagates outward from the point of impact in the form of waves. When the waves reach the twig, some of the pebble's energy is transferred to it, causing the twig to bob up and down. In this way, both energy and *information*—the fact that the pebble entered the water—are transferred from the place where the pebble landed to the location of the twig. We could tell just by observing the twig that a pebble (or some small object) had entered the water. With a little additional physics, we could even estimate the pebble's energy.

A wave is *not* a physical object. No water traveled from the point of impact of the pebble to the twig; at any location on the surface, the water surface simply moved up and down as the wave passed. What, then, *does* move across the pond surface? The answer is that the wave is the *pattern* of up-and-down motion, and it is this pattern that is transmitted from one point to the next as the disturbance moves across the water.

Figure 2.3 shows how wave properties are quantified. The **wave period** is the number of seconds needed for the wave to repeat itself at some point in space. The **wavelength** is the number of meters needed for the wave to repeat itself at a given moment in time. It can be measured as the distance between two adjacent wave *crests*, two adjacent wave *troughs*, or any other two similar points on adjacent wave cycles (for example, the points marked "×" in Figure 2.3). The maximum departure of the wave from the undisturbed state—still air, say, or a flat pond surface—is called its **amplitude.**

The number of wave crests passing any given point per unit time is called the wave's **frequency.** If a wave of a given wavelength moves at high speed, then many crests pass by per second and the frequency is high. Conversely, if the same wave moves slowly, then its frequency is low. The frequency of a wave is just one divided by the wave's period:

$$\text{frequency} = \frac{1}{\text{period}}.$$

Waves ripple out from where a pebble hit the water . . .

. . . to where a twig is floating.

This insert shows a series of "snapshots" of the pond surface as the wave passes by.

Undisturbed pond surface

Direction of wave motion

INTERACTIVE ▲ **FIGURE 2.2 Water Wave** The passage of va wave across a pond causes the surface of the water to bob up and down, but there is no movement of water from one part of the pond to another. **MA**

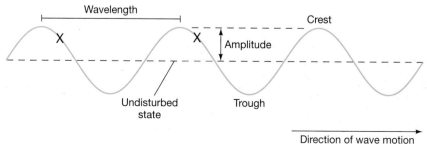

Wavelength

Crest

Amplitude

X   X

Undisturbed state

Trough

Direction of wave motion

INTERACTIVE ◄ **FIGURE 2.3 Wave Properties** A typical wave has a direction of motion, wavelength, and amplitude. In one wave period, the entire pattern shown here moves one wavelength to the right. **MA**

Since frequency counts events per unit time, it is expressed in units of inverse time (cycles per second), called hertz (Hz) in honor of the 19th-century German scientist Heinrich Hertz, who studied the properties of radio waves. For example, a wave with a period of 5 seconds (5 s) has a frequency of (1/5) cycles/s = 0.2 Hz meaning that one wave crest passes a given point in space every 5 seconds.

A wave moves a distance equal to one wavelength in one wave period. The product of wavelength and frequency therefore equals the *wave velocity*:

$$\text{wavelength} \times \text{frequency} = \text{velocity}.$$

Thus, if the wave in our earlier example had a wavelength of 0.5 m, its velocity is (0.5 m) × (0.2 Hz) = 0.1 m/s. Wavelength and wave frequency are *inversely* related—doubling one halves the other.

CONCEPT CHECK ▶ What is a wave? What four basic properties describe a wave, and what relationships, if any, exist among them?

2.2 Waves in What?

Scientists know that radiation travels as a wave because light displays characteristic behavior common to *all* kinds of wave motion. For example, light waves tend to "bend around corners," just like sound waves or ocean waves passing a breakwater (Figure 2.4a). This is called **diffraction.** In addition, the crests and troughs of waves coming from different sources can reinforce or partly cancel one another (Figure 2.4b). This is known as **interference.** We are not normally aware of this behavior in the case of electromagnetic radiation because the effects are generally too small to be noticeable in everyday life. However, they are easily measured in the laboratory and are very important considerations when designing and building telescopes, as we will see in Chapter 3. Their discovery early in the 19th century provided strong evidence in favor of the theory that light moves as a wave.

Waves of radiation do differ in one fundamental respect from water waves, sound waves, or any other waves that travel through a material medium—radiation needs *no* medium. When light travels from a distant cosmic object, it moves through the virtual vacuum of space. Sound waves, by contrast, cannot do this, despite what you hear in sci-fi movies! If we were to remove all the air from a room, conversation would be impossible (even with suitable breathing apparatus to keep our test subjects alive), because sound waves cannot exist without air or some other physical medium to support them. Communication by flashlight or radio, however, would be entirely possible.

The ability of light to travel through empty space was once a great mystery. The idea that light, or any other kind of radiation, could move as a wave through nothing at all seemed to violate common sense, yet it is now a cornerstone of modern physics.

In diffraction, a wave bends around an obstacle, such as a breakwater.

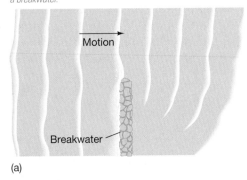

(a)

In interference, two waves combine to produce a wave equal to the sum of the two.

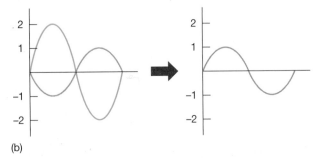

(b)

▲ **FIGURE 2.4 Wave Behavior** Diffraction (a) and interference (b) are two basic types of behavior shared by all waves, including electromagnetic radiation.

Interactions Between Charged Particles

To understand more about the nature of light, consider an *electrically charged* particle, such as an **electron** or a **proton.** Electrons and protons are elementary particles—fundamental components of matter—that carry the basic unit of charge. Electrons are said to carry a *negative* charge, while protons carry an equal and opposite *positive* charge. Just as a massive object exerts a gravitational force on any other massive object, an electrically charged particle exerts an *electrical* force on every other charged particle in the universe. ∞ *(Sec. 1.4)* Unlike gravity, however, which is always attractive, electrical forces can be either attractive or repulsive. Particles having like charges (both negative or both positive) repel one another; particles having unlike charges attract (Figure 2.5a).

Extending outward in all directions from our charged particle is an **electric field,** which determines the electric force exerted by the particle on other charged particles (Figure 2.5b). The strength of the electric field, like that of the gravitational field, decreases with increasing distance from the source according to an inverse-square law—doubling the distance reduces the strength of the field by a factor of 4. ∞ *(Sec. 1.4)* By means of the electric field, the particle's presence is "felt" by other charged particles, near and far.

Now suppose our particle begins to vibrate, perhaps because it becomes heated or collides with some other particle. Its changing position causes its associated electric field to change, and this changing field in turn causes the electrical force exerted on other charges to vary (Figure 2.5c). If we measure the changes in the forces on these other charges, we learn about our original particle. Thus, *information about our particle's motion is transmitted through space via a changing electric field.* This *disturbance* in the particle's electric field travels through space as a wave.

Electromagnetism

The laws of physics tell us that a **magnetic field** must accompany every changing electric field. Magnetic fields govern the influence of *magnetized* objects on one another, much as electric fields govern interactions between charged particles. The fact that one end of a compass needle always points to magnetic north is the result of the interaction between the magnetized needle and Earth's magnetic field (Figure 2.6). Magnetic fields also exert forces on moving electric charges (that is, electric currents)—electric meters and motors rely on this basic fact. Conversely, moving charges create magnetic fields (the electromagnets found in loudspeakers and electric motors are familiar examples).

Electric and magnetic fields are inextricably linked to one another. A change in either one *necessarily* creates the other. For this reason, the disturbance produced by our moving charge actually consists of oscillating electric *and* magnetic fields, always oriented perpendicular to one another and moving together through space (Figure 2.7). These fields do not exist as independent entities. Rather, they are different aspects of a single physical phenomenon: **electromagnetism**. Together they constitute an *electromagnetic wave* that carries energy and information from one part of the universe to another.

Now consider a distant cosmic object—a star. It is made up of charged particles, mainly protons and electrons, in constant motion. As these charged contents move around, their electric fields change, and electromagnetic waves are produced. These waves travel outward into space, and eventually some reach Earth. Other charged particles, either in the molecules in our eyes or in our experimental apparatus (a telescope or radio receiver, for example), respond to the electromagnetic field changes by vibrating in tune with the received radiation. This response is how we "see" the radiation—with our eyes or with our detectors.

How quickly does one charge feel the change in the electromagnetic field when another begins to move—that is, how fast does an electromagnetic wave travel? Both theory and experiment tell us that all electromagnetic waves move at a very specific speed—the **speed of light** (always denoted by the letter c). Its value is 299,792.458 km/s in a vacuum (and somewhat less in material substances, such as air or water). In this text, we round this value off to $c = 3.00 \times 10^5$ km/s. This is an extremely high speed. In the time needed to snap your fingers—about a tenth of a second—light can travel three quarters of the way around our planet! According to the theory of relativity (see *More Precisely 13-1*), the speed of light is the fastest speed possible.

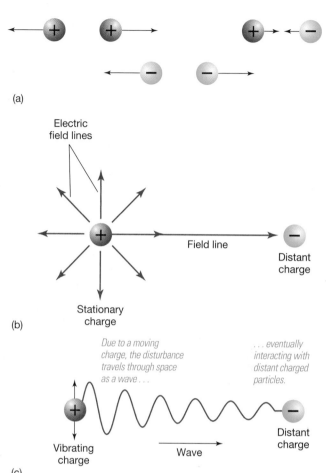

▲ **FIGURE 2.5 Charged Particles** (a) Particles carrying like electrical charges repel one another; particles with unlike charges attract. (b) A charged particle is surrounded by an electric field, which determines the particle's influence on other charged particles. (c) If a charged particle begins to vibrate, its electric field changes.

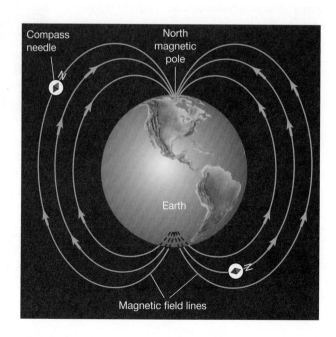

▶ **FIGURE 2.6 Magnetism** Earth's magnetic field interacts with a magnetic compass needle, causing the needle to become aligned with the field—that is, to point toward Earth's north (magnetic) pole.

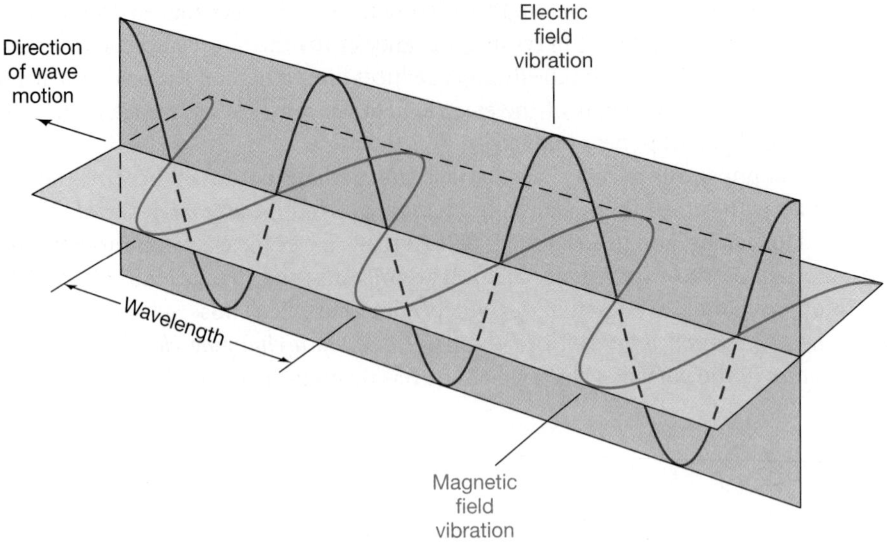

▶ **FIGURE 2.7 Electromagnetic Wave** Electric and magnetic fields vibrate perpendicular to each other. Together they form an electromagnetic wave that moves through space at the speed of light.

| CONCEPT | What is light? List some similarities |
| CHECK ▶ | and differences between light waves |

and waves on water or in air.

2.3 The Electromagnetic Spectrum

White light is a mixture of colors, which we conventionally divide into six major hues—red, orange, yellow, green, blue, and violet. As shown in Figure 2.8, we can separate a beam of white light into a rainbow of these basic colors—called a *spectrum* (plural: spectra)—by passing it through a prism. This experiment was first reported by Isaac Newton more than 300 years ago. In principle, the original beam of white light could be recovered by passing the spectrum through a second prism to recombine the colored beams.

Components of Visible Light

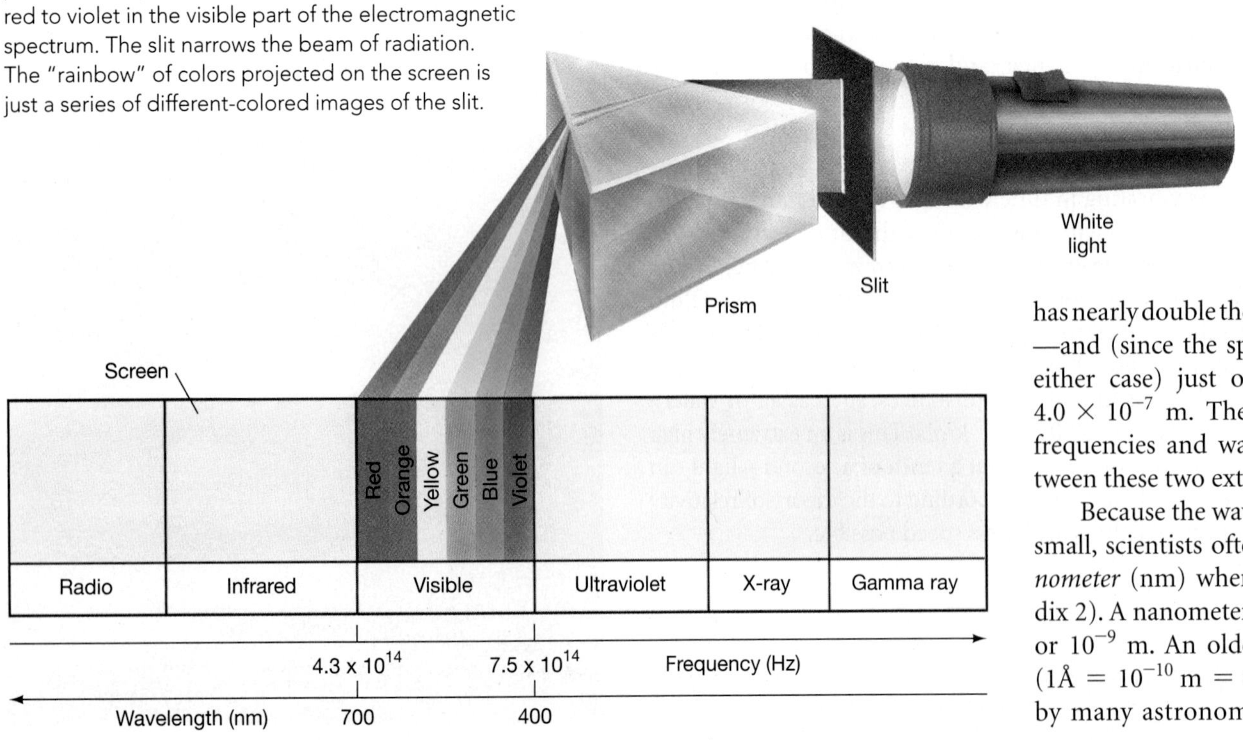

▼ **FIGURE 2.8 Visible Spectrum** When passed through a prism, white light splits into its component colors, spanning red to violet in the visible part of the electromagnetic spectrum. The slit narrows the beam of radiation. The "rainbow" of colors projected on the screen is just a series of different-colored images of the slit.

What determines the color of a beam of light? The answer is its frequency (or, equivalently, its wavelength). We see different colors because our eyes react differently to electromagnetic waves of different frequencies. Red light has a frequency of roughly 4.3×10^{14} Hz, corresponding to a wavelength of about 7.0×10^{-7} m. Violet light, at the other end of the visible range, has nearly double the frequency—7.5×10^{14} Hz —and (since the speed of light is the same in either case) just over half the wavelength—4.0×10^{-7} m. The other colors we see have frequencies and wavelengths intermediate between these two extremes.

Because the wavelength of visible light is so small, scientists often use a unit called the *nanometer* (nm) when describing it (see Appendix 2). A nanometer is one-billionth of a meter, or 10^{-9} m. An older unit called the *angstrom* ($1Å = 10^{-10}$ m = 0.1 nm) is also widely used by many astronomers and atomic physicists, although the nanometer is now preferred. Thus, the visible spectrum covers the wavelength range from 400 to 700 nm (4000 to 7000 Å). The radiation to which our eyes are most sensitive has a wavelength near the middle of this range, at about 550 nm (5500 Å), in the yellow-green region of the spectrum.

The Full Range of Radiation

Figure 2.9 plots the entire range of electromagnetic radiation. To the low-frequency, long-wavelength side of visible light lies radio and infrared radiation. Radio frequencies include radar, microwave radiation, and the familiar AM, FM, and TV bands. We perceive infrared radiation as heat. To the high-frequency, short-wavelength side of visible light lies ultraviolet, X-ray, and gamma-ray radiation. Ultraviolet radiation, lying just

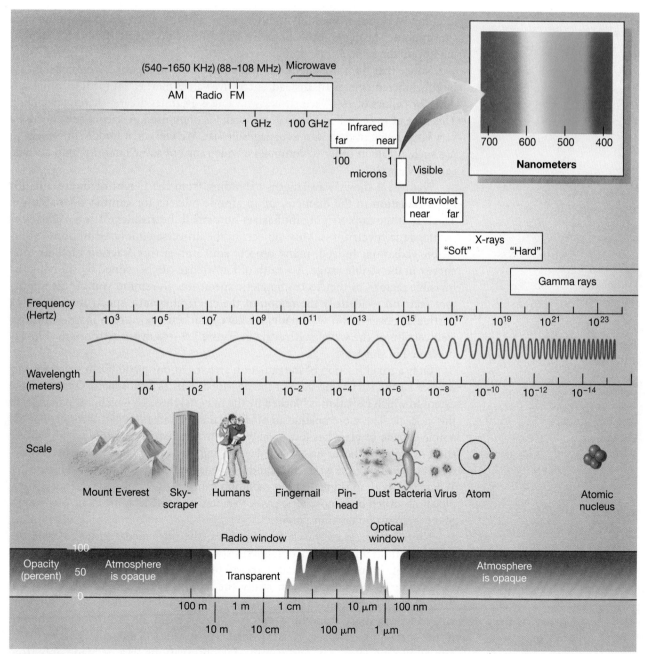

INTERACTIVE ▲ FIGURE 2.9 **Electromagnetic Spectrum**   The entire electromagnetic spectrum, running from long-wavelength, low-frequency radio waves to short-wavelength, high-frequency gamma rays.   **MA**

beyond the violet end of the visible spectrum, is responsible for suntans and sunburns. X-rays are perhaps best known for their ability to penetrate human tissue and reveal the state of our insides without our resorting to surgery. Gamma rays have the shortest wavelengths. They are often associated with radioactivity and are invariably damaging to any living cells they encounter.

All these spectral regions, including the visible, collectively make up the **electromagnetic spectrum**. Remember that despite their greatly differing wavelengths and the very different roles they play in everyday life on Earth, all types of electromagnetic radiation are basically the same and all move at the same speed—the speed of light c.

Figure 2.9 is worth studying carefully, as it contains a great deal of information. Note that wave frequency (in hertz) increases from left to right, while wavelength (in meters) increases from right to left. Scientists often disagree on the "correct" way to display wavelengths and frequencies in diagrams of this type, and it is common for astronomers working in different parts of the spectrum to adopt opposite conventions. Throughout this book we will consistently display frequency increasing toward the *right*.

Notice that the wavelength and frequency scales in Figure 2.9 do not increase by equal increments of 10. Instead, successive values marked on the horizontal axis differ by factors of 10—each successive value is 10 times greater than its neighbor. This type of scale (called a *logarithmic* scale) is often used in science to condense a very large range of data into a manageable size. We will often find it convenient to use such a scale in order to compress a wide range of some quantity onto a single easy-to-view plot.

Figure 2.9 shows wavelengths extending from the height of mountains for radio radiation to the diameter of an atomic nucleus for gamma-ray radiation. The box at the upper right emphasizes how small the visible portion of the electromagnetic spectrum is. Most objects in the universe emit large amounts of invisible radiation. Indeed, many objects emit only a tiny fraction of their total energy in the visible range. A wealth of knowledge can be gained by studying the invisible regions of the electromagnetic spectrum. To remind you of this important fact and to identify the region of the electromagnetic spectrum in which a particular observation was made, we have attached a spectrum icon—an idealized version of the wavelength scale in Figure 2.9—to every astronomical image presented in this text.

Only a small fraction of the radiation arriving at our planet from space actually reaches Earth's surface because of the *opacity* of Earth's atmosphere. **Opacity** is the extent to which radiation is blocked by the material through which it is passing—in this case, air. The more opaque an object is, the less radiation gets through. For example, a pane of glass has low opacity (i.e., is *transparent*) to visible light, while a sheet of paper or the air on a foggy day has high opacity. Earth's atmospheric opacity is plotted along the wavelength and frequency scales at the bottom of Figure 2.9. Where the shading is greatest, no radiation can get in or out—the energy is completely absorbed by atmospheric gases. Where there is no shading at all, our atmosphere is almost totally transparent.

Note that there are just a few *windows* (the unshaded regions) at well-defined locations in the electromagnetic spectrum, where Earth's atmosphere is transparent. In much of the radio and all of the visible portion of the spectrum, the opacity is low, and we can study the universe at those wavelengths from ground level. In parts of the infrared range, the atmosphere is partially transparent, so we can make some infrared observations from the ground. Over the rest of the spectrum, however, the atmosphere is opaque. As a result, most infrared, and all ultraviolet, X-ray, and gamma-ray observations, can be made only from high-flying balloons or (more commonly) above the atmosphere, from orbiting satellites (see Section 3.5).

CONCEPT CHECK ▶ In what sense are radio waves, visible light, and X-rays one and the same phenomenon?

2.4 Thermal Radiation

All macroscopic objects—fires, ice cubes, people, stars—emit radiation at all times. They radiate because the microscopic charged particles in them are in constant random motion, and whenever charges change their state of motion, electromagnetic radiation is emitted. The **temperature** of an object is a direct measure of the amount of microscopic motion within it (see *More Precisely 2-1*). The hotter the object, the higher its temperature, the faster its constituent particles move and the more energy they radiate.

The Blackbody Spectrum

Intensity is a term often used to specify the amount or strength of radiation at any point in space. Like frequency and wavelength, intensity is a basic property of electromagnetic radiation. No natural object emits all of its radiation at just one frequency. Instead, the energy is often spread out over a range of frequencies. By studying the way in which the intensity of this radiation is distributed across the electromagnetic spectrum, we can learn much about the object's properties.

Figure 2.10 illustrates schematically the distribution of radiation emitted by any object. Note that the curve peaks at a single, well-defined frequency (marked)

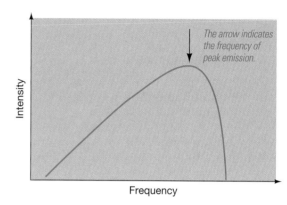

▲ **FIGURE 2.10 Ideal Blackbody Curve** The blackbody curve represents the spread of the intensity of electromagnetic radiation emitted across the electromagnetic spectrum.

2-1 MORE PRECISELY

The Kelvin Temperature Scale

The atoms and molecules that make up any piece of matter are in constant random motion. This motion represents a form of energy known as *thermal energy*. The quantity we call temperature is a direct measure of this internal motion: the higher an object's temperature, the faster the random motion of its constituent particles. More precisely, the temperature of a piece of matter specifies the average thermal energy of the particles it contains.

The temperature scale probably most familiar to you, the Fahrenheit scale, is now a peculiarity of American society. Most of the rest of the world uses the Celsius temperature scale, in which water freezes at 0 degrees (0°C) and boils at 100 degrees (100°C), as illustrated in the accompanying figure.

There are, of course, temperatures below the freezing point of water. Temperatures can in principle reach as low as −273.15°C (although we know of no matter anywhere in the universe that is actually this cold); this is the temperature at which, theoretically, all atomic and molecular motion effectively ceases. It is convenient to construct a temperature scale based on this lowest possible temperature, which is called *absolute zero*. Scientists commonly use such a scale, called the Kelvin scale in honor of the 19th-century British physicist Lord Kelvin. Since it takes absolute zero as its starting point, the Kelvin scale differs from the Celsius scale by 273.15°. In this book, we round off the decimal places and simply use the relationship

$$\text{kelvins} = \text{degrees Celsius} + 273.$$

Thus,

- Thermal motion of atoms and molecules ceases at 0 kelvins (0 K).
- Water freezes at 273 kelvins (273 K).
- Water boils at 373 kelvins (373 K).

Note, by the way, that the unit is "kelvins," or "K," not "degrees kelvin" or "°K."

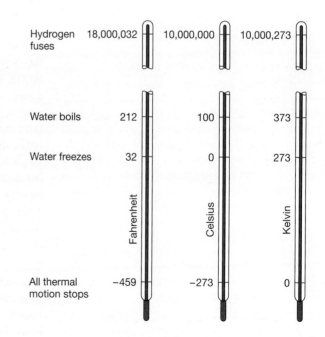

and falls off to lesser values above and below that frequency. However, the curve is not symmetrical about the peak—the falloff is more rapid on the high-frequency side of the peak than it is toward lower frequencies. This overall shape is characteristic of the electromagnetic radiation emitted by *any* object, regardless of its size, shape, composition, or temperature.

The curve drawn in Figure 2.10 refers to a mathematical idealization known as a *blackbody*—an object that absorbs all radiation falling upon it. In a steady state (that is, if the temperature remains constant), a blackbody must reemit the same amount of energy as it absorbs. The **blackbody curve** shown in the figure describes the distribution of that reemitted radiation. No real object absorbs or radiates as a perfect blackbody. However, in many cases the blackbody curve is a very good approximation to reality, and the properties of blackbodies provide important insights into the behavior of real objects, including stars.

SELF-GUIDED TUTORIAL   Continuous Spectra (MA) and Blackbody Radiation

The Radiation Laws

As illustrated in Figure 2.11, the entire blackbody curve shifts toward higher frequencies (shorter wavelengths) and greater intensities as an object's temperature increases. Even so, the *shape* of the curve remains the same. This shifting of radiation's peak frequency with temperature is familiar to us all. Very hot glowing objects, such as lightbulb filaments or stars, emit visible light because their blackbody curves peak in or near the visible range. Cooler objects, such as warm rocks or household radiators, produce invisible radiation—they are warm to the touch but are not glowing hot to the eye. These objects emit most of their radiation in the lower-frequency infrared portion of the electromagnetic spectrum.

There is a very simple connection between the frequency or wavelength at which most radiation is emitted and the absolute temperature (that is, temperature measured in kelvins—see *More Precisely 2-1*) of the emitting object: The peak frequency is *directly proportional* to the temperature. This relationship, known as **Wien's law** after Wilhelm Wien, the German scientist who formulated it in 1897, is usually expressed as:

$$\text{wavelength of peak emission} \propto \frac{1}{\text{temperature}}.$$

The symbol "$\propto$" just means "is proportional to." Wien's law tells us that the hotter the object, the bluer its radiation. For example, an object with a temperature of 6000 K (Figure 2.11c) emits most of its energy in the visible part of the spectrum, with a peak wavelength of 480 nm. At 600 K (Figure 2.11b), the object's emission would peak at 4800 nm, well into the infrared. At a temperature of 60,000 K (Figure 2.11d), the peak would move all the way through the visible range to a wavelength of 48 nm, in the ultraviolet.

It is also a matter of everyday experience that as the temperature of an object increases, the *total* amount of energy it radiates (summed over all frequencies) increases rapidly. For example, the heat given off by an electric heater increases sharply as the heater warms up and begins to emit visible light. In fact, the total amount of energy radiated per unit time is proportional to the *fourth power* of an object's temperature:

$$\text{total energy radiated per second} \propto \text{temperature}^4.$$

This relationship is called **Stefan's law,** after the 19th-century Austrian physicist Josef Stefan. It implies that the energy emitted by a body rises dramatically as the body's temperature increases. Doubling the temperature, for example, causes the total energy radiated to increase by a factor of 16. The radiation laws are presented in more detail in *More Precisely 2-2.*

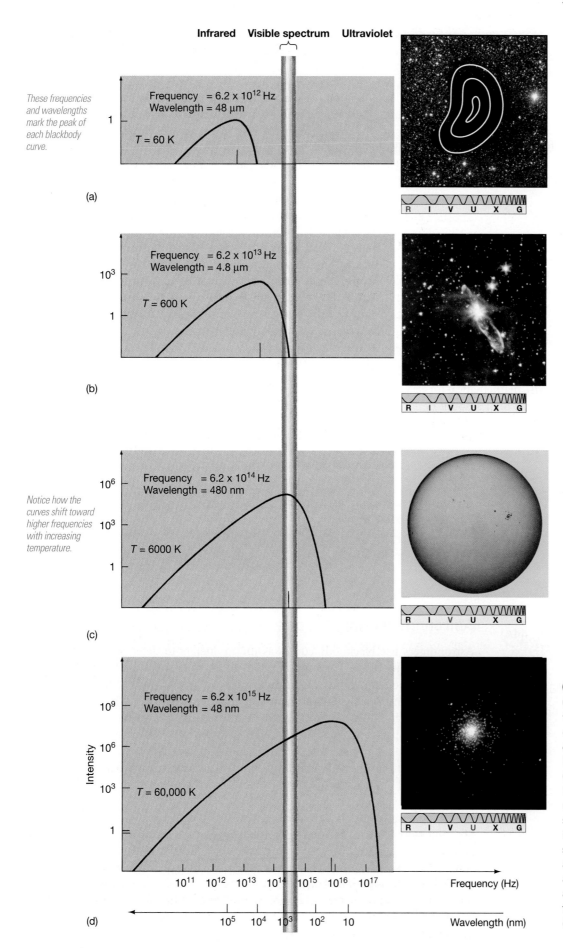

Infrared   Visible spectrum   Ultraviolet

These frequencies and wavelengths mark the peak of each blackbody curve.

Frequency   = 6.2 x 10^{12} Hz
Wavelength = 48 μm

T = 60 K

(a)

Frequency   = 6.2 x 10^{13} Hz
Wavelength = 4.8 μm

T = 600 K

(b)

Notice how the curves shift toward higher frequencies with increasing temperature.

Frequency   = 6.2 x 10^{14} Hz
Wavelength = 480 nm

T = 6000 K

(c)

Frequency   = 6.2 x 10^{15} Hz
Wavelength = 48 nm

T = 60,000 K

Intensity

Frequency (Hz)

Wavelength (nm)

(d)

INTERACTIVE ◄ FIGURE 2.11 **Blackbody Curves** Comparison of blackbody curves for four cosmic objects. (a) A cool, dark galactic cloud called Barnard 68. At a temperature of 60 K, it emits mostly radio radiation, shown here as overlaid contours. (b) A dim, young star (shown white in the inset photograph) called Herbig-Haro 46. The star's atmosphere, at 600 K, radiates mainly in the infrared. (c) The Sun's surface, at approximately 6000 K, is brightest in the visible region of the electromagnetic spectrum. (d) Some very hot, bright stars in a cluster called Messier 2, as observed by an orbiting space telescope above Earth's atmosphere. At a temperature of 60,000 K, these stars radiate strongly in the ultraviolet. (MA) *(ESO; AURA; SST; GALEX)*

2-2 MORE PRECISELY

More on the Radiation Laws

Wien's law relates the temperature T of an object to the wavelength λ_{max} at which the object's blackbody radiation spectrum peaks. (The Greek letter λ—lambda—is conventionally used to denote wavelength.) Mathematically, if we measure T in kelvins, we find that

$$\lambda_{max} = \frac{0.29 \text{ cm}}{T}.$$

Thus, at 6000 K (the approximate surface temperature of the Sun), the wavelength of maximum intensity is 0.29/6000 cm, or 480 nm, corresponding to the yellow-green part of the visible spectrum. A cooler star with a temperature of 3000 K has a peak wavelength of 970 nm, just longward of the red end of the visible spectrum, in the near infrared. The blackbody curve of a star with a temperature of 12,000 K peaks at 242 nm, in the near ultraviolet, and so on.

We can also give Stefan's law a more precise mathematical formulation. With T again measured in kelvins, the total amount of energy emitted per square meter of surface per second (a quantity known as the *energy flux F*) is given by

$$F = \sigma T^4.$$

energy per unit area | temperature to the fourth power

constant

This equation is usually referred to as the *Stefan-Boltzmann* equation. (Stefan's student, Ludwig Boltzmann, was an Austrian physicist who played a central role in the development of the laws of thermodynamics during the late 19th and early 20th centuries.) The constant σ (the Greek letter sigma) is known as the Stefan-Boltzmann constant.

Notice just how rapidly the energy flux increases with increasing temperature. A piece of metal in a furnace, when at a temperature of 3500 K, radiates energy at a rate of about 850 W for every square centimeter of its surface area. Doubling its temperature to 7000 K (so that it becomes yellow to white hot, by Wien's law) increases the energy emitted by a factor of 16 (four "doublings"), to 13.6 *kilo*watts (kW) (13,600 W) per square centimeter. Notice also that the law relates to energy emitted *per unit area*. The flame of a blowtorch is considerably hotter than a bonfire, but the bonfire emits far more energy *in total*, because it is much larger.

The SI unit of energy is the joule (J). Probably more familiar is the closely related unit called the watt (W), which measures power—the rate at which energy is emitted or expended by an object. One watt is the emission of 1 joule per second; for example, a 100-W lightbulb emits energy (mostly in the form of infrared and visible light) at a rate of 100 J/s. In these units, the Stefan-Boltzmann constant has the value $\sigma = 5.67 \times 10^{-8} \ W/m^2 \ K^4$.

CONCEPT CHECK ▶ Describe, in terms of the radiation laws, how the appearance of an incandescent lightbulb changes as you turn a dimmer switch to increase its brightness from "off" to "maximum."

Astronomical Applications

Astronomers use blackbody curves as thermometers to determine the temperatures of distant objects. For example, study of the solar spectrum makes it possible to measure the temperature of the Sun's surface. Observations of the electromagnetic radiation from the Sun at many frequencies yield a curve shaped somewhat like that shown in Figure 2.10. The Sun's curve peaks in the visible part of the electromagnetic spectrum. The Sun also emits a lot of infrared and a little ultraviolet radiation. Applying Wien's law to the blackbody curve that best fits the solar spectrum, we find that the temperature of the Sun's surface is approximately 6000 K (Figure 2.11c). A more precise measurement, based on the detailed solar spectrum (Section 2.6), yields a temperature of 5800 K.

Other cosmic objects have surfaces very much cooler or hotter than the Sun's, emitting most of their radiation in invisible parts of the spectrum. For example, the relatively cool surface of a very young star might measure 600 K and emit mostly infrared radiation (Figure 2.11b). Cooler still is the interstellar gas cloud from which the star formed. At a temperature of 60 K, such a cloud would emit mainly long-wavelength radiation in the radio and infrared parts of the spectrum (Figure 2.11a). The brightest stars, by contrast, have surface temperatures as high as 60,000 K and, hence, emit mostly ultraviolet radiation (Figure 2.11d).

2.5 Spectroscopy

Radiation can be analyzed with an instrument known as a **spectroscope**. In its most basic form, this device consists of an opaque barrier with a slit in it (to form a narrow beam of light), a prism (to split the beam into its component colors), and either a detector or a screen (to allow the user to view the resulting spectrum). Figure 2.12 shows such an arrangement.

Emission Lines

The spectra encountered in the previous section are examples of **continuous spectra.** A lightbulb, for instance, emits radiation of all wavelengths (but mostly in the visible and near-infrared ranges), with an intensity distribution that is well described by the blackbody curve corresponding to the bulb's temperature. Viewed through a spectroscope, the spectrum of the light from the bulb would show the familiar rainbow running from red to violet without interruption, as presented in Figure 2.12.

Not all spectra are continuous, however. For instance, if we took a glass jar containing pure hydrogen gas and passed an electrical discharge through it (a little like a lightning bolt arcing through Earth's atmosphere), the gas would begin to glow—that is, it would emit radiation. If we were to examine that radiation with our spectroscope, we would find that its spectrum consisted of only a few bright lines on an otherwise dark background, quite unlike the continuous spectrum of the lightbulb. Figure 2.13 shows this schematically (the lenses have been removed for clarity), and a more detailed rendering of the spectrum of hydrogen appears in the top panel of Figure 2.14. The light produced by the hydrogen in this experiment does *not* consist of all possible colors but instead includes only a few narrow, well-defined **emission lines**, narrow "slices" of the continuous spectrum. The black background represents all the wavelengths *not* emitted by hydrogen.

After some experimentation, we would also find that although we could alter the intensity of the lines (for example, by changing the amount of hydrogen in the jar or the strength of the electrical discharge), we could not alter their color (in other words, their frequency or wavelength). The particular pattern of spectral emission lines shown in Figure 2.13 is a property of the element hydrogen—whenever we perform this experiment, the same characteristic **emission spectrum** is the result. Other elements yield different emission spectra. Depending on which element is involved, the pattern of lines can be fairly simple or very complex. Always, though, it is *unique* to that element. The emission spectrum of a gas thus provides a "fingerprint" that allows scientists to deduce its presence by spectroscopic means. An analogy is a bar code that specifies uniquely the cost of a supermarket item. Examples of the emission spectra of some common substances are shown in Figure 2.14.

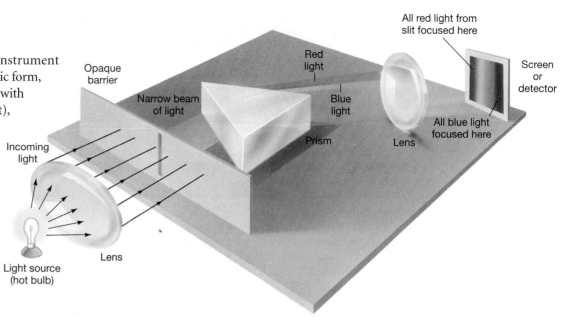

Opaque barrier
Narrow beam of light
Incoming light
Lens
Light source (hot bulb)
Prism
Red light
Blue light
Lens
All red light from slit focused here
All blue light focused here
Screen or detector

▲ **FIGURE 2.12 Spectroscope** A simple spectroscope allows a narrow beam of light to pass through a thin slit and then into a prism where it is split into its component colors. A lens then focuses the light into a sharp image that is either projected onto a screen, as shown here, or analyzed as it strikes the detector.

(MA) **SELF-GUIDED TUTORIAL** Emission Spectra

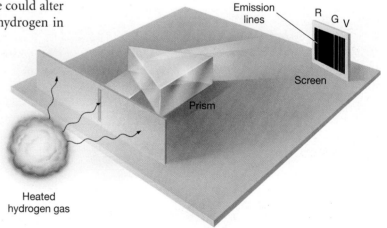

Emission lines
R G V
Prism
Screen
Heated hydrogen gas

INTERACTIVE ▶ **FIGURE 2.13 Emission Spectrum** Instead of a continuous spectrum, the light from excited (heated) hydrogen gas consists of a series of distinct spectral lines. (For simplicity, the focusing lens has been omitted.) (MA)

ANALOGY: Spectra are analogous to supermarket bar codes that uniquely specify the type and cost of a product.

US $24.00 / $36.50 CAN
ISBN 0-465-07835-4

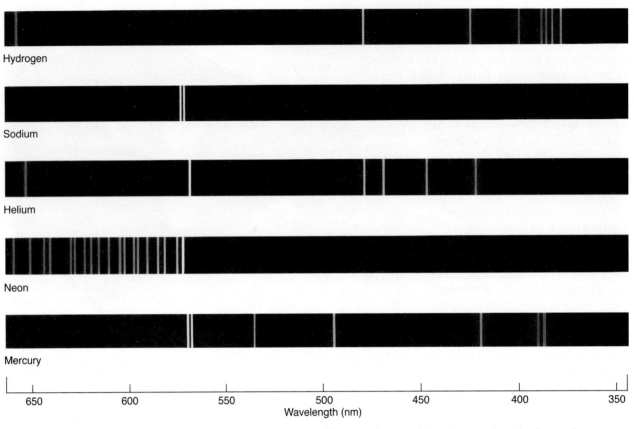

▲ **FIGURE 2.14 Elemental Emission** The emission spectra of some well-known elements. Note that wavelengths shorter than 400 nm, shown here in shades of purple, are in the ultraviolet part of the spectrum and thus invisible to the human eye. *(Wabash Instrument Corporation)*

CONCEPT
CHECK ► What are absorption and emission lines? What do they tell us about the properties of the gas producing them?

The scale of this spectrum extends from long wavelengths in red . . .

. . . to short wavelengths in blue.

Absorption Lines

When sunlight is split by a prism, at first glance it appears to produce a continuous spectrum. However, closer scrutiny shows that the solar spectrum is interrupted by a large number of narrow dark lines, as shown in Figure 2.15. These lines represent wavelengths of light that have been removed (absorbed) by gases present either in the outer layers of the Sun or in Earth's atmosphere. These gaps in the spectrum are called **absorption lines**. The absorption lines in the solar spectrum are referred to collectively as *Fraunhofer lines*, after the 19th-century German physicist Joseph Fraunhofer, who measured and cataloged more than 600 of them.

At around the time solar absorption lines were discovered, scientists found that absorption lines could also be produced in the laboratory by passing a beam of

◄ **FIGURE 2.15 Solar Spectrum** The Sun's visible spectrum shows hundreds of dark absorption lines superimposed on a bright continuous spectrum. This high-resolution spectrum is displayed in a series of 48 horizontal strips stacked vertically; each strip covers a small portion of the entire spectrum from left to right. If the strips were placed side by side, the full spectrum would be some 6 meters (20 feet) across! *(AURA)*

light from a continuous source through a cool gas, as shown in Figure 2.16. They quickly observed a connection between emission and absorption lines: The absorption lines associated with a given gas occur at precisely the *same* wavelengths as the emission lines produced when the gas is heated. Both sets of lines therefore contain the *same* information about the composition of the gas.

The study of the ways in which matter emits and absorbs radiation is called **spectroscopy**. The observed relationships between the three types of spectra—continuous, emission line, and absorption line—are illustrated in Figure 2.17 and may be summarized as follows:

1. A luminous solid or liquid, or a sufficiently dense gas, emits light of all wavelengths and so produces a continuous spectrum of radiation (Figure 2.12).
2. A low-density hot gas emits light whose spectrum consists of a series of bright emission lines. These lines are characteristic of the chemical composition of the gas (Figure 2.13).
3. A low-density cool gas absorbs certain wavelengths from a continuous spectrum, leaving dark absorption lines in their place, superimposed on the continuous spectrum. These lines are characteristic of the composition of the intervening gas. They occur at precisely the same wavelengths as the emission lines produced by the gas at higher temperatures (Figure 2.16).

These rules are collectively known as **Kirchhoff's laws**, after the German physicist Gustav Kirchhoff, who published them in 1859.

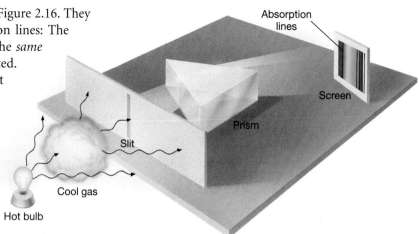

INTERACTIVE ▲ FIGURE 2.16 **Absorption Spectrum** When a cool gas is placed between a source of continuous radiation (such as a hot lightbulb) and the detector/screen, the resulting color spectrum is crossed by a series of dark absorption lines. These lines are formed when the intervening cool gas absorbs certain wavelengths (colors) from the original beam of light. The absorption lines appear at precisely the same wavelengths as the emission lines that would be produced if the gas **(MA)** were heated to high temperatures (see Figure 2.13).

(MA) **SELF-GUIDED TUTORIAL** Absorption Spectra

Astronomical Applications

Once astronomers realized that spectral lines were indicators of chemical composition, they set about identifying the observed lines in the Sun's spectrum. Almost all of the lines observed from extraterrestrial sources could be attributed to known elements. For example, many of the Fraunhofer lines in sunlight are associated with the element iron. However, some unfamiliar lines also appeared in the solar spectrum. In 1868 astronomers realized that those lines must correspond to a previously unknown element. It was given the name helium, after the Greek word *helios*, meaning "Sun." Only in 1895, almost three decades after its detection in sunlight, was helium discovered on Earth.

The development of spectroscopy is another example of the scientific method in action. ∞ *(Sec. 0.5)* As technology evolved and experimental measurement techniques improved, scientists realized that spectra were unique fingerprints

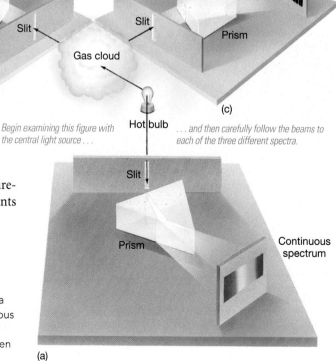

Begin examining this figure with the central light source . . .

. . . and then carefully follow the beams to each of the three different spectra.

▶ FIGURE 2.17 **Kirchhoff's Laws** A source of continuous radiation, here represented by a lightbulb, is used to illustrate Kirchhoff's laws of spectroscopy. (a) The unimpeded beam shows the familiar continuous spectrum of colors. (b) When the source is viewed through a cloud of hydrogen gas, a series of dark hydrogen absorption lines appears in the continuous spectrum. These lines are formed when the gas absorbs some of the bulb's radiation and reemits it in random directions. (c) When the gas is viewed from the side, a fainter hydrogen emission spectrum is seen, consisting of reemitted radiation.

of matter. They then went on to use this knowledge as a means of determining the composition of objects that were otherwise impossible to reach.

Yet for all the information that 19th-century astronomers could extract from observations of stellar spectra, they still lacked a theory explaining *how* those spectra arose. Despite their sophisticated spectroscopic equipment, they knew little more about the physics of stars than did Galileo or Newton. The next step in the circle of scientific progress was an explanation of Kirchhoff's empirical laws in terms of fundamental physical principles, just as Newtonian mechanics explained Kepler's empirical laws of planetary motion. ∞ *(Sec. 1.4)* And as with Newton's laws, the new theory—called *quantum mechanics*—opened the door to an explosion of scientific understanding far beyond the context in which it was originally conceived.

To understand how scientists use spectroscopy to obtain detailed information about astronomical objects from the light they emit, we must delve more deeply into the processes that produce line spectra.

2.6 Formation of Spectral Lines

By the start of the 20th century, physicists had accumulated substantial evidence that light sometimes behaves in a manner that simply cannot be explained by the wave theory of radiation. As we have just seen, absorption and emission lines are produced only at certain, very specific wavelengths. This would not be the case if light behaved only as a continuous wave and matter always obeyed the laws of Newtonian mechanics. It became clear that when light interacts with matter on very small scales, it does so not in a smooth, continuous way but in a discontinuous, stepwise manner. The challenge was to find an explanation for this unexpected behavior. The solution revolutionized our view of nature and now forms the foundation not just for physics and astronomy, but for virtually all of modern science.

Atomic Structure

To explain the formation of spectral lines, we must understand not just the nature of light but also something of the structure of **atoms**—the microscopic building blocks from which all matter is constructed. Let us start with the simplest atom, hydrogen, which consists of a single negatively charged electron orbiting a positively charged proton. The proton forms the central **nucleus** (plural: nuclei) of the atom. Because the positive charge on the proton exactly cancels the negative charge on the electron, the hydrogen atom as a whole is electrically neutral.

How does this picture of the hydrogen atom relate to the characteristic emission and absorption lines associated with hydrogen gas? If an atom absorbs some energy in the form of radiation, that energy must cause some internal change. And if the atom emits energy, it must come from somewhere within the atom. The energy absorbed or emitted by the atom is associated with changes in the motion of the orbiting electron.

In 1913, Danish physicist Niels Bohr developed a model of the atom that provided the first explanation of hydrogen's observed spectral lines. His work earned him the 1922 Nobel Prize in Physics. Now known simply as the **Bohr model** of the atom, its essential features are as follows. First, there is a state of lowest energy—the **ground state**—which represents the "normal" condition of the electron as it orbits the nucleus. Second, there is a maximum energy that the electron can have and still be part of the atom. If the electron acquires more than that maximum energy, it is no longer bound to the nucleus, and the atom is said to be *ionized*. An atom having fewer (or more) than its normal complement of electrons, and, hence, a net electrical charge, is called an **ion**. Third, and most important (and also least intuitive), between those two energy levels, the electron can exist only in certain sharply defined energy states, often referred to as *orbitals*.

An atom is said to be in an **excited state** when an electron occupies an orbital other than the ground state. The electron then lies at a greater than normal distance from its parent nucleus, and the atom has a greater than normal amount of energy. The excited state with the lowest energy (that is, the one closest to the ground state) is called the *first excited state*, that with the second-lowest energy the *second excited state*, and so on.

In Bohr's model, each electron orbital was pictured as having a specific radius, like a planetary orbit in the solar system, as shown in Figure 2.18. The modern view is not so simple. Although each orbital *does* have a precise energy, the electron is now envisioned as being smeared out in an *electron cloud* surrounding the nucleus, as illustrated in Figure 2.19. It is common to speak of the average distance from the cloud to the nucleus as the radius of the electron's orbital. When a hydrogen atom is in its ground state, the radius of the orbital is about 0.05 nm (0.5 Å). As the orbital energy increases, the radius increases, too. For clarity the diagrams below use solid lines to represent electron orbitals, but bear in mind that the fuzziness in Figure 2.19 is a more accurate depiction of reality.

An atom can become excited by absorbing some light energy from a source of electromagnetic radiation or by colliding with some other particle—another atom, for example. However, the atom cannot stay in that state forever. After about 10^{-8} s (second), it returns to its ground state.

The Particle Nature of Radiation

Here now is the crucial point that links atoms to radiation and allows us to interpret atomic spectra. Because electrons may exist only in orbitals having specific energies, atoms can absorb only specific amounts of energy as their electrons are boosted into excited states. Likewise, atoms can emit only specific amounts of energy as their electrons fall back to lower energy states. Thus, the amount of light energy absorbed or emitted in these processes *must correspond precisely to the energy difference between two orbitals*. This requires that light must be absorbed and emitted in the form of little "packets" of electromagnetic radiation, each carrying a very specific amount of energy. We call these packets **photons.** A photon is, in effect, a "particle" of electromagnetic radiation.

The idea that light sometimes behaves not as a continuous wave but as a stream of particles was first suggested by Albert Einstein in 1905. By that time, the experimental evidence clearly indicated that, on microscopic scales, electromagnetic radiation often displayed particle properties. Scientists understood that the notion of radiation as a wave was incomplete, but they did not know how to reconcile its two seemingly contradictory natures. Einstein's great breakthrough (for which he won the 1919 physics Nobel Prize) came when he realized that all of the puzzling experiments could be explained by a simple, but critically important, connection between the particle and wave aspects of light. He found that the energy contained within a photon had to be proportional to the frequency of the radiation:

$$\text{photon energy} \propto \text{radiation frequency}.$$

Thus, for example, a "red" photon having a frequency of 4×10^{14} Hz (corresponding to a wavelength of about 750 nm, or 7500 Å) has 4/7 the energy of a "blue" photon, of frequency of 7×10^{14} Hz. Because it connects the *energy* of a photon with the *color*

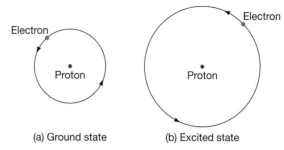

(a) Ground state   (b) Excited state

▲ **FIGURE 2.18 Classical Atom**   An early-20th-century conception of the hydrogen atom—the Bohr model—pictured its electron orbiting the central proton in a well-defined orbit, much like a planet orbiting the Sun. Two electron orbits of different energies are shown: (a) the ground state and (b) an excited state.

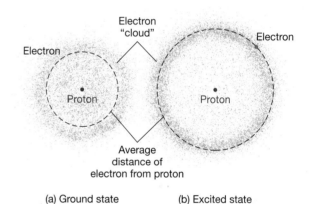

(a) Ground state   (b) Excited state

▲ **FIGURE 2.19 Modern Atom**   The modern view of the hydrogen atom sees the electron as a "cloud" surrounding the nucleus.

MA ANIMATION/VIDEO Classical Hydrogen Atom I

MA ANIMATION/VIDEO Classical Hydrogen Atom II

of the light it represents, this relationship is the final piece in the puzzle of how to understand the spectra we see.

Environmental conditions ultimately determine which description—wave or stream of particles—better fits the behavior of electromagnetic radiation. As a general rule of thumb, in the macroscopic realm of everyday experience, radiation is more usefully described as a wave, and in the microscopic domain of atoms, it is best characterized as a stream of particles.

The Spectrum of Hydrogen

Figure 2.20 illustrates the absorption and emission of photons by a hydrogen atom. In part (a), the atom absorbs a photon of radiation and makes a transition from the ground state to the first excited state, then emits a photon of precisely the same energy and drops back to the ground state. The energy difference between the two states corresponds to an ultraviolet photon of wavelength 121.6 nm (1216 Å).

Figure 2.20(b) depicts the absorption of a more energetic (higher frequency, shorter wavelength) ultraviolet photon, this one having a wavelength of 102.6 nm

ANIMATION/VIDEO Photon Emission **MA**

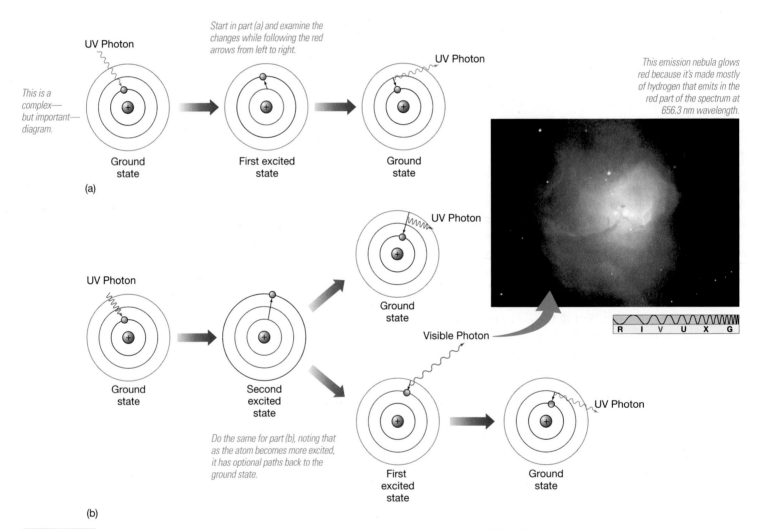

(a)

(b)

INTERACTIVE ▲ **FIGURE 2.20 Atomic Excitation** (a) Absorption of an ultraviolet (UV) photon (left) by a hydrogen atom causes the momentary excitation of the atom into its first excited state (center). Eventually, the atom returns to its ground state (right), in the process emitting a photon having exactly the same energy as the original photon. (b) Absorption of a higher-energy UV photon may boost the atom into a higher excited state, from which there are several possible paths back to the ground state. At the top, the electron falls immediately back to the ground state, emitting a photon identical to the one it absorbed. At the bottom, the electron initially falls into the first excited state, producing visible radiation of wavelength 656.3 nm—the characteristic red glow of excited hydrogen. (Inset: NASA) **MA**

(1026 Å), causing the atom to jump to the *second* excited state. From that state the electron may return to the ground state via either one of two alternate paths.

1. It can proceed directly back to the ground state, in the process emitting an ultraviolet 102.6-nm photon identical to the one that excited the atom in the first place.
2. Alternatively, it can *cascade* down one orbital at a time, emitting *two* photons: one having an energy equal to the difference between the second and first excited states and the other having an energy equal to the difference between the first excited state and the ground state.

The second step in the cascade produces a 121.6-nm ultraviolet photon, just as in Figure 2.20(a). However, the first part of the cascade—the one from the second to the first excited state—produces a photon of wavelength 656.3 nm (6563 Å), which is in the visible part of the electromagnetic spectrum. This photon is seen as red light. An individual atom—if one could be isolated—would emit a momentary red flash. The inset in Figure 2.20 shows an astronomical object whose red color is the result of precisely this process.

Absorption of still more energy can boost the electron to even higher orbitals within the atom. As the excited electron cascades back down to the ground state, the atom may emit many photons, each with a different energy and, hence, a different color. In this case, the resulting spectrum shows many distinct spectral lines. For hydrogen, all transitions ending at the ground state produce ultraviolet photons. However, downward transitions ending at the *first* excited state give rise to spectral lines in or near the visible portion of the electromagnetic spectrum (Figure 2.14). Because they form the most easily observable part of the hydrogen spectrum and were the first to be discovered, these lines (also known as *Balmer lines*) are often referred to simply as the "hydrogen series" and are denoted by the letter H. Individual transitions are labeled with Greek letters, in order of increasing energy (decreasing wavelength): The Hα (H alpha) line corresponds to the transition from the second to the first excited state and has a wavelength of 656.3 nm (red); Hβ (H beta; third to first) has wavelength 486.1 nm (green); Hγ (H gamma; fourth to first) has wavelength 434.1 nm (blue); and so on. We will use these designations (especially Hα and Hβ) frequently in later chapters.

PROCESS OF ▶ SCIENCE CHECK In what ways was the wave theory of radiation unable to account for detailed observations of atomic spectra? Describe some key aspects of the theory of quantum mechanics, and explain how the new theory explains spectral lines.

Kirchhoff's Laws Explained

Let's reconsider our earlier discussion of emission and absorption lines in terms of the model just presented. In Figure 2.17(b) a beam of radiation shines through a cloud of gas. The beam contains photons of all energies, but most of them do not interact with the gas because the gas can absorb only photons having precisely the right energy to cause an electron to jump from one orbital to another. Photons having energies that cannot produce such a jump pass through the gas unhindered. Photons having the right energies are absorbed, excite the gas, and are removed from the beam. This is the cause of the dark absorption lines in the spectrum. These lines are direct indicators of the energy differences between orbitals in the atoms making up the gas.

The excited gas atoms rapidly return to their original states, each emitting one or more photons in the process. Most of these reemitted photons leave at angles that do *not* take them through the slit and on to the detector. A second detector looking at the cloud from the side (Figure 2.17c) would record the reemitted energy as an emission spectrum. (This is what we are seeing in the inset to Figure 2.20.) Like the absorption spectrum, the emission spectrum is characteristic of the gas, not of the original beam.

The case of a *continuous* spectrum in Figure 2.17(c) is shown for emitted photons escaping from the bulb without further interaction with matter. Actually, the situation in a denser source of radiation (a thick gas cloud or in a liquid or solid body) is more complex. There, a photon is likely to interact with atoms, free electrons, and ions in the body many times before finally escaping, exchanging some

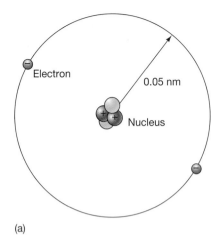

(a)

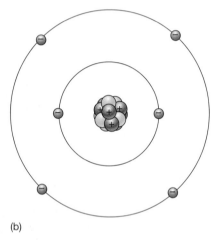

(b)

▲ **FIGURE 2.21 Helium and Carbon** (a) A helium atom in its ground state has two electrons within the lowest-energy orbital around a nucleus containing two protons and two neutrons. (b) A carbon atom in its ground state has six electrons orbiting around a six-proton, six-neutron nucleus—two of the electrons in an inner orbital and the other four at a greater distance from the center.

CONCEPT CHECK ▶ How does the structure of an atom determine the atom's emission and absorption spectra?

▼ **FIGURE 2.22 Hydrogen Spectra** The emission spectrum of molecular hydrogen (a) differs greatly from that of simpler atomic hydrogen (b). *(Bausch & Lomb Inc.)*

energy with the matter at each encounter. The net result is that the emitted radiation displays a continuous spectrum, in accordance with Kirchhoff's first law. The spectrum is approximately that of a blackbody with the same temperature as the source.

More Complex Spectra

All hydrogen atoms have the same structure—a single electron orbiting a single proton—but, of course, there are many other kinds of atoms, each having a unique internal structure. The number of protons in the nucleus of an atom determines which **element** the atom represents. That is, just as all hydrogen atoms have a single proton, all oxygen atoms have 8 protons, all iron atoms have 26 protons, and so on.

The next simplest element after hydrogen is helium. Its nucleus is made up of two protons and two **neutrons** (another elementary particle with a slightly larger mass than a proton, but carrying no electrical charge). Two electrons orbit the nucleus. As with hydrogen and all other atoms, the "normal" condition for helium is to be electrically neutral, with the negative charge of the orbiting electrons exactly canceling the positive charge of the nucleus (Figure 2.21a).

More complex atoms contain more protons (and neutrons) in the nucleus and have correspondingly more orbiting electrons. For example, an atom of carbon (Figure 2.21b) consists of six electrons orbiting a nucleus containing six protons and six neutrons. As we progress to heavier and heavier elements, the number of orbiting electrons increases, and consequently the number of possible electronic transitions rises rapidly. The number increases further as the temperature increases and atoms become excited or even ionized. The result is that very complicated spectra can be produced. The complexity of atomic spectra generally reflects the complexity of the source atoms. A good example is the element iron, which contributes several hundred of the Fraunhofer absorption lines seen in the solar spectrum. The many possible transitions of its 26 orbiting electrons yield an extremely rich line spectrum.

Even more complex spectra are produced by **molecules.** A molecule is a tightly bound group of atoms held together by interactions among their orbiting electrons—interactions called *chemical bonds.* Much like atoms, molecules can exist only in certain well-defined energy states, and they produce emission or absorption spectral lines when they make a transition from one state to another. Because molecules are more complex than atoms, the rules of molecular physics are also much more complex. Nevertheless, as with atomic spectral lines, painstaking experimental work over many decades has determined the precise frequencies at which millions of molecules emit and absorb radiation. These lines are molecular fingerprints, just like their atomic counterparts, enabling researchers to identify and study one kind of molecule to the exclusion of all others.

Molecular lines usually bear little resemblance to the spectral lines associated with their component atoms. For example, Figure 2.22(a) shows the emission spectrum of the simplest molecule known—molecular hydrogen. Notice how different it is from the spectrum of atomic hydrogen shown in part (b).

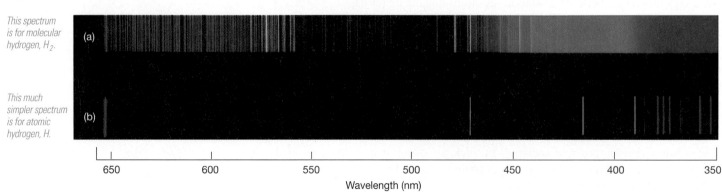

This spectrum is for molecular hydrogen, H_2.

(a)

This much simpler spectrum is for atomic hydrogen, H.

(b)

650 600 550 500 450 400 350

Wavelength (nm)

2.7 The Doppler Effect

Most of us have had the experience of hearing the pitch of a train whistle change from high shrill (high frequency, short wavelength) to low blare (low frequency, long wavelength) as the train approaches and then recedes. This motion-induced change in the observed frequency of a wave is known as the **Doppler effect**, in honor of Christian Doppler, the 19th-century Austrian physicist who first explained it. Applied to cosmic sources of electromagnetic radiation, it has become one of the most important observational tools in all of 20th-century astronomy. Here's how it works.

Imagine a wave moving from the place where it is generated toward an observer who is not moving with respect to the wave source (Figure 2.23a). By noting the distances between successive wave crests, the observer can determine the wavelength of the emitted wave. Now suppose that the wave source begins to move (Figure 2.23b). Because the source moves between the times of emission of one wave crest and the next, wave crests in the direction of motion of the source are seen to be *closer together* than normal, while crests behind the source are more widely spaced. Thus, an observer in front of the source measures a *shorter* wavelength than normal, while one behind sees a *longer* wavelength.

The greater the relative speed of source and observer, the greater the observed shift. In terms of the net velocity of *recession* between source and observer, the apparent wavelength and frequency (measured by the observer) are related to the true quantities (emitted by the source) by

$$\frac{\text{apparent wavelength}}{\text{true wavelength}} = \frac{\text{true frequency}}{\text{apparent frequency}}$$

$$= 1 + \frac{\text{recession velocity}}{\text{wave speed}}.$$

A positive recession velocity means that the source and the observer are moving apart; a negative value means that they are approaching. The wave speed is the speed of light c in the case of electromagnetic radiation. For most of this text, the recession velocity will be small compared to the speed of light. Only when we come to discuss the properties of black holes (Chapter 13) and the structure of the universe on the largest scales (Chapter 17) will we have to reconsider this formula.

Note that, in the figure, the source is shown in motion. However, the same general statements hold whenever there is any *relative* motion between source and observer. Note also that only motion along the line joining source and observer, known as *radial* motion, appears in the above equation. Motion *transverse* (perpendicular) to the line of sight has no significant effect.

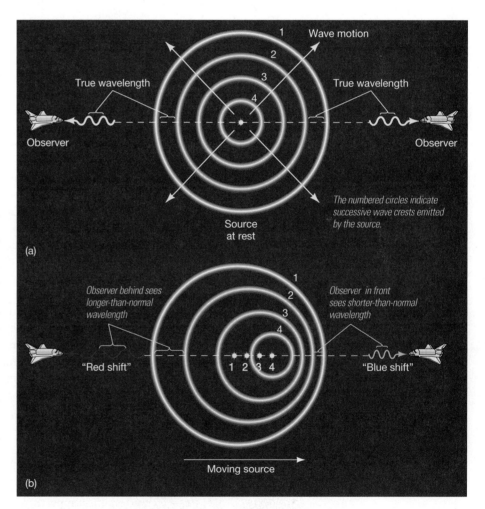

(a)

(b)

INTERACTIVE ▶ FIGURE 2.23 Doppler Effect   (a) Wave motion from a source toward an observer at rest with respect to the source. As seen by the observer, the source is not moving, so the wave crests are just concentric spheres (shown here as circles). (b) Waves from a moving source tend to "pile up" in the direction of motion and be "stretched out" on the other side. As a result, an observer situated in front of the source measures a shorter-than-normal wavelength—a blueshift—while an observer behind the source sees a redshift.

ANALOGY: Police use handheld radar guns—and the Doppler effect—to catch speeders.

Reflected radiation (blue crests) from the oncoming car shifts toward shorter wavelengths.

These are realistic spectra emitted by objects moving at different speeds.

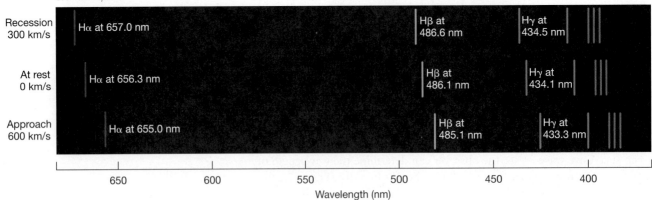

The middle spectrum is unshifted for an object at rest.

Recession 300 km/s — Hα at 657.0 nm · Hβ at 486.6 nm · Hγ at 434.5 nm

At rest 0 km/s — Hα at 656.3 nm · Hβ at 486.1 nm · Hγ at 434.1 nm

Approach 600 km/s — Hα at 655.0 nm · Hβ at 485.1 nm · Hγ at 433.3 nm

Wavelength (nm) — 650 · 600 · 550 · 500 · 450 · 400

INTERACTIVE ▲ **FIGURE 2.24 Doppler Shift** The Doppler effect shifts the entire spectrum of a moving object to higher or lower frequencies. The spectrum at top shows the redshift of the hydrogen lines from an object moving at a speed of 300 km/s away from the observer; that at bottom shows those same lines blueshifted while the object moves at double that speed toward the observer.

SELF-GUIDED TUTORIAL Doppler Effect

 CONCEPT CHECK ▶ How might the Doppler effect be used in determining the mass of a distant star?

In astronomical jargon, the wave measured by an observer situated in front of a moving source is said to be *blueshifted*, because blue light has a shorter wavelength than red light. Similarly, an observer situated behind the source will measure a longer-than-normal wavelength—the radiation is said to be *redshifted*. This terminology holds even for invisible radiation, for which "red" and "blue" have no meaning. Any shift toward shorter wavelengths is called a blueshift, and any shift toward longer wavelengths is called a redshift.

The Doppler effect is not normally noticeable in visible light on Earth—the speed of light is so large that the wavelength change is far too small to be noticeable for everyday terrestrial velocities. However, using spectroscopic techniques, astronomers routinely use the Doppler effect to measure the line-of-sight velocity of cosmic objects by determining the extent to which known spectral lines are shifted to longer or shorter wavelengths. For example, suppose that an astronomer observes the red Hα line in the spectrum of a star to have a wavelength of 657 nm instead of the 656.3 nm measured in the laboratory. (How does she know it is the same line? Because, as illustrated in Figure 2.24, she realizes that *all* the hydrogen lines are shifted by the same fractional amount—the characteristic pattern of lines identifies hydrogen as the source.) Using the above equation, she calculates that the star's radial velocity is 657/656.3 − 1 = 0.0056 times the speed of light. In other words, the star is receding from Earth at a rate of 320 km/s.

The motions of nearby stars and distant galaxies—even the expansion of the universe itself—have all been measured in this way. Motorists stopped for speeding on the highway have experienced another, much more down-to-earth, application. As illustrated in the inset to Figure 2.23, police radar measures speed by means of the Doppler effect, as do the radar guns used to clock the velocity of a pitcher's fastball or a tennis player's serve.

2.8 Spectral-Line Analysis

Astronomers apply the laws of spectroscopy in analyzing radiation from beyond Earth. A nearby star or a distant galaxy takes the place of the lightbulb in our previous examples, an interstellar cloud or a stellar (or even planetary) atmosphere plays the role of the intervening cool gas, and a spectrograph attached to a telescope replaces our simple prism and detector. We list below some of the properties of emitters and absorbers that can be determined by careful analysis of radiation received on (or near) Earth. We will encounter other important examples as our study of the cosmos unfolds.

1. The *composition* of an object is determined by matching its spectral lines with the laboratory spectra of known atoms and molecules.

2. The *temperature* of an object emitting a continuous spectrum can be measured by matching the overall distribution of radiation with a blackbody curve. Temperature may also be determined from detailed studies of spectral lines. In Section 10.3 we will see how stellar temperatures are most accurately measured by spectroscopic means.

3. The *(line-of-sight) velocity* of an object is measured by determining the Doppler shift of its spectral lines.

4. An object's *rotation rate* can be determined by measuring the *broadening* (smearing out over a range of wavelengths) produced by the Doppler effect in emitted or reflected spectral lines.

5. The *pressure* of the gas in the emitting region of an object can be measured by its tendency to broaden spectral lines. The greater the pressure, the broader the line.

6. The *magnetic field* of an object can be inferred from a characteristic splitting it produces in many spectral lines, when a single line divides into two. (This is known as the *Zeeman effect*.)

Given sufficiently sensitive equipment, there is almost no end to the wealth of data contained in starlight. However, deciphering the extent to which each of many competing factors influences a spectrum can be a very difficult task. Typically, the spectra of many elements are superimposed on one another, and several physical processes are occurring simultaneously, each modifying the spectrum in its own way. The challenge facing astronomers is to unravel the extent to which each mechanism contributes to spectral line profiles and therefore obtain meaningful information about the source of the lines.

CONCEPT CHECK ▶ Why is it important for astronomers to analyze spectral lines in detail?

CHAPTER REVIEW

SUMMARY

LO1 **Electromagnetic radiation** (p. 44) travels through space in the form of a **wave** (p. 44). Any electrically charged object is surrounded by an **electric field** (p. 46) that determines the force it exerts on other charged objects. When a charged particle moves, information about that motion is transmitted via the particle's changing **electric** and **magnetic fields** (pp. 46, 47). The information travels at the **speed of light** (p. 47) as an electromagnetic wave.

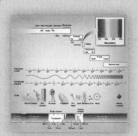

LO2 The **electromagnetic spectrum** (p. 50) consists of (in order of decreasing wavelength, or increasing frequency) **radio waves**, **infrared radiation**, **visible light**, **ultraviolet radiation**, **X-rays**, and **gamma rays** (p. 44). The **opacity** (p. 50) of Earth's atmosphere—the extent to which it absorbs radiation—varies greatly with the wavelength of the radiation. Only radio waves, some infrared wavelengths, and visible light can penetrate the atmosphere and reach the ground.

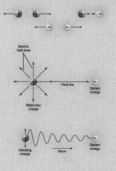

LO3 The **temperature** (p. 51) of an object is a measure of the speed with which its constituent particles move. The **intensity** (p. 51) of radiation emitted by an object has a characteristic distribution, called a **blackbody curve** (p. 52), that depends only on the temperature of the object. **Wien's law** (p. 52) tells us that the wavelength at which the object's radiation peaks is inversely proportional to its temperature. **Stefan's law** (p. 52) states that the total amount of energy radiated is proportional to the fourth power of the temperature.

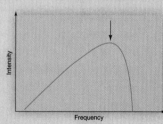

LO4 Many hot objects emit a **continuous spectrum** (p. 55) of radiation, containing light of all wavelengths. A hot gas may instead produce an **emission spectrum** (p. 55), consisting only of a few well-defined **emission lines** (p. 55) of specific frequencies, or colors. Passing a continuous beam of radiation through cool gas will produce **absorption lines** (p. 56) at precisely the same frequencies as would be present in the gas's emission spectrum.

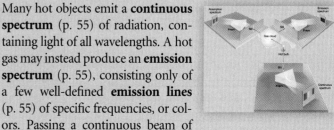

L05 **Atoms** (p. 58) are made up of negatively charged electrons orbiting a positively charged **nucleus** (p. 58) consisting of positively charged protons and electrically neutral **neutrons** (p. 62). The number of protons determines the type of **element** (p. 62) the atom represents. In the **Bohr model** (p. 58), a hydrogen atom has a minimum-energy **ground state** (p. 58), representing its "normal" condition. When the electron has a higher-than-normal energy, the atom is in an **excited state** (p. 59). For any given atom, only certain, well-defined energies are possible. In the modern view, the electron is envisaged as being spread out in a "cloud" around the nucleus but still having a sharply defined energy.

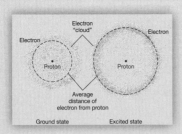

L06 When an electron moves from one energy state to another in an atom, the difference in the energy between the states is emitted or absorbed in the form of "packets" of electromagnetic radiation—**photons** (p. 59). Because the energy levels have definite energies,

the photons also have definite energies that are characteristic of the type of atom involved. The energy of a photon determines the frequency, and hence the color, of the light emitted or absorbed.

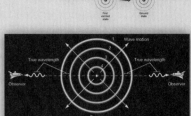

L07 Our perception of the wavelength of a beam of light can be altered by the source's velocity relative to us. This motion-induced change in the observed frequency of a wave is called the **Doppler effect** (p. 63). Any net motion of the source away from the observer causes a redshift—a shift to lower frequencies—in the received beam. Motion toward the observer causes a blueshift. The extent of the shift is directly proportional to the source's radial velocity relative to the observer.

MasteringAstronomy®   For instructor-assigned homework go to www.masteringastronomy.com

Problems labeled **POS** explore the process of science.   **VIS** problems focus on reading and interpreting visual information.   **LO** connects to the introduction's numbered Learning Outcomes.

REVIEW AND DISCUSSION

1. Define the following wave properties: period, wavelength, amplitude, frequency.

2. Compare and contrast the gravitational and electric forces.

3. **LO1** Describe how light radiation leaves a star, travels through the vacuum of space, and finally is seen by someone on Earth.

4. **LO2** What do radio waves, infrared radiation, visible light, ultraviolet radiation, X-rays, and gamma rays have in common? How do they differ?

5. In what regions of the electromagnetic spectrum is the atmosphere transparent enough to allow observations from the ground?

6. What is a blackbody? Describe the radiation it emits.

7. **POS** If Earth were completely blanketed with clouds and we couldn't see the sky, could we learn about the realm beyond the clouds? What forms of radiation might penetrate the clouds and reach the ground?

8. **LO3** In terms of its blackbody curve, describe what happens as a red-hot glowing coal cools.

9. **LO6** **POS** Explain how astronomers might use spectroscopy to determine the composition and temperature of a star.

10. **LO4** What are continuous, emission, and absorption spectra? How are they produced?

11. What is the normal condition for atoms? What is an excited atom? What are orbitals?

12. **LO5** Why do atoms absorb and reemit radiation at characteristic frequencies?

13. **POS** Suppose a luminous cloud of gas is discovered emitting an emission spectrum. What can be learned about the cloud from this observation?

14. What is the Doppler effect, and how does it alter the way in which we perceive radiation?

15. **LO7** How do astronomers use the Doppler effect to determine the velocities of astronomical objects?

CONCEPTUAL SELF-TEST: TRUE OR FALSE?/MULTIPLE CHOICE

1. The wavelength of green light is about the size of an atom. (T/F)

2. Two otherwise identical objects have temperatures of 1000 K and 1200 K, respectively. The object at 1200 K emits roughly twice as much radiation as the object at 1000 K. (T/F)

3. As you drive away from a radio transmitter, the radio signal you receive from the station is shifted to longer wavelengths. (T/F)

4. Imagine an emission spectrum produced by a container of hydrogen gas. Changing the amount of hydrogen in the container will change the colors of the lines in the spectrum. (T/F)

5. In the previous question, changing the gas in the container from hydrogen to helium will change the colors of the lines occurring in the spectrum. (T/F)

6. The energy of a photon is inversely proportional to the wavelength of the radiation. (T/F)

7. An electron moves to a higher energy level in an atom after absorbing a photon of a specific energy. (T/F)

8. Compared with ultraviolet radiation, infrared radiation has a greater (a) wavelength; (b) amplitude; (c) frequency; (d) energy.

9. An X-ray telescope located in Antarctica would not work well because of (a) the extreme cold; (b) the ozone hole; (c) continuous daylight; (d) Earth's atmosphere.

10. A star much cooler than the Sun would appear (a) red; (b) blue; (c) smaller; (d) larger.

11. The blackbody curve of a star moving toward Earth would have its peak shifted (a) to lower intensity; (b) toward higher energies; (c) toward longer wavelengths; (d) toward lower energies.

12. The visible spectrum of sunlight reflected from Saturn's cold moon Titan would be expected to be (a) continuous; (b) an emission spectrum; (c) an absorption spectrum.

13. Astronomers analyze starlight to determine a star's (a) temperature; (b) composition; (c) motion; (d) all of the above.

14. VIS According to Figure 2.11 (Blackbody Curves), an object having a temperature of 1000 K emits mostly (a) infrared light; (b) visible light; (c) ultraviolet light; (d) X-rays.

15. VIS In Figure 2.20 (Atomic Excitation), the total energy of the two photons emitted in the branch at the lower right is (a) greater than, (b) less than, (c) approximately equal to, (d) exactly equal to, the energy of the UV photon absorbed at the left.

PROBLEMS

The number of squares preceding each problem indicates its approximate level of difficulty.

1. ■ A sound wave moving through water has a frequency of 256 Hz and a wavelength of 5.77 m. What is the speed of sound in water?

2. ■ What is the wavelength of a 100-MHz (FM 100) radio signal?

3. ■ What would be the frequency of an electromagnetic wave having a wavelength equal to Earth's diameter (12,800 km)? In what part of the electromagnetic spectrum would such a wave lie?

4. ■■ The blackbody emission spectrum of object A peaks in the ultraviolet region of the electromagnetic spectrum at a wavelength of 200 nm. That of object B peaks in the red region, at 650 nm. Which object is hotter, and, according to Wien's law, how many times hotter is it? According to Stefan's law, how many times more energy per unit area does the hotter body radiate per second?

5. ■ Normal human body temperature is about 37°C. What is this temperature in kelvins? What is the peak wavelength emitted by a person with this temperature? In what part of the spectrum does this lie?

6. ■■ According to the Stefan-Boltzmann law, how much energy is radiated into space per unit time by each square meter of the Sun's surface (see *More Precisely 2-2*)? If the Sun's radius is 696,000 km, what is the total power output of the Sun?

7. ■ By what factor does the energy of a 1-nm X-ray photon exceed that of a 10-MHz radio photon? How many times more energy has a 1-nm gamma ray than a 10-MHz radio photon?

8. ■■ How many different photons (that is, photons of different frequencies) can be emitted as a hydrogen atom in the second excited state falls back, directly or indirectly, to the ground state? What are their wavelengths? What about a hydrogen atom in the third excited state?

9. ■ The Hα line (Section 2.6) of a star is received on Earth at a wavelength of 656 nm. What is the star's radial velocity relative to Earth?

10. ■■■ You are observing a spacecraft moving in a circular orbit of radius 100,000 km around a distant planet. You happen to be located in the plane of the spacecraft's orbit. You find that the spacecraft's radio signal varies periodically in wavelength between 2.99964 m and 3.00036 m. Assuming that the radio is broadcasting at a constant wavelength, what is the mass of the planet?

ACTIVITIES

Collaborative

1. Find a spectrum of the Sun with a wavelength scale on it. Google is a good place to start. Select some absorption lines and determine their wavelengths by interpolation. Now, try to identify the element that produced these lines. Use a reference such as Moore's *A Multiplet Table of Astrophysical Interest,* available on the NASA Astrophysics Data System. Work with the darkest lines before trying the fainter ones. How many elements can you find?

2. Stand near (but not too near!) a train track or busy highway and wait for a train or traffic to pass by. Can you notice the Doppler effect in the pitch of the engine noise or whistle blowing? How does the sound frequency depend on the train's (a) speed and (b) motion toward or away from you? Divide your group into two. One subgroup should time the train's motion and hence calculate approximately its speed. The other (consisting of the more musically inclined!) should estimate the perceived frequency change of the whistle when the train is moving first toward you and then away from you.

Individual

1. Locate the constellation Orion. Its two brightest stars are Betelgeuse and Rigel. Which is hotter? How can you tell? Which of the other stars scattered across the night sky are hot, and which are cool?

2. Obtain a handheld spectroscope, available from your school science lab or online. In the shade, point the spectroscope at a white cloud or white piece of paper that is in direct sunlight. Look for the absorption lines in the Sun's spectrum. Note their wavelength from the scale inside the spectroscope. How many of the lines can you identify by comparing your list with the Fraunhofer lines given in many astronomy reference books or on Wikipedia?

3

Telescopes

The Tools of Astronomy

At its heart, astronomy is an observational science. Painstaking observations of cosmic phenomena almost always precede any clear theoretical understanding of their nature. As a result, our detecting instruments—our telescopes—have evolved to observe as broad a range of wavelengths as possible. Until the middle of the 20th century, telescopes were limited to visible light. Since then, however, technological advances have expanded our view of the universe to all parts of the electromagnetic spectrum. Some telescopes are sited on Earth, others must be placed in space, and designs vary widely from one region of the spectrum to another. Whatever the details of its construction, however, the basic purpose of a telescope is to collect electromagnetic radiation and deliver it to a detector for detailed study.

MasteringAstronomy® Visit www.masteringastronomy.com for quizzes, animations, videos, interactive figures, and self-guided tutorials.

◀ One of astronomy's most advanced telescopes today is actually a group of many radio antennas that work together as a team. Seen here spread across 16 km on the Chajnantor plateau in the Atacama Desert high atop the Chilean Andes, each dish of ALMA (for Atacama Large Millimeter/submillimeter Array) captures radiation from cosmic sources with a slightly different perspective, thereby achieving superb resolution. With its 66 antennas, each one 12 m in diameter, ALMA is opening up a whole new window on the universe. This device, which has been years in the making, is expected to revolutionize our knowledge of many areas of astronomy and astrophysics. *(ESO/NAOJ/NRAO)*

3.1 Optical Telescopes

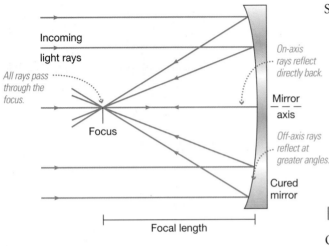

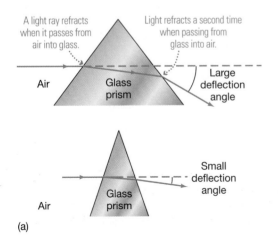

ANALOGY: Telescopes are "light buckets" that collect photons, just as actual buckets collect raindrops falling into them.

▲ **FIGURE 3.1 Reflecting Mirror** Curved mirrors focus to a single point all rays of light arriving parallel to the mirror axis. The arrows indicate the directions of the incoming and reflected rays.

Simply put, a **telescope** is a "light bucket"—a device whose primary function is to capture as much radiation as possible from a given region of the sky and concentrate it into a focused beam for analysis. We begin our study of astronomical hardware with *optical* telescopes, designed to collect wavelengths visible to the human eye. Later, we will look at telescopes designed to capture and analyze radiation in other, *invisible* regions of the electromagnetic spectrum. Throughout this text we will see how advances in our understanding of the universe have always gone hand in hand with technological improvements in both the sensitivity and the spectral range of our detectors. This combination of technology and science is as vital today as it was when Galileo first turned his telescope toward the skies. ∞ *(Sec. 1.2)*

Reflecting and Refracting Telescopes

Optical telescopes fall into two basic categories—*reflectors* and *refractors*. Figure 3.1 shows how a **reflecting telescope** uses a curved mirror to gather and concentrate a beam of light. This mirror, usually called the *primary mirror* because telescopes generally contain more than one, is constructed so that all light rays arriving parallel to its axis (the imaginary line through the center of and perpendicular to the mirror) are reflected to pass through a single point, called the *focus*. The distance between the primary mirror and the focus is the *focal length*. In astronomical contexts, the focus of the primary mirror is referred to as the **prime focus.**

A **refracting telescope** uses a lens instead of a mirror to focus the incoming light, relying on refraction rather than reflection to achieve its purpose. **Refraction** is the bending of a beam of light as it passes from one transparent medium (for example, air) into another (such as glass). Figure 3.2(a) illustrates how a prism can be used to change the direction of a beam of light. As illustrated in Figure 3.2(b), we can think of a lens as a series of prisms combined in such a way that all light rays striking the lens parallel to the axis are refracted to pass through the focus.

Astronomers often use telescopes to make **images** of their fields of view. Figure 3.3 illustrates how this is accomplished, in this case by the primary mirror in a reflecting telescope. Light from a distant object reaches the telescope as nearly parallel rays. A ray of light entering the instrument parallel to the mirror axis is reflected through the focus. Light from a slightly different direction—that is, at a slight angle to the axis—is focused to a slightly different point. In this way, an image is formed near the focus. Each point in the image corresponds to a different angle in the field of view. Often, a lens known as an *eyepiece* is used to magnify the image before it

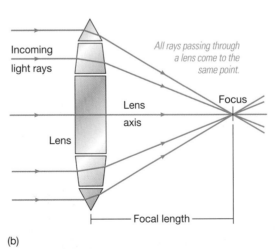

(b)

◀ FIGURE 3.2 **Refracting Lens** (a) Refraction by a prism changes the direction of a light ray. (b) A lens can be thought of as a series of prisms.

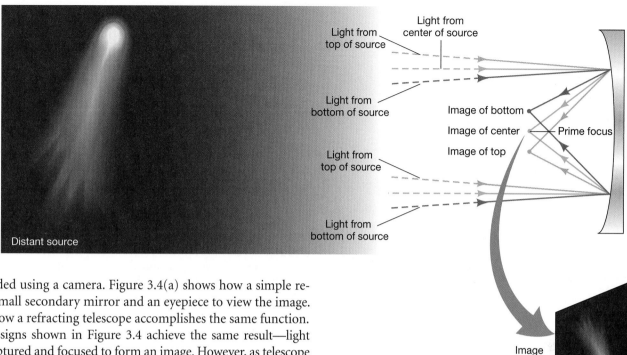

is viewed by eye or recorded using a camera. Figure 3.4(a) shows how a simple reflecting telescope uses a small secondary mirror and an eyepiece to view the image. Figure 3.4(b) illustrates how a refracting telescope accomplishes the same function.

The two telescope designs shown in Figure 3.4 achieve the same result—light from a distant object is captured and focused to form an image. However, as telescope *size* has steadily increased over the years (as discussed in Section 3.2) astronomers have come to favor reflecting instruments over refractor for a number of reasons:

1. Just as a prism separates white light into its component colors, the lens in a refracting telescope focuses red and blue light differently (the blue focus lying closer to the lens). This deficiency is known as *chromatic aberration.* Careful design and choice of materials can largely correct this problem, but it is difficult to eliminate it entirely, and it requires the use of very high-quality glass for the body of the lens. Mirrors do not suffer from this defect.

2. As light passes through the lens, some of it is absorbed by the glass. This absorption is a relatively minor problem for visible radiation, but it can be severe for infrared and ultraviolet observations because glass blocks most of the radiation

▲ **FIGURE 3.3 Image Formation** An image is formed by a mirror as rays of light coming from different points on a distant object focus to slightly different locations. Notice that the image is inverted (that is, upside down).

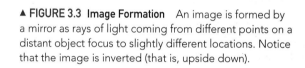

SELF-GUIDED TUTORIAL Chromatic Aberration

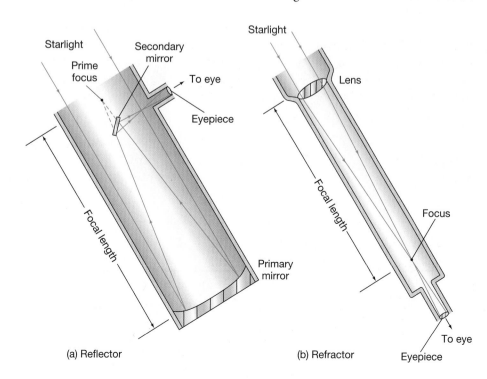

◄ **FIGURE 3.4 Reflectors and Refractors** Comparison of (a) reflecting and (b) refracting telescopes. Both types are used to gather and focus electromagnetic radiation. The image is viewed with a small magnifying lens called an *eyepiece.*

in those regions of the electromagnetic spectrum. This problem obviously does not affect mirrors.

3. A large lens can be quite heavy. Because it can be supported only around its edge (so as not to block the incoming radiation), the lens tends to deform under its own weight. A mirror does not have this drawback because it can be supported over its entire back surface.

4. A lens has two surfaces that must be accurately machined and polished, which can be a difficult task. A mirror has only one surface.

For these reasons, *all* large modern optical telescopes are reflectors.

Types of Reflecting Telescopes

SELF-GUIDED TUTORIAL Reflecting Telescopes

SELF-GUIDED TUTORIAL The Optics of a Simple Lens

Figure 3.5 shows some basic reflecting telescope designs. Radiation from a star enters the instrument, passes down the main tube, strikes the primary mirror, and is reflected back toward the prime focus, near the top of the tube. Sometimes astronomers place their recording instruments at the prime focus. However, it can be very inconvenient, or even impossible, to suspend bulky pieces of equipment there. More often, the light is intercepted on its path to the focus by a secondary mirror and redirected to a more convenient location, as in Figure 3.5(b–d).

In a **Newtonian telescope** (named after Isaac Newton, who invented this design), the light is intercepted by a flat secondary mirror before it reaches the prime focus and deflected 90°, usually to an eyepiece at the side of the instrument (Figure 3.5b). This is a popular design for smaller reflecting telescopes, such as those used by amateur astronomers.

Alternatively, astronomers may choose to work on a rear platform where they can use detecting equipment too heavy or delicate to hoist to the prime focus. In this case, the light reflected by the primary mirror toward the prime focus is intercepted by a convex secondary mirror, which reflects the light back down the tube and through a small hole at the center of the primary mirror (Figure 3.5c). This arrangement is known as a **Cassegrain telescope** (after Guillaume Cassegrain, a French lensmaker). The point behind the primary mirror where the light from the star finally converges is called the *Cassegrain focus*. A well-known example of a Cassegrain telescope is the *Hubble Space Telescope (HST; see Discovery 3-1)*, named

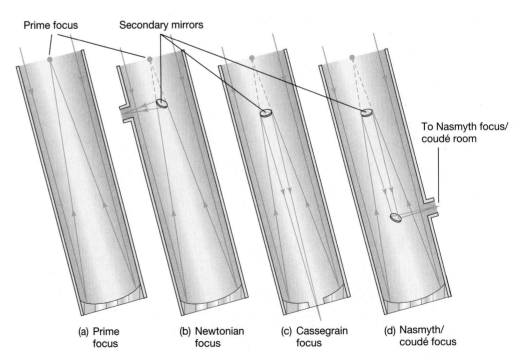

▶ FIGURE 3.5 **Reflecting Telescopes** Four reflecting telescope designs: (a) prime focus, (b) Newtonian focus, (c) Cassegrain focus, and (d) Nasmyth or coudé focus. Each design uses a primary mirror at the bottom of the telescope to capture radiation, which is then directed along different paths for analysis.

Prime focus   Secondary mirrors

To Nasmyth focus/ coudé room

(a) Prime focus   (b) Newtonian focus   (c) Cassegrain focus   (d) Nasmyth/ coudé focus

for one of America's most notable astronomers, Edwin Hubble. The telescope's detectors all lie directly behind the primary mirror and are capable of making measurements in the optical, infrared, and ultraviolet parts of the spectrum.

A more complex observational configuration requires starlight to be reflected by several mirrors. As in the Cassegrain design, light is first reflected by the primary mirror toward the prime focus and is then reflected back down the tube by a secondary mirror. Next, a third, much smaller, mirror reflects the light out of the telescope, where (depending on the details of the telescope's construction) the beam may be analyzed by a detector mounted alongside, at the *Nasmyth focus,* or it may be directed via a further series of mirrors into an environmentally controlled laboratory known as the *coudé* room (from the French word for "bent"). This laboratory is separate from the telescope itself, enabling astronomers to use very heavy and finely tuned equipment that cannot be placed at any of the other foci (all of which necessarily move with the telescope). The arrangement of mirrors is such that the light path to the coudé room does not change as the telescope tracks objects across the sky.

Figure 3.6(a) shows the twin 10-m-diameter optical/infrared telescopes of the Keck Observatory on Mauna Kea in Hawaii, operated jointly by the California Institute of Technology and the University of California. The diagram in part (b) illustrates the light paths and some of the foci. Observations may be made at the Cassegrain, Nasmyth, or coudé focus, depending on the needs of the user. As the size of the person in part (c) indicates, this is indeed a very large telescope—in fact, the two mirrors are currently among the very largest on Earth. We will see numerous examples throughout this text of Keck's many important discoveries.

◀ **FIGURE 3.6 Keck Telescope** (a) The two 10-m telescopes of the Keck Observatory. (b) Artist's illustration of the telescope; the blue arrows show some of the paths starlight can take within the telescope and some of the locations where instruments may be placed. (c) One of the 10-m mirrors. Note the technician in orange coveralls at center. (*W. M. Keck Observatory*)

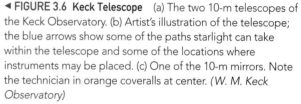
(c)

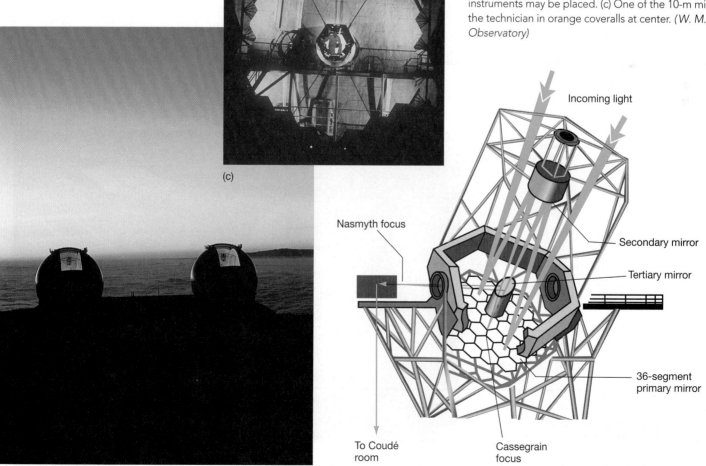
(a)

(b)

3-1 DISCOVERY

The Hubble Space Telescope

The *Hubble Space Telescope (HST)* is the largest, most complex, most sensitive observatory ever deployed in space. At more than $8 billion (including the cost of several missions to service and refurbish the system), it is also the most expensive scientific instrument ever built and operated. Built jointly by NASA and the European Space Agency, *HST* was designed to allow astronomers to probe the universe with at least 10 times finer resolution and with some 30 times greater sensitivity to light than existing Earth-based devices. The first figure shows the telescope being lifted out of the cargo bay of the space shuttle *Discovery* in the spring of 1990.

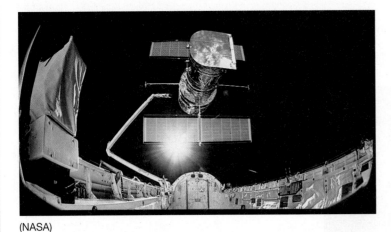

(NASA)

The accompanying "see-through" illustration displays *HST*'s main features. The telescope's Cassegrain design (Section 3.1) reflects light from its 2.4-m-diameter primary mirror (the large bluish disk at center) back to a smaller, 0.3-m secondary mirror, which sends the light through a small opening in the center of the primary mirror, from where it can be directed to any of several instruments (shown here in various colors at left) arrayed in the aft bay of the spacecraft. The large red objects are sensors that guide the pointing of the telescope, and the huge blue panels collect light from the Sun to power everything onboard. For scale, each of the instruments is about the size of a refrigerator. They were designed to be maintained by NASA astronauts, and indeed most of the telescope's original instruments have been upgraded or replaced since *HST* was launched. The current detectors on the telescope span the visible, near-infrared, and near-ultraviolet regions of the electromagnetic spectrum, from about 100 nm (UV) to 2200 nm (IR).

Soon after launch, astronomers discovered that the telescope's primary mirror had been polished to the wrong shape. The mirror is too flat by 2 μm, about 1/50 the width of a human hair, making it impossible to focus light as well as expected. In 1993, shuttle astronauts replaced *Hubble*'s gyroscopes to help the telescope point more accurately, installed sturdier versions of the solar panels that power the telescope's electronics, and—most important—inserted an intricate set of small mirrors to compensate for the faulty primary mirror. *Hubble*'s resolution is now close to the original design specifications, and the telescope has regained much of its lost sensitivity. Additional service missions were performed in 1997, 1999, 2002, and 2009 to replace instruments and repair faulty systems.

ANIMATION/VIDEO *Hubble Space Telescope* in Orbit

ANIMATION/VIDEO Deployment of the James Webb Space Telescope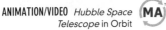

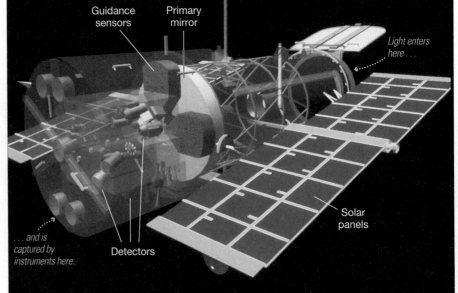

Guidance sensors

Primary mirror

Light enters here . . .

. . . and is captured by instruments here.

Detectors

Solar panels

(D. Berry)

A good example of *Hubble*'s scientific capabilities can be seen by comparing the two images of the spiral galaxy M100, shown here. On the left is one of the best ground-based photographs of this beautiful galaxy, showing rich detail and color in its spiral arms. On the right, to the same scale and orientation, is an *HST* image showing improvements in both resolution and sensitivity. (The chevron-shaped field of view is caused by the corrective optics inserted into the telescope in 1993; an additional trade-off is that *Hubble*'s field of view is smaller than those of ground-based telescopes.) The inset shows *Hubble*'s exquisite resolution of small fields of view.

During its decade and a half of operation, *Hubble* has literally revolutionized our view of the sky, helping to rewrite some theories of the universe along the way. By measuring the properties of stars and supernovae within distant galaxies, it has helped establish the size and the expansion rate of the cosmos and has provided insights into the past and future evolution of the universe. *Hubble* has studied newborn galaxies almost at the limit of the observable universe with unprecedented clarity, allowing astronomers to see the interactions and collisions that may have shaped the evolution of our own Milky Way. Turning its gaze to the hearts of galaxies closer to home, *Hubble* has provided strong evidence for supermassive black holes in their cores. Within our own Galaxy, it has given astronomers stunning new insights into the physics of star formation and the evolution of

stellar systems and stars of all sizes, from superluminous giants to objects barely more massive than planets. Finally, in our solar system, *Hubble* has given scientists new views of both the planets and their moons, and of the tiny fragments from which they formed long ago. Many spectacular examples of the telescope's remarkable capabilities appear throughout this book.

With *Hubble* now ending its second decade of highly productive service, NASA has formally approved plans for the telescope's successor. The *James Webb Space Telescope* (*JWST*), named after the administrator who led NASA's *Apollo* program during the 1960s and 1970s, will dwarf *Hubble* in both scale and capability. Sporting a 6.5-m segmented mirror with seven times *Hubble*'s collecting area and containing a formidable array of detectors optimized for use at visible and infrared wavelengths, *JWST* will orbit the Sun some 1.5 million km outside Earth's orbit, far beyond the Moon. NASA plans to launch the telescope in 2018. Its primary mission is to study the formation of the first stars and galaxies, measure the large-scale structure of the universe, and investigate the evolution of planets, stars, and galaxies.

Note that the *JWST* launch schedule likely means an interruption of a few years in space-based astronomical observations at near-optical wavelengths. Fortunately, the 2008 servicing mission extended *Hubble*'s operating lifetime beyond 2012, but few expect that the telescope will survive until *JWST* becomes fully operational.

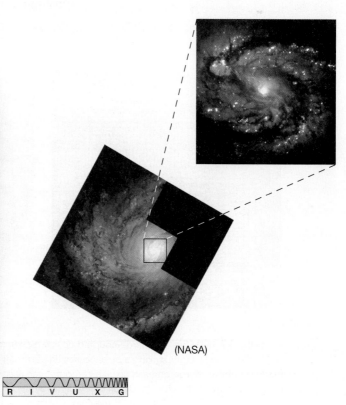

(NASA)

(D. Malin/Anglo-Australian Telescope)

R I V U X G

Detectors and Image Processing

Most modern telescopes use electronic detectors known as **charge-coupled devices**, or **CCDs,** to record and store their data. A CCD (Figure 3.7) consists of a wafer of silicon divided into a two-dimensional array of many tiny picture elements, known as *pixels*. When light strikes a pixel, an electric charge builds up on the device. The amount of charge is directly proportional to the number of photons striking each pixel—in other words, to the intensity of the light at that point. The charge buildup is monitored electronically, and a two-dimensional image is obtained (see Figures 3.13c and d). A CCD is typically a few square centimeters in area and may contain several million pixels, generally arranged on a square grid. As technology continually improves, both the areas of CCDs and the number of pixels they contain are steadily increasing. Incidentally, the technology is not limited to astronomy—home video and digital cameras contain CCD chips similar in basic design to those in use at the great astronomical observatories of the world.

CCDs have two important advantages over photographic plates, which were the staple of astronomers for over a century. First, CCDs are much more *efficient* than photographic plates, recording as many as 75 percent of the photons striking them, compared with less than 5 percent for photographic methods. This means that a CCD instrument can image objects 10 to 20 times fainter—or the same object 10 to 20 times faster—than can a photographic plate. Second, CCDs produce a faithful representation of an image in a digital format that can be manipulated by software, stored on disk, or sent across the Internet to an observer's home institution for detailed analysis.

Astronomers use computer processing on digital CCD images to compensate for known instrumental defects and even partially remove unwanted background noise in the signal, allowing them to see features in their data that would otherwise remain hidden. In addition, the computer can often carry out many of the tedious and time-consuming chores that must be performed before an image or spectrum reaches

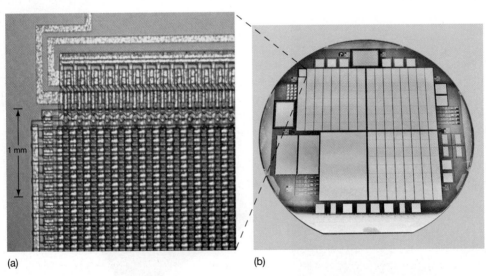

(a)   (b)

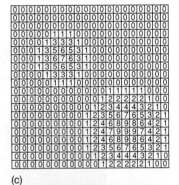

(c)

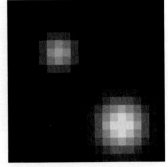

(d)

▲ **FIGURE 3.7 CCD Chip**   A charge-coupled device (CCD) consists of millions of tiny light-sensitive cells called pixels. Light striking a pixel causes an electrical charge to build up on it. By electronically reading out the charge on each pixel, a computer can reconstruct the pattern of light—the image—falling on the chip. (a) Detail of a CCD array. (b) A CCD chip mounted for use at the focus of a telescope. (c) Typical data from the chip consist of an array of numbers, running from 0 to 9 in this simplified example. Each number represents the intensity of the radiation striking that particular pixel. (d) The resulting image. *(MIT Lincoln Lab; AURA)*

(a) (b) (c) (d)

its final "clean" form. Figure 3.8 illustrates how computerized image-processing techniques were used to correct for known instrumental problems in the *Hubble Space Telescope,* allowing much of the planned resolution of the telescope to be recovered even before its repair in 1993.

▲ FIGURE 3.8 **Image processing** (a) Ground-based view of the star cluster R136, a group of stars in the Large Magellanic Cloud (a nearby galaxy). (b) The "raw" image of this same region as seen by the *Hubble Space Telescope* in 1990, before the first repair mission. (c) The same image after computer processing that partly compensated for imperfections in the mirror. (d) The same region as seen by the repaired *HST* in 1994, here observed at a somewhat bluer wavelength. (*AURA/NASA*)

3.2 Telescope Size

Large telescopes have two main advantages over small ones—*light-gathering power,* the ability to capture light, and *resolving power,* the ability to distinguish fine detail.

Light-Gathering Power

Astronomers spend much of their time observing very distant—and, hence, very *faint*—cosmic sources. In addition, they often want to obtain *spectra,* which entail spreading the incoming light out into its component colors. ∞ *(Sec. 2.5)* In order to make such detailed observations, astronomers must collect as much light as possible for analysis by their detectors.

The two factors determining a telescope's ability to collect light are *exposure time* and *collecting area.* **Exposure time** is simply the time spent gathering light from a source. Doubling the exposure time doubles the total amount of radiation captured. In Chapter 16 (p. 440) we will see an extreme example of a long-exposure image: constructed in 2003 from more than 300 hours of observations of the same small region of the sky, the so-called Hubble Ultra-Deep Field continues to provide astronomers with volumes of information on some of the most distant galaxies in the universe. However, such long exposures are rare. Access to telescopes and the sophisticated instruments attached to them is a precious commodity, and observatories are reluctant to allocate large amounts of time to a single observation.

The second factor controlling a telescope's light-gathering power is **collecting area,** the area capable of intercepting and focusing radiation—the "size of the bucket" (recall Figure 3.1). The larger the telescope's reflecting mirror (or refracting lens), the more light it collects and the easier it is to measure and study an object's radiative properties. The observed brightness of an astronomical object is directly proportional to the area of the telescope's mirror and, hence, to the *square* of the mirror diameter (Figure 3.9). For example, a 5-m telescope produces an image 25 times

(a)

(b)

▶ FIGURE 3.9 **Sensitivity** Telescope size affects the image of a cosmic source, in this case the Andromeda galaxy. Both of these photographs had the same exposure time, but image (b) was taken with a telescope twice the size of that used to make image (a). Fainter detail can be seen as the diameter of the telescope mirror increases because larger telescopes are able to collect more photons per unit time. (*Adapted from AURA*)

▶ **FIGURE 3.10 Mauna Kea Observatory** (a) The world's highest ground-based observatory, at Mauna Kea, Hawaii, is perched atop an extinct volcano more than 4 km (nearly 14,000 feet) above sea level. Among the domes visible in the picture are those housing the Canada–France–Hawaii 3.6-m telescope, the 8.1-m Gemini North instrument, the 2.2-m telescope of the University of Hawaii, Britain's 3.8-m infrared facility, and the twin 10-m Keck telescopes. To the right of the twin Kecks is the Japanese 8.3-m Subaru telescope. (b) The mirror in the Subaru telescope. (*R. Wainscoat; NAOJ*)

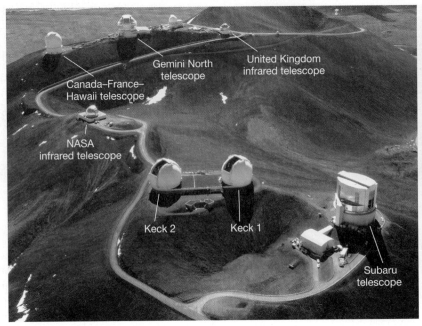

(a)

(b)

brighter than a 1-m instrument because a 5-m mirror has $5^2 = 25$ times the collecting area of a 1-m mirror. We can also think of this relationship in terms of the *time* required for a telescope to collect enough energy to create a recognizable image. A 5-m telescope produces an image 25 times faster than a 1-m device because it gathers energy at a rate 25 times greater. Put another way, a 1-hour time exposure

CONCEPT CHECK ▶ Why do the largest modern telescopes use mirrors to gather and focus light?

▶ **FIGURE 3.11 VLT Observatory** Located at the Paranal Observatory in Atacama, Chile, the European Southern Observatory's Very Large Telescope (VLT) is the world's largest optical telescope. Four 8.2-m reflecting telescopes are used in tandem to create the effective area of a single 16-m mirror. (*ESO*)

with a 1-m telescope is roughly equivalent to a 2.4-minute (1/25 of an hour) time exposure with a 5-m instrument.

The twin Keck telescopes, shown in Figure 3.6 and in a larger view in Figure 3.10(a), are a case in point. Each telescope combines 36 hexagonal 1.8-m mirrors into the equivalent collecting area of a single 10-m reflector. The high altitude and large size of these telescopes makes them particularly well suited for detailed spectroscopic studies of very faint objects. To the right of the Keck domes in Figure 3.10(a) is the 8.3-m Subaru (the Japanese name for the Pleiades) telescope. Its mirror, shown in Figure 3.10(b), is one of the largest single mirrors (as opposed to the segmented design used in Keck) yet built. In terms of total available collecting area, the largest telescope system now operating is the European Southern Observatory's Very Large Telescope (VLT), located at Cerro Paranal, in Chile (Figure 3.11). Its four separate 8.2-m mirrors can function as a single instrument of equivalent diameter 16.4 m. Both Keck and the VLT are designed to operate in the optical and near-infrared parts of the electromagnetic spectrum.

Resolving Power

The second advantage of large telescopes over smaller instruments is their superior **angular resolution.** In general, *resolution* refers to the ability of any device, such as a camera or a telescope, to form distinct, separate images of objects lying close together in the field of view. The finer the resolution, the better we can distinguish the objects and the more detail we can see. In astronomy, where we are always concerned with angular measurement, "close together" means "separated by a small angle on the sky," so angular resolution is the factor that determines our ability to see fine structure (see *More Precisely 0-1* for more detail on angular measure). Astronomers typically need to resolve objects only a few arc seconds (″) across. Figure 3.12 illustrates the result of increasing resolving power with views of the Andromeda galaxy at several different resolutions.

One important factor limiting a telescope's resolution is *diffraction,* the tendency of light—and all waves, for that matter—to "bend" around corners. ∞ *(Sec. 2.2)* Diffraction introduces a certain fuzziness, or loss of resolution, into the system. The degree of fuzziness—the minimum angular separation that can be distinguished—determines the angular resolution of the telescope.

As discussed in *More Precisely 3-1,* the amount of diffraction is directly proportional to the wavelength of the radiation and inversely proportional to the diameter of the telescope mirror. For light of any given wavelength, large telescopes produce less diffraction than small ones. The resolution given by the formula in *More Precisely 3-1* is known as the **diffraction-limited resolution** of the telescope. Thus, a 5-m telescope observing in blue light has a diffraction-limited resolution of about 0.02″. A 1-m telescope would have a diffraction limit of 0.1″ at the same wavelength, and so on. For comparison, the angular resolution of the human eye in the middle of the visual range is about 0.5′.

(a)

(b)

(c)

(d)

R I V U X G

▶ **FIGURE 3.12 Resolution** Detail becomes clearer in the Andromeda galaxy as the angular resolution is improved some 600 times, from (a) 10′, to (b) 1′, (c) 5″, and (d) 1″. *(Adapted from AURA)*

3-1 MORE PRECISELY

Diffraction and Telescope Resolution

The resolution of a telescope is ultimately controlled by *diffraction*, the process where light spreads out as it passes a corner or through an opening. ∞ *(Sec. 2.2)* Because of diffraction, it is impossible to focus a parallel beam of light to a sharp point, even with a perfectly constructed mirror. As illustrated in the accompanying figure, a wave passing through a gap is diffracted, creating a "fuzzy" shadow on the screen at right. Light and dark gradations represent crests and troughs of the wave, which define the wavelength (see Figure 2.3). In the absence of any diffraction, the shadow would be perfectly sharp—but that never happens in reality.

The amount of diffraction—that is, the amount of "fuzziness" introduced into an instrument—depends both on the *wavelength* of the radiation and the *size* of the opening (in our case, the *diameter* of the primary mirror). For a circular mirror and otherwise perfect optics, the angular resolution of a telescope (in convenient units) is

$$\text{angular resolution (arc seconds)} = 0.25 \frac{\text{wavelength (μm)}}{\text{mirror diameter (m)}},$$

where 1 μm (micrometer) $= 10^{-6}$ m $= 1000$ nm. (Recall that the symbol μ is the Greek letter mu.)

Thus, the best possible angular resolution that can be obtained using a 1-m telescope in blue light (with a wavelength of 400 nm $= 0.4$ μm) is about 0.25 $(0.4/1)'' = 0.1''$. But if we were to use our 1-m telescope to make observations in the near infrared, at a wavelength of 10 μm (10,000 nm), the best resolution we could obtain would be only 0.25 $(10/1)'' = 2.5''$. Observations in the infrared or radio range are often limited by the effects of diffraction. A 1-m radio telescope operating at a wavelength of 1 cm would have an angular resolution of just under 1°.

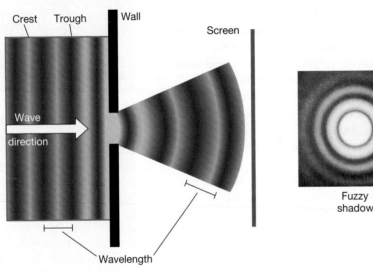

Crest Trough   Wall

Screen

Wave direction

Wavelength

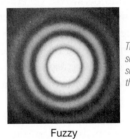

This is what is seen on the screen when the light hits it.

Fuzzy shadow

3.3 High-Resolution Astronomy

Even large telescopes have limitations. For example, according to the discussion in the preceding section, the 10-m Keck telescope should have an angular resolution of around 0.01″ in blue light. In practice, however, without the technological advances discussed in this section, it could not do better than about 1″. In fact, apart from instruments using special techniques developed to examine some particularly bright stars, *no* ground-based optical telescope built before 1990 can resolve astronomical objects to much better than 1″. The reason is *turbulence* in Earth's atmosphere—small-scale eddies of swirling air all along the line of sight, which blur the image of a star even before the light reaches our instruments.

Atmospheric Blurring

As we observe a star, atmospheric turbulence produces continuous small changes in the optical properties of the air between the star and our telescope (or eye). The light from the star is refracted slightly, again and again, and the stellar image dances around on the detector (or retina). This is the cause of the well-known twinkling of stars. The same basic process makes objects appear to shimmer when viewed across a hot roadway on a summer day, because turbulent hot air just above the road surface constantly deflects and distorts the light rays reaching us.

On a typical night at a good observing site, the maximum deflection produced by the atmosphere is around 1″. Consider taking a photograph of a star under such conditions. After a few minutes' exposure time (long enough for the intervening atmosphere to have undergone many small, random changes), the dancing sharp image of the star has been smeared out over a roughly circular region 1″ or so in diameter (Figure 3.13). Astronomers use the term **seeing** to describe the effects of atmospheric turbulence. The circle over which a star's light is spread is called the **seeing disk**. In this jargon, "good seeing" means that the air is relatively stable and the seeing disk is small—as little as a few tenths of an arc second across in some exceptional cases.

To achieve the best possible seeing, telescopes are sited on mountaintops (to get above as much of the atmosphere as possible) in locations where the atmosphere is known to be fairly stable and relatively free of dust, moisture, and light pollution from cities. In the continental United States, these sites tend to be in the desert Southwest. The U.S. National Observatory for optical astronomy in the Northern Hemisphere is located high on Kitt Peak near Tucson, Arizona. The site was chosen because of its many dry, clear nights, and typical seeing of around 1″. Even better conditions are found on Mauna Kea in Hawaii (Figure 3.10) and in the Andes Mountains of Chile (Figure 3.11), which is why so many large telescopes have recently been constructed at those two exceptionally clear sites.

CONCEPT CHECK ▶ Give two reasons why astronomers need to build very large telescopes.

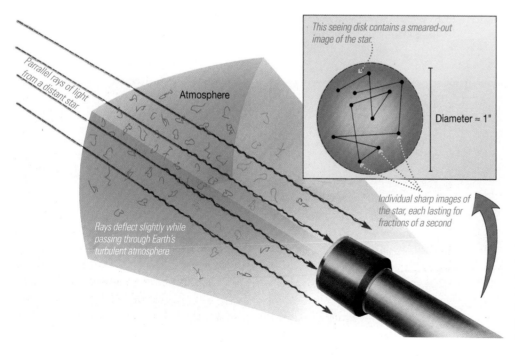

This seeing disk contains a smeared-out image of the star.

Parallel rays of light from a distant star

Atmosphere

Diameter ≈ 1″

Rays deflect slightly while passing through Earth's turbulent atmosphere.

Individual sharp images of the star, each lasting for fractions of a second

◄ **FIGURE 3.13 Atmospheric Turbulence** Light rays from a distant star strike a telescope detector at slightly different locations because of turbulence in Earth's atmosphere. Over time, the light covers a roughly circular region on the detector, and even the point-like image of a star is recorded as a small disk, called the seeing disk.

A telescope placed above the atmosphere, in Earth orbit, can achieve resolution close to the diffraction limit, subject only to the engineering restrictions of building and placing large structures in space. The *Hubble Space Telescope* (*Discovery 3-1*) has a 2.4-m mirror and a diffraction limit of 0.05″ (in blue light).

New Telescope Design

ANIMATION/VIDEO Adaptive Optics **(MA)**

In addition to using high-quality optics and selecting the best observing sites, astronomers have developed other techniques to produce the sharpest possible images. By analyzing the image *while the light is still being collected,* it is possible to adjust the telescope from moment to moment to reduce the effects of mirror distortion, temperature changes, and bad seeing. By these means, an increasing number of large telescopes, including Keck and the VLT, can now achieve resolutions close to their theoretical diffraction limits.

The collection of techniques aimed at controlling environmental and mechanical fluctuations in the properties of the telescope itself is known as **active optics**. Active optics systems often include improved dome design to control airflow, precise control of the mirror temperature, and the use of actuators (pistons) behind the mirror to maintain its precise shape at all times (Figure 3.14a). Figure 3.14(b) illustrates dramatically how active optics can improve image resolution.

An even more ambitious approach is known as **adaptive optics**. This technique deforms the shape of a mirror's surface under computer control while the image is being exposed in order to undo the effects of atmospheric turbulence. In the system shown in Figure 3.15(a), lasers probe the atmosphere above the telescope, returning information about the air's swirling motion to a computer that modifies the mirror thousands of times per second to compensate for poor seeing. Other adaptive optics systems monitor standard stars in the field of view, constantly adjusting the mirror's shape to preserve those stars' appearance. Figure 3.15(b) presents an example of the improvement in image quality that can be obtained by these means.

Many of the world's largest telescopes now incorporate sophisticated adaptive optics systems, and resolutions as fine as a few hundredths of an arc second have been reported for observations in the near-infrared, significantly better than the resolution of the much smaller *HST* at the same wavelengths. Adaptive optics systems are giving astronomers the best of both worlds, achieving with large ground-based optical telescopes the kind of resolution once attainable only from space.

CONCEPT CHECK ▶ Why is Earth's atmosphere a problem for optical astronomers? What can they do about it?

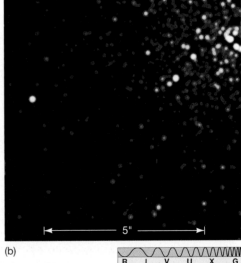

▶ **FIGURE 3.14 Active Optics** Infrared images of part of the star cluster R136 (see also Figure 3.8) contrast the resolution obtained (a) without and (b) with an active-optics system. *(ESO)*

(a)   (b)

5″

R   I   V   U   X   G

(a)

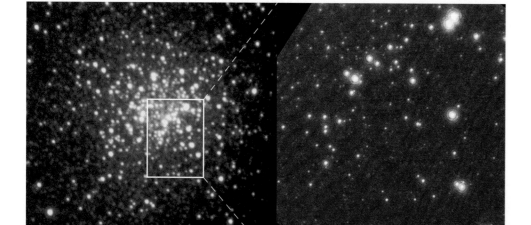

(b)

◄ **FIGURE 3.15 Adaptive Optics** (a) In this test at the Lick Observatory's 3-m Shane telescope in California, a laser is used to create an "artificial star" to improve guiding. The laser beam probes the atmosphere above the telescope, allowing tiny computer-controlled changes to be made to the shape of the mirror surface thousands of times each second. (b) Clarity of view is very important in astronomy. The uncorrected, visible-light image (left) of the star cluster NGC 6934 is resolved to a little less than 1". With adaptive optics applied (right), the resolution in the infrared is improved to nearly 10 times better, and more stars are seen more clearly. (*L. Hatch/Lick Observatory; NOAO*)

3.4 Radio Astronomy

In addition to the visible radiation that penetrates Earth's atmosphere on a clear day, radio radiation also reaches the ground, and astronomers have built many ground-based **radio telescopes** capable of detecting cosmic radio waves. ∞ *(Sec. 2.3)* These devices have all been constructed since the 1950s—radio astronomy is a much younger field than optical astronomy.

Essentials of Radio Telescopes

Figure 3.16 shows the world's largest steerable radio telescope, the large 105-m-diameter (340-foot-diameter) telescope located at the National Radio Astronomy Observatory in West Virginia. Conceptually, the operation of a radio telescope is the same as a prime-focus optical reflector (Figure 3.5a). It has a large, horseshoe-shaped mount supporting a huge, curved metal dish that serves as the collecting area. The dish captures cosmic radio waves and reflects them to the focus, where a receiver detects the signals and channels them to a computer.

(a)   (b)

▲ FIGURE 3.16 Radio Telescope   (a) The world's largest fully steerable radio telescope, the 105-m-diameter device at the National Radio Astronomy Observatory in Green Bank, West Virginia, is 150 m tall—taller than the Statue of Liberty and nearly as tall as the Washington Monument. (b) A schematic diagram shows the path taken by an incoming beam of radio radiation (colored blue). *(NRAO)*

Radio telescopes are built large in part because cosmic radio sources are extremely faint. Many sources simply don't emit many radio photons (see for example Figure 2.11), the photons don't carry much energy, and the sources themselves are often very distant. ⟳ *(Sec. 2.6)* In fact, the total amount of radio energy received by Earth's entire surface is less than a trillionth of a watt. Compare this with the roughly 10 *million* watts our planet's surface receives in the form of infrared and visible light from any of the bright stars visible in the night sky. In order to capture enough radio energy to allow detailed measurements to be made, a large collecting area is essential.

Because of diffraction, the angular resolution of radio telescopes is generally quite poor compared with that of their optical counterparts. Typical wavelengths of radio waves are about a million times longer than those of visible light, and even the enormous sizes of radio dishes only partly offset this effect. The 105-m telescope shown in Figure 3.16 can achieve resolution of about 1′ at a wavelength of 3 cm, although the instrument was designed to be most sensitive to wavelengths around 1 cm, where the resolution is approximately 20′. The best angular resolution obtainable with a single radio telescope is about 10″ (for the largest instruments operating at millimeter wavelengths), about 100 times coarser than the capabilities of the best optical mirrors.

Figure 3.17 shows the world's largest radio telescope, built in 1963 in Arecibo, Puerto Rico. Approximately 300 m in diameter, the telescope's reflecting surface lies in a natural depression in a hillside and spans nearly 20 acres. The receiver is strung among several limestone hills. Its enormous collecting area makes this telescope the most sensitive on Earth. The huge but fixed dish creates one distinct disadvantage, however. The Arecibo telescope cannot be pointed to follow cosmic objects across the sky. Its observations are limited to those objects that happen to pass within about 20° of overhead as Earth rotates.

Receivers at the focus are suspended nearly 150 m (about 45 stories) above the dish's center.

Detector

Collecting area

▲ FIGURE 3.17 Arecibo Observatory The 300-m-diameter dish at the National Astronomy and Ionospheric Center near Arecibo, Puerto Rico. The two insets show close-ups of the receivers hanging high above the dish (left) and technicians adjusting the dish surface to make it smoother (right). (D. Parker/T. Acevedo/NAIC; Cornell University)

The Value of Radio Astronomy

Despite the inherent disadvantage of relatively poor angular resolution, radio astronomy enjoys some advantages too. Unlike visible light, radio waves are not deflected or scattered by Earth's atmosphere, and the Sun itself is a relatively weak source of radio energy. As a result, radio observations can cover almost the entire sky, 24 hours a day. Only within a few degrees of the Sun does solar radiation swamp radio observations of more distant objects. In addition, radio observations can often be made through cloudy skies, and radio telescopes can detect the longest-wavelength radio waves even during rain or snowstorms.

However, perhaps the greatest value of radio astronomy—and all invisible astronomies—is that it opens up a whole new window on the universe. There are two main reasons for this. First, just as objects that are bright in the visible part of the spectrum (the Sun, for example) are not necessarily strong radio emitters, many of the strongest radio sources in the universe emit little or no visible light. Second, visible light may be strongly absorbed by interstellar dust along the line of sight to a source. Radio waves, on the other hand, are generally unaffected by intervening matter. Many parts of the universe cannot be seen at all by optical means but are easily detectable at radio wavelengths. Radio observations therefore allow us to see whole new classes of objects that would otherwise be completely unknown.

Figure 3.18 shows a visible-light photograph of a distant galaxy called Centaurus A. Superimposed on the optical image is a radio map of the same region. The radio image is represented in **false color,** a technique commonly used for displaying images taken in nonvisible light. The colors do not represent the actual wavelength

CONCEPT CHECK ► Cosmic radio waves are very weak, and the resolution of radio telescopes is often poor, so what can astronomers hope to learn from radio astronomy?

▶ **FIGURE 3.18 Radio Galaxy** This is a composite image of the Centaurus galaxy, showing its optical view at center and its radio emission in lobes well beyond (here shown in false color, with red indicating greatest radio intensity, blue the least). *(J. Burns)*

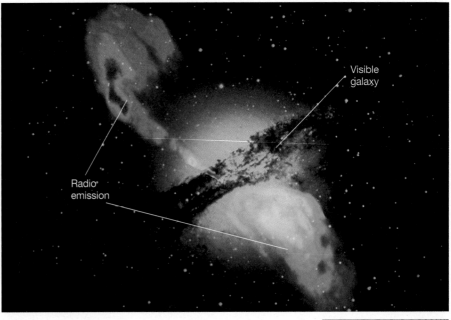

of the radiation emitted, but instead some other property of the source, in this case intensity, descending from red to yellow, green, and blue. Notice just how different the galaxy looks at radio and optical wavelengths. The visible galaxy emits little or no radio energy, while the large blobs (called *radio lobes,* and thought to have been ejected from the center of the galaxy by explosive events long ago) are completely invisible. Centaurus A is an example of a *radio galaxy* (to be studied in more detail in Chapter 15). It actually emits far more energy in the form of radio waves than it does as visible light, but without radio astronomy we would be entirely ignorant of this fact, and of the galaxy's violent past.

Interferometry

Radio astronomers can sometimes improve angular resolution by using a technique known as **interferometry,** making it possible to produce radio images of much higher angular resolution than can be achieved with even the best optical telescopes, on Earth or in space. In interferometry, two or more radio telescopes are used in tandem to observe the same object at the same wavelength and at the same time. The combined instruments together make up an **interferometer** (Figure 3.19). By means of electronic cables or radio links, the signals received by each antenna in the array making up the interferometer are sent to a central computer that combines and stores the data as the antennas track their target.

Interferometry works by analyzing how the signals *interfere* with each other when added together. ∞ *(Sec. 2.2)* Consider an incoming wave striking two detectors

(a)

(b)

◀ **FIGURE 3.19 VLA interferometer** (a) This large interferometer, called the Very Large Array, or VLA for short, is located on the plain of San Augustin in New Mexico. It comprises 27 dishes spread along a Y-shaped pattern about 30 km across. (b) The dishes are mounted on railroad tracks so that they can be repositioned easily. (Another powerful interferometer is shown in the Chapter Opener photo on page 68.) *(NRAO)*

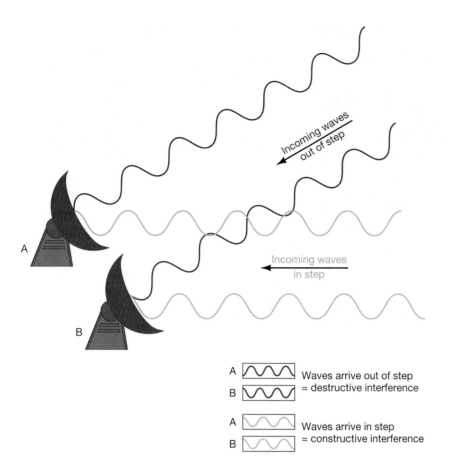

◀ **FIGURE 3.20 Interferometry**   Two telescopes, A and B, record different signals from the same incoming wave because of the time it takes the radiation to traverse the distance between them. When the signals are combined, the amount of interference depends on the wave's direction of motion, providing a means of measuring the position of the source in the sky. Here, the dark blue waves come from a source high in the sky and are out of phase, thus causing destructive interference. But when the same source has moved because of Earth's rotation (light-blue waves), the interference can be constructive.

(Figure 3.20). Because the detectors lie at different distances from the source, the signals they record will, in general, be out of step with one another and, when combined, will interfere destructively, partly canceling each other out. Notice that the amount of interference depends on the *direction* in which the wave is traveling relative to the line joining the detectors. As Earth rotates and the antennae track their target, careful computer analysis of the combined signal results in a high-resolution image of the distant object.

As far as resolving power is concerned, the effective diameter of an interferometer is the distance between its outermost dishes. In other words, two small dishes can act as opposite ends of an imaginary but huge single radio telescope, dramatically improving the angular resolution. Large interferometers made up of many dishes, like the Very Large Array (VLA) shown in Figure 3.19, now routinely attain radio resolution of a few arc seconds, comparable to that of many ground-based optical instruments (without adaptive optics). Figure 3.21 compares an interferometric radio map of a nearby galaxy with a photograph of that same galaxy made using a large optical telescope. (The radio image in Figure 3.18 was also obtained by interferometric means.) The larger the distance (or *baseline*) separating the telescopes, the better the resolution attainable. Very-long-baseline interferometry (VLBI), using instruments separated by thousands of kilometers, can achieve resolution in the milliarcsecond (0.001″) range.

Although the technique was originally developed by radio astronomers, interferometry is no longer restricted to the radio domain. Radio interferometry became feasible when electronic equipment and computers achieved speeds great enough to combine and analyze radio signals from separate radio detectors without loss of data. As the technology has improved, it has become possible to apply the same methods to higher-frequency radiation. Millimeter-wavelength interferometry has already become an established and important observational technique, and both the Keck and VLT telescopes are designed to be used for infrared—and eventually optical—interferometric work.

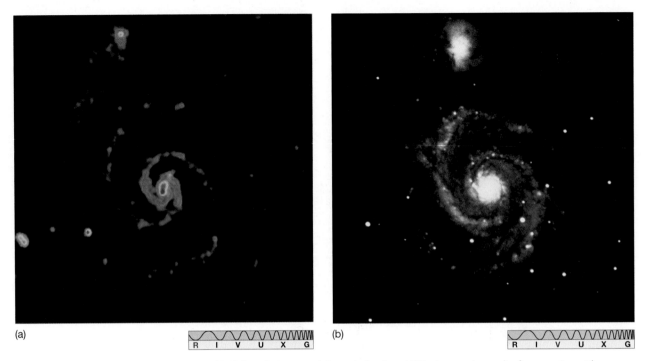

(a)

(b)

R I V U X G

R I V U X G

▲ FIGURE 3.21 Radio-optical Comparison   (a) VLA radio image of the spiral galaxy M51 observed at radio frequencies with an angular resolution of a few arc seconds. (b) Visible-light image of the same galaxy, made with the 4-m Kitt Peak optical telescope and displayed on the same scale as (a). (NRAO; AURA)

Perhaps the highest-resolution interferometric instrument currently in operation is the six-telescope optical array operated by the Center for High Angular Resolution Astronomy (CHARA) on Mount Wilson in California (Figure 3.22). Although each telescope is only 1 m in diameter, the placement of the array over the mountain results in a combined light beam having resolution equivalent to a single telescope 300 m across. CHARA is not designed to produce images of the stars it studies; however, it can resolve details as small as 0.0002' across, allowing the positions, orbits, and even radii of some stars to be measured with exquisite accuracy.

▶ FIGURE 3.22 Optical Interferomer   This aerial photo shows the CHARA array intermingled with the existing equipment of the historic Mount Wilson Observatory, in California. The small 1-m telescopes in the array are numbered. (E. Simison/Sea West Enterprises, Inc.)

3.5 Space-Based Astronomy

As we saw in Chapter 2, Earth's atmosphere is opaque to electromagnetic radiation outside of the radio, infrared, and optical windows. ∞ *(Sec. 2.3)* Most other wavelengths can only be studied from space. The rise of these "other astronomies" has therefore been closely tied to the development of the space program.

Infrared and Ultraviolet Astronomy

Infrared studies are a very important component of modern observational astronomy. Much of the gas between the stars has a temperature between a few tens and a few hundreds of kelvins, and Wien's law tells us that the infrared domain is the natural portion of the electromagnetic spectrum in which to study this material. ∞ *(Sec. 2.4)* Generally, **infrared telescopes** resemble optical telescopes, but the infrared detectors are sensitive to longer-wavelength radiation. Although most infrared radiation is absorbed by the atmosphere (primarily by water vapor), there are a few windows in the high-frequency part of the infrared spectrum where the opacity is low enough to permit ground-based observations. ∞ *(Fig. 2.9)* As we have seen, some of the most useful infrared observing is done from the ground using large telescopes equipped with adaptive optics systems.

For most infrared observations, astronomers must place their instruments above most or all of Earth's atmosphere. Improvements in balloon-, aircraft-, rocket-, and satellite-based telescope technologies have made infrared research a powerful tool for studying the universe (Figure 3.23). As might be expected, the infrared telescopes that can be carried above the atmosphere are considerably

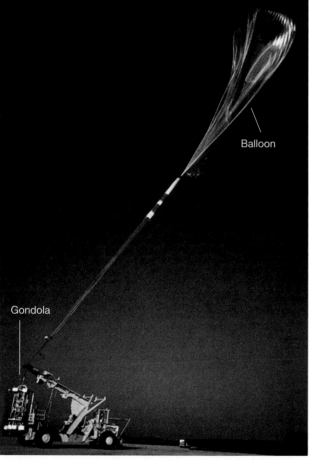

Balloon

Gondola

(a)

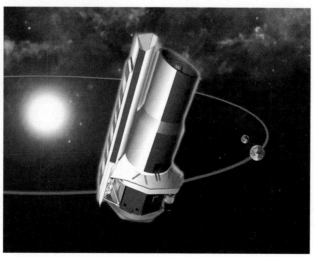

(b)

◄ FIGURE 3.23 Infrared Telescopes (a) A gondola containing a 1-m infrared telescope is readied for its balloon-borne ascent to an altitude of about 30 km, where it will capture infrared radiation that cannot penetrate Earth's atmosphere. (b) This artist's conception shows the *Spitzer Space Telescope* orbiting in an Earth-trailing orbit around the Sun while surveying the infrared sky at wavelengths ranging from 3 to 200 μm. *(SAO; JPL)*

► **FIGURE 3.24 Infrared Images**
(a) An optical photograph taken near San Jose, California, and (b) an infrared photo of the same area taken at the same time. Infrared radiation can penetrate smog much better than short-wavelength visible light. The same advantage pertains to astronomical observations (c). An optical view of an especially dusty part of the central region of the Orion Nebula is more clearly revealed (d) in this infrared image showing a cluster of stars behind the obscuring dust. (*Lick Observatory; NASA*)

These foggy optical views . . .   *. . . are much clearer when viewed in the infrared.*

smaller than the massive instruments found in ground-based observatories. Nevertheless, their infrared view penetrates the clouds of dust and gas surrounding many cosmic objects, allowing astronomers to study regions of space that are completely obscured at visible wavelengths and revolutionizing our understanding of the universe. Figure 3.24 shows some terrestrial and astronomical examples of advantages gained by viewing the infrared part of the spectrum.

In August 2003, NASA launched the 0.85-m *Spitzer Space Telescope (SST,* shown in Figure 3.23b), named in honor of Lyman Spitzer, Jr., a renowned astrophysicist and the first person to propose (in 1946) that a large telescope be located in space. The facility's detectors are designed to operate at wavelengths between 3.6 and 160 μm. Unlike previous space-based observatories, *SST* does not orbit Earth, but instead follows our planet in its orbit around the Sun, trailing millions of kilometers behind to minimize Earth's heating effect on its detectors. The spacecraft is currently drifting away from Earth at the rate of 0.1 AU per year. The detectors were cooled to near absolute zero in order to observe infrared signals from space without interference from the telescope's own heat. Figure 3.25 shows some examples of the spectacular imagery obtained from NASA's latest eye on the universe; many more of its images are seen throughout this book.

Unfortunately, the liquid helium keeping *SST*'s detectors cool could not be confined indefinitely, and it slowly leaked away into space. In early 2009, *Spitzer* entered a new "warm" phase of operation as its temperature increased to roughly 30K—still very cool by Earth standards, but warm enough for the telescope's own thermal emission to overwhelm the long-wavelength detectors on board. (Recall how Wien's law tells us that the thermal emission of a 30-K object peaks at a wavelength of roughly 100 μm.) ∞ *(Sec. 2.4)* However, the craft's shorter-wavelength

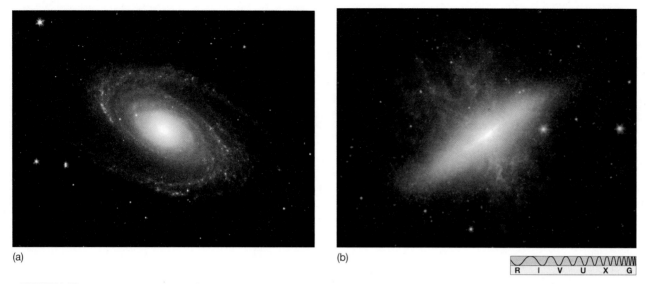

(a)

(b)

R I V U X G

▲ FIGURE 3.25 Spitzer Images   Images from the *Spitzer Space Telescope* show its camera's superb capabilities. (a) The magnificent spiral galaxy M81 is about 12 million light-years away. (b) Its companion, M82, is not so serene, rather resembling a smoking hot cigar. *(JPL)*

detectors (at around 3.6 and 4.5 μm) are expected to remain operational for a further 3–4 years, and the telescope will continue to be an important astronomical resource during that time.

To the short-wavelength, higher-energy side of the visible spectrum lies the ultraviolet domain. This region of the spectrum, extending in wavelength from 400 nm (blue light) down to a few nanometers (the soft, or low-energy, end of the X-ray region), has only recently begun to be explored. Because Earth's atmosphere is partially opaque to radiation below 400 nm and is totally opaque to radiation below about 300 nm, astronomers cannot conduct any useful ultraviolet observations from the ground, not even from the highest mountaintop. Rockets, balloons, or satellites are therefore essential to any **ultraviolet telescope**—a device designed to capture and analyze this high-frequency radiation.

Figure 3.26(a) shows an image of a supernova remnant—the remains of a violent stellar explosion that occurred some 12,000 years ago (see Chapter 12)—obtained by

▼ FIGURE 3.26 Ultraviolet Images   (a) A camera on board the *Extreme Ultraviolet Explorer* satellite captured this image of the Cygnus Loop supernova remnant, the result of a massive star having exploded. The glowing field of debris lies some 1500 light-years from Earth. Based on the velocity of the outflowing debris, astronomers estimate that the explosion occurred about 12,000 years ago. (b) This false-color image of the galaxies M81 and M82 (the same ones as in Figure 3.25), made here by the *Galaxy Evolution Explorer* satellite, reveals stars forming in the blue arms well away from the galaxy's center. *(NASA/GALEX)*

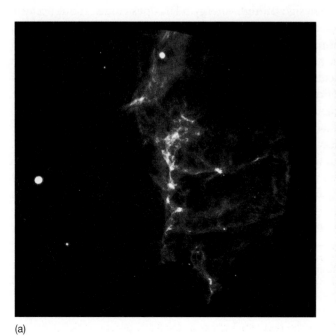

(a)

(b)

R I V U X G

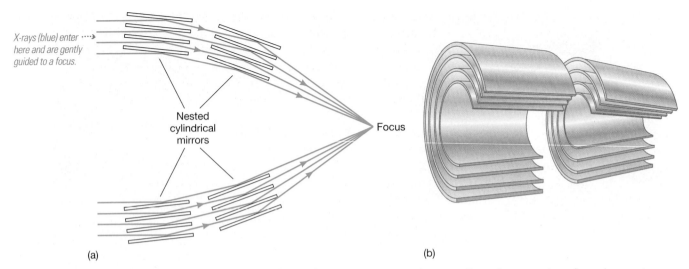

X-rays (blue) enter here and are gently guided to a focus.

Nested cylindrical mirrors

Focus

(a)

(b)

▲ FIGURE 3.27 X-Ray Telescope   (a) The arrangement of nested mirrors in an X-ray telescope allows the rays to be reflected at grazing angles to form an image. (b) A cutaway 3D rendition of the mirrors, showing their shape more clearly.

PROCESS OF ▶
SCIENCE CHECK
Describe some of the kinds of observations that astronomers make across the entire electromagnetic spectrum. Why are such observations necessary?

ANIMATION/VIDEO *Chandra* Light and Data Paths   **MA**

the *Extreme Ultraviolet Explorer (EUVE)* satellite, launched in 1992. Since its launch, *EUVE* has mapped out our local cosmic neighborhood as it presents itself in the far ultraviolet and has radically changed astronomers' conception of interstellar space in the vicinity of the Sun. The *Hubble Space Telescope*, best known as an optical telescope, is also a superb ultraviolet instrument, as is the *Galaxy Evolution Explorer (GALEX)* satellite, launched in 2003, one of whose images is seen in Figure 3.26(b).

High-Energy Astronomy

High-energy astronomy studies the universe as it presents itself to us in X-rays and gamma rays—the types of radiation whose photons have the highest frequencies and, hence, the greatest energies. How do we detect radiation of such short wavelengths? First, it must be captured high above Earth's atmosphere because none of it reaches the ground. Second, its detection requires the use of equipment basically different from that used to capture the relatively low-energy radiation discussed up to this point.

The difference in the design of **high-energy telescopes** comes about because X-rays and gamma rays cannot be reflected easily by any kind of surface. Rather, these rays tend to either pass straight through or else be absorbed by any material they strike. When X-rays barely graze a surface, however, they can be reflected from it in a way that yields an image, although the mirror design is fairly complex (Figure 3.27). High-quality data from imaging X-ray telescopes have driven major advances in our understanding of high-energy phenomena throughout the universe.

In July 1999, NASA launched the *Chandra X-Ray Observatory* (named in honor of the Indian astrophysicist Subramanyan Chandrasekhar and shown in Figure 3.28). With greater sensitivity, a wider field of view, and better resolution than any previous X-ray telescope, *Chandra* is providing high-energy astronomers with new levels of observational detail. Figure 3.29 shows the first image returned by *Chandra*—a supernova remnant known as Cas A, all that now remains of a star in the constellation Cassiopeia that was observed to explode about 320 years ago. The false-color

◀ FIGURE 3.28 Chandra Observatory   The *Chandra* X-ray telescope is shown here during the final stages of its construction. Its effective angular resolution is 1″, allowing this spacecraft to produce images of quality comparable to that of optical photographs. *Chandra* now occupies an elliptical orbit high above Earth; its farthest point from our planet, 140,000 km out, reaches almost one-third of the way to the Moon. (*NASA*)

image shows 50 million-kelvin gas in the wisps of ejected stellar material; the bright white point at the very center of the debris may be a black hole. The European *X-Ray Multi-Mirror* satellite (now known as *XMM-Newton*) was launched in December 1999. *XMM-Newton* is more sensitive than *Chandra* (that is, it can detect fainter X-ray sources), but it has significantly poorer angular resolution (5′, compared with 0.5″ for *Chandra*), making the two missions complementary to one another.

Gamma-ray astronomy is the youngest entrant into the observational arena. For gamma rays, no method of producing a true image (in the sense of Section 3.1) has yet been devised—present-day gamma-ray telescopes simply point in a specified direction and count photons received. As a result, only fairly coarse (1° resolution) observations can be made. Nevertheless, even at this resolution, there is much to be learned. Cosmic gamma rays were originally detected in the 1960s by the U.S. *Vela* series of satellites, whose primary mission was to monitor illegal nuclear detonations on Earth. Since then, several X-ray telescopes have also been equipped with gamma-ray detectors.

Gamma-ray astronomy traces the most violent events in the universe, on scales ranging from stars to galaxies. NASA's *Compton Gamma-Ray Observatory (CGRO)*, placed in orbit in 1991, scanned the sky and studied individual objects in much greater detail than ever before. The *CGRO* mission ended in June 2000, when, following a failure of one of the spacecraft's three gyroscopes, NASA opted for a controlled reentry and dropped the satellite into the Pacific Ocean. In August 2008, NASA launched the *Fermi Gamma-Ray Space Telescope*. With greater sensitivity to a broader range of gamma-ray energies than *CGRO*, *Fermi's* capabilities

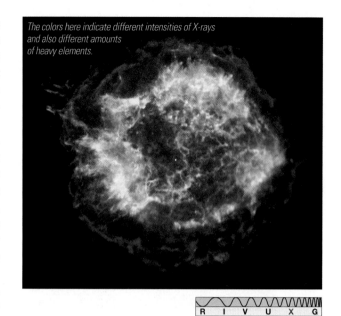

The colors here indicate different intensities of X-rays and also different amounts of heavy elements.

R I V U X G

▲ FIGURE 3.29 X-Ray Image   A false-color *Chandra* X-ray image of the supernova remnant Cassiopeia A, a debris field of scattered, hot gases that were once part of a massive star. Roughly 10,000 light-years from Earth and barely visible in the optical part of the spectrum, Cas A is awash in brilliantly glowing X-rays spread across some 10 light-years. *(CXC/SAO)*

TABLE 3.1	Astronomy at Many Wavelengths	
Radiation	**General Considerations**	**Common Applications (Chapter Reference)**
Radio	Can penetrate dusty regions of interstellar space. Earth's atmosphere largely transparent to radio wavelengths. Can be observed day or night. High resolution at long wavelengths requires very large telescopes.	Radar studies of planets (1, 6) Planetary magnetic fields (7) Interstellar gas clouds (11) Center of Milky Way Galaxy (14) Galactic structure (14, 15) Active galaxies (15) Cosmic background radiation (17)
Infrared	Can penetrate dusty regions of interstellar space. Earth's atmosphere only partially transparent to IR radiation, so some observations must be made from space.	Star formation (11) Cool stars (11, 12) Center of Milky Way Galaxy (14) Active galaxies (15) Large-scale structure of the universe (16, 17)
Visible	Earth's atmosphere transparent to visible light.	Planets (6, 7) Stars and stellar evolution (9, 10, 12) Galactic structure (14, 15) Large-scale structure of the universe (16, 17)
Ultraviolet	Earth's atmosphere opaque to UV radiation, so observations must be made from space.	Interstellar medium (11) Hot stars (12)
X-ray	Earth's atmosphere opaque to X-rays, so observations must be made from space. Special mirror configurations needed to form images.	Stellar atmospheres (9) Neutron stars and black holes (13) Active galactic nuclei (15) Hot gas in galaxy clusters (16)
Gamma ray	Earth's atmosphere opaque to gamma rays, so observations must be made from space. Cannot form images.	Neutron stars (13) Active galactic nuclei (15)

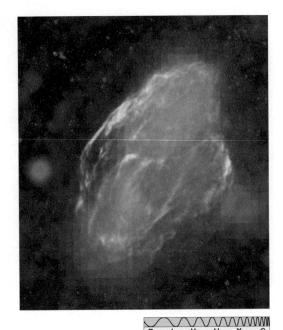

◄ **FIGURE 3.30 Gamma-Ray Image**   A false-color gamma-ray image from the *Fermi* telescope, showing the remains of a violent event (a supernova) in a region named W44. *(NASA)*

are greatly expanding astronomers' view of the high-energy universe. Figure 3.30 shows the aftermath of a vast stellar explosion, the kind of violent event that emits much of its energy in the form of gamma rays. An early all-sky image from *Fermi* appears in Figure 3.31(e).

Full-Spectrum Coverage

Table 3.1 lists the basic regions of the electromagnetic spectrum and describes objects typically studied in each frequency range. Bear in mind that the list is far from exhaustive and that many astronomical objects are now routinely observed at many different electromagnetic wavelengths. As we proceed through the text, we will discuss more fully the wealth of information that high-precision astronomical instruments can provide us.

As an illustration of the sort of comparison that full-spectrum coverage allows, Figure 3.31 shows a series of images of the Milky Way Galaxy. They were made by several instruments, at wavelengths ranging from radio to gamma ray, over a period of about 5 years. By comparing the features visible in each, we immediately see how multiwavelength observations can complement each other, greatly extending our perception of the universe around us.

CONCEPT CHECK ► List two scientific benefits of placing telescopes in space. What might be some drawbacks?

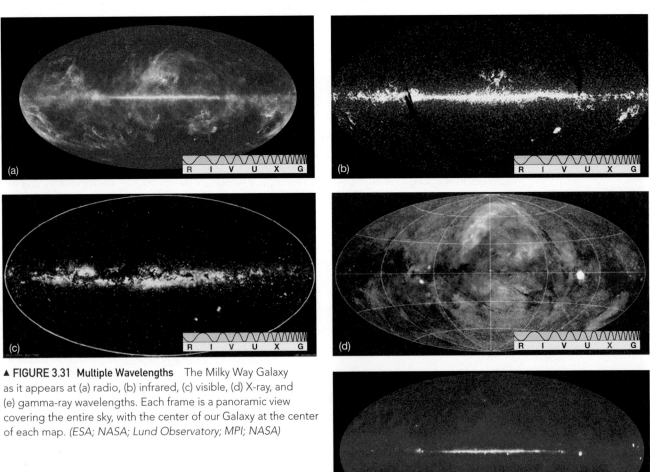

▲ **FIGURE 3.31 Multiple Wavelengths**   The Milky Way Galaxy as it appears at (a) radio, (b) infrared, (c) visible, (d) X-ray, and (e) gamma-ray wavelengths. Each frame is a panoramic view covering the entire sky, with the center of our Galaxy at the center of each map. *(ESA; NASA; Lund Observatory; MPI; NASA)*

CHAPTER REVIEW

SUMMARY

L01 A **telescope** (p. 70) is a device designed to collect as much light as possible from some distant source and deliver it to a detector for detailed study. Astronomers prefer **reflecting telescopes** (p. 70) because large mirrors are lighter and much easier to construct than large lenses, and they also suffer from fewer optical defects. Instruments may be placed inside the telescope at the **prime focus** (p. 70), or a secondary mirror may be used to reflect the light to an external detector. Most modern telescopes use **charge-coupled devices** (p. 76) to collect and store data in digital form for later analysis.

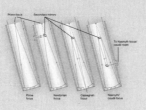

L02 The light-gathering power of a telescope depends on its **collecting area** (p. 77), which is proportional to the square of the mirror diameter. To study the faintest sources of radiation, astronomers must use large telescopes. Large telescopes also suffer least from the effects of diffraction and, hence, can achieve better **angular resolution** (p. 78) once the blurring effects of Earth's atmosphere are overcome. The amount of diffraction is proportional to the wavelength of the radiation under study and inversely proportional to the size of the mirror.

L03 The resolution of most ground-based optical telescopes is limited by **seeing** (p. 81)—the blurring effect of Earth's turbulent atmosphere, which smears the pointlike images of a star into a **seeing disk** (p. 81) a few arc seconds in diameter. Astronomers can improve a telescope's resolution by using **active optics** (p. 82), in which a telescope's environment and focus are carefully monitored and controlled, and **adaptive optics** (p. 82), in which the blurring effects of atmospheric turbulence are corrected for in real time.

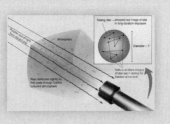

L04 **Radio telescopes** (p. 83) are conceptually similar in construction to optical reflecting telescopes, although they are generally much larger than optical instruments, in part because so little radio energy reaches Earth from space. Their main disadvantage is that diffraction of long-wavelength radio

waves limits their resolution. Their principal advantage is that they allow astronomers to explore a new part of the electromagnetic spectrum and of the universe—many astronomical radio emitters are completely undetectable in visible light. In addition, radio observations are largely unaffected by Earth's atmosphere, weather, and the location of the Sun.

L05 To increase the effective area of a telescope, and, hence, improve its resolution, several instruments may be combined into an **interferometer** (p. 86), in which the interference pattern of radiation received by two or more detectors is used to reconstruct a high-precision map of the source. Using **interferometry** (p. 86), radio telescopes can produce images much sharper than those from the best optical equipment.

L06 **Infrared** (p. 89) and **ultraviolet telescopes** (p. 91) are similar in basic design to optical systems. **High-energy telescopes** (p. 92) study the X-ray and gamma-ray regions of the electromagnetic spectrum. X-ray telescopes can form images of their field of view, although the mirror design is more complex than for lower-energy instruments. Gamma-ray telescopes simply point in a certain direction and count photons received. Infrared studies in some parts of the infrared range can be carried out using large ground-based systems. However, because of atmospheric opacity, most infrared, and all ultraviolet, X-ray, and gamma-ray observations must be made from space.

L07 Different physical processes can produce very different kinds of electromagnetic radiation, and the image of a given object in long-wavelength, low-energy radio waves may bear little resemblance to its appearance in high-energy X-rays or gamma rays. Observations at wavelengths spanning the electromagnetic spectrum are essential to a complete understanding of astronomical events.

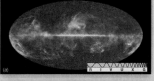

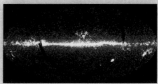

MasteringAstronomy®   For instructor-assigned homework go to www.masteringastronomy.com

Problems labeled **POS** explore the process of science. **VIS** problems focus on reading and interpreting visual information. **LO** connects to the introduction's numbered Learning Outcomes.

REVIEW AND DISCUSSION

1. **LO1** List three advantages of reflecting telescopes over refracting telescopes.

2. **LO2** What are the largest optical telescopes in use today? Why do astronomers want their telescopes to be as large as possible?

3. How does Earth's atmosphere affect what is seen through an optical telescope?

4. What advantages does the *Hubble Space Telescope* have over ground-based telescopes? List some disadvantages.

5. What are the advantages of a CCD over a photograph?

6. **LO3** Is the resolution of a 2-m telescope on Earth's surface limited more by atmospheric turbulence or by the effects of diffraction?

7. Why do radio telescopes have to be very large?

8. **LO4** Which astronomical objects are best studied with radio techniques?

9. **LO5** What is interferometry, and what problem in radio astronomy does it address?

10. Compare the highest resolution attainable with optical telescopes to the highest resolution attainable with radio telescopes (including interferometers).

11. What special conditions are required to conduct observations in the infrared?

12. In what ways do the mirrors in X-ray telescopes differ from those found in optical instruments?

13. **LO6** Why was *Fermi* placed in space, rather than on the ground?

14. **LO7 POS** What are the main advantages of studying objects at many different wavelengths of radiation?

15. **POS** Our eyes can see light with an angular resolution of about 1′—equivalent to about a third of a millimeter at arm's length. Suppose our eyes detected only infrared radiation, with 1° angular resolution. Would we be able to make our way around on Earth's surface? To read? To sculpt? To create technology?

CONCEPTUAL SELF-TEST: TRUE OR FALSE/MULTIPLE CHOICE

1. The main advantage to using the *HST* is the increased amount of "nighttime" viewing it affords. (T/F)

2. The term "seeing" is used to describe how faint an object can be detected by a telescope. (T/F)

3. One of the primary advantages of CCDs over photographic plates is the former's high efficiency in detecting light. (T/F)

4. Radio telescopes are large in part to improve their angular resolution, which is poor because of the long wavelengths at which they are used to observe the skies. (T/F)

5. Infrared astronomy can only be done from space. (T/F)

6. Gamma-ray telescopes employ the same basic design that optical instruments use. (T/F)

7. Because gamma rays have very short wavelengths, gamma-ray telescopes can achieve extremely high angular resolution. (T/F)

8. The main reason that most professional research telescopes are reflectors is that (a) mirrors produce sharper images than lenses do; (b) their images are inverted; (c) they do not suffer from the effects of seeing; (d) large mirrors are easier to build than large lenses.

9. If telescope mirrors could be made in odd sizes, the one with the *most* light-gathering power would be (a) a triangle with 1-m sides; (b) a square with 1-m sides; (c) a circle 1 m in diameter; (d) a rectangle with two 1-m sides and two 2-m sides.

10. The primary reason professional observatories are built on the highest mountaintops is to (a) get away from city lights; (b) be above the rain clouds; (c) reduce atmospheric blurring; (d) improve chromatic aberration.

11. When multiple radio telescopes are used for interferometry, resolving power is most improved by increasing (a) the distance between telescopes; (b) the number of telescopes in a given area; (c) the diameter of each telescope; (d) the electrical power supplied to each telescope.

12. The *Spitzer Space Telescope* is stationed far from Earth because (a) this increases the telescope's field of view; (b) the telescope is sensitive to electromagnetic interference from terrestrial radio stations; (c) doing so avoids the obscuring effects of Earth's atmosphere; (d) Earth is a heat source and the telescope must be kept very cool.

13. The best way to study young stars hidden behind interstellar dust clouds would be to use (a) X-rays; (b) infrared light; (c) ultraviolet light; (d) blue light.

14. **VIS** The image shown in Figure 3.12 (Resolution) is sharpest when the ratio of wavelength to telescope size is (a) large; (b) small; (c) close to 1; (d) none of these.

15. **VIS** Table 3.1 (Astronomy at Many Wavelengths) suggests that the best frequency range in which to study the hot (million-kelvin) gas found among the galaxies in the Virgo cluster would be (a) at radio frequencies; (b) in the infrared; (c) in X-rays; (d) in gamma rays.

PROBLEMS

The number of squares preceding each problem indicates its approximate level of difficulty.

1. ■ A certain telescope has a $10' \times 10'$ field of view that is recorded using a CCD chip having 2048×2048 pixels. What angle on the sky corresponds to 1 pixel? What would be the diameter of a typical seeing disk ($1''$ radius), in pixels?

2. ■ The *SST*'s planned operating temperature is 5.5 K. At what wavelength (in micrometers, μm) does the telescope's own blackbody emission peak? How does this wavelength compare with the wavelength range in which the telescope is designed to operate? ∞ *(More Precisely 2-2)*

3. ■ A 2-m telescope can collect a given amount of light in 1 hour. Under the same observing conditions, how much time would be required for a 6-m telescope to perform the same task? A 12-m telescope?

4. ■ A space-based telescope can achieve a diffraction-limited angular resolution of $0.05''$ for red light (wavelength 700 nm). What would the resolution of the instrument be (a) in the infrared, at wavelength 3.5 μm, and (b) in the ultraviolet, at wavelength 140 nm?

5. ■■ Two identical stars are moving in a circular orbit around one another, with an orbital separation of 2 AU. The system lies 200 light-years from Earth. If we happen to view the orbit head-on, how large a telescope would we need to resolve the stars, assuming diffraction-limited optics at a wavelength of 2 μm?

6. ■■ What is the greatest distance at which *HST*, operating in blue light (400 nm), could resolve the stars in the previous question?

7. ■ The photographic equipment on a telescope is replaced by a CCD. If the photographic plate records 5 percent of the light reaching it, but the CCD records 90 percent, how much time will the new system take to collect as much information as the old detector recorded in a 1-hour exposure?

8. ■ The Andromeda galaxy lies about 2.5 million light-years away. To what distances do the angular resolutions of *SST* ($3''$), *HST* ($0.05''$), and a radio interferometer ($0.001''$) correspond at that distance?

9. ■ What would be the equivalent single-mirror diameter of a telescope constructed from two separate 10-m mirrors? Four separate 8-m mirrors?

10. ■ Estimate the angular resolutions of (a) a radio interferometer with a 5000-km baseline, operating at a frequency of 5 GHz, and (b) an infrared interferometer with a baseline of 50 m, operating at a wavelength of 1 μm.

ACTIVITIES

Collaborative

1. Determine the maximum size interferometer your group could build if you placed 2-m radio telescopes at each of your homes. What would be its resolution at a wavelength of 1 cm?

2. Your group has been assigned to observe the region of the sky around Orion to look for hot, bright young stars hidden in molecular clouds. Explain which of the telescopes described in the text would be your best choice, and estimate the level of detail you might expect to see.

Individual

1. Take some photographs of the night sky. You will need a location with a clear, dark sky, a good digital camera that lets you control the exposure time, a tripod and cable release, and a watch with a seconds display visible in the dark. Set your camera to the "manual" setting for the exposure and attach the cable release so you can control it. Set the focus on infinity. Point the camera at the desired constellation, seen through the viewfinder, and take a 20- to 30-second exposure. Don't touch any part of the camera or hold on to the cable release during the exposure to minimize all vibration. Keep a log of your shots.

2. For some variation, vary your exposure times, take hours-long exposures for star trails, use different lenses such as wide-angle or telephoto, or place the camera piggyback on a telescope tracking an object in the sky so you can take exposures that are a few minutes long. Experiment and have fun!

3. Which image of the Milky Way Galaxy in Figure 3.31 (Multiple Wavelengths) provides the most interesting information? Explain your reasoning.

Our Planetary System

As our study of astronomy expands, we embark on the next in a series of increasingly larger steps that will ultimately take us to the limits of the observable universe. This journey begins with Earth, progresses to the study of other planets, and continues on to the magnificent Sun. The dimension of our planetary family is so large that more than a million Earths could be stacked side by side across the whole solar system.

Part 2 relates our present understanding of our extended, complex home in space. That understanding is still a work in progress, but the results thus far have revolutionized our knowledge both of our present cosmic neighborhood and of Earth's rich natural history.

The images here illustrate the range of sizes encountered in Part 2, from the scattered debris of asteroids and comets to the central Sun.

Venus ~ 10^7 m

Moon ~ 10^6 m

Asteroid ~ 10^4 m

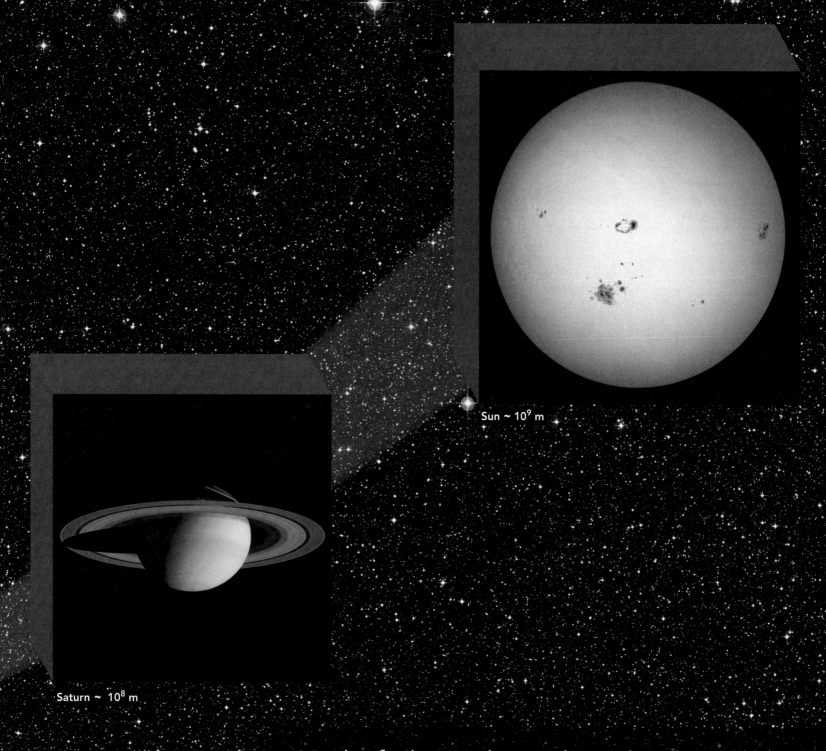

Sun ~ 10^9 m

Saturn ~ 10^8 m

Much of what we know about the planets was discovered in the last few decades.

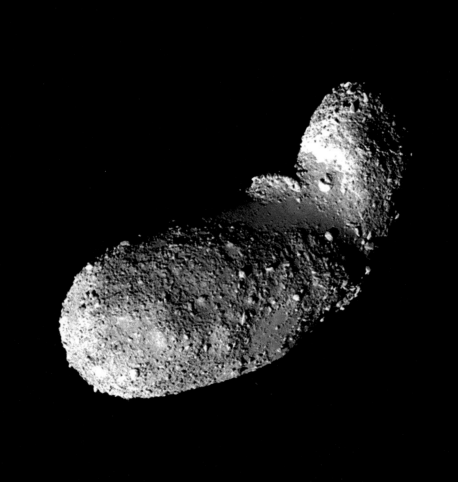

4

The Solar System

Interplanetary Matter and the Birth of the Planets

In less than a generation, we have learned more about our solar system—the Sun and everything that orbits it—than in all previous centuries. By studying the planets, their moons, and the countless fragments of material that orbit in interplanetary space, astronomers have gained a richer outlook on our own home in space. The discoveries of the past few decades have revolutionized our understanding not only of the present state of our cosmic neighborhood, but also of its history, for our solar system is filled with clues to its own origin and evolution. Paradoxically, the richest sources of information about the earliest days of the solar system are not the large bodies orbiting the Sun, but rather the asteroids, comets, and meteoroids that pepper interplanetary space. Far more than the planets themselves, these fragments hold a record of the formative stages of our planetary system and have much to teach us about the origin of our world.

MasteringAstronomy® Visit www.masteringastronomy.com for quizzes, animations, videos, interactive figures, and self-guided tutorials.

◄ Only within the past few decades have scientists taken seriously the idea that life on Earth has been disrupted over the course of billions of years by asteroid and comet impacts. It is very much in our own interest to monitor such stray objects that occasionally glide by Earth—and sometimes hit us! This image shows a close-up of the asteroid Itokawa, which is only 0.5 km long—about five soccer fields across. It was photographed as the Japanese spacecraft, *Hayabusa*, having launched from Earth in 2003, slowly approached the asteroid in 2005. The craft then soft-landed, scooped up some rocky debris, and took off for Earth, landing back home in 2010. This mission also scientifically proved that asteroids like this one are the source of most meteorites—the oldest matter in the solar system. (*JAXA*)

LEARNING OUTCOMES

Studying this chapter will enable you to:

LO1 Describe the scale and structure of the solar system, and list the basic differences between terrestrial and jovian planets.

LO2 Summarize the orbital and physical properties of asteroids.

LO3 Describe the composition and structure of a typical comet, and explain what a comet's orbit tells us about its probable origin.

LO4 Summarize the orbital and physical properties of meteoroids, and explain how these bodies are related to asteroids and comets.

LO5 List the major facts and exceptions that any theory of solar system formation must explain.

LO6 Outline the condensation theory of planetary formation, and indicate how it accounts for the major features of the solar system.

LO7 Explain how the condensation theory accounts for the terrestrial and jovian planets, as well as the smaller bodies scattered throughout the solar system.

LO8 Describe the main methods by which astronomers have detected planets beyond the solar system.

LO9 Outline the properties of the known extrasolar planets and relate to current theories of solar system formation.

4.1 An Inventory of the Solar System

Our **solar system** contains one star (the Sun), eight planets orbiting that star, 169 moons (at last count) orbiting those planets, five *dwarf planets,* seven *asteroids* and more than 100 *Kuiper belt objects* larger than 300 km (200 miles) in diameter, tens of thousands of smaller (but well-studied) asteroids and Kuiper belt objects, myriad *comets* a few kilometers in diameter, and countless *meteoroids* less than 100 m across. This list will undoubtedly grow as we continue to explore our cosmic neighborhood. Table 4.1 lists some basic orbital and physical properties of the eight planets, with a few other solar system objects included for comparison.

ANIMATION/VIDEO An Astronomical Ruler

Planetary Properties

The arrangement of the major bodies in the solar system is shown in Figure 4.1. Mercury is the planet closest to the Sun, followed by Venus, Earth, Mars, Jupiter, Saturn, Uranus, and Neptune. The asteroids lie mainly in a broad belt between the orbits of Mars and Jupiter, the Kuiper belt beyond Neptune.

The planets' distances from the Sun are known from their periods and Kepler's laws once the scale of the solar system is set by radar ranging on Venus. ∞ *(Sec. 1.3)* The distance from the Sun to Neptune is about 30 AU, some 700,000 times Earth's radius and roughly 12,000 times the distance from Earth to the Moon. Yet despite this vast extent, the planets all lie very close to the Sun, astronomically speaking. The diameter of Neptune's orbit is less than 1/3000 of a light-year, while the next nearest star to the Sun is several light-years distant. Note that the planetary orbits are not evenly spaced. Roughly speaking, the spacing between adjacent orbits doubles as we move outward from the Sun.

All the planets orbit the Sun counterclockwise as seen from above Earth's North Pole and in nearly the same plane as Earth (the ecliptic plane). ∞ *(Sec. 1.1)* Mercury deviates slightly from this rule—its orbital plane lies at 7° to the ecliptic. Still, we can think of the solar system as being quite flat. If we were to view the planets' orbits from a vantage point in the ecliptic plane about 50 AU from the Sun, no planet's orbit would be noticeably tilted. Figure 4.2 is a photograph of the planets Mercury, Venus, Mars, Jupiter, and Saturn, taken during an April 2002 planetary alignment. These five planets can (occasionally) be found in the same region of the sky in large part because their orbits lie nearly in the same plane in space.

Earth's radius has long been known through conventional surveying techniques and, more recently, by satellite measurements. ∞ *(Discovery 0-1)* Other planetary radii are found by measuring the planets' angular sizes and then employing elementary geometry, as discussed in *More Precisely 4-1.* Figure 4.3 illustrates the sizes of the planets relative to the Sun.

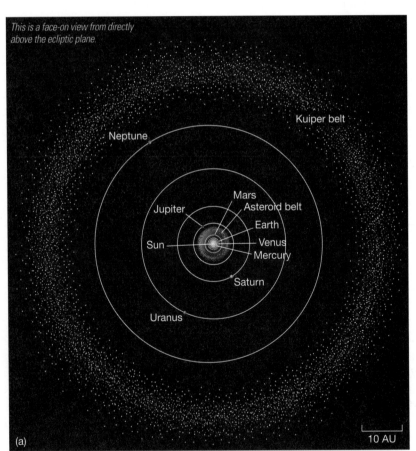

This is a face-on view from directly above the ecliptic plane.

Kuiper belt

Neptune

Jupiter

Mars
Asteroid belt

Earth

Sun

Venus

Mercury

Saturn

Uranus

(a)   10 AU

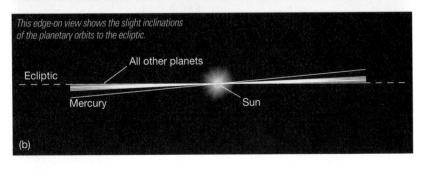

This edge-on view shows the slight inclinations of the planetary orbits to the ecliptic.

All other planets

Ecliptic

Mercury   Sun

(b)

◀ **FIGURE 4.1 Solar System** Major bodies of the solar system include the Sun, planets, and asteroids. Except for Mercury, the orbits of the planets are almost circular (a) and lie nearly in the same plane (b). The entire solar system, including the Kuiper belt, spans nearly 100 AU.

4-1 MORE PRECISELY

Measuring Sizes with Geometry

We saw in Chapter 1 how astronomers use parallax, radar ranging, and Kepler's laws to determine the distances to objects in the solar system. ∞ *(Sec. 1.3)* Knowing the distance, we can convert a body's angular size into its physical size using a simple argument first made by the Greek geometer Euclid.

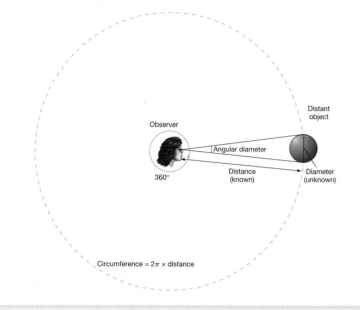

The accompanying figure shows an observer measuring the angular diameter of an object of known distance; we have added a large circle centered on the observer and passing through the object. To compute the object's size, we note that the ratio of its actual diameter to the circumference of the circle (2π times the distance to the object) must be equal to the ratio of the observed angular diameter to one full revolution, 360°:

$$\frac{\text{diameter}}{2\pi \times \text{distance}} = \frac{\text{angular diameter}}{360°}.$$

Note the basic similarity of this reasoning to that presented in Chapter 0. ∞ *(Discovery 0-1)* From the above equation we find

$$\text{diameter} = \text{distance} \times \frac{\text{angular diameter}}{57.3°}.$$

For example, radar ranging on the Moon tells us that its distance is 384,000 km. The Moon's angular diameter is measured to be about 31 arc minutes—a little over half a degree. It therefore follows that the Moon's actual diameter is 384,000 km × (31/60)°/57.3° = 3460 km. A more precise measurement gives 3476 km.

Although the observations are straightforward and the geometrical reasoning elementary, simple measurements such as these form the basis for almost every statement made in this book about size and scale in the universe.

A planet's mass is determined by observing its gravitational influence on some nearby object and applying Newton's laws of motion and gravity. ∞ *(More Precisely 1-1)* Before the Space Age, astronomers calculated planetary masses either by tracking the orbits of the planets' moons (if any) or by measuring the small but detectable distortions the planets produce in each other's orbits. Today astronomers can accurately determine the masses of most of the objects listed in Table 4.1 through their gravitational effects on artificial satellites and space probes launched from Earth. We discuss measurements of the orbits, masses, and radii of asteroids and comets in more detail in Section 4.2. With nearly 99.9 percent of the total mass, the Sun is clearly the "senior partner" in the solar system. Its gravity dominates the motion of everything else.

The final column in Table 4.1 lists a quantity called **density,** a measure of the "compactness" of an object, computed by dividing the object's mass (in kilograms) by its volume (in cubic meters). Earth's average density is 5500 kg/m³. For comparison, the density of water is 1000 kg/m³, rocks on Earth's surface have densities in the range 2000–3000 kg/m³, and iron has a density of about 8000 kg/m³. Earth's atmosphere at sea level has a density of only a few kilograms per cubic meter.

▶ **FIGURE 4.2 Planetary Alignment** Six planets—Mercury, Venus, Mars, Jupiter, Saturn, and Earth—are shown during a planetary alignment in April 2002. The Sun and Moon are just below the horizon. *(J. Lodriguss)*

TABLE 4.1 **Properties of Some Solar System Objects**

Object	Orbital Semimajor Axis (AU)	Orbital Period (Earth Years)	Mass (Earth Masses)	Radius (Earth Radii)	Number of Known Moons	Average Density (kg/m3)	(Earth = 1)
Mercury	0.39	0.24	0.055	0.38	0	5400	0.98
Venus	0.72	0.62	0.82	0.95	0	5200	0.95
Earth	1.0	1.0	1.0	1.0	1	5500	1.00
Moon	—	—	0.012	0.27	—	3300	0.60
Mars	1.5	1.9	0.11	0.53	2	3900	0.71
Ceres (asteroid)	2.8	4.7	0.00015	0.073	0	2700	0.49
Jupiter	5.2	11.9	318	11.2	64	1300	0.24
Saturn	9.5	29.4	95	9.5	62	700	0.13
Uranus	19.2	84	15	4.0	27	1300	0.24
Neptune	30.1	164	17	3.9	13	1600	0.29
Pluto (Kuiper-belt object)	39.5	249	0.002	0.2	4	2100	0.38
Comet Hale-Bopp	180	2400	1.0×10^{-9}	0.004	—	100	0.02
Sun	—	—	332,000	109	—	1400	0.25

Terrestrial and Jovian Planets

Astronomers can infer a planet's bulk composition from a combination of density measurements, spacecraft data, and theoretical calculations (see Chapters 5–8). Based on all of these, we can draw a clear distinction between the inner and outer members of our planetary system. Simply put, the inner planets—Mercury, Venus, Earth, and Mars—are small, dense, and *rocky* in composition. The outer worlds—Jupiter, Saturn, Uranus, and Neptune—are large, of low density, and *gaseous.*

Because the physical and chemical properties of Mercury, Venus, and Mars are somewhat similar to Earth's, the four innermost planets are called the **terrestrial planets.** (The word *terrestrial* derives from the Latin word *terra,* meaning "land" or "earth.") The larger outer planets—Jupiter, Saturn, Uranus, and Neptune—are all similar to one another chemically and physically (and very different from the terrestrial worlds). They are labeled the **jovian planets,** after Jupiter, the largest member of the group. (The word *jovian* comes from *Jove,* another name for the Roman god Jupiter.) The jovian worlds are all much larger than the terrestrials and quite different from them in both composition and structure. Table 4.2 compares and contrasts some key properties of these two distinct planetary classes.

There are important differences among planets within each category. For example, when we take into account how the weight of the outer layers squeezes the interiors of the terrestrial planets to different extents (greatest for

ANIMATION/VIDEO Size and Scale of the Terrestrial Planets I

ANIMATION/VIDEO The Gas Giants

◀ **FIGURE 4.3 Sun and Planets** Relative sizes of the planets and our Sun. Notice that Jupiter, Saturn, Uranus, and Neptune are much larger than Earth and the other inner planets. However, even these large planets are dwarfed by the still larger Sun.

Earth, least for Mercury), we find that the average *uncompressed densities* of the terrestrial worlds—that is, the densities they would have in the absence of any gravitational compression—decrease steadily as we move farther from the Sun. This decrease in density indicates that, despite their other similarities, the overall compositions of these planets differ significantly one from the other. Another trend is found among the jovian worlds. All four are thought to have large, dense "terrestrial" cores up to 10 or 15 times the mass of Earth, but these cores account for an increasing fraction of the planet's total mass as we move outward from the Sun.

Solar System Debris

One of the principal goals of planetary science is to understand how the solar system formed and to explain the physical conditions now found on Earth and elsewhere in our planetary system. The last two sections of this chapter are devoted to this topic. Ironically, studies of the most accessible planet—Earth itself—do not help us much in our quest because information about our planet's early stages was obliterated long ago by atmospheric erosion and geological activity, such as earthquakes and volcanoes. Much the same holds true for the other planets (with the possible exception of Mercury, where a near-airless and geologically inactive surface still retains some imprint of the distant past). The problem is that the large bodies of the solar system have all *evolved* significantly since they formed, making it difficult to decipher the circumstances of their births.

A much better place to look for hints of conditions in the early solar system is on its *small* bodies—the planetary moons and the asteroids, meteoroids, Kuiper belt objects, and comets that make up interplanetary debris—for nearly all such fragments contain traces of solid and gaseous matter from the earliest times. They represent truly ancient material—many have scarcely changed since they formed along the rest of the solar system billions of years ago. Therefore, as a prologue to our study of the formation of the solar system, we first examine in more detail the contents of interplanetary space.

TABLE 4.2	Comparison Between the Terrestrial and Jovian Planets	
Terrestrial	**Jovian**
close to the Sun	far from the Sun
closely spaced orbits	widely spaced orbits
small masses	large masses
small radii	large radii
predominantly rocky	predominantly gaseous
solid surface	no solid surface
high density	low density
slower rotation	faster rotation
weak magnetic fields	strong magnetic fields
no rings	many rings
few moons	many moons

CONCEPT CHECK ▶ Why do astronomers draw such a clear distinction between the inner and the outer planets?

4.2 Interplanetary Matter

In the vast space among the eight known planets move countless small chunks of matter, ranging in size from a few hundred kilometers in diameter down to tiny grains of dust. The major constituents of this cosmic debris are asteroids, comets, Kuiper belt objects, and meteoroids. **Asteroids** and **meteoroids** are fragments of rocky material, somewhat similar in composition to the outer layers of the terrestrial planets. The distinction between the two is simply their size—anything larger than 100 m in diameter (corresponding to a mass exceeding 10,000 tons) is conventionally called an asteroid; anything smaller is a meteoroid. **Comets** are predominantly icy rather than rocky in composition (although they do contain some rocky material) and have typical diameters in the 1- to 10-km range. The **Kuiper belt** is an "outer asteroid belt" of sorts, also consisting of icy bodies, including the former planet Pluto. Comets and Kuiper belt objects are quite similar in chemical makeup to some of the icy moons of the outer planets—and may very well be the progenitors of those small bodies.

In 2006, the International Astronomical Union, the organization that oversees the rules for astronomical terminology, introduced a new category of solar system object: A **dwarf planet** is a body that orbits the Sun and is massive enough for its gravity to have pulled it into a spherical shape, but not so massive that it has cleared the region around its orbit of smaller bodies. (We will return to the reasons for this somewhat controversial decision in Chapter 8.) By this definition,

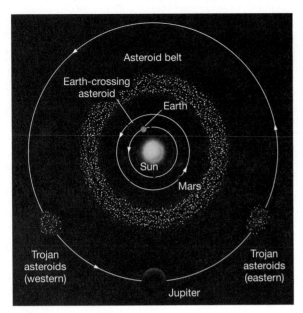

▲ **FIGURE 4.4 Inner Solar System** The main asteroid belt, along with the orbits of Earth, Mars, and Jupiter. Note the Trojan asteroids clumped at two locations in Jupiter's orbit. The orbits of the main-belt asteroids generally lie within the belt region of the figure. The red ellipse shows the orbit of an Earth-crossing asteroid.

INTERACTIVE ▲ **FIGURE 4.5 Asteroids, from Earth** The Earth-crossing asteroid Icarus has an orbit that passes within 0.2 AU of the Sun, well within Earth's orbit. Icarus occasionally comes close to Earth, making it one of the best-studied asteroids in the solar system. Its motion relative to the stars makes it appear as a streak (marked) in this long-exposure photograph. (*Palomar Observatory/Caltech*) **MA**

the three largest "small bodies" in the solar system—the asteroid Ceres, three Kuiper belt objects, including Pluto, and the "trans-Neptunian object" Eris—are all dwarf planets.

Taken together, these small bodies account for a negligible fraction of the total mass of the solar system—probably only a few millionths the mass of the Sun. Their combined gravitational effect is miniscule compared even with that of our Moon, and they play no important role in the present-day workings of the planets or their moons. Why, then, do we study them in any detail? The answer is that they are of crucial importance to our studies, for they are the keys to answering some very fundamental questions about our planetary environment.

Asteroid Orbits

The name *asteroid* comes from the Greek word *asteroeides*, which means "like a star," but asteroids are definitely not stars. They are too small even to be classified as planets. Astronomers often refer to them as minor planets. Researchers have so far cataloged well over 100,000 asteroids with well-determined orbits. The total number of known asteroids (that is, including those whose orbits are not yet known with sufficient accuracy to make them "official") is now approaching half a million. As sketched in Figure 4.4, the vast majority are found in a region of the solar system known as the **asteroid belt,** located between 2.1 and 3.3 AU from the Sun, roughly midway between the orbits of Mars (at 1.5 AU) and Jupiter (at 5.2 AU). All but one of the known asteroids revolve about the Sun in prograde orbits—in the same sense (direction) as Earth and the other planets. Again like the planets, most asteroids have orbits fairly close to the ecliptic plane (with inclinations less than 10–20°). Unlike the almost circular paths of the major planets, however, asteroid orbits are generally noticeably elliptical in shape.

In addition to the main-belt asteroids, a few hundred **Trojan asteroids** share an orbit with Jupiter, remaining a constant 60° ahead of or behind that planet as it circles the Sun. This peculiar orbital behavior is not a matter of chance—the Trojan asteroids are held in place by a stable balance between the gravitational fields of Jupiter and the Sun. Calculations first performed by the 18th-century French mathematician Joseph Louis Lagrange show that interplanetary matter that strays into one of the two regions of space now occupied by the Trojan asteroids can remain there indefinitely, perfectly synchronized with Jupiter's motion.

The orbits of most asteroids have eccentricities lying in the range 0.05–0.3, ensuring that they always remain between the orbits of Mars and Jupiter. However, some **Earth-crossing asteroids** move on orbits that intersect Earth's orbit, most likely because the gravitational field of Mars or Jupiter has deflected these small bodies into the inner solar system. Currently, some 9000 Earth-crossing asteroids are known. Most have been discovered since the late 1990s, when systematic searches for such objects began. The mile-wide Earth-crossing asteroid Icarus is shown (as the white streak) in Figure 4.5.

Roughly 1300 Earth crossers are officially designated "potentially hazardous," meaning that they are more than about 150 m in diameter (three times the size of the impactor responsible for the Barringer crater shown in Figure 4.19) and move in orbits that could bring them within 0.05 AU (7.5 million km) of our planet. All told, between 2000 and 2010, some 200 asteroids (that we know of!) passed within 0.05 AU of our planet, a couple of them coming within the orbit of the Moon. Similar numbers are expected during the next decade. None of the currently known potentially hazardous asteroids is expected to impact Earth during the next century—the closest predicted near miss will occur in April 2029, when the 350-m asteroid Apophis will pass approximately 30,000 km above our planet's surface.

Calculations indicate that most Earth-crossing asteroids *will* eventually collide with Earth and that during any given million-year period, roughly three asteroids strike our planet. Several dozen large basins and eroded craters on Earth are suspected to be sites of ancient asteroid collisions. The many large impact craters on the Moon, Venus, and Mars are direct evidence of similar events on other worlds. Such an impact could be catastrophic by human standards. Even a 1-km object carries more than 100 times more energy than all the nuclear weapons currently in existence on Earth and would devastate an area hundreds of kilometers in diameter. Should a sufficiently large asteroid hit our planet, it might even cause the extinction of an entire species. Indeed, many scientists think that the extinction of the dinosaurs was the result of just such an impact ∞ *(Discovery 4-1)*.

Asteroid Properties

Most asteroids are too small to resolve with even the largest Earth-based telescopes. Their images appear as featureless points, with no discernible surface features. ∞ *(Sec. 3.2)* However, astronomers can estimate the sizes of these tiny worlds by measuring the amount of sunlight they reflect and the heat they radiate. In addition, from time to time an asteroid happens to pass directly in front of a star, as seen from Earth, allowing astronomers to determine the asteroid's size and shape with great accuracy. The largest resident of the asteroid belt, the dwarf planet Ceres, has a diameter of 940 km. Only about two dozen asteroids are larger than 200 km across, and most are far smaller. Large asteroids are roughly spherical because (as with planets) gravity is the dominant force determining their shape. Smaller bodies can be highly irregular in shape.

Because the gravitational effect of an asteroid on its neighbors is very small and hard to measure accurately, astronomers have measured the masses of only a few of them. The mass of Ceres is just 1/10,000 the mass of Earth. The total mass of all the known asteroids probably amounts to less than one-tenth the mass of the Moon.

Asteroid compositions are inferred from spectroscopic and other measurements in the infrared, visible, and ultraviolet parts of the spectrum. The darkest (least reflective) asteroids contain significant amounts of water ice and other volatile substances and are rich in organic (carbon-based) molecules. They are known as *carbonaceous* asteroids. The more reflective *silicate* asteroids are composed primarily of rocky material. The few asteroid densities known from direct measurement of masses and radii are generally consistent with compositions inferred by spectroscopic means. Silicate asteroids predominate in the inner portions of the asteroid belt, but carbonaceous asteroids are more common overall, and their fraction increases as we move outward.

The first close-up views of asteroids were provided by the Jupiter probe *Galileo*, which made close encounters with asteroid Gaspra in October 1991 and Ida in August 1993 (Figures 4.6a, b). Both asteroids are irregularly shaped bodies a few tens of kilometers across, pitted with craters ranging in size from a few hundred meters to 2 km wide, and covered with layers of dust of variable thickness. They are thought to be fragments of larger objects that broke up following violent collisions hundreds of millions of years ago. To the surprise of most astronomers, close inspection of the Ida image (Figure 4.6b) revealed the presence of a tiny moon (now named Dactyl), just 1.5 km across, orbiting the asteroid. From the *Galileo* images, astronomers were able

▶ **FIGURE 4.6 Asteroids, Close-up** (a) The asteroid Gaspra as seen from a distance of 1600 km by the space probe *Galileo*. (b) The asteroid Ida photographed by *Galileo* from a distance of 3400 km. (Ida's moon, Dactyl, is visible at right.) The resolution in these photographs is about 100 m. True-color images show the surfaces of both bodies to be a fairly uniform gray. (c) Asteroid Mathilde, imaged by the *NEAR* spacecraft on its way to the asteroid Eros. The largest craters in this image are about 20 km across—much larger than those seen on Ida or Gaspra. The reason may be Mathilde's low density and rather soft composition. *(NASA)*

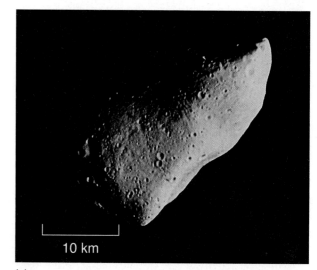

10 km

(a)

30 km

(b)

30 km

(c)

R I V U X G

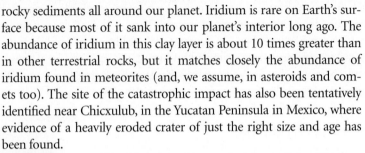

4-1 DISCOVERY

What Killed the Dinosaurs?

The name *dinosaur* derives from the Greek words *deinos* (terrible) and *sauros* (lizard), but dinosaurs were no ordinary reptiles. In their prime, they were the rulers of Earth. Their fossilized remains have been found on every continent. They dominated Earth for well over 100 million years (for comparison, humans have been around for a little over 2 million years). Yet according to the fossil record, these creatures vanished from Earth quite suddenly about 65 million years ago. What happened to them?

Many explanations have been offered for the extinction of these creatures. Devastating plagues, magnetic field reversals, increased geological activity, severe climate change, and supernova explosions (see Chapter 12) have all been proposed. In the 1980s, it was suggested that a large extraterrestrial object—a 10- to 15-km-wide asteroid or comet—struck Earth 65 million years ago, and this is now (arguably) the leading explanation for the demise of the dinosaurs. Sketched in the first figure, the impact released millions of times more energy than the largest nuclear bombs ever constructed by humans, and kicked huge quantities of dust high into the atmosphere. The dust may have shrouded our planet for years, virtually extinguishing the Sun's rays. On the darkened surface, plants could not survive. The entire food chain was disrupted, and the dinosaurs, at the top of that chain, became extinct.

(D. Hardy)

Although we have no direct astronomical evidence for or against this idea, we can estimate the chances that a large asteroid or comet will strike Earth today on the basis of the number of objects presently on Earth-crossing orbits. The second figure shows the likelihood of an impact as a function of the size of the impacting body. The horizontal scale indicates the energy released by the collision, measured in *megatons* of TNT. The megaton—4.2×10^{16} joules, the explosive yield of a large nuclear warhead—is the only common terrestrial measure of energy adequate to describe the violence of these events. ∞ *(More Precisely 2-2)*

We see that 100 million-megaton impacts, like the planet-wide catastrophe thought to have wiped out the dinosaurs, are very rare, occurring only once every 10 million years or so. However, smaller impacts, equivalent to "only" a few tens of kilotons of TNT (roughly equivalent to the bomb that destroyed Hiroshima in 1945), could happen every few years. The most recent large impact was the Tunguska explosion in Siberia in 1908, which packed a roughly 1-megaton punch (see Figure 4.21).

The main geological evidence supporting this theory is a layer of clay enriched with the element iridium, found in 65 million-year-old rocky sediments all around our planet. Iridium is rare on Earth's surface because most of it sank into our planet's interior long ago. The abundance of iridium in this clay layer is about 10 times greater than in other terrestrial rocks, but it matches closely the abundance of iridium found in meteorites (and, we assume, in asteroids and comets too). The site of the catastrophic impact has also been tentatively identified near Chicxulub, in the Yucatan Peninsula in Mexico, where evidence of a heavily eroded crater of just the right size and age has been found.

The idea that extraterrestrial events could cause catastrophic change on Earth was rapidly accepted by most astronomers, but it was controversial among paleontologists and geologists. Opponents argued that the amount of iridium in the clay layer varies greatly from place to place across the globe, and there is no complete explanation of why that should be so. Perhaps, they suggested, the iridium was produced by volcanoes and had nothing to do with an extraterrestrial impact at all.

Still, in the decades since the idea was first suggested, the focus of the debate seems to have shifted. As is often the case in science, the debate has evolved, sometimes erratically, as new data have been obtained, but the reality of a major impact 65 million years ago has become widely accepted. ∞ *(Sec. 0.5)* Much of the argument now revolves around the question of whether that event actually caused the extinction of the dinosaurs or merely accelerated a process already under way. The realization that catastrophic impacts can and do occur marked an important milestone in our understanding of planetary evolution (see also *Discovery 7-2*).

As a general rule, we can expect that global catastrophes are bad for the dominant species on a planet. As the dominant species on Earth, we are the ones who now stand to lose the most.

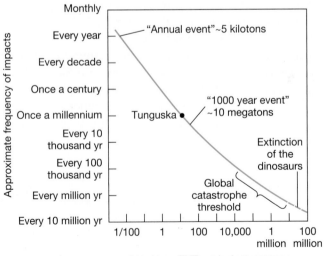

▶ FIGURE 4.7 **Asteroid Eros**   The asteroid Eros, as seen by the *NEAR-Shoemaker* probe. Craters of all sizes, ranging from 50 m (the resolution of the image) to 5 km, pit the surface. The inset shows a close-up image of a "younger" section of the surface, where loose material from recent impacts has apparently filled in and erased all trace of older craters. *(JHU/NASA)*

to estimate Dactyl's orbit around Ida and hence (using Newton's laws) measure Ida's mass at about 4–5×10^{16} kg and infer a density of 2100–3100 kg/m^3. ∞ *(Sec. 1.4)* For comparison, the density of surface rock on Earth is around 3000 kg/m^3.

The *Near Earth Asteroid Rendezvous (NEAR)* spacecraft visited the asteroid Mathilde in 1997 on its way to the mission's main target, the asteroid Eros. Shown in Figure 4.6(c), Mathilde is some 60 km across. By sensing its gravitational pull, *NEAR* measured Mathilde's mass to be about 10^{17} kg, implying a density of just 1300 kg/m^3. To account for this low density, scientists speculate that the asteroid's interior must be quite porous. Indeed, many asteroids seem to be more like loosely bound "rubble piles" than pieces of solid rock. On arrival at Eros on February 14, 2000, *NEAR* (by then renamed *NEAR-Shoemaker*) went into orbit around the asteroid, sending back high-resolution images of Eros and making detailed measurements of its size, shape, gravitational and magnetic fields, composition, and structure (Figure 4.7). Eros is a heavily cratered, rocky body, with mass 7×10^{15} kg and a roughly uniform density of around 2400 kg/m^3, and it is extensively fractured due to innumerable impacts in the past.

In July 2011, NASA's *Dawn* probe entered orbit around Vesta, the second largest asteroid in the solar system. Vesta's most striking surface feature (Figure 4.8) is a set of deep troughs girdling its equator. Overall, the surface appears much rougher than most asteroids in the main asteroid belt, and age estimates based on cratering (see Section 5.6) indicate that Vesta's southern hemisphere is only 1–2 billion years old, much younger than the north. The south polar region also contains one of the largest mountains in the solar system, some 22 km high. The origins of these features are not well understood, but they may have formed when an impact with another large body fractured Vesta's interior. The asteroid belt is a violent place!

After orbiting Vesta for a year, *Dawn* moved on in July 2012 to the dwarf planet Ceres, where it should arrive in 2015.

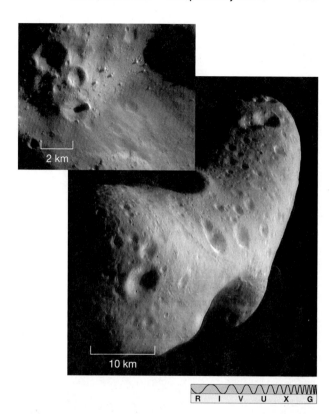

R I V U X G

MA ANIMATION/VIDEO Orbiting Eros

CONCEPT CHECK ▶   Describe some basic similarities and differences between asteroids and the inner planets.

MA ANIMATION/VIDEO NEAR Descent

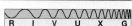
R I V U X G

◀ FIGURE 4.8 **Asteroid Vesta**   NASA's *Dawn* spacecraft snapped this shot of Vesta, the solar system's second largest asteroid. It is 500 km across and orbits between Mars and Jupiter. *Dawn* is now on its way to Ceres, the largest known asteroid. *(NASA)*

Comets

Comets are usually discovered as faint, fuzzy patches of light on the sky while still several astronomical units away from the Sun. Traveling in a highly elliptical orbit, a comet brightens and develops an extended **tail** as it nears the Sun. (The name "comet" derives from the Greek word *kome*, meaning "hair.") As the comet departs from the Sun's vicinity, its brightness and its tail diminish until it once again becomes a faint point of light receding into the distance.

Probably the most famous comet of all is Halley's Comet (Figure 4.9a), whose appearance at 76-year intervals has been documented at every passage since 240 B.C. Its most recent appearance was in 1986. A spectacular show, the tail of Halley's Comet can reach almost a full astronomical unit in length, stretching many tens of degrees across the sky. More recently, in 1997, comet Hale-Bopp, an unusually large and bright comet with a 40°-long tail, was almost certainly the most-watched comet in history (see Figure 4.10).

The various parts of a typical comet are shown in Figure 4.9(b). Like the planets, comets emit no visible light of their own—they shine by reflected (or reemitted) sunlight. The **nucleus,** or main solid body, of a comet is only a few kilometers in diameter. During most of the comet's orbit, far from the Sun, only this frozen nucleus exists. But if a comet comes within a few astronomical units of the Sun, its icy surface becomes too warm to remain stable. Part of it becomes gaseous and expands into space, forming a diffuse **coma** (halo) of dust and evaporated gas around the nucleus. The coma gets larger and brighter as the comet nears the Sun. At maximum size, it can measure 100,000 km in diameter—as large as Jupiter or Saturn. Engulfing the coma, an invisible **hydrogen envelope** stretches across millions of kilometers of space. The comet's tail, most pronounced when the comet is closest to the Sun, is larger still, sometimes spanning as much as 1 AU. From Earth, only the coma and tail of a comet are visible to the naked eye. Most of the comet's light comes from the coma. However, most of the mass resides in the nucleus.

Comets can have two types of tail. An **ion tail** is approximately straight, and often made of glowing, linear streamers like those shown toward the top of Figure 4.10. Its emission spectrum indicates numerous ionized atoms and molecules that have lost some of their normal complement of electrons. ∞ *(Sec. 2.6)* A **dust tail** is usually broad, diffuse, and gently curved, as shown clearly at the center of Figure 4.10. It is rich in microscopic dust particles that reflect sunlight, making the tail visible from afar.

Both types of tails are in all cases directed away from the Sun by the **solar wind,** an invisible stream of matter and radiation escaping from the Sun. (In fact, astronomers first inferred the existence of the solar wind from observations of comet tails.) Consequently, as depicted in Figure 4.11, the tail, be it ion or dust, always lies outside a comet's orbit and actually *leads* the comet during the portion of the orbit that is outbound from the Sun. The light particles that make up the ion tails are more strongly influenced by the solar wind than by the Sun's gravity, so those tails always point directly away from the Sun. The heavier dust particles are less affected by the pressure of the solar wind and tend to follow the comet's orbit around the Sun, giving rise to the slight curvature of the dust tails.

Based on the best available observations, experts now consider the nucleus of a comet to be composed of dust particles plus some small rocky fragments all trapped within a loosely packed mixture of methane, ammonia, carbon dioxide, and ordinary water ice, the density of the mixture being about 100 kg/m³. Comets are often described as "dirty snowballs." Even as atoms, molecules, and dust particles boil off into space, creating a comet's coma and tail, the nucleus remains frozen at a temperature of only a few tens of kelvins. Estimates of typical cometary

▼ FIGURE 4.9 **Halley's Comet** (a) Halley's Comet in 1986, about 1 month before it rounded the Sun. (b) Diagram of a typical comet, showing the nucleus, coma, hydrogen envelope, and tail, approximately to the same scale as in part (a). The tail is not a sudden streak in time across the sky, as in the case of meteors or fireworks. Instead, it travels along with the rest of the comet, always pointing away from the Sun. *(NOAO)*

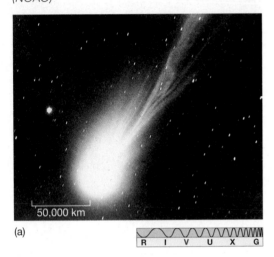

50,000 km

(a)

R   I   V   U   X   G

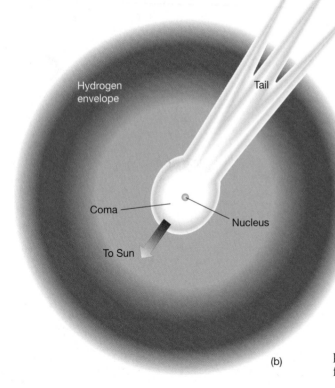

Hydrogen envelope

Tail

Coma

Nucleus

To Sun

(b)

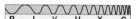

◄ FIGURE 4.10 **Comet Tails** In 1997, comet Hale-Bopp displayed both an ion tail (dark blue) and a dust tail (white blue), showing the gentle curvature and inherent fuzziness of the dust. At the comet's closest approach to the Sun, its tail stretched nearly 40° across the sky. *(W. Pacholka)*

ANIMATION/VIDEO Comet Hale-Bopp Nucleus Animation

masses range from 10^{12} to 10^{16} kg, comparable to the masses of small asteroids. Figure 4.12 shows the nucleus of Halley's comet, as seen by the European *Giotto* spacecraft, which approached within 600 km of the nucleus in 1986.

The past few years have seen several new missions to study comets at close range. In January 2004, NASA's *Stardust* mission approached within 150 km of the nucleus of comet P/Wild 2 (*Wild* is German, pronounced "Vilt"), collecting cometary particles in a specially designed foamlike aerogel dust detector (see Figure 4.13). The comet was chosen because it is a relative newcomer to the inner solar system, having been deflected onto its present orbit by an encounter with Jupiter in 1974. It therefore has not been subject to much solar heating or loss of mass by evaporation since it formed long ago and probably has not changed significantly since our solar system formed. On its return to Earth in January 2006, *Stardust* provided researchers with the first ever samples of cometary material. Detailed chemical analysis revealed evidence for organic material apparently formed in deep space, as well as the unexpected presence of silicate (rocky) materials that should only have formed at high temperatures, challenging astronomers' models of solar system formation.

The second mission had a much more violent end. On July 4, 2005, a 400-kg projectile from NASA's *Deep Impact* spacecraft crashed into comet Tempel 1 at more than 10 km/s (23,000 mph), blasting gas and debris from the comet's interior out into interplanetary space, while the spacecraft itself watched from a distance of 500 km. Figure 4.14 shows the mothership's spectacular view of the explosion about 1 minute after impact. Spectroscopic analysis of the ejected gas has provided scientists with their clearest view yet of the internal composition of a comet, and, hence, of the primordial matter of the early solar system, confirming the presence of water ice and many organic molecules. ∞ *(Sec. 2.8)* Observations of the crater suggest a low-density "fluffy" internal composition, consistent with the "snowball" picture of cometary structure just described.

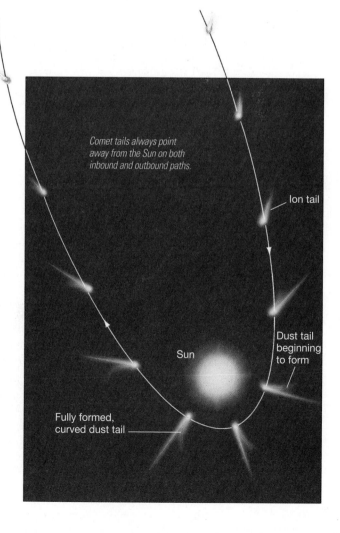

▶ FIGURE 4.11 **Comet Trajectory** As it approaches the Sun, a comet develops an ion tail, which is directed away from the Sun. Closer in, the dust tail displays marked curvature and tends to lag behind the ion tail. Compare this to comet Hale-Bopp (Figure 4.10).

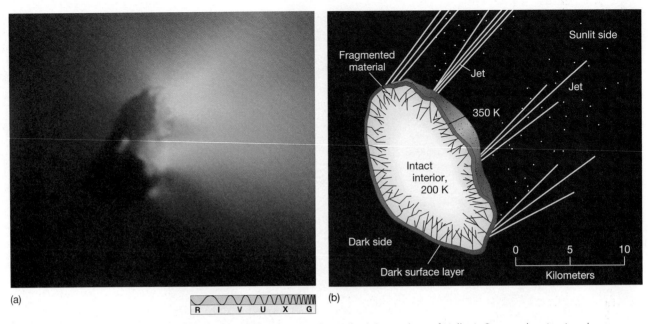

(a)

(b)

R I V U X G

▲ **FIGURE 4.12 Halley's Comet Close-up** (a) The *Giotto* spacecraft resolved the nucleus of Halley's Comet, showing it to be very dark. The Sun is toward the right in this image. The brightest areas are jets of evaporated gas and dust spewing from the comet's nucleus. (b) A diagram of Halley's nucleus, showing its size, shape, jets, and other physical and chemical properties. *(ESA/Max Planck Institute)*

Comet Orbits

ANIMATION/VIDEO Deep Impact Simulation

ANIMATION/VIDEO Anatomy of a Comet Part I

INTERACTIVE ▼ **FIGURE 4.13 Interactive Stardust at Wild-2** (a) The *Stardust* spacecraft captured this image (a) of comet P/Wild-2 in 2004, just before the craft passed through the comet's coma. (b) Onboard is a detector made of a foam-like jelly (called *aerogel*) that is 99.8% air, yet is strong enough to stop and store cometary dust particles as they hit the spacecraft. (c) Upon return of the craft to Earth in 2006, analysis began of the minute tracks in the aerogel, the ends of which contain captured comet dust fragments. *(NASA)*

Their highly elliptical orbits take most comets far beyond Pluto, where, in accordance with Kepler's second law, they spend most of their time. ⚬ *(Sec. 1.3)* The majority take hundreds of thousands, even millions, of years to complete a single orbit around the Sun. Unlike the orbits of other solar system objects, most comet orbits are *not* confined to within a few degrees of the ecliptic plane. Rather, they exhibit all inclinations and orientations, both prograde and retrograde, roughly uniformly distributed in all directions from the Sun.

A few *short-period* comets (conventionally defined as those having orbital periods of less than 200 years) are exceptions to the above general statements. According to Kepler's third law, these bodies never venture far beyond the orbit of Pluto. Their orbital orientations also differ from the norm—they tend to be prograde and lie close to the ecliptic plane, like the orbits of the planets and most asteroids.

Short-period comets originate beyond the orbit of Neptune, in a region of the outer solar system called the **Kuiper belt** (after Gerard Kuiper, a pioneer in infrared and planetary astronomy). Rather like the asteroids in the inner solar system, most Kuiper belt comets move in roughly circular orbits between about 30 and 100 AU

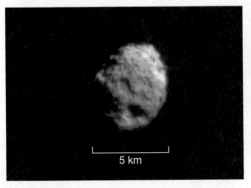

5 km

(a)

(b)

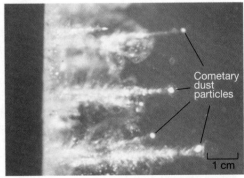

Cometary dust particles

1 cm

(c)

R I V U X G

from the Sun, always remaining outside the orbits of the jovian planets. Occasionally, however, either a chance encounter between two comets or (more likely) the gravitational influence of Neptune "kicks" a Kuiper belt comet into an eccentric orbit that brings it into the inner solar system and into our view. The observed orbits of these comets reflect the flattened structure of the Kuiper belt.

More than 1000 *Kuiper belt objects* are currently known. Because they are so small and distant, only a tiny fraction of the total has so far been observed. Researchers estimate that the total mass of the Kuiper belt may exceed that of the asteroid belt by a factor of 100 or more (although it is probably still less than the mass of Earth). The most prominently known Kuiper belt object is the former planet Pluto, but larger bodies orbit beyond Neptune. In 2005, astronomers discovered an object (since named *Eris*) some 10 percent larger and 30 percent more massive than Pluto, moving on an eccentric orbit with a semimajor axis of 70 AU. The discovery of Eris was the main reason for Pluto's demotion from major planet status in 2006.

So how do we account for the apparently random orbital orientations of the long-period comets? Only a tiny portion of a typical long-period cometary orbit lies within the inner solar system, so it follows that for every comet we see, there must be many more similar objects far from the Sun. On these general grounds, astronomers reason that there must be a huge cloud of comets lying far beyond the orbit of Pluto, completely surrounding the Sun. It is named the **Oort cloud,** after the Dutch astronomer Jan Oort, who first wrote in the 1950s of the possibility of such a vast and distant reservoir of inactive, frozen comets. The Kuiper belt and the orbits of some typical Oort cloud comets are sketched in Figure 4.15.

Based on the observed orbital properties of long-period comets, researchers think that the Oort cloud may be up to 100,000 AU (0.5 light-years) in diameter. Like their Kuiper belt counterparts, however, most Oort cloud comets never come anywhere near the Sun. Indeed, Oort cloud comets rarely approach even the orbit of Pluto. Only when the gravitational field of a passing star happens to deflect a comet into an extremely eccentric orbit that passes through the inner solar system do we get to see it at all. Because the Oort cloud surrounds the Sun in all directions instead of being confined to the ecliptic plane like the Kuiper belt, the long-period comets we see can come from any direction in the sky.

▲ FIGURE 4.14 **Deep Impact** The 5-km-sized nucleus of Comet Tempel 1 is shown at top before impact in 2005 and at bottom shortly after collision with a small projectile launched from the *Deep Impact* robot spacecraft. *(NASA)*

ANIMATION/VIDEO Anatomy of a Comet Part II

CONCEPT CHECK ▶ In what sense are the comets we see very unrepresentative of comets in general?

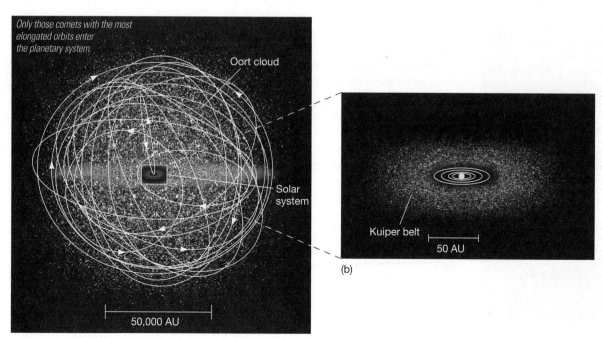

Only those comets with the most elongated orbits enter the planetary system.

Oort cloud

Solar system

50,000 AU

(a)

Kuiper belt

50 AU

(b)

◀ FIGURE 4.15 **Comet Reservoirs** (a) Diagram of the Oort cloud, showing a few cometary orbits. The solar system is smaller than the overlaid box at the center of the figure. (b) The Kuiper belt, the source of short-period comets, whose orbits hug the ecliptic plane.

(a)

(b)

R   I   V   U   X   G

▲ **FIGURE 4.16 Meteor Trails** A bright streak of light called a meteor is produced when a fragment of interplanetary debris plunges into the atmosphere, heating the air to incandescence. (a) A small meteor photographed against a backdrop of stars and the Northern Lights. (b) These meteors (one with a red smoke trail) streaked across the sky during the height of the Leonid meteor storm in November 2001. *(P. Parviainen, J. Lodriguss)*

Meteoroids

On a clear night it is possible to see a few *meteors*—shooting stars—every hour. A **meteor** is a sudden streak of light in the night sky caused by friction between air molecules in Earth's atmosphere and an incoming piece of asteroid, meteoroid, or comet. This friction heats and excites the air molecules, which then emit light as they return to their ground states, producing the characteristic bright streak shown in Figure 4.16. Note that the brief flash that is a meteor is in no way similar to the broad, steady swath of light associated with a comet's tail. A meteor is a fleeting event in Earth's atmosphere, whereas a comet tail exists in deep space and can remain visible in the sky for weeks or even months.

Before encountering the atmosphere, the chunk of debris causing a meteor was almost certainly a meteoroid, simply because these small interplanetary fragments are far more common than either asteroids or comets. Any piece of interplanetary debris that survives its fiery passage through our atmosphere and finds its way to the ground is called a **meteorite.**

The smallest meteoroids are mainly the rocky remains of broken-up comets. Each time a comet passes near the Sun, some fragments dislodge from the main body. The fragments initially travel in a tightly knit group of dust or pebble-sized objects called a **meteoroid swarm,** moving in nearly the same orbit as the parent comet. Over the course of time, the swarm gradually disperses along the orbit, so that eventually the **micrometeoroids,** as these small meteoroids are known, become more or less smoothly spread around the parent comet's orbit. If Earth's orbit happens to intersect the orbit of such a young cluster of meteoroids, a spectacular *meteor shower* can result. Earth's motion takes it across a given comet's orbit at most twice a year (depending on the precise orbit of each body). Because the intersection occurs at the same time each year (Figure 4.17), the appearance of certain meteor showers is a regular and (fairly) predictable event (Table 4.3).

Comet debris continues to disperse. 3

Fragments continue along the comet orbit as it begins to break up.

Earth

2

Sun

Meteor showers occur when cometary debris hits Earth. 4

Comet breaks up as it rounds the Sun. 1

◄ **FIGURE 4.17 Meteor Showers** A meteoroid swarm associated with a given comet intersects Earth's orbit at specific locations, giving rise to meteor showers at specific times of the year. If the comet's path happens to intersect Earth's, the result is a meteor shower each time Earth passes through the intersection (point 4).

TABLE 4.3 Some Prominent Meteor Showers

Morning of Maximum Activity	Shower Name/Radiant	Rough Hourly Count	Parent Comet
Jan. 3	Quadrantid/Bootes	40	—
Apr. 21	Lyrid/Lyra	10	1861I (Thatcher)
May 4	Eta Aquarid/Aquarius	20	Halley
June 30	Beta Taurid/Taurus	25	Encke
July 30	Delta Aquarid/Aquarius/Capricorn	20	—
Aug. 12	Perseid/Perseus	50	1862III (Swift-Tuttle)
Oct. 9	Draconid/Draco	up to 500	Giacobini-Zimmer
Oct. 20	Orionid/Orion	30	Halley
Nov. 7	Taurid/Taurus	10	Encke
Nov. 16	Leonid/Leo	12[1]	1866I (Tuttle)
Dec. 13	Geminid/Gemini	50	3200 Phaeton[2]

[1]Every 33 years (most recently in 1999 and 2000), as Earth passes through the densest region of this meteoroid swarm, we see intense showers, reaching 1000 meteors per minute for brief periods of time.
[2]Phaeton is an asteroid with no cometary activity, but its orbit matches the meteoroid paths very well.

Meteor showers are named for their *radiant,* the constellation from whose direction they appear to come (Figure 4.18). For example, the Perseid shower appears to emanate from the constellation Perseus. It can last for several days but reaches its maximum every year on the morning of August 12, when upward of 50 meteors per hour can be observed. Astronomers use the speed and direction of a meteor's flight to compute its interplanetary trajectory. This is how some meteoroid swarms have come to be identified with well-known cometary orbits.

Larger meteoroids—those more than a few centimeters in diameter—are usually *not* associated with comets. They are more likely small bodies that have strayed from the asteroid belt, possibly as the result of asteroid collisions. Their orbits can sometimes be reconstructed in a manner similar to that used to determine the orbits of meteor showers. In most cases, their computed orbits do indeed intersect the asteroid belt, providing the strongest evidence we have that this is where they originated. These objects are responsible for most of the cratering on the surfaces of the Moon, Mercury, Venus, Mars, and some of the moons of the jovian planets. A few meteorites have been identified as originating on the Moon or Mars, blasted off the surfaces of those bodies by some impact long ago.

Meteoroids smaller than about a meter across (roughly a ton in mass) generally burn up in Earth's atmosphere. Larger bodies reach the surface, where they can cause significant damage, such as the kilometer-wide Barringer Crater shown in Figure 4.19. From the size of this crater, we can estimate that the meteoroid responsible must have had a mass of about 200,000 tons and a diameter of perhaps 50 m. Only 25 tons of iron meteorite fragments have been found at the crash site. The remaining mass must have been scattered by the explosion at impact, broken down by subsequent erosion, or buried in the ground. Currently, Earth is scarred with nearly 100 craters larger than 0.1 km in diameter. Most of these are so heavily eroded by weather and geological activity that they can be identified only in satellite photography, as in Figure 4.20. Fortunately, such major collisions between Earth and large meteoroids are thought to be rare events now. On average, they occur only once every few hundred thousand years (see *Discovery 4-1*).

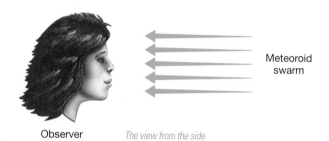

Meteoroid swarm

Observer   *The view from the side*

(a)

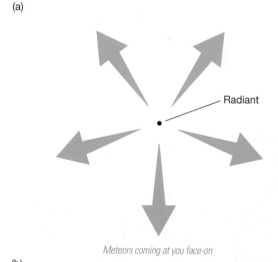

Radiant

Meteors coming at you face-on

(b)

ANALOGY: Meteor swarms coming at you face-on are analogous to parallel train tracks that seem to converge in the distance.

▲ **FIGURE 4.18 Radiant** (a) A group of meteoroids approaches an observer, all of them moving in the same direction at the same speed. (b) From an observer's viewpoint, the trajectories of the meteoroids (and the meteor shower they produce) appear to spread out from a central point, called the radiant.

▲ **FIGURE 4.19 Barringer Crater** The Barringer Meteor Crater, near Winslow, Arizona, 1.2 km in diameter and 0.2 km deep, resulted from a meteorite impact about 25,000 years ago. The meteoroid was probably about 50 m across and likely weighed around 200,000 tons. *(U.S. Geological Survey)*

▲ **FIGURE 4.20 Manicouagan Reservoir** This photograph, taken from orbit by the U.S. *Skylab* space station, shows the ancient impact basin that forms Quebec's Manicouagan Reservoir. A large meteorite landed there about 200 million years ago. *(NASA)*

One of the most recent documented meteoritic events occurred in central Siberia on June 30, 1908 (Figure 4.21). The presence of only a shallow depression as well as a complete lack of fragments implies that this intruder exploded several kilometers above the ground, leaving a blasted depression at ground level but no well-formed crater. Recent calculations suggest that the object in question was a rocky meteoroid about 30 m across. The explosion, estimated to have been equal in energy to a 10-megaton nuclear detonation, was heard hundreds of kilometers away and produced measurable increases in atmospheric dust levels all across the Northern Hemisphere.

Most meteorites are rocky (Figure 4.22a), although a few are composed mainly of iron and nickel (Figure 4.22b). Their composition is much like that of the rocky inner planets or the Moon, except that some of their lighter elements—such as hydrogen and oxygen—are depleted. Scientists think that the rocky meteorites are associated with the rocky, silicate asteroids and that their light elements boiled away when the bodies on which they originated were wholly or partially molten. Some meteorites do show evidence of strong heating in the past, most likely during the collisions that liberated them from their parent asteroids. Others show no evidence of past heating and may date back to the formation of the solar system. Oldest of all are the *carbonaceous* meteorites, black or dark gray in color and very likely related to the carbonaceous asteroids.

Finally, almost all meteorites are *old*. Radioactive dating shows most of them to be between 4.4 and 4.6 billion years old—roughly the age of the oldest Moon rocks brought back to Earth by *Apollo* astronauts (Chapter 5). Meteorites, along with asteroids, comets, and some lunar rocks, provide essential clues to the original state of matter in the solar neighborhood and to the birth of our planetary system.

 Why are astronomers so interested in interplanetary matter?

◄ **FIGURE 4.21 Tunguska Debris** The Tunguska event of 1908 leveled trees over a vast area. Although the impact of the blast was tremendous and its sound audible for hundreds of kilometers, this Siberian site is so remote that little was known about the event until scientific expeditions arrived to study it many years later. *(Sovfoto/Eastfoto)*

(a)

(b)

▲ FIGURE 4.22 **Meteorite Samples** (a) A stony (silicate) meteorite often has a dark crust, created when its surface is melted by the tremendous heat generated during its passage through the atmosphere. The coin at the bottom is for scale. (b) Iron meteorites are much rarer than the stony variety. Most, like the one in this photo, show characteristic crystalline patterns when their surfaces are cut, polished, and etched with acid. *(Science Graphics)*

4.3 The Formation of the Solar System

You might be struck by the vast range of physical and chemical properties found in the solar system. Our astronomical neighborhood may seem more like a great junkyard than a smoothly running planetary system. Is there some underlying principle that unifies the facts just outlined? Remarkably, the answer is yes. The origin of the solar system is a complex and as yet incompletely solved puzzle, but (we think!) the basic outlines are at last becoming understood.

A comprehensive theory of the formation of the solar system has been a dream of astronomers for centuries. Its development is a case study in the scientific method, on a par with the sweep of ideas presented in Chapters 1 and 2, although in this case the final form of the theory remains to be determined. ∞ *(Sec. 0.5)* As we will see, competing hypotheses have risen, fallen, and in some cases risen again in response to improved observations of our planetary environment. Nevertheless, we now seem to have a coherent picture of how our solar system formed.

Until the mid-1990s, theories of planetary system formation concentrated almost exclusively on our own solar system, for the very good reason that astronomers had no other examples of planetary systems on which to test their ideas. However, all that has now changed. We currently know of about 700 **extrasolar planets**—planets orbiting stars other than the Sun (Figure 4.23)—to challenge our theories. And challenge them they do! As we will see in Section 4.4, the new planetary systems discovered so far seem to have properties quite different from our own and may require us to rethink in some radical ways our concept of how stars and planets form.

Still, we currently have only limited information on these systems—little more than orbits and mass estimates for the largest planets. Accordingly, we begin our study by outlining the comprehensive theory that accounts, in detail, for most of the observed properties of our own planetary system: the solar system. Bear in mind, though, that no part of the scenario we describe here is in any way unique to our own system. The same basic processes should have occurred during the formative stages of many of the stars in our Galaxy. Later we will see how our theories hold up in the face of the new extrasolar data.

R I V U X G

▲ FIGURE 4.23 **Extrasolar Planet** Most known extrasolar planets are too faint to be detectable against the glare of their parent stars. However, in this system, called 2M1207, the parent itself (centered) is very faint—a so-called *brown dwarf* (see Chapter 12)—allowing the planet (lower left) to be detected in the infrared. This planet has a mass about five times that of Jupiter and orbits 55 AU from the star, which is 230 light-years away. *(VLT/ESO)*

Model Requirements

Based on the measured ages of the oldest meteorites, as well as the ages of Earth and lunar rocks, planetary scientists think that the age of the solar system is 4.6 billion years—an enormous period of time (although still far less than the 14 billion-year age of the universe—see Chapter 17). What happened so long ago to create the planetary system we see today? And what evidence survives from those early times to constrain our theories? Any theory of the origin and architecture of our planetary system must adhere to these 10 known facts:

1. *Each planet is relatively isolated in space.* The planets orbit at progressively larger distances from the central Sun; they are not bunched together.
2. *The orbits of the planets are nearly circular.* With the exception of Mercury, which we will argue is a special case, each planetary orbit closely describes a perfect circle.
3. *The orbits of the planets all lie in nearly the same plane.* The planes swept out by the planets' orbits are accurately aligned to within a few degrees. Again, Mercury is a slight exception.
4. *The planets all orbit the Sun in the same direction (counterclockwise as viewed from above Earth's North Pole).* Virtually all the large-scale motions in the solar system (other than cometary orbits) are in the same plane and in the same sense. The plane is that of the Sun's equator, and the sense is that of the Sun's rotation.
5. *Most planets rotate on their axis in roughly the same sense as the Sun.* Exceptions are Venus and Uranus.
6. *Most of the known moons revolve about their parent planet in the same direction as the planet rotates on its axis.* Of the large moons in the solar system, only Neptune's moon Triton is an exception.
7. *Our planetary system is highly differentiated.* The terrestrial planets are characterized by high densities, moderate atmospheres, slow rotation rates, and few or no moons. The jovian planets have low densities, thick atmospheres, rapid rotation rates, and many moons.
8. *Asteroids are very old and exhibit a range of properties not characteristic of either the terrestrial or the jovian planets or their moons.* Asteroids share, in rough terms, the bulk orbital properties of the planets. However, they appear to be made of primitive, unevolved material, and the meteorites that strike Earth are the oldest rocks known.
9. *The Kuiper belt is a collection of asteroid-sized icy bodies orbiting beyond Neptune.* Pluto is now considered such a Kuiper belt object.
10. *The Oort cloud comets are primitive, icy fragments that do not orbit in the plane of the ecliptic and reside primarily at large distances from the Sun.* While similar to the Kuiper belt in composition, the Oort cloud is a completely distinct part of the outer solar system.

These 10 facts strongly suggest a high degree of order in our solar system. The large-scale architecture is too neat, and the ages of the components too uniform, to be the result of random chaotic events. The overall organization points toward a single formation, an ancient but one-time event 4.6 billion years ago.

In the next few chapters we will see that some planetary properties (such as atmospheric composition and interior structure) have gradually *evolved* into their present states during the billions of years since the planets formed. However, *no* such evolutionary explanation exists for the items in the preceding list. For example, Newton's laws imply that the planets must move in elliptical orbits with the Sun at one focus, but they offer no explanation of why the observed orbits should be roughly circular, coplanar, and prograde. We know of no way in which the planets could have started off in random paths, then later evolved into the orbits we see today. Their basic orbital properties must have been established at the outset.

In addition to its many regularities, our solar system also has many notable *irregularities,* some of which we have already mentioned. Far from threatening our theory, however, these irregularities are important facts for us to consider in shaping our explanations. For example, any theory explaining solar system formation must not insist that *all* planets rotate in the same sense or have only prograde moons, because that is not what we observe. Instead, the theory should provide strong reasons for the observed planetary characteristics yet be flexible enough to allow for and explain the deviations, too. And, of course, the existence of the asteroids and comets that tell us so much about our past must be an integral part of the picture. That's a tall order, yet many researchers now think we are close to that level of understanding.

Nebular Contraction

Modern theory holds that planets are by-products of the process of star formation (Chapter 11). Imagine a large cloud of interstellar dust and gas—called a **nebula**—a light-year or so across. Now suppose that due to some external influence, such as a collision with another interstellar cloud or perhaps the explosion of a nearby star, the nebula starts to contract under the influence of its own gravity. As it contracts, it becomes denser and hotter, eventually forming a star—the Sun—at its center.

In 1796 the French mathematician-astronomer Pierre Simon de Laplace showed mathematically that conservation of angular momentum (objects spin faster as they shrink—see *More Precisely 4-2*) demands that our hypothetical nebula must spin faster as it contracts. The increase in rotation speed, in turn, causes the nebula's *shape* to change as it shrinks. Centrifugal forces (the outward "push" due to rotation) tend to oppose the contraction in directions perpendicular to the rotation axis, with the result that the nebula collapses most rapidly along the rotation axis. As shown in Figure 4.24, by the time it has shrunk to about 100 AU, the cloud has flattened into a pancake-shaped disk. This swirling mass destined to become our solar system is usually referred to as the **solar nebula.**

If we now suppose that planets form out of this spinning disk, we can begin to understand the origin of much of the architecture observed in our planetary system today, such as the near-circularity of the planets' orbits and the fact that they move in the same sense in almost the same plane. The idea that planets form from such a disk is called the **nebular theory.**

Astronomers are fairly confident that the solar nebula formed such a disk because we can see similar disks around other stars. Figure 4.25(a) shows a visible-light image of the region around the star Beta Pictoris, which lies about 50 light-years from the Sun. When the light from Beta Pictoris itself is removed and the resulting image enhanced by a computer, a faint disk of matter (viewed almost edge-on here) can be seen. This particular disk is roughly 1000 AU across—about 10 times the diameter of the Kuiper belt. Astronomers think that Beta Pictoris is a very young star, perhaps only 100 million years old, and that we are witnessing it pass through an evolutionary stage similar to that experienced by our own Sun 4.6 billion years ago. Figure 4.25(b) shows an artist's conception of the disk.

As we saw in Chapter 0, scientific theories must continually be tested and refined as new data become available. ∞ *(Sec. 1.5)* Unfortunately for the original nebular theory, while Laplace's description of the collapse and flattening of the solar nebula was essentially correct, we now know that a disk of gas would *not* form clumps of matter that would subsequently evolve into planets. In fact, modern computer calculations predict just the opposite: Any clumps in the gas would tend to disperse, not contract further. However, the model currently favored by most astronomers, known as the **condensation theory,** rests squarely on the old nebular

**PROCESS OF ►
SCIENCE CHECK** Why is it important that a theory of solar system formation make clear statements about how planets arose yet not be too rigid in its predictions?

MA ANIMATION/VIDEO Beta Pictoris Warp

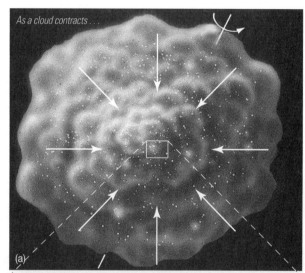
As a cloud contracts . . .
(a)

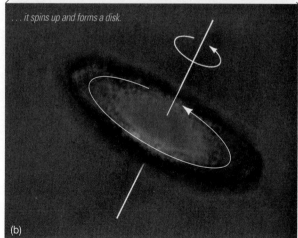
. . . it spins up and forms a disk.
(b)

INTERACTIVE ▲ FIGURE 4.24 Angular Momentum
(a) Conservation of angular momentum demands that a contracting, rotating cloud must spin faster as its size decreases. (b) Eventually, a small part of it destined to become the solar system came to resemble a gigantic pancake. The large blob at the center ultimately became the Sun. **MA**

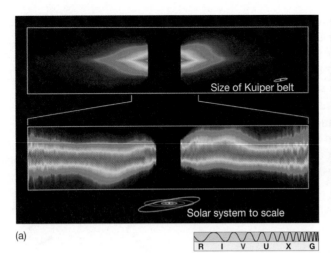

(a)

R I V U X G

(b)

▲ FIGURE 4.25 Beta Pictoris (a) A computer-enhanced view of a disk of warm matter surrounding the star Beta Pictoris. False color is used here to accentuate the details. The bottom image is a close-up of the inner part of the disk, whose warp is possibly caused by the gravitational pull of unseen companions. In both images, the overwhelmingly bright central star has been covered to let us see the much fainter disk surrounding it. (b) An artist's conception of the disk of clumped matter, showing the warm disk with a young star at the center and several comet-sized or larger bodies already forming. Mottled dust is seen throughout—such protoplanetary regions are probably very "dirty." (NASA; D. Berry)

theory, combining its basic physical reasoning with new information about interstellar chemistry to avoid most of the original theory's problems.

The key new ingredient is *interstellar dust* in the solar nebula. Astronomers now recognize that the space between the stars is strewn with microscopic dust grains, an accumulation of the ejected matter of many long-dead stars. These dust particles probably formed in the cool atmospheres of old stars, then grew by accumulating more atoms and molecules from the interstellar gas. The end result is that interstellar space is littered with tiny chunks of icy and rocky matter, having typical diameters of about 10^{-5} m. Figure 4.26 shows one of many such dusty regions in the vicinity of the Sun.

ANIMATION/VIDEO The Formation of the Solar System

Dust grains play an important role in the evolution of any gas cloud. Dust helps to cool warm matter by efficiently radiating heat away in the form of infrared radiation. When the cloud cools, its molecules move around more slowly, reducing the internal pressure and allowing the nebula to collapse more easily. ⚭ *(More Precisely 2-1)* Furthermore, the dust grains greatly speed up the process of collecting enough atoms to form a planet. They act as **condensation nuclei**—microscopic platforms to which other atoms can attach, forming larger and larger balls of matter. This is similar to the way raindrops form in Earth's atmosphere; dust and soot in the air act as condensation nuclei around which water molecules cluster.

Planet Formation

According to the condensation theory, the planets formed from the solar nebula (Figure 4.27a) in three distinct stages. The first two apply to all planets, the third only to the giant jovian worlds.

The first stage of planet formation began when dust grains in the solar nebula formed condensation nuclei around which matter began to accumulate (Figure 4.27b). This vital step greatly hastened the critical process of forming the first small clumps of matter. Once these clumps

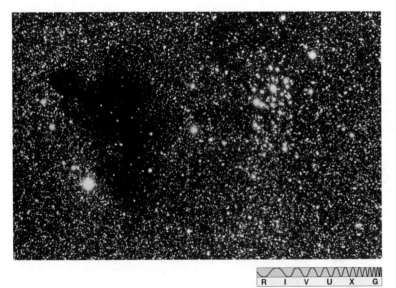

R I V U X G

▲ FIGURE 4.26 Dark Cloud Interstellar gas and dark dust lanes mark this region of star formation. The cloud known as Barnard 86 (dark, empty space at left) flanks a cluster of young blue stars called NGC 6520 (right). (D. Malin/Anglo-Australian Observatory)

formed, they grew rapidly by sticking to other clumps. (Imagine a snowball thrown through a fierce snowstorm, growing bigger as it encounters more snowflakes.)

As the clumps grew larger, their surface areas increased, and consequently, the rate at which they swept up new material accelerated. They gradually formed larger

▶ FIGURE 4.27 **Solar System Formation** The condensation theory of planet formation.
(a) The solar nebula after it has contracted and flattened to form a spinning disk (Figure 4.24b).
The large red blob in the center will become the Sun. Smaller blobs in the outer regions may
become jovian planets (see Figure 4.29). (b) Dust grains act as condensation nuclei, forming
clumps of matter that collide, stick together, and grow into moon-sized planetesimals.
(c) After a few million years, strong winds from the still-forming Sun begin expelling nebular
gas, and some massive planetesimals in the outer solar system have already captured gas
from the nebula. (d) With the gas ejected, planetesimals continue to collide and grow.
(e) Over the course of a hundred million years or so, planetesimals are accreted or ejected,
leaving a few large planets that travel in roughly circular orbits.

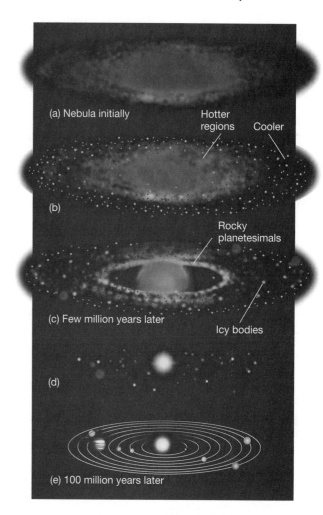

(a) Nebula initially

Hotter regions   Cooler

(b)

Rocky planetesimals

(c) Few million years later

Icy bodies

(d)

(e) 100 million years later

and larger objects, the size of pebbles, baseballs, basketballs, and boulders. Figure 4.28
shows an infrared view of a relatively nearby star whose protostellar disk is thought to
be in just this state. Eventually, this process of **accretion**—the gradual growth of small
objects by collision and sticking—created objects a few hundred kilometers across.

At the end of the first stage of planet formation, the solar system was made up
of hydrogen and helium gas and millions of **planetesimals**—objects the size of small
moons, having gravitational fields just strong enough to affect their neighbors. By
then, their gravity was strong enough to sweep up material that would otherwise not
have collided with them, so their rate of growth became faster still. During this second
phase of planet formation, gravitational forces between planetesimals caused them
to collide and merge, forming larger and larger objects. Because larger bodies exert
stronger gravitational pulls, eventually almost all the planetesimal material was swept
up into a few large **protoplanets**—accumulations of matter that would eventually
evolve into the planets we know today.

As the protoplanets grew, a competing process became important. Their strong
gravitational fields produced many high-speed collisions between planetesimals
and protoplanets. These collisions led to **fragmentation** as small objects broke into
still smaller chunks that were then swept up by the protoplanets. Only a relatively
small number of 10- to 100-km fragments escaped capture by a planet or a moon
and became the asteroids and comets.

After about 100 million years, the primitive solar system had evolved into eight
protoplanets, dozens of protomoons, and a glowing **protosun** at the center. Roughly
a billion more years were required to sweep the system clear of interplanetary trash.
This was a period of intense meteoritic bombardment whose effects on the Moon
and elsewhere are still evident today (see Chapter 5).

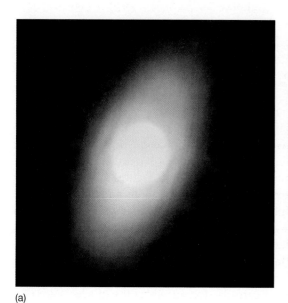

(a)

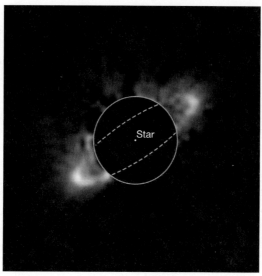

Star

(b)

R I V U X G

◀ FIGURE 4.28 **Newborn Solar Systems?**
(a) This infrared image, taken by the *Spitzer
Space Telescope,* of the bright star Fomalhaut,
some 25 light-years from Earth, shows a
circumstellar disk in which the process of
accretion is underway. The star itself is well
inside the yellowish blob at center. (b) This
Hubble Space Telescope image blocks out
the central (circled) parts of another such
disk around a more distant star HR4796A, but
shows the edges more clearly. *(NASA)*

4-2 MORE PRECISELY

The Concept of Angular Momentum

Most celestial objects rotate. Planets, moons, stars, and galaxies all have some *angular momentum,* which we may define loosely as the tendency of a body to keep spinning or moving in a circle, or, equivalently, how much effort must be expended to stop it. Angular momentum is a fundamental property of any rotating object, as important a quantity as its mass or its energy.

Intuitively, we know that the more massive an object, or the larger it is, or the faster it spins, the harder it is to stop. In fact, angular momentum depends on the object's *mass, rotation rate* (measured in, say, revolutions per second), and *radius,* in a very specific way:

$$\text{angular momentum} \propto \text{mass} \times \text{rotation rate} \times \text{radius}^2.$$

Recall that the symbol ∝ means "is proportional to." The constant of proportionality depends on the details of how the object's mass is distributed.

According to Newton's laws of motion, angular momentum is *conserved* at all times—that is, it must remain constant before, during, and after a physical change in any object, so long as no external forces act. This allows us to relate changes in size to changes in rotation rate. As illustrated in the accompanying figure, if a spherical object having some spin begins to contract, the above relationship demands that it spin faster, so that the product

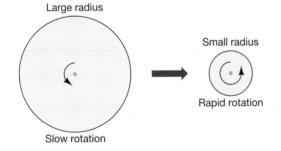

Large radius

Slow rotation

Small radius

Rapid rotation

mass × angular speed × radius² remains constant. The sphere's mass does not change during the contraction, but its radius clearly decreases. The rotation speed must therefore increase in order to keep the total angular momentum unchanged. Figure skaters use this principle to spin faster by drawing in their arms and slow down by extending them (as shown in the second accompanying figure). The mass of the human body remains the same, but its overall radius changes, causing the body's rotation rate to change in order to keep the angular momentum constant.

To take an example from the text, suppose that the interstellar gas cloud from which our solar system formed was initially 1 light-year across and rotating very slowly—once every 10 million years, say. The cloud's mass doesn't change (we assume) during the contraction, but its radius decreases. The rotation rate must therefore *increase* in order to keep the total angular momentum unchanged. By the time it has collapsed to a radius of 100 AU, its radius has shrunk by a factor of (1 light-year/100 AU) ≈ 9.5×10^{12} km/1.5×10^{10} ≈ 630. Conservation of angular momentum then implies that its (average) spin rate increases by a factor of $630^2 \approx$ 400,000, to roughly one revolution every 25 years, close to the orbital period of Saturn.

Incidentally, the law of conservation of angular momentum also applies to planetary orbits (where now "radius" is the distance from the planet to the Sun). In fact, Kepler's second law *is* just conservation of angular momentum, expressed another way. *(Sec. 1.3)*

(M. Powell/Getty Images)

Making the Jovian Planets

ANIMATION/VIDEO Protoplanetary Disk Destruction (MA)

ANIMATION/VIDEO Protoplanetary Disks in the Orion Nebula

The two-stage accretion picture just described has become the accepted model for the formation of the terrestrial planets in the inner solar system. However, the origin of the giant jovian worlds is less clear. Two somewhat different views, with important consequences for our understanding of extrasolar planets, have emerged.

In the first, more conventional scenario depicted in Figures 4.27(c and d), the four largest protoplanets in the outer solar system grew rapidly and became massive enough to enter a third phase of planetary development—their strong gravitational fields swept up large amounts of gas directly from the solar nebula. As we will see, there was a lot more raw material available for planet building in the outer solar system, so protoplanets grew much faster there. In this scenario, the cores of the jovian planets reached the point where they could capture nebular gas in less than a few million years. Compare this to the inner solar system, where the smaller, terrestrial protoplanets never reached this stage. Their growth was slow, taking 100 million years to form the planets we know today, and their masses remained relatively low.

In the second scenario, the giant planets formed through instabilities in the cool outer regions of the solar nebula—not so different from Laplace's original idea—mimicking on small scales the collapse of the initial interstellar cloud. In this view, the jovian protoplanets formed directly and very rapidly, skipping the initial accretion stage and perhaps taking less than a thousand years to acquire much of their mass. These first protoplanets had gravitational fields strong enough to scoop up gas and dust from the solar nebula, allowing them to grow into the giants we see today. Figure 4.29 illustrates this alternative formation path.

Ongoing studies of planetary composition and internal structure (for example, the size of the protoplanetary core), as well as observations of extrasolar planets, may one day allow us to distinguish between these competing theories. In either case, once the jovian protoplanets reached the critical size at which they could capture nebular gas, they grew rapidly. Their growth rate increased further as their gravitational fields intensified, and they reached their present masses in just a few million years.

Many of the jovian moons probably also formed by accretion, but on a smaller scale, in the gravitational fields of their parent planets. Once nebular gas began to flow onto the large jovian protoplanets, conditions probably resembled a miniature solar nebula, with condensation and accretion continuing to occur. The large moons of the outer planets almost certainly formed in this way. Some of the smaller moons may be captured planetesimals.

What of the gas that made up most of the original cloud? Why don't we see it today throughout the planetary system? All young stars apparently experience a highly active evolutionary stage known as the *T-Tauri phase* (see Chapter 11), during which their radiation emission and stellar winds are very intense. When our Sun entered this phase, a few million years after the solar nebula formed, any gas remaining between the planets was blown away into interstellar space by the solar wind and the Sun's radiation pressure. Afterward, all that remained were protoplanets and planetesimal fragments, ready to continue their long evolution into the solar system we know today. Note that the evolution of the Sun sets the time frame for the formation of the jovian planets. The outer planets *must* have formed before the nebula dispersed.

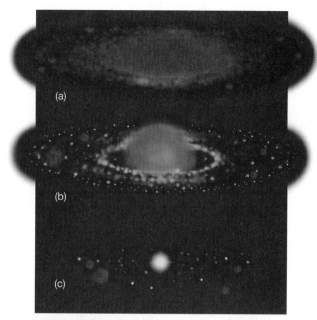

INTERACTIVE ▲ **FIGURE 4.29 Jovian Condensation** As an alternative to the growth of massive protoplanetary cores followed by accretion of nebular gas, some or all of the giant planets may have formed directly by instabilities in the cool gas of the outer solar nebula. Part (a) shows the same instant as Figure 4.27(a).(b) Only a few thousand years later, four gas giants have formed (red blobs), circumventing the accretion process sketched in Figure 4.27. With the nebula gone (c), the giant planets have taken their place in the outer **MA** solar system.

Differentiation of the Solar System

The condensation theory explains the basic composition differences among the terrestrial and jovian planets and the smaller bodies that constitute the solar system. Indeed, it is in this context that the term *condensation* derives its true meaning.

As the solar nebula contracted under the influence of gravity, it heated up as it flattened into a disk. The density and temperature were greatest near the central protosun and much lower in the outlying regions. In the hot inner regions, dust grains broke apart into molecules, which in turn split into atoms. Most of the original dust in the inner solar system disappeared at this stage, although the grains in the outermost parts probably remained largely intact.

The destruction of the dust in the inner solar nebula introduced an important new ingredient into the theoretical mix, one that we omitted from our earlier discussion. With the passage of time, the gas radiated away its heat, and the temperature decreased at all locations except in the very core, where the Sun was forming. Everywhere beyond the protosun new dust grains began to condense out, much as raindrops, snowflakes, and hailstones condense from moist, cooling air here on Earth. It may seem strange that although there was plenty of interstellar dust early on, it was mostly destroyed, only to form again later. However, a critical change had occurred. Initially, the nebular gas was uniformly peppered with dust grains. When the dust re-formed later, the distribution of grains was very different.

Figure 4.30 shows the temperature in various parts of the primitive solar system just before the onset of accretion. At any given location, the only materials to condense out were those able to survive the temperature there. In the innermost

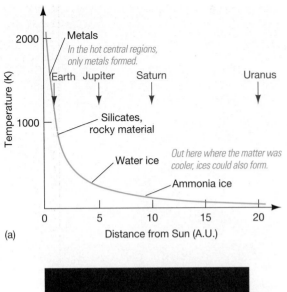

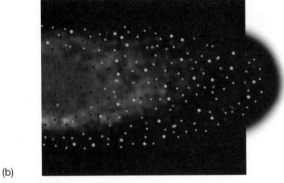

▲ **FIGURE 4.30 Temperature in the Early Solar Nebula**
(a) Theoretically computed variation of temperature across the primitive solar nebula illustrated in part (b), which shows half of the disk in Figure 4.27(b).

regions, around Mercury's present orbit, only metallic grains could form—it was simply too hot for anything else to exist. Farther out, at about 1 AU, it was possible for rocky, silicate grains to form, too. Beyond about 3 or 4 AU, water ice could exist, and so on. More and more material could condense out at greater and greater distances from the Sun. The composition of the grains at any given distance from the Sun determined the type of planetesimal—and ultimately planet—that formed there. The present-day structure of the asteroid belt, with rocky silicates more common in the inner regions and carbonaceous bodies, descendents of icier planetesimals, most prevalent at larger radii, still reflects these early conditions.

Beyond about 5 AU from the center, the temperature was low enough to allow several abundant gases—water vapor, ammonia, and methane—to condense into solid form. As we will see, these compounds are still important constituents of jovian atmospheres. Consequently, the planetesimals destined to become the cores of the jovian planets were formed under cold conditions out of low-density, icy material. Because more material could condense out of the solar nebula at these radii than in the inner regions near the protosun, accretion began sooner, with more resources to draw on. If they hadn't already formed through instabilities in the cold nebular gas, the outer planets grew rapidly to the point where they could accrete not just grains, but nebular gas also, and the eventual result was the hydrogen-rich jovian worlds we see today.

In the inner regions of the primitive solar system, the environment was too hot for ices to survive. Many of the abundant heavier elements, such as silicon, iron, magnesium, and aluminum, combined with oxygen to produce a variety of rocky materials. Planetesimals in the inner solar system were therefore rocky or metallic, as were the protoplanets and planets they ultimately formed. Here is an additional reason why the jovian planets grew so much bigger than the terrestrial worlds. The inner regions of the nebula had to wait for the temperature to drop so that a few rocky grains could appear and begin the accretion process, whereas accretion in the outer solar system began almost as soon as the solar nebula collapsed to a disk.

The heavier materials condensed into grains in the outer solar system, too, of course. However, there they would have been vastly outnumbered by the far more abundant light elements. The outer solar system is not deficient in heavy elements—rather, the inner solar system is *underrepresented in light material*.

Asteroids and Comets

In the inner solar system, most rocky planetesimals collided with or were ejected by the growing terrestrial planets. Only a few remain today as the asteroid belt. Planetesimals beyond the orbit of Mars failed to accumulate into a protoplanet because the huge gravitational field of nearby Jupiter continuously disturbed their motion, nudging and pulling at them, preventing them from aggregating into a planet. Many of the Trojan asteroids have likely been locked in their odd orbits by the combined gravitational pulls of Jupiter and the Sun since those earliest times.

Once the jovian planets formed, they exerted strong gravitational forces on the planetesimals in the outer solar system. Over a period of hundreds of millions of years, and after repeated gravitational "kicks" from the giant planets, particularly Jupiter and Saturn, most of the interplanetary fragments in the outer solar system were flung into orbits taking them far from the Sun (Figure 4.31). Those fragments now make up the Oort cloud. Interactions with Uranus and Neptune were gentler, and generally did not eject planetesimals to large distances. Rather, these encounters tended to deflect small bodies into eccentric orbits that carried them into the inner solar system, where they collided with a planet or were kicked into the Oort cloud by Jupiter or Saturn. Most of the original planetesimals formed beyond the orbit of Neptune are still there, making up the Kuiper belt.

The interactions that cleared the outer solar system of comets also caused significant changes in the orbits of the planets themselves. Computer simulations indicate that, during the ejection process, Jupiter moved slightly closer to the Sun, its orbital semimajor

CONCEPT CHECK ▶ Would you expect to find asteroids and comets orbiting other stars?

axis decreasing by a few tenths of an astronomical unit, while the other giant planets migrated outward—possibly by as much as 10 AU in the case of Neptune (Figure 4.31b).

The deflection of icy planetesimals into the inner solar system played an important role in the evolution of the terrestrial planets too. A long-standing puzzle in the condensation theory's account of the formation of the inner planets was the origin of the water and other volatile gases on Earth and elsewhere. At formation, the inner planets' surface temperatures were far too high, and their gravity too low, to capture and retain those gases. The answer seems to be that the comets from the outer solar system bombarded the newly born inner planets, supplying them with water *after* their formation.

Solar System Regularities and Irregularities

The condensation theory accounts for the 10 facts listed at the start of this section. The growth of planetesimals throughout the solar nebula, with each protoplanet ultimately sweeping up the material near it, accounts for the planets' wide spacing (point 1, although the theory does not adequately explain the regularity of the spacing). That the planets' orbits are circular (2), in the same plane (3), and in the same direction as the Sun's rotation on its axis (4) is a direct consequence of the solar nebula's shape and rotation. The rotation of the planets (5) and the orbits of the moon systems (6) are due to the tendency of smaller structures to inherit the overall sense of rotation of their parent. The heating of the nebula and the Sun's ignition resulted in the observed differentiation (7), while the debris from the accretion–fragmentation stage naturally accounts for the asteroids (8) and comets (9).

We mentioned earlier that an important aspect of any theory of solar system formation is its ability to accommodate deviations. In the condensation theory, that capacity is provided by the randomness inherent in the encounters that combined planetesimals into protoplanets. As the numbers of large bodies decreased and their masses increased, individual collisions acquired greater and greater importance. The effects of these collisions can still be seen today in many parts of the solar system.

We will discuss many solar system irregularities in the next few chapters; most can be explained as the results of random collisions. A case in point is the anomalously slow and retrograde rotation of Venus (Chapter 6), which can be explained if we assume that the last major collision in the formation history of that planet just happened to involve a near head-on encounter between two protoplanets of comparable mass. A similar encounter, this time involving Earth, may have formed our Moon (Chapter 5). Scientists usually do not like to invoke random events to explain observations. However, there seem to be many instances where pure chance has played an important role in determining the present state of the universe.

4.4 Planets Beyond the Solar System

The condensation theory was developed to explain just one planetary system—our own. But a critical test of any scientific theory is its applicability and predictive power outside the context in which it was originally conceived. ∞ *(Sec. 0.5)* The discovery in recent years of numerous planets orbiting other stars presents astronomers with the opportunity—indeed, the scientific obligation—to confront their theories of solar system formation against a new body of observational data.

Detecting Extrasolar Planets

The past few years have seen enormous strides in the search for extrasolar planets. These advances have been achieved through steady improvements in both telescope and detector technology and computerized data analysis. With few exceptions it is not yet possible to obtain images of these newly discovered worlds. The techniques

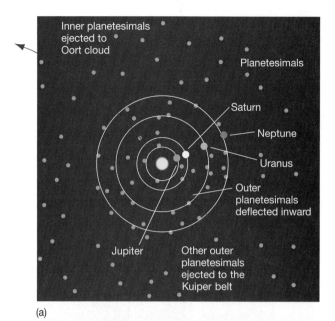

(a)

(b)

INTERACTIVE ▲ FIGURE 4.31 **Planetesimal Ejection** The ejection of icy planetesimals to form the Oort cloud and Kuiper belt. (a) Initially, once the giant planets had formed, leftover planetesimals were found throughout the solar system. Interactions with Jupiter and Saturn apparently "kicked" planetesimals out to very large radii (the Oort cloud). (b) After hundreds of millions of years and as a result of the inward and outward "traffic," the orbits of all four giant planets were significantly modified by the time the planetesimals inside Neptune's orbit had been ejected. As depicted here, Neptune was affected most and may have moved outward by as much as 10 AU.

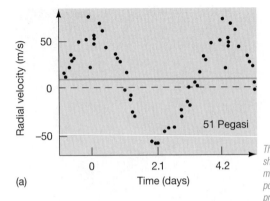

(a)

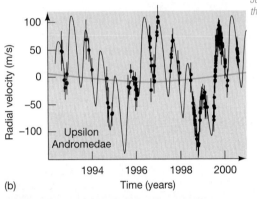

The blue lines show the maximum possible signal produced by Jupiter orbiting the Sun.

(b)

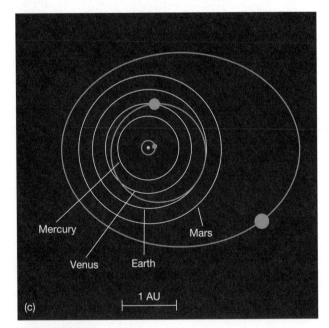

(c)

INTERACTIVE ▲ FIGURE 4.32 **Planets Revealed** (a) Discovery data of the Doppler shift of the star 51 Pegasi reveal a clear periodic signal indicating the presence of a planetary companion of mass at least half the mass of Jupiter. (b) Radial velocity data for Upsilon Andromedae are more complex, but are well fit (solid line) by a three-planet system orbiting the star. (c) A sketch of the inferred orbits of the three planets in the Upsilon Andromedae system (in orange) with the orbits of the terrestrial planets superimposed for comparison (in white). (MA)

ANIMATION/VIDEO Survey for Transiting Extrasolar Planets

used to find them are generally indirect, based on analysis of light from the parent star, not from the unseen planet.

Most planets discovered to date have been detected via their gravitational effect on their parent star. As a planet orbits a star, gravitationally pulling one way and then the other, the star "wobbles" slightly. The more massive the planet or the closer its orbit to the star, the greater its gravitational pull and hence the star's movement. If the wobble happens to occur along our line of sight to the star, then we see small fluctuations in the star's radial velocity, which astronomers can measure using the Doppler effect. ∞ *(Sec. 2.7)*

Figure 4.32 shows two sets of radial velocity data that betray the presence of planets orbiting other stars. Part (a) shows the line-of-sight velocity of the star 51 Pegasi, a near-twin to our Sun some 40 light-years away. These data were acquired in 1994 by Swiss astronomers using the 1.9-m telescope at Haute-Provence Observatory in France, and they were the first firm evidence for an extrasolar planet orbiting a Sun-like star. The regular 50 m/s fluctuations in the star's velocity have since been confirmed by several groups of astronomers and imply that a planet of at least half the mass of Jupiter orbits 51 Pegasi in a circular orbit with a period of 4.2 days.

Figure 4.32(b) shows another set of Doppler data, this time revealing one of the most complex systems of planets discovered to date—a triple-planet system orbiting another nearby Sun-like star named Upsilon Andromedae. The three planets have minimum masses of 0.7, 2.1, and 4.3 times the mass of Jupiter, and orbital semimajor axes of 0.06, 0.83, and 2.6 AU, respectively. Figure 4.32(c) sketches their orbits, with the orbits of the solar system terrestrial planets shown for scale.

The Doppler technique suffers from the limitation that the angle between the line of sight and the planet's orbital plane cannot be determined. Simply put, we cannot distinguish between a low-speed orbit seen edge-on and a high-speed orbit seen almost face-on (so only a small component of the orbital motion contributes to the line-of-sight Doppler effect). However, in some systems this is not the case. For example, observations of the solar-type star HD 209458, which lies some 150 light-years from Earth, reveal a small (1.7 percent) but clear drop in brightness each time its 0.6-Jupiter-mass companion, orbiting at a distance of just 7 million km (0.05 AU), passes between the star and Earth (Figure 4.33).

Such **planetary transits** are relatively rare because they require us to see the orbit almost exactly edge-on, but when they do occur, they allow an unambiguous determination of the planet's mass and radius. In this case, the resulting density of the planet is just 200 kg/m³, consistent with a high-temperature gas giant planet orbiting very close to its parent star. Some 250 extrasolar planets—about one-third of the total currently known—produce measurable fluctuations in brightness in their partner stars.

Only a small fraction of planetary systems will be oriented in just the right way to show transits, so planet hunters adopt the strategy of repeatedly surveying thousands of stars in the hope of detecting a transit should one occur. Space-based telescopes are particularly well suited to this task, as they can stare continuously at a given region of the sky, making simultaneous, high-precision observations of the target stars. Orbiting instruments can measure the tiny brightness changes (less than 1 part in 10⁴) needed to detect an Earth-like planet orbiting a Sun-like star. The European *CoRoT* mission (short for **Co**nvection, **Ro**tation, and planetary **T**ransits), launched in 2006, and NASA's *Kepler* spacecraft, launched in 2009, are monitoring hundreds of thousands of Sun-like stars, searching for brightness fluctuations due to planetary transits.

Exoplanet Properties

To date (as of mid-2012), astronomers have confirmed almost 800 extrasolar planets orbiting more than 600 stars. Most lie within 500 light years of the Sun, although many of the planets discovered by *CoRoT* and *Kepler* lie at much greater distances. About 10 percent of the nearby stars surveyed to date have been found to host planets. In most cases only a single planet has been detected, but roughly 100 *multiple-planet systems* are also known. Most astronomers expect both the fraction of stars with planets

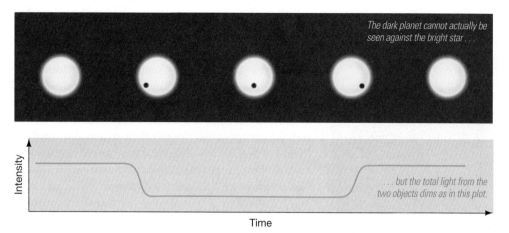

◄ FIGURE 4.33 **Extrasolar Transit**
As an extrasolar planet passes between us and its parent star, the light from the star dims in a characteristic way. The planet orbiting a Sun-like star known as HD 209458 is 200,000 km across and transits every 3.5 days, blocking about 2 percent of the star's light each time it does so.

and the number of planets per star to increase as detection and data analysis technology continue to improve. In 2012 the *Kepler* team published a list of more than 2300 planet candidates—stars with brightness variations consistent with planetary transits. Follow-up observations will undoubtedly show that some of these are "false alarms," but many of them are likely to turn out to be real, greatly expanding the exoplanet database. Already, more than 60 of the Kepler candidates have been verified. These include some of the smallest exoplanets yet discovered.

Figure 4.34 presents the observed masses and semimajor axes of roughly 400 extrasolar planets, and introduces some jargon used in the field. Each dot in the figure represents a planet, and we have added points corresponding to Earth, Neptune, and Jupiter in our own solar system. Massive exoplanets are often referred to as *Jupiters,* while less massive, but still "jovian" planets are *Neptunes.* The dividing line between Jupiters and Neptunes is somewhat arbitrary, but is often taken at about twice the mass of Neptune, or 0.1 Jupiter masses. The terminology is intended to distinguish between planets that are mostly gas, like Jupiter, and those that have substantial rocky cores, like Neptune, but bear in mind that the practical division is based only on mass—we have no information on most of these planets' composition or internal structure.

Planets having masses between roughly 2 and 10 Earth masses (about half the mass of Neptune) are known as **super-Earths.** The upper limit in this case is significant, as theorists think that 10 Earth masses represents the minimum mass of a planetary core needed for it to accrete large amounts of nebular gas and begin to form a gas giant (Section 4.3). Below 2 Earth masses, exoplanets are simply called *Earths.*

Exoplanets are further subdivided depending on their distances from their parent stars. Planets with orbital semimajor axes less than 0.1 AU are said to be *hot,* while those on wider orbits are called *cold.* Again, the dividing line is arbitrary—the actual temperature of a planet depends not just on its orbit, but also on the composition of the planet's atmosphere and the temperature and brightness of the central star.

Most of the planets observed so far fall into the "cold Jupiter" or "cold Neptune" categories, like the jovian planets in our own solar system, although their orbits are generally somewhat smaller than those of the jovian planets—less than a few astronomical units across—and much more eccentric. Fewer than 20 percent have eccentricities less than 0.1, whereas no jovian planet in our solar system has an eccentricity greater than 0.06. Figure 4.35 plots the actual orbits of some of these planets, with Earth's orbit superimposed for comparison. A sizable minority—about a third—of all observed exoplanets move in "hot" orbits very close to their parent stars and have surface temperatures as high as 1000–2000 K. The most massive ones were the first to be discovered, and they were quickly dubbed **hot Jupiters.** They represent a new class of planet and have no counterparts in our own solar system.

About 60 super-Earths are currently known. They are found in both hot and cold orbits. Some, especially the lower-mass ones, might be large terrestrial planets.

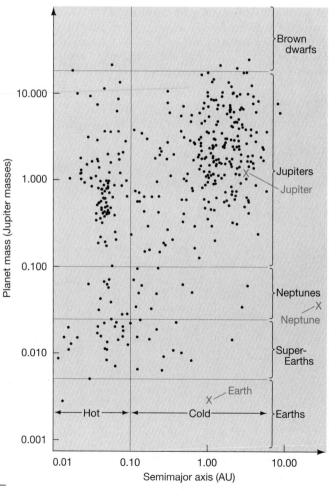

▲ FIGURE 4.34 **Extrasolar Planetary Parameters** Masses and orbital semimajor axes of several hundred extrasolar planets. Each point represents one planetary orbit. Corresponding points for Earth, Jupiter, and Neptune in our solar system are also shown. Planets are classified by familiar solar system names, depending on mass, and as hot or cold, depending on distance from their parent star.

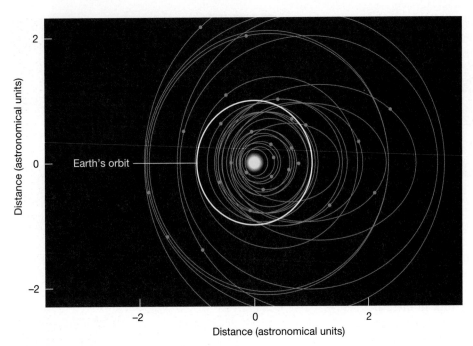

Distance (astronomical units) (vertical axis)

Earth's orbit

Distance (astronomical units) (horizontal axis)

▲ **FIGURE 4.35 Extrasolar Orbits**   The orbits of many extrasolar planets residing more than 0.15 AU from their parent star, superimposed on a single plot, with Earth's orbit shown for comparison (as dashed white). All the extrasolar planets shown are comparable in mass to Jupiter. A plot of all known extrasolar planets would be too cluttered to show.

Others could be icy planetary cores that never managed to accrete significant amounts of nebular gas. Still others may have substantial atmospheres of light gases but never grew to "Neptune" status—they are sometimes referred to as *gas dwarfs*. These latter two categories, if real, would be new classes of planet unknown in the solar system. So far, 5 exo-Earths have been found, all but one moving on hot orbits close to their parent stars, and unlikely to resemble anything we'd want to call home. However, all these numbers are likely to change very soon. The *Kepler* candidate list contains dozens of potential Earths and hundreds of potential super-Earths; we can expect a substantial fraction of these to be confirmed.

The fact that we don't see many low-mass planets or more massive planets on wide orbits (i.e., toward the right or lower part of Figure 4.34) is not too surprising. It is what astronomers call a **selection effect.** Lightweight and/or distant planets simply don't produce large enough velocity fluctuations for them to be easily detectable. Generally speaking, the methods employed so far are biased toward finding large and/or massive objects orbiting close to their parent stars. Those systems would be expected to give the strongest signal, and they are precisely what have been observed.

Exoplanet Composition

If an exoplanet happens to transit its star, we can often determine both its mass and its radius, and hence estimate its density and possibly even its composition. More than 200 transiting hot Jupiters have been measured in this way. But when astronomers calculate their densities the numbers they obtain are generally much lower than predicted by theory. The computed densities range from around 1500 kg/m³ (slightly more than the density of Jupiter) down to just 200 kg/m³ (roughly the density of styrofoam) and are generally *inconsistent,* by a wide margin, with theoretical models, even assuming the lightest possible composition of pure hydrogen and helium. The leading explanation for this discrepancy is that the heat of the nearby parent star has puffed these planets up far beyond their normal sizes. However, no current model can account for the observed range in densities, and for now a complete explanation of this phenomenon eludes astronomers.

Currently, some 25 transiting Earths and super-Earths are known (although many more await confirmation in the *Kepler* candidate list). Their densities range from 500 to 9000 kg/m³. The low end of this range suggests planets harboring large amounts of light gases, probably with rocky/icy cores and hydrogen/helium atmosphere—gas dwarfs. The high end suggests mainly rocky composition—compressed Earths. The intermediate densities suggest planets composed of water and/or other ices. Figure 4.36 presents a visual comparison of Earth, Neptune, and two super-Earths whose physical properties are relatively well known. CoRoT 7b is 4.8 times more massive and 1.7 times larger than Earth, implying a mean density of 5300 kg/m³—very similar to that of Earth, although since this planet orbits just 0.02 AU from its parent Sun-like star, surface conditions are much more extreme than those on Earth. Meanwhile, the planet GJ 1214b has a mass 5.7 times that of Earth and a radius of 2.7 Earth radii, for a mean density of 1600 kg/m³. Definitely not rocky, this planet may be composed mainly of water and/or ice, possibly surrounding a small rocky core, with an atmosphere of hydrogen and helium.

The composition, structure, and history of super-Earths are of great interest to astronomers as they seek to understand how these bodies form—and why they are absent in our solar system. Transit times, once measured, can be accurately predicted, and astronomers can time their observations to obtain spectroscopic and other

▶ FIGURE 4.36 Super-Earth Comparison Two transiting super-Earths whose masses and radii are accurately known are shown here compared with Earth and Neptune. Based on their average densities, these two new worlds seem to be very different from one another—one is rocky, somewhat like Earth or Neptune's core, but the other may be composed predominantly of water and ice. Also shown is the recently discovered "Earth" KOI 961d (see Figure 4.39), whose mass (and hence composition) is currently unknown.

information on the starlit face of the planet during its "quarter" phase, allowing them to probe the planet's atmospheric composition and dynamics. ∞ *(Sec. 2.5)* Since planets—even the hot ones—are cooler than their parent stars, their reflected light is most easily distinguished from the background starlight at infrared wavelengths, and the *Spitzer Space Telescope* has played a vital role in these studies. ∞ *(Secs. 2.4, 3.5)* So far, hydrogen, sodium, methane (CH_4), carbon dioxide (CO_2), and water vapor have been detected. Observations such as these will be critical in pinning down the content and structure of exoplanet atmospheres.

Spectroscopic observations of the parent stars reveal what may be a crucial piece in the puzzle of extrasolar planet formation. Stars having compositions similar to that of the Sun are statistically *much more likely* to have planets orbiting them than are stars containing smaller fractions of the key elements carbon, nitrogen, oxygen, silicon, and iron. Because the elements found in a star reflect the composition of the nebula from which it formed, and the elements just listed are the main ingredients of interstellar dust, this finding provides strong support for the condensation theory (Section 4.3). Dusty disks really are more likely to form planets.

Is Our Solar System Unusual?

Not so long ago, many astronomers argued that the condensation scenario described earlier in this chapter was in no way unique to our own system. Today we know that planetary systems *are* common, but, by and large, the ones we see don't really look much like ours! We can legitimately ask whether our solar system really is unusual, and whether the observations described above undermine the current theory of solar system formation.

Most cold extrasolar Jupiters and Neptunes move in orbits that are considerably more eccentric than the orbits of the jovian planets in the solar system. Does this make our system fundamentally different from the others? Probably not. To a large extent, this discrepancy can be explained by the selection effect just described—eccentric orbits tend to produce larger velocities and hence are more readily discovered. As search techniques improve, astronomers are finding more and more Jupiter-mass (and lower-mass) planets on wider and less eccentric orbits. Figure 4.37 shows evidence for one of the most "Jupiter-like" planets yet detected: a 0.95-Jupiter-mass object moving on a roughly circular orbit around a near-twin of our own Sun. The planet's period is 9.1 years. It is too early to say whether cold Jupiters on nearly circular orbits will turn out to be unusual or common among exoplanets, but they clearly exist among the systems already observed.

Are the eccentric extrasolar orbits we see consistent with the condensation theory? The answer is yes. The theory actually allows many ways in which massive planets can end up in eccentric orbits. Indeed, an important aspect of solar system formation not mentioned in our earlier discussion is the fact that many theorists worried about how Jupiter could have remained in a circular orbit after it formed in the protosolar disk! Jupiter-sized planets may be knocked into eccentric orbits by interactions with other Jupiter-sized planets or by the effects of nearby stars. And if they formed by gravitational instability, they could have had eccentric orbits right from the start, and then we must explain how those orbits circularized in the case of the solar system.

What of the hot Jupiters, which have no counterparts among the planets of the solar system? Here, too, there is an explanation that fits within the condensation theory!

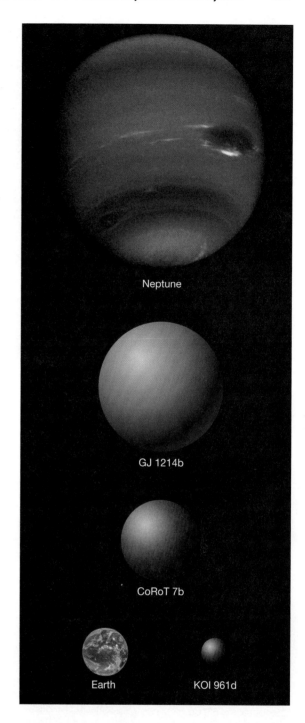

Neptune

GJ 1214b

CoRoT 7b

Earth

KOI 961d

▶ FIGURE 4.37 An Exo-Jupiter? Velocity "wobbles" in the star HD 154345 reveal the presence of the extrasolar planet with the most "Jupiter-like" orbit yet discovered. The parent star is almost identical to the Sun, and the 0.95-Jupiter-mass planet orbits at a distance of 4.2 AU with an orbital eccentricity of 0.04.

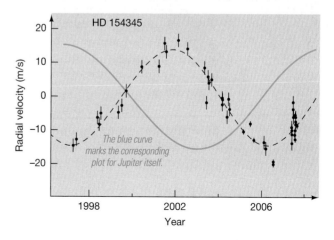

HD 154345

The blue curve marks the corresponding plot for Jupiter itself.

Radial velocity (m/s)

1998   2002   2006

Year

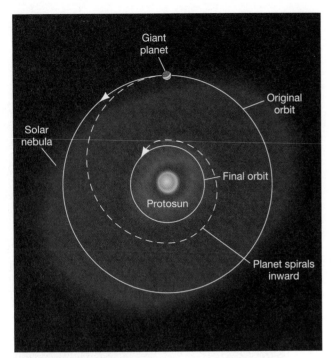

▲ **FIGURE 4.38 Sinking Planet** Friction between a giant plant and the nebular disk in which it formed tends to make the planet spiral inward. The process continues until the disk is dispersed by the wind from the central star, possibly leaving the planet in a "hot" orbit.

(MA) ANIMATION/VIDEO Hot Jupiter Extrasolar Planet Evaporating

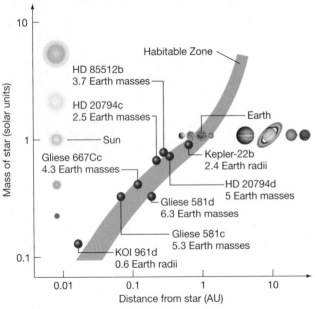

▲ **FIGURE 4.39 Habitable Zones** Every star is surrounded by a habitable zone, within which an Earth-like planet could have liquid water on its surface. The planets of the solar system, seven extrasolar super-Earths, and one possible exo-Earth lying in or near their habitable zones are marked.

In the mid-1980s theorists realized that friction between the giant planets and the nebula in which they moved should have caused the planets to drift inward quite rapidly. Even before the first hot Jupiter was observed, theorists knew that, depending on how long the disk survived before being dispersed by the newborn Sun, the process could easily have deposited planets in an orbit very close to the parent star, as illustrated in Figure 4.38. The theorists were right, and the observed hot Jupiters may provide a much-needed connection between extrasolar planetary systems and our own.

Curiously, a Jupiter-sized planet sinking inward through the disk in which planetesimals are forming is not detrimental to the formation of terrestrial planets. The hot Jupiters must have reached their scorched orbits before the gas disk dispersed—that is, within a few million years of the formation of the solar nebula and long *before* the formation of the terrestrial planets. Computer simulations indicate that the inward migration stirs up the planetesimals as the giant planet moves through, but for the most part does not disrupt or eject them. The main consequence is to mix in more icy material from farther out, possibly resulting in planets more massive and more water-rich than would otherwise be the case.

Searching for Earths

Giant planets are interesting objects, but to many astronomers the real goal of extrasolar planet research is the detection of terrestrial planets with conditions similar to those found on Earth—and, by extension, the possible discovery of life elsewhere in the universe. This latter goal is increasingly becoming the motivation behind solar system exploration, too. No Earth-mass planets have yet been found, but the growing numbers of super-Earths in the exoplanet catalog suggest that detection of an Earth is just a matter of time.

If Earth-like bodies are our ultimate goal, what parameters are of greatest interest to astronomers? As we saw earlier, many astronomers think that a key requirement for the development of life as we know it is the existence of liquid water on (or under) a planet's surface, implying a surface temperature roughly in the 0–100° C range. The planet's temperature depends both on its distance from its parent star and on the star's intrinsic brightness. Figure 4.39 illustrates how a **habitable zone** (really, a three-dimensional shell) surrounds any given star. Within that zone, liquid water can exist on the planet, making it a possible abode for life.

For low-mass, faint stars, the habitable zone is small and lies close to the star. For more massive, brighter stars it lies much farther out and can be 1 AU or more wide. Three terrestrial planets—Venus, Earth, and Mars—lie in or near the Sun's habitable zone. Any or all of them might have seen the development of life, given the right circumstances (see Chapter 6). Earth-like planets orbiting within the habitable zone of their parent stars are regarded as the best candidates for harboring life as we know it beyond the solar system.

At the time of writing, seven super-Earths (more than 10 percent of the current total) and one exo-Earth are known to orbit in or near the habitable zones of their parent stars. They are indicated on Figure 4.39. All orbit relatively low-mass stars, mostly near (within 40 light years of) the Sun, although two (Kepler-22b and KOI 961d) are much more distant *Kepler* objects. Unfortunately, none of the nearby planets transit their parent star, so their radii are not known, and the masses of the transiting *Kepler* planets are undetermined, so in all cases the densities, and hence compositions, of these worlds are unknown. Most of these bodies orbit close to the "hot" edge of the habitable zone, but this is yet another aspect of the selection effect described earlier—of all small planets, those in close orbits are the most likely to be detected. The *Kepler* candidate list contains a further 27 Earths and Super Earths that may soon be added to this figure.

Many planet hunters are confident that within the next decade (or sooner), observational techniques will reach the level of sophistication at which jovian and terrestrial planets similar to those in our solar system should be readily detectable—if they exist. Advances during the next decade will either bring numerous detections of extrasolar planets in "solar system" orbits, or allow astronomers to conclude that systems like our own really are a small minority. Either way, the consequences are profound.

CONCEPT CHECK ▶ Why is it not too surprising that the extrasolar planetary systems detected thus far have properties quite different from our solar system?

CHAPTER REVIEW

SUMMARY

L01 The planets of our **solar system** (p. 102) all orbit the Sun counterclockwise, as viewed from above Earth's North Pole, on roughly circular orbits that lie close to the ecliptic plane. The orbit of Mercury is the most eccentric and also has the greatest inclination. The orbital spacing increases as we move outward from the Sun. The inner **terrestrial planets** (p. 104)—Mercury, Venus, Earth, Mars—are all of comparable density and generally rocky, whereas all the outer **jovian planets** (p. 104)—Jupiter, Saturn, Uranus, and Neptune— have much lower densities and are composed mostly of gaseous or liquid hydrogen and helium.

L02 Most **asteroids** (p. 105) orbit in a broad band called the **asteroid belt** (p. 106), between the orbits of Mars and Jupiter. The **Trojan asteroids** (p. 106) share Jupiter's orbit, remaining 60° ahead or behind the planet as it circles the Sun. A few **Earth-crossing asteroids** (p. 106) have orbits that intersect Earth's orbit and will eventually collide with our planet. The largest asteroids are a few hundred kilometers across. Most are much smaller. Asteroids in the inner part of the asteroid belt are predominantly rocky; asteroids farther out contain larger fractions of water ice and organic material.

L03 **Comets** (p. 105) are icy and rocky fragments that normally orbit far from the Sun. We see a comet by the sunlight reflected from the dust and vapor released if its orbit happens to bring it near the Sun. Stretching behind the kilometer-size **nucleus** (p. 110) of a comet is a long **tail** (p. 110) formed by the interaction between the cometary material and the **solar wind** (p. 110). Most comets reside in the **Oort cloud** (p. 113), a vast reservoir of cometary material tens of thousands of astronomical units across, completely surrounding the Sun. Short-period comets, having orbital periods of less than about 200 years, originate in the **Kuiper belt** (p. 112), a broad band of icy material beyond the orbit of Neptune.

L04 A **meteoroid** (p. 105) is a piece of rocky interplanetary debris smaller than 100 m across. A meteoroid that enters Earth's atmosphere produces a **meteor** (p. 114), a bright streak of light across the sky. If part of it reaches the ground, the remnant is called a **meteorite** (p. 114). Each time a comet rounds the Sun, some cometary material becomes dislodged, forming a **meteoroid swarm** (p. 114)—a group of small **micrometeoroids** (p. 114) that travel in the comet's orbit. Larger meteoroids are pieces of material chipped off asteroids following collisions in the asteroid belt. Most meteorites are between 4.4 and 4.6 billion years old.

L05 Any theory of solar system formation must explain the planets' roughly circular, coplanar, widely spaced orbits in the same direction around the Sun, with terrestrial planets near the Sun and jovian planets farther out—as well as the existence of the asteroids, Kuiper belt, and comets, and irregularities such as retrograde rotation of planets or moons.

L06 According to the **condensation theory,** as the **solar nebula** (p. 119) contracted under its own gravity, it began to spin faster, eventually forming a disk. **Protoplanets** (p. 121) formed in the disk and became planets, while the central **protosun** (p. 121) became the Sun. Particles of interstellar dust helped cool the solar nebula and acted as **condensation nuclei** (p. 120) that began the planet-building process. Small clumps of matter grew by **accretion** (p. 121), sticking together and growing into moon-size **planetesimals** (p. 121) whose gravitational fields accelerated the accretion. As planetesimals collided and merged, a few planet-size objects remained.

L07 Planets in the outer solar system became so large that they captured hydrogen and helium from the solar nebula, forming the jovian worlds. When the Sun became a star, its strong winds blew away any remaining nebular gas. The terrestrial planets are rocky because they formed in the hot inner regions of the solar nebula. Farther out, the nebula was cooler, so water and ammonia ice could also form. Many leftover planetesimals were ejected into the Oort cloud by the outer planets, leaving the Kuiper belt behind. The planetesimals in the asteroid belt never formed a planet, probably because of Jupiter's gravity.

L08 Astronomers have identified more than 700 **extrasolar planets** (p. 117). Most have been discovered by observing their parent star wobble back and forth as the planet orbits. However, a significant and growing fraction has been observed passing in front of the star, slightly reducing its brightness.

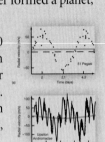

L09 Known exoplanetary masses range from that of Earth to many times that of Jupiter. Some planets move on "hot" orbits close to their parent star, while others move on wide, "cold" orbits similar to those of the jovian planets in the solar system. **Hot Jupiters** (p. 127) and **super-Earths** (p. 127) are new classes of planet not known in the solar system. It is not known whether our solar system is unusual among planetary systems. The new exoplanetary classes are compatible with the condensation theory. A few super-Earths have been found in or near the **habitable zones** (p. 130) of their parent stars.

MasteringAstronomy®   For instructor-assigned homework go to www.masteringastronomy.com

Problems labeled POS explore the process of science. VIS problems focus on reading and interpreting visual information. LO connects to the introduction's numbered Learning Outcomes.

REVIEW AND DISCUSSION

1. **L01** Name three important differences between the terrestrial planets and the jovian planets.

2. **L02 POS** Why are asteroids, comets, and meteoroids important to planetary scientists?

3. Are all asteroids found in the asteroid belt?

4. Describe the consequences of a 10-km asteroid striking Earth?

5. **L03** What are comets like when they are far from the Sun? What happens when they enter the inner solar system?

6. Why can comets approach the Sun from any direction but asteroids generally orbit close to the plane of the elliptic?

7. **L04** Distinguish between a meteor, a meteoroid, and a meteorite.

8. What causes a meteor shower?

9. **L05 POS** Describe the basic features of the nebular theory of solar system formation, and give three examples of how this theory explains some observed features of the present-day solar system.

10. **L06 POS** What key ingredient in the modern condensation theory was missing or unknown in the nebula theory?

11. Why are jovian planets so much larger than terrestrial planets?

12. **L07** How did the temperature structure of the solar nebula determine planetary composition?

13. Where did Earth's water come from?

14. **L08** How do astronomers set about looking for extrasolar planets?

15. **L09 POS** In what ways do observed extrasolar planetary systems differ from our own solar system?

CONCEPTUAL SELF-TEST: TRUE OR FALSE?/MULTIPLE CHOICE

1. The largest planets also have the largest densities. (T/F)

2. The total mass of all the planets is much less than the mass of the Sun. (T/F)

3. Asteroids were recently formed by the collision and breakup of an object orbiting in the asteroid belt. (T/F)

4. Most comets have short periods and orbit close to the ecliptic plane. (T/F)

5. The solar system is of largely uniform composition. (T/F)

6. Asteroids, meteoroids, and comets are remnants of the early solar system. (T/F)

7. Astronomers have no theoretical explanation for the "hot Jupiters" observed orbiting some other stars. (T/F)

8. If we were to construct an accurate scale model of the solar system on a football field with the Sun at one end and Pluto at the other, the planet closest to the center of the field would be (a) Earth; (b) Jupiter; (c) Saturn; (d) Uranus.

9. In the leading theory of solar system formation, the planets (a) were ejected from the Sun following a close encounter with another star; (b) formed from the same flattened, swirling gas cloud that formed the Sun; (c) are much younger than the Sun; (d) are much older than the Sun.

10. The solar system is differentiated because (a) the heavy elements in the outer solar system sank to the center; (b) the light elements in the inner solar system fell into the Sun; (c) the light elements in the inner solar system were carried off as comets; (d) only rocky and metallic particles could form close to the Sun.

11. Water on Earth was (a) transported here by comets; (b) accreted from the solar nebula; (c) produced by volcanoes in the form of steam; (d) created by chemical reactions involving hydrogen and oxygen shortly after Earth formed.

12. The number of confirmed exoplanets is (a) less than 10; (b) roughly 50; (c) more than 500; (d) more than 5000.

13. Astronomers have not yet reported any Earth-like planets orbiting other stars because (a) there are none; (b) they are not detectable

with current technology; (c) no nearby stars are of the type expected to have Earth-like planets; (d) the government is preventing us from reporting their discovery.

14. **VIS** In Figure 4.39 (Habitable Zones), the habitable zone of a star twice as massive as the Sun (a) is centered at roughly 3 AU from the star; (b) is more than 10 AU wide; (c) lies entirely within 1 AU of the star; (d) is the same size as the Sun's habitable zone.

15. **VIS** According to Figure 4.30, the temperature in the solar nebula at the location now at the center of the asteroid belt is (a) 2000 K; (b) 900 K; (c) 400 K; (d) 100 K.

PROBLEMS

The number of squares preceding each problem indicates its approximate level of difficulty.

1. ■ Suppose the average mass of each of 20,000 asteroids in the solar system is 10^{17} kg. Compare the total mass of these asteroids to the mass of Earth. Assuming a spherical shape and a density of 3000 kg/m^3, estimate the diameter of an asteroid having this average mass.

2. ■■ The asteroid Icarus (Figure 4.5) has a perihelion distance of 0.19 AU and an orbital eccentricity of 0.83. What is its semimajor axis and aphelion distance from the Sun?

3. ■ The largest asteroid, Ceres, has a radius 0.073 times the radius of Earth and a mass of 0.0002 Earth masses. How much would an 80-kg astronaut weigh on Ceres?

4. ■■ It is observed that the number of asteroids (or meteoroids) of a given diameter is roughly inversely proportional to the square of the diameter. Assuming a density of 3000 kg/m^3 and approximating the actual distribution of asteroids and meteoroids as a single 1000-km body (Ceres), one hundred 100-km bodies, 10 thousand 10-km bodies, and so on, calculate the total mass in the form of 1000-km bodies, 100-km bodies, 10-km bodies, and 1-km bodies.

5. ■■ (a) Using the version of Kepler's laws of planetary motion from Section 1.5, calculate the orbital period of an Oort cloud comet if the semimajor axis of the comet's orbit is 50,000 AU. (b) What is the maximum possible aphelion distance for a short-period comet with an orbital period of 125 years?

6. ■■ The planet orbiting star HD187123 has a semimajor axis of 0.042 AU. If the star's mass is 1.06 times the mass of the Sun, calculate how many times the planet has orbited its star since the paper announcing its discovery was published on December 1, 1998.

7. ■ Given the data provided in the text, calculate the gravitational acceleration at the surfaces of the two transiting super-Earths (CoRoT 7b and GJ 1214B) discussed in Section 4.4.

8. ■■ Stellar radial velocity variations as small as 1 m/s can be detected with current technology. For a Jupiter-mass planet orbiting a Sun-like star, this corresponds to a planetary orbital velocity of approximately 1 km/s. What is the radius of the widest circular orbit on which Jupiter could currently be detected orbiting the Sun?

9. ■ A typical comet contains 10^{13} kg of water ice. How many comets would be needed to account for the 2×10^{21} kg of water presently found on our planet? If this amount of water accumulated over a period of 0.5 billion years, how frequently must Earth have been struck by comets during that time?

10. ■■■ The amount of energy reaching a planet's surface per unit time is proportional to the luminosity of its parent star divided by the square of the planet's distance from the star. If the luminosity is proportional to the fourth power of the star's mass, estimate the orbital distance (in AU) at which an Earth-like exoplanet orbiting a 0.5 solar mass star would receive the same amount of energy from its star as Earth does from the Sun.

ACTIVITIES

Collaborative

1. Which of the "irregular" characteristics of the solar system listed in Section 4.3. do you think are most and least likely to have occurred by pure chance? Can you think of other irregular features not on the list?

2. Do you think our solar system is unusual? Justify your opinion using data obtained from *The Extrasolar Planets Encyclopedia*, which you can find online at http://exoplanet.eu. Use the *Encyclopedia* to find examples of (1) a hot Jupiter, (2) a cold Jupiter on an orbit like Jupiter's, (3) a hot super-Earth, (4) a super-Earth in the habitable zone.

3. What should be the U.S. government's policy on the mining of minerals from asteroids? Justify the policy.

Individual

1. You can begin to visualize the ecliptic—the plane of the planets' orbits—just by noticing the path of the Sun throughout the day and of the full Moon in the course of a single night. It helps if you watch from one spot, such as your backyard or a rooftop. It's also good to have a general notion of direction. (West is where the Sun sets!) The movements of the Sun, Moon, and planets are confined to a narrow pathway across our sky. This pathway reflects the plane of the solar system, the ecliptic.

2. The only way to tell an asteroid from a star is to watch it over several nights. The magazines *Sky & Telescope* and *Astronomy* often publish charts for especially prominent asteroids. Look for Ceres, Pallas, or Vesta, the brightest asteroids. Use the chart to locate the appropriate star field and aim binoculars at that location in the sky. You may be able to pick out the asteroid from the chart. If you can't, make a rough drawing of the entire field. Come back a night or two later and look again. The "star" that has moved is the asteroid.

3. There are a number of major meteor showers every year, but if you plan to watch one, be sure to note the phase of the Moon. Bright moonlight (or city lights) can obliterate a meteor shower. A common misconception is that most meteors are seen in the direction of the shower's radiant. If you trace the paths of the meteors backward in the sky, they do all come from the radiant, but most meteors don't become visible until they are 20° or 30° from the radiant—they can appear in all parts of the sky! Just relax and let your eyes rove among the stars. You will generally see many more meteors in the hours before dawn than in the hours after sunset. Why do meteors have different brightnesses? Can you detect their variety of colors? Watch for meteors that appear to "explode" as they fall and vapor trails that linger after the meteor itself has disappeared.

5

Earth and Its Moon

Our Cosmic Backyard

If we are to appreciate the universe, we must first come to know our own home. By cataloging Earth's properties and attempting to explain them, we set the stage for a comparative study of all the other planets. From an astronomical perspective, this is a compelling reason to study the structure and history of our own world. What about the Moon? It is by far our closest neighbor in space, yet despite its nearness, it is a world very different from our own. It has no air, no sound, no water, no weather. Boulders and pulverized dust litter the landscape. Why, then, do we study it? In part, simply because it is our nearest neighbor and dominates our night sky. Beyond that, however, we study this body because it holds important clues to our own past. The very fact that it hasn't changed much since its formation means the Moon is a crucial key to unlocking the secrets of the solar system.

MasteringAstronomy® Visit www.masteringastronomy.com for quizzes, animations, videos, interactive figures, and self-guided tutorials.

◀ America's manned exploration of the Moon was arguably the greatest engineering feat of the 20th century, indeed one of the greatest of all time. Nine crewed missions were launched to the Moon, a dozen astronauts were landed, and all returned safely to Earth. The *Apollo* program ended in 1972, as quickly as it had begun a decade earlier—largely because of political posturing at the height of the Cold War. Here, an *Apollo 15* astronaut near Mount Hadley is adjusting some instruments designed to measure the composition of the soil and detect any "moonquakes" that might occur. Given the lack of wind and water on the Moon, the boot prints in the foreground are destined to survive for more than a million years. *(NASA)*

5.1 Earth and the Moon in Bulk

 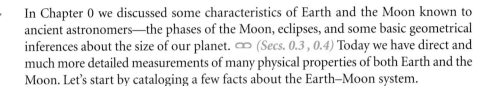

In Chapter 0 we discussed some characteristics of Earth and the Moon known to ancient astronomers—the phases of the Moon, eclipses, and some basic geometrical inferences about the size of our planet. ∞ *(Secs. 0.3 , 0.4)* Today we have direct and much more detailed measurements of many physical properties of both Earth and the Moon. Let's start by cataloging a few facts about the Earth–Moon system.

Physical Properties

Table 5.1 presents some basic data on Earth and the Moon. The first five columns list mass, radius, and average density for Earth and the Moon, determined as described in Chapter 4. ∞ *(Sec. 4.1)* The next two columns present important measures of a body's gravitational field. **Surface gravity** is the strength of the gravitational force at the body's surface. ∞ *(More Precisely 1-1)* **Escape speed** is the speed required for any object—an atom, a baseball, or a spaceship—to escape forever from the body's gravitational pull (see *More Precisely 5-1,* p. 140). By either measure, the Moon's gravitational pull is much weaker than Earth's. As we will see, this fact has played an important role in determining the very different evolutionary paths followed by the two worlds. The final column in Table 5.1 lists (sidereal) rotation periods, which have long been accurately known from Earth-based observations.

These data will form the basis for our comparative study of our home planet and its nearest neighbor. We will expand our catalog of "planetary" characteristics and peculiarities in the next three chapters as we study in turn the other members of the solar system.

One important quantity not listed in Table 5.1 is the *distance* from Earth to the Moon. Parallax, with Earth's diameter as a baseline, has long provided astronomers with reasonably accurate measurements of the Earth–Moon distance. ∞ *(Sec. 0.4)* However, radar and laser ranging[1] now yield far more accurate measurements of the Moon's orbit. ∞ *(Sec. 1.3)* Today the Moon's precise distance from Earth at any instant is known to within about 2 cm. The semimajor axis of the Moon's orbit around Earth is 384,000 km.

Overall Structure

Figure 5.1 compares and contrasts the main regions of these two very dissimilar worlds. As indicated in Figure 5.1(a), our planet may be divided into six main regions. In Earth's interior a thick **mantle** surrounds a smaller, two-part **core.** At the surface we have a relatively thin **crust,** comprising the solid continents and the seafloor; and the **hydrosphere,** comprising rivers, lakes, and the liquid oceans. An **atmosphere** of air lies just above the surface. At much greater altitudes a zone of charged particles trapped by our planet's magnetic field forms Earth's **magnetosphere.**

[1] This technique is conceptually similar to radar ranging. It determines distance by measuring how long a pulse of laser light fired from Earth takes to return after reflection from the lunar surface.

TABLE 5.1	Some Properties of Earth and the Moon						
	Mass		**Radius**		**Average Density**	**Surface Gravity**	**Escape Speed**	
	(kg)	**(Earth = 1)**	**(km)**	**(Earth = 1)**	**(kg/m³)**	**(Earth = 1)**	**(km/s)**	**Rotation Period**
Earth	6.0×10^{24}	1.00	6400	1.00	5500	1.00	11	23ʰ 56ᵐ
Moon	7.4×10^{22}	0.012	1700	0.27	3300	0.17	2.4	27.3 days

▶ **FIGURE 5.1 Earth and Moon** (a) Earth's inner core of radius 1300 km is surrounded by a 2200 km thick liquid outer core. Most of the rest of Earth's interior is taken up by the mantle, topped by a thin crust only a few tens of kilometers thick. The liquid portions of Earth's surface make up the hydrosphere. Above the hydrosphere and solid crust lies the atmosphere, most of it within 50 km of the surface. Earth's outermost region is the magnetosphere, extending thousands of kilometers into space. (b) The Moon's rocky outer mantle is about 900 km thick. Its inner mantle is a semisolid layer somewhat similar to the upper regions of Earth's mantle.

The Moon lacks a hydrosphere, an atmosphere, and a magnetosphere. Its internal structure (Figure 5.1b) is not as well studied as that of Earth, for the very good reason that the Moon is much less accessible. Nevertheless, as we will see, the basic interior regions on Earth—*crust, mantle,* and *core*—also exist on the Moon, although their properties differ somewhat from those of their Earthly counterparts.

5.2 The Tides

Earth is unique among the planets in that it has large quantities of liquid water on its surface. Approximately three-quarters of Earth's surface is covered by water, to an average depth of some 3.6 km. A familiar hydrospheric phenomenon is the daily fluctuation in ocean level known as the **tides.** At most coastal locations on Earth there are two low tides and two high tides each day. The "height" of the tides—the magnitude of the variation in sea level—can range from a few centimeters to many meters, depending on the location and time of year, averaging about a meter on the open sea. An enormous amount of energy is contained in the daily tidal motion of the oceans.

Gravitational Deformation

What causes the tides? A clue comes from the fact that they exhibit daily, monthly, and yearly cycles. In fact, the tides are a direct result of the gravitational influence of the Moon and the Sun on Earth. We have already seen how gravity keeps Earth and the Moon in orbit about one another, and both in orbit around the Sun. For simplicity, let's first consider just the interaction between Earth and the Moon.

Recall that the strength of the gravitational force depends on the inverse square of the distance separating any two objects. ∞ *(Sec. 1.4)* Thus, the Moon's gravitational attraction is greater on the side of Earth that faces the Moon than on the opposite side, some 12,800 km (Earth's diameter) farther away. This difference in the gravitational force is small, only about 3 percent, but it produces a noticeable deformation—a stretching along the line joining Earth to the Moon called a **tidal bulge.** The effect is greatest in Earth's oceans, because liquid can most easily move around on our planet's surface. As illustrated in Figure 5.2, the ocean becomes a little deeper in some places (along the Earth–Moon line) and shallower in others (in the perpendicular direction). The daily tides we see result as Earth rotates beneath this deformation.

This *differential force,* the variation of the Moon's (or the Sun's) gravity across Earth, is called a **tidal force.** The *average* gravitational interaction between two bodies determines their orbit around one another. However, the tidal force, superimposed on that average, tends to deform the bodies. We will see many situations in this book where such forces are critically important in understanding astronomical

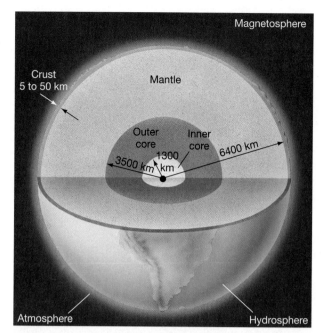

(a)

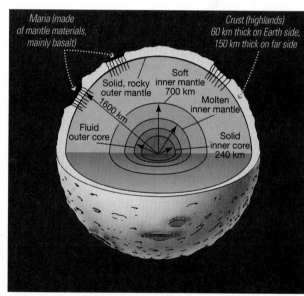

(b)

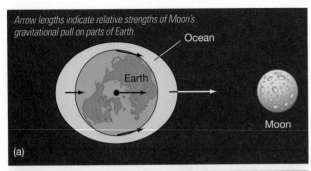

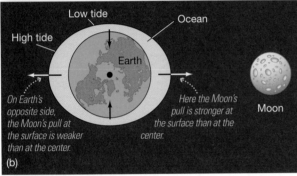

◄ **FIGURE 5.2 Lunar Tides**   This exaggerated illustration shows how the Moon induces tides on both the near and far sides of Earth. (a) The lunar gravitational force is greatest on the side nearest the Moon and smallest on the opposite side. (b) The differences between the lunar forces experienced at the locations shown in part (a) and the force exerted by the Moon on Earth's center. Closest to the Moon, the oceans tend to be pulled away from Earth; on the far side, Earth tends to be pulled away from the oceans. The net result is a tidal bulge.

phenomena. Note that the tidal force is not in any sense a "new" force. It is still the familiar force of gravity, just viewed in a different context—its effect on extended bodies such as planets and moons. We still use the word *tidal* in these other contexts, even though we are not discussing oceanic tides, and sometimes not even planets at all.

Notice in Figure 5.2 that the side of Earth opposite the Moon also experiences a tidal bulge. The different gravitational pulls—greatest on that part of Earth closest to the Moon, weaker at Earth's center, and weakest of all on Earth's opposite side—cause average tides on opposite sides of our planet to be approximately equal in height. On the side nearer the Moon, the ocean water is pulled slightly toward the Moon. On the opposite side, the ocean water is left behind as Earth is pulled closer to the Moon. Thus, high tide occurs *twice,* not once, every day.

Both the Moon and the Sun exert tidal forces on our planet. The tidal force drops off very rapidly with distance, but even though the Sun is around 375 times farther away from Earth than is the Moon, the Sun's mass is so much greater (by about a factor of 27 million) that its tidal influence is still significant—about half that of the Moon. Thus, instead of one tidal bulge being created on Earth, there are two: one pointing toward the Moon, the other toward the Sun. The interaction between them accounts for the changes in the height of the tides over the course of a month or a year. When Earth, Moon, and Sun are roughly lined up—at new or full Moon—the gravitational effects reinforce one another, and the highest tides occur (Figure 5.3a). These tides are known as *spring tides.* When the Earth–Moon line is perpendicular to the Earth–Sun line (at the first and third quarters), the daily tides are smallest (Figure 5.3b). These are termed *neap tides.*

Tidal Locking

The Moon rotates once on its axis in 27.3 days—exactly the same time as it takes to complete one revolution around Earth. ⚭ *(Sec. 0.3)* As a result, the Moon presents the same face toward Earth at all times; that is, the Moon has a near side, which is always visible from Earth, and a far side, which never is (Figure 5.4a). This condition, in which the rotation period of a body is precisely equal to (or *synchronized* with) its orbital period around another body, is known as a **synchronous orbit.**

The fact that the Moon is in a synchronous orbit around Earth is no accident. It is an inevitable consequence of the tidal gravitational interaction between those two bodies. To see how this can occur, let's continue our discussion of tides on Earth. Because of Earth's rotation, the tidal bulge raised by the Moon does not point directly at the Moon, as indicated in Figure 5.2. Instead, through the effects of friction, both between the crust and the oceans and within Earth's interior, Earth's rotation tends to drag the tidal bulge around with it, causing the bulge to be displaced slightly ahead of the Earth–Moon line (Figure 5.4b). The Moon's asymmetrical gravitational pull on this slightly offset bulge acts to slow our planet's rotation.

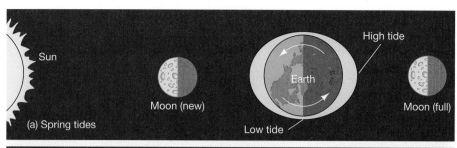

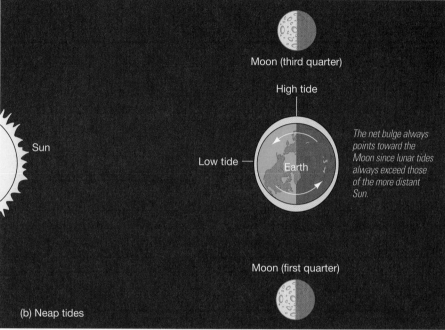

(a) Spring tides

(b) Neap tides

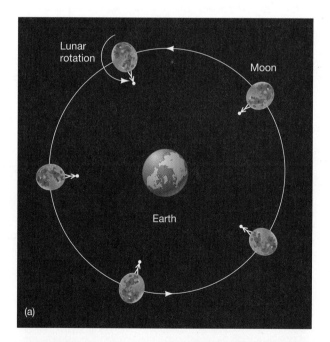

INTERACTIVE ◀ FIGURE 5.3 **Solar and Lunar Tides** The combined effects of the Sun and the Moon produce variations in high and low tides. (a) When the Moon is either full or new, Earth, Moon, and Sun are approximately aligned, and the tidal bulges raised in Earth's oceans by the Moon and the Sun reinforce one another. (b) When the Moon is in either its first or its third quarter, the tidal effects of the Moon and the Sun partially cancel each other, and the tides are smallest. **MA**

According to fossil measurements, the rate of decrease is just 2 milliseconds every century—not much on the scale of a human lifetime, but it means that half a billion years ago Earth's day was just over 21 hours long and its year contained 410 days. At the same time, the Moon spirals slowly away from Earth, increasing its average distance from our planet by about 4 cm per century. This process will continue until (billions of years from now) Earth rotates on its axis at exactly the same rate as the Moon orbits Earth. Earth's rotation will become synchronized, or **tidally locked,** with the Moon's motion. At that time the Moon will always be above the same point on Earth and will no longer lag behind the bulge it raises.

Basically the same process is responsible for the Moon's synchronous orbit. Earth and the Moon have evolved together. Just as the Moon raises tides on Earth, Earth also produces a tidal bulge in the Moon. Indeed, because Earth is so much more massive than the Moon, the lunar tidal bulge is considerably larger, and the

INTERACTIVE ▶ FIGURE 5.4 **Tidal Locking** (a) As the Moon orbits Earth, it keeps one face permanently pointed toward our planet. To the astronaut shown here, Earth is always directly overhead. In fact, because of Earth's tides, the Moon is slightly elongated in shape, with its long axis perpetually pointing toward Earth. (b) The tidal bulge raised in Earth by the Moon does not point directly at the Moon. Instead, because of the effects of friction, the bulge points slightly ahead of the Moon, in the direction of Earth's rotation. (Both figures are greatly exaggerated for clarity.) **MA**

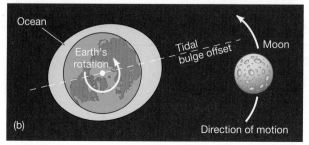

5-1 MORE PRECISELY

Why Air Sticks Around

Why does Earth have an atmosphere but the Moon does not? Why doesn't our atmosphere disperse into space? The answer is that *gravity* holds it down. However, gravity is not the only influence acting. If it were, all of Earth's air would have fallen to the surface long ago. *Heat* competes with gravity to keep the atmosphere buoyant. Our atmosphere stays in place because the buoyancy force balances the force of gravity.

Let's explore this competition between gravity and heat in a little more detail. All gas molecules are in constant random motion. The temperature of any gas is a direct measure of this motion—the hotter the gas, the faster the molecules are moving. ∞ *(More Precisely 2-1)* The rapid movement of heated molecules creates *pressure* that tends to oppose the force of gravity, preventing our atmosphere from collapsing under its own weight.

An important measure of the strength of a body's gravity is its *escape speed*—the speed needed for any object to escape forever from its surface. This speed increases with increased mass or decreased radius of the parent body. In fact, the escape speed is proportional to the speed of a circular orbit at the body's surface. In convenient units, it can be expressed as

$$\text{escape speed (in km/s)} =$$
$$11.2 \sqrt{\frac{\text{mass of body (in Earth masses)}}{\text{radius of body (in Earth radii)}}}.$$

In other words, you need high speed to escape the gravitational pull of a very massive or very small body, but you can escape from a less massive or larger body at lower speeds. If the *mass* of the parent body is quadrupled, the escape speed doubles. If the parent body's *radius* quadruples, then the escape speed is halved.

To determine whether or not a planet will retain an atmosphere, we must compare the planet's escape speed with the average speed of the gas particles making up the atmosphere. This speed depends not only on the temperature of the gas but also on the mass of the individual molecules—the hotter the gas or the smaller the molecular mass, the higher the average speed of the molecules:

$$\text{average molecular speed (in km/s)} =$$
$$0.157 \sqrt{\frac{\text{gas temperature (K)}}{\text{molecular mass (hydrogen atom masses)}}}.$$

Thus, increasing the absolute temperature of a sample of gas by a factor of four—from, for example, 100 K to 400 K—doubles the average speed of its constituent molecules. And at a given temperature, molecules of hydrogen in air move, on average, four times faster than molecules of oxygen, which are 16 times heavier.

At any instant, a tiny fraction of the molecules in any gas have speeds much greater than average, and some molecules are always moving fast enough to escape, even when the average molecular speed is much less than the escape speed. As a result, all planetary atmospheres slowly leak away into space. Don't be alarmed, however—the leakage is usually very gradual. As a rule of thumb, if the escape speed from a planet or moon exceeds the average speed of a given type of molecule by a factor of 6 or more, then molecules of that type will not have escaped from the planet's atmosphere in significant quantities since the solar system formed.

For air on Earth (with a temperature of about 300 K), the mean molecular speed of oxygen (mass = 32 hydrogen masses) and nitrogen (mass = 28) is about 0.6 km/s, comfortably below one-sixth of the escape speed (11.2 km/s). As a result, Earth is able to retain its atmosphere. However, if the Moon originally had an Earth-like atmosphere, the lunar escape speed of just 2.4 km/s means that any original atmosphere long ago dispersed into interplanetary space. The same reasoning can be used to understand atmospheric composition. For example, hydrogen molecules (mass = 2) move, on average, at about 2 km/s in Earth's atmosphere at sea level. Consequently, they have had plenty of time to escape since our planet formed, which is why we find very little hydrogen in Earth's atmosphere today.

CONCEPT CHECK ▶ In what ways do tidal forces differ from the familiar inverse-square force of gravity?

synchronization process correspondingly faster. The Moon's rotation became tidally locked to Earth long ago. Most of the moons in the solar system are similarly locked by the tidal forces of their parent planets.

5.3 Atmospheres

Earth's atmosphere is much more than just air for us to breathe or a "blanket" to keep us warm. As we will see, this thin layer of gas clinging to the edge of a planet or a moon also shields that body from potentially damaging solar and cosmic radiation and may radically alter conditions on the surface.

Earth's Atmosphere

Our planet's atmosphere is a mixture of gases, the most abundant of which are nitrogen (78 percent by volume), oxygen (21 percent), argon (0.9 percent), and carbon dioxide (0.03 percent). Water vapor is a variable part of the atmosphere, making up anywhere from 0.1 to 3 percent, depending on location and climate. The presence of a large amount of free oxygen makes our atmosphere unique in the solar system—Earth's oxygen is a direct consequence of the emergence of life on our planet. We will discuss the formation and evolution of planetary atmospheres in more detail in Chapter 6, when we consider the terrestrial planets as a whole.

Figure 5.5 shows a cross-section of Earth's atmosphere. Compared with Earth's overall dimensions, the extent of the atmosphere is not great. Half of it lies within 5 km of the surface, and all but 1 percent of it is found below 30 km. The region below about 12 km is called the **troposphere.** Everything on Earth's surface lies within it—even Mount Everest, the tallest mountain on our planet, rises only 9 km above sea level. Above the troposphere, extending up to an altitude of 40–50 km, lies the **stratosphere.** Between 50 and 80 km from the surface lies the **mesosphere.** Above about 80 km, in the **ionosphere,** the atmosphere is kept partly ionized by solar ultraviolet radiation. These various atmospheric regions are distinguished from one another by the behavior of the temperature (decreasing or increasing with altitude) in each. Atmospheric pressure decreases steadily with increasing altitude.

The troposphere is the region of Earth's (or any other planet's) atmosphere where **convection** occurs. Convection is the constant upwelling of warm air and the concurrent downward flow of cooler air to take its place. In Figure 5.6, part of Earth's surface is heated by the Sun. The air immediately above the warmed surface is heated, expands a little, and becomes less dense. As a result, it becomes buoyant and starts to rise. At higher altitudes the opposite effect occurs: The air gradually cools, grows denser, and sinks back to the ground. The cool air at the surface rushes in to replace the hot, buoyant air that has risen. In this way, a circulation pattern is established. These **convection cells** of rising and falling air contribute to atmospheric heating and are responsible for surface winds and all the weather we experience. Above the troposphere the atmosphere is stable and the air is calm.

Within the stratosphere lies the **ozone layer,** where atmospheric oxygen, ozone, and nitrogen absorb incoming solar ultraviolet radiation. Ozone is a form of oxygen. Most atmospheric oxygen consists of two oxygen atoms combined into a molecule; ozone consists of three oxygen atoms combined into each molecule. Ozone is

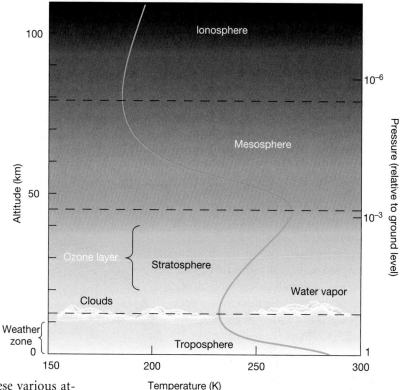

▲ FIGURE 5.5 **Earth's Atmosphere** Diagram of Earth's atmosphere, showing the changes in temperature (blue curve, bottom axis) and pressure (right-hand axis) from the planet's surface to the bottom of the ionosphere. Pressure decreases steadily with increasing altitude, but the temperature may fall or rise, depending on height above the ground.

▶ FIGURE 5.6 **Convection** Convection occurs whenever cool matter overlies warm matter. The resulting circulation currents are familiar to us as the winds in Earth's atmosphere, caused by the solar-heated ground. Over and over, hot air rises, cools, and falls back to Earth. Eventually, steady circulation patterns are established and maintained, provided that the source of heat (the Sun, in the case of Earth) remains intact.

5-1 DISCOVERY

Earth's Growing Ozone "Hole"

During the last two centuries, human technology has begun to produce measurable—and possibly permanent—changes in our planet. You can probably think of many instances, mostly negative, of such changes, from the threat of nuclear war to the reality of air and water pollution worldwide. One example very much in the news and relevant to the discussion in this chapter is the depletion of Earth's ozone layer. (See *Discovery 5–2* for an unrelated, but even more serious, threat to our environment.)

A particularly undesirable by-product of human technology is a group of chemicals known as chlorofluorocarbons (CFCs), relatively simple compounds once widely used for a variety of purposes— propellant in aerosol cans, dry-cleaning products, and in air conditioners and refrigerators. In the 1970s scientists discovered that, instead of quickly breaking down after use as had previously been thought, CFCs accumulate in the atmosphere and are carried high into the stratosphere by convection. There they are broken down by sunlight, releasing chlorine, which quickly reacts with ozone (O_3; see Section 5.3), turning it into oxygen (O_2). In chemical terms the chlorine is said to act as a *catalyst*—it is not consumed in the reaction, and so it survives to react with many more ozone molecules. A single chlorine atom can destroy up to 100,000 ozone molecules before being removed by other, less frequent chemical reactions. In this way,

even a small quantity of CFCs turns out to be extraordinarily efficient at destroying atmospheric ozone.

As we saw in the text, ozone is part of the protective "blanket" of gases in our atmosphere that protect us from the harsh realities of outer space—in this case, solar ultraviolet radiation. Thus, the result of CFC emission is a substantial increase in ultraviolet radiation levels at Earth's surface, with detrimental effects to most living organisms. The accompanying figure shows the development of an ozone "hole" (colored magenta) over the Antarctic—a region where atmospheric circulation and low temperatures conspire each Antarctic spring to create a vast circumpolar cloud of ice crystals, which act to promote the ozone-destroying reactions, resulting in ozone levels about 50 percent below normal for the region.

Since the hole (really a relative lack of ozone) was discovered in the 1980s, its depth and area have grown each year. Its peak size is now larger than North America. Ozone depletion is not confined to the Antarctic, although the effect is greatest there. Smaller holes have been observed in the Arctic, and occasional ozone depletions of up to 20 percent have been reported at lower northern latitudes.

In the late 1980s, when the effect of CFCs on the atmosphere was realized, the world moved remarkably rapidly to curtail their production and use, with the goal of phasing them out entirely by 2030. Substantial cuts have already been made. Still, scientists think that even if all remaining CFC emissions were to stop today, it would still take several decades for CFCs to leave the atmosphere completely.

South America

Antarctica

Center of ozone hole

formed in the atmosphere when solar ultraviolet radiation interacts with oxygen molecules. Its concentration is greatest at an altitude of around 25 km. The ozone layer is one of the insulating layers that protect life on Earth from the harsh environment of outer space. By absorbing potentially dangerous high-frequency radiation, it acts as a planetary umbrella. Without it, advanced life (at least on Earth's surface) would be at best impaired and at worst impossible. *Discovery 5-1* discusses how human activities may now be affecting this vital part of our atmosphere.

The Greenhouse Effect

SELF-GUIDED TUTORIAL The Greenhouse Effect

ANIMATION/VIDEO Seasonal Changes on the Earth

The Sun emits most of its energy in the visible and near-infrared (that is, wavelengths only slightly longer than red light) regions of the electromagnetic spectrum. Because Earth's atmosphere is largely transparent to radiation of this type, almost all of the solar radiation not absorbed by or reflected from clouds in the upper atmosphere shines directly onto Earth's surface, heating it up. ∞ *(Sec. 2.3)*

Earth's surface reradiates the absorbed energy. As the temperature rises, the amount of radiated energy increases rapidly, according to Stefan's law, and eventually our planet radiates as much energy back into space as it receives from the Sun. In the absence of any complicating effects, this balance would be achieved at an average surface temperature of about 250 K (−23°C). At that temperature, Wien's law tells us that most of the reemitted energy is in the form of far-infrared (roughly 10 μm) radiation. ∞ *(More Precisely 2-2)*

However, there is a complication. Long-wavelength infrared radiation is partially blocked by Earth's atmosphere, mainly because of carbon dioxide and water vapor, both of which absorb very efficiently in the infrared portion of the spectrum. Even though these two gases account for only a tiny fraction of our atmosphere, they manage to absorb a large fraction of all the infrared radiation emitted from the surface. Consequently, only some of that radiation escapes back into space. The rest is radiated back to the surface, causing the temperature to rise.

This partial trapping of solar radiation is known as the **greenhouse effect.** The name comes about because a similar process operates in a greenhouse—sunlight passes relatively unhindered through glass panes, but much of the infrared radiation reemitted by the plants is blocked by the glass and cannot get out.[2] Consequently, the interior of the greenhouse heats up, and flowers, fruits, and vegetables can grow even on cold winter days. The radiative processes that determine the temperature of Earth's atmosphere are illustrated in Figure 5.7. Earth's greenhouse effect makes our planet about 40 K hotter than would otherwise be the case—a very important difference, as it raises the average temperature *above* the freezing point of water.

The magnitude of the greenhouse effect is very sensitive to the concentration of *greenhouse gases* (that is, gases that absorb infrared radiation efficiently) in the atmosphere. As just mentioned, water vapor and carbon dioxide are both important greenhouse gases. Carbon dioxide is of particular concern today, because its concentration in Earth's atmosphere is increasing, largely as a result of the burning of fossil fuels (principally oil and coal) in the industrialized world. Carbon dioxide levels have increased by more than 20 percent in the last century, and they continue to rise at a present rate of 4 percent per decade. Many scientists think that this increase, if left unchecked, may result in global temperature increases of several kelvins over the next half-century, enough to melt much of the polar ice caps and cause dramatic, perhaps even catastrophic, changes in Earth's climate (see *Discovery 5-2*).

Lunar Air?

What about the Moon's atmosphere? That's easy—for all practical purposes, there is none! All of it escaped long ago. More massive objects have a better chance of retaining their atmospheres because the more massive an object, the greater the speed needed for atoms and molecules to escape (see *More Precisely 5-1*). The Moon's escape speed is only 2.4 km/s, compared with 11.2 km/s for Earth (Table 5.1). Simply put, the Moon has a lot less pulling power—any atmosphere it might once have had is gone forever.

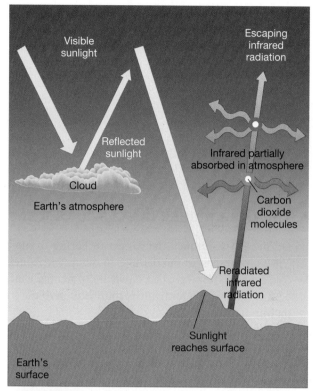

INTERACTIVE ▲ **FIGURE 5.7 Greenhouse Effect** Sunlight not reflected by clouds reaches Earth's surface, warming it up. Infrared radiation reradiated from the surface is partially absorbed by carbon dioxide in the atmosphere, causing the overall surface temperature to rise.

SELF-GUIDED TUTORIAL Atmospheric Lifetimes

ANIMATION/VIDEO Ozone Hole Over the Antarctic

---

[2]Actually, although this process does contribute to warming the interior of a greenhouse, it is not the most important effect. A greenhouse works mainly because its glass panes prevent convection from carrying heat up and away from the interior. Nevertheless, the name "greenhouse effect" to describe the heating effect due to Earth's atmosphere has stuck.

Lacking the moderating influence of an atmosphere, the Moon experiences wide variations in surface temperature. Noontime temperatures can reach 400 K, well above the boiling point of water (373 K). ∞ *(More Precisely 2-1)* At night (which lasts nearly 14 Earth days) or in the shade, temperatures fall to about 100 K, well below water's freezing point (273 K). With such high daytime temperatures and no atmosphere to help retain it, any water that ever existed on the Moon's surface has most likely evaporated and escaped. Not only is there no lunar hydrosphere, but all the lunar

5-2 DISCOVERY

The Greenhouse Effect and Global Warming

Discovery 5-1 outlined the danger to Earth's ozone layer posed by CFCs—products of modern technology with unexpected global consequences. However, an older, and potentially far more serious, hazard comes from humanity's contribution to the *greenhouse effect* on our planet.

We saw the greenhouse effect in Section 5.3: so-called greenhouse gases in Earth's atmosphere, notably water vapor and carbon dioxide (CO_2), tend to trap heat leaving the surface, raising our planet's temperature by several tens of degrees Celsius. The greenhouse effect is not a bad thing—it is the reason that water exists in the liquid state on Earth's surface, and thus it is crucial to the existence and survival of life on our planet (see Chapter 18). However, if atmospheric greenhouse gas levels rise unchecked, the consequences can be catastrophic (for a particularly extreme example, see Section 6.8).

Since the Industrial Revolution in the 18th century, and particularly over the past few decades, human activities on Earth have steadily raised the level of carbon dioxide in our atmosphere, as illustrated in the first figure. Fossil fuels (coal, oil, and gas), still the dominant energy source of modern industry, all release CO_2 when burned. At the same time, the extensive forests that once covered much of our planet are being systematically destroyed to make room for human expansion. Forests play an important role in this problem because vegetation absorbs carbon dioxide, thus providing a natural control mechanism for atmospheric CO_2. Deforestation therefore also tends to increase the level of greenhouse gases in Earth's atmosphere.

Global warming is the slow rise in Earth's surface temperature caused by the increased greenhouse effect resulting from higher levels of atmospheric carbon dioxide. As shown in the second figure, average global temperatures have risen by about 0.5°C during the past century. This may not seem like much, but climate models predict that if CO_2 levels continue to rise, a further increase of as much as 5°C may be possible by the end of the 21st century. Such a rise would be enough to cause serious climatic change on a global scale.

We saw in *Discovery 5-1* how, once the environmental impact of CFCs was identified, rapid steps were taken to curb their use. Curiously, given the more serious nature of the potential damage, a concerted response to global warming has been much slower in coming. Most scientists see the human-enhanced greenhouse effect as a real threat to Earth's climate and urge prompt and deep reductions

in CO_2 emissions, along with steps to slow and ultimately reverse deforestation. Some, however, particularly those connected with the industries most responsible for the production of greenhouse gases, argue that Earth's long-term response to increased greenhouse emissions is too complex for simple conclusions to be drawn and that immediate action is unnecessary. They suggest that the current temperature trend may be part of some much longer cycle or that natural environmental factors may in time stabilize, or even reduce, the level of CO_2 in the atmosphere without human intervention.

Given the stakes, it is perhaps not surprising that these arguments have become far more political than scientific in tone—not at all like the deliberative scientific method presented elsewhere in this text! The basic observations and much of the basic science are generally not seriously questioned, but the interpretation, long-term consequences, and proper response are all hotly debated. Separating the issues sometimes is not easy, but the outcome may be of vital importance to life on Earth.

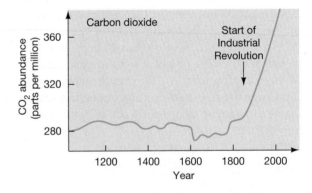

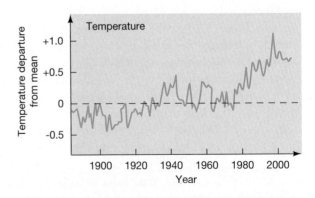

samples returned by the U.S. and Soviet Moon programs were absolutely bone dry. Lunar rock doesn't even contain minerals with water molecules locked within their crystal structure. Terrestrial rocks, conversely, are almost always 1 or 2 percent water.

The Moon is not entirely devoid of water, however. In November 1996, radar echoes from the U.S. *Clementine* spacecraft suggested the presence of water ice at the lunar poles. Since the Sun never rises more than a few degrees above the horizon, as seen from these regions, temperatures on the permanently shaded floors of craters near the poles never exceed about 100 K. Consequently, scientists theorized, ice could have remained permanently frozen there since being dumped on the Moon by comets during the very early days of the solar system, never melting or vaporizing and hence never escaping into space. ∞ *(Sec. 4.3)*

In March 1998, NASA announced that the *Lunar Prospector* mission had confirmed *Clementine's* findings, detecting large amounts of water ice—possibly totaling trillions of tons—at both lunar poles, lying perhaps half a meter below the surface.

In 2009, in an attempt to gain more information about lunar ice, NASA placed the *Lunar Reconnaissance Orbiter (LRO)* spacecraft into a polar orbit just 50 km above the Moon's surface. During its 1-year mission, *LRO* collected detailed information on the lunar surface, with particular emphasis on the polar regions. In late 2009, mission scientists carried out a very direct experiment in their search for ice. The Centaur rocket that boosted both *LRO* and its sister spacecraft, *LCROSS* (the *Lunar CRater Observation and Sensing Satellite*), into lunar orbit was crashed into a deeply shadowed crater near the lunar south pole. *LCROSS* followed and watched from a few thousand kilometers behind, radioing detailed spectroscopic data back to Earth via *LRO* until it too impacted the Moon minutes later. Detailed analysis of the *LCROSS* data confirmed the presence of water molecules in the ejecta. The amount of water was not great—only about 1 part in 100,000, less than in desert sand on Earth—but was more than enough to corroborate the earlier reports.

NASA's *Gravity Recovery and Interior Laboratory (GRAIL)* mission, launched in September 2011, placed two spacecraft in coordinated orbits around the moon for several months. They measured the lunar gravity field in unprecedented detail, affording scientists a clearer understanding of the Moon's subsurface structure and, indirectly, of how Earth and the other terrestrial planets formed.

5.4 Internal Structure of Earth and the Moon

Although we reside on Earth, we cannot easily explore our planet's interior, because drilling gear can penetrate rock only so far before breaking. No substance—not even diamond, the hardest known material—can withstand the conditions below a depth of about 10 km. That's rather shallow compared with Earth's 6400-km radius. Fortunately, geologists have developed techniques that can indirectly probe the deep recesses of our planet.

Seismology

A sudden dislocation of rocky material in Earth's crust—an **earthquake**—causes the entire planet to vibrate a little. It literally rings like a giant bell (but one tuned so low that human ears cannot detect the sound). These vibrations are not random. They are systematic waves, called **seismic waves** (after the Greek word for "earthquake"), that move outward from the site of the quake. Like all waves, they carry information. This information can be detected and recorded using sensitive equipment—a *seismograph*—designed to monitor Earth tremors.

Decades of earthquake research have demonstrated the existence of many kinds of seismic waves. Two are of particular importance to the study of Earth's internal structure (Figure 5.8). First to arrive at a monitoring site after a distant earthquake are the

CONCEPT CHECK ▶ Why is the greenhouse effect important for life on Earth? What might be the consequences if the effect continues to strengthen?

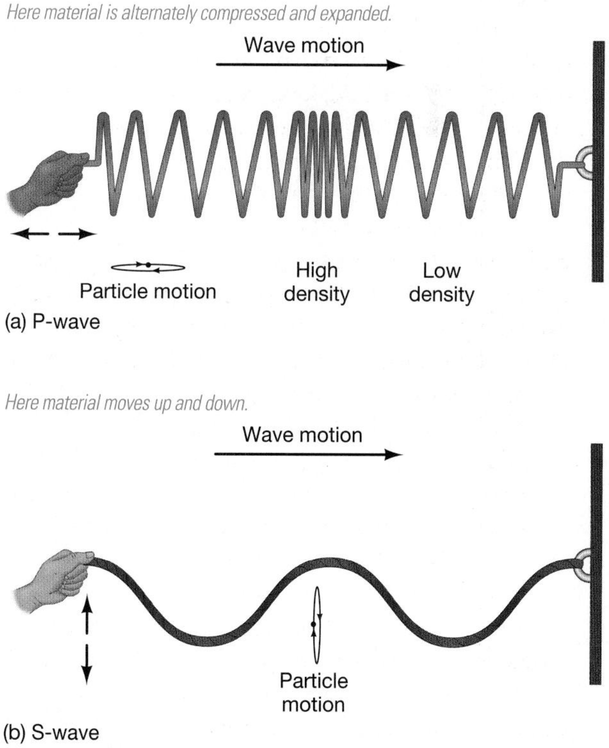

Here material is alternately compressed and expanded.

Wave motion

Particle motion   High density   Low density

(a) P-wave

Here material moves up and down.

Wave motion

Particle motion

(b) S-wave

▲ **FIGURE 5.8 P- and S-waves**   (a) Pressure (P) waves traveling through Earth's interior cause material to vibrate in a direction parallel to the direction of motion of the wave. (b) Shear (S) waves produce motion perpendicular to the direction in which the wave travels, pushing material from side to side.

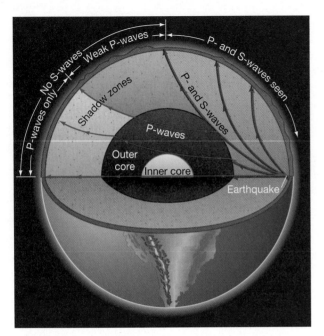

▲ **FIGURE 5.9 Seismic Waves** Earthquakes generate pressure (P, or primary) and shear (S, or secondary) waves that can be detected at seismographic stations around the world. The waves bend while moving through Earth's interior because of the variation in density and temperature within our planet. S-waves (colored red) are not detected by stations "shadowed" by the liquid core of Earth. P-waves (colored green) do reach the side of Earth opposite the earthquake, but their interaction with Earth's core produces another shadow zone, where almost no P-waves are registered.

PROCESS OF ▶ Describe how scientists combine
SCIENCE CHECK theory and observation to create
models of Earth's interior. Give examples of some
Earth properties that we actually observe and some
that are derived purely from the models.

primary waves, or *P-waves.* These are *pressure* waves, a little like ordinary sound waves in air, that alternately expand and compress the medium (the core or mantle) through which they move. Seismic P-waves usually travel at speeds ranging from 5 to 6 km/s and can travel through both liquids and solids. Some time later (the actual delay depends on the distance from the earthquake site), secondary waves, or *S-waves,* arrive. These are *shear* waves causing side-to-side motion, like waves in a guitar string. S-waves normally travel through Earth's interior at 3–4 km/s; however, they cannot travel through liquid, which absorbs them. The speed of each type depends on the density and physical state of the matter through which it is traveling. Consequently, by measuring the time taken for waves to move from the site of an earthquake to one or more monitoring stations on Earth's surface, geologists can infer the density of matter in the interior.

Modeling Earth's Interior

Because earthquakes occur often and at widespread places across the globe, geologists have accumulated a large amount of data about seismic-wave properties. They have used these data, along with direct knowledge of surface rocks, to model Earth's interior. Our knowledge of the deepest recesses of our planet is based almost entirely on modeling and indirect observation. We will find many more examples of this powerful combination throughout the text.

Figure 5.9 illustrates some paths followed by seismic waves from the site of an earthquake. Monitoring stations on the side of Earth opposite a quake never detect S-waves (colored red in the figure)—they are blocked by material within Earth's interior. Furthermore, while P-waves (green) always arrive at stations diametrically opposite the quake, other parts of Earth's surface receive almost none. Most geologists think that S-waves are absorbed by a liquid core at Earth's center and that P-waves are refracted at the core boundary, much as light is refracted by a lens. The result is the S- and P-wave "shadow zones" we observe. The fact that every earthquake exhibits these shadow zones is the best evidence that the core of our planet is molten. The radius of this **outer core,** as determined from seismic data, is about 3500 km. There is also evidence that the P-waves, which can pass through the liquid outer core, are reflected off the surface of a solid **inner core,** of radius 1300 km.

Figure 5.10 presents a model that most scientists accept. Earth's outer core is surrounded by a thick mantle and topped with a thin crust. The mantle is about 3000 km thick and accounts for the bulk (80 percent) of our planet's volume. The crust has an average thickness of only 15 km—a little less (around 8 km) under the oceans and somewhat more (20–50 km) under the continents. Note just how thin the crust is relative to the size of our planet: 15 km, about twice the height of the tallest mountains on our planet, is a mere one quarter of 1 percent of Earth's radius. The average density of crust material is around 3000 kg/m^3. Density and temperature both increase with depth. From Earth's surface to its very center, the density increases from roughly 3000 kg/m^3 to a little more than 12,000 kg/m^3, while the temperature rises from just under 300 K to well over 5000 K. Much of the mantle has a density midway between the densities of the core and crust: about 5000 kg/m^3.

The high central density suggests to geologists that Earth's core must be rich in nickel and iron. Under the heavy pressure of the overlying layers, these metals (whose densities under surface conditions are around 8000 kg/m^3) can be compressed to the high densities predicted by the model. The sharp density increase at the core–mantle boundary results from the difference in composition between these two regions. Unlike the dense metallic core, the mantle is composed of less dense, rocky material—compounds of silicon and oxygen. Note that there is no comparable jump in density or temperature at the inner core boundary—the material there simply changes from the liquid to the solid state.

Because geologists have been unable to drill deeper than about 10 km, no experiment has yet recovered a sample of the mantle. However, we are not entirely ignorant of the mantle's properties. In a **volcano,** molten rock upwells from below the crust, bringing a little of the mantle to us in the form of lava and providing some inkling of the composition of Earth's interior. The composition of the upper mantle is probably quite similar to the dark gray material known as *basalt* often found near volcanoes. With a density between 3000 and 3300 kg/m³, it contrasts with the lighter *granite* (density 2700–3000 kg/m³) that constitutes much of Earth's crust.

Differentiation

Earth, then, is not a homogeneous ball of rock. Instead, it has a layered structure, with a low-density rocky crust at the surface, intermediate-density rocky material in the mantle, and a high-density metallic core. (As we will see, the other terrestrial planets have similar internal structures.) This variation in density and composition is known as **differentiation.** Why isn't our planet just one big, rocky ball of uniform density? The answer is that at some time in the distant past, much of Earth was *molten,* allowing the higher-density matter to sink to the core, displacing lower-density material toward the surface. A remnant of this ancient heating exists today: Earth's central temperature is nearly equal to that of the Sun's surface. Two processes played important roles in heating Earth to the point where differentiation could occur. First, very early in its history, our planet experienced a violent bombardment by interplanetary debris, an essential part of the process by which Earth and the other planets formed and grew. ∞ *(Sec. 4.3)* This bombardment probably generated enough heat to melt much of our planet.

The second process that heated Earth after its formation and contributed to differentiation was **radioactivity**—the release of energy by certain unstable elements, such as uranium and thorium, that were present in the solar nebula when Earth formed. ∞ *(Sec. 4.3)* These elements emit energy as their complex, heavy nuclei break up into simpler, lighter ones. Rock is such a poor conductor of heat that the energy released through radioactivity took a very long time to reach the surface and leak away into space. As a result, the heat built up in the interior, adding to the energy left there by Earth's formation. Geologists think that enough radioactive elements were originally spread throughout the primitive planet, like raisins in a cake, that the entire planet—from crust to core—could have melted and remained molten, or at least semisolid, for about a billion years.

The Lunar Interior

The Moon's average density, about 3300 kg/m³, is quite similar to the density of the lunar surface rock obtained by U.S. and Soviet missions. This similarity all but eliminates any chance that the Moon has a large, massive, dense nickel–iron core like that within Earth. The low average lunar density suggests that the Moon contains substantially fewer heavy elements (such as iron) than does Earth.

Most of our detailed information on the Moon's interior comes from seismic data obtained from equipment left on the lunar surface by astronauts. These measurements indicate only very weak *moonquakes* deep within the lunar interior. Even if you stood directly above one of these quakes, you would not feel the vibrations. The average moonquake releases about as much energy as a firecracker, and no large quakes have ever been detected. This barely perceptible seismic activity confirms the idea that the Moon is geologically dead. Nevertheless, researchers can use these weak lunar vibrations to obtain information about the Moon's interior.

Combining all available data with mathematical models of the lunar interior indicates that most of the Moon is of almost uniform density, but chemically differentiated—that is, the chemical properties change from core to surface. As noted in Figure 5.1(b), the models suggest a central core about 330 km in radius, surrounded by a 400-km-thick inner mantle of semisolid rock having properties similar to those of Earth's upper mantle. Above these regions lies a 900-km-thick outer mantle of solid rock, topped by a 60- to 150-km-thick crust.

Data from the gravity experiment aboard the *Lunar Prospector* probe (Section 5.3), combined with magnetic measurements made as the Moon passed through Earth's

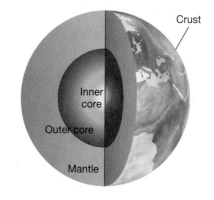

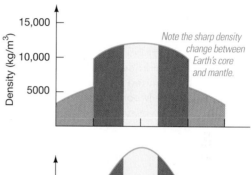

Note the sharp density change between Earth's core and mantle.

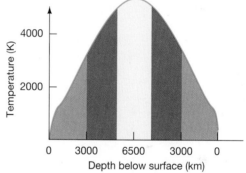

▲ **FIGURE 5.10 Earth's Interior**   Computer models of Earth's interior imply that the density and temperature vary greatly through the mantle and the core.

magnetic "tail" (see Figure 5.22), imply that the lunar core is denser and more iron-rich than the rest of the Moon. Theoretical models predict a central temperature as low as 1500 K, too cool to melt rock or iron, but seismic measurements suggest that parts of the core may be molten, indicating a higher temperature, perhaps due to heating by radioactive elements. A reanalysis of the *Apollo* seismic data in 2011 broadly confirmed all these findings. As best we can tell, the Moon has a solid, mainly iron inner core roughly 240 km in radius. The rest of the core, as well as the innermost 150 km of the surrounding inner mantle, is liquid. However, our knowledge of the Moon's deep interior is still quite limited. Scientists hope that the *GRAIL* mission (see Section 5.3) will reveal more detail on the Moon's internal structure.

The crust on the Moon's far side is considerably thicker (150 km) than the crust on the side nearer Earth (60 km). The reason is most likely related to Earth's gravitational pull. Just as heavier material tries to sink to the center of Earth, the denser far-side lunar mantle tended to sink below the lighter far-side crust in the presence of Earth's gravitational field. In other words, while the Moon was cooling and solidifying, the far-side lunar mantle was pulled a little closer to Earth than the far-side crust. In this way, the crust and mantle became slightly off-center with respect to one another. The result was the thicker crust on the Moon's far side.

CONCEPT CHECK ▶ How would our knowledge of Earth's interior be changed if our planet were geologically dead, like the Moon?

5.5 Surface Activity on Earth

Earth is geologically alive today. Its interior seethes, and its surface constantly changes. Many clear indicators of geological activity, in the form of earthquakes and volcanic eruptions, are scattered across our globe. Erosion by wind and water has obliterated much of the evidence from ancient times, but the sites of more recent activity are well documented.

Continental Drift

Figure 5.11 is a map of the currently active areas of our planet. Nearly all these sites have experienced surface activity during this century. The intriguing aspect

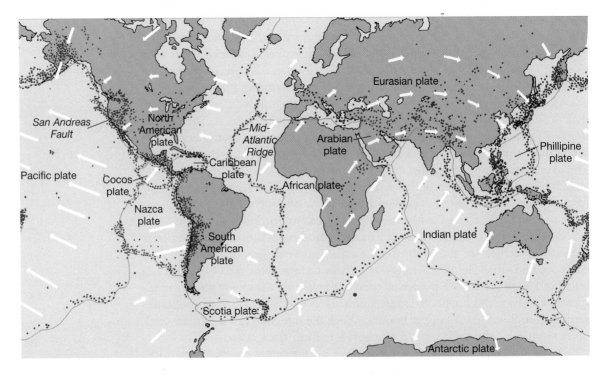

▶ **FIGURE 5.11 Global Plates** The red dots represent active sites where major volcanoes or earthquakes occurred in the 20th century. Taken together, the sites outline vast "plates," indicated in dark blue, that drift around on the surface of our planet. The white arrows show the general directions and speeds of the plate motions.

of the figure is that the active sites are not spread evenly across our planet. Instead, they trace well-defined lines of activity, where crustal rocks shift (resulting in earthquakes) or mantle material upwells (in volcanoes).

In the mid-1960s, scientists realized that these lines are the outlines of gigantic *plates,* or slabs of Earth's surface and that these plates are slowly drifting around the surface of our planet. These plate motions, popularly known as *continental drift,* have created the mountains, oceanic trenches, and many other large-scale features across the face of planet Earth. The technical term for the study of plate movement and the reasons for it is **plate tectonics.** Some plates are made mostly of continental landmasses, some are made of a continent plus a large part of an ocean floor, and some contain no continental land at all and are made solely of ocean floor. For the most part, the continents are just passengers riding atop much larger plates.

The plates move at an extremely slow rate. Typical speeds amount to only a few centimeters per year—about the same speed as your fingernails grow. Nevertheless, over the course of Earth history, the plates have had plenty of time to move large distances. For example, a drift rate of 2 cm per year can cause two continents (for example, Europe and North America) to separate by 4000 km—the width of the Atlantic Ocean—in 200 million years. That's a long time by human standards, but it represents only about 5 percent of Earth's age.

As the plates drift around, we might expect collisions to be routine. Indeed, plates do collide, but they are driven by enormous forces and do not stop easily. Instead, they just keep crunching into one another. Figure 5.12 shows the result of a collision between ancient continents in what is now western America. As Earth's rocky crust crumbles and folds, mountains are formed—in this case, the Rocky Mountains that range from New Mexico to British Columbia, where the surface has been lifted nearly 5 km by tectonic forces.

Not all plates collide head-on. As indicated by the arrows on Figure 5.11, sometimes plates slide or shear past one another. A good example is the most famous active region in North America—the San Andreas Fault in California (Figure 5.13), which forms part of the boundary between the Pacific and North American plates. Along the fault, these two plates are not moving in quite the same direction, nor at quite the same speed. Like parts in a poorly oiled machine, their motion is neither steady nor smooth. Instead, they tend to stick, then suddenly lurch forward as surface rock gives way. The resulting violent, jerky movements have been the cause of many major earthquakes along the fault line.

At still other locations, such as the seafloor under the Atlantic Ocean, the plates are moving apart. As they recede, new mantle material wells up between them, forming midocean ridges. Today, hot mantle material is rising through a crack all along the Mid-Atlantic Ridge, which extends, like a seam on an enormous baseball, all the way from the North Atlantic to the southern tip of South America. Radioactive dating indicates that material has been upwelling along the ridge more or less steadily for the past 200 million years. The Atlantic seafloor is slowly growing as the North and South American plates move away from the Eurasian and African plates.

What Drives the Plates?

What is responsible for the enormous forces that drag plates apart in some locations and ram them together in others? The answer (Figure 5.14) is *convection*—the same process we encountered earlier in our study of Earth's atmosphere (Section 5.3). Each plate is made up of crust plus a small portion of upper mantle. Below the plates, at a depth of perhaps 50 km, the temperature is sufficiently high that the mantle at that depth is soft enough to flow, very slowly, although it is not molten.

This is a perfect setting for convection—warm matter underlying cool matter. The warm mantle rock rises, just as hot air rises in our atmosphere, and large

▲ **FIGURE 5.12 Rocky Mountains** Mountain building results mostly from plate collisions. Here, the folding and jostling of rock is visible high in the northern Rockies, a mountain range upthrust about 70 million years ago. For scale, note at bottom right the evergreen trees, which are about 10 m tall. *(M. Chaisson)*

▲ **FIGURE 5.13 Californian Fault** A small portion of the San Andreas fault in California. The fault is the result of the North American and Pacific plates sliding past one another. The Pacific plate, which includes a large slice of the California coast, is drifting to the northwest relative to the North American plate. *(D. Parker/Science Photo Library)*

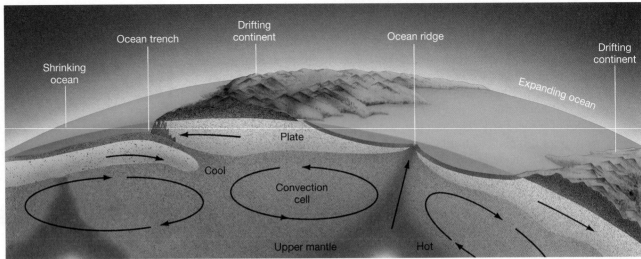

INTERACTIVE ▲ FIGURE 5.14 **Plate Drift** The motion of Earth's plates is probably caused by convection—in this case, giant circulation patterns in the upper mantle that drag the plates across the surface.

circulation patterns become established. Riding atop these convection patterns are the plates. The circulation is extraordinarily sluggish. Semisolid rock takes millions of years to complete one convection cycle. Although the details are far from certain and remain controversial, many researchers suspect that it is large-scale convection patterns near plate boundaries that cause the plates to move.

Figure 5.15 illustrates how all the continents nearly fit together like pieces of a puzzle. Geologists think that sometime in the past a single gargantuan landmass dominated our planet. This ancestral supercontinent, known as *Pangaea* (meaning "all lands"), is shown in Figure 5.15(a). The rest of the planet was covered with water. The present locations of the continents, along with measurements of their current drift rates, suggest that Pangaea was the major land feature on Earth approximately 200 million years ago. Dinosaurs, which were then the dominant form of life, could have sauntered from Russia to Texas via Boston without getting their feet wet. The other frames in Figure 5.15 show how Pangaea split apart, its separate pieces drifting across Earth's surface, eventually becoming the familiar continents we know today. Very probably there has been a long series of "Pangaeas" stretching back in time over much of Earth's history, as tectonic forces have continually formed, destroyed, and re-formed our planet's landmasses. Probably there will be many more.

PROCESS OF ▶
SCIENCE CHECK Go to a library or the Web and research the work of Alfred Wegener on the theory of continental drift—the forerunner of modern plate tectonics. What seemingly unrelated facts did the theory connect and explain? Why did the theory encounter such resistance in the scientific community?

CONCEPT ▶
CHECK Describe the causes and some consequences of plate tectonics on Earth.

Plate Tectonics on the Moon

There is no evidence for plate tectonics on the Moon today—no obvious extensive fault lines, no significant seismic activity, no ongoing mountain building. Plate

▼ FIGURE 5.15 **Pangaea** Given the current estimated drift rates and directions of the plates, we can trace their movements back into the past. About 200 million years ago, they would have been at the approximate positions shown in (a). The continents' current positions are shown in (d).

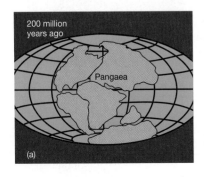

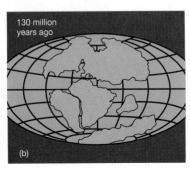

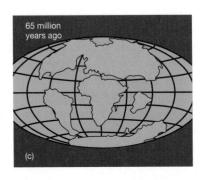

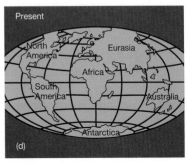

tectonics requires both a relatively thin outer rocky layer, which is easily fractured into continent-sized pieces, and a soft, convective region under it, to make the pieces move. On the Moon neither of these ingredients exists. The Moon's thick crust and solid upper mantle make it impossible for pieces of the surface to move relative to one another. There simply isn't enough energy left in the lunar interior for plate tectonics to work.

5.6 The Surface of the Moon

On Earth the combined actions of air, water, and geological activity erode our planet's surface and reshape its appearance almost daily. As a result, most of the ancient history of our planet's surface is lost to us. The Moon, though, has no air, no water, and no ongoing volcanic or other geological activity (although, as we will see, there is strong evidence that the young Moon underwent a period of intense geological activity before settling into its current inactive state). Consequently, features dating back almost to the formation of the Moon are still visible today. For this reason, studies of the lunar surface are of great importance to Earth geologists and have played a major role in shaping theories of the early development of our planet.

Large-Scale Features

The first observers to point their telescopes at the Moon noted large, roughly circular, dark areas that resemble (they thought) Earth's oceans. They called these regions **maria,** a Latin word meaning "seas" (singular: *mare*). The largest of them (Mare Imbrium, the Sea of Showers) is about 1100 km in diameter. Today we know that the maria are actually extensive flat plains that resulted from the spread of lava during an earlier, volcanic period of lunar evolution (Section 5.8). In a sense, then, the maria *are* oceans—ancient seas of molten lava, now solidified.

Early observers also saw light-colored areas that resembled Earth's continents. Originally dubbed *terrae,* from the Latin word for "land," these regions are now known to be elevated several kilometers above the maria. Accordingly, they are usually called the lunar **highlands.** Both types of region are visible in Figure 5.16, a photographic mosaic of the full Moon. These light and dark surface features are also evident to the naked eye, creating the face of the familiar "man-in-the-Moon."

We have already seen numerous examples of how science progresses hand in hand with technology, and studies of the Moon are no exception. However, the Moon (for now, at least) is unique in astronomy in that manned and unmanned space probes have actually explored parts of its surface and returned samples to Earth. Based on analyses of lunar rock brought back to Earth by *Apollo* astronauts and robot Soviet landers, geologists have found important differences in both *composition* and *age* between the highlands and the maria.

The highlands are made largely of rocks rich in aluminum, making them lighter in color and lower in density (2900 kg/m^3). The maria's basaltic matter contains more iron, giving it a darker color and greater density (3300 kg/m^3). Loosely speaking, the highlands represent the Moon's crust, while the maria are made of mantle material. Maria rock is quite similar to terrestrial basalt, and geologists think that it arose on the Moon much as basalt did on Earth, through the upwelling of molten material through the crust. Radioactive dating indicates ages of more than 4 billion years for highland rocks and from 3.2 to 3.9 billion years for those from the maria.[3]

---

[3]Radioactive dating compares the rates at which different radioactive elements in a sample of rock decay into lighter elements. The age returned by this technique is the time since the rock solidified.

▲ FIGURE 5.16 **Full Moon, Near Side** A photographic mosaic of the full Moon, north pole at the top. Because the Moon emits no visible radiation of its own, we can see it only by the reflected light of the Sun. *(UC/Lick Observatory)*

 ANIMATION/VIDEO One Small Step

 ANIMATION/VIDEO *Full Rotation of the Moon*

 ANIMATION/VIDEO *Ranger Spacecraft Descent to Moon*

 CONCEPT CHECK ▶ Describe three important ways in which the lunar maria differ from the highlands.

▲ **FIGURE 5.17 Full Moon, Far Side** This image of the far side of the Moon, which we can never see from home, is actually a collection of more than 15,000 small photos taken by the *Lunar Reconnaissance Orbiter,* a NASA robot that repeatedly orbits the Moon. Only a few small maria exist on the heavily cratered far side. *(NASA)*

Until spacecraft flew around the Moon, no one on Earth had any idea what the Moon's hidden half looked like. To the surprise of most astronomers, when the far side of the Moon was mapped, first by Soviet and later by U.S. spacecraft, no major maria were found. The lunar far side (Figure 5.17) is composed almost entirely of highlands.

Cratering

Because the smallest lunar features we can distinguish with the naked eye are about 200 km across, we see little more than the maria and highlands when we gaze at the Moon. Through a telescope, however (Figure 5.18), we find that the lunar surface is scarred by numerous bowl-shaped **craters** (after the Greek word for "bowl").

Most craters formed eons ago as the result of meteoritic impact. ⚬⚬ *(Sec. 4.2)* (Compare the ancient craters in Figures 5.18c and 5.20 with the much more recent Barringer Crater in Figure 4.19.) Meteoroids generally strike the Moon at speeds of several kilometers per second. At these speeds even a small piece of matter carries an enormous amount of energy—for example, a 1-kg object hitting the Moon's surface at 10 km/s would release as much energy as the detonation of 10 kg of TNT. As illustrated in Figure 5.19, impact by a meteoroid causes sudden and tremendous pressures to build up on the lunar surface, heating the normally brittle rock and deforming the ground. The ensuing explosion pushes previously flat layers of rock up and out, forming a crater.

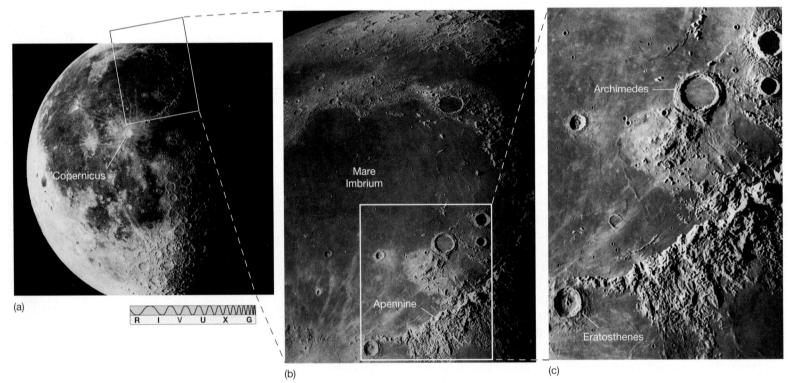

(a)

(b)

(c)

▲ **FIGURE 5.18 Moon, Close-up** (a) The Moon near third quarter makes visible surface features near the *terminator,* where sunlight strikes at a sharp angle and the lit lunar globe changes to dark. (b) Magnified view of a region near the terminator, as seen from Earth through a large telescope. (c) Enlargement of a portion of (b). The smallest craters visible here have diameters of about 2 km, about twice the size of the Barringer Crater shown in Figure 4.19. *(UC/Lick Observatory; Caltech)*

The material thrown out by the explosion surrounds the crater in a layer called an *ejecta blanket,* the ejected debris ranging in size from fine dust to large boulders. The larger pieces of ejected material may themselves create secondary craters. Many of the rock samples brought back by the *Apollo* astronauts show patterns of repeated shattering and melting—direct evidence of the violent shock waves and high temperatures produced in meteoritic impacts.

Lunar craters come in all sizes, reflecting the range in sizes of the impactors that create them. The largest craters are hundreds of kilometers in diameter, the smallest microscopic. Because the Moon has no protective atmosphere, even tiny interplanetary fragments can reach the lunar surface unimpeded. Figure 5.20(a) shows the result of a large meteoritic impact on the Moon, Figure 5.20(b) a crater formed by a micrometeoroid.

Craters are found everywhere on the Moon's surface, but the older highlands are much more heavily cratered than the younger maria. Knowing the ages of the highlands and maria, researchers can estimate the rate of cratering in the past. They conclude that the Moon, and presumably the entire inner solar system, experienced a sudden sharp drop in meteoritic bombardment rate about 3.9 billion years ago. The rate of cratering has been roughly constant since that time.

This time—3.9 billion years in the past—is taken to represent the end of the accretion process through which planetesimals became planets. ∞ *(Sec. 4.3)* The lunar highlands solidified and received most of their craters before that time. The great basins that formed the maria are thought to have been created during the final stages of heavy meteoritic bombardment between about 4.1 and 3.9 billion years ago. The largest impacts were so violent that they cracked the lunar crust (see Figure 5.1b). Molten lava from the mantle welled up through these cracks to fill the large basins, then solidified as it cooled, resulting in the maria we see today.

Lunar Erosion

Meteoritic impact is the only important source of erosion on the Moon. Over billions of years, collisions with meteoroids, large and small, have scarred, cratered, and sculpted the lunar landscape. At the present average rates, one new 10-km-diameter lunar crater is formed every 10 million years, one new 1-m-diameter crater is created about once a month, and 1-cm-diameter craters appear every few minutes. In addition, a steady "rain" of micrometeoroids also eats away at the lunar surface (see Figure 5.21). The accumulated dust from countless impacts (called the lunar *regolith*) covers the lunar surface to an average depth of about 20 m, thinnest on the maria (about 10 m) and thickest on the highlands (more than 100 m in places).

Despite this barrage from space, the Moon's present-day erosion rate is still very low—about 10,000 times less than on Earth. For example, the Barringer Meteor Crater (Figure 4.19) in the Arizona desert, one of the largest meteor craters on Earth, is only 25,000 years old, but it is already decaying. It will probably disappear completely in a mere million years, quite a short time geologically. If a crater that size had formed on the Moon even a billion years ago, it would still be plainly visible today.

INTERACTIVE ▶ FIGURE 5.19 **Meteoroid Impact**   Stages in the formation of a crater by meteoritic impact. (a) The meteoroid strikes the surface, releasing a large amount of energy. The resulting explosion ejects material from the impact site and (c) sends shock waves through the underlying surface. (d) Eventually, a characteristic crater results, surrounded by a field of ejected material.

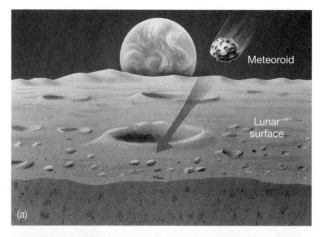

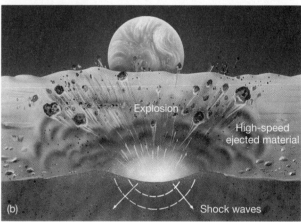

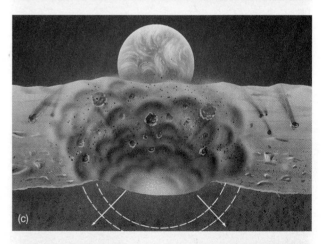

▶ FIGURE 5.20 **Lunar Craters** (a) Two smaller craters called Reinhold and Eddington sit amid the secondary cratering resulting from the impact that created the 90-km-wide Copernicus Crater (near the horizon) about a billion years ago. (b) Craters of all sizes litter the lunar landscape. Some shown here, embedded in glassy beads retrieved by *Apollo* astronauts, measure only 0.01 mm across. (The scale at the top is in millimeters.) The beads themselves were formed during the explosion following a meteoroid impact, when surface rock was melted, ejected, and rapidly cooled. *(NASA)*

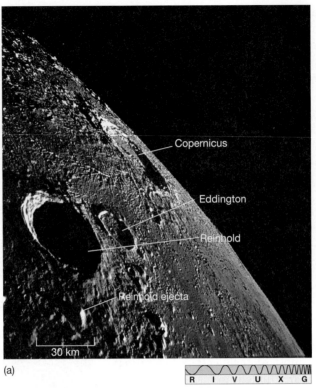

(a)

R I V U X G

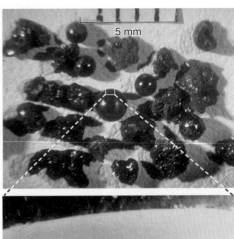

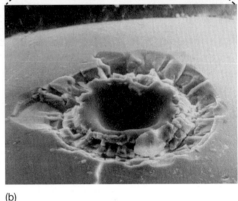

(b)

R I V U X G

▲ FIGURE 5.21 **Lunar Surface** Despite the complete lack of wind and water, the lunar surface is slowly eroded by the constant "rain" of impacting meteoroids, especially micrometeoroids. Note the soft edges of the background hills of this image. The *Apollo* astronaut's boot prints in the foreground lunar dust are only a few centimeters deep, but they will survive for more than a million years. *(NASA)*

5.7 Magnetospheres

Simply put, the magnetosphere is the region around a planet that is influenced by that planet's magnetic field. It forms a buffer zone between the planet and the high-energy particles of the solar wind. It can also provide important insights into the planet's interior structure.

Earth's Magnetosphere

Shown in Figure 5.22, Earth's magnetic field extends far above the atmosphere, completely surrounding our planet. The *magnetic field lines,* which indicate the strength and direction of the field at any point in space, run from south to north, as indicated by the white arrowheads in the figure. The north and south *magnetic poles,* where the axis of an imaginary bar magnet within our planet (Figure 5.23) intersects Earth's surface, are very roughly aligned with Earth's spin axis.

Earth's inner magnetosphere contains two doughnut-shaped zones of high-energy charged particles, one located about 3000 km and the other 20,000 km above Earth's surface. These zones are named the **Van Allen belts**, after the American physicist whose instruments on board some early rocket flights during the late 1950s first detected them. We call them *belts* because they are most pronounced near Earth's equator and because they completely surround the planet. Figure 5.23 shows how these invisible regions envelop Earth except near the North and South Poles.

The particles that make up the Van Allen belts originate in the solar wind. Traveling through space, neutral particles and electromagnetic radiation are unaffected by Earth's magnetism, but electrically charged particles are strongly influenced. As illustrated in the inset to Figure 5.23, a magnetic field exerts a force on a moving charged particle, causing the particle to spiral around the magnetic field lines. In this way, charged particles—mainly electrons and protons—from the solar wind can become trapped by Earth's magnetism. Earth's magnetic field

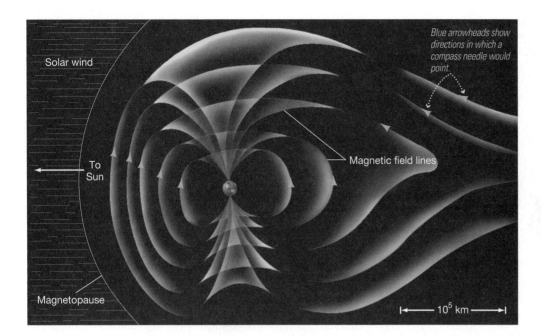

Blue arrowheads show directions in which a compass needle would point.

Solar wind

To Sun

Magnetopause

Magnetic field lines

◄ **FIGURE 5.22 Earth's Magnetosphere** The magnetosphere is the region surrounding a planet wherein particles from the solar wind are trapped by the planet's magnetic field. Far from Earth, the magnetosphere is greatly distorted by the solar wind, with a long "tail" extending from the nighttime side of Earth (here, at right) far into space. The magnetopause is the boundary of the magnetosphere in the sunward direction.

10^5 km

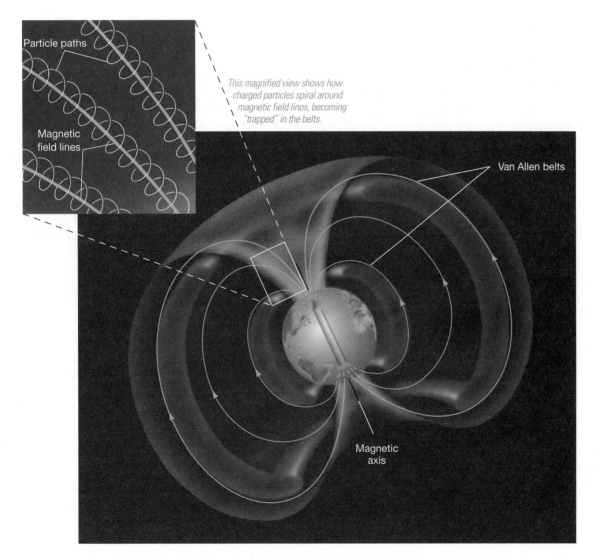

Particle paths

Magnetic field lines

This magnified view shows how charged particles spiral around magnetic field lines, becoming "trapped" in the belts.

Van Allen belts

Magnetic axis

◄ **FIGURE 5.23 Van Allen Belts** Earth's magnetic field resembles somewhat the field of an enormous bar magnet buried inside our planet. High above Earth's atmosphere, the magnetosphere (light blue-green area) contains two doughnut-shaped regions (grayish areas) of magnetically trapped charged particles. These are the Van Allen belts.

▲ FIGURE 5.24 Aurorae   A colorful aurora rapidly flashes across the sky, resembling huge wind-blown curtains glowing in the dark. *(NCAR)*

ANIMATION/VIDEO   Northern and Southern Lights

CONCEPT CHECK ▶   What does the existence of a planetary magnetic field tell us about the planet's interior?

exerts electromagnetic control over these particles, herding them into the Van Allen belts. The outer belt contains mostly electrons; the much heavier protons accumulate in the inner belt.

Particles from the Van Allen belts often escape from the magnetosphere near Earth's north and south magnetic poles, where the field lines intersect the atmosphere. Their collisions with air molecules create a spectacular light show called an **aurora** (Figure 5.24). This colorful display results when atmospheric atoms, excited upon collision with the charged particles, fall back to their ground states and emit visible light. ∞ *(Sec. 2.6)* Aurorae are most brilliant at high latitudes, especially inside the Arctic and Antarctic circles. In the north the spectacle is called the *aurora borealis,* or *Northern Lights.* In the south it is called the *aurora australis,* or *Southern Lights.*

Just as Earth's magnetosphere influences the charged particles of the solar wind, the stream of incoming solar wind particles also affects our magnetosphere. As shown in Figure 5.22, the sunward (daytime) side is squeezed toward Earth's surface, while the opposite side has a long "tail" often extending hundreds of thousands of kilometers into space.

Earth's magnetosphere plays an important role in controlling many of the potentially destructive charged particles that venture near our planet. Without the magnetosphere, Earth's atmosphere—and perhaps the planet surface, too—would be bombarded by harmful particles, possibly damaging many forms of life. Some researchers have even suggested that had the magnetosphere not existed in the first place, life might never have arisen on planet Earth.

Earth's magnetic field is not an intrinsic part of our planet. Instead, it is continuously generated in Earth's core and exists only because the planet is rotating. As in the dynamos that generate electrical power in automobiles and power stations, Earth's magnetism is produced by the spinning, electrically conducting, liquid metal core deep within our planet. Indeed, the theory describing the creation of planetary magnetic fields is called **dynamo theory.** Both rapid rotation and a conducting liquid core are needed for this mechanism to work. As we will see, this link between magnetism and internal structure is very important to studies of other planets, as we have few other probes of planetary interiors.

Lunar Magnetism

No Earth-based observation or spacecraft measurement has ever detected any lunar magnetic field. Based on our current understanding of how Earth's magnetic field is created, this is not surprising. As we have just seen, researchers think that planetary magnetism requires a rapidly rotating liquid metal core. Because the Moon rotates slowly and because its core is probably neither molten nor particularly rich in metals, the absence of a lunar magnetic field is exactly what we expect.

5.8 History of the Earth–Moon System

Given all the data, can we construct a reasonably consistent history of Earth and the Moon? The answer seems to be yes. Many specifics are still debated, but a consensus now exists.

Formation of the Moon

Sometime around 4.6 billion years ago Earth formed by accretion in the solar nebula. ∞ *(Sec. 4.3)* The formation of the Moon is somewhat less certain. The basic problem is that Earth and the Moon are too *dissimilar* in both density and composition for them simply to have formed together out of the same preplanetary matter (the *coformation* theory). However, there are enough *similarities,* particularly between their mantles, to make it unlikely that Earth and the Moon formed entirely independently and subsequently became bound to one another, presumably after a close encounter (the *capture* theory). In addition, there are good theoretical arguments against each scenario—the former because it is hard to reconcile it with the theory of solar system formation outlined in Chapter 4, the latter because of the extreme improbability of the event.

Today many astronomers favor a scenario often called the *impact* theory, which supposes a glancing collision between a large, Mars-sized object and a youthful, molten Earth. As we have seen, such collisions were probably quite frequent in the early solar system. ∞ *(Sec. 4.3)* Computer simulations of such a catastrophic event (Figure 5.25) show that most of the bits and pieces of splattered Earth could have recombined into a stable orbit, forming the Moon. If Earth had already formed an iron core by the time the collision occurred, the Moon could indeed have ended up with a composition similar to Earth's mantle. During the collision, any iron core in the impacting object would have been left behind, eventually to become part of Earth's core. Thus, both the Moon's overall similarity to Earth's mantle and its lack of a dense central core are explained.

The quest to understand the origin of the Moon highlights the interplay between theory and observation that characterizes modern science. ∞ *(Sec. 0.5)* Detailed data from generations of unmanned and manned lunar missions have allowed astronomers to discriminate between competing theories of the formation of the Moon, discarding some and modifying others. At the same time, the condensation theory of solar system formation provides a natural context in which the currently favored impact theory can occur. ∞ *(Sec. 4.3)* Without the idea that planets formed by collisions of smaller bodies, such an impact might well have been viewed as so improbable that the theory would never have gained ground.

Lunar Evolution

The approximate age of the oldest rocks discovered in the lunar highlands is 4.4 billion years, so we know that at least part of the lunar crust must already have solidified by that time. The oldest known Earth rocks are of roughly the same age, although samples more than 4 billion years old are rare, suggesting that erosion on Earth has been quite effective at hiding the details of our planet's distant past. At formation, the Moon was already depleted in heavy metals relative to Earth.

Earth was at least partially molten during most of its first billion years of existence. Denser matter sank toward the core while lighter material rose to the surface—Earth became differentiated. The intense meteoritic bombardment that helped melt Earth at early times subsided about 3.9 billion years ago. Radioactive heating in the interior continued even after Earth's surface cooled and solidified, but it, too, diminished with time.

As our planet cooled, it did so from the outside in, because regions closest to the surface could most easily lose their excess heat to space. In this way, the surface developed a crust, and the differentiated interior attained the layered structure now implied by seismic studies. Today, radioactive heating continues throughout Earth, but there is probably not enough of it to melt any part of our planet. The high temperatures in the core are mainly the trapped remnant of a much hotter Earth that existed eons ago.

The lunar interior evolved quite differently from Earth's, in large part because of the Moon's smaller size. Small objects cool more quickly than large ones

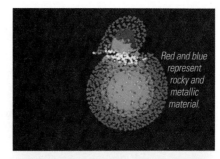

Red and blue represent rocky and metallic material.

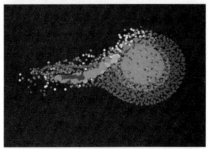

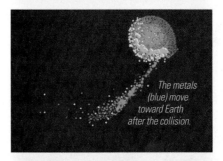

The metals (blue) move toward Earth after the collision.

This is the newly formed Moon.

▶ **FIGURE 5.25 Moon Formation**   This sequence shows a simulated collision between Earth and an object the size of Mars. (The sequence proceeds from top to bottom and zooms out dramatically.) Note how most of the impactor's metallic core becomes part of Earth, leaving the Moon composed primarily of rocky material. *(W. Benz)*

(a) 4 billion years ago   (b) 3 billion years ago   (c) Today

▲ **FIGURE 5.26 Lunar Evolution**   Paintings of the Moon (a) about 4 billion years ago, after much of the meteoritic bombardment had subsided and the surface had somewhat solidified; (b) about 3 billion years ago, after molten lava had made its way up through surface fissures to fill the low-lying impact basins and create the smooth maria; and (c) today, with much of the originally smooth maria now heavily pitted with craters formed at various times within the past 3 billion years. (*U.S. Geological Survey*)

R   I   V   U   X   G

▲ **FIGURE 5.27 Large Lunar Crater**   A large lunar crater, the Orientale Basin. The meteoroid that produced this crater upthrust much surrounding matter, which can be seen as concentric rings of cliffs called the Cordillera Mountains. Notice the smaller, sharper, younger craters that have impacted this ancient basin more recently. (*NASA*)

CONCEPT CHECK ►   What single factor is most responsible for the evolutionary differences between Earth and the Moon?

(basically because heat from the interior has less distance to go to reach the surface), and so the Moon rapidly lost its internal heat to space. During the earliest phases of the Moon's existence—roughly the first half-billion years—the meteoritic bombardment was violent enough to heat, and keep molten, most of the surface layers, perhaps to a depth of 400 km in places. However, the intense heat derived from these collisions probably did not penetrate into the deep lunar interior, because rock is a very poor conductor of heat. As on Earth, radioactivity probably heated the Moon, but (because the heat could escape more easily) not sufficiently to transform it from a warm, semi-solid object to a completely liquid one. The Moon must have differentiated during this period. Its small iron core also formed at this time.

About 3.9 billion years ago, when the heaviest bombardment ceased, the Moon was left with a solid crust dented with numerous large basins (Figure 5.26a). The crust ultimately became the highlands, and the basins soon flooded with lava and became the maria. Between 3.9 and 3.2 billion years ago, lunar volcanism filled the basins with the basaltic material we see today. The age of the youngest maria—3.2 billion years—apparently indicates the time when this volcanic activity finally subsided (Figure 5.26b). The maria are the sites of the last extensive lava flows on the Moon. Their smoothness compared with the more rugged highlands disguises their great age.

Not all these great craters became flooded with lava, however. One of the youngest is the Orientale Basin (Figure 5.27), which formed about 3.9 billion years ago. It did not undergo much subsequent volcanism, so we can recognize it as an impact crater rather than a mare. Similar "unflooded" basins can be seen on the lunar far side.

Because of Earth's gravitational pull, the lunar crust became thicker on the far side than on the near side. Therefore, lava from the interior had a shorter route through the crust to the surface on the Moon's Earth-facing side. As a result, relatively little volcanic activity occurred on the far side, and no large maria were created—the crust was simply too thick to allow that to occur.

As the Moon continued to cool, volcanic activity ended as the thickness of the solid surface layer increased. The crust is now far too thick for volcanism or plate tectonics to occur. With the exception of a few meters of surface erosion from eons of meteoritic bombardment (Figure 5.26c), the lunar landscape has remained more or less structurally frozen for the past 3 billion years. The Moon is dead now, and it has been dead for a long time.

CHAPTER REVIEW

SUMMARY

LO1 The six main regions of Earth are (from inside to outside) a central metallic **core** (p. 136), which is surrounded by a thick rocky **mantle** (p. 136) and topped with a thin **crust** (p. 136). Above the surface is the **atmosphere** (p. 136), composed primarily of nitrogen and oxygen. Surface winds and weather in the lower atmosphere, or **troposphere** (p. 141) are caused by **convection** (p. 141), whereby heat moves from one place to another by rising or sinking streams of air. Higher still lies the **magnetosphere** (p. 136), where charged particles from the Sun are trapped by Earth's magnetic field. The Moon has a partially differentiated interior consisting of a central core, a mantle, and a thick crust. It has no atmosphere, because its gravity is too weak to retain one, and no magnetosphere.

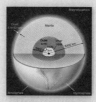

LO2 The daily **tides** (p. 137) in Earth's oceans are caused by the gravitational effects of the Moon and the Sun, which raise **tidal bulges** (p. 137) in the oceans. Their size depends on the orientations of the Sun and the Moon relative to Earth. A differential gravitational force is called a **tidal force** (p. 137), even when no oceans or even planets are involved. The tidal interaction between Earth and the Moon is causing Earth's spin to slow and is responsible for the Moon's **synchronous orbit** (p. 138), in which the same side of the Moon always faces our planet.

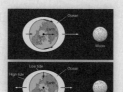

LO3 At high altitudes, in the **ionosphere** (p. 141), the atmosphere is kept ionized by the absorption of high-energy radiation and particles from the Sun. Between the ionosphere and the troposphere lies the **ozone layer** (p. 141), where incoming solar ultraviolet radiation is absorbed. Both these layers help protect us from dangerous radiation from space. The **greenhouse effect** (p. 143) is the absorption and trapping of infrared radiation emitted by Earth's surface by atmospheric gases (primarily carbon dioxide and water vapor). It makes our planet's surface some 40 K warmer than would otherwise be the case.

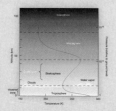

LO4 We study Earth's interior by observing how **seismic waves** (p. 145) produced by **earthquakes** (p. 145) travel through the mantle. Earth's iron core consists of a solid **inner core** (p. 146) surrounded by a liquid **outer core** (p. 146). The process by which dense material sinks

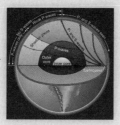

to the center of a planet while lighter material rises to the surface is called **differentiation** (p. 147). The differentiation of Earth implies that our planet must have been at least partially molten in the past, because of bombardment by material from interplanetary space and heat released by **radioactivity** (p. 147) in Earth's interior.

LO5 Earth's surface is made up of enormous slabs, or plates. The slow movement of these plates across the surface is known as continental drift, or **plate tectonics** (p. 149). Earthquakes, volcanism, and mountain building are associated with plate boundaries, where plates may collide, move apart, or rub against one another. The "fit" of the continents to one another and the ages of rocks near oceanic ridges argue in favor of this theory. The motion of the plates is thought to be driven by convection in Earth's mantle. On the Moon, the crust is too thick and the mantle too cool for plate tectonics to occur.

LO6 The main surface features on the Moon are the dark **maria** (p. 151) and the lighter-colored **highlands** (p. 151). **Craters** (p. 152) of all sizes, caused by impacting meteoroids, are found everywhere on the lunar surface. The highlands are older than the maria and are much more heavily cratered. Meteoritic impacts are the main source of erosion on the lunar surface. There is no volcanic activity on the Moon because all volcanism was stifled by the Moon's cooling mantle shortly after extensive lava flows formed the maria more than 3 billion years ago. Astronomers can use the amount of cratering to deduce the ages of surfaces on the Moon and elsewhere in the solar system.

LO7 Charged particles from the solar wind are trapped by Earth's magnetic field lines to form the **Van Allen belts** (p. 154). When particles from the Van Allen belts hit Earth's atmosphere, they heat and ionize the atoms there, causing the atoms to glow in an **aurora** (p. 156). Planetary magnetic fields are produced by the motion of rapidly rotating, electrically conducting fluid (such as molten iron) in a planet's core. The Moon rotates slowly and lacks a conducting liquid core, which accounts for the absence of a lunar magnetic field.

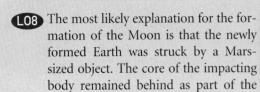 

LO8 The most likely explanation for the formation of the Moon is that the newly formed Earth was struck by a Mars-sized object. The core of the impacting body remained behind as part of the core of our planet, and debris splattered into space formed the Moon. Lunar volcanism was stifled by the Moon's cooling mantle shortly after the extensive lava flows formed the maria more than 3 billion years ago.

MasteringAstronomy® For instructor-assigned homework go to www.masteringastronomy.com

Problems labeled **POS** explore the process of science. **VIS** problems focus on reading and interpreting visual information. **LO** connects to the introduction's numbered Learning Outcomes.

REVIEW AND DISCUSSION

1. **LO2** Explain how the Moon produces tides in Earth's oceans.

2. What does it mean to say the Moon is in a synchronous orbit around Earth? How did the Moon come to be in such an orbit?

3. What is convection? What effect does it have on (a) Earth's atmosphere and (b) Earth's interior?

4. In contrast to Earth, the Moon undergoes extremes in temperature. Why?

5. Use the concept of escape speed to explain why the Moon has no atmosphere.

6. **LO3** Is the greenhouse effect operating in Earth's atmosphere helpful or harmful? Give examples. What are the consequences of an enhanced greenhouse effect?

7. **LO1** The density of water in Earth's hydrosphere and the density of rocks in the crust are both lower than the average density of the planet as a whole. What does this fact tell us about Earth's interior?

8. **LO4 POS** Give two reasons why geologists think that part of Earth's core is liquid.

9. **POS** What clue does Earth's differentiation provide to our planet's history?

10. **LO5** What process is responsible for the surface mountains, oceanic trenches, and other large-scale features on Earth's surface?

11. In what sense were the lunar maria once "seas"?

12. What is the primary source of erosion on the Moon? Why is the average rate of lunar erosion so much less than on Earth?

13. **LO6 POS** Name two pieces of evidence indicating that the lunar highlands are older than the maria.

14. **LO7** Give a brief description of Earth's magnetosphere. Why does the Moon have no magnetosphere?

15. **LO8 POS** Describe the theory of the Moon's origin currently favored by many astronomers.

CONCEPTUAL SELF-TEST: TRUE OR FALSE?/MULTIPLE CHOICE

1. Because of tidal forces, the Moon is in a synchronous orbit around Earth. (T/F)

2. Lunar maria are extensive lava-flow regions. (T/F)

3. Except for the layer of air closest to Earth's surface, the ozone layer is the warmest part of the atmosphere. (T/F)

4. Earth's magnetic field is the result of our planet's large, permanently magnetized iron core. (T/F)

5. Motion of the crustal plates is driven by convection in Earth's upper mantle. (T/F)

6. Volcanic activity continues today on the surface of the Moon. (T/F)

7. Like Earth, the Moon has a molten metal core. (T/F)

8. If you were making a scale model of Earth, representing our planet by a 12-inch basketball, the inner core would be about the size of (a) a half-inch ball bearing; (b) a 2-inch golf ball; (c) a 4-inch tangerine; (d) a 7-inch grapefruit.

9. Earth's average density is about the same as that of (a) a glass of water; (b) a heavy iron meteorite; (c) an ice cube; (d) a chunk of black volcanic rock.

10. Sunlight absorbed by Earth's surface is reemitted in the form of (a) microwave, (b) infrared, (c) visible, (d) ultraviolet radiation.

11. The deepest that geologists have drilled into Earth is about the same as (a) the height of the Statue of Liberty; (b) the altitude most commercial jet airplanes fly; (c) the distance between New York and Los Angeles; (d) the distance between the United States and China.

12. If Earth had no Moon, then tides would (a) not occur; (b) occur more often and with more intensity; (c) still occur, but not really be measurable; (d) occur with the same frequency, but would not be as strong.

13. The most likely theory of the formation of the Moon is that it (a) was formed by the gravitational capture of a large asteroid; (b) formed simultaneously with Earth's formation; (c) was created from a

collision scooping out the Pacific Ocean; (d) formed from a collision of Earth with a Mars-sized object.

14. VIS According to Figure 5.5 (Earth's Atmosphere), commercial jet airplanes flying at 10 km are in (a) the troposphere; (b) the stratosphere; (c) the ozone; (d) the mesosphere.

15. VIS The first figure in *Discovery 5-2* shows that CO_2 levels in Earth's atmosphere began to rise rapidly (a) in the middle ages; (b) in 1600; (c) in the mid-19th century; (d) in the late 20th century.

PROBLEMS

The number of squares preceding each problem indicates its approximate level of difficulty.

1. ■ What would Earth's surface gravity and escape speed be if the entire planet had a density equal to that of the crust (3000 kg/m³, say)?

2. ■ The Moon's mass is 1/80 that of Earth, and the lunar radius is 1/4 Earth's radius. Based on these figures, calculate the total weight on the Moon of a 100-kg astronaut with a 50-kg spacesuit and backpack, relative to his or her weight on Earth.

3. ■■ Most of Earth's ice is found in Antarctica, where permanent ice caps cover approximately 0.5 percent of Earth's surface area and are 3 km thick, on average. Earth's oceans cover roughly 71 percent of our planet to an average depth of 3.6 km. Assuming that water and ice have roughly the same density, estimate by how much sea level would rise if global warming were to cause the Antarctic ice caps to melt.

4. ■■ You are standing on Earth's surface during a total eclipse, and both the Moon and the Sun are directly overhead. What fraction is your weight decreased due to their combined tidal gravitational force?

5. ■ Based on the data presented in the text, estimate the fractions of Earth's volume represented by (a) the inner core, (b) the outer core, (c) the mantle, and (d) the crust.

6. ■■ Approximating Earth's atmosphere as a layer of gas 7.5 km thick, with a uniform density of 1.3 kg/m³, calculate the atmosphere's total mass of the atmosphere. Compare your result with Earth's mass.

7. ■■■ As discussed in the text, without the greenhouse effect, Earth's average surface temperature would be about 250 K. With the greenhouse effect, it is some 40 K higher. Use this information and Stefan's law to calculate the fraction of infrared radiation leaving Earth's surface that is absorbed by greenhouse gases in the atmosphere. ∞ *(Sec. 2.4)*

8. ■ Following an earthquake, how long would it take a P-wave moving in a straight line with a speed of 5 km/s to reach the opposite side of Earth?

9. ■■ Using the rate given in the text for the formation of 10-km craters on the Moon, estimate how long it would take to cover the Moon with new craters of this size. How much higher must the cratering rate have been in the past to cover the entire lunar surface with such craters in the 4.6 billion years since the Moon formed?

10. ■ The *Hubble Space Telescope* has a resolution of about 0.05 arc second. What is the smallest object it could see on the surface of the Moon? Give your answer in meters.

ACTIVITIES

Collaborative

1. Go online and read about global warming. How much carbon dioxide is produced each year by human activities? How does this compare with the total amount of carbon dioxide in Earth's atmosphere? Do all—or most—scientists agree that global warming is an inevitable consequence of carbon dioxide production? What political initiatives are currently under way to address the problem? As a group, which if any do you think are likely to succeed?

2. The estimated cost of transporting a gallon of water from Earth to the Moon is about $100,000. By determining how much water each group member uses in a single day, estimate the cost of taking a single day's supply of water for your group to the Moon.

Individual

1. Go to a sporting goods store and get a tide table; many stores near the ocean provide them free. Choose a month, and plot the height of one high and one low tide versus the day of the month. Now mark the dates when the primary phases of the Moon occur. How well does the phase of the Moon predict the tides?

2. Observe the Moon during an entire cycle of phases. When does the Moon rise, set, and appear highest in the sky at each major phase? What is the interval of time between each phase?

3. If you have binoculars, turn them on the Moon when it appears at twilight and when it appears high in the sky. Draw pictures of what you see. What differences do you notice in your two drawings? What color is the Moon when seen near the horizon? What color is the Moon when seen high in the sky? Why is there a difference?

4. Watch the Moon over a period of hours on a night when you can see one or more bright stars near it. Estimate how many Moon diameters it moves per hour, relative to the stars. Knowing the Moon is about 0.5° in diameter, how many degrees per hour does it move? Based on this, what is your estimate of the Moon's orbital period?

6

The Terrestrial Planets

A Study in Contrasts

With Earth and the Moon as our guides, we now expand our field of view to study the other terrestrial planets. As we explore these worlds and seek to understand the similarities and differences among them, we begin our comparative study of the only planetary system we know. Mercury in many ways is kin to Earth's Moon, and much can be learned by comparing Mercury with our own satellite. Venus and Mars both have properties more like Earth's, and we learn about these two terrestrial worlds by drawing parallels with our planet. It is possible that Venus, Earth, and Mars had many similarities when they formed, yet Earth today is vibrant, teeming with life, while Venus is an uninhabitable inferno and Mars is a dry, dead world. What were the factors leading to these present conditions? In answering this question, we will discover that a planet's environment, as well as its composition, can play a critical role in determining its future.

MasteringAstronomy®   Visit www.masteringastronomy.com for quizzes, animations, videos, interactive figures, and self-guided tutorials.

LEARNING OUTCOMES

Studying this chapter will enable you to:

LO1 Explain how Mercury's rotation has been influenced by its orbit around the Sun.

LO2 Describe how the atmospheres of Venus and Mars differ from one another and from Earth's.

LO3 Compare the surface of Mercury with that of the Moon, and describe how Mercury's surface features formed.

LO4 Compare the surfaces of Venus and Mars with that of Earth.

LO5 Explain why many scientists think that Mars once had running water and a thick atmosphere.

LO6 Identify the major similarities and differences in the internal structures and geological histories of the four terrestrial planets.

LO7 List the basic factors influencing atmospheric evolution, and explain why the atmospheres of Venus, Mars, and Earth are now so different from one another.

◄ The search for life on Mars continues unabated. "Follow the water" is a good guide when prospecting for life—where there's water, there may well be life. Although Mars today seems as dry as any desert on Earth, there is growing evidence for a wetter Mars billions of years ago when the Martian climate was perhaps warmer. This true-color mosaic of many images was made in 2008 by the *Phoenix* spacecraft (bottom left), which landed in Mars' arctic region. Amid the many scattered rocks, which hold "memories" of the ancient events that formed them, this view shows ground patterns similar to those in permafrost areas on Earth. Subsurface ice was later detected within the trenches shown that were dug by the robot's arm, which scooped up soil and tested it in an onboard mini-chemistry lab. *(NASA)*

6.1 Orbital and Physical Properties

Mercury, the innermost planet, lies close to the Sun and is visible above the horizon for at most 2 hours before the Sun rises or after it sets. Orbiting somewhat farther from the Sun (but still within Earth's orbit), the next planet, Venus, is visible for a little longer—up to 3 hours, depending on the time of year (Figure 6.1). The interior orbits of these planets ensure that we never see them far from the Sun in the sky. Contrast this with the exterior orbit of the planet Mars. From our earthly viewpoint, Mars appears to traverse the entire sky, always keeping close to the ecliptic (and occasionally executing retrograde loops, as we saw in Chapter 1). ∞ *(Sec. 1.1)*

Venus is the third brightest object in the entire sky (only the Sun and the Moon are brighter). Like all the planets, it shines by reflected sunlight. It is so bright because almost all the sunlight reaching it is reflected from thick clouds that envelop the planet. You can even see Venus in the daytime if you know where to look. The much fainter Mercury is visible to the naked eye only when the Sun's light is blotted out—just before dawn, just after sunset, and during a total solar eclipse. Orange-red Mars is also quite easy to spot in the night sky. Because of its less reflective surface, smaller size, and greater distance from the Sun, Mars does not appear as bright as Venus, as seen from Earth. However, at its brightest—at closest approach, as marked on Figure 6.1(a)—Mars is still brighter than any star.

Table 6.1 expands Table 5.1 to include Mercury, Venus, and Mars and adds two additional properties: surface temperature and surface atmospheric pressure. In seeking to understand the other terrestrial planets, astronomers are guided by our much more detailed knowledge of Earth and the Moon. Mercury's high average density, for instance, tells us that this planet must have a large iron core, consistent with the condensation theory's account of its formation. ∞ *(Sec. 4.3)* However, in many other respects Mercury is similar to Earth's Moon, leading us to use the Moon as a model for understanding Mercury's past. Venus and, to a lesser extent, Mars are more similar to Earth, so Earth provides the natural starting point for studies of those planets. For example, despite the lack of seismic data, we nevertheless assume that Venus has a metallic core and rocky mantle similar to those shown for Earth in Figure 5.1. Even the widely different atmospheres of these two worlds can be explained in familiar earthly terms.

6.2 Rotation Rates

In principle, astronomers can determine the rate at which a planet spins simply by observing the motion of some prominent surface feature across the planet's disk. Unfortunately, for Mercury and Venus this turns out not to work well because surface markings are hard or impossible to discern. Astronomers had to develop other techniques to probe the rotations of these bodies.

Mercury's Curious Spin

From Earth, even through a large telescope, we see Mercury only as a slightly pinkish, almost featureless disk. The largest ground-based telescopes can resolve features on the surface of Mercury about as well as we can perceive features on our Moon

(a)

(b)

R   I   V   U   X   G

◀ FIGURE 6.1 **Terrestrial Planet Orbits**   (a) The orbits of Mercury and Venus mean that neither planet is ever found far from the Sun, as seen from Earth. Mercury's greatest angular distance from the Sun is 28°, that of Venus 47°. Mars, on the other hand, lies outside Earth's orbit and can appear anywhere on the ecliptic plane, depending on the two planets' positions in their respective orbits. (b) Mercury, Venus, and Mars in the evening sky, along with Jupiter and the Moon. *(J. Sanford/Science Photo Library)*

TABLE 6.1   Some Properties of the Terrestrial Planets and Earth's Moon

	(kg)	(Earth = 1)	(km)	(Earth = 1)	Average Density (kg/m³)	Surface Gravity (Earth = 1)	Escape Speed (km/s)	Surface Rotation Period (solar days)	Surface Temperature (K)	Atmospheric Pressure (Earth = 1)
Mercury	3.3×10^{23}	0.055	2400	0.38	5400	0.38	4.2	59	100–700	—
Venus	4.9×10^{24}	0.82	6100	0.95	5300	0.91	10	-243^1	730	90
Earth	6.0×10^{24}	1.00	6400	1.00	5500	1.00	11	1.00	290	1.0
Mars	6.4×10^{23}	0.11	3400	0.53	3900	0.38	5.0	1.03	180–270	0.007
Moon	7.3×10^{22}	0.012	1700	0.27	3300	0.17	2.4	27.3	100–400	—

[1]The minus sign indicates retrograde rotation.

with our unaided eyes. Figure 6.2 is one of the few photographs of Mercury taken from Earth that shows any indication of surface features. In the days before close-up images were obtainable from space, astronomers could only speculate about the faint, dark markings this photograph reveals.

In the mid-19th century, an Italian astronomer named Giovanni Schiaparelli attempted to measure Mercury's rotation rate by watching surface features move around the planet. He concluded that Mercury always keeps one side facing the Sun, much as our Moon always presents only one face to Earth. The explanation suggested for this synchronous rotation was the same as for the Moon: The tidal bulge raised in Mercury by the Sun had modified the planet's rotation rate until the bulge always points directly at the Sun. ∞ *(Sec. 5.2)* Although the surface features could not be seen clearly, the combination of Schiaparelli's observations and a plausible physical explanation was enough to convince most astronomers. The belief that Mercury rotated synchronously with its revolution about the Sun (once every 88 Earth days) persisted for almost a century.

In 1965 astronomers making radar observations of Mercury from the Arecibo radio telescope in Puerto Rico discovered that this long-held view was in error. ∞ *(Sec. 3.4)* The technique they used is illustrated in Figure 6.3, which shows a radar signal reflecting from the surface of a hypothetical planet. Because the planet is rotating, a pulse of outgoing radiation of a single frequency is broadened—"smeared out"—by the Doppler effect by an amount that depends on the planet's rotation speed. ∞ *(Sec. 2.7)* The radiation reflected from the side moving toward us returns at a slightly higher frequency than does the radiation reflected from the receding side. (Think of the two hemispheres as separate sources of radiation moving at slightly different velocities, one toward us and one away from us.) By measuring the extent of this broadening, we can determine the rate at which the planet rotates.

In this way, astronomers found that the rotation period of Mercury is not 88 days, as had previously been thought, but just under 59 days—in fact, exactly two-thirds of Mercury's year. This odd state of affairs surely did not occur by chance. In fact, 19th-century astronomers were correct in thinking that Mercury's rotation was governed by the tidal effect of the Sun. However, the combination of the Sun's gravity and Mercury's eccentric orbit has caused the planet's rotation to be more complicated than that of the Moon. Unable to come into a state of precisely synchronous rotation (because Mercury's orbital speed changes significantly from place to place in its orbit), Mercury did the next best thing. It presents the same

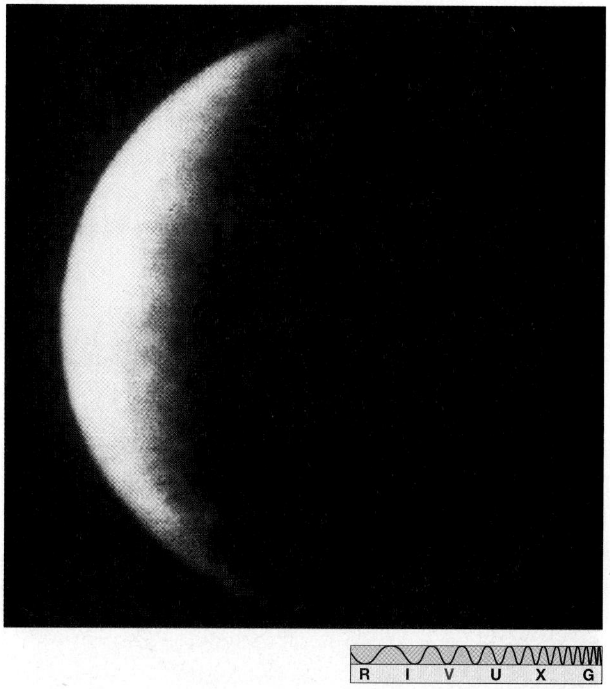

▲ FIGURE 6.2 **Mercury** Photograph of Mercury taken from Earth with a large ground-based optical telescope. Only a few faint surface features are discernible. *(Palomar Observatory/Caltech)*

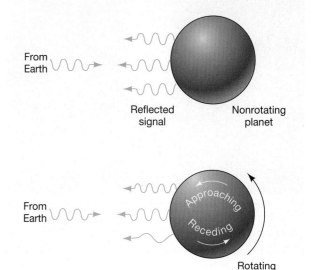

▶ FIGURE 6.3 **Planetary Radar** A radar beam (blue waves) reflected from a rotating planet yields information about both the planet's line-of-sight motion and its rotation rate.

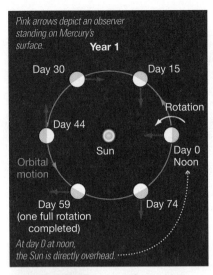
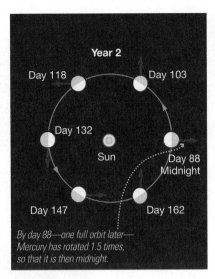
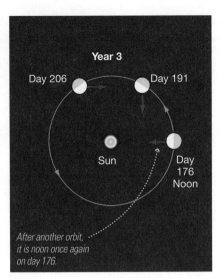

Pink arrows depict an observer standing on Mercury's surface.

Year 1

Day 30   Day 15

Rotation

Day 44

Sun   Day 0 Noon

Orbital motion

Day 59 (one full rotation completed)

Day 74

At day 0 at noon, the Sun is directly overhead.

Year 2

Day 118   Day 103

Day 132

Sun   Day 88 Midnight

Day 147   Day 162

By day 88—one full orbit later—Mercury has rotated 1.5 times, so that it is then midnight.

Year 3

Day 206   Day 191

Sun   Day 176 Noon

After another orbit, it is noon once again on day 176.

INTERACTIVE ▲ **FIGURE 6.4 Mercury's Rotation**
Mercury's orbital and rotational motions combine to produce a solar day that is 2 Mercury years long. **MA**

ANIMATION/VIDEO Transit of Mercury **MA**

CONCEPT CHECK ▶ How has the Sun's gravity influenced Mercury's rotation?

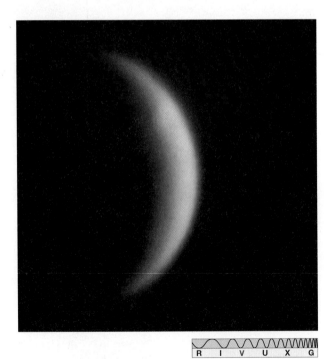

▲ **FIGURE 6.5 Venus** This photograph, taken from Earth, shows Venus with its creamy yellow mask of clouds. No surface detail can be seen because the clouds completely obscure our view of what lies beneath them. *(NOAO)*

face to the Sun not every time around, but every other time. Figure 6.4 illustrates the implications of this odd rotation for a hypothetical inhabitant of Mercury. The planet's solar day—the time from one noon to the next—is 2 Mercury years long!

The Sun also influences the tilt of Mercury's spin axis. Because of the Sun's tides, Mercury's rotation axis is almost exactly perpendicular to its orbital plane. Thus, the noontime Sun is always directly overhead for someone standing on the equator and always on the horizon for someone standing at either pole.

Venus and Mars

The same clouds whose reflectivity make Venus so easy to see in the night sky also make it impossible for us to discern any surface features, at least in visible light. Figure 6.5, one of the best photographs of Venus taken with an Earth-based telescope, shows an almost uniform yellow-white disk, with rare hints of clouds. Because of the cloud cover, astronomers did not know Venus's rotation period until the development of radar techniques in the 1960s, when Doppler broadening of returning radar echoes indicated an unexpectedly sluggish 243-day rotation period. Furthermore, Venus's spin was found to be *retrograde*—that is, opposite that of Earth and most other solar system objects and in the direction opposite Venus's orbital motion. The planet's rotation axis is almost exactly perpendicular to its orbital plane, just as Mercury's is.

We have no "evolutionary" explanation for Venus's anomalous rotation. It is not the result of any known interaction with the Sun, Earth, or any other solar system body. At present, the best explanation astronomers can offer is that during the final stages of Venus's formation in the early solar system, the planet was struck by a large body, much like the one that may have hit Earth and formed the Moon. ∞ *(Secs. 4.3, 5.8)* That impact was sufficient to reduce the planet's spin almost to zero, leaving Venus rotating as we now observe.

In contrast to Mercury and Venus, surface markings are easily seen on Mars (Figure 6.6), allowing astronomers to track the planet's rotation. Mars rotates once on its axis every 24.6 hours—close to 1 Earth day. The planet's equator is inclined to the orbital plane at an angle of 24.0°, very similar to Earth's inclination of 23.5°. Thus, as Mars orbits the Sun, we find both daily and seasonal cycles, just as on Earth. However, the Martian seasons are complicated somewhat by variations in solar heating due to the planet's eccentric orbit. Figure 6.7 summarizes the rotations and orbits of the four terrestrial planets.

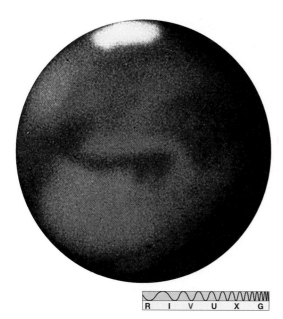

R I V U X G

▲ **FIGURE 6.6 Mars** This deep-red (800-nm) image of Mars was taken at Pic du Midi Observatory, an exceptionally clear site in the French Alps. One of the planet's polar caps appears at the top, and a few other surface markings are visible. *(CNRS and Université Paul Sabatier)*

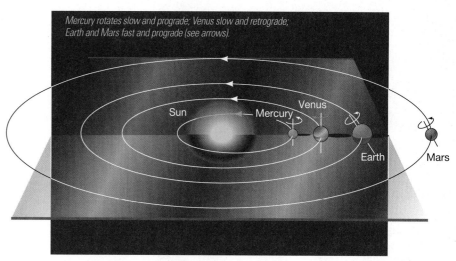

Mercury rotates slow and prograde; Venus slow and retrograde; Earth and Mars fast and prograde (see arrows).

▲ **FIGURE 6.7 Terrestrial Planets' Spin** The inner planets of the solar system—Mercury, Venus, Earth, and Mars—display widely different rotational properties. All orbit the Sun in the same direction and in nearly the same plane, but Venus rotates clockwise as seen from above the plane of the ecliptic, whereas Mercury, Earth, and Mars all spin counterclockwise. This is a perspective view, roughly halfway between a flat edge-on view and a direct overhead view.

6.3 Atmospheres

In this section we outline what is known of the present-day atmospheres of Mercury, Venus, and Mars. We will return to this subject at the end of the chapter, when we discuss how the terrestrial atmospheres evolved into their present states.

ANIMATION/VIDEO Transit of Venus

Mercury

To date, only two spacecraft have visited Mercury. NASA's *Mariner 10* made three flybys during 1974 and 1975. (A "flyby," in NASA parlance, is any space mission in which a probe passes relatively close to a planet—within a few planetary radii, say—but does not go into orbit around it.) Those close encounters were the source of virtually all the detailed information we had on Mercury until NASA's next mission, *Messenger*, made the first of three flybys in 2008, before going into orbit around the planet in 2011.

Mercury has no appreciable atmosphere. *Mariner 10* did find a trace of what was at first thought to be an atmosphere, but the gas is now known to be mostly temporarily trapped hydrogen and helium from the solar wind. Mercury holds this material for just a few weeks before it leaks away again into space. *Messenger* measured the composition of the gas during its first flyby and found that, while it is indeed composed largely of hydrogen and helium like the Sun, it also contains atoms that must have been kicked off the planet's surface by interactions with the solar wind.

The absence of any significant atmosphere on Mercury is readily explained by the planet's high surface temperature (up to 700 K at noon on the equator) and low mass (only 4.5 times the mass of the Moon). Any atmosphere Mercury might once have had escaped long ago. ∞ *(More Precisely 5-1)* With no atmosphere to retain heat, Mercury's surface temperature falls to about 100 K during the planet's long night. Mercury's 600 K temperature range is the largest of any planet or moon in the solar system. Near the poles, where the Sun's light arrives almost parallel to the surface, the temperature remains low at all times. Recent Earth-based radar studies suggest that Mercury's polar temperature could be as low as 125 K and that the poles may be permanently covered with extensive thin sheets of water ice.

(a)

(b)

◄ **FIGURE 6.8 Venus, Up Close** (a) Venus as photographed in the ultraviolet by the *Pioneer* spacecraft's cameras 200,000 km away from the planet. This image was made by capturing solar radiation reflected from the planet's clouds, which are composed mostly of sulfuric acid droplets, much like the corrosive acid in a car battery. (b) Infrared images of Venus's south pole, taken by *Venus Express* a few hours apart, allow us to see the polar vortex deeper into the planet's clouds. *(NASA/ESA)*

Venus

The atmosphere of Venus provides an excellent illustration of how theories are only as good as the data on which they are based. ∞ *(Sec. 0.5)* In the 1930s, scientists used spectroscopy to measure the temperature of Venus's upper atmosphere and found it to be about 240 K, not much different from that of Earth's stratosphere. ∞ *(Sec. 5.3)* Taking into account the cloud cover and the planet's nearness to the Sun, and assuming that Venus had an atmosphere much like our own, researchers concluded that Venus might have an average surface temperature only a few degrees higher than Earth's. In the 1950s, however, when radio observations of the planet penetrated the cloud layer and gave the first indication of conditions near the surface, they revealed a temperature exceeding 600 K! Almost overnight, the popular conception of Venus changed from lush tropical jungle to arid, uninhabitable desert.

Since then, spacecraft data have revealed the full extent of the differences between the atmospheres of Venus and Earth. Venus's atmosphere is much more massive than our own, and it extends to a much greater height above the planet's surface. The surface pressure on Venus is about 90 times the pressure at sea level on Earth, equivalent to an (Earth) underwater depth of about 1 km. (Unprotected humans cannot dive much below 100 m.) The surface temperature is a sizzling 730 K.

The main component (96.5 percent) of Venus's atmosphere is carbon dioxide. Almost all of the remaining 3.5 percent is nitrogen. Given Venus's similarity to Earth in mass, radius, and location in the solar system, it is commonly assumed that Venus and Earth must have started off looking somewhat alike. However, there is no sign of the large amount of water vapor that would be present if a volume of water equivalent to Earth's oceans had once existed on Venus and later evaporated. If Venus started off with Earth-like composition, something happened to its water, for the planet is now an exceedingly dry place. Even the highly reflective clouds are composed not of water vapor, as on Earth, but of sulfuric acid droplets.

Venus's atmospheric patterns are much more evident when examined in the ultraviolet. Some of Venus's upper-level clouds absorb this high-frequency radiation, thereby increasing the contrast. Figure 6.8(a) is an ultraviolet image taken in 1979 by the U.S. *Pioneer Venus* spacecraft from a distance of 200,000 km (compare the optical image shown in Figure 6.5). The large, fast-moving cloud patterns lie between 50 and 70 km above the surface. Upper-level winds reach speeds of 400 km/h relative to the planet. Below the clouds, extending down to an altitude of 30 km, is a layer of haze. Below 30 km, the air is clear.

Figure 6.8(b) is a mosaic of infrared images taken in 2006 by the European *Venus Express* orbiter, whose cameras could partially penetrate the planet's thick haze. It shows a **polar vortex**—a relatively stable, long-lived wind flow circling the planet's south pole. Polar vortices are well known to atmospheric scientists and are expected in any rotating body (planet or moon) with an atmosphere. Earth's south polar vortex plays an important role in confining and concentrating the gases responsible for our planet's Antarctic ozone hole. ∞ *(Sec. 7.2)* By watching how the vortex structure changes in time, scientists hope to understand the forces governing the global circulation of Venus's atmosphere.

CONCEPT CHECK ► Describe some important differences between the atmospheres of Venus and Earth.

Mars

Well before the arrival of spacecraft, astronomers knew from Earth-based spectroscopy that the Martian atmosphere is quite thin and composed primarily of carbon dioxide. Spacecraft measurements revealed that the atmospheric pressure is only about 1/150 the pressure of Earth's atmosphere at sea level. The Martian atmosphere is 95.3 percent carbon dioxide, 2.7 percent nitrogen, and 1.6 percent argon, plus small amounts of oxygen, carbon monoxide, and water vapor. While there is some superficial similarity in composition between the atmospheres of Mars and Venus, the two planets clearly must have had very different atmospheric histories. Average surface temperatures on Mars are about 70 K cooler than on Earth.

6.4 The Surface of Mercury

Figure 6.9 shows a picture of Mercury taken during the *Messenger* flyby in 2008 when the spacecraft was about 30,000 km from the planet. Figure 6.10 shows a higher-resolution photograph of the planet that demonstrates striking similarities to our Moon. Indeed, much of Mercury's cratered surface bears a strong resemblance to the Moon's highlands. The crater walls are generally not as high as on the Moon, the craters are not as deep, and the ejected material landed closer to the impact site, exactly as we would expect given Mercury's greater surface gravity (which is a little more than twice that of the Moon). Mercury, however, shows no extensive lava flow regions akin to the lunar maria.

As on the Moon, Mercury's craters are the result of meteoritic bombardment. The craters are not so densely packed as their lunar counterparts, however, and there are extensive **intercrater plains.** Following *Mariner 10*'s visit, the leading explanation for Mercury's relative lack of craters was that the older impact craters had been erased by volcanic activity, in much the same way as the Moon's maria filled in craters as they formed. More detailed observations by *Messenger* appear to confirm that conclusion. Still, Mercury's intercrater plains do not look much like mare material. They are much lighter in color and not as flat.

Mercury has at least one type of surface feature not found on the Moon. Figure 6.11 shows a **scarp,** or cliff, that does not appear to be the result of volcanic or other familiar geological activity. The scarp cuts across several craters, indicating that

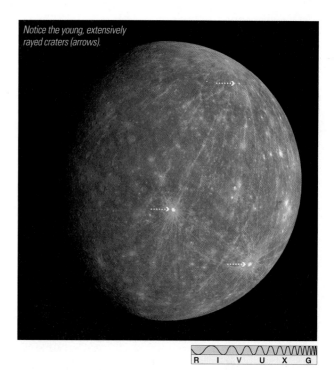

Notice the young, extensively rayed craters (arrows).

R I V U X G

▲ FIGURE 6.9 **Mercury, Up Close** Mercury is imaged here as a mosaic of several visible-light photographs—a composite image constructed from many individual images—taken by the *Messenger* spacecraft in 2008 as it bypassed the planet. (*NASA*)

CONCEPT CHECK ▶ How do scarps on Mercury differ from geological faults on Earth?

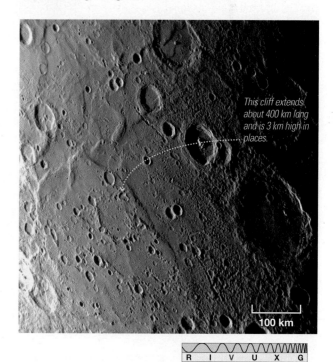

This cliff extends about 400 km long and is 3 km high in places.

100 km

R I V U X G

▲ FIGURE 6.11 **Mercury's Surface** Scarps, or ridges, on Mercury's surface as photographed by *Messenger*. This cliff appears to have formed when the planet's crust cooled and shrank early in its history, causing a crease in the surface. (*NASA*)

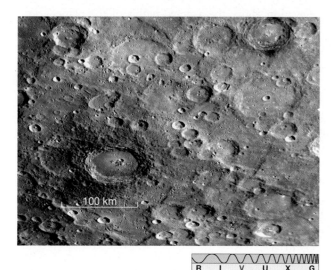

100 km

R I V U X G

▲ FIGURE 6.10 **Mercury, Very Close** Another photograph of Mercury by *Messenger,* this one at higher resolution. The dark material around the crater at lower left is typical of many large craters on Mercury. Its origin is not yet understood. (*NASA*)

◀ **FIGURE 6.12 Caloris Basin** Mercury's most prominent geological feature—the Caloris Basin—measures about 1400 km across and is ringed by concentric mountain ranges that reach more than 3 km high in places. This huge circular basin, shown here in orange in this false-colored visible image from *Messenger*, is similar in size to the Moon's Mare Imbrium and spans more than half of Mercury's radius. *(NASA)*

whatever produced it occurred after most of the meteoritic bombardment was over. Numerous scarps have been found in the *Mariner* and *Messenger* images. Mercury shows no evidence of crustal motions, so the scarps did not form by tectonic processes such as those responsible for fault lines on Earth. Instead, the scarps probably formed when the crust cooled, shrank, and split long ago. Afterwards, one side of the crack moved upward relative to the other side, forming the scarp cliff face. If we can apply to Mercury the cratering age estimates we use for the Moon, the scarps appeared about 4 billion years ago.

Much of the discussion in Chapter 5 about the surface of Earth's Moon applies equally well to Mercury. ∞ *(Sec. 5.6)* Figure 6.12 shows the result of what was probably the last great event in the geological history of Mercury—an immense bull's-eye crater called the Caloris Basin, formed eons ago by the impact of a large asteroid. ∞ *(Sec. 5.8)* Compare this basin with the Orientale Basin on the Moon (Figure 5.27). *Messenger* studies of this basin and the surrounding surface have been instrumental in demonstrating that volcanism was mainly responsible for the intercrater plains.

6.5 The Surface of Venus

Although the clouds of Venus are thick and the surface is totally shrouded, we are by no means ignorant of the planet's surface. Radar astronomers have bombarded Venus with radio signals, and analysis of the radar echoes yields a map of surface features. Except for Figure 6.17, all the views of Venus in this section are "radargraphs" (not photographs) created in this way. Most recently, the U.S. *Magellan* spacecraft has provided very high-resolution radar images of the planet.

Large-Scale Topography

Figure 6.13(a) shows a relatively low-resolution map of Venus made by *Pioneer Venus* in 1979. Surface elevation above the average radius of the planet's surface is indicated by color, with white representing the highest elevations, blue the lowest. (Note that the blue has nothing to do with oceans!) For comparison, Figure 6.13(b) shows a map of Earth to the same scale and at the same spatial resolution. Figure 6.14 is a 1995 mosaic of *Magellan* images of Venus. The orange color is based on optical data returned from spacecraft that landed on the planet.

Venus's surface appears to be mostly smooth, resembling rolling plains with modest highlands and lowlands. Only two continent-sized features, called Ishtar Terra and Aphrodite Terra, adorn the landscape, and these contain mountains comparable in height to those on Earth. The highest peaks rise some 14 km above the level of the deepest surface depressions. (The highest point on Earth, the summit of Mount Everest, lies about 20 km above the deepest section of the ocean floor.) The elevated "continents" occupy only 8 percent of Venus's total surface area whereas continents on Earth make up about 25 percent of our planet's surface.

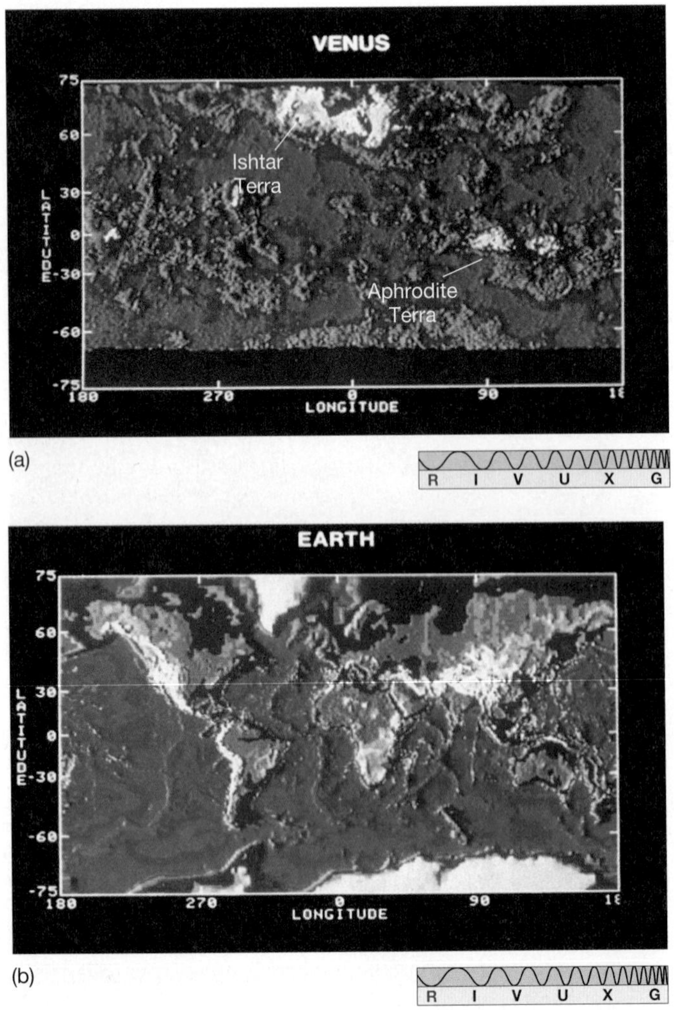

◀ **FIGURE 6.13 Venus Radar Map** (a) Radar map of the surface of Venus, based on *Pioneer Venus* data. Color represents elevation, with white the highest areas and blue the lowest. (b) A similar map of Earth, at the same spatial resolution. *(NASA)*

► FIGURE 6.14 **Venus Magellan Map** A planetwide mosaic of Venus made from *Magellan* images, colored in roughly the same way as Figure 6.13 to represent altitude: blue is lowest, white highest. *(NASA)*

This large, dragon-shaped continent is called Aphrodite Terra.

The larger continent-sized formation, Aphrodite Terra, is located on Venus's equator. It is comparable in size to Africa. Before *Magellan*'s arrival, some researchers had speculated that Aphrodite Terra might have been the site of something equivalent to seafloor spreading on Earth, where two tectonic plates moved apart and molten rock rose to the surface in the gap between them, forming an extended ridge. This is just what is happening today at Earth's Mid-Atlantic Ridge, which is clearly visible in Figure 6.13(b). However, the *Magellan* images seem to rule out any plate-tectonic activity on Venus, and the Aphrodite region shows no signs of spreading. The crust appears buckled and fractured, suggesting large compressive forces, and there seem to have been numerous periods when extensive lava flows occurred.

Volcanism and Cratering

Although erosion by the planet's atmosphere may play some part in obliterating surface features, the most important factor is volcanism (volcanic activity), which appears to resurface the planet every few hundred million years. Many areas of Venus have volcanic features. Figure 6.15(a) shows a *Magellan* image of seven pancake-shaped lava domes, each about 25 km across—a little bigger than Washington, D.C. They probably formed when lava oozed out of the surface, formed the dome, then withdrew, leaving the crust to crack and subside. Lava domes such as these are found in several locations on Venus. Figure 6.15(b) shows a computer-generated three-dimensional view of the domes.

The most common volcanoes on the planet are of the type known as **shield volcanoes.** Those on Earth are associated with lava welling up through a "hot spot" in the crust (like the Hawaiian Islands). They are built up over long periods of time by successive eruptions and lava flows. A characteristic of shield volcanoes is the formation of a *caldera,* or crater, at the summit when the underlying lava withdraws and the surface collapses. A large shield volcano, called Gula Mons, is shown (again as a computer-generated view) in Figure 6.15(c).

The largest volcanic structures on Venus are huge, roughly circular regions known as **coronae.** A large corona, called Aine, can be seen in Figure 6.16. Coronae are unique to Venus. They appear to be the result of upwelling motions in the mantle that caused the surface to bulge outward, but never developed into full-fledged convection as on Earth. Coronae generally have volcanoes both in and

 ANIMATION/VIDEO Topography of Venus

 ANIMATION/VIDEO Flight Over Alpha Regio

▼ FIGURE 6.15 **Venus Surface Features** (a) These dome-shaped structures on Venus resulted when molten rock bulged out of the ground and then retreated, leaving behind a thin, solid crust that later cracked and subsided. (b) A computer-generated three-dimensional representation of four of the lava domes. (c) A *Magellan* view of the large shield volcano known as Gula Mons, whose volcanic caldera at the summit is about 100 km across and 4 km high. *(NASA)*

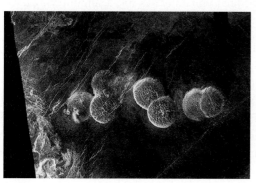

(a)

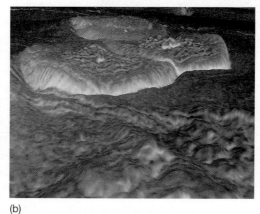

(b)

(c)

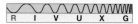

▶ FIGURE 6.16 **Venus Corona** This corona, called Aine, lies in the plains south of Aphrodite Terra and is about 300 km across. Note the pancake-shaped lava domes at top, the many fractures in the crust around the corona, and the large impact craters with their surrounding white (rough) ejecta blankets that stud the region. *(NASA)*

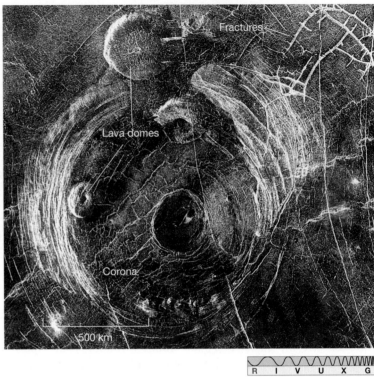

ANIMATION/VIDEO Flight Over Sif Mons Volcano

▼ FIGURE 6.17 **Venus in Situ** One of the first true-color views of the surface of Venus, radioed back to Earth from the Russian *Venera 14* spacecraft, which made a soft landing in 1975. The amount of sunlight penetrating Venus's cloud cover is about the same as that reaching Earth's surface on a heavily overcast day. *(Russian Space Agency)*

around them, and their rims usually show evidence of extensive lava flows into the plains below.

Two pieces of indirect evidence suggest that volcanism on Venus continues today. First, the level of sulfur dioxide above Venus's clouds shows large and fairly frequent fluctuations, possibly as the result of volcanic eruptions on the surface. Second, orbiting spacecraft have observed bursts of radio energy from the planet's surface similar to those produced by the lightning discharges that often occur in the plumes of erupting volcanoes on Earth. However, while quite persuasive, these pieces of evidence are still only circumstantial. No "smoking gun" (or erupting volcano) has yet been seen, so the case for active volcanism is not yet complete.

A few Soviet *Venera* spacecraft have landed on Venus's surface. Each survived for about an hour before being destroyed by the intense heat, its electronic circuitry melting in this planetary oven. Figure 6.17 shows one of the first photographs of the surface of Venus radioed back to Earth. The flat rocks visible in this image show little evidence of erosion and are apparently quite young, supporting the idea of ongoing surface activity. Later Soviet landers performed simple chemical analyses of the surface. Some of the samples studied were found to be predominantly basaltic, again implying a volcanic past. Others resembled terrestrial granite.

Not all the craters on Venus are volcanic in origin. Some were formed by meteoritic impact. The largest impact craters on Venus are generally circular, but those less than about 15 km in diameter can be quite asymmetric. Figure 6.18(b) shows a *Magellan* image of a relatively small impact crater, about 10 km across, in Venus's southern hemisphere. Geologists think that the light-colored region is an ejecta blanket—material thrown from the crater following the impact. Its irregular shape may be the result of a large meteorite that broke up just before impact, with the separate pieces hitting the surface near one another. This

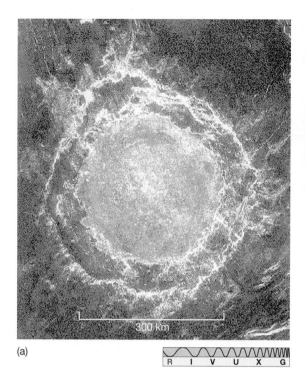

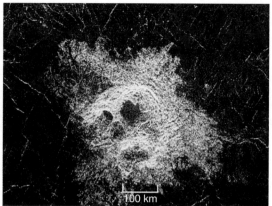

(b)

(a)

R I V U X G

▲ **FIGURE 6.18 Impact Craters on Venus**
(a) Venus's largest crater, named after anthropologist
Margaret Mead, has a double-ringed structure.
(b) A *Magellan* image of a multiple-impact crater
in Venus's southern hemisphere displays irregularly
shaped light-colored ejected debris that was
probably caused by a meteorite that fragmented
just prior to impact; dark regions in the crater may
be pools of solidified lava. *(NASA)*

seems to be a fairly common fate for medium-sized bodies (1 km or so in diameter) that plow through Venus's dense atmosphere.

Figure 6.18(a) shows the largest-known impact feature on Venus, the 280-km-diameter crater Mead. Its double-ringed structure is in many ways similar to the Moon's Mare Orientale. ⟱ *(Sec. 5.8)* Numerous impact craters, again identifiable by their ejecta blankets, can also be seen in Figure 6.16.

6.6 The Surface of Mars

Earth-based observations of Mars at closest approach can distinguish surface features as small as 100 km across—about the same resolution as the unaided human eye can achieve when viewing the Moon. However, when Mars is closest to us and most easily observed, it is also full, so the angle of the Sun's rays does not permit us to see any topographical detail, such as craters or mountains. Even through a large telescope Mars appears only as a reddish disk with some light and dark patches and prominent polar caps (Figure 6.6).

Mars's surface features undergo slow seasonal changes over the course of a Martian year—a consequence of Mars's axial tilt and somewhat eccentric orbit. The polar caps grow or shrink according to the seasons, almost disappearing during the Martian summer. The dark features also vary in size and shape. To fanciful observers around the start of the 20th century, these changes suggested the annual growth of vegetation—and much more (see *Discovery 6-1*)—but, as with Venus, these speculations were not confirmed. The changing polar caps are mostly frozen carbon dioxide, not water ice as at Earth's North and South Poles (although smaller "residual" caps of water ice, which persist even through the warmer summer months, also exist). The dark regions are just highly cratered and eroded areas on the surface. During summer in the Martian southern hemisphere, planetwide dust storms sweep up the dry dust and carry it aloft, sometimes for months at a time, eventually depositing it elsewhere on the planet. Repeated covering and uncovering of the Martian landscape gives the impression from a distance of surface variability, but it is only the thin dust cover that changes.

 SELF-GUIDED TUTORIAL SuperSpaceship—Voyage to Venus

CONCEPT CHECK ▶ Are the volcanoes on Venus mainly associated with the movement of tectonic plates, as on Earth?

 ANIMATION/VIDEO *Hubble* View of Mars

6-1 DISCOVERY

Martian Canals?

The year 1877 was an important one in the human study of the planet Mars. The Red Planet came unusually close to Earth, affording astronomers an especially good view. Of particular note was the discovery, by U.S. Naval Observatory astronomer Asaph Hall, of the two moons circling Mars. But most exciting was the report of the Italian astronomer Giovanni Schiaparelli on his observation of a network of linear markings that he termed *canali*. In Italian, *canali* can simply mean "grooves" or "channels," but it can also mean "canals." As far as we know, Schiaparelli did not intend to imply that the canali were anything other than natural, but the word was translated into English as *canals*, suggesting that the grooves had been constructed by intelligent beings. The world's press (especially in the United States) sensationalized these observations, and some astronomers began drawing elaborate maps of Mars, showing oases and lakes where canals met in desert areas.

Percival Lowell (see photo), a successful Boston businessman (and brother of the poet Amy Lowell and Harvard president Abbott Lawrence Lowell), became fascinated by these reports. He abandoned his business and purchased a clear-sky site at Flagstaff, Arizona, where he built a major observatory. He devoted his life to achieving a better understanding of the Martian "canals." In doing

so, he championed the idea that Mars was drying out and that an intelligent society had constructed the canals to transport water from the wet poles to the arid equatorial deserts.

Alas, the Martian valleys and channels photographed by robot spacecraft during the 1970s are far too small to be the canali that Schiaparelli, Lowell, and others thought they saw on Mars. The entire episode represents a classic case in the history of science—a case in which well-intentioned observers, perhaps obsessed with the notion of life on other worlds, let their personal opinions and prejudices seriously affect their interpretations of reasonable data. The accompanying figures of Mars show how surface features (which were probably genuinely observed by astronomers at the turn of the century) might have been imagined to be connected. The figure on the left is a photograph of how Mars actually looked in a telescope at the end of the 19th century. The sketch at right is an interpretation (done at the height of the canal hoopla) of the pictured view. The human eye, under physiological stress, tends to connect dimly observed yet distinctly separated features. Humans saw patterns and canals where none in fact existed.

The chronicle of the Martian canals illustrates how the scientific method requires scientists to acquire new data to sort out sense from nonsense, fact from fiction. Rather than simply believing the claims about the Martian canals, other scientists demanded further observations to test Lowell's hypothesis. Eventually, improved observations, climaxing in the *Mariner* and *Viking* exploratory missions to the Red Planet nearly a century after all the fuss began, totally disproved the existence of canals. It often takes time, but the scientific method does eventually lead to progress in understanding reality.

(Lowell Observatory)

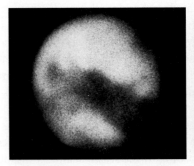
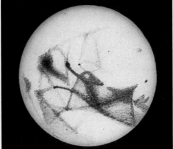

(Lowell Observatory)

Large-Scale Topography

There have been far more space missions to Mars than to any other planet, and many of these missions have played important roles in our understanding of the planet. Figure 6.19 shows a detailed mosaic of Mars, constructed from images made by one of the U.S. *Viking* spacecraft in orbit around the planet.

A striking feature of the terrain of Mars is the marked difference between the northern and southern hemispheres (Figure 6.20). The northern hemisphere is

▶ **FIGURE 6.19 Mars Globe**   This highly detailed mosaic of Mars is based on images from a *Viking* spacecraft in orbit around the planet. Mars's Tharsis region, 5000 km across, bulges out from the equator, rising to a height of about 10 km. The two large volcanoes on the left mark the approximate peak of the Tharsis bulge. Dominating the center of the field of view is a vast "canyon" known as Valles Marineris—the Mariner Valley. *(NASA)*

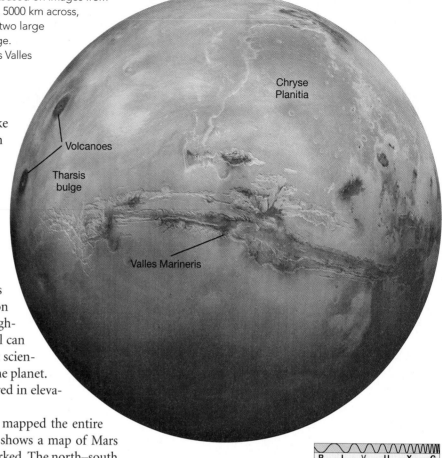

made up largely of rolling volcanic plains, somewhat like the lunar maria but much larger than any plains found on Earth or the Moon. They were apparently formed by eruptions involving enormous volumes of lava. The plains are strewn with blocks of volcanic rock, as well as with boulders blasted out of impact areas by infalling meteoroids. The southern hemisphere consists of heavily cratered highlands lying several kilometers above the level of the lowland north.

The northern plains are cratered much less than the southern highlands, suggesting that the northern surface is younger—perhaps 3 billion years old, compared with 4 billion in the south. In places, the boundary between the southern highlands and the northern plains is quite sharp. The surface level can drop by as much as 4 km in a distance of 100 km or so. Most scientists assume that the southern terrain is the original crust of the planet. How most of the northern hemisphere could have been lowered in elevation and flooded with lava remains a mystery.

In the late 1990s, NASA's *Mars Global Surveyor* satellite mapped the entire Martian surface to an accuracy of a few meters. Figure 6.21 shows a map of Mars based on these data, with some prominent surface features marked. The north–south asymmetry is very obvious in this map. The major geological feature on the planet is the *Tharsis bulge,* visible at the left of Figure 6.19 and marked in Figure 6.21. Roughly the size of North America, Tharsis lies on the Martian equator and rises some 10 km above the rest of the Martian surface. To the east and west of Tharsis lie wide depressions, hundreds of kilometers across and up to 3 km deep. If we wished to extend the idea of "continents" from Earth and Venus to Mars, we would say that Tharsis is the

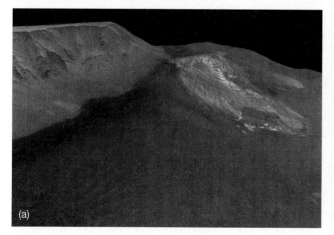
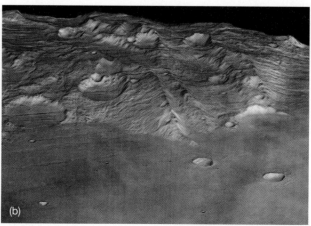

▲ **FIGURE 6.20 Mars, Up Close**   (a) Mars's northern hemisphere consists of rolling volcanic plains. (b) The southern Martian highlands are heavily cratered. Both of these *Mars Express* photographs are in true color and show roughly the same scale, nearly 1000 km across. *(ESA)*

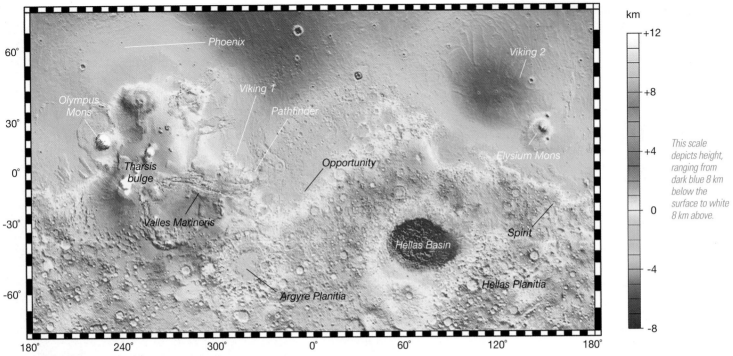

INTERACTIVE ▲ FIGURE 6.21 **Mars Map**   A computer-generated map of planet Mars, based on detailed measurements made by *Mars Global Surveyor*. Color represents height of the surface according to the scale at right; note the great difference in elevation between the northern and southern hemispheres. Some surface features are labeled, such as the Mariner Valley and the Hellas impact basin, as are the *Viking*, **MA** *Pathfinder*, *Exploration Rover*, and *Phoenix* robot landing sites. *(NASA)*

ANIMATION/VIDEO Flight Over Columbia Hills **MA**

ANIMATION/VIDEO Flight Over Tharsis **MA**

ANIMATION/VIDEO Flight Over Mariner Valley **MA**

only continent on the Martian surface. However, as on Venus, there is no sign of plate tectonics—the continent of Tharsis is not drifting like its Earth counterparts. ∞ *(Sec. 5.5)* Tharsis appears to be even less heavily cratered than the northern hemisphere, making it the youngest region on the planet. It is estimated to be only 2–3 billion years old.

Associated with the Tharsis bulge is a great canyon known as Valles Marineris (the Mariner Valley). Shown in its entirety in Figure 6.21 and cutting across the center of Figure 6.19, this enormous crack in the Martian surface is not really a canyon in the terrestrial sense, because running water played no part in its formation. Geologists think it was formed by the same crustal forces that pushed the Tharsis region upward, causing the surface to split and crack. Cratering studies suggest that Valles Marineris is at least 2 billion years old. Similar (but smaller) cracks, formed in a similar way, have been found in the Aphrodite Terra region of Venus.

Valles Marineris runs for almost 4000 km along the Martian equator and extends about one-fifth of the way around the planet. At its widest, it is 120 km across, and it is as deep as 7 km in places. Earth's Grand Canyon in Arizona would easily fit into one of its side "tributary" cracks. It is so large that it can even be seen from Earth. This Martian geological feature is not a result of Martian plate tectonics, however. For some reason, the crustal forces that formed it never developed into full-fledged plate motion as on Earth.

Almost diametrically opposite Tharsis lies the Hellas Basin (Figure 6.21). Located in the southern highlands, it paradoxically contains the lowest point on the Martian surface. Some 3000 km across, the floor of the basin lies nearly 9 km below its rim and over 6 km below the average level of the planet's surface. Its shape and structure identify it as an impact feature. Its formation must have caused a major redistribution of the young Martian crust—maybe even enough to account for a substantial portion of the highlands, according to some researchers. Its heavily cratered floor indicates that the impact occurred very early in Martian history—perhaps 4 billion years ago.

The giant Borealis Basin around the Martian north pole—most of the blue region at the top of Figure 6.21 (see also Figure 6.26)—may be the result of one of the largest known impacts in the solar system. Recent research, based on computer simulations of the collision and detailed data from the *Mars Global Surveyor* and *Mars Reconnaissance Orbiter* spacecraft, suggests that the basin could have formed when a giant impactor some 2000 km across— twice the size of the largest asteroid, Ceres—struck the planet a grazing blow during the formative stages of the solar system. ∞ *(Sec. 4.3)* These ideas are controversial and still hotly debated by planetary scientists, but proponents claim that the resulting impact feature would be comparable in size to the observed basin, and the collision might explain why the northern hemisphere of Mars is so much lower than and differs so radically from the south.

Martian Volcanism

Mars contains the largest known volcanoes in the solar system. Four particularly large ones are found on the Tharsis bulge, two of them visible in Figure 6.19. Biggest of all is Olympus Mons (Figure 6.22), which lies on the northwestern slope of Tharsis, just over the left horizon of Figure 6.19, but marked on Figure 6.21. Olympus Mons is 700 km in diameter at its base—only slightly smaller than the state of Texas—and rises to a height of 25 km above the surrounding plains.

Like the volcanoes on Venus, those on Mars are not associated with plate motion, but instead are shield volcanoes, sitting atop hot spots in the Martian mantle. Spacecraft images of the Martian surface reveal many hundreds of volcanoes. Most of the largest are associated with the Tharsis bulge, but many smaller ones are also found in the northern plains. It is not known whether any of them are still active. However, from the extent of impact cratering on their slopes, it appears that some of them erupted as recently as 100 million years ago.

The great height of Martian volcanoes is a direct consequence of the planet's low surface gravity. ∞ *(Sec. 5.1)* As a shield volcano forms and lava flows and spreads, the new mountain's height depends on its ability to support its own weight. The lower the surface gravity, the less the lava weighs and the higher the mountain can be. Mars has a surface gravity only 40 percent that of Earth, and its volcanoes rise roughly 2.5 times as high.

Evidence for Past Water on Mars

Although the Valles Marineris was not formed by running water, photographic evidence reveals that liquid water once existed in great quantity on the surface of Mars. Two types of flow features were seen by the *Viking* orbiters: **runoff channels** and **outflow channels.**

The runoff channels (Figure 6.23) are found in the southern highlands. They are extensive systems—sometimes hundreds of kilometers in total length—of interconnecting, twisting channels that merge into larger, wider channels. They bear a strong resemblance to river systems on Earth, and geologists think that this is just what they are: the dried-up beds of long-gone rivers that once carried rainfall on

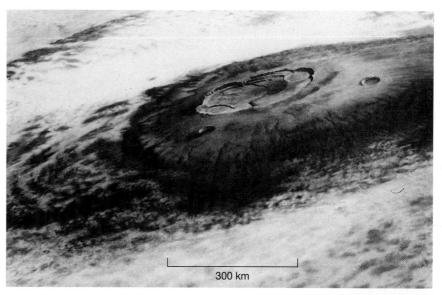

300 km

R I V U X G

▲ **FIGURE 6.22 Martian Volcano**   Olympus Mons, the largest volcano (now dormant) known on Mars or anywhere else in the solar system, is nearly three times taller than Mount Everest on Earth, measuring about 700 km across its base and rising 25 km high at its peak. By comparison, the largest volcano on Earth, Hawaii's Mauna Loa, measures a mere 120 km across and peaks just 9 km above the Pacific Ocean floor. *(NASA)*

 SELF-GUIDED TUTORIAL Comparative Planetology: Mars

PROCESS OF ►
SCIENCE CHECK   Use examples drawn from the study of Mercury, Venus, or Mars to illustrate how improvements in observational data can radically change the interpretation of a scientific theory.

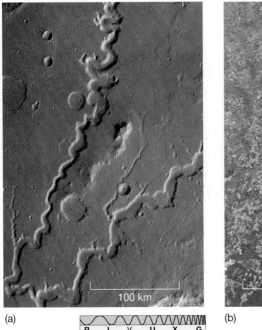

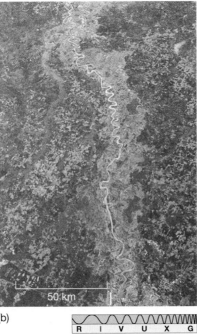

(a)

(b)

R I V U X G

R I V U X G

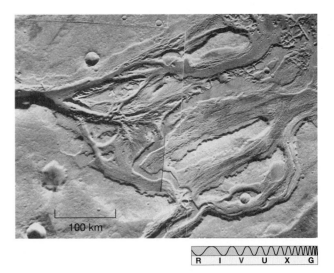

100 km

R I V U X G

▲ **FIGURE 6.24 Martian Outflow** This outflow channel near the Martian equator bears witness to a catastrophic flood that occurred about 3 billion years ago. *(NASA)*

▲ **FIGURE 6.23 Martian Channel** (a) This runoff channel on Mars is about 400 km long and up to 5 km wide in places. (b) The Red River on Earth runs from the Texas Panhandle to the Mississippi River. The two differ mainly in that there is currently no liquid water in this, or any other, Martian valley. *(ESA/NASA)*

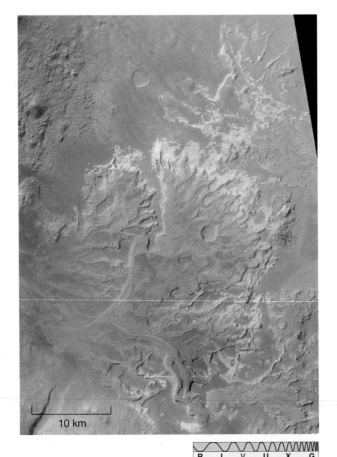

10 km

R I V U X G

Mars from the mountains down into the valleys. The systems of runoff channels are often referred to as *valley networks*. They speak of a time 4 billion years ago (the age of the Martian highlands) when the atmosphere was thicker, the surface warmer, and liquid water widespread.

The outflow channels (Figure 6.24) are probably relics of catastrophic flooding on Mars long ago. They appear only in equatorial regions and generally do not form extensive interconnected networks like the runoff channels. Instead, they are probably the paths taken by huge volumes of water draining from the southern highlands into the northern plains. Judging from the width and depth of the channels, the flow rates must have been truly enormous—perhaps as much as a hundred times greater than the 10^5 tons per second carried by the Amazon River, the largest river system on Earth. Flooding shaped the outflow channels about 3 billion years ago, about the same time as the northern volcanic plains formed.

Scientists speculate that Mars may have enjoyed an extended early period during which rivers, lakes, and perhaps even oceans adorned its surface. Figure 6.25 is a *Mars Global Surveyor* image showing what mission specialists think may be a delta—a fan-shaped network of channels and sediments where a river once flowed into a larger body of water, in this case a lake filling a crater in the southern highlands. Other researchers go even farther, suggesting that the data provide evidence for large open expanses of water on the early Martian surface. Figure 6.26(a) is a computer-generated view of the Martian north polar region, showing the extent of what may have been an ancient ocean covering much of

◀ **FIGURE 6.25 Martian River Delta** Did this fan-shaped region of twisted streams form as a river flowed into a larger sea? If it did, the *Mars Global Surveyor* image supports the idea that Mars once had large bodies of liquid water on its surface. Not all scientists agree with this interpretation, however. *(NASA)*

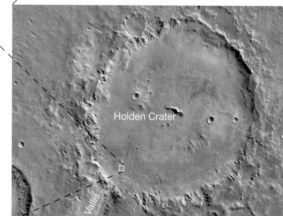

► FIGURE 6.26 **Ancient Ocean?** (a) A possible ancient Martian ocean might once have spanned the polar regions. The blue areas in this computer-generated map indicate depth below the average radius of the planet, thus approximate possible ocean depth. (Color scale is the same as in Figure 6.21.) (b) This high-resolution image shows tentative evidence for erosion by standing water in the floor of Holden Crater, about 140 km across. (*NASA*)

(a)

the northern lowlands. The Hellas basin (Figure 6.21) is another candidate for an ancient Martian sea.

These ideas remain controversial. Proponents point to features such as the terraced "beaches" shown in Figure 6.26(b), which could conceivably have been left behind as a lake or ocean evaporated and the shoreline receded. But detractors maintain that the terraces could also be due to geological activity, perhaps related to tectonic forces that depressed the northern hemisphere far below the level of the south, in which case they have nothing whatever to do with Martian water. Furthermore, *Mars Global Surveyor* data seem to indicate that the Martian surface contains too few *carbonate* rock layers—layers containing compounds of carbon and oxygen—that should have been formed in abundance in an ancient ocean. Their absence supports the picture of a cold, dry Mars, which never experienced the extended mild period required to form lakes and oceans. However, data from the most recent *NASA* landers (see below) imply that at least some parts of the planet did in fact experience long periods in the past during which liquid water existed on the surface. Obviously the debate is far from over.

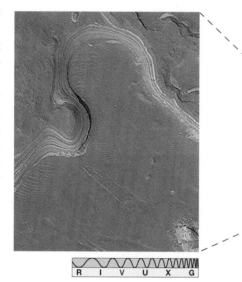

R I V U X G

(b)

Where Is the Water Today?

Astronomers have no direct evidence for liquid water anywhere on the Martian surface today, and the amount of water vapor in the Martian atmosphere is tiny. Yet the extent of the outflow channels and the other evidence just described indicates that a huge total volume of water existed on Mars in the past. Where did all that water go? The most likely answer is that much of Mars's original water is now locked in a layer of **permafrost**, a layer of water ice lying just below the planet's surface, much like that found in Earth's arctic regions, with more contained in the polar caps.

Figure 6.27 shows indirect evidence for the permafrost layer in the form of a fairly typical Martian impact crater named Yuty. Unlike the lunar craters discussed in Chapter 5, Yuty's ejecta blanket gives the distinct impression of a liquid that has splashed or flowed out of the crater. Most likely, the explosive impact heated and liquefied the permafrost, resulting in the fluid appearance of the ejecta. ∞ *(Sec. 5.4)* More direct evidence for subsurface ice came in 2002, when the *Mars Odyssey* orbiter detected extensive deposits of water ice crystals (actually, the hydrogen they contain) mixed with the Martian surface layers. In some locations ice appears to make up as much as 50 percent by volume of the planet's soil.

As noted earlier, Mars's polar caps actually consist of two distinct components. The *seasonal caps* vary in size as atmospheric carbon dioxide alternately freezes and

▲ **FIGURE 6.27 Martian Crater**   The ejecta from Mars's crater Yuty evidently was once liquid. This type of crater is sometimes called a "splosh" crater. *(NASA)*

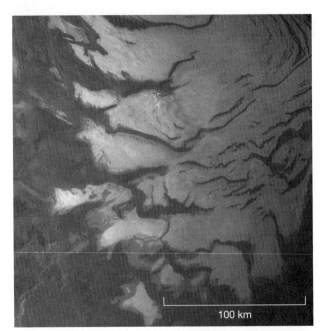

▲ **FIGURE 6.28 Martian Polar Cap**   The residual (permanent) polar caps on Mars may be major storehouses for water on the planet. This is the smaller southern cap, roughly 350 km in diameter. *(ESA)*

evaporates during the winter and summer months. However, the permanently frozen *residual caps* (Figure 6.28) are now known to be composed of water ice. The presence of ice in the larger northern cap has long been known from spectroscopic observations of water vapor above them as some small fraction of the cap evaporates in the Sun's heat during the summer months. The composition of the southern residual cap was finally established only in 2004 by spectroscopic imaging observations made by the European Space Agency's *Mars Express* orbiter. The thickness of the caps is uncertain, but it is quite possible that they represent a major storehouse for water on Mars.

Prior to the arrival of *Mars Global Surveyor*, astronomers thought that all the water on and below the Martian surface existed in the form of ice. However, in 2000, *Surveyor* mission scientists reported the discovery of numerous small-scale "gullies" in Martian cliffs and crater walls that apparently were carved by running water in the relatively recent past. These features are too small to have been resolved by *Viking* cameras. Figure 6.29(a) shows one such gully, found in the inner rim of a Martian impact crater in the southern highlands. Its structure has many similarities to the channels carved by flash floods on Earth. The ages of these intriguing features are uncertain, and might be as great as a million years in some cases, but some *Surveyor* data suggest that some of them may still be active today, implying that liquid water might exist in some regions of Mars at depths of less than 500 meters.

Some scientists dispute this interpretation, however, arguing that the "fluid" responsible for the gullies could have been solid (granular) or even liquid carbon dioxide, expelled under great pressure from the Martian crust.

Further data taken from orbit have deepened the mystery of the Martian gullies. Figure 6.29(b) shows two images of an unnamed impact crater in the southern highlands. The white streak in the second image is thought (by some) to be a frozen mudslide, where liquid water briefly flowed down the inside of the crater wall, carrying rocky debris with it, then froze on the chilly Martian surface. Its composition is uncertain, but one thing is clear—whatever it is, it formed recently, demonstrating that the production of these features is an ongoing process. It seems that a lot more study, perhaps even a human visit, will be needed before this issue is settled.

Scientists think that 4 billion years ago, as the Martian climate changed, the running water that formed the runoff channels began to freeze, forming the permafrost and drying out the river beds. Mars remained frozen for about a billion years, until volcanic (or some other) activity heated large regions of the surface, melting the permafrost and causing the flash floods that created the outflow channels. Subsequently, volcanic activity subsided, the water refroze, and Mars once again became a dry world. The present level of water vapor in the Martian atmosphere is the maximum possible given the atmosphere's present density and temperature. Estimates of the total amount of water stored as permafrost and in the polar caps are still quite uncertain, but it is likely that if all the water on Mars were to become liquid, it would cover the surface to a depth of roughly 10 meters.

Exploration by Martian Landers

Remote sensing—taking images and other measurements from orbit—is very important, but in many cases there is no substitute for a close-up look. Six U.S. spacecraft have successfully landed on the Martian surface. Their landing sites, marked on Figure 6.21, spanned a variety of Martian terrains. Their goals included detailed geological and chemical analyses of Martian surface rocks, the search for life, and search for water (an essential requirement for life as we know it—see Chapter 18).

▶ FIGURE 6.29 **Running Water on Mars?** (a) This high-resolution *Mars Global Surveyor* view of a crater wall near the Mariner Valley shows evidence of "gullies" apparently formed by running water in the relatively recent past. (b) Comparison of two images taken 6 years apart of another Martian impact crater shows that something—the white streak (lower right), possibly water—flowed across the surface. *(NASA)*

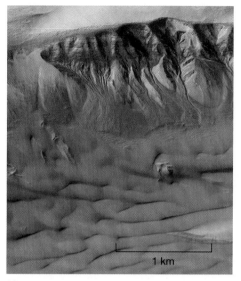
(a)

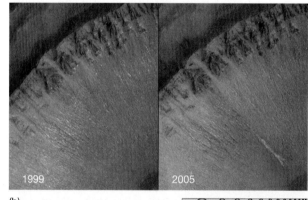
(b)

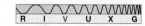

Both U.S. *Viking* missions dispatched landers to the Martian surface. Figure 6.30(a) is the view from *Viking 1*, which touched down near the planet's equator just east of Tharsis. The photograph shows a windswept, gently rolling, desolate plain littered with rocks of all sizes, not unlike a high desert on Earth. This view may be quite typical of the low-latitude northern plains. The *Viking* landers performed numerous chemical analyses of the rock on Mars's surface and performed several experiments designed specifically to test for life (see *Discovery 6-2*). One important finding of these studies was the high iron content of the planet's crust. Chemical reactions between the iron-rich surface soil and trace amounts of oxygen in the atmosphere are responsible for the iron oxide (rust) that gives Mars its characteristic red color.

The next successful mission to the Martian surface was *Mars Pathfinder.* During the unexpectedly long lifetime of its mission in 1997 (it lasted almost 3 months instead of the anticipated 1 month), the lander performed measurements of the Martian atmosphere and atmospheric dust while its robot rover *Sojourner* (Figure 6.30b) carried out chemical analyses of the soil and rocks within about 50 m of the parent craft. In addition, more than 16,000 images of the region were returned to Earth. The soil chemistry in the vicinity of the landing site was similar to that found by the *Viking* landers. Interestingly, analyses of the nearby rocks revealed chemical makeups different from that of the Martian meteorites found on Earth (see *Discovery 6-2*).

The landing site for the *Pathfinder* mission had been carefully chosen to lie near the mouth of an outflow channel, and the size distribution and composition of the many rocks and boulders surrounding the lander were consistent with their having been deposited there by flood waters. In addition, the presence of numerous

(a)

(b)

▲ FIGURE 6.30 **Martian Landscape** (a) Wide-angle view from the perspective of the *Viking 1* spacecraft. The fine-grained soil and the reddish rock-strewn terrain stretching toward the horizon contain substantial amounts of iron ore; the surface of Mars is literally rusting away. The sky is a pale pink, the result of airborne dust. (b) The Mars rover *Sojourner* is seen at left inspecting a large Martian rock, nicknamed Yogi by mission scientists. In the foreground is *Sojourner's* mother ship, *Pathfinder,* with its solar panels that provide power and airbags that aided landing. *(NASA)*

6-2 DISCOVERY

Life on Mars?

Even before the *Viking* missions reached Mars in 1976, astronomers had abandoned hope of finding life on the planet. Scientists knew there were no large-scale canal systems, no surface water, almost no oxygen in the atmosphere, and no seasonal vegetation changes. The present lack of liquid water on Mars especially dims the chances for life there now. However, running water and possibly a dense atmosphere in the past may have created conditions suitable for the emergence of life long ago. In the hope that some form of microbial life might have survived to the present day, the *Viking* landers carried out experiments designed to detect biological activity. The accompanying pair of photographs show the robot arm of one of the landers digging a shallow trench.

All three *Viking* biological experiments assumed some basic similarity between hypothetical Martian bacteria and those found on Earth. A *gas-exchange* experiment offered a nutrient broth to any residents of a sample of Martian soil and looked for gases that would signal metabolic activity. A *labeled-release* experiment added compounds containing radioactive carbon to the soil, then waited for results signaling that Martian organisms had either eaten or inhaled this carbon. Finally, a *pyrolitic-release* experiment added radioactively tagged carbon dioxide to a sample of Martian soil and atmosphere, waited awhile, then removed the gas and tested the soil (by heating it) for signs that something had absorbed the tagged gas. In all cases, contamination by terrestrial bacteria was a major concern. Indeed, any release of Earth organisms would have invalidated these and all future such experiments on Martian soil. Both *Viking* landers were carefully sterilized prior to launch.

Initially, all three experiments appeared to be giving positive signals! However, subsequent studies showed that the results could all be explained by inorganic (that is, nonliving) chemical reactions. Thus, we have no evidence for even microbial life on the Martian surface. The *Viking* robots detected peculiar reactions that mimic in some ways the basic chemistry of living organisms, but they did not detect life itself.

One criticism of the *Viking* experiments is that they searched only for life now living. Today, Mars seems locked in an ice age—the kind of numbing cold that would prohibit sustained life as we know it. If bacterial life did arise on an Earth-like early Mars, however, then we might be able to find its fossilized remains preserved on or near the Martian surface. This is one reason that scientists are

so eager to land more spacecraft on the planet, particularly at the poles, whose ice caps may offer the best environment for finding Martian life (or its remains).

Surprisingly, one alternative place to look for evidence of life on Mars is right here on Earth. The accompanying figure shows ALH84001, a blackened, 2-kg meteorite about 17 cm across, found in 1984 in Antarctica. Chemical analysis strongly indicates that it originated on Mars. It was apparently blasted off that planet long ago by a meteoritic impact of some sort, thrown into space, and eventually captured by Earth's gravity. Based on the estimated cosmic-ray exposure it received before reaching Earth, it left Mars about 16 million years ago.

On the basis of all the data accumulated from studies of ALH84001, in 1996 a group of scientists argued that they had discovered fossilized evidence for life on Mars. They pointed to globular structures similar to those produced by bacteria on Earth (top left inset), the presence of chemical compounds sometimes associated with Earth biology, and curved, rodlike structures (right image) resembling Earth bacteria, which the researchers interpreted as fossils of primitive Martian organisms. In 1999 the team released an

Robot arm

Trench

(NASA)

| R | I | V | U | X | G |

(Before)   (After)

rounded pebbles strongly suggested the erosive action of running water at some time in the past.

In 2004 the twin landers *Spirit* and *Opportunity* of the Mars Exploration Rover mission arrived at their targets on opposite sides of the planet (see Figure 6.21). The two robots were designed to operate for just 3 months, but each continued to send back vital data for several years. During their lifetime on Mars, the landers roamed the surface within several kilometers of their landing sites, making chemical and

analysis of a second meteorite also thought to have come from Mars, again reporting evidence for microbial life having similarities in size, shape, and arrangement to known nanobacteria on Earth.

These claims were and remain very controversial. Many experts categorically disagree that evidence of Martian life has been found—not even fossilized life. They maintain that, as in the case of the *Viking* experiments, the evidence could all be due to chemical reactions not requiring any kind of biology. The scale of the supposed fossil structures is also important. They are only about 0.5 μm across, 1/30 the size of ancient bacterial cells found fossilized on Earth. Furthermore, several key experiments have not yet been done, such as testing the suspected fossils for evidence of cell walls or of any internal cavities where body fluids would have resided. Nor has anyone yet found in either meteorite any amino acids, the basic building blocks of life as we know it (see Chapter 18). In addition, there is the huge problem of contamination—after all, ALH84001 was found on Earth and apparently sat in the Antarctic ice fields for 13,000 years before being picked up by meteorite hunters.

As things now stand, it's a matter of interpretation—at the frontiers of science, issues are usually not as clear-cut as we would hope. Only additional analysis and new data—perhaps in the form of samples returned directly from the Martian surface—will tell for sure if primitive Martian life existed long ago. Most workers in the field seem to have concluded that, taken as a whole, the results do not support the claim of ancient life on Mars. Still, even some skeptics concede that as much as 20 percent of the organic material in ALH84001 could have originated on the Martian surface—although that is a far cry from proving the existence of life there.

Whether or not ALH84001 actually turns out to contain evidence of ancient Martian life, the scientific uproar it has caused has forced scientists to think much more carefully about the kinds of life that might be able to survive in extreme environments—those with temperatures, pressures, or chemical makeup not normally regarded as hospitable to life as we know it on Earth. These studies have fueled the emerging field of *exobiology*—the search for and study of life on other planets, in our own solar system and beyond. As we will see in Chapters 8 and 18, the subsurface oceans of Jupiter's moon Europa and the hydrocarbon seas of Saturn's frigid moon Titan have joined Mars as prime candidates in the search for extraterrestrial life.

Should the claim of life on Mars hold up against the weight of healthy skepticism in the scientific community, these findings may go down in history as one of the greatest scientific discoveries of all time. We are—or at least were—not alone in the universe! Maybe . . .

Carbonate globule

ANIMATION/VIDEO
Meteorites Ejected From Mars

ANIMATION/VIDEO
Microscopic Martian Fossils?

5 cm

(NASA)

Structure inside globules

geological studies of rocks they encountered. The primary goal of the mission was to look for evidence of liquid water on Mars at any time in the past, and in that they succeeded. The rovers' findings provide the best evidence yet for standing water on the ancient Martian surface and have changed the minds of many skeptical scientists on the subject of water on Mars.

Spirit's landing site was rocky and similar in many ways to the terrains encountered by earlier landers, although closer study by the science team reveals that most of

▲ FIGURE 6.31 **Mars Panorama** A panoramic view of the terrain where NASA's *Opportunity* rover landed on Mars in 2004. This is Endurance crater, roughly 130 meters across. *(NASA)*

ANIMATION/VIDEO Mars Exploration Rover Landing

ANIMATION/VIDEO Flight Over Opportunity at Gustav Crater

CONCEPT CHECK ► Why do astronomers think that the Martian climate was once quite different from today's?

the rocks in the lander's vicinity do appear to have been extensively altered by water long ago. Halfway around the planet, however, *Opportunity* appears to have hit the jackpot in its quest, finding itself surrounded by rocks showing every chemical and geological indication of having been very wet—possibly immersed in salt water—in the distant past (Figure 6.31). Several indicators suggest that the rocks near *Opportunity*'s landing site have been alternately underwater and dry for extended periods of time, possibly as a shallow lake alternately filled and evaporated numerous times during the course of Martian history. If life ever did exist at this site, the sorts of rocks found there might have preserved a fossil record very well, but *Opportunity* was not equipped to carry out such studies.

The most recent mission to Mars is NASA's *Phoenix* mission, which landed in the planet's north polar region (see Figure 6.21) in May 2008. Its objectives included determining if the Martian arctic is or was capable of supporting life, looking for ice or other evidence of water, and exploring the Martian polar climate. Scientists had long been eager to send a spacecraft to the Martian polar regions because of the expectation of finding water ice there. The spacecraft was not a rover, but instead contained a sophisticated array of equipment to collect and analyze the surrounding soil.

Phoenix confirmed the presence of subsurface water ice at the landing site (Figure 6.32; see also the chapter-opening photo on page 162) and found clay and carbonates in the soil, both indicators of a wet environment at some time in the past, although scientists do not know if the water resulted from seasonally melting ice or is a feature of the much more distant past. Initially, the overall composition of the soil seemed quite Earth-like, but later analysis suggested some chemical differences that might make it less friendly to life, at least as we know it. *Phoenix* touched down in late fall in the Martian northern hemisphere, and during its final weeks, its sensors reported the first snow as winter closed in. The mission ended when diminishing sunlight and extreme low temperatures shut down the lander's power supply.

◄ FIGURE 6.32 **Martian Exploration** The *Phoenix* lander's robotic arm is shown here with a surface sample in its scoop, just before delivering it to an onboard miniature chemistry laboratory. The circular hardware at bottom is one of its solar panels. *(NASA)*

6.7 Internal Structure and Geological History

As with Earth and the Moon, planetary scientists can combine measurements of a planet's bulk properties with detailed observations of its gravity and magnetic field to build a model of the planetary interior. ∞ *(Secs. 5.4, 5.7)* Our understanding of the internal structure and evolution of Mercury, Venus, and Mars has steadily grown as models have improved and spacecraft have radioed back more and more detailed information on these worlds.

Mercury

Mercury's magnetic field, discovered by *Mariner 10*, is about 1/100 of Earth's field. The discovery that Mercury has a magnetic field came as a surprise to planetary scientists, who, having detected no magnetic field in the Moon, expected Mercury to have none either. In Chapter 5 we saw how a combination of liquid metal core and rapid rotation is necessary for the production of a planetary magnetic field. ∞ *(Sec. 5.7)* Mercury certainly does not rotate rapidly and it may also lack a liquid metal core, yet a magnetic field undeniably surrounds it. Although weak, the field is strong enough to deflect the solar wind and create a small magnetosphere around the planet.

Before *Messenger*'s arrival, scientists thought it most likely that Mercury's magnetic field was a "fossil remnant" dating back to the distant past when the planet's core solidified. However, detailed observations now suggest that the field is in fact generated, but curiously, the field is significantly offset from the planet's center, making the field at the north pole much stronger than at the south. The details of how such a relatively strong and asymmetric field can be produced by a slowly rotating planet remain to be resolved.

Mercury's magnetic field and high average density (roughly (5400 kg/m^3) taken together imply that most of the planet's interior is dominated by a large, iron-rich core having a radius of perhaps 1800 km and accounting for some 60 percent of the planet's mass. The ratio of core volume to total planet volume is greater for Mercury than for any other object in the solar system. Figure 6.33 illustrates the relative sizes and internal structures of Earth (and perhaps also Venus), the Moon, Mercury, and Mars.

Like the Moon, Mercury seems to have been geologically dead for roughly the past 4 billion years. Again as on the Moon, the lack of present-day geological activity on Mercury results from the mantle's being solid, preventing volcanism or tectonic motion. Largely on the basis of studies of the Moon, scientists have pieced together the following outline of Mercury's early history.

When Mercury formed 4.6 billion years ago, its location in the hot inner region of the early solar system ensured a dense, largely metallic overall composition.

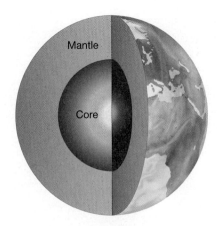

Mantle

Core

Earth

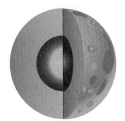

Mars

Mercury

Moon

▲ FIGURE 6.33 **Terrestrial Planet Interiors**   The internal structures of Earth, the Moon, Mercury, and Mars, drawn to the same scale. Note how large a fraction of Mercury's interior is core. The interior structure of Venus is largely unknown, but is thought to be similar to that of Earth.

∞ *(Sec. 4.3)* During the next half-billion years, Mercury melted and differentiated, just as the other terrestrial worlds did. It suffered the same intense meteoritic bombardment as the Moon. Being more massive than the Moon, Mercury cooled more slowly, so its crust was thinner than the Moon's and volcanic activity more common. Lava erased more craters, leading to the intercrater plains found by *Mariner 10*.

As the planet's large iron core formed and then cooled, the planet began to shrink, causing the surface to contract. This compression produced the scarps seen on Mercury's surface and may have prematurely terminated volcanic activity by squeezing shut the cracks and fissures on the surface. Thus, Mercury did not experience the subsequent extensive volcanic outflows that formed the lunar maria. Despite its larger mass and greater internal temperature, Mercury has probably been geologically inactive for even longer than the Moon.

Venus

Both U.S. and Soviet spacecraft failed to detect any magnetosphere around Venus. Given that Venus's average density is similar to Earth's, it seems likely that Venus has an Earth-like overall composition and a partially molten iron-rich core. The lack of any detectable magnetic field, then, is almost surely the result of the planet's extremely slow rotation. ∞ *(Sec. 5.7)*

Because none of the *Venera* landers carried seismic equipment, no direct measurements of the planet's interior have ever been made, and theoretical models of the interior have very little hard data to constrain them. The physical similarities between Venus and Earth suggest that the core/mantle structure of Venus resembles that of our own planet. However, to many geologists the surface of Venus resembles that of the young Earth, at an age of perhaps a billion years. At that time, volcanic activity had already begun on Earth, but the crust was still relatively thin and the convective processes in the mantle that drive plate tectonic motion were not yet established.

Why has Venus remained in that immature state and not developed plate tectonics as Earth did? That question remains to be answered. Some planetary geologists have speculated that the high surface temperature on Venus has inhibited evolution by slowing the planet's cooling rate. Possibly the high surface temperature has made the crust too soft for Earth-style plates to develop. Or perhaps the high temperature and soft crust led to more volcanism, tapping the energy that might otherwise have gone into convective motion.

Mars

In September 1997 the orbiting *Mars Global Surveyor* succeeded in detecting a very weak Martian field, about 1/800 that of Earth. However, this is most likely a local anomaly, not a global field. Because Mars rotates rapidly, the weakness of the magnetic field is taken to mean that the planet's core is nonmetallic, or nonliquid, or both. ∞ *(Sec. 5.7)*

Mars's small size means that any internal heat would have been able to escape more easily than in a larger planet like Earth or Venus. The evidence of ancient surface activity, especially volcanism, suggests that at least parts of the Martian interior must have melted at some time in the past, but the lack of current activity and the absence of any significant magnetic field indicate that the melting was never as extensive as on Earth. The latest data indicate that Mars's core has a diameter of about 2500 km, is composed largely of iron sulfide (a compound about twice as dense as surface rock), and is still at least partly molten.

Mars appears to be a planet where large-scale tectonic activity almost started but was stifled by the planet's rapidly cooling outer layers. On a larger, warmer planet, the upwelling that formed the Tharsis bulge might have developed into full-fledged plate tectonics, but the Martian mantle became too rigid, the crust too thick, for that to occur. Instead, the upwelling fired volcanic activity, perhaps even up to the present day, but geologically much of the planet died 2 billion years ago.

CONCEPT CHECK ▶ Why do Venus and Mars lack magnetic fields?

6.8 Atmospheric Evolution on Earth, Venus, and Mars

ANIMATION/VIDEO Terrestrial Planets II

Now that we have studied the structure and history of the three terrestrial worlds that still have atmospheres today—Venus, Earth, and Mars—let's return to the question of why their atmospheres are so different from one another.

The Runaway Greenhouse Effect on Venus

Given the distance of Venus from the Sun, the planet was not expected to be such a pressure cooker. Why is Venus so hot? And if, as seems likely, Venus started off like Earth, then why is its atmosphere now so different from Earth's?

The answer to the first question is easy. Given the present composition of its atmosphere, Venus is hot because of the *greenhouse effect.* ∞ *(Sec. 5.3)* Venus's dense atmosphere is made up almost entirely of a prime greenhouse gas, carbon dioxide (Figure 6.34). This thick blanket absorbs about 99 percent of all the infrared radiation released from the surface of Venus and is the immediate cause of the planet's sweltering 730 K surface temperature. The answer to the second question is much more complex, however, and requires us to consider in detail how the atmospheres of the terrestrial planets formed and what happened to the greenhouse gases they contained.

The terrestrial atmospheres did not come into being along with the planets themselves. Instead, they developed over many millions of years as **secondary atmospheres,** made up of gases released from the planets' interiors by volcanic activity—a process called *outgassing.* Volcanic gases are rich in water vapor, carbon dioxide, sulfur dioxide, and compounds containing nitrogen. On Earth, as the surface temperature fell and the water vapor condensed, oceans formed. Most of the carbon dioxide and sulfur dioxide dissolved in the oceans or combined with surface rocks. Solar ultraviolet radiation liberated nitrogen from its chemical bonds with other elements, and a nitrogen-rich atmosphere slowly appeared.

The basic mechanism controlling the level of carbon dioxide in Earth's atmosphere is the competition between its ongoing production by volcanic (and perhaps human) activity and its absorption by the rocks and oceans that make up our planet's surface. ∞ *(Sec. 5.4)* Illustrated in simplified form in Figure 6.35(a), this constant recycling of atmospheric carbon dioxide is known as the **carbon cycle.** The presence of liquid water greatly accelerates the absorption process—carbon dioxide dissolves in water, eventually reacting with surface material to form carbonate rocks. At the same time, plate tectonics steadily releases carbon dioxide back into the air. The roughly stable level of carbon dioxide in Earth's atmosphere today is the result of a balance between these opposing forces.

The initial stages of atmospheric development on Venus probably took place in much the same way as just described for our own planet. The real difference between Earth and Venus is that the greenhouse gases in Venus's secondary atmosphere never left the atmosphere as they did on Earth. Indeed, if all the dissolved or chemically combined carbon dioxide on Earth were released back into our present-day atmosphere, its new composition would be 98 percent carbon dioxide and 2 percent nitrogen, and it would have a pressure about 70 times its current value. In other words, apart from the presence of oxygen (which is a direct consequence of life on Earth) and water (whose absence on Venus will be explained in a moment), Earth's atmosphere would look a lot like that of Venus.

To understand what happened on Venus, imagine taking Earth from its present orbit and placing it in that of Venus. Being closer to the Sun, our planet would warm up. More water would evaporate from the oceans, leading to an increase in

▶ **FIGURE 6.34 Venus's Atmosphere**   Because Venus's atmosphere is much deeper and denser than Earth's, a much smaller fraction of the infrared radiation leaving the planet's surface escapes into space. The result is a much stronger greenhouse effect than on Earth and a correspondingly hotter planet.

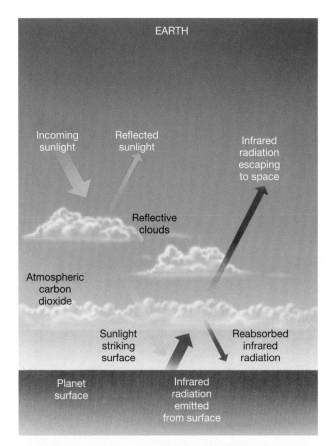

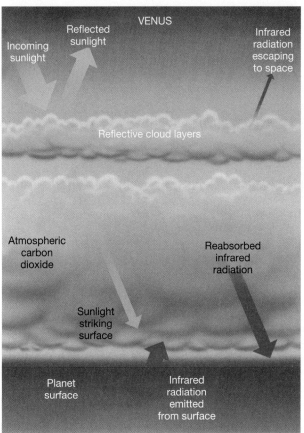

EARTH

Atmospheric CO_2 absorbed by rocks, water

CO_2 released by volcanism

Planet surface

(a)

VENUS

Little atmospheric CO_2 absorbed by surface rocks

CO_2 released by volcanism

Planet surface

(b)

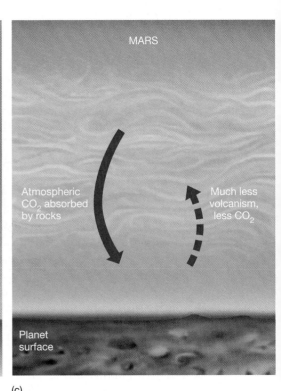

MARS

Atmospheric CO_2 absorbed by rocks

Much less volcanism, less CO_2

Planet surface

(c)

▲ FIGURE 6.35 **Carbon Cycle**   (a) On Earth, competition between volcanic activity and absorption into surface rocks and water keeps atmospheric carbon dioxide at a modest level, leading to a small but beneficial greenhouse effect. (b) On Venus, the temperature is so high that relatively little carbon can be absorbed, leading to the runaway greenhouse effect. (c) On Mars, there is little volcanic activity, and so most of the carbon dioxide now resides in the surface rocks.

atmospheric water vapor. Because the ability of both the oceans and surface rocks to hold carbon dioxide diminishes with increasing temperature, more carbon dioxide would enter the atmosphere. The additional greenhouse heating would warm our planet still further, leading to a further increase in atmospheric greenhouse gases, and so on. This **runaway greenhouse effect** would eventually lead to the complete evaporation of the oceans, restoring all the original greenhouse gases to the atmosphere. Essentially the same thing must have happened on Venus long ago (Figure 6.35b), leading to the planetary inferno we see today.

The greenhouse effect on Venus was even more extreme in the past, when the atmosphere also contained water vapor. By adding to the blanketing effect of the carbon dioxide, the water vapor helped push the surface of Venus to temperatures perhaps twice as hot as at present. At those high temperatures, the water vapor was able to rise high into the planet's upper atmosphere—so high that solar ultraviolet radiation broke it up into its components, hydrogen and oxygen. The light hydrogen rapidly escaped, the reactive oxygen quickly combined with other atmospheric gases, and all water on Venus was lost forever.

Evolution of the Martian Atmosphere

The runaway greenhouse effect on Venus defines the "hot" edge of the habitable zone that features so prominently in discussions of Earth-like exoplanetary systems. ∞ *(Sec. 4.4)* The "cold" edge of the zone is governed by the history of our other planetary neighbor, Mars.

Presumably, Mars also had an outgassed secondary atmosphere early in its history. As we saw in Section 6.6, around 4 billion years ago Mars may have had a relatively dense atmosphere rich in carbon dioxide, perhaps even with blue skies and rain. Planetary scientists estimate that the greenhouse effect could have kept average surface temperatures above 0°C (Figure 6.36). Sometime during the next billion years, however, most of the Martian atmosphere disappeared. Possibly some of it escaped due to impacts with large bodies in the early solar system, and a large part may have leaked away into space because of the planet's weak gravity, but most of the remainder probably became unstable and was lost in a kind of "reverse runaway greenhouse effect." ∞ *(More Precisely 5-1)* The following scenario is accepted by many planetary scientists (although

▲ FIGURE 6.36 **Ancient Mars**   Artist's conception of Mars some 4 billion years ago, with a dwindling atmosphere and some lingering surface water. (*Kees Veenenbos*)

not all—those who discount the evidence presented earlier for liquid water and a thick early atmosphere on Mars obviously require no explanation for their absence today).

As we have just seen, Mars cooled faster than Earth and apparently never developed large-scale plate tectonic motion. As a result, even taking into consideration the large volcanoes discussed earlier in this chapter, Mars has had on average far less volcanism than Earth does, so the processes depleting carbon dioxide were much more effective than those replenishing it. The level of atmospheric carbon dioxide steadily declined, its greenhouse effect diminished, and the planet cooled, causing still more carbon dioxide to be absorbed in the surface layers and leave the atmosphere. The carbon cycle turned into a "one-way street" of ever lower temperatures and decreasing levels of carbon dioxide in the atmosphere. Calculations suggest that much of the Martian atmospheric carbon dioxide could have been lost in this way in a relatively short period of time, perhaps as quickly as a few hundred million years.

As the temperature continued to fall, water froze out of the atmosphere, lowering still further the level of atmospheric greenhouse gases and accelerating the cooling. Eventually, even carbon dioxide began to freeze out, particularly at the poles, and Mars reached the frigid state we see today—a cold, dry planet with most of its original complement of atmospheric gases now residing in or under the barren surface.

CONCEPT CHECK ▶ If Venus and Mars had formed at Earth's distance from the Sun, what might their climates be like today?

CHAPTER REVIEW

SUMMARY

LO1 Mercury's rotation rate is strongly influenced by the tidal effect of the Sun, which causes the planet to rotate exactly one and a half times for every one revolution around the Sun.

LO2 Mercury has no permanent atmosphere. The atmospheres of Venus and Mars are mainly carbon dioxide. Venus's atmosphere is extremely hot and 90 times denser than Earth's. The density of the cool Martian atmosphere is only 0.7 percent that of Earth's.

LO3 Mercury's surface is heavily cratered, much like the lunar highlands. Mercury lacks lunar-like maria but has extensive **intercrater plains** (p. 169) and cliff-forming cracks, or **scarps** (p. 169), in its crust. The plains were caused by lava flows early in Mercury's history. The scarps were apparently formed when the planet's core cooled and shrank, causing the surface to crack.

LO4 Because of its thick cloud cover, Venus's surface cannot be seen in visible light from Earth, although it has been thoroughly mapped by radar. Many lava domes and **shield volcanoes** (p. 171) have been found. No eruptions have been observed, although there is indirect evidence that Venus is still volcanically active today. The northern hemisphere of Mars consists mainly of rolling plains, while the south is

rugged highlands. The reason for this asymmetry is unknown. Mars's major surface feature is the Tharsis bulge. Associated with the bulge is the largest-known volcano in the solar system and a huge crack, called the Valles Marineris, in the planet's surface.

LO5 There is clear evidence that water once existed in great quantity on Mars. The **runoff channels** (p. 177) are the remains of ancient Martian rivers. The **outflow channels** (p. 177) are the paths taken by floods that cascaded from the southern highlands into the northern plains. Today, a large amount of that water may be locked up in the polar caps and in a layer of **permafrost** (p. 179) lying under the Martian surface. Images from orbiting spacecraft seem to show what may be ancient coastlines; others suggest that there may still be liquid water below the surface in some places. The most recent craft to land on the surface have found strong geological evidence for standing water.

LO6 Mercury's weak magnetic field seems to be a fossil remnant from the time when the planet's iron core solidified. Geological activity on Mercury ceased long ago. Venus does not have a detectable magnetic field even though, like Earth, it probably has a molten core. This is due to the planet's slow rotation. The surface of Venus shows no sign of plate tectonics. Features called **coronae** (p. 171) are thought to have been caused by an upwelling of mantle material that never

developed into full convective motion. Mars rotates rapidly, but has a very weak magnetic field, implying that its core is nonmetallic, nonliquid, or both. Mars also appears to have had some form of tectonic activity long ago, but plate tectonics never developed because the planet cooled too rapidly.

LO7 Any initial atmospheres of light gases that Earth, Venus, and Mars had at formation rapidly escaped. All the terrestrial planets probably developed **secondary**

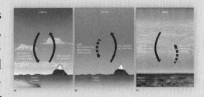

atmospheres (p. 187), outgassed from volcanoes, early in their lives. On Mercury the atmosphere escaped, as on the Moon. On Earth much of the outgassed material became absorbed in surface rocks or dissolved in the oceans. On Venus the **runaway greenhouse effect** (p. 188) has resulted in all the planet's carbon dioxide being left in the atmosphere, leading to the extreme conditions we observe today. On Mars part of the secondary atmosphere escaped into space. Of the remainder, most of the carbon dioxide is now locked up in surface rock, and most of the water vapor is now stored in the permafrost and the polar caps.

MasteringAstronomy®   For instructor-assigned homework go to www.masteringastronomy.com

Problems labeled **POS** explore the process of science. **VIS** problems focus on reading and interpreting visual information. **LO** connects to the introduction's numbered Learning Outcomes.

REVIEW AND DISCUSSION

1. In contrast to Earth, Mercury undergoes extremes in temperature. Why?

2. What do Mercury's magnetic field and large average density imply about the planet's interior?

3. **LO3** How is Mercury's evolutionary history like that of the Moon? How is it different?

4. **LO1 POS** Mercury used to be called "the Moon of the Sun." Why do you think it had this name? What piece of scientific evidence proved the name inappropriate?

5. Why does Venus appear so bright to the naked eye?

6. **LO6** Venus probably has a molten iron-rich core like Earth. Why doesn't it also have a magnetic field?

7. **LO4 POS** How did radio observations of Venus made in the 1950s change our conception of that planet?

8. What are the main constituents of Venus's atmosphere? What are clouds in the upper atmosphere made of?

9. What is the runaway greenhouse effect, and how might it have altered the climate of Venus?

10. Why is Mars red?

11. **LO5 POS** What is the evidence that water once flowed on Mars? Is there water on Mars today?

12. Why were Martian volcanoes able to grow so large?

13. **LO2** Since Mars has an atmosphere and it is composed mostly of a greenhouse gas, why isn't there a significant greenhouse effect to warm its surface?

14. Do you think that sending humans to Mars in the near future is a reasonable goal? Why or why not?

15. **LO7** Compare and contrast the evolution of the atmospheres of Mars, Venus, and Earth. Include a discussion of the importance of volcanoes and water.

CONCEPTUAL SELF-TEST: TRUE OR FALSE/MULTIPLE CHOICE

1. Mercury's solar day is longer than its solar year. (T/F)

2. Numerous surface features on Venus can be seen in Earth-based images made in the ultraviolet part of the spectrum. (T/F)

3. Evidence of lava flows is common on the surface of Venus. (T/F)

4. There is strong circumstantial evidence that active volcanism continues on Venus. (T/F)

5. Mars has the largest volcanoes in the solar system. (T/F)

6. There are many indications of past plate tectonics on Mars. (T/F)

7. Valles Marineris is comparable in size to Earth's Grand Canyon. (T/F)

8. Mercury's large overall density suggests that the planet (a) has an interior structure similar to that of Earth's Moon; (b) has a dense

metal core; (c) has a weaker magnetic field than the Moon; (d) is younger than the Moon.

9. Venus's surface is permanently obscured by clouds. As a result, the surface has been studied primarily by (a) robotic landers; (b) orbiting satellites using radar; (c) spectroscopy; (d) radar signals from Earth.

10. Compared with Earth, Venus has a level of plate tectonic activity that is (a) much more rapid; (b) somewhat less rapid; (c) about the same; (d) virtually nonexistent.

11. Venus's atmospheric temperature is (a) about the same as Earth's; (b) cooler than temperatures on Mercury; (c) hotter than temperatures on Mercury; (d) high due to the presence of sulfuric acid.

12. In terms of area, the extinct Martian volcano Olympus Mons is about the size of (a) Mt. Everest; (b) Colorado; (c) North America; (d) Earth's Moon.

13. In comparison to the atmosphere of Venus, the vastly different atmospheric character of Mars is likely due to a(n) (a) ineffective greenhouse effect; (b) a reverse greenhouse effect; (c) absence of greenhouse gases in the Martian atmosphere; (d) greater distance from the Sun.

14. VIS In Figure 6.3 (Planetary Radar), part (b), the highest-frequency reflected radiation is received from (a) the upper part of the planet; (b) the center of the planet; (c) the lower part of the planet.

15. VIS Of the following, the best evidence for the existence of liquid water on an ancient Mars is Figure (a) 6.19; (b) 6.22; (c) 6.23; (d) 6.30.

PROBLEMS

The number of squares preceding each problem indicates its approximate level of difficulty.

1. ■ How long does it take a radar signal to travel from Earth to Mercury and back when Mercury is at its closest point to Earth?

2. ■■ Assume that a planet will have lost its initial atmosphere by the present time if the average (daytime) molecular speed exceeds one-sixth of the escape speed (see *More Precisely 5-1*). What would Mercury's mass have to be in order for it still to have a nitrogen (molecular weight 28) atmosphere?

3. ■■ Approximating Venus's atmosphere as a layer of gas 50 km thick with a uniform density of 21 kg/m^3, calculate its total mass. Compare your answer with the mass of Earth's atmosphere (Chapter 5, Problem 6) and with the mass of Venus.

4. ■ *Pioneer Venus* observed high-level clouds moving around Venus's equator in 4 days. What was their speed in kilometers per hour? In miles per hour?

5. ■■ (a) What is the size of the smallest feature that can be distinguished on the surface of Venus (at closest approach) by the Arecibo radio telescope at an angular resolution of $1'$? (b) Could an infrared telescope with an angular resolution of $0.1''$ distinguish impact craters on the surface of Venus?

6. ■■■ Calculate the orbital period of the *Magellan* spacecraft, moving around Venus on an elliptical orbit with a minimum altitude of 294 km and a maximum altitude of 8543 km above the planet's surface. In 1993 the spacecraft's orbit was changed to have minimum and maximum altitudes of 180 km and 541 km, respectively. What was the new period?

7. ■ Verify that the surface gravity on Mars is 40 percent that of Earth. What would you weigh on Mars?

8. ■■ The outflow channel shown in Figure 6.24 is about 10 km across and 100 m deep. If it carried 10^7 metric tons (10^{10} kilograms) of water per second, as stated in the text, estimate the speed at which the water must have flowed.

9. ■■ Calculate the total mass of a uniform layer of water covering the entire Martian surface to a depth of 2 m (see Section 6.8). Compare it with the mass of Mars and with the mass of Venus's atmosphere (Problem 3).

10. ■■ Mars has two small, irregular moons, Phobos and Deimos, which orbit at distances of 9400 km and 23,500 km, respectively, from the planet. Their longest diameters are 28 km (Phobos) and 16 km (Deimos). Calculate their maximum angular diameters, as seen by an observer on the Martian surface directly under their orbits. Would the observer ever see a total solar eclipse?

ACTIVITIES

Collaborative

1. If Mars is visible in the night sky, observe it with as large a telescope as is available to you; binoculars will not be of much use. Use your almanac (or go online) to find out which Martian season is occurring at the time of your observation, which hemisphere is tilted in Earth's direction, and what longitude is pointing toward Earth. Sketch what you see. Look carefully and take your time. Repeat this observation several times over the course of a night. Take turns making the sketches. Afterward, try to identify the various features you have seen by referring to known objects on Mars (e.g., as shown in Figure 6.6). You should also be able to see the planet's rotation by watching the surface features move. Do the same the next night. Because Mars's rotation period is so similar to Earth's, you should see the same surface features again.

Individual

Consult an almanac to determine where Mercury, Venus, and Mars are in the sky this year.

1. Try to spot Mercury in the morning or evening twilight—not an easy task! From the Northern Hemisphere, the best evening sightings of the planet take place in the spring; the best morning sightings take place in the fall.

2. Find out when Venus will next pass between Earth and the Sun. How many days before and after this event can you glimpse the planet with the naked eye? Using binoculars or a small telescope, examine Venus as it goes through its phases. Note the phase and the relative size (you can compare its size to the field of view in a telescope; always use the same eyepiece). Observe it every few days or once a week. Make a table of its shape, size, and relative brightness. Can you see the correlations between these properties first recognized by Galileo?

3. Several months before opposition, Mars begins retrograde motion. Chart the planet's motion relative to the stars to determine when it stops moving eastward and begins moving toward the west. Notice the increase in Mars's brightness as it approaches opposition.

7

The Jovian Planets
Giants of The Solar System

Beyond the orbit of Mars, the solar system is a very different place. In sharp contrast to the small, rocky terrestrial bodies found near the Sun, the outer solar system presents us with a totally unfamiliar environment—huge gas balls, peculiar moons, planet-circling rings, and a wide variety of physical and chemical properties, many of which are only poorly understood even today. U.S. spacecraft visiting the outer planets over the past two decades have revealed these worlds in detail that was only dreamed of by centuries of Earth-bound astronomers. The jovian planets—Jupiter, Saturn, Uranus, and Neptune—differ from Earth and from each other in many ways, but they have much in common, too. As with the terrestrial planets, we learn from their differences as well as from their similarities.

MasteringAstronomy® Visit www.masteringastronomy.com for quizzes, animations, videos, interactive figures, and self-guided tutorials.

◄ Engineering feats of the modern space age allow us to scientifically explore many of the diverse worlds of our solar system. Here, we see in great detail the spectacular ring system around the giant planet Saturn, a huge ball of lightweight gas surrounded by an intricate disk of orbiting rocky debris—a planet far different from Earth. This image combines visible, infrared, and ultraviolet data and was taken by cameras aboard the *Cassini* robot spacecraft as it cruised behind the planet and through its shadow in 2005. Looking back toward the inner parts of the solar system in this view, we can see not only part of the bright Sun (bottom center), but also a dim and distant speck of reflected light—our home planet Earth floating in space (at the 8 o'clock position). *(JPL)*

7.1 Observations of Jupiter and Saturn

The View from Earth

Named after the most powerful god of the Roman pantheon, Jupiter is the third-brightest object in the night sky (after the Moon and Venus), making it easy to locate and study from Earth. Ancient astronomers could not have known the planet's true size, but their name choice was very apt—Jupiter is by far the largest planet in the solar system. Figure 7.1(a) is a photograph of Jupiter taken through a small telescope on Earth, showing the planet's overall reddish coloration and the alternating light and dark bands paralleling its equator. Figure 7.1(b) is a *Hubble Space Telescope* image of Jupiter, showing the colored bands in much more detail and displaying numerous oval structures within them (note in particular the large orange oval at lower right). These atmospheric features are quite unlike anything found on the terrestrial planets. In further contrast to the inner planets, Jupiter has many moons that vary greatly in size and other properties. The four largest, shown in Figure 7.1(a), are visible from Earth with a small telescope (and, for a few people, with the naked eye). They are known as the *Galilean moons* after Galileo Galilei, who discovered them in 1610. ∞ *(Sec. 1.2)*

Saturn (Figure 7.2), the next major body we encounter as we continue outward beyond Jupiter, was the most distant planet known to Greek astronomers. Named after the father of Jupiter in Greco-Roman mythology, Saturn orbits at almost twice Jupiter's distance from the Sun, making it considerably fainter (as seen from Earth) than either Jupiter or Mars. The planet's banded atmosphere is somewhat similar to Jupiter's, but Saturn's atmospheric bands are much less distinct. Overall, the planet has a rather uniform butterscotch hue. Again, like Jupiter and unlike the inner planets, Saturn has many moons orbiting it. Saturn's best-known feature, its spectacular *ring system*, is clearly visible in Figure 7.2. The moons and rings of the jovian planets are the subject of Chapter 8.

Spacecraft Exploration

Most of our detailed knowledge of the jovian planets has come from NASA spacecraft visiting those worlds. Of particular importance are the two *Voyager* probes and the more recent *Galileo* and *Cassini* missions.

The *Voyager* spacecraft left Earth in 1977, reaching Jupiter in March (*Voyager 1*) and July (*Voyager 2*) of 1979. Both craft subsequently used Jupiter's strong gravity to send them on to Saturn in a maneuver called a *gravity assist* (see *Discovery 7-1*). *Voyager 2* took advantage of a rare planetary configuration that enabled it to use Saturn's gravity to propel it to Uranus and then on to Neptune in a spectacularly successful "grand tour" of the outer planets. Each craft carried equipment to study planetary magnetic fields and magnetospheres, as well as radio, visible-light, and infrared sensors to analyze reflected and emitted radiation from the jovian planets and their moons. The data they returned revolutionized our knowledge of all the jovian worlds and remain our primary source of detailed information on Uranus and Neptune. The two *Voyager* craft are now headed out of the solar system, still returning data as they race toward interstellar space.

Galileo was launched in 1989 and reached Jupiter in December 1995, after a circuitous route that took it three times through the inner solar system, receiving

(a)

(b)

R I V U X G

▲ FIGURE 7.1 **Jupiter** (a) Photograph of Jupiter made with an Earth-based telescope, showing the planet and its four Galilean moons. (b) A *Hubble Space Telescope* image of Jupiter, in true color—that is, as the human eye would see it. *(AURA/NASA)*

R I V U X G

◀ FIGURE 7.2 **Saturn** Saturn as seen by the *Hubble Space Telescope* in 2003 when the planet was at its maximum tilt (27°) toward Earth in its 30-year orbit about the Sun. The colors in this photo are natural—that is, as we would see it if we had telescopic vision. (See also the full-page opening photo on page 192.) *(NASA)*

gravity assists from both Venus and Earth before finally reaching its target. The mission had two components: an atmospheric probe and an orbiter. The probe, slowed by a heat shield and a parachute, made measurements and chemical analyses as it descended into Jupiter's atmosphere. The orbiter executed a complex series of gravity-assisted maneuvers through Jupiter's moon system, returning to some moons studied by *Voyager* and visiting others for the first time. Many of *Galileo*'s findings are described in Chapter 8.

The mission was scheduled to end in December 1997 but was so successful that NASA extended its lifetime three times, for a total of 6 additional years, to obtain even more detailed data on Jupiter's inner moons. Eventually, however, the spacecraft began to run out of the fuel needed to fine-tune its orbit and keep its antenna pointed toward Earth. Knowing that they would ultimately lose control of the probe, and wishing to avoid at all costs any possible contamination of Jupiter's moon Europa (see Section 8.1), the *Galileo* team elected to end the mission by steering the spacecraft into Jupiter's atmosphere in September 2003.

In 1997, NASA launched the *Cassini* mission to Saturn. The craft reached Saturn in 2004 after four planetary gravity assists—two from Venus and one each from Earth and Jupiter—en route. Figure 7.3 shows *Cassini*'s best close-up view of Jupiter during its encounter with that planet in 2001. In December 2004 the craft released a probe built by the European Space Agency into the atmosphere of Titan, Saturn's largest moon (see Section 8.2). *Cassini* is now orbiting among Saturn's moons, much as *Galileo* did at Jupiter, returning detailed data on the moons, the planet, and its famous ring system. The mission was originally scheduled to last for 4 years after arrival at Saturn, but the gravity-assisted maneuver that placed the probe in orbit around the planet was so accurate that almost no fuel had to be used to fine-tune the trajectory, extending *Cassini*'s lifetime by several years. The current mission, now renamed *Cassini Solstice*, is scheduled to end in 2017.

7.2 The Discoveries of Uranus and Neptune

British astronomer William Herschel discovered Uranus in 1781. Herschel was mapping the faint stars in the sky when he came across an unusual object that he described as "a curious either nebulous star or perhaps a comet." Repeated observations showed that it was neither. The object appeared as a disk in Herschel's 6-inch telescope and moved relative to the stars, but it traveled too slowly to be a comet. Herschel soon realized that he had found the seventh planet in the solar system.

Because this was the first new planet discovered in recorded history, the event caused quite a stir. The story goes that Herschel's first instinct was to name the new planet *Georgium Sidus* (Latin for "George's star") after his king, George III of England. The world was saved from a planet named George by the wise advice of another astronomer, Johann Bode, who suggested instead that the tradition of using names from ancient mythology be continued and that the planet be named after Uranus, the father of Saturn.

Uranus is just barely visible to the naked eye, under the best possible observing conditions. It looks like a faint, undistinguished star. Even through a large optical telescope, Uranus appears hardly more than a tiny, pale, slightly greenish disk. Figure 7.4 shows a close-up visible-light image taken by *Voyager 2* in 1986. Uranus's apparently featureless atmosphere contrasts sharply with the bands and spots visible in the atmospheres of the other jovian planets.

ANIMATION/VIDEO *Galileo* Mission to Jupiter

▲ FIGURE 7.3 **The View from *Cassini*** This *Cassini* spacecraft image of Jupiter shows intricate atmospheric clouds of different heights, thicknesses, and chemical composition. The image is approximately true color. *(NASA)*

▶ FIGURE 7.4 **Uranus** A close-up view of Uranus sent back to Earth by *Voyager 2* while the spacecraft was whizzing past the planet at 10 times the speed of a rifle bullet. Color is natural and the upper atmosphere is nearly featureless, except for a few wispy clouds in the northern hemisphere. *(NASA)*

7-1 DISCOVERY

Spaceflight in the Solar System

Since the 1960s, dozens of uncrewed space missions have traveled throughout the solar system. All of the planets have been probed at close range, and robot spacecraft have also visited numerous comets and asteroids. The impact of these missions on our understanding of our planetary system has been revolutionary. The five chapters of Part 2 of this text display many examples of marvelous images radioed back to Earth by the remarkable robots that have explored our nearby environment in space.

(Sovfoto/Eastfoto)

Mercury has been visited by two spacecrafts to date—in the mid-1970s U.S. *Mariner 10* bypassed (but did not orbit or land on) the planet while snapping thousands of images, revealing it to be almost as heavily cratered as the Moon. More than 30 years went by before NASA's *Messenger* probe arrived at Mercury in 2011, where it is now actively mapping the surface of this peculiar place. By contrast, more robots have arrived at Venus than any other planet. The Soviet Union took the lead in the 1960s when nearly a dozen *Venera* probes orbited (and some actually landed on) Venus—see the first figure showing Russian engineers building the craft. Since then the United States (*Pioneer* and *Magellan*—see the second figure showing its launch in 1989 from the Space Shuttle) and Europe (*Express*) have sent several more craft

to spy on this hellish world, especially using radar. The new data have taught us much about Venus's surface and atmosphere, and the results have helped us better understand weather on our own planet Earth.

Mars has been the target of very active robotic exploration. The United States has sent more than a dozen probes to orbit the "red planet" and often to land on its surface; Russia and Europe have also aimed craft at Mars, but most have missed the planet or crash landed. Several U.S. *Mariner* craft paved the way in the 1960s, showing the planet to be surprisingly inhospitable yet geologically intriguing. *Viking* in the 1970s was one of NASA's finest missions, not only safely landing two craft but also searching for life (which it did not find). During the past decade, a series of miniaturized robots have been roving over Mars's surface, sampling the air and dirt, drilling into stones, and digging for ice. The true-color panoramic view below was taken in 2005 from the rover, *Spirit* whose mother ship is shown in the foreground. This and other rovers, such as *Phoenix*, which landed on Mars

(NASA)

(NASA)

in 2008, have mainly been searching for water, as "follow the water" is a popular mantra when searching for extraterrestrial life—but they have not yet found either liquid water or life of any kind.

Celestial mechanics—the study of the motions of gravitationally interacting objects—is an essential tool for scientists and engineers wishing to navigate manned and unmanned spacecraft throughout the solar system. Robot probes can now be sent on stunningly accurate trajectories, expressed in the trade with such slang phrases as "sinking a corner shot on a billion-kilometer pool table." Near-flawless rocket launches, aided by occasional midcourse changes in flight paths, now enable interplanetary navigators to steer remotely controlled spacecraft through an imaginary window of space just a few kilometers wide and billions of kilometers away.

However, sending a spacecraft to another planet requires a lot of energy—often more than can be conveniently provided by a rocket launched from Earth or safely transported in a shuttle for launch from orbit. Faced with these limitations, mission scientists often use their knowledge of celestial mechanics to make "slingshot" maneuvers to boost an interplanetary probe to a more energetic orbit and also aid navigation toward the target, all at no additional cost!

The figure at right illustrates a gravitational slingshot, or *gravity assist,* in action. A spacecraft approaches a planet, passes close by, and then escapes along a new trajectory. Obviously, the spacecraft's *direction* of motion is changed by the encounter. Less obviously, the spacecraft's *speed* is also altered as the planet's gravity propels the spacecraft in the direction of the planet's motion. By careful choice of incoming trajectory, the craft can either speed up (by passing "behind" the planet, as shown) or slow down (by passing in front) by as much as twice the planet's orbital speed. Of course, there is no free lunch—the spacecraft gains energy from, or loses it to, the planet's

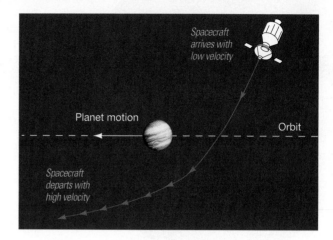

motion, causing its orbit to change ever so slightly. However, since planets are so much more massive than spacecraft, the effect is tiny.

Such a slingshot maneuver has been used many times in missions to both the inner and the outer planets, as illustrated in the last figure. Shown here are the trajectories of the two U.S. *Voyager* spacecraft, which were perhaps humankind's greatest solar system venture to date. The gravitational pulls of these giant worlds whipped each craft around at each visitation in the 1970s and 1980s, enabling flight controllers to get considerable extra "mileage" out of the probes. *Voyager 1* is now high above the plane of the solar system, having been deflected up and out following its encounter with Saturn. *Voyager 2* continued on for a "Grand Tour" of the four jovian planets and is now well beyond former planet Pluto in the Kuiper belt. NASA's current *New Horizons* mission used a gravity assist from Jupiter to propel the spacecraft out to Pluto's orbit.

Most of what we know about the outer planets has been gained from spacecraft propelled in this way. As described in the text, the U.S. *Galileo* mission to Jupiter (a 7-year flight that arrived in 1995) and the U.S./European *Cassini* mission to Saturn (arriving in 2004) used multiple gravity assists, both to reach the target planet and subsequently to navigate among that planet's moons. Every encounter with a moon had a slingshot effect—sometimes accelerating and sometimes slowing the probe, but each time moving it into a different orbit—and every one was carefully calculated long before the spacecraft ever left Earth. As noted in this and the next chapter, both of these robotic missions have returned spectacular images and detailed data about these giant planets and their bizarre moons.

1: Sept 5, 1977
2: Aug 20, 1977

Jupiter at launch

1: Mar 3, 1979
2: Aug 9, 1979

Earth

Saturn at launch

2: Aug 27, 1981

1: Nov 13, 1980

Voyager 1

2: Aug 15, 1989

Voyager 2

Neptune at launch

2: Jan 30, 1986

Uranus at launch

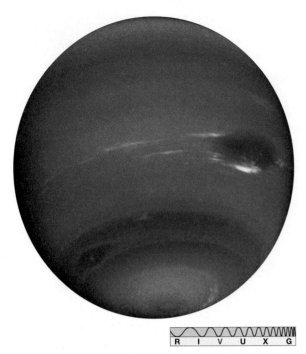

▲ **FIGURE 7.5 Neptune** Neptune as seen in natural color by *Voyager* 2 from a distance of 1,000,000 km. Its cloud streaks range in width from 50 km to 200 km. *(NASA)*

CONCEPT CHECK ▶ How did observations of Uranus lead to the discovery of Neptune?

Once Uranus was discovered, astronomers set about charting its orbit. They quickly discovered a discrepancy between the planet's predicted and observed positions. Try as they might, they could not find an elliptical orbit that fit the planet's path. By the early 19th century the discrepancy had grown to a quarter of an arc minute, far too large to be explained away as observational error. Uranus seemed to be violating—slightly—Kepler's laws of planetary motion. ∞ *(Sec. 1.3)*

Astronomers realized that although the Sun's gravitational pull dominates the planet's orbital motion, the small deviation meant that some unknown body was exerting a much weaker but still measurable gravitational force on Uranus. There had to be *another* planet in the solar system influencing Uranus. In September 1845, after almost 2 years of work, an English mathematician named John Adams solved the problem of determining the new planet's mass and orbit. In June 1846 a French mathematician, Urbain Leverrier, independently came up with essentially the same answer. Later that year a German astronomer named Johann Galle found the new planet within one or two degrees of the predicted position. The new planet was named Neptune, and Adams and Leverrier are now jointly credited with its discovery.

Unlike Uranus, distant Neptune cannot be seen with the naked eye, although it can be seen through binoculars or a small telescope. In fact, according to his notes, Galileo may actually have seen Neptune, although he surely had no idea what it really was at the time. Through a large telescope on Earth, the planet appears only as a small, bluish disk, with few features visible. Even under the best observing conditions, only a few markings, suggestive of multicolored cloud bands, can be seen. With *Voyager 2*'s arrival, much more detail emerged (Figure 7.5)—in fact, the bulk of our knowledge of the two outermost jovian worlds has come from just this one spacecraft. Superficially, at least, Neptune resembles a blue-tinted Jupiter, with atmospheric bands and spots clearly evident.

7.3 Bulk Properties of the Jovian Planets

Table 7.1 extends Table 6.1 to include the four jovian planets. Figure 7.6 shows them to scale, along with Earth for comparison.

Physical Characteristics

The large masses and radii and relatively low average densities of these gigantic worlds imply that they differ radically in both composition and structure from the terrestrial planets. ∞ *(Sec. 4.1)* Saturn's density is actually less than that of water—the planet would float, if only we could find a bathtub big enough! Hydrogen and helium make up most of the mass of Jupiter and Saturn and about half the mass of Uranus and Neptune. These light gases have densities on Earth (at room temperature and sea level) of 0.08 kg/m^3 and 0.16 kg/m^3, respectively, but they are compressed enormously by the jovian planets' strong gravitational fields. The abundance of hydrogen and helium on these worlds is a consequence of the strong jovian gravity. The jovian planets are massive enough to have retained even the lightest gas—hydrogen—and very little of their original atmospheres have escaped since the birth of the solar system 4.6 billion years ago. ∞ *(More Precisely 5-1)*

None of the jovian planets has a solid surface of any kind. Their gaseous atmospheres just become hotter and denser with depth because of the pressure of the overlying layers, eventually becoming liquid in the interior. What we see from Earth is simply the outermost layer of atmospheric clouds.

At the heart of each jovian planet lies a dense, compact core many times more massive than Earth. As we will see in Section 7.6, three of the four (Jupiter, Saturn,

► **FIGURE 7.6 Jovian Planets**   Jupiter, Saturn, Uranus, and Neptune, showing their relative scales compared to Earth. Note the very different sizes and atmospheric features of these five worlds. *(NASA)*

Jupiter

Saturn

Earth

Uranus

Neptune

and Neptune) have significant internal heating, which impacts the behavior and appearance of their atmospheres.

Rotation Rates

Because the jovian planets have no solid surfaces to "tie down" the gas flow, different parts of their atmospheres can and do move at different speeds. This state of affairs—when the rotation rate is not constant from one location to another—is called **differential rotation.** It is not possible in solid objects like the terrestrial worlds but is normal for fluid bodies such as the jovian planets (and the Sun—see Chapter 9).

The amount of differential rotation on Jupiter is small—the equatorial regions rotate once every $9^h 50^m$, while the higher latitudes take about 6 minutes longer. On Saturn the difference between the equatorial and polar rotation rates is 26 minutes, again slower at the poles. On Uranus and Neptune the differences are greater still—more than 2 hours for Uranus and 6 for Neptune—with the poles rotating more rapidly in both cases. The differential rotation observed in the clouds reflects large-scale wind flows in the planets' atmospheres.

More meaningful measurements of overall planetary rotation rates are provided by observations of their magnetospheres. All four jovian worlds have strong magnetic fields and emit radiation at radio wavelengths. The strength of this radio emission varies with time and repeats periodically. Scientists assume that this period matches the rotation of the planet's deep interior, where (as on Earth) the magnetic field arises. ⚬ *(Sec. 5.7)* The rotation periods listed in Table 7.1 were obtained in this way. There is no clear relationship between the interior and atmospheric rotation

ANALOGY: Water in a very big bathtub illustrates two useful facts:

Saturn is less dense than water and would therefore float.

Water swirling down the drain rotates differentially—faster at the center than at the edges.

(MA)™ **SELF-GUIDED TUTORIAL** Jupiter—Differential Rotation

TABLE 7.1   Planetary Properties

	Mass		Radius		Average Density (kg/m³)	Surface[1] Gravity (Earth = 1)	Escape Speed (km/s)	Rotation Period (solar days)	Axial Tilt (degrees)	Surface[1] Temperature (K)	Surface[1] Magnetic Field (Earth = 1)
	(kg)	(Earth = 1)	(km)	(Earth = 1)							
Mercury	3.3×10^{23}	0.550	2,400	0.38	5400	0.38	4.3	59	0	100–700	0.01
Venus	4.9×10^{24}	0.82	6,100	0.95	5300	0.91	10	-243^2	179	730	0.0
Earth	6.0×10^{24}	1.00	6,400	1.00	5500	1.00	11	1.00	23	290	1.0
Mars	6.4×10^{23}	0.11	3,400	0.53	3900	0.38	5.0	1.03	24	180–270	0.0
Jupiter	1.9×10^{27}	320	71,000	11	1300	2.5	60	0.41	3	120	14
Saturn	5.7×10^{26}	95	60,000	9.5	710	1.1	36	0.43	27	97	0.7
Uranus	8.7×10^{25}	15	26,000	4.0	1200	0.91	21	-0.69^2	98	58	0.7^3
Neptune	1.0×10^{26}	17	25,000	3.9	1700	1.1	24	0.72	30	59	0.4^3

[1]For the jovian planets, "surface" refers to the cloud tops.

[2]The minus sign indicates retrograde rotation.

[3]Average values. Because of the planet's asymmetric magnetic field geometry (see Section 7.6), the field strength varies substantially from one point on the surface to another.

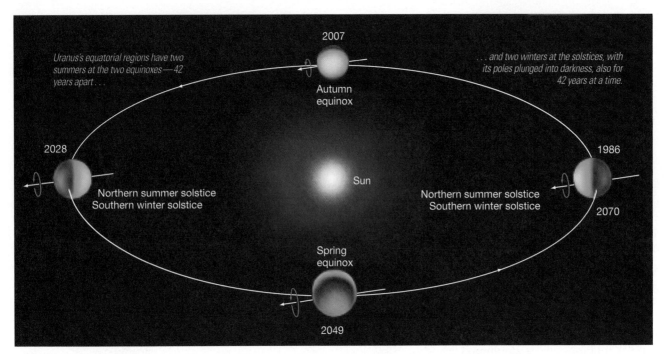

Uranus's equatorial regions have two summers at the two equinoxes—42 years apart . . .

2007

Autumn equinox

. . . and two winters at the solstices, with its poles plunged into darkness, also for 42 years at a time.

2028

Northern summer solstice
Southern winter solstice

Sun

1986

Northern summer solstice
Southern winter solstice

2070

Spring equinox

2049

▲ **FIGURE 7.7 Seasons on Uranus**   Because of Uranus's axial tilt of 98°, the planet experiences the most extreme seasons known in the solar system.

rates of the jovian planets. In Jupiter and Saturn the interior rotation rate matches the polar rotation. In Uranus the interior rotates more slowly than any part of the atmosphere. In Neptune the reverse is true.

Curiously, the *Cassini* measurement of Saturn's rotation period using the planet's magnetic field is about 6 minutes longer than the corresponding result obtained by *Voyager* more than 20 years earlier. Scientists don't think that Saturn's actual rotation rate has changed during this relatively short time. Rather, it seems that the planet's magnetic field may not be as good an indicator of interior rotation as was previously thought.

The observations used to measure a planet's rotation period also let us determine the orientation of its rotation axis. Jupiter's rotation axis is almost perpendicular to the orbit plane, the axial tilt being only 3°, quite small compared with Earth's tilt of 23°. The tilts of Saturn and Neptune (Table 7.1) are similar to those of Earth and Mars.

Each planet in our solar system seems to have some outstanding peculiarity. In the case of Uranus, it is its rotation axis. Unlike the other major planets, which have their spin axes (very) roughly perpendicular to the ecliptic plane, Uranus's rotation axis lies almost within that plane. We might say that Uranus is "tipped over" onto its side, for its axial tilt is 98°. (Because the north pole lies below the ecliptic plane, the planet's rotation, like that of Venus, is classified as retrograde.) As a result, the north pole[1] of Uranus, at some point in the planet's orbit, points almost directly toward the Sun. Half a Uranian year later the south pole faces the Sun, as illustrated in Figure 7.7. When *Voyager 2* encountered the planet in 1986, the north pole happened to be pointing nearly directly at the Sun, so it was midsummer in the northern hemisphere. At the same time, the south polar regions were halfway through a 42-year period of permanent darkness.

No one knows why Uranus is tilted in this way. Astronomers speculate that a catastrophic event, such as a grazing collision between the planet and another planet-size body, might have altered the planet's spin axis. ⚲ *(Sec. 4.3)* However, there is no direct evidence for such an occurrence and no theory to tell us how we should seek to confirm it.

CONCEPT CHECK ▶   Why do observations of a planet's magnetosphere allow astronomers to measure the rotation rate of the interior?

---

[1]We adopt here the convention that a planet's rotation is counterclockwise as seen from above the north pole (that is, planets always rotate from west to east).

7.4 Jupiter's Atmosphere

Just as Earth has been our yardstick for the terrestrial worlds, Jupiter will be our guide to the outer planets. We therefore begin our study of jovian atmospheres with Jupiter itself.

Overall Appearance and Composition

Visually, Jupiter is dominated by two atmospheric features (both clearly visible in Figure 7.1): a series of ever-changing atmospheric cloud bands arranged parallel to the equator and an oval atmospheric blob called the **Great Red Spot.** The cloud bands display many colors—pale yellows, light blues, deep browns, drab tans, and vivid reds, among others. The Red Spot is one of many features associated with Jupiter's weather. This egg-shaped region, the long diameter of which is approximately twice Earth's diameter, seems to be a hurricane that has persisted for hundreds of years.

What is the cause of Jupiter's colors? The most abundant gas in Jupiter's atmosphere is molecular hydrogen (roughly 86 percent of all molecules), followed by helium (nearly 14 percent). Small amounts of atmospheric methane, ammonia, and water vapor are also found. None of these gases can, by itself, account for Jupiter's observed coloration. For example, frozen ammonia and water vapor would produce white clouds, not the many colors we see. Scientists think that complex chemical processes occurring in Jupiter's turbulent atmosphere are responsible for these colors, although the details are not fully understood. The trace elements sulfur and phosphorus may play important roles in influencing the cloud colors—particularly the reds, browns, and yellows. The energy that powers the reactions comes in many forms: the planet's own internal heat, solar ultraviolet radiation, aurorae in the planet's magnetosphere, and lightning discharges within the clouds. ∞ *(Sec. 5.7)*

Astronomers describe the banded structure of Jupiter's atmosphere as consisting of a series of lighter-colored **zones** and darker **belts** crossing the planet. The zones and belts vary in both latitude and intensity during the year, but the general pattern is always present. These variations appear to be the result of convective motion in the planet's atmosphere. (Recall from Chapter 5 that convection can occur whenever cool gas overlies hotter material, as is the case here.) ∞ *(Sec. 5.3) Voyager* sensors indicated that the zones lie above upward-moving convective currents, while the belts are the downward part of the cycle, where material is generally sinking, as illustrated in Figure 7.8. Thus, because of the upwelling material below them, the zones are regions of high pressure; the belts, conversely, are low-pressure regions. However, observations made during the *Cassini* flyby have challenged this standard view, suggesting that upward convection is actually confined to the belts. For now, planetary scientists have no clear resolution of this contradiction between the *Voyager* and *Cassini* results.

The belts and zones are Jupiter's equivalents of the familiar high- and low-pressure systems that control the weather on Earth. A major difference from Earth is that Jupiter's rapid differential rotation causes these systems to wrap all the way around the planet, instead of forming localized circulating storms as on our own world.

▶ **FIGURE 7.8 Jupiter's Convection** The colored bands in Jupiter's atmosphere are associated with vertical convective motion. As on Earth, winds tend to blow from high-pressure regions to low-pressure regions. Jupiter's rapid rotation channels these winds into an east–west flow pattern, as indicated by the three yellow-red arrows drawn atop the belts and zones. (*NASA*)

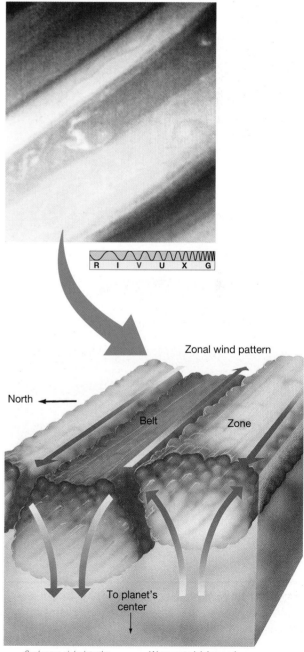

This is a real Voyager *photo of Jupiter's clouds, showing its actual banded structure.*

R I V U X G

Zonal wind pattern

North

Belt

Zone

To planet's center

Cooler gas sinks into the atmosphere, creating darker bands atop lower-pressure regions.

Warm material rises and creates lighter-colored zones.

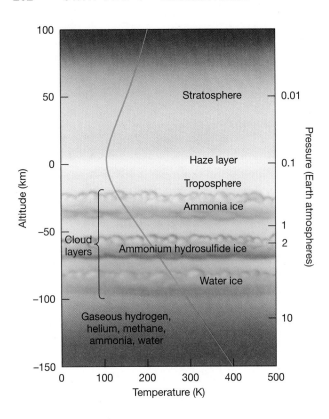

◄ FIGURE 7.9 **Jupiter's Atmosphere**   Models of the vertical structure of Jupiter's atmosphere indicate that the planet's clouds are arranged in three main layers, each with quite different colors and chemistry. The white regions are the tops of the upper ammonia clouds. The yellows, reds, and browns are associated with the second cloud layer, which is composed of ammonium hydrosulfide ice. The lowest (bluish) cloud layer is water ice. The blue curve shows how Jupiter's atmospheric temperature depends on altitude. (For comparison with Earth, see Figure 5.5)

Because of the pressure difference between them, the belts and zones lie at slightly different heights in Jupiter's atmosphere. Cloud chemistry is very sensitive to temperature, and the temperature difference between the two levels is the main reason for the different colors we see.

Underlying the bands is an apparently stable pattern of east-west winds called Jupiter's **zonal flow.** The equatorial regions of Jupiter's atmosphere rotate faster than the planet as a whole, with an average flow speed of about 500 km/h in the easterly direction, somewhat similar to the jet stream on Earth. At higher latitudes there are alternating regions of westward and eastward flow, the alternations corresponding to the pattern of belts and zones. The flow speed decreases toward the poles. Near the poles, where the zonal flow disappears, the band structure vanishes also.

Atmospheric Structure

Jupiter's clouds are arranged in several layers, with white ammonia clouds generally overlying the colored layers, whose composition we will discuss in a moment. Above the ammonia clouds lies a thin, faint layer of haze created by chemical reactions similar to those that cause smog on Earth. When we observe Jupiter's colors, we are looking down to many different depths in the planet's atmosphere.

Figure 7.9 is a diagram of Jupiter's atmosphere, based on observations and computer models. Because the planet lacks a solid surface to use as a reference level for measuring altitude, scientists conventionally take the top of the troposphere (the turbulent region containing the clouds we see) to lie at 0 km. With this level as the zero point, the troposphere's colored cloud layers all lie at negative altitudes in the diagram. As on other planets, weather on Jupiter is the result of convection in the troposphere. The haze layer lies at the upper edge of Jupiter's troposphere. The temperature at this level is about 110 K. Above the troposphere, as on Earth, the temperature rises as the atmosphere absorbs solar ultraviolet light.

Below the haze layer, at an altitude of −30 km, lie white, wispy clouds of ammonia ice. The temperature at this level is 125–150 K. A few tens of kilometers below the ammonia clouds, the temperature is a little warmer—above 200 K—and the clouds are made up mostly of droplets or crystals of ammonium hydrosulfide, produced by reactions between ammonia and hydrogen sulfide. At deeper levels the ammonium hydrosulfide clouds give way to clouds of water ice or water vapor. The top of this lowest cloud layer, which is not seen in visible-light images of Jupiter, lies some 80 km below the top of the troposphere.

In December 1995 the *Galileo* atmospheric probe arrived at Jupiter. The probe survived for about an hour before being crushed by atmospheric pressure at an altitude of −150 km (that is, just at the bottom of Figure 7.9). The probe's entry location (Figure 7.10) was in Jupiter's equatorial zone and, as luck would have it, coincided with an atypical hole almost devoid of upper-level clouds, and with abnormally low water content. However, once these factors were taken into account, *Galileo*'s findings on wind speed, temperature, and composition were in good agreement with the description presented above.

R I V U X G

▲ FIGURE 7.10 *Galileo's* **Entry Site**   The arrow on this image shows where the *Galileo* atmospheric probe plunged into Jupiter's cloud deck on December 7, 1995. Until its demise, the probe took numerous weather measurements, transmitting those signals to the orbiting mother ship, which then relayed them to Earth. (*NASA*)

Weather on Jupiter

In addition to the large-scale zonal flow, Jupiter has many small-scale weather patterns. The Great Red Spot (Figure 7.11) is a prime example. Observational records indicate that it has existed continuously in one form or another for more than 300 years, and it may well be much older. *Voyager* and *Galileo* observations show the Spot to be a region of swirling, circulating winds, like a whirlpool or a terrestrial hurricane, a persistent and vast atmospheric storm. The Spot varies in length, averaging about 25,000 km. It rotates around Jupiter at a rate similar to that of the planet's interior, suggesting that its roots lie far below the atmosphere. Astronomers think that the Spot is somehow sustained by Jupiter's large-scale atmospheric motion. The zonal motion north of the Spot is westward, while that to the south is eastward (Figure 7.11), supporting the idea that the Spot is confined and powered by the zonal flow. Turbulent eddies form and drift away from the spot's edge. The center, however, remains tranquil in appearance, like the eye of a hurricane on Earth.

The *Voyager* mission discovered many smaller spots that are also apparently circulating storm systems. Examples of these smaller systems are the **white ovals** seen in many images of Jupiter (see, for example, the one just south of the Great Red Spot in Figure 7.11). Their high cloud tops give these regions their white color. The white oval in Figure 7.11 is known to be at least 40 years old. The inset shows a **brown oval**—a long-lived hole in the overlying clouds that allows us to look down into the lower atmosphere. Many more oval storm systems—dark and light—can be seen in the *Cassini* view of Jupiter shown in Figure 7.3.

The simplified picture presented in Figure 7.8, with wind direction alternating between adjacent bands as Jupiter's rotation deflects the winds into eastward or westward streams, is really too crude to describe the many details observed in Jupiter's atmosphere. More likely, the interaction between convection in Jupiter's atmosphere and the planet's rapid differential rotation channels the largest eddies into the observed zonal pattern, but smaller eddies—like the Red Spot and white ovals—tend to cause localized irregularities in the flow.

In the late 1990s astronomers noted with interest the collision and merger of three relatively small white ovals in Jupiter's atmosphere (Figure 7.12a). For several years the resultant larger system remained a white oval, but in early 2006 it changed from white to brown to red, becoming in a matter of months a smaller version of the Great Red Spot! Scientists think that the color of the Great Red Spot may be due to that storm's enormous size and strength, which lifts its cloud tops high above the surrounding clouds where solar ultraviolet radiation causes chemical reactions that produce the color. The reddening of the smaller spot may indicate that that storm is intensifying, and might even rival the Great Red Spot some day.

Merger and growth may well be the mechanism by which large storms on the jovian planets form and strengthen. Small storms that approach too close

A typical break ("brown oval") in Jupiter's atmosphere

5,000 km

Note the complex turbulence to the left of both the Red Spot and the smaller white oval below it.

Direction of gas flow

This is Earth, drawn to scale relative to Jupiter's clouds.

10,000 km

R I V U X G

INTERACTIVE ▲ **FIGURE 7.11 Jupiter's Red Spot** *Voyager 1* took this photograph of Jupiter's Great Red Spot (upper right) from a distance of about 100,000 km. The arrows indicate direction of gas flow above, below, and inside the Great Red Spot. *(NASA)* MA

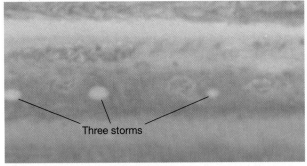

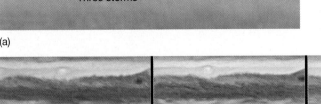

Three storms

(a)

◄ FIGURE 7.12 **Red Spot Evolution** (a) Between 1997 and 2000, astronomers watched as three white ovals in Jupiter's southern hemisphere merged to form a single large storm. Each oval, captured here by the *Cassini* spacecraft cameras, is about half the size of Earth. (b) In mid-2008, the *Hubble* telescope recorded this sequence of images at monthly intervals (left to right), showing a "baby red spot" (arrows) approaching the Great Red Spot and being destroyed by it. *(NASA)*

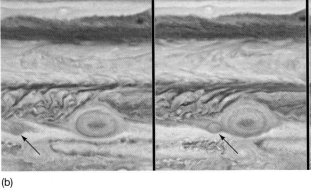

(b)

to larger ones are swept up and absorbed. Figure 7.12(b) shows a time sequence of *Hubble Space Telescope* images of the Great Red Spot and a smaller, "baby red spot" that appeared in 2008 at almost the same latitude as the bigger storm, only to be shredded by the spot's swirling winds within a few months.

Although we cannot fully explain their formation, we can at least offer a partial explanation for the longevity of storm systems on Jupiter. On Earth a large storm, such as a hurricane, forms over the ocean and may survive for many days, but it dies quickly once it encounters land because the land disrupts the energy supplied and flow patterns that sustain the storm. Jupiter has no continents, so once a storm is established and has reached a size at which other storm systems cannot destroy it, apparently little affects it. The larger the system, the longer its lifetime.

7.5 Atmospheres of the Outer Jovian Worlds

Composition of Saturn's Atmosphere

Saturn is not as colorful as Jupiter. Figure 7.13(a) shows yellowish and tan cloud bands that parallel the equator, but these regions display less atmospheric structure than do the bands on Jupiter. No obvious large spots or ovals adorn Saturn's cloud decks. Storms do exist, but the color changes that distinguish them on Jupiter are largely absent on Saturn. Figure 7.13(b) is the highest-resolution full-global image ever made of Saturn—a mosaic of more than a hundred photographs taken in true color—showing clarity yet subtlety in the structure of its cloud deck.

Spectroscopic studies show that Saturn's atmosphere consists of molecular hydrogen (92.4 percent) and helium (7.4 percent), with traces of methane and ammonia. As on Jupiter, hydrogen and helium dominate. However, the fraction of helium on Saturn is substantially less than on Jupiter. It is unlikely that the processes that created the outer planets preferentially stripped Saturn of nearly half its helium or that the missing helium somehow escaped from the planet while the hydrogen remained behind. Instead, astronomers theorize that at some time in Saturn's past the helium liquefied and then sank toward the center of the planet, reducing the abundance of that gas in the outer layers. The reason for this gas-to-liquid transformation is discussed in Section 7.6.

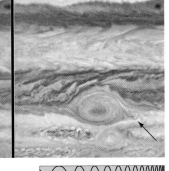

(a)

(b)

◄ FIGURE 7.13 **Saturn, Up Close** (a) The banded structure of Saturn's atmosphere appears clearly in this *Voyager* 2 image. (Three of Saturn's moons appear at bottom; a fourth, not visible here, casts a black shadow on the cloud tops.) (b) This *Cassini* image is actually a mosaic of many images taken in true color. *(NASA)*

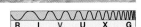
R I V U X G

Figure 7.14 illustrates the vertical structure of Saturn's atmosphere. In many respects it is similar to Jupiter's (Figure 7.9), except that Saturn's overall atmospheric temperature is a little lower because of its greater distance from the Sun. As on Jupiter, the troposphere contains clouds arranged in three distinct layers, composed (in order of increasing depth) of ammonia ice, ammonium hydrosulfide ice, and water ice. Above the clouds lies a layer of haze.

The total thickness of the three cloud layers in Saturn's atmosphere is roughly 250 km—about three times the cloud thickness on Jupiter—and each layer is thicker than its counterpart on Jupiter. The reason for this difference is Saturn's weaker gravity. At the haze level, Jupiter's gravitational field is nearly 2.5 times stronger than Saturn's, so Jupiter's atmosphere is pulled much more powerfully toward the center of the planet and the cloud layers are squeezed more closely together.

The colors of Saturn's cloud layers, as well as the planet's overall yellowish coloration, are likely due to the same basic photochemistry as on Jupiter. However, because Saturn's clouds are thicker, there are fewer holes and gaps in the top layer, with the result that we rarely glimpse the more colorful levels below. Instead, we see only the topmost layer, which accounts for Saturn's less-varied appearance.

Weather on Saturn

Computer-enhanced images of Saturn clearly show bands, oval storm systems, and turbulent flow patterns, many looking very much like those on Jupiter. In addition, there is an apparently quite stable east–west zonal flow, although the wind speed is considerably greater than on Jupiter and shows fewer east–west alternations. The equatorial eastward jet stream reaches a brisk 1500 km/h. Astronomers do not fully understand the reasons for the differences between Jupiter's and Saturn's flow patterns.

Figure 7.15(a) shows a sequence of *Hubble Space Telescope* images of a storm system on Saturn. First detected in September 1990 as a large white spot

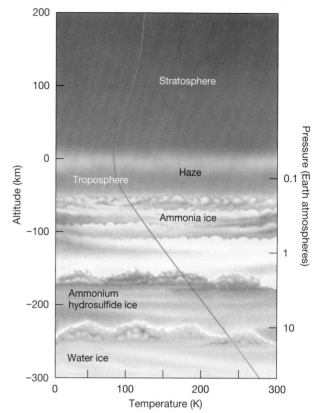

▲ **FIGURE 7.14 Saturn's Atmosphere** The vertical structure of Saturn's atmosphere contains several cloud layers, like Jupiter, but Saturn's weaker gravity results in thicker clouds and a more uniform appearance.

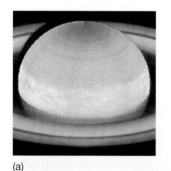
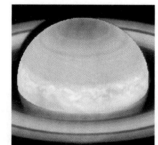

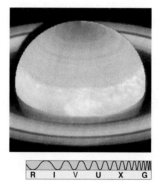

(a)

▲ **FIGURE 7.15 Storm on Saturn** (a) Circulating and evolving cloud systems on Saturn, imaged by the *Hubble Space Telescope* at approximately 2-hour intervals (left to right). (b) This huge storm was observed by the *Cassini* spacecraft in 2011. While churning its way through the northern hemisphere, it seems to leave behind a "tail" wrapped around the planet. *(NASA)*

 ANIMATION/VIDEO Saturn Cloud Rotation

(b)

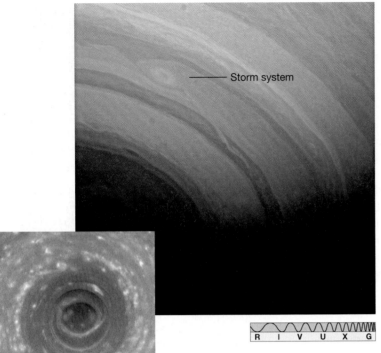

Storm system

This is a close-up of the polar vortex
at Saturn's south pole.

R I V U X G

◀ FIGURE 7.16 Saturn's Storm Systems   This false-color infrared image from *Cassini* shows what is thought to be a vast thunderstorm on Saturn. This storm also generated bursts of radio waves resembling the static created by lightning on Earth. *(NASA)*

just south of Saturn's equator, by November of that year the spot had developed into a band of clouds completely encircling the planet's equator. Figure 7.15(b) shows the same stormy bands as observed by the *Cassini* spacecraft some 20 years later. Astronomers think that the white coloration arose from crystals of ammonia ice that formed when an upwelling plume of warm gas penetrated the cool upper cloud layers. Because the crystals were freshly formed, they had not yet been affected by the chemical reactions that color the planet's other clouds. The turbulent flow patterns seen in and around the spot had many similarities to the flow around Jupiter's Great Red Spot. Routine observations of temporary atmospheric phenomena on the outer worlds are giving scientists greater insight into the dynamics of planetary atmospheres.

Since *Cassini*'s arrival at Saturn, scientists have been able to study the planet's storm systems in much more detail. Figure 7.16 shows a particularly large and complex system that appeared in 2006. These storms may be Saturn's equivalent of thunderstorms on Earth (although on a scale comparable to our entire planet). *Cassini*'s detectors have measured strong bursts of radio waves associated with them, most likely produced by intense lightning discharges deep below the cloud tops. The lightning is powered by convection and precipitation (water and ammonia "rain"), as on Earth, but the bursts are millions of times stronger than anything ever witnessed here at home. Astronomers suspect that this storm may actually be a long-lived phenomenon rooted deep in Saturn's atmosphere—perhaps a little like Jupiter's Great Red Spot—but normally hidden below the upper cloud layers. Occasionally it flares up, producing a bright plume visible from outside.

The inset to Figure 7.16 shows *Cassini*'s view of an Earth-sized *polar vortex* at Saturn's south pole. The orbiter's close-up imagery has allowed scientists to study the system with unprecedented clarity. As mentioned in Chapter 6, *Venus Express* has mapped a similar vortex on Venus in great detail. ∞ *(Sec. 6.3)* The opportunity to compare polar vortices in planets with such different atmospheres, orbits, and rotational properties as Venus, Earth, and Saturn offers astronomers a unique opportunity for comparative planetology.

Atmospheric Composition of Uranus and Neptune

The atmospheres of Uranus and Neptune are very similar in composition to that of Jupiter—the most abundant gas is molecular hydrogen (84 percent), followed by helium (about 14 percent) and methane, which is more abundant on Neptune (about 3 percent) than on Uranus (2 percent). Ammonia is not observed in any significant quantity in the atmosphere of either planet, most likely because, at the low atmospheric temperatures of those planets, the ammonia exists in the form of ice crystals, which are much harder to detect spectroscopically than gaseous ammonia.

The blue color of the outer jovian planets is the result of their relatively high percentages of methane. Because methane absorbs long-wavelength red light quite efficiently, sunlight reflected from these planets' atmospheres is deficient in red and yellow photons and appears blue-green or blue. The greater the concentration of methane, the bluer the reflected light. Therefore Uranus, having less methane, looks blue-green, while Neptune, having more methane, looks distinctly blue.

Weather on Uranus and Neptune

Voyager 2 detected only a few atmospheric features on Uranus (Figure 7.17), and even they became visible only after extensive computer enhancement. Few clouds exist in Uranus's cold upper atmosphere. Clouds of the kind found on Jupiter and Saturn are found only at lower, warmer atmospheric levels. The absence of high-level clouds means we must look deeply into the planet's atmosphere to see any structure, and the bands and spots that characterize flow patterns on the other jovian worlds are largely washed out on Uranus by intervening stratospheric haze. Series of computer-processed images such as Figures 7.17(a–c) have shown astronomers that Uranus's atmospheric clouds and flow patterns move around the planet in the same direction as the planet's rotation, with wind speeds ranging from 200 to 500 km/h.

Although Neptune lies farther from the Sun, its internal heat (see Section 7.6) means that its upper atmosphere is actually slightly warmer than that of Uranus. This is probably the reason why Neptune's atmospheric features are more easily visible than those of Uranus (Figure 7.18). The extra warmth causes Neptune's stratospheric haze to be thinner than the haze on Uranus and also makes the cloud layers less dense than they would otherwise be, with the result that they lie at higher levels in the atmosphere.

Voyager 2 detected numerous white methane clouds on Neptune (some of them visible in Figure 7.5) lying about 50 km above the main cloud tops. Neptune's equatorial winds blow from east to west with speeds of over 2000 km/h relative to the planet's interior. Why such a cold planet should have such rapid winds, and why they should be retrograde relative to the west-to-east rotation of the interior, remains a mystery.

Neptune sports several storm systems similar in appearance to those seen on Jupiter, and probably produced and sustained by the same basic processes. The largest storm, the **Great Dark Spot** (Figure 7.18a), was discovered by *Voyager 2* in 1989. Comparable in size to Earth, the Spot was located some 20°

CONCEPT CHECK ► List some similarities and differences between Jupiter's belts and zones and weather systems on Earth.

CONCEPT CHECK ► Why are atmospheric features much less evident on Uranus than on the other jovian planets?

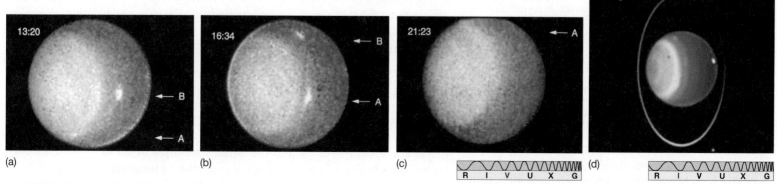

▲ **FIGURE 7.17 Uranus's Rotation** (a), (b), and (c) These computer-enhanced *Hubble Space Telescope* images, taken at roughly 4-hour intervals, show the motion of a pair of bright clouds (labeled A and B) in Uranus's southern hemisphere. (The numbers at the top give the time of each photo.) (d) An infrared image of Uranus shows that planet's ring system, as well as a number of clouds (pink and red regions) in the upper atmosphere. *(NASA)*

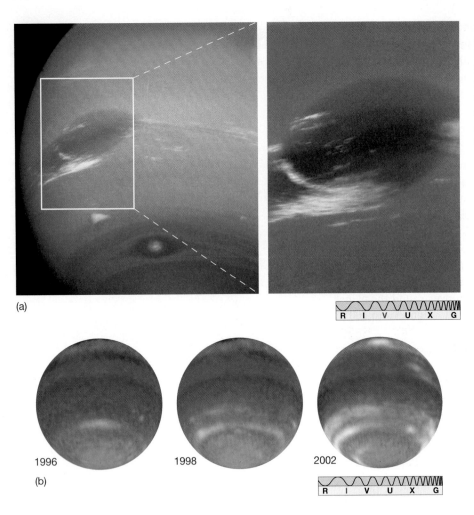

(a)

R I V U X G

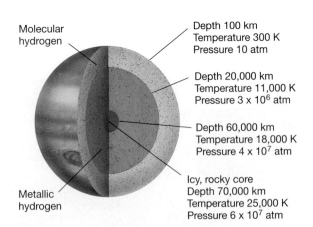

1996   1998   2002

(b)

R I V U X G

◄ FIGURE 7.18 **Neptune's Dark Spot**   (a) Close-up views, taken by *Voyager 2* of the Great Dark Spot of Neptune, show a large storm system in the planet's atmosphere, possibly similar in structure to Jupiter's Great Red Spot. The entire dark spot is roughly the size of Earth. (b) These *Hubble Space Telescope* views of Neptune were taken years apart (as marked). The cloud features (mostly methane ice crystals) are tinted pink here because they were imaged in the infrared, but they are white in visible light. *(NASA)*

south of the planet's equator—quite similar to the Great Red Spot on Jupiter. The Dark Spot also exhibited other characteristics in common with Jupiter's Red Spot. The flow around it was counterclockwise, and there appeared to be turbulence where the winds associated with the Dark Spot interacted with the zonal flow to its north and south. Astronomers did not have long to study the Dark Spot's properties, however. By the time detailed *Hubble Space Telescope* observations of the planet became available in the mid-1990s, the Spot had disappeared (Figure 7.18b). At around the same time, a new Dark Spot of comparable size became visible in Neptune's northern hemisphere. The origins and lifetimes of large storm systems on the jovian planets are currently not well understood.

7.6 Jovian Interiors

The visible cloud layers of the jovian planets are all less than a few hundred kilometers thick. How, then, do we determine the conditions beneath the clouds, in the unseen interiors of these distant worlds? We have no seismographic information on these regions (but see *Discovery 7-2*), nor do we happen to live on a more or less similar planet that we can use for comparison. Instead, we must use all available bulk data on each planet, along with our knowledge of physics and chemistry, to construct a model of its interior that agrees with observations. ⚬⚬ *(Sec. 0.5)* Our statements about the interiors of the jovian worlds, then, are really statements about the models that best fit the facts. As we will see throughout this book, astronomers rely heavily on model building, often aided by powerful computers, as a tool to interpret data about objects too far away for direct exploration.

Internal Structure

As illustrated in Figure 7.19, both the temperature and the pressure of Jupiter's atmosphere increase with depth. At a depth of a few thousand kilometers, the gas makes a gradual transition into the liquid state. By a depth of about 20,000 km the pressure is about 3 million times greater than atmospheric pressure on Earth. Under these conditions the hot liquid hydrogen is compressed so much that it undergoes another transition, this time to a metallic state, with properties in many ways similar to some liquid metals on Earth. Of particular importance for Jupiter's magnetic field, this metallic hydrogen is an excellent conductor of electricity.

Molecular
hydrogen

Metallic
hydrogen

Depth 100 km
Temperature 300 K
Pressure 10 atm

Depth 20,000 km
Temperature 11,000 K
Pressure 3×10^6 atm

Depth 60,000 km
Temperature 18,000 K
Pressure 4×10^7 atm

Icy, rocky core
Depth 70,000 km
Temperature 25,000 K
Pressure 6×10^7 atm

▲ FIGURE 7.19 **Jupiter's Interior**   Jupiter's internal structure, as deduced from spacecraft measurements and theoretical modeling. Pressure and temperature increase with depth, and the atmosphere gradually liquefies over the outermost few thousand kilometers. Below 20,000 km, the hydrogen behaves like a liquid metal. At the center of the planet lies a rocky core (red region), itself larger than any of the terrestrial planets.

Jupiter's rotation produces a strong outward push that causes a pronounced bulge at the planet's equator—Jupiter's equatorial radius (Table 7.1) exceeds its polar radius by nearly 7 percent. However, careful calculations indicate that if Jupiter were composed of hydrogen and helium alone, it should be *more* flattened than is observed. The actual degree of flattening implies a dense *core* having a mass perhaps 10 times the mass of Earth. The core's precise composition is unknown, but many scientists think it consists of materials similar to those found in the terrestrial worlds: molten, perhaps semisolid, rock. Because of Jupiter's enormous central pressure— approximately 50 million times that on Earth's surface, or 10 times that at Earth's center—the core must be compressed to very high densities (perhaps twice the core density of Earth). It is probably not much more than 20,000 km in diameter.

Saturn has the same basic internal parts as Jupiter, but their relative proportions are different: Its metallic hydrogen layer is thinner, and the central dense core (again inferred from studies of the planet's polar flattening) is larger—around 15 Earth masses. Because of its lower mass, Saturn has a less extreme core temperature, density, and pressure than Jupiter. The central pressure is around one-tenth of Jupiter's— not too different from the pressure at the center of Earth.

The interior pressures of Uranus and Neptune are sufficiently low that hydrogen stays in its molecular form all the way into the planets' cores. Currently, astronomers theorize that deep below the cloud layers, Uranus and Neptune may have high-density, "slushy" interiors containing thick layers of highly compressed water clouds. It is also possible that much of the planets' ammonia could be dissolved in the water, producing a thick, electrically conducting layer that could conceivably explain the planets' magnetic fields. At the present time, however, we don't know enough about the interiors to assess the correctness of this picture.

The cores of Uranus and Neptune are inferred mainly from considerations of those planets' densities. Because of their lower masses, we would not expect the hydrogen and helium in the interiors of Uranus and Neptune to be compressed as much as in the two larger jovian worlds, yet the average densities of Uranus and Neptune are actually *greater* than the density of Saturn and similar to that of Jupiter. Astronomers interpret this to mean that Uranus and Neptune each have Earth-sized cores some 10 times more massive than our planet, with compositions similar to the cores of Jupiter and Saturn.

ANIMATION/VIDEO The Gas Giants Part II

Our current knowledge of the internal structures of the four jovian worlds is summarized in Figure 7.20. The next space mission to significantly expand our knowledge of jovian interiors will be NASA's *Juno* probe, launched in August 2011 and due to arrive at Jupiter in 2016. From its polar orbit, *Juno* will map Jupiter's gravity and magnetism to study the planet's internal structure. It will also observe the composition and circulation of the planet's atmosphere.

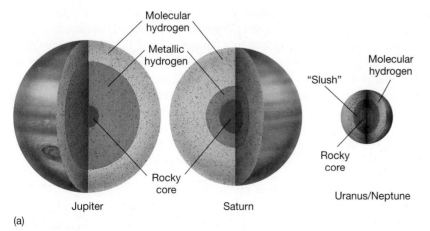

(a)

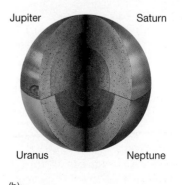

(b)

INTERACTIVE ▲ FIGURE 7.20 Jovian Interiors A comparison of the interior structures of the four jovian planets. (a) The planets drawn to scale. (b) The relative proportions of the various internal zones.

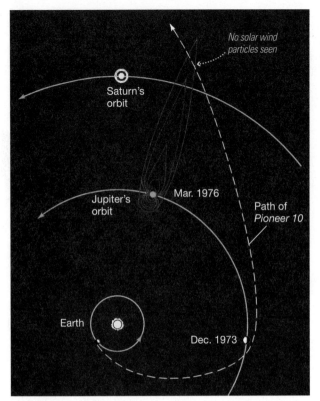

▲ FIGURE 7.21 *Pioneer* Mission   The *Pioneer 10* spacecraft (a forerunner to the *Voyager* missions) did not detect any solar particles while moving far behind Jupiter in March 1976, indicating that the tail of Jupiter's magnetosphere (drawn in red) extends beyond the orbit of Saturn.

Magnetospheres

The combination of rapid overall rotation and an extensive region of highly conductive fluid in its interior gives Jupiter by far the strongest planetary magnetic field in the solar system. ∞ *(Sec. 5.7)* The field at the cloud tops is some 14 times that of Earth, but taking the planet's radius into account implies an intrinsic field strength some 20,000 times Earth's. Jupiter is surrounded by a vast sea of energetic charged particles (mostly electrons and protons), somewhat similar to Earth's Van Allen belts but very much larger. Direct spacecraft measurements show Jupiter's magnetosphere to be almost 30 million km across (north to south), roughly a million times more voluminous than Earth's magnetosphere and far larger than the entire Sun. All four Galilean moons lie within it.

On the sunward side the boundary of Jupiter's magnetic influence on the solar wind lies about 3 million km from the planet. On the opposite side, however, the magnetosphere has a long "tail" extending away from the Sun at least as far as Saturn's orbit (some 4 AU, or 600 million kilometers), as sketched in Figure 7.21. Near the planet the magnetic field channels particles from the magnetosphere into the upper atmosphere, forming aurorae vastly larger and more energetic than those observed on Earth (Figure 7.22). ∞ *(Sec. 5.7)*

Saturn's electrically conducting interior and rapid rotation also produce a strong magnetic field and an extensive magnetosphere. However, because of the considerably smaller mass of Saturn's metallic hydrogen zone, the magnetic field strength at the planet's cloud tops is only about 1/20 that of Jupiter. Saturn's magnetosphere extends about 1 million kilometers toward the Sun and is large enough to contain the planet's ring system and most of its moons.

Voyager 2 found that both Uranus and Neptune have fairly strong magnetic fields, comparable to those of Saturn and Earth. Taking the planets' radii into account, we find intrinsic field strengths a few percent that of Saturn, or 30–40 times Earth's. Both planets have substantial magnetospheres, populated largely by electrons and protons either captured from the solar wind or created from hydrogen gas escaping from the planets below. Figure 7.23 compares the magnetic field structures of Earth and the four jovian planets. The locations and orientations of the bar magnets represent the observed planetary fields, and the sizes of the bars indicate overall magnetic field strength. Note that the fields of Uranus and Neptune are not aligned with the planets' rotation axes and are significantly offset from the planets' centers. Astronomers do not yet understand why this should be so.

Internal Heating

On the basis of Jupiter's distance from the Sun, astronomers expected to find the temperature of the cloud tops to be around 105 K. At that temperature, they reasoned, Jupiter would radiate back into space the same amount of energy as it receives from the Sun. However, radio and infrared observations of the planet revealed a black-body spectrum corresponding to a temperature of 125 K, implying (from Stefan's law) that Jupiter emits about twice—$(125/105)^4$—as much energy as reaches the planet in the form of sunlight. ∞ *(Sec. 2.4)*

Astronomers theorize that the source of Jupiter's excess energy is heat left over from the planet's formation. ∞ *(Sec. 4.3)* In the distant past, Jupiter's interior was probably very much hotter than it is today (although not nearly hot enough for Jupiter ever to have become a star like the Sun). That early heat is slowly leaking out through the planet's heavy atmospheric blanket, resulting in the excess emission we observe today.

◀ FIGURE 7.22 Aurorae on Jupiter   The main (underlying) image was taken by the *Hubble* telescope in visible (true-color) light, but the two insets at the poles were taken in the ultraviolet part of the spectrum. The oval-shaped aurorae, extending hundreds of kilometers above Jupiter's surface, result from charged particles escaping the jovian magnetosphere and colliding with the atmosphere, causing the gas to glow. (See Figure 5.24.) *(NASA)*

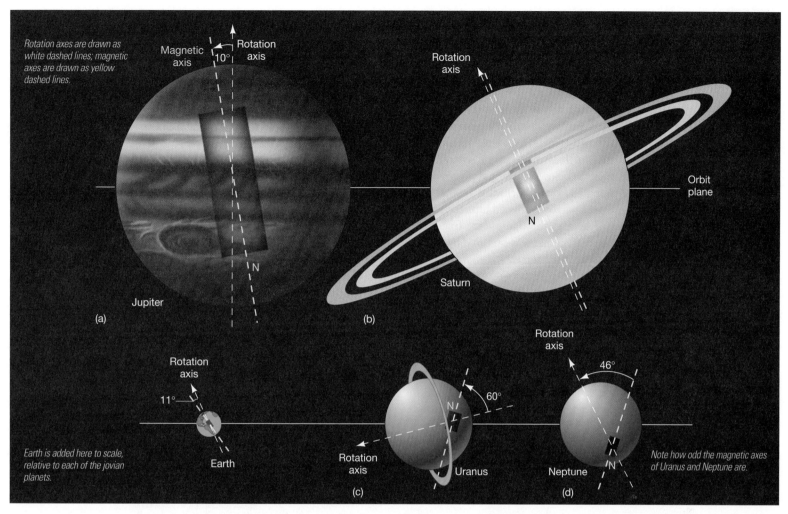

Rotation axes are drawn as white dashed lines; magnetic axes are drawn as yellow dashed lines.

Magnetic axis — Rotation axis 10°

Jupiter (a)

Rotation axis — Saturn (b) — Orbit plane — N

Rotation axis 11° — Earth (c) — Rotation axis — 60° N — Uranus

Rotation axis — 46° — Neptune (d) N

Earth is added here to scale, relative to each of the jovian planets.

Note how odd the magnetic axes of Uranus and Neptune are.

▲ FIGURE 7.23 **Jovian Magnetic Fields** Comparison of magnetic field strengths, orientations, and offsets in the four jovian planets. In each case the magnetic field is drawn as that of a simple bar magnet. The size and location of each magnet represent the strength and orientation of the planetary field.

Infrared measurements indicate that Saturn, too, has an internal energy source. Saturn's cloud-top temperature is 97 K, substantially higher than the temperature at which Saturn would reradiate all the energy it receives from the Sun. In fact, Saturn radiates away almost *three times* more energy than it absorbs. However, the explanation for Jupiter's excess emission doesn't work for Saturn. Saturn is smaller than Jupiter and so must have cooled more rapidly—fast enough, in fact, that its original supply of heat should have been used up long ago. What then is happening inside Saturn to produce this extra heat?

The explanation for this strange state of affairs also explains Saturn's atmospheric helium deficit. At the temperatures and high pressures found in Jupiter's interior, liquid helium *dissolves* in liquid hydrogen. In Saturn, where the internal temperature is lower, the helium doesn't dissolve so easily and tends to form droplets instead. (The basic phenomenon is familiar to cooks who know that it is generally much easier to dissolve ingredients in hot liquids than in cold ones.) Saturn probably started out with a fairly uniform mix of hydrogen and helium, but the helium tended to condense out of the surrounding hydrogen, much as water vapor condenses out of Earth's atmosphere to form a mist. The amount of helium condensation was greatest in the planet's cool outer layers, where the mist turned to rain about 2 billion years ago. A light shower of liquid helium has been falling through Saturn's interior ever since. This helium precipitation is responsible for depleting the outer layers of their helium content.

As the helium sinks toward the center, the planet's gravitational field compresses it and heats it up. The energy thus released is the source of Saturn's internal heating. In the distant future the helium rain will stop and Saturn will cool until its outermost layers radiate only as much energy as they receive from the Sun. When that happens, the temperature at Saturn's cloud tops will be 74 K.

PROCESS OF SCIENCE CHECK ▶ Why is mathematical modeling essential to understanding the interior structures of the jovian planets?

7-2 DISCOVERY

A Cometary Impact

In 1994, astronomers were granted a rare alternative means of studying Jupiter's atmosphere and interior when a comet, called Shoemaker-Levy 9, collided with the planet. Discovered in 1993, the comet had an unusual flattened nucleus made up of several large pieces, as shown in the first image (a) below. The largest piece was no larger than 1 km across. Tracing the orbit backward in time, researchers found that the comet had narrowly missed Jupiter in 1992. They realized that the objects shown in the figure were fragments produced when a previously "normal" comet was captured by Jupiter and torn apart by the planet's strong tidal forces. ∞ *(Sec. 5.2)* The data also revealed an even more remarkable fact—the comet would collide with Jupiter on its next approach!

Between July 16 and July 22, 1994, as most of the telescopes on (and in orbit around) Earth watched, the comet fragments struck Jupiter's upper atmosphere at speeds of over 60 km/s, causing a series of enormous explosions. Each impact created, for a period of a few minutes, a brilliant fireball hundreds of kilometers across and having a temperature of many thousands of kelvins. The energy released in each explosion was comparable to a billion terrestrial nuclear detonations. One of the largest pieces of the comet, fragment G, produced the spectacular fireball shown in the second image (b). The effects on the planet's atmosphere and the vibrations produced throughout Jupiter's interior were observable for days after the impact.

As far as we can determine, none of the cometary fragments breached the jovian clouds. Only *Galileo* had a direct view of the impacts on the back side of Jupiter, and in every case the explosions seemed to occur high in the atmosphere, above the uppermost cloud layer. Spectral lines from water vapor, silicon, magnesium, and iron were observed, but all apparently came from the comet itself—which really did resemble a loosely packed snowball. ∞ *(Sec. 4.2)* Debris from the impacts, like the "black eye" shown in image (c), spread slowly around Jupiter and eventually encircled the planet. It took years for all the cometary matter to settle into the interior.

The Shoemaker-Levy 9 impact has scientific significance far beyond its direct effect on Jupiter. Since the 1960s, planetary scientists have become increasingly aware that collisions between small bodies—asteroids, comets, and meteoroids—and the major planets are relatively commonplace events in the solar system. As we have seen, the consequences for life can be catastrophic. ∞ *(Discovery 4-1)* The fact that we witnessed such an event during the few decades since it became technologically feasible for us to do so underscores just how frequent these encounters probably are and has convinced many scientists of the importance of planetary impacts to solar system evolution.

The final figure (d) shows a 2009 comet collision with Jupiter, comparable in violence to the 1994 event, discovered by pure chance on the 15th anniversary of the Shoemaker-Levy 9 impact! Such collisions may well become an important future source of information about planetary interiors.

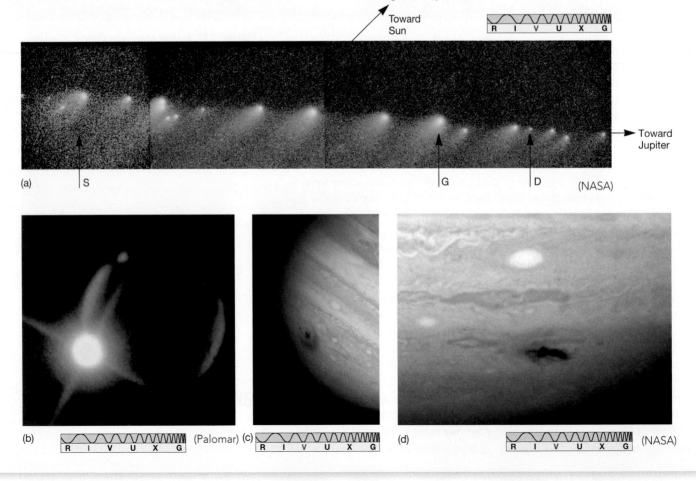

(a)   S   Toward Sun   R I V U X G   Toward Jupiter   G   D   (NASA)

(b)   R I V U X G   (Palomar)   (c)   R I V U X G   (d)   R I V U X G   (NASA)

Uranus apparently has no internal energy source, radiating into space as much energy as reaches it from the Sun. It never experienced helium precipitation, and its initial heat was lost to space long ago. Surprisingly, given its other similarities to Uranus, Neptune *does* have an internal heat source—it emits 2.7 times more energy than it receives from the Sun. The reason is still uncertain. Some scientists think that Neptune's relatively high concentration of methane insulates the planet, helping to maintain its initially high internal temperature.

CONCEPT CHECK ▶ Why do all the jovian planets have strong magnetic fields?

CHAPTER REVIEW

SUMMARY

LO1 Jupiter and Saturn were the outermost planets known to ancient astronomers. Uranus was discovered in the 18th century, by chance. Neptune was discovered after mathematical calculations of Uranus's slightly non-Keplerian orbit revealed the presence of an eighth planet.

LO2 The four jovian planets are all much more massive than the terrestrial worlds and are of much lower density. They are composed primarily of hydrogen and helium. However, all have large, dense cores, each roughly 10 times Earth's mass and of chemical composition similar to that of the terrestrial planets. The jovian planets all rotate more rapidly than Earth. Having no solid surfaces, they display **differential rotation** (p. 199)—the rotation rate varies with latitude and with depth below the cloud tops. For unknown reasons, Uranus's spin axis lies nearly in the ecliptic plane, leading to extreme seasonal variations in solar heating on the planet as it orbits the Sun.

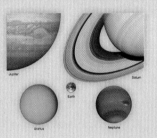

LO3 The cloud layers on all the jovian worlds are arranged into bands of light-colored zones and darker belts crossing the planets parallel to their equators. The bands are the result of convection in the planets' interiors and the planets' rapid rotation. The lighter zones are the tops of upwelling, warm currents, and the darker bands are cooler regions, where gas is sinking. Underlying them is a stable pattern of eastward or westward wind flow called a **zonal flow.** The wind direction alternates as we move north or south away from the equator. Jupiter's atmosphere consists of three main cloud layers. The colors we see are the result of chemical reactions at varying depths below the cloud tops.

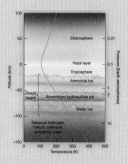

LO4 A major weather pattern on Jupiter is the **Great Red Spot** (p. 201), a huge hurricane that apparently has been raging for at least three centuries. Other, smaller, atmospheric features—**white ovals** (p. 203) and **brown ovals** (p. 203)—are also observed. Similar systems are found on the other jovian planets, although they are less distinct on Saturn and Uranus. Long-lived storms on Saturn may lie hidden below the planet's clouds. The **Great Dark Spot** (p. 207) on Neptune had many similarities to Jupiter's Red Spot. It disappeared within a few years of its discovery by *Voyager 2*. Storms in the jovian atmospheres are long-lived because there is no solid surface to dissipate their energy.

LO5 The atmospheres of the jovian planets become hotter and denser with depth, eventually becoming liquid. In Jupiter and Saturn the interior pressures are so high that the hydrogen near the center exists in a liquid-metallic form rather than as molecular hydrogen. In Uranus and Neptune this change from molecular hydrogen to metallic hydrogen apparently does not occur. Radio waves emitted from the planets' magnetospheres provide a measure of their interior rotation rates. The combination of rapid rotation and conductive interiors means that all four jovian planets have strong magnetic fields and extensive magnetospheres.

LO6 Three jovian planets radiate more energy than they receive from the Sun. On Jupiter and Neptune the source of this energy is most likely heat left over from the planet's formation. On Saturn the heating is the result of helium precipitation in the interior, where helium liquefies and forms droplets that then fall toward the center of the planet. This process is also responsible for reducing the amount of helium observed in Saturn's outer layers.

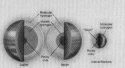

MasteringAstronomy®   For instructor-assigned homework go to www.masteringastronomy.com

Problems labeled **POS** *explore the process of science.* **VIS** *problems focus on reading and interpreting visual information.* **LO** *connects to the introduction's numbered Learning Outcomes.*

REVIEW AND DISCUSSION

1. How was Uranus discovered?

2. **LO1 POS** Why did astronomers suspect an eighth planet beyond Uranus?

3. Why has Jupiter retained most of its original atmosphere?

4. What is differential rotation, and how is it observed on Jupiter?

5. Why does Saturn have a less varied appearance than Jupiter?

6. **LO2** How (and why) do the thicknesses of Saturn's various layers (clouds, molecular and metallic hydrogen, core) compare to those in Jupiter?

7. **LO4** What is the Great Red Spot? What is the source of its energy?

8. Describe the cause of the colors in the jovian planets' atmospheres.

9. **LO3** Compare and contrast the atmospheres of the jovian planets, describing how the differences affect the appearance of each.

10. Explain the theory that accounts for Jupiter's internal heat source.

11. **LO6** Describe the theoretical model that accounts for Saturn's missing helium and the surprising amount of heat the planet radiates.

12. What is unusual about the rotation of Uranus?

13. What is responsible for Jupiter's enormous magnetic field?

14. **LO5** How are the interiors of Uranus and Neptune thought to differ from those of Jupiter and Saturn?

15. **POS** How much do you think we might know about the outer solar system without the *Voyager*, *Galileo*, and *Cassini* missions?

CONCEPTUAL SELF-TEST: TRUE OR FALSE?/MULTIPLE CHOICE

1. The solid surface of Jupiter lies just below the cloud layers that are visible from Earth. (T/F)

2. Storms in the jovian planets' atmospheres are generally much longer lived than storms on Earth. (T/F)

3. The element helium plays an important role in producing Saturn's atmospheric colors. (T/F)

4. Jupiter emits more energy than it receives from the Sun. (T/F)

5. Although often referred to as a gaseous planet, Jupiter is mostly liquid in its interior. (T/F)

6. Neptune probably does not have a rocky core. (T/F)

7. Around northern midsummer on Uranus, an observer near the north pole would observe the Sun high and almost stationary in the sky. (T/F)

8. The magnetic fields of Uranus and Neptune are highly tilted relative to their rotation axes and offset from the planets' centers. (T/F)

9. The main constituent of Jupiter's atmosphere is (a) hydrogen; (b) helium; (c) ammonia; (d) carbon dioxide.

10. Saturn's cloud layers are much thicker than those of Jupiter because Saturn has (a) more mass; (b) lower density; (c) a weaker magnetic field; (d) weaker gravity.

11. Compared to Earth, Saturn's core is roughly (a) half the mass; (b) the same mass; (c) twice as massive; (d) 10 times more massive.

12. Uranus was discovered about the same time as (a) Columbus reached North America; (b) the U.S. Declaration of Independence; (c) the American Civil War; (d) the Depression.

13. Compared with Uranus, the planet Neptune is (a) much smaller; (b) much larger; (c) roughly the same size; (d) closer to the Sun.

14. **VIS** According to Figure 7.9 (Jupiter's Atmosphere), if ammonia and ammonium hydrosulfide ice were transparent to visible light, Jupiter would appear (a) bluish; (b) red; (c) tawny brown; (d) exactly as it does now.

15. **VIS** According to Figure 7.21, Jupiter's magnetosphere extends about (a) 1 AU; (b) 5 AU; (c) 10 AU; (d) 20 AU beyond the planet.

PROBLEMS

The number of squares preceding each problem indicates its approximate level of difficulty.

1. ■ The *Hubble Space Telescope* has an angular resolution of 0.05″. What is the size of the smallest feature visible on (a) Jupiter, at 4.2 AU, and (b) Neptune, at 29 AU?

2. ■ What is the gravitational force on Uranus from Neptune, at closest approach? Compare with the Sun's gravitational force on Uranus.

3. ■ What is the round-trip travel time of light from Earth to Jupiter (at the minimum Earth–Jupiter distance of 4.2 AU)? How far would a spacecraft orbiting Jupiter at a speed of 20 km/s travel in that time?

4. ■■ What would be the mass of Jupiter if it were composed entirely of hydrogen at a density of 0.08 kg/m^3 (the density of hydrogen at sea level on Earth)?

5. ■ Verify the statement made in the text that the force of gravity at Jupiter's cloud tops is nearly 2.5 times that at Saturn's cloud tops.

6. ■ Given Jupiter's current atmospheric temperature, what is the smallest possible mass the planet could have and still have retained its hydrogen atmosphere? ∞ *(More Precisely 5-1)*

7. ■ How long does it take for Saturn's equatorial flow, moving at 1500 km/h, to encircle the planet?

8. ■■ If the temperature at Saturn's cloud tops is 97 K and the planet radiates three times more energy than it receives from the Sun, use Stefan's law to calculate what the temperature would be in the absence of any internal heat source. ∞ *(More Precisely 2-2)*

9. ■■ If the core of Uranus has a radius twice that of planet Earth and an average density of 8000 kg/m³, calculate the mass of Uranus outside the core. What fraction of the planet's total mass is its core?

10. ■ From Wien's law, at what wavelength does Neptune's thermal emission peak? In what part of the electromagnetic spectrum does this wavelength lie? ∞ *(More Precisely 2-2, Fig. 2.9)*

ACTIVITIES

Collaborative

1. Jupiter's rapid (10-hour) rotation and ever-changing atmospheric features make it a highly dynamic telescopic object. Even a small telescope should reveal several of the planet's cloud belts, as well as its four brightest moons. Can you see the Great Red Spot? Make a series of sketches of the planet. You'll have to work fast because of the planet's rotation! Try to complete each sketch within 20–30 minutes, and prepare in advance a series of sheets of paper with a circle to represent the planet. Note the date and time of each sketch. By carefully timing some easily recognizable surface feature, such as the Great Red Spot, measure Jupiter's rotation period. (You may have to come back on another night to be able to conveniently observe the chosen feature cross the entire planetary disk.) Repeat your observations a week or two later. Have any of the surface features, or their relation to one another, changed?

Individual

1. Consult a sky chart and find Jupiter in the night sky. It's an easy-to-find naked-eye object. Do any stars look as bright as Jupiter? What other differences do you notice between Jupiter and the stars?

2. Use a telescope to look at Saturn. Does the planet appear flattened? Can you see any atmospheric features?

3. Locate Uranus in the night sky. It may be barely visible to the naked eye, but binoculars will make the search much easier. (Hint: Uranus shines more steadily than the background stars.) Can you detect the planet's color?

4. The search for Neptune requires a much more determined effort! A telescope is best, but high-powered binoculars mounted on a steady support will also do. Comparing Uranus and Neptune, which planet appears bluer? Through a telescope, does either planet appear as a disk, or do they look more like points of light?

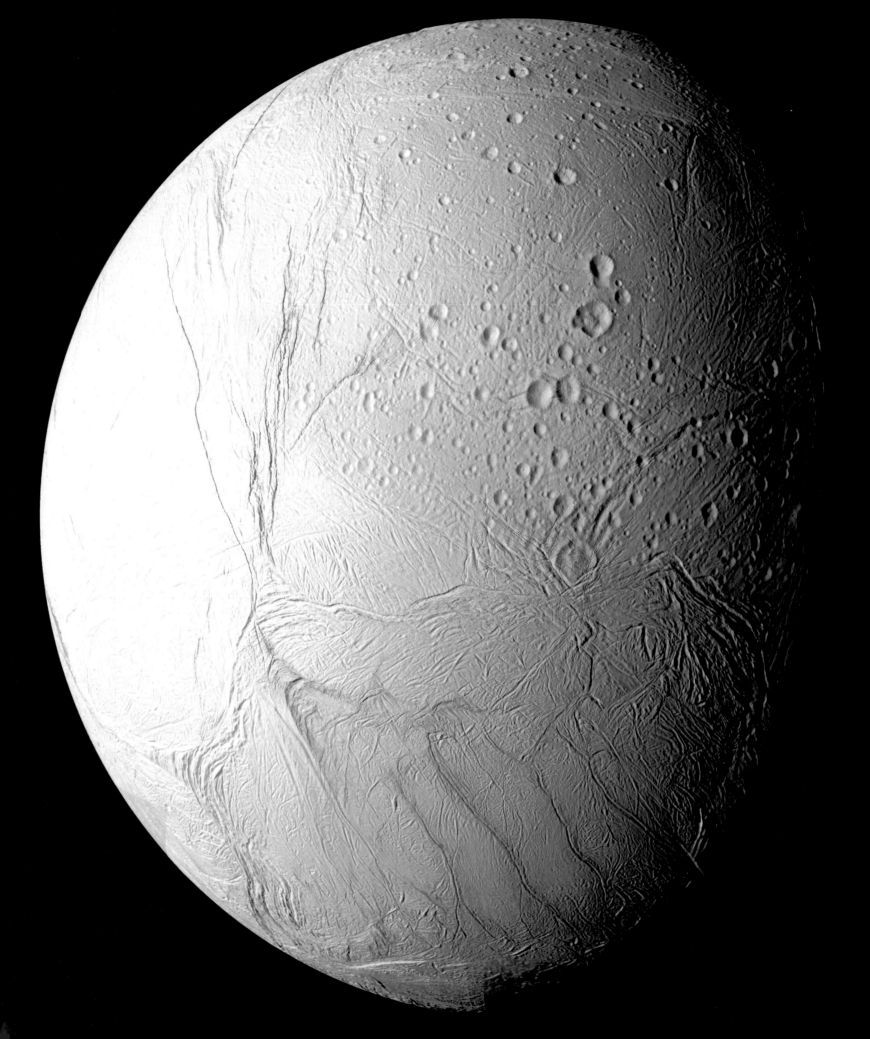

8

Moons, Rings, and Plutoids

Small Worlds Among Giants

All four jovian planets have systems of moons and rings that exhibit fascinating variety and complexity. The six largest jovian moons have many planetary features, providing us with insight into the terrestrial worlds. Two of these moons, Jupiter's Europa and Saturn's Titan, have become prime candidates for study in the search for life elsewhere in the solar system. Like the moons, the rings of the jovian planets also differ greatly. The rings of Saturn—the best-known ring system among the jovian planets—are one of the most spectacular displays in the sky. Plutoids are dwarf planets outside the orbit of Neptune—small icy bodies having much in common with the large jovian moons and little similarity to the eight major planets. For these reasons we study them here, along with the moons and rings of their giant neighbors.

MasteringAstronomy® Visit www.masteringastronomy.com for quizzes, animations, videos, interactive figures, and self-guided tutorials.

◄ The number of known moons in the solar system increased rapidly during the late 1990s, when bigger telescopes and spaceprobes enabled astronomers to better inventory objects in our cosmic neighborhood. But the more closely we examine these alien worlds, the more complex and bizarre they seem to be. In this puzzling image taken by the *Cassini* spacecraft, Saturn's tiny ice-covered moon Enceladus (only 500 km in diameter) shows evidence of a youthful terrain in the south where craters are mostly absent. The long blue streaks (about 1 km wide) are likely fractures in the ice through which gas escapes to form a thin but real atmosphere. *(JPL)*

8.1 The Galilean Moons of Jupiter

All four jovian planets have extensive moon systems—Jupiter has at least 64 natural satellites orbiting it, Saturn has 53, Uranus 27, and Neptune 13. These moons range greatly in size and other physical properties, but they fall into three natural groups.

The first group consists of the 6 "large" jovian moons, each comparable in size to Earth's Moon. They are listed in Table 8.1, along with our own Moon for comparison. All have diameters greater than 2500 km and are large enough to have their own distinctive geologies and histories. Next are the 12 "medium-sized" moons—spherical bodies with diameters ranging from about 400 to 1500 km. Their heavily cratered surfaces offer tantalizing clues to the past and present states of the environment in the outer solar system. Finally, there are the many "small" moons—irregularly shaped chunks of ice, all less than 300 km across—that exhibit a bewildering variety of complex and fascinating motions. Most are probably little more than captured interplanetary debris. ∞ *(Sec. 4.3)* The number of known small moons has increased rapidly in recent years (and will probably continue to rise) thanks to steadily improving ground and space-based technology.

In this chapter we will focus our attention on the large and midsize jovian moons, beginning with the Galilean satellites of Jupiter. ∞ *(Sec. 1.2)*

A "Miniature Solar System"

Jupiter's four large moons travel on nearly circular, prograde orbits in Jupiter's equatorial plane. Moving outward from the planet, they are named Io, Europa, Ganymede, and Callisto, after the mythical attendants of the Roman god Jupiter. The two *Voyager* spacecraft and, more recently, the *Galileo* mission returned many remarkable images of these small worlds, allowing scientists to discern much fine surface detail on each. Figure 8.1 is a *Voyager 1* image of Io and Europa, with Jupiter providing a spectacular backdrop. Figure 8.2 shows the four Galilean moons to scale, along with their relative orbits.

The Galilean moons range in size from slightly smaller than Earth's Moon (Europa) to slightly larger than Mercury (Ganymede). Their densities decrease with increasing distance from Jupiter in a manner very reminiscent of the falloff in density of the terrestrial planets with increasing distance from the Sun. ∞ *(Sec. 4.1)* Indeed, many astronomers think that the formation of Jupiter and its Galilean satellites may have mimicked on a small scale the formation of the Sun and the inner planets, with generally denser moons forming closer to the intense heat of the young Jupiter. ∞ *(Sec. 4.3)*

ANIMATION/VIDEO Galilean Moons Transit Jupiter **(MA)**

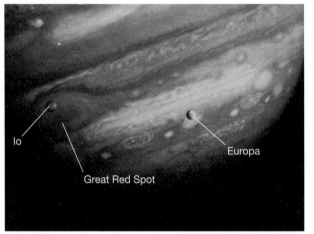

▲ **FIGURE 8.1 Jupiter, Up Close** *Voyager 1* took this photograph of Jupiter with ruddy Io on the left and pearl-like Europa toward the right. Note the scale of objects here: Io and Europa are both comparable in size to our Moon; the Great Red Spot is roughly twice as big as Earth. *(NASA)*

TABLE 8.1	The Large Moons of the Solar System						
Name	Parent Planet	Distance from (km)	Parent Planet (planet radii)	Orbit Period (Earth days)	Radius (km)	Mass (Earth's Moon = 1)	Density (kg/m^3)
Moon	Earth	384,000	60.2	27.3	1740	1.00	3300
Io	Jupiter	422,000	5.90	1.77	1820	1.22	3500
Europa	Jupiter	671,000	9.38	3.55	1570	0.65	3000
Ganymede	Jupiter	1,070,000	15.0	7.15	2630	2.02	1900
Callisto	Jupiter	1,880,000	26.3	16.7	2400	1.46	1800
Titan	Saturn	1,220,000	20.3	16.0	2580	1.83	1900
Triton	Neptune	355,000	14.3	−5.88[1]	1350	0.29	2100

[1]The minus sign indicates a retrograde orbit.

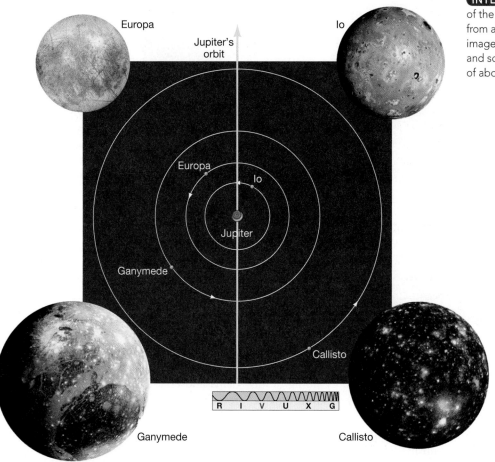

INTERACTIVE ◄ FIGURE 8.2 Galilean Moons   The orbits of the four Galilean moons of Jupiter, drawn to scale, as seen from above the planet's north pole. The four insets show actual images of those moons, taken by the *Galileo* spacecraft and scaled here as they would appear from a distance of about 1 million kilometers. *(NASA)*

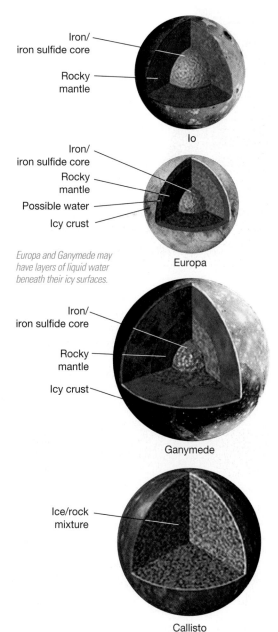

Europa and Ganymede may have layers of liquid water beneath their icy surfaces.

Detailed gravity measurements made by the *Galileo* orbiter support this picture of the origin of the Jupiter system. The data suggest that Io and Europa have large, iron-rich cores surrounded by thick mantles of generally rocky composition, perhaps similar to the crusts of the terrestrial planets. Europa has a water/ice outer shell between 100 and 200 km thick. Ganymede and Callisto are less dense overall; low-density materials, such as water ice, may account for as much as half their total mass. Ganymede has a relatively small metallic core topped by a rocky mantle and a thick icy outer shell. Callisto seems to be a largely undifferentiated mixture of rock and ice. Figure 8.3 summarizes our knowledge of the interiors of these bodies.

Io: The Most Active Moon

Io is the most geologically active object in the entire solar system. In mass and radius it is fairly similar to Earth's Moon, but there the resemblance ends. Shown in Figure 8.4, Io's uncratered surface is a collage of oranges, yellows, and blackish browns.

As *Voyager 1* sped past Io, it made an outstanding discovery: Io has active volcanoes! The spacecraft photographed eight erupting volcanoes. Six of them were still

▶ FIGURE 8.3 **Galilean Moon Interiors**   Cutaway diagrams of the interior structure of the four Galilean satellites. Moving outward from Io to Callisto, densities steadily decrease as the composition shifts from rocky mantles and metallic cores in Io and Europa, to a thick icy crust and smaller core in Ganymede, to an almost uniform rock and ice mix in Callisto.

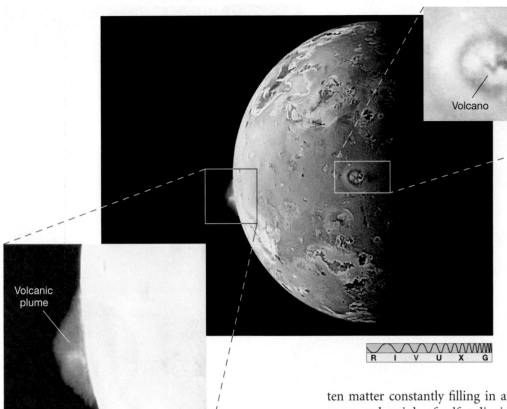

Volcanic plume

Surface

(a)

R I V U X G

◀ FIGURE 8.4 Io   Jupiter's innermost moon, Io, differs from the other three Galilean satellites. Its surface is kept smooth and brightly colored by constant volcanism, shown here as dark, circular features. The two inserts (b) show Io's umbrella-like eruptions as Galileo flew past in 1997; both volcanoes are about 300 km across, and the plume at the left is about 150 km high. (NASA)

Volcano   Plume

ANIMATION/VIDEO Jupiter Icy Moons Orbiter Mission

ANIMATION/VIDEO Io Cutaway

ANIMATION/VIDEO Galileo's View of Europa

erupting when *Voyager 2* passed by 4 months later. Inset (a) in Figure 8.4 shows one volcano, called Prometheus, ejecting matter at speeds of up to 2 km/s to an altitude of about 150 km. The orange color immediately surrounding the volcanoes most likely results from sulfur compounds in the ejected material. Io's exceptionally smooth surface is apparently the result of molten matter constantly filling in any "dents and cracks." Io has a thin atmosphere composed mainly of sulfur dioxide, produced by volcanic activity and temporarily retained by the moon's gravity. ∞ *(More Precisely 5-1)*

By the time *Galileo* arrived, some of the volcanoes observed by *Voyager 2* had subsided, but many new ones were seen—in fact, *Galileo* found that Io's surface features can change significantly in as little as a few weeks. In all, more than 80 active volcanoes have been identified on Io. The largest, known as Loki (located on the "back" side of Io in Figure 8.4), is larger than the state of Maryland and emits more energy than all of Earth's volcanoes combined. According to *Galileo*'s instruments, lava temperatures on Io generally range from 650 to 900 K, although temperatures as high as 2000 K—far hotter than any earthly volcano— have been measured at some locations. Mission scientists speculate that these superhot volcanoes may be similar to those that occurred on Earth more than 3 billion years ago.

Io is far too small to have geological activity similar to that on Earth. It should be long dead, like our Moon, its internal heat lost to space billions of years ago. Instead, the source of Io's energy is *external*—Jupiter's gravity. Because Io orbits very close to Jupiter, the planet's huge gravitational field produces strong tidal forces on the moon, resulting in a large (100 m) tidal bulge. ∞ *(Sec. 5.2)* If Io were Jupiter's only satellite, it would long ago have come into a state of synchronous rotation with the planet. In that case, Io would move in a perfectly circular orbit with one face permanently turned toward Jupiter. The tidal bulge would be stationary with respect to the moon, and there would be no internal stresses and no volcanism.

But Io is not alone. As it orbits, it is constantly tugged by the gravity of its nearest large neighbor, Europa. These tugs are small and not enough to cause any great tidal effect in and of themselves, but they are sufficient to make Io's orbit slightly noncircular, preventing the moon from settling into a precisely synchronous state. The moon's orbital speed varies from place to place as it revolves around the planet, but its rate of rotation on its axis remains constant. Thus, it cannot keep one face permanently turned toward Jupiter. Instead, Io rocks or wobbles slightly from side to side on its axis as it moves. The tidal bulge, however, always points directly toward Jupiter, so it moves back and forth across Io's surface as the moon wobbles. The conflicting forces acting on Io result in enormous tidal stresses that continually flex and squeeze its interior.

Just as repeated back-and-forth bending of a wire can produce heat through friction, this ever-changing distortion continually energizes Io. Researchers estimate that the total amount of heat generated within Io as a result of tidal flexing is about 100 million megawatts—five times the total power consumption of all the nations on Earth. The large amount of heat generated within Io causes huge jets of gas and molten rock to squirt out of the surface. It is likely that much of Io's interior is soft or molten, with only a relatively thin solid crust overlying it.

Europa: Liquid Water Locked in Ice

Europa (Figure 8.5) is a world very different from Io. Europa has relatively few craters on its surface, again suggesting that recent activity has erased the scars of ancient meteoric impacts. However, the surface displays a vast network of lines crisscrossing bright, clear fields of water ice. Many of these linear bands extend halfway around the satellite.

Based on the *Voyager* images, planetary scientists theorized that Europa is covered by an ocean of liquid water whose surface layers are frozen. Later images from *Galileo* seem to support this idea. Figure 8.5(b) is a high-resolution image of this weird moon, showing ice-filled surface cracks similar to those seen in Earth's polar ice floes. Figure 8.5(c) shows what appear to be icebergs—flat chunks of ice that have been broken apart and reassembled, perhaps by the action of water below. Mission scientists speculate that Europa's ice may be several kilometers thick and that the ocean below it may be 100 km deep. As on Io, the ultimate source of the energy that keeps the ocean liquid and also drives the internal motion leading to the cracks in the icy surface is Jupiter's gravity. The heating mechanism is essentially similar to that on Io, but the effects are less extreme on Europa because of its greater distance from the planet.

Other detailed images of the surface support this hypothesis. Figure 8.5(d) shows a region where Europa's icy crust appears to have been pulled apart and new material has filled in the gaps between the separating ice sheets. Elsewhere on the surface, *Galileo* found what appeared to be the icy equivalent of lava flows on Earth—regions where water apparently erupted through the surface and flowed for many kilometers before solidifying. The scarcity of impact craters on Europa implies that the processes responsible for these features did not stop long ago. Rather, they must be ongoing.

Further evidence comes from studies of Europa's magnetic field. *Galileo* found that Europa has a weak magnetic field that constantly changes strength and direction. This finding is consistent with the idea that the field is generated by the action of Jupiter's magnetism on a shell of electrically conducting fluid about 100 km below Europa's surface—in other words, the salty liquid water layer suggested by the surface observations. These results convinced quite a few skeptical scientists of the reality of Europa's deep ocean.

The likelihood that Europa has an extensive layer of liquid water below its surface ice opens up many interesting avenues of speculation about the possible development of life there. In the rest of the solar system, only Earth has liquid water on or near its surface, and most scientists agree that water played a key role in the appearance of life here (see Chapter 18). Europa may well contain more liquid water than exists on our entire planet! Of course, the existence of water does not *necessarily* imply the emergence of life. Europa, even with its liquid ocean, is still a hostile environment compared to Earth. Nevertheless, the

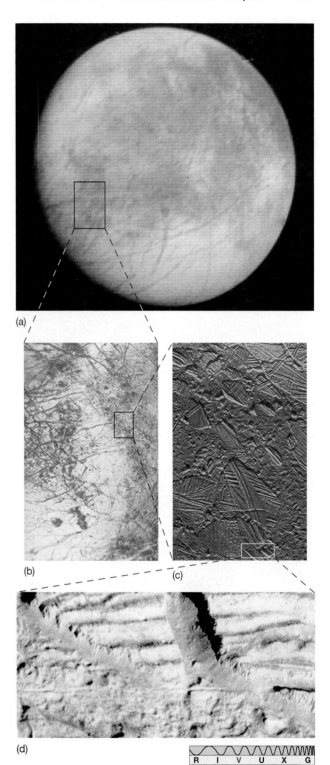

▲ **FIGURE 8.5 Europa** (a) A *Voyager 1* mosaic of Europa. (b) Its icy surface is only lightly cratered, indicating that some ongoing process must be obliterating impact craters soon after they form. (c) This image from the *Galileo* spacecraft shows a smooth yet tangled surface, called Conamara Chaos, resembling the huge ice floes that cover Earth's polar regions. (d) This detailed *Galileo* image shows "pulled apart" terrain that suggests liquid water upwelling from the interior and freezing, filling in the gaps between separating surface ice sheets. *(NASA)*

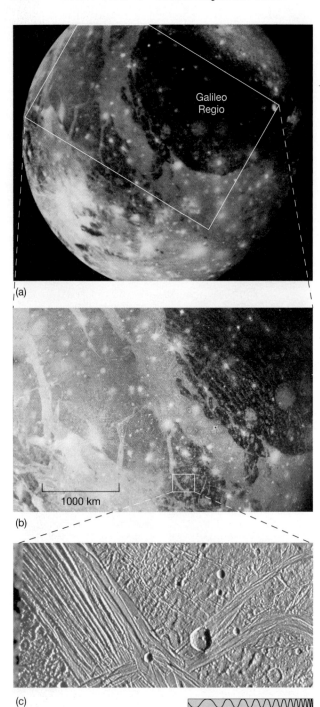

(a)

1000 km

(b)

(c)

R I V U X G

▲ FIGURE 8.6 Ganymede (a) and (b) *Voyager 2* images of Ganymede. The dark regions are the oldest parts of the moon's surface and may represent its original icy crust. Galileo Regio spans some 3200 km. The lighter, younger regions are the result of flooding and freezing that occurred within a billion years or so of Ganymede's formation. (c) Grooved terrain on Ganymede may have been caused by a process similar to plate tectonics on Earth. The area shown in this *Galileo* image spans about 50 km and reveals a multitude of ever-smaller ridges, valleys, and craters, down to the resolution of a football field. *(NASA)*

ANIMATION/VIDEO: *Galileo's* View of Ganymede

possibility, even a remote one, of life on Europa was an important motivating factor in the decision to extend the *Galileo* mission for a total of 6 additional years. ∞ *(Sec. 7.1)* No new visits are currently planned to the Jupiter system, although planetary scientists are lobbying hard for a new mission dedicated to this fascinating world.

Ganymede and Callisto: Fraternal Twins

Ganymede, shown in Figure 8.6, is the largest moon in the solar system, exceeding not only Earth's Moon, but also the planet Mercury in size. It has many impact craters on its surface, and light and dark patterns reminiscent of features on Earth's Moon. In fact, Ganymede's history has many parallels with that of the Moon (with water ice replacing lunar rock).

As with the inner planets and Earth's Moon, we can estimate the ages of Ganymede's surface features by counting craters. We learn that the darker regions are the oldest parts of Ganymede's surface. They are the original icy surface, just as the ancient highlands on our Moon are its original crust. Ganymede's surface has darkened with age as micrometeorite dust has slowly covered it. The light-colored regions are much less heavily cratered, and so must be younger. They are Ganymede's "maria" and probably formed in a manner similar to the way the maria on the Moon were created. ∞ *(Sec. 5.6)* Intense meteoritic bombardment caused liquid water—Ganymede's counterpart to the Moon's molten lava—to upwell from the interior, flooding the impacting regions before solidifying. Not all of Ganymede's surface features follow the lunar analogy, however. Ganymede has a system of grooves and ridges (Figure 8.6b, c) that may have resulted from crustal tectonic motion, much as Earth's surface undergoes mountain building and faulting at plate boundaries. ∞ *(Sec. 5.5)* The process apparently stopped about 3 billion years ago when the cooling crust became too thick.

Callisto, shown in Figure 8.7, is in many ways similar in appearance to Ganymede, although Callisto is more heavily cratered and has few fault lines. Callisto's most obvious feature is a huge series of concentric ridges surrounding two large basins, one clearly visible in Figure 8.7. The ridges resemble the ripples made as a stone hits water, but on Callisto they probably resulted from a cataclysmic impact with a meteorite. The upthrust ice was partially melted, but it resolidified quickly, before the ripples had a chance to subside. Today both the ridges and the rest of the crust are frigid ice and show no obvious signs of geological activity. Apparently, Callisto froze before plate tectonic or other activity could start. The density of impact craters on the large impact basins indicates that they formed long ago, perhaps 4 billion years in the past.

Ganymede's internal differentiation indicates that the moon was largely molten at some time in the past, yet Callisto is undifferentiated and hence apparently never melted (Figure 8.3). ∞ *(Sec. 5.4)* Researchers are uncertain why two such similar bodies evolved so differently. Complicating things further, the *Galileo* orbiter found that Ganymede has a weak magnetic field, about 1 percent that of Earth, suggesting that this moon, too, may have liquid or "slushy" water below its surface. ∞ *(Sec. 5.7)* If that is so, then Ganymede's heating and differentiation must have happened relatively recently—less than a billion years ago, based on recent estimates of how rapidly the moon's heat escapes into space.

Scientists have no clear explanation for how this could have occurred, although some speculate that interactions among the inner moons may have caused Ganymede's orbit to have changed significantly about 1 billion years ago and that prior tidal heating by Jupiter helped melt the moon's interior. These interactions may conceivably also be related to the near-perfect 1:2:4 ratio among the orbital periods of the three inner Galilean moons (see Table 8.1), which astronomers cannot easily explain but are unwilling to dismiss as simple coincidence.

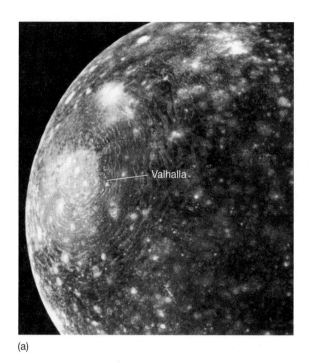

Valhalla

(a)

(b)

◄ FIGURE 8.7 **Callisto** (a) Callisto resembles Ganymede in overall composition but is more heavily cratered. Extending nearly 1500 km from the Valhalla basin center, its concentric ridges formed when "ripples" from a large meteoritic impact froze before they could disperse completely. (b) This higher-resolution *Galileo* image of Callisto's equatorial region, about 200 km across, displays more clearly its heavy cratering. (*NASA*)

8.2 The Large Moons of Saturn and Neptune

The other two jovian moons, Titan and Triton, are very different worlds. Titan's thick atmosphere continues to evolve even today, while distant Triton, the smallest of the six "large" moons, may well be a recent acquisition from the Kuiper belt. Both pose new riddles and offer unique insights into the formation and evolution of our planetary system.

Titan: A Moon with an Atmosphere

Titan is the largest and most intriguing of Saturn's moons. From Earth, it is visible only as a barely resolved orange disk. Long before the *Voyager* missions, astronomers knew from spectroscopic observations that the moon's ruddy coloration is caused by something quite special—an atmosphere. So anxious were mission planners to obtain a closer look that they programmed *Voyager 1* to pass very close to Titan, even though that meant the spacecraft could not then use Saturn's gravity to continue on to Uranus and Neptune.

Unfortunately, despite this close encounter, the moon's surface remained a mystery. The spacecraft's view was completely obscured by a thick, uniform haze layer, in some ways similar to the smog found over many cities on Earth, that envelops the moon (Figure 8.8). More detailed, close-up information on Titan is one of the principal goals of the ongoing *Cassini* mission. ⚬ *(Discovery 7-1)* The spacecraft's carefully planned trajectory has included some 70 visits to Titan so far, with another 50 planned during the *Cassini Solstice* mission extension now underway and due to end in 2017.

CONCEPT CHECK ▶ What is the source of energy for all the activity observed on Jupiter's Galilean moons?

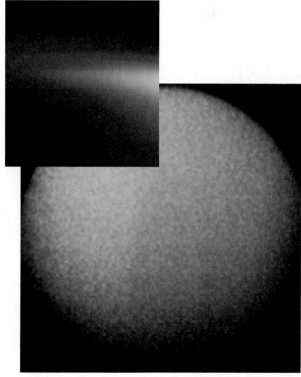

► FIGURE 8.8 **Titan** Larger than Mercury and roughly half the size of Earth, Titan was photographed from only 4000 km away as the *Voyager 1* spacecraft passed by in 1980. The inset shows a contrast-enhanced true color image of the haze layers in Titan's upper atmosphere, taken by the *Cassini* spacecraft in 2005. (*NASA*)

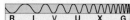

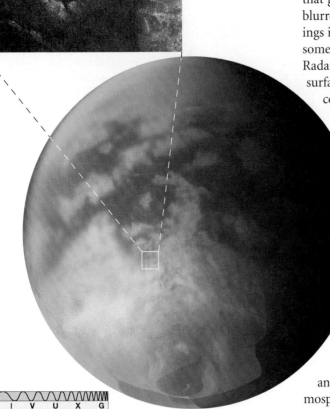

R I V U X G

▲ FIGURE 8.9 **Titan Revealed**   *Cassini's* telescopes revealed this infrared, false-color view of Titan's surface in late 2004. The inset shows a circular surface feature thought to be an icy volcano. *(NASA/ESA)*

Figure 8.9 shows how *Cassini's* infrared instruments penetrated the moon's foggy atmosphere, revealing details of the surface. The image shows light and dark regions near the center of the field of view, thought to be icy plateaus, apparently smeared with hydrocarbon tar. Ridges and cracks on the moon's surface suggest that geological activity, in the form of "titanquakes," may be common. The rather blurred boundaries between the light and dark regions, the peculiar surface markings in the light-colored region, and the absence of extensive cratering suggest that some sort of erosion is occurring, perhaps as a result of wind or volcanic activity. Radar imaging has revealed few large (10–100 km diameter) craters on the moon's surface, but fewer small craters than expected given Titan's location in Saturn's congested ring plane (see Section 8.4). The insert in Figure 8.9 shows what appears to be an icy volcano, supporting the view that the moon's surface is geologically active.

Titan's atmosphere is thicker and denser than Earth's and is certainly more substantial than that of any other moon. Radio and infrared observations made by *Voyager 1* and *Cassini* show that the atmosphere is composed mostly of nitrogen (more than 98 percent), with methane and trace amounts of other gases making up the rest. The surface temperature is a frigid 94 K, roughly what we would expect on the basis of Titan's distance from the Sun. Under these conditions, as on Ganymede and Callisto, water ice plays the role of rock on Earth, and liquid water the role of lava. The presence of an atmosphere allows us to extend this analogy further—at the temperatures typical of the lower atmosphere, methane and ethane behave like water on Earth, raising the possibility of methane rain, snow, and fog and ethane rivers and oceans!

On January 14, 2005, the *Huygens* probe, transported to Saturn by *Cassini* and released 3 weeks earlier, arrived at Titan and parachuted through the thick atmosphere to the moon's surface. Figure 8.10(a) shows an intriguing image radioed

▶ FIGURE 8.10 **Titan's Surface**   (a) This photograph of the surface was taken from an altitude of 8 km as the *Huygens* probe descended. It shows a network of dark channels reminiscent of streams (at center) or rivers draining from the light-shaded uplifted terrain into darker, low-lying regions (at bottom). (b) *Huygens*'s view of its landing site, in approximately true color. The foreground icy "rocks" are only a few centimeters across. *(NASA/ESA)*

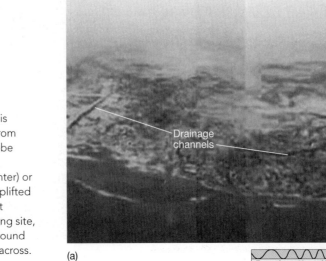

Drainage channels

(a)

R I V U X G

(b)

back from *Huygens* during its descent. It appears to show a network of drainage channels leading to a shoreline, but this interpretation is uncertain. The probe landed on solid ground, and for the next hour transmitted images and instrument readings to *Cassini* as it passed overhead. The hazy view from the surface (Figure 8.10b) reveals an icy landscape. The rocks in the foreground are a few centimeters across and show evidence of erosion by liquid of some sort.

Figure 8.11 shows the probable structure of Titan's atmosphere, based on *Voyager 1* and *Huygens* measurements. Despite Titan's low mass (a little less than twice that of Earth's Moon) and hence low surface gravity (one-seventh of Earth's), the atmospheric pressure at ground level is 60 percent greater than that on Earth. Titan's atmosphere contains about 10 times more gas than Earth's atmosphere and, because of Titan's weaker gravitational pull, extends 10 times farther into space than does our own. The top of the main haze layer lies about 200 km above the surface (Figure 8.8 inset), although there are additional layers, seen primarily through their absorption of ultraviolet radiation, at 300 km and 400 km.

Below the haze the atmosphere is clear, although rather gloomy because so little sunlight gets through. *Cassini* detected low-lying methane clouds at roughly the altitude (20 km) predicted by the theoretical models, but the clouds were less common than scientists had expected, with the result that the rain supposedly responsible for the channels in Figure 8.10(a) is not as widespread as was previously thought. Current models suggest that the methane rain may be a seasonal phenomenon, falling onto the poles in winter and evaporating again in summer. *Huygens* reached Titan almost at the end of winter at its landing site.

In 2003, radio astronomers using the Arecibo telescope reported the detection of liquid hydrocarbon lakes on Titan's surface. ∞ *(Sec. 3.4)* In 2007, *Cassini* mission specialists confirmed this finding, presenting radar images showing numerous lakes, some many tens of kilometers in length, near Titan's north polar regions (Figure 8.12). As with the *Magellan* radar images of Venus, the darkest regions are extremely smooth, in Titan's case implying that they are composed of liquid; the shapes of the regions also strongly suggest liquid bodies. ∞ *(Sec. 6.5)*

Subsequently, *Cassini* confirmed the presence of liquid ethane in the lakes, although their exact composition remains uncertain. Methane is surely present—the methane rain feeds the lakes, but it is much more volatile than ethane under the conditions found on Titan, so it probably evaporates rapidly, leaving heavier hydrocarbons behind. Computer models indicate that the lakes are mostly ethane (75 percent), with methane (10 percent) and propane (7 percent) making up much of the remainder. No waves have been detected on the lakes, hinting that they may also contain heavier tar-like hydrocarbons, increasing their viscosity. The number of lakes observed near Titan's south pole is significantly smaller than the number seen in the north, suggesting that the lakes grow during the rainy winter season. Scientists hope that repeated observations on future *Cassini* passes will reveal changes in the size and structure of the lakes as spring on Titan gives way to summer.

Titan's atmosphere seems to act like a gigantic chemical factory. Powered by the energy of sunlight, it is undergoing a complex series of chemical reactions that ultimately result in the observed composition. The upper atmosphere is thick with smoggy haze, and the surface may be covered with hydrocarbon sediment that has settled from the clouds. Spectrometers aboard *Cassini* have detected organic molecules below the haze layers. Exploration of the lakes and

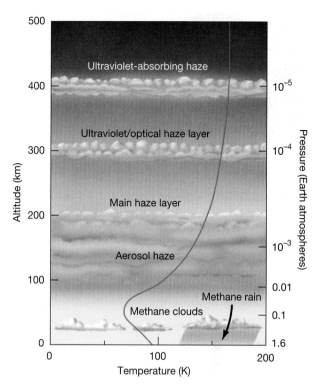

▲ FIGURE 8.11 **Titan's Atmosphere** The structure of Titan's atmosphere, as deduced from *Voyager 1* observations. The blue curve represents temperature at different altitudes.

 ANIMATION/VIDEO *Huygens* Landing on Titan

▶ FIGURE 8.12 **Titan's Lakes** Radar aboard *Cassini* detected many smooth regions (colored dark blue in this false-colored radio image) thought to be lakes of liquid methane, near Titan's north pole. This scene spans about 200 km vertically, with the largest features much smaller than any of the Great Lakes on Earth. *(NASA/ESA)*

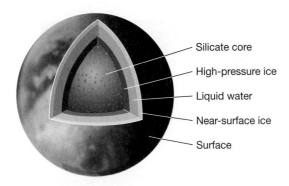

▲ **FIGURE 8.13 Titan's Interior** Based on measurements of Titan's gravitational field during numerous flybys, Titan's interior seems to be largely a rocky silicate mix. Most intriguing is the subsurface layer of liquid water, similar to that hypothesized on Jupiter's Europa and Ganymede.

other surface features by *Cassini* and future missions may afford scientists the opportunity to study the kind of chemistry thought to have occurred billions of years ago on Earth—the prebiotic chemical reactions that eventually led to life on our own planet (see Chapter 18).

Why does Titan have a thick atmosphere when Ganymede and Callisto do not? The answer lies largely in Titan's 94 K surface temperature, which is considerably lower than that of Jupiter's moons, making it easier for the moon to retain an atmosphere. *(More Precisely 5-1)* Also, gases may have become trapped in the surface ice when the moon formed. As Titan's internal radioactivity warmed it, the ice released the trapped gases, forming a thick methane–ammonia atmosphere. Sunlight split the ammonia into hydrogen, which escaped into space, and nitrogen, which remained in the atmosphere. The methane, which was less easily broken apart, survived intact. Detailed studies by *Cassini* scientists of the composition of Titan's atmosphere suggest that the atmosphere is steadily escaping, in large part due to the constant buffeting by Saturn's harsh magnetosphere, and may have been much thicker—perhaps 5–10 times denser—in the distant past.

Each pass of *Cassini* allows scientists to probe the gravity of Titan, and repeated passes, coupled with knowledge of the properties of Titan's likely constituents, have allowed construction of a remarkably detailed model of the moon's interior (Figure 8.13). The model indicates that Titan has a rocky core surrounded by a thick mantle of water ice, but intriguingly, also predicts the presence of a thick layer of liquid water a few tens of kilometers below the surface. Thus, Titan joins Europa, Ganymede, Earth, and (as we will see) Enceladus on the list of solar system objects containing large bodies of liquid water, with all that implies for the prospects of life developing there.

Triton: Captured from the Kuiper Belt?

Neptune's large moon, Triton, is the smallest of the seven large moons in the solar system, with about half the mass of the next smallest, Europa. Lying 4.5 billion km from the Sun and possessing an icy surface that reflects much of the solar radiation reaching it, Triton has a temperature of just 37 K. *Voyager 2* found that this moon has an extremely thin nitrogen atmosphere and a solid, frozen surface probably composed primarily of water ice.

A *Voyager 2* mosaic of Triton's south polar region is shown in Figure 8.14. The moon's low temperatures produce a layer of nitrogen frost that forms, evaporates, and reforms seasonally over the polar caps. The frost is visible as the pinkish region at the bottom right of Figure 8.14. Overall, there is a marked lack of cratering on Triton, presumably indicating that surface activity has obliterated the evidence of most impacts. There are many other signs of an active past. Triton's face is scarred by large fissures similar to those seen on Ganymede, and the moon's odd cantaloupe-like terrain may indicate repeated faulting and deformation over the moon's lifetime. In addition, Triton has numerous frozen lakes of water ice, thought to be volcanic in origin.

Triton's surface activity is not just a thing of the past. As *Voyager 2* passed the moon, its cameras detected great jets of

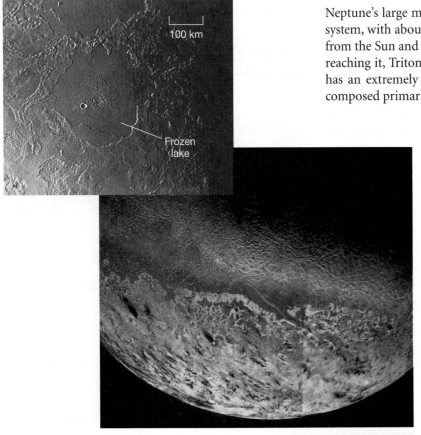

◀ **FIGURE 8.14 Triton** The south polar region of Triton, showing a variety of terrains ranging from deep ridges and gashes to what appear to be frozen water lakes, all indicative of past surface activity. The pinkish region at the lower right is nitrogen frost, forming the moon's polar cap. The roughly circular lakelike feature shown in the inset may have been caused by the eruption of an ice volcano. *(NASA)*

nitrogen gas erupting several kilometers into the sky. These geysers may result when liquid nitrogen below Triton's surface is heated and vaporized by some internal energy source, or perhaps even by the Sun's feeble light. Scientists conjecture that nitrogen geysers may be very common on Triton and are perhaps responsible for much of the moon's thin atmosphere.

Triton is unique among the large moons of the solar system in that its orbit around Neptune is *retrograde*. Also, with an orbital inclination of about 20°, Triton is the only large jovian moon not to orbit in its parent planet's equatorial plane. The event(s) that placed Triton on a tilted, retrograde orbit are the subject of considerable speculation. Triton's peculiar orbit and surface features suggest to many astronomers that the moon did not form as part of the Neptune system but instead was captured from the Kuiper belt, probably not too long ago. The surface deformations on Triton certainly suggest violent and relatively recent events in the moon's past, but the details remain uncertain. The absence of a "normal" midsize moon system around Neptune (see Section 8.3) is also often cited as evidence for some catastrophic event in the planet's history.

Whatever Triton's past, its future seems clear. Because of its retrograde orbit, the tidal bulge that Triton raises on Neptune tends to make the moon spiral *toward* the planet rather than away from it (as our Moon moves away from Earth). ⟳ *(Sec. 5.2)* Calculations indicate that Triton is doomed to be torn apart by Neptune's tidal gravitational field, probably in no more than about 100 million years.

CONCEPT CHECK ▶ Why was *Voyager 1* unable to photograph the surface of Titan? Why do we now have a much better understanding of the moon's surface?

8.3 The Medium-Sized Jovian Moons

Table 8.2 lists the 12 medium-sized satellites of the solar system, conventionally defined as bodies having radii between about 200 and 800 km. The 6 that orbit Saturn are shown in Figure 8.15. The remainder (5 orbiting Uranus, 1 orbiting Neptune) are shown in Figure 8.16 at the same scale as Figure 8.15. Their densities suggest that all 12 are composed largely of rock and water ice. All move on nearly circular paths and are tidally locked by their parent planet's gravity, and so have permanently "leading" and "trailing" faces. ⟳ *(Sec. 5.2)*.

Most of these moons have old, heavily cratered surfaces, with no evidence for extensive geological activity. Prior to *Cassini*'s arrival, scientists thought that the light-colored *wispy terrain* on Saturn's Rhea and Dione (see Figure 8.15) was the result of events in the distant past during which water was released from the moons' interiors and condensed on their surfaces. However, *Cassini* images reveal that the markings are in fact bright ice cliffs created by *tectonic fractures,* where stresses in the moons' icy interiors as they cooled and contracted caused the surface layers to crack and buckle. Dione also has "maria" of sorts, where flooding from ice volcanism appears to have obliterated the older craters. Both Tethys in the Saturn system and Ariel orbiting Uranus have extensive cracks on their surfaces, which may also be tectonic fractures, or possibly the results of violent ancient impacts.

The Uranian moons Titania, Oberon, and Umbriel are heavily cratered and show few signs of geological activity—they are comparable to Rhea in appearance (and probable history), except they lack Rhea's wispy streaks. All the moons of Uranus are darker than those of Saturn, indicating much less reflective surfaces. The most likely explanation is **radiation darkening**, where solar radiation and high-energy particles break up surface molecules, leading to chemical reactions that slowly form a layer of dark, organic material. However, our knowledge of these distant moons is quite limited—our only source of detailed information is the 1988 *Voyager 2* flyby. ⟳ *(Discovery 7-1)*

Saturn's moon Iapetus has a distinctly "two-faced" appearance. Its leading hemisphere is very dark, while the other is very bright. For years astronomers puzzled over how the dark markings could adorn only the leading face, as there seemed to be no

ANIMATION/VIDEO Saturn Satellite Transit

▶ FIGURE 8.15 **Saturnian Moons** Saturn's six medium-sized satellites, to scale, as imaged by the *Cassini* spacecraft. All are heavily cratered and all are shown here in natural color. For scale, part of Earth's Moon is shown at left. *(UC/Lick Observatory; NASA)*

Iapetus shows strong contrast between its icy (top) and cratered (bottom) hemispheres.

Enceladus is volcanically active.

Mimas

Enceladus

Iapetus

Tethys

Rhea and Dione have icy cliffs caused by surface cracking.

Rhea

Dione

Earth's Moon

R I V U X G

nearby material for the moon to sweep up as it orbited the planet. However, in 2009 the *Spitzer Space Telescope* discovered a new huge, but very diffuse, ring more than 6 million kilometers from Saturn, apparently associated with the small moon Phoebe. ⚬⚬ *(Sec. 3.5)* Iapetus lies at the inner edge of this ring, and the steady accumulation of ring particles over billions of years, coupled with the effects of solar heating and radiation darkening, could naturally account for the moon's asymmetric appearance.

TABLE 8.2	The Medium-Sized Moons of the Solar System						
Name	Parent Planet	Distance from (km)	Parent Planet (planet radii)	Orbit Period (Earth days)	Radius (km)	Mass (Moon = 1)	Density (kg/m3)
Mimas	Saturn	186,000	3.10	0.94	200	0.00051	1100
Enceladus	Saturn	238,000	3.97	1.37	250	0.00099	1000
Tethys	Saturn	295,000	4.92	1.89	530	0.0085	1000
Dione	Saturn	377,000	6.28	2.74	560	0.014	1400
Rhea	Saturn	527,000	8.78	4.52	760	0.032	1200
Iapetus	Saturn	3,560,000	59.3	79.3	720	0.022	1000
Miranda	Uranus	130,000	5.09	1.41	240	0.00090	1200
Ariel	Uranus	191,000	7.48	2.52	580	0.018	1700
Umbriel	Uranus	266,000	10.4	4.14	590	0.016	1400
Titania	Uranus	436,000	17.1	8.71	790	0.048	1700
Oberon	Uranus	583,000	22.8	13.5	760	0.041	1600
Proteus	Neptune	118,000	4.76	1.12	210	—	—

◀ FIGURE 8.16 **Moons of Uranus and Neptune** The five medium-sized moons of Uranus and the single medium-sized moon of Neptune, Proteus, shown at the same scale as Figure 8.15. All six moons are heavily cratered, and most show no clear evidence of any present or past geological activity. *(NASA)*

R I V U X G

Iapetus's other prominent surface feature is a giant 20-km-high, 1400-km-long ridge spanning half of the moon's circumference. Discovered by *Cassini* in 2005, it is clearly visible cutting across the bottom third of the moon in Figure 8.15. It is unique in the solar system, and so far defies explanation.

One surprising finding of the *Cassini* mission has been that Saturn's moon Enceladus (see the chapter-opening photo on page 216) has ongoing geological activity. Its surface is so bright and shiny—reflecting virtually 100 percent of the sunlight falling on it—that astronomers think it is completely coated with fine crystals of pure ice, the icy "ash" of water volcanoes. Much of the surface is devoid of impact craters, which have been erased by "lava flows" composed of liquid water emerging from the moon's interior and freezing on the surface. The processes involved are similar to the geothermal activity found in volcanic regions on Earth, and many astronomers prefer to describe these features as geysers (recall the nitrogen geysers on Triton, discussed in Section 8.2). *Cassini* detected icy jets emerging from geysers near Enceladus's south pole and a transient water vapor atmosphere surrounding the moon. The geysers steadily replenish the atmosphere as it escapes from the moon's weak gravity to form Saturn's E ring (see Section 8.4). Interestingly, some of the jets were found to contain organic material, fueling speculation about the possibility of life in the moon's liquid water interior.

Why is there so much activity on such a small moon? The best explanation seems to be that the internal heating is the result of tidal stresses, much like those operating in Io and Europa (Section 8.1). The orbit of Enceladus is slightly noncircular due to the gravity of the other moons, especially Dione. Saturn's tidal force on Enceladus is only one-quarter of the force exerted by Jupiter on Io, but this is still enough to partially liquefy the interior and cause the activity observed.

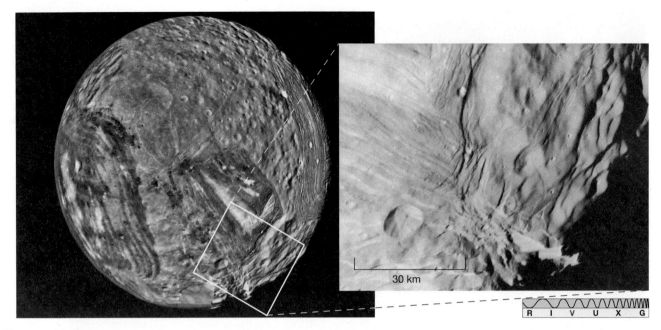

▲ FIGURE 8.17 **Miranda** The asteroid-sized innermost moon of Uranus, photographed by *Voyager 2*. Miranda has a fractured surface suggestive of a violent past, but the cause of the grooves and cracks is unknown. *(NASA)*

CONCEPT CHECK ► What surface property of the medium-sized moons tells us that the outer solar system was once a very violent place?

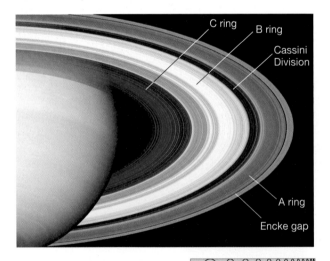

▲ FIGURE 8.18 **Saturn's Rings** Saturn as observed by the *Cassini* spacecraft in 2005. The main ring features are marked and are shown here in false color in order to enhance contrast. *(NASA/ESA)*

Most of the midsized moons show heavy cratering, indicating the cluttered and violent planetary environment that existed just after the outer planets formed, as the remaining planetesimals were accreted by larger bodies or ejected from the vicinity of the jovian planets to form the Oort cloud. ∞ *(Sec. 4.3)* The innermost midsize moons, Mimas (Saturn) and Miranda (Uranus), show perhaps the clearest evidence of violent meteoritic impacts. The impact that caused the large crater visible on the image of Mimas in Figure 8.15 must have come very close to shattering the moon. Miranda (Figure 8.17) displays a wide range of surface terrains, including ridges, valleys, faults, and many other geological features. To explain why Miranda seems to combine so many types of terrain, some researchers have hypothesized that it has been catastrophically disrupted several times, with the pieces falling back together in a chaotic, jumbled way. However, it will be a long time before we can obtain more detailed information to test this theory.

8.4 Planetary Rings

All four jovian worlds have planetary ring systems girdling their equators. The properties and appearance of each planet's rings are intimately connected with the planet's small and medium-sized moons, with many of the inner jovian moons orbiting close to (or even within) the parent planet's rings. Saturn's rings are by far the best known and best observed, and we study them here in greatest detail. However, the rings of all the jovian planets have much to tell us about the environment in the outer solar system.

Saturn's Spectacular Ring System

Figure 8.18 shows the main features of Saturn's rings. From Earth, astronomers identified three rings, which they simply labeled A, B, and C. The **A ring** lies farthest from the planet and is separated from the inner **B** and **C rings** by a dark gap called the **Cassini Division**, named for the 17th century French–Italian astronomer

Giovanni Domenico Cassini. A smaller gap, known as the **Encke Gap,** is found in the outer part of the A ring. Its width is 270 km. No finer ring details are visible from Earth. Of the three main rings, the B ring is brightest, followed by the somewhat fainter A ring, and then by the almost translucent C ring. The additional rings listed in Table 8.3 cannot be seen in Figure 8.18.

Because Saturn's rings lie in the planet's equatorial plane and the planet's rotation axis is tilted with respect to the ecliptic, the appearance of the rings (as seen from Earth) changes seasonally, as shown in Figure 8.19. As Saturn orbits the Sun, the angle at which the rings are illuminated varies. When the planet's north or south pole is tipped toward the Sun (which means it is either summer or winter on Saturn), the highly reflective rings are at their brightest. In Saturn's spring and fall, the rings are close to being edge-on both to the Sun and to observers on Earth, so they seem to disappear. One important deduction we can make from this simple observation is that the rings are very *thin.* In fact, we now know that their thickness is less than a few hundred meters, even though they are more than 200,000 km in diameter.

Saturn's rings are not solid objects. In 1857 Scottish physicist James Clerk Maxwell realized that they must be composed of a great number of small solid *particles,* all independently orbiting Saturn like so many tiny moons. Measurements of the Doppler shift of sunlight reflected from the rings reveal that the orbital speeds decrease with increasing distance from the planet in precise accordance with Kepler's laws and Newton's law of gravity. ∞ *(Sec. 1.4)* The highly reflective nature of the rings suggests that the particles are made of ice, and infrared observations in the 1970s confirmed that water ice is in fact a prime ring constituent. *Cassini* also detected substantial amounts of "mud"—small rocky particles and dust—mixed in. Radar observations and later *Voyager* and *Cassini* studies of scattered sunlight showed that the sizes of the icy ring particles range from fractions of a millimeter to tens of meters in diameter, with most particles being about the size (and composition) of large, dirty snowballs. The rings are thin because collisions between ring particles tend to keep them moving in circular orbits in a single plane. Any particle straying from this orderly configuration soon runs into other ring particles and is eventually forced back into the fold.

The Roche Limit

To understand why rings form, consider the fate of a small moon orbiting close to a massive planet such as Saturn (Figure 8.20). As we bring our hypothetical moon closer to the planet, the tidal force exerted on it by the planet increases, stretching it along the direction to the planet. ∞ *(Sec. 5.2)* The tidal force increases rapidly as the distance to the planet decreases. Eventually, a point is reached where the planet's tidal force becomes greater than the internal forces holding the moon together, and the moon is torn apart by the planet's gravity. The pieces of the satellite move on their own individual orbits around that planet, eventually spreading all the way around it in the form of a ring.

For any given planet and any given moon, this critical distance inside which the moon is destroyed is known as the **Roche limit,** after the 19th-century French mathematician Edouard Roche, who first calculated it. If our hypothetical moon is held together by its own gravity, and if its average density is similar to that of the parent planet (both reasonably good assumptions for Saturn's larger moons), then the Roche limit is approximately 2.4 times the radius of the planet. For example, in the case of Saturn (of radius 60,000 km), the Roche limit

TABLE 8.3	The Rings of Saturn				
	Inner Radius		Outer Radius		
Ring	(km)	(planet radii)	(km)	(planet radii)	Width (km)
D	60,000	1.00	74,000	1.23	14,000
C	74,000	1.23	92,000	1.53	18,000
B	92,000	1.53	117,600	1.96	25,600
A	122,200	2.04	136,800	2.28	14,600
F	140,500	2.34	140,600	2.34	100
E	210,000	3.50	300,000	5.00	90,000

(MA) ANIMATION/VIDEO Saturn Ring Plane Crossing

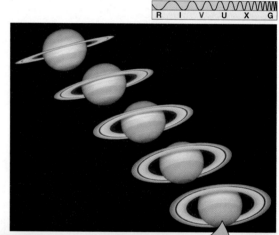

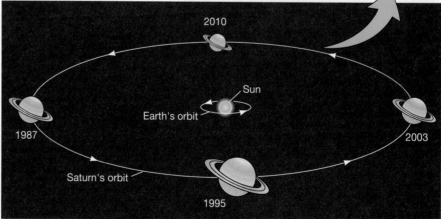

▲ **FIGURE 8.19 Saturn's Tilt** (a) Over time Saturn's rings change their appearance to terrestrial observers as the tilted ring plane orbits the Sun. The roughly true-color images (inset) span a period of several years from 2003 (bottom) to nearly the present (top) and show how the rings change from our perspective on Earth, from almost edge-on to more nearly face-on. (*NASA*)

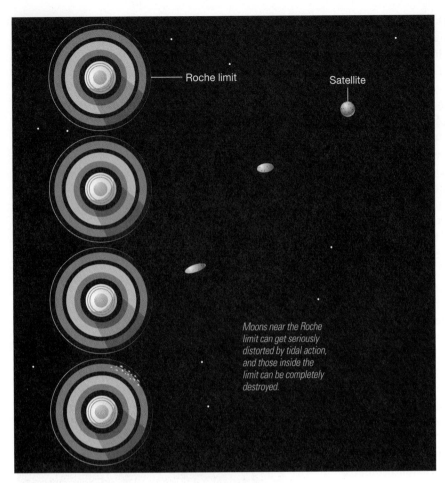

Moons near the Roche limit can get seriously distorted by tidal action, and those inside the limit can be completely destroyed.

◄ FIGURE 8.20 **Roche Limit**   The tidal field of a planet first distorts (top) and then destroys (bottom) a moon that strays too close.

lies roughly 144,000 km from the planet's center—just outside the outer edge of the A ring. The rings of Saturn occupy the region inside Saturn's Roche limit. These considerations apply equally well to the other jovian worlds. As indicated in Figure 8.21, all four ring systems in our solar system are found within (or close to) the Roche limit of their parent planet.

Note that the calculation of the Roche limit applies only to moons massive enough for their own gravity to be the dominant binding force. Space probes and sufficiently small moons can survive within the Roche limit because they are held together not by gravity but rather by interatomic (electromagnetic) forces, so the above arguments do not apply.

Fine Structure in Saturn's Rings

In 1979 the *Voyager* probes found that Saturn's main rings are actually composed of tens of thousands of narrow **ringlets**. A *Cassini* view of this fine ring structure is shown in Figure 8.22. This intricate structure is not fixed, but instead varies both with time and with location in the rings. The ringlets form when the self-gravity of the ring particles, possibly in combination with the effects of Saturn's inner moons, creates waves of matter—regions of high and low density—that move in the plane of the rings like ripples on the surface of a pond. Ring particles bunch up in some places and move apart in others, creating the ringlets we see.

The narrower gaps in the rings—about 20 in all—are the result of small *moonlets* embedded in them. Because these moonlets are much larger (perhaps 10

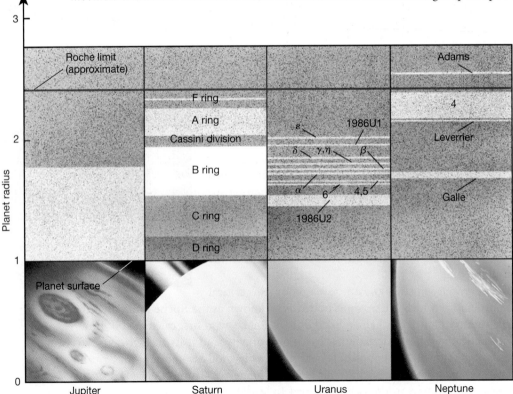

► FIGURE 8.21 **Jovian Ring Systems**   All of the distances to the rings of Jupiter, Saturn, Uranus, and Neptune are expressed in planetary radii. The red line represents the Roche limit, and all the rings lie within (or very close to) this limit of the parent planet.

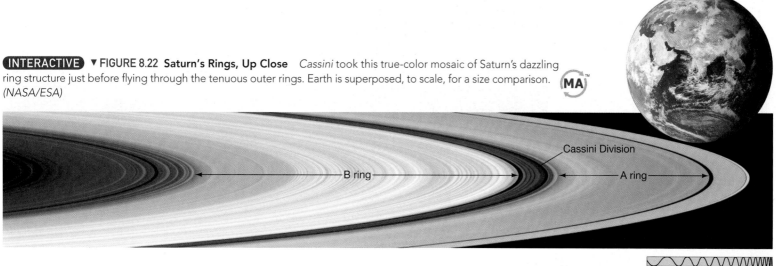

INTERACTIVE ▼ FIGURE 8.22 **Saturn's Rings, Up Close** *Cassini* took this true-color mosaic of Saturn's dazzling ring structure just before flying through the tenuous outer rings. Earth is superposed, to scale, for a size comparison. (*NASA/ESA*)

MA ANIMATION/VIDEO *Voyager* Ring Spokes Animation

or 20 km across) than the largest ring particles, they sweep up ring material as they go. Prior to *Cassini*'s arrival, only one of these moonlets had been found, despite many careful searches of the *Voyager* images. *Cassini* has now identified numerous small moonlets (a few kilometers across) in the gaps between the main rings.

The largest gap—the Cassini Division—has a different origin. It owes its existence to the gravitational influence of Saturn's innermost medium-sized moon, Mimas. Over time, particles orbiting within the Division have been deflected by Mimas's gravity into eccentric orbits that cause them to collide with other ring particles, effectively moving them into new orbits at different radii. The net result is that the number of ring particles in the Cassini Division is greatly reduced.

Outside the A ring lies perhaps the strangest ring of all: the faint, narrow **F ring** (Figure 8.23). It was discovered by *Pioneer 11* in 1979, but its full complexity became evident only when the *Voyager* spacecraft took a closer look. Its oddest feature is that it looks as though it is made up of several strands braided together. The ring's complex structure, as well as its thinness, arises from the influence of two small moons, known as **shepherd satellites**, that orbit about 1000 km on either side of it. Their gravitational effect gently guides any particle straying too far out of the F ring back into it. *Cassini* has discovered several more braided rings within the main ring system. The braids are the result of waves created in the rings by the moons. However, the details of how the braids are produced, and why the shepherd moons are there at all, remain subjects of active research.

Voyager 2 also found a faint series of rings, now known collectively as the **D ring**, inside the inner edge of the C ring, stretching down almost to Saturn's cloud tops. The D ring contains relatively few particles and is so dark that it is completely invisible from Earth. The faint **E ring,** also a *Voyager 2* discovery, lies well outside the main ring structure. It is thought to be the result of volcanism on the moon Enceladus, discussed earlier. Figure 8.24 is a *Cassini* image showing the rings from a perspective never before seen—behind the planet looking back toward the eclipsed Sun. Just as diffuse airborne dust is most easily seen against the light streaming in through a sunlit window, the planet's faint rings show up clearly in this remarkable back-lit view. In addition, this image reveals several more faint rings, some of them also associated with the orbits of various small moons.

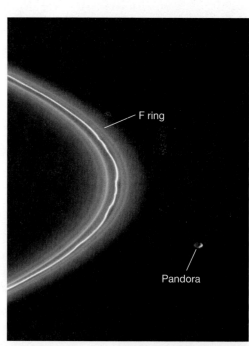

▶ FIGURE 8.23 **Shepherd Moon** Saturn's narrow F ring appears to contain kinks and braids, making it unlike any of Saturn's other rings. Its thinness is caused by two shepherd satellites that orbit the ring, one a few hundred kilometers inside the ring and the other a similar distance outside it. One of the potato-shaped shepherding satellites, called Pandora and roughly 100 km across, can be seen at right. (*NASA*)

ANALOGY: Similar to the way the combined efforts of shepherd and sheepdogs contain sheep within a somewhat wavy line, shepherd moons confine loose material within the braided strands of Saturn's rings.

▶ FIGURE 8.24 Back-Lit Rings   *Cassini* took this spectacular image of Saturn's rings as it passed through Saturn's shadow. The normally hard to see F, G, and E outer rings are clearly visible in this contrast-enhanced image. The inset shows the moon Enceladus orbiting within the E ring; its eruptions give rise to the ring's icy particles. *(NASA)*

▼ FIGURE 8.25 Jupiter's Ring   Jupiter's faint ring, as photographed nearly edge-on by *Voyager 2*. Made of dark fragments of rock and dust possibly chipped off the innermost moons by meteorites, the ring was unknown before the two *Voyager* spacecraft arrived at the planet. *(NASA)*

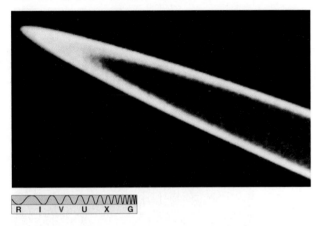

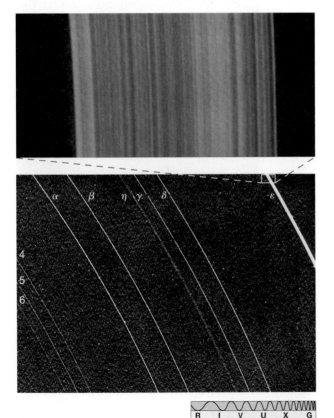

The Rings of Jupiter, Uranus, and Neptune

One of the many remarkable findings of the 1979 *Voyager* missions was the discovery of a faint ring of matter encircling Jupiter in the plane of the planet's equator (Figure 8.25). This ring lies roughly 50,000 km above the top cloud layer, inside the orbit of the innermost moon and is probably composed of dark fragments of rock and dust chipped off the inner moons by meteorites. The outer edge of the ring is quite sharply defined, although a thin sheet of material may extend all the way down to Jupiter's cloud tops. In the direction perpendicular to the equatorial plane, the ring is only a few tens of kilometers thick.

The ring system surrounding Uranus was discovered in 1977, when astronomers observed the planet passing in front of a bright star, momentarily dimming the star's light. Such a **stellar occultation** happens a few times per decade and allows astronomers to measure planetary structures that are too small and faint to be detected directly. The ground-based observations revealed 9 thin rings encircling Uranus. Two more were discovered by *Voyager 2*. The locations of all 11 rings are indicated in Figure 8.21.

The nine rings known from Earth-based observation are shown in Figure 8.26, a *Voyager 2* image that makes it readily apparent that Uranus's rings are quite different from those of Saturn. While Saturn's rings are bright and wide, with relatively narrow, dark gaps between them, those of Uranus are dark, narrow, and widely spaced. However, like Saturn's rings, the rings of Uranus are all less than a few tens of meters thick (measured in the direction perpendicular to the ring plane). Like the F ring of Saturn, Uranus's narrow rings require shepherd satellites to keep them in place. During its 1986 flyby, *Voyager 2* detected the shepherds of the Epsilon ring (Figure 8.27). Many other, undetected shepherd satellites must also exist.

As shown in Figure 8.28, Neptune is surrounded by four dark rings. Three are quite narrow, like the rings of Uranus, and one is quite broad and diffuse, more like Jupiter's ring. The outermost ring is noticeably clumped in places. From Earth we see not a complete ring, but only partial arcs; the unseen parts of the ring are simply too thin (unclumped) to be detected. The connection between the rings and the planet's small inner satellites has not yet been firmly established, although many astronomers now think the clumping is caused by shepherd satellites.

◀ FIGURE 8.26 Uranus's Rings   The main rings of Uranus, as imaged by *Voyager 2*. All nine of the rings known before the spacecraft's arrival can be seen in this photo. The inset at top shows a closeup of the Epsilon ring, revealing some internal structure. *(NASA)*

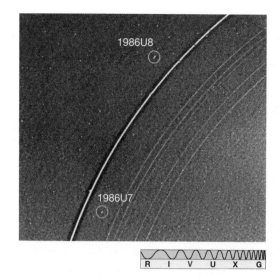

▶ FIGURE 8.27 **More Shepherd Moons** These two small moons, now named Cordelia and Ophelia, were discovered by *Voyager 2* in 1986. They shepherd Uranus's Epsilon ring, keeping it from diffusing away. *(NASA)*

The Formation of Planetary Rings

The dynamic behavior—waves, collisions, and interactions with moons—observed in the rings of Saturn and the other jovian planets suggests to many researchers that the rings are quite young—perhaps no more than 50 million years old, just 1 percent of the age of the solar system. If the rings are indeed so young, then either they are replenished from time to time, perhaps by fragments of moons chipped off by meteoritic impact, or they are the result of a relatively recent possibly catastrophic event in the planet's system—a moon torn apart by tidal forces, or destroyed in an impact with a large comet or even with another moon.

Astronomers estimate that the total mass of Saturn's rings is enough to make a satellite about 250 km in diameter. If such a satellite strayed inside Saturn's Roche limit or was destroyed (perhaps by a collision) near that radius, a ring system could have resulted. While we do not know if Saturn's rings originated in this way, this is the likely fate of Neptune's large moon Triton. As mentioned earlier (in Section 8.2), Triton will be torn apart by Neptune's tidal gravitational field within 100 million years or so. By that time it is quite possible that Saturn's ring system will have disappeared, and Neptune will then be the planet in the solar system with spectacular rings!

CONCEPT CHECK ▶ What is the Roche limit, and what is its relevance to planetary rights?

8.5 Beyond Neptune

The inventory of known Kuiper belt objects has expanded greatly since the early 1990s. Ground-based telescopes have led the way in the painstaking effort to capture the meager amounts of sunlight reflected from these dark inhabitants of the outer solar system. However, one resident of the Kuiper belt has been known for decades—Pluto. Let's begin our study of this distant region by reviewing what is known about its most prominent member.

The Discovery of Pluto

In Chapter 7 we saw how irregularities in Uranus's orbit led to the discovery of Neptune in the mid-19th century. ∞ *(Sec. 7.2)* By the end of that century, observations of the orbits of Uranus and Neptune suggested that Neptune's influence was not sufficient to account for all of the observed irregularities in Uranus's motion. Furthermore, it seemed that Neptune itself was being affected by some other unknown body. Astronomers hoped to pinpoint the location of this new planet using techniques similar to those that had guided them to the discovery of Neptune.

One of the most ardent searchers was Percival Lowell, a capable and persistent observer and one of the best-known astronomers of his day. ∞ *(Discovery 6-1)* Basing his investigation primarily on the motion of Uranus (Neptune's orbit was still relatively poorly determined at the time), Lowell set about calculating where the supposed ninth planet should be. He searched for it, without success, during the decade preceding his death in 1916. In 1930 the American astronomer Clyde Tombaugh, working with improved equipment and photographic techniques at the Lowell Observatory, finally succeeded in finding Lowell's ninth planet, only 6° away from Lowell's predicted position. The new object was named Pluto for the Roman god of the dead, who reigns over eternal darkness (and also because its first two letters and its astronomical symbol PL are Lowell's initials).

On the face of it, the discovery of Pluto looks like another spectacular success for Newtonian mechanics. However, the supposed irregularities in the motions of Uranus and Neptune did not in fact exist and the mass of Pluto, not measured accurately until the 1980s, is far too small to have caused them anyway. The discovery of Pluto owed much more to simple luck than to complex mathematics!

▲ FIGURE 8.28 **Neptune's Faint Rings** In this long-exposure image, Neptune (center) is heavily overexposed and has been artificially obscured (by an instrument) to make the rings easier to see. One of the two fainter rings lies between the inner bright ring and the planet. The others lie between the two bright rings. *(NASA)*

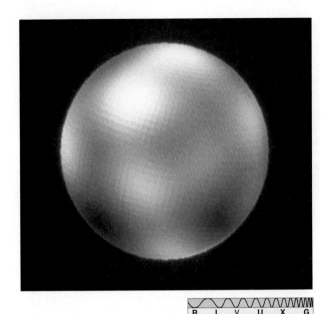

R I V U X G

▲ FIGURE 8.29 Pluto   A surface map of Pluto—not a photograph, but rather a "modeled" view created by carefully combining 24 *Hubble Space Telescope* images with a mathematical description of the surface. *(NASA)*

At nearly 40 AU from the Sun, Pluto is often hard to distinguish from the background stars. It is never visible to the naked eye. Pluto is among the largest objects in the solar system not studied at close range by unmanned spacecraft, although NASA has launched a 2006 mission to Pluto and the Kuiper belt, currently scheduled to reach the outer solar system in 2015. For now, though, every new discovery about this remote world is the result of painstaking observations either from Earth or from Earth-orbiting instruments. Figure 8.29 is a composite map of Pluto's surface obtained by combining many *Hubble Space Telescope* images. It shows surface detail at about the same level as can be seen on Mars with a small telescope. Other than the bright polar caps, none of the surface features has been conclusively identified. Some of the dark regions may be craters or impact basins, as on Earth's Moon.

The Pluto–Charon System

In 1978 astronomers at the U.S. Naval Observatory discovered that Pluto has a satellite. It is now named Charon, after the mythical boatman who ferried the dead across the river Styx into Hades, Pluto's domain. The discovery photograph of Charon is shown in Figure 8.30(a). Charon is the large white bump at the top right of the image. Figure 8.30(b) shows a 2005 *Hubble Space Telescope* image that clearly resolves the two bodies. Based on studies of Charon's orbit, astronomers have measured the mass of Pluto to great accuracy. It is just 0.0021 Earth masses (or roughly 10^{22} kg)—or about 0.17 times the mass of Earth's Moon.

By pure chance, Charon's orbit over the 6-year period from 1985 to 1991 produced a series of *eclipses* as Pluto and Charon repeatedly passed in front of one another, as seen from Earth. With more good fortune, these eclipses took place while Pluto was closest to the Sun, making for the best possible Earth-based observations. Basing their calculations on the variations in light as Pluto and Charon periodically hid each other, astronomers computed the masses and radii of both bodies and determined their orbit plane. The mass of Charon is about 0.12 that of Pluto, giving the Pluto–Charon system by far the largest satellite-to-object mass ratio in the solar system. Pluto's radius is 1150 km, about one-fifth that of Earth. Charon is about 600 km in radius and orbits 19,600 km from Pluto.

The masses and radii of Pluto and Charon imply average densities of 2100 kg/m³—just what we would expect for bodies of that size made up mostly of water ice, like

▶ FIGURE 8.30 Pluto and Charon   (a) The discovery photograph of Pluto's moon Charon. The moon is the small blotch of light on the top right portion of the image. (b) The Pluto–Charon system, as seen by the *Hubble Space Telescope*. *(NASA)*

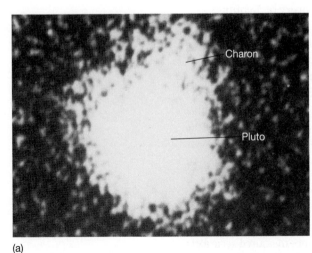

(a)

(b)

R I V U X G

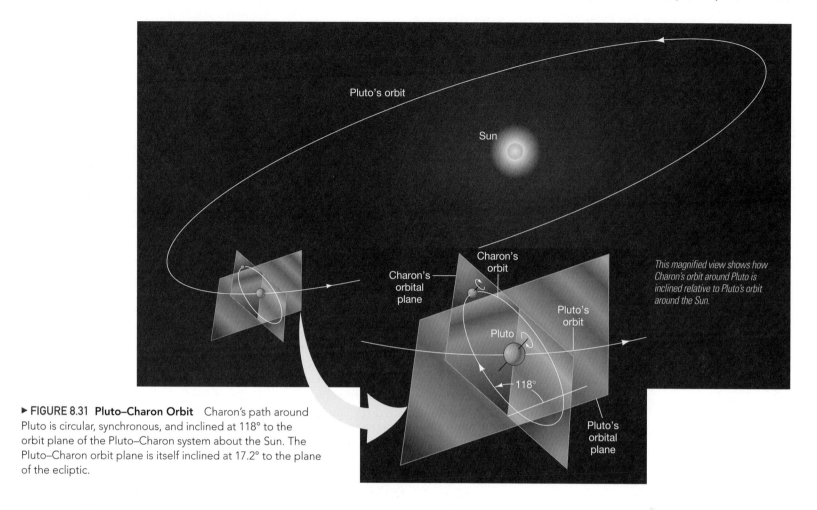

This magnified view shows how Charon's orbit around Pluto is inclined relative to Pluto's orbit around the Sun.

▶ **FIGURE 8.31 Pluto–Charon Orbit** Charon's path around Pluto is circular, synchronous, and inclined at 118° to the orbit plane of the Pluto–Charon system about the Sun. The Pluto–Charon orbit plane is itself inclined at 17.2° to the plane of the ecliptic.

several of the large moons of the outer planets. In fact, Pluto is very similar in both mass and radius to Neptune's large moon, Triton—which, as we have just seen, is thought to be a captured Kuiper belt object. By studying sunlight reflected from their surfaces, astronomers also found that Pluto and Charon are tidally locked as they orbit each other. Their orbital period, and the rotation period of each, is 6.4 days. As shown in Figure 8.31, Charon's orbit and the spin axes of both objects are inclined at an angle of 118° to the plane of the ecliptic.

In late 2005, scientists using *HST* reported the discovery of two additional small moons, perhaps 100–200 km in diameter, orbiting Pluto at about twice the distance of Charon (see Figure 8.30). They have since been named Nix (for the goddess of darkness and mother of Charon) and Hydra (a mythical nine-headed monster). A fourth moon, tentatively termed S/2011/P1, was spotted in 2011.

 What do Pluto and Triton have in common?

Plutoids and the Kuiper Belt

Most Kuiper belt objects are not nearly as well observed as Pluto, which happens to be the largest known member of the class and orbits near the inner edge of the belt, making it appear relatively bright as seen from Earth. Objects beyond Neptune—even the large ones—are very faint. Nevertheless, as observational techniques have improved, the number of objects detected beyond Neptune has risen rapidly, and astronomers have pieced together a better understanding of their sizes and orbits. For example, we now know that about 25 percent of all Kuiper belt objects have exactly the same orbital period as Pluto. This is not a coincidence, of course—it is caused by gravitational interactions between these bodies and the planet Neptune, which maintains all their orbital periods at precisely 1.5 times that of Neptune.

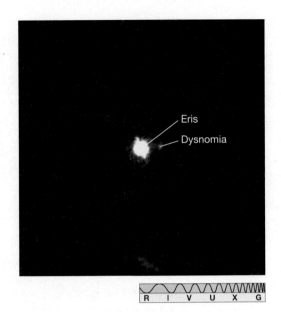

▲ FIGURE 8.32 **Kuiper Belt Object** The trans-Neptunian object Eris and its small moon Dysnomia (named after the Greek goddess of discord and her daughter of chaos and lawlessness) were imaged in the infrared at the Keck Observatory in Hawaii. *(Keck)*

More than 1200 "trans-Neptunian" objects are currently known. Most orbit in the Kuiper belt, conventionally defined to lie between 40 and 50 AU from the Sun. Because they are so small and distant, researchers reason that only a tiny fraction of the total has so far been observed. The total number of Kuiper belt objects larger than 100 km across is estimated to be more than 100,000. If so, the combined mass of all the debris beyond Neptune could be hundreds of times larger than the mass of the inner asteroid belt (although still less than the mass of Earth).

As the details were filled in and the number of known objects increased, it became increasingly clear to astronomers that Pluto was not distinctly different from the other small bodies in the outer solar system, as had once been supposed. The Kuiper belt object Quaoar (pronounced "Kwah-o-ar," and named for a native American creation god), discovered in 2002, is roughly 1200 km across—larger than the largest asteroid, Ceres, and more than half the size of Pluto. The Kuiper belt objects Haumea, discovered in 2003, and Makemake, in 2005, are larger still—both some 1500–2000 km across. (Both names are drawn from Hawaiian mythology.) But the final blow came with the discovery in 2005 of the (non-Kuiper belt)[1] object Eris (appropriately named for the Greek goddess of discord, and shown in Figure 8.32), whose diameter, measured in 2006 by *HST*, is 2350 km—bigger than Pluto.

Figure 8.33 shows the sizes of some of the largest bodies orbiting beyond Neptune. It also includes yet another intriguing object called Sedna, discovered in 2003 and thought to have a diameter of about 1500 km. It is the farthest known object in the solar system. Its highly elliptical orbit takes it out to almost 1000 AU from the Sun—almost to the (theoretical) inner edge of the Oort cloud. There may well be more Pluto-sized (or larger) objects still out there waiting to be discovered—the possibility has not been conclusively ruled out.

Even before the discovery of Eris, many astronomers had already concluded that Pluto was simply a large Kuiper belt object, playing much the same role in the Kuiper belt as Ceres does among the asteroids. Once it became clear that Eris was larger than Pluto, pressure mounted to find a classification that reflected astronomers' new understanding of the outer solar system. As mentioned in Chapter 4, the result was the invention of a new category of solar system object—*dwarf planet*—containing Pluto, Eris, Ceres, and now also Haumea and Makemake. ⚭ *(Sec. 4.1)*

Some astronomers are unhappy at Pluto's demotion from the solar system "A list." They argue (probably correctly) that the new dwarf planet category was concocted largely to exclude Pluto and Eris from planetary status. Others applaud the redefinition as long-overdue recognition of Pluto's true identity. The arguments over terminology are not over, but it seems unlikely that future changes will restore Pluto to its former status. Few astronomers doubt that if Pluto were discovered today, its classification as a

---

[1]Technically, because its eccentric orbit extends well beyond the Kuiper belt, Eris is not classified as a Kuiper belt object. However, like the members of the Kuiper belt, it most likely formed as an icy planetesimal in the outer solar system and was subsequently kicked out by Neptune, so in terms of composition and history it probably differs little from "true" Kuiper belt objects. ⚭ *(Sec. 4.3)*

▼ FIGURE 8.33 **Trans-Neptunian Objects** Some large trans-Neptunian objects, including Pluto and the largest known, called Eris, with part of Earth and the Moon added for scale. Most diameters are approximate, as they are estimated from each object's observed brightness. *(NASA; Caltech)*

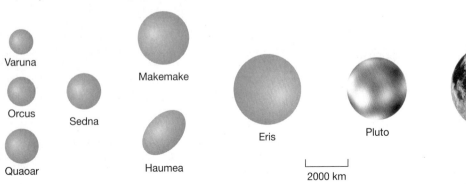

Varuna

Orcus

Sedna

Quaoar

Makemake

Haumea

Eris

Pluto

Moon

2000 km

Earth

member (the biggest yet, the headlines would say!) of the Kuiper belt would be assured. As if for consolation, in 2008 the *International Astronomical Union* decided that the icy dwarf planets beyond Neptune would collectively be known as **plutoids.**

Pluto's reclassification illustrates the way in which science evolves. Our conception of the cosmos has undergone many changes—some radical, others more gradual—since the time of Copernicus. ⚭ *(Sec. 1.2)* As our understanding grows, our terminology and classifications change. The situation with Pluto has a close parallel in the discovery of the first asteroids in the early 19th century. They, too, were initially classified as planets; indeed, in the 1840s, leading astronomy texts listed no fewer than 11 planets in our solar system, including as numbers 5 through 8 the asteroids Vesta, Juno, Ceres, and Pallas. Within a couple of decades, however, the discovery of several dozen more asteroids had made it clear that these small bodies represented a new class of solar system objects, separate from the planets, and the number of planets fell to eight (including the newly discovered Neptune).

Much of the observational work on the Kuiper belt began as a search for a tenth planet. It is ironic that the end result of these efforts has been a reduction in the number of "true" (that is, major) solar system planets back to eight.

**PROCESS OF ►
SCIENCE CHECK** Why is Pluto no larger regarded as a major planet?

CHAPTER REVIEW

SUMMARY

LO1 Six large moons in the outer solar system are comparable to or larger than Earth's Moon in size and mass. The four Galilean moons of Jupiter have densities that decrease with increasing distance from the planet. Io has active volcanoes powered by the constant flexing of the moon by Jupiter's tidal forces. Europa's cracked, icy surface may well conceal an ocean of liquid water. Ganymede and Callisto have heavily cratered surfaces. Ganymede's magnetic field implies relatively recent geological activity, but Callisto shows none.

LO2 Saturn's large moon Titan has a thick atmosphere that obscures the moon's surface and may be the site of complex cloud and surface chemistry. The moon's surface is so cold that water behaves like rock and liquid methane and ethane flow like water. Sensors aboard *Cassini* have mapped the moon's surface, revealing evidence for ongoing erosion and volcanic activity. The *Huygens* probe landed on the icy surface and photographed what may be channels carved by flowing methane.

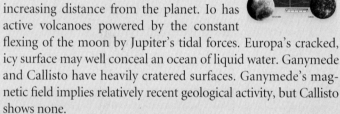

LO3 Neptune's moon Triton has a fractured surface of water ice and a thin atmosphere of nitrogen, probably produced by nitrogen "geysers" on its surface. Triton is the only large moon in the solar system to have a retrograde orbit around its parent planet. This orbit is unstable and will eventually cause Triton to be torn apart by Neptune's gravity.

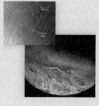

LO4 The medium-sized moons of Saturn and Uranus are made up predominantly of rock and water ice. Many are heavily cratered, and some must have come close to being destroyed by the impacts whose craters we now see. Some show evidence of recent geological activity; in most cases the cause of the activity is not known.

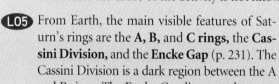

LO5 From Earth, the main visible features of Saturn's rings are the **A, B,** and **C rings,** the **Cassini Division,** and the **Encke Gap** (p. 231). The Cassini Division is a dark region between the A and B rings. The Encke Gap lies near the outer edge of the A ring. The rings are made up of trillions of individual particles, ranging in size from dust grains to boulders. Their total mass is comparable to that of a small moon. Interactions between the ring particles and the planet's inner moons are responsible for tens of thousands of narrow **ringlets** (p. 232). The tenuous **E-ring** (p. 233) is associated with volcanism on the moon Enceladus. Saturn's narrow **F ring** (p. 233) lies just outside the A ring. Its kinked, braided structure is due to two small **shepherd satellites** (p. 233) that orbit close to it and prevent it from breaking up.

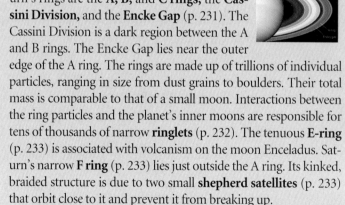

LO6 The **Roche limit** (p. 231) of a planet is the distance within which the planet's tidal field overwhelms the internal gravity of an orbiting moon, tearing the moon apart and forming a ring. All the rings of the four jovian planets lie inside (or close to) their parent planets' Roche limits.

LO7 Jupiter has a faint, dark ring extending down to the planet's cloud tops. Uranus has a series of dark, narrow rings first detected from Earth by **stellar occultation** (p. 234), which occurs when a body passing between Earth and a distant star obscures the light from that star. Shepherd satellites keep Uranus's rings from breaking apart. Neptune has three narrow rings like Uranus's and one broad ring, like Jupiter's.

LO8 Pluto was discovered in 1930 after a laborious search for a planet that was supposedly affecting Uranus's orbital motion. We now know that Pluto is far too small to have had any detectable influence on Uranus's path. Pluto has a large moon, Charon, and two smaller moons. Studies of Charon's orbit around Pluto have allowed the masses and radii of both bodies to be accurately determined. Several bodies beyond Neptune's orbit have masses comparable to or greater than that of Pluto; Eris is the most massive. Eris, Pluto, and two other Kuiper belt objects are examples of **plutoids** (p. 239).

MasteringAstronomy®   For instructor-assigned homework go to www.masteringastronomy.com

Problems labeled POS explore the process of science. VIS problems focus on reading and interpreting visual information. LO connects to the introduction's numbered Learning Outcomes.

REVIEW AND DISCUSSION

1. **LO1** How do the density and composition of the Galilean moons vary with increasing distance from Jupiter?

2. What is unusual about Jupiter's moon Io?

3. **POS** Why is there speculation that the Galilean moon Europa might be an abode for life?

4. **POS** Why do scientists think that Ganymede's interior may have been heated as recently as a billion years ago?

5. **LO2** What properties of Saturn's largest moon, Titan, make it of particular interest to astronomers?

6. **LO3** What is the predicted fate of Triton?

7. **LO4** What is the evidence for geological activity on Saturn's mid-sized moons?

8. **LO5** Seen from Earth, Saturn's rings sometimes appear broad and bright but at other times seem to disappear. Why?

9. **LO6 POS** Why do many astronomers think Saturn's rings formed quite recently?

10. What effect does Mimas have on the rings of Saturn?

11. Describe the behavior of shepherd satellites.

12. **LO7** How do the rings of Neptune differ from those of Uranus and Saturn?

13. **LO8 POS** Why don't astronomers regard Pluto as a major planet?

14. In what respect is Pluto more like a moon than a planet?

15. How was the mass of Pluto determined?

CONCEPTUAL SELF-TEST TRUE OR FALSE?/MULTIPLE CHOICE

1. Scientists speculate that Europa may have liquid water below its frozen surface. (T/F)

2. Saturn's ring particles are made up mainly of water ice. (T/F)

3. Saturn's rings lie within the planet's Roche limit. (T/F)

4. Two small shepherd satellites are responsible for the unusually complex form of Saturn's F ring. (T/F)

5. Titan's surface is obscured by thick clouds of ammonia ice. (T/F)

6. Uranus has no large moons. (T/F)

7. Triton's orbit is unusual because it is retrograde. (T/F)

8. Pluto was discovered via its gravitational effect on Neptune. (T/F)

9. Io's surface appears very smooth because it (a) is continually re-surfaced by volcanic activity; (b) is covered with ice; (c) has been shielded by Jupiter from meteorite impacts; (d) is liquid.

10. The Galilean moons of Jupiter are sometimes described as a min-iature inner solar system because (a) there are the same number of Galilean moons as there are terrestrial planets; (b) the moons have generally "terrestrial" composition; (c) the moons' densities decrease with increasing distance from Jupiter; (d) the moons all move on circular, synchronous orbits.

11. Compared to Earth's surface, the pressure at the surface of Titan is (a) less than; (b) about the same as; (c) about 1.5 times greater than that of Earth.

12. The rings of Uranus are (a) broad and bright; (b) narrow and dark; (c) narrow and bright; (d) undetectable from Earth.

13. The solar system object most similar to Pluto is (a) Mercury; (b) the Moon; (c) Titan; (d) Triton.

14. VIS In Figure 8.11 (Titan's Atmosphere), the atmospheric temper-ature at the top of the main haze layer is approximately (a) 50 K; (b) 100 K; (c) 150 K; (d) 200 K.

15. VIS From Figure 8.19 (Saturn's Tilt), the next time Saturn's rings will appear roughly edge-on as seen from Earth will be around (a) 2025; (b) 2015; (c) 2020; (d) 2035.

PROBLEMS

The number of squares preceding each problem indicates its approximate level of difficulty.

1. ■ What is the orbital speed, in kilometers per second, of ring particles at the inner edge of Saturn's B ring? Compare the particles speed with that of a satellite in low Earth orbit (500 km altitude, say). Why are these speeds so different?

2. ■■ Io orbits Jupiter at a distance of six planetary radii from the planet's center once every 42 hours. If Jupiter completes one rotation every 10 hours, use Kepler's third law to calculate how far from the center a satellite must orbit in order to appear stationary above the planet. ∞ *(Sec. 1.3)*

3. ■■■ Calculate the difference between the acceleration due to Jupiter's gravity at Io's center and the acceleration on Io's surface at the point closest to the planet. This is the tidal acceleration exerted by Jupiter on the moon. ∞ *(Sec. 5.2)* Compare this with Io's surface gravity. Repeat the comparison for Earth and Earth's Moon.

4. ■ Show that Titan's surface gravity is about one-seventh that of Earth's. What is Titan's escape speed? ∞ *(More Precisely 5-1)*

5. ■■ Compare the apparent sizes of the Galilean moons, as seen from Jupiter's cloud tops, with the angular diameter of the Sun at Jupiter's distance. Would you expect ever to see a total solar eclipse from Jupiter's cloud tops? ∞ *(More Precisely 4-1)*

6. ■■ Assuming a spherical shape and a uniform density of 2000 kg/m^3, calculate how small an icy moon of one of the outer planets would have to be before a 40 m/s (about 90 mph) fastball could escape.

7. ■ The total mass of material in Saturn's rings is about 10^5 tons (10^8 kg). Suppose the average ring particle is 6 cm in radius (the size of a large snowball) and has a density of 1000 kg/m^3. How many ring particles are there?

8. ■ How close is Charon to Pluto's Roche limit?

9. ■ What would be your weight on Pluto? On Charon?

10. ■■■ What is the round-trip travel time of light from Earth to Pluto (at a distance of 40 AU)? How far would a spacecraft moving in a circular orbit 500 km above Pluto's surface travel during that time? To how many orbits does this correspond?

ACTIVITIES

Collaborative

1. Observe the orbits of the Galilean moons of Jupiter. You'll need a telescope and a tripod or other mount to keep the image steady. Center the planet in the field of view. You should be able to see its red-and-tan cloud bands, as well as three or four points of light along a line from Jupiter's center. These are the Galilean moons. Carefully sketch what you see, or if you can attach a camera, make an image of the field of view. In your sketch, show Jupiter, the nearby moons, any bright background stars, and the full field of view of the telescope. Keep track of the time at which the observation was made. Repeat this procedure roughly every hour over the next 3–4 nights—work in shifts so no one has to lose too much sleep! The inner two moons will complete at least one orbit during this time. Your sketches should allow you to visualize the orbits of the moons. Estimate their orbital periods and radii relative to Jupiter's radius and compare your results with the numbers in Table 8.1.

Individual

1. Use binoculars to look at Jupiter. Can you see any of the four large moons? If you come back the following evening, the moons' positions will have changed. Have some changed more than others?

2. Through a telescope, you should have a much better view of Jupiter's moons. Before observing, look up the positions of the Galilean moons in a current magazine such as *Astronomy* or *Sky & Telescope*. Identify each of the moons. Watch Io over a period of an hour. Can you see its motion? Do the same for Europa.

3. Use a telescope to look at Saturn's rings. Can you see a dark line in them? This is the Cassini Division. Can you see the shadow of the rings on the planet's surface? Can you see any moons? They line up with the rings; Titan is often the farthest out and is always the brightest. Use an almanac to identify each moon you find.

The Stars

All stars evolve—they are born, mature, and die. They form in huge clouds of gas and dust, their nuclear fires ignite much as in the Sun, and eventually they end their lives by returning to space some of the heavy elements created within their cores. The poetic notion that we are made of stardust is a romantic one—but it also happens to be true!

Part 3 describes many diverse stellar and interstellar regions that resemble a "galactic ecosystem"—an evolutionary balance nearly as intricate and delicate as life in a tide pool or a tropical forest. The theory of stellar evolution gives astronomers a powerful, unifying theme with which to study all stars.

The images here illustrate a sampling of the stars and stellar systems of Part 3, all residents of an even larger Milky Way Galaxy.

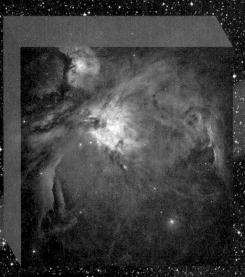
Nebula ~ 10^{16} m

Dusty Disk ~ 10^{14} m

Red Giant ~ 10^{11} m

Galaxy ~ 10^{21} m

Star Cluster ~ 10^{18} m

Nearly a trillion trillion stars inhabit our
universe; the Sun is but one of these stars.

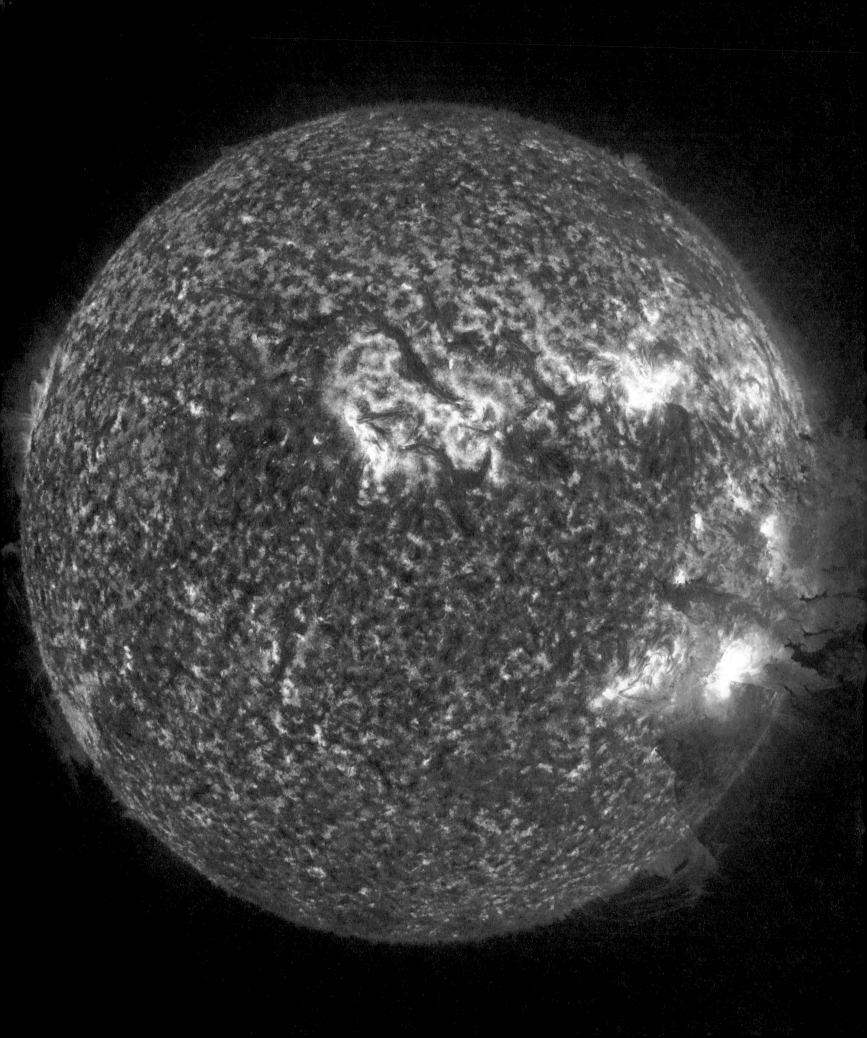

9

The Sun

Our Parent Star

Living in the solar system, we have the chance to study, at close range, perhaps the most common type of cosmic object—a star. Our Sun is a star, and a fairly average star at that, but with one unique feature: It is very close to us— some 300,000 times closer than the next nearest star, Alpha Centauri. Whereas Alpha Centauri is 4.3 light-years away, the Sun is only 8 light-minutes from us. Consequently, astronomers know far more about the properties of the Sun than about any of the other distant points of light in the universe. Indeed, a good fraction of all our astronomical knowledge is based on modern studies of the Sun. Just as we studied our parent planet, Earth, to set the stage for our exploration of the solar system, we now study our parent star, the Sun, as the next step in our exploration of the universe.

MasteringAstronomy®   Visit www.masteringastronomy.com for quizzes, animations, videos, interactive figures, and self-guided tutorials.

◄ The Sun is our star—the main source of energy that powers weather, climate, and life on Earth. Humans simply would not exist without the Sun. Although we take it for granted each and every day, the Sun is of great importance to us in the cosmic scheme of things. This spectacular image, taken through a filter in the ultraviolet part of the spectrum to emphasize contrast on the solar surface, shows moderate activity (in white), including a large solar flare on the northern limb. The image was observed in 2011 by the *Solar Dynamics Observatory*—a 2-ton robot that orbits Earth but stares at the Sun unblinkingly 24 hours a day, eavesdropping on this gas ball's surface, atmosphere, and interior. *(NASA)*

R I V U X G

▲ FIGURE 9.1 The Sun   The inner part of this composite, filtered image of the Sun shows a sharp edge, although our star, like all stars, is composed of a gradually thinning gas. The edge appears sharp because the solar photosphere is so thin. The outer portion of the image is the solar corona, normally too faint to be seen, but visible during an eclipse, when the light from the solar disk is blotted out. (Note the blemishes; they are sunspots.) *(NOAO)*

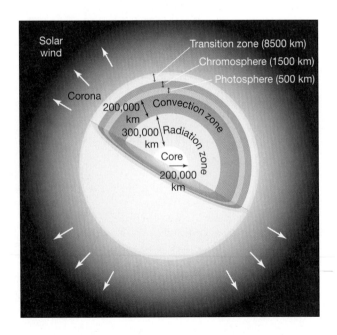

9.1 The Sun in Bulk

The Sun is the sole source of light and heat for the maintenance of life on Earth. It is a **star**, a glowing ball of gas held together by its own gravity and powered by nuclear fusion (a process in which two or more atomic nuclei combine and release energy—see Section 9.5) at its center. In its physical and chemical properties, the Sun is very similar to most other stars, regardless of when and where they formed. Far from detracting from our interest in the Sun, this very mediocrity is one of the main reasons we study it—most of what we know about the Sun also applies to many other stars in the sky. The Sun is the best-studied star in the universe, and it is no exaggeration to say that our knowledge of the cosmos rests squarely on our understanding of solar physics.

Overall Structure

Table 9.1 lists some basic solar data. With a radius larger than 100 Earth radii, a mass greater than 300,000 Earth masses, and a surface temperature well above the melting point of any known material, the Sun is clearly very different from any other body we have encountered so far.

The Sun has a surface of sorts—not a solid surface (the Sun contains no solid material), but rather that part of the brilliant gas ball we perceive with our eyes or view through a heavily filtered telescope. The part of the Sun that emits the radiation we see is called the **photosphere** (after the Greek word *photos*, meaning "light"). Its radius (the radius of the Sun in Table 9.1) is about 700,000 km. However, as we will see, its thickness is probably no more than 500 km—less than 0.1 percent of the radius—which is why we perceive the Sun as having a well-defined, sharp edge (Figure 9.1).

The main regions of the Sun are illustrated in Figure 9.2. Just above the photosphere is the Sun's lower atmosphere, called the **chromosphere**, about 1500 km thick. Above that lies a region called the **transition zone**, where the temperature rises dramatically. Above 10,000 km, and stretching far beyond, is a thin, hot upper atmosphere, the solar **corona**. At still greater distances the corona turns into the **solar wind**, which flows away from the Sun and permeates the entire solar system. ∞ *(Sec. 4.2)*

Below the photosphere, extending down some 200,000 km, is the **convection zone**, a region where the material of the Sun is in constant convective motion. ∞ *(Sec. 5.3)* Below the convection zone lies the **radiation zone**, where solar energy is transported toward the surface by radiation rather than by convection. The term *solar interior* is often used to mean both the radiation and convection zones. The central **core**, roughly 200,000 km in radius, is the site of powerful nuclear fusion reactions that generate the Sun's enormous energy output.

Solar rotation can be measured by timing sunspots and other surface features as they cross the solar disk. ∞ *(Sec. 1.2)* These observations indicate that the Sun completes one rotation in about a month, but it does not do so as a solid body. Instead, it spins differentially—faster at the equator and slower at the poles, like Jupiter and Saturn ∞ *(Sec. 7.3)*. The equatorial rotation period is about 25 days. Sunspots are never seen above latitude 60° (north or south), but at that latitude they indicate a 31-day period. Other measurement techniques, such as those described in Section 9.2, reveal that the rotation period continues to increase as we approach the poles. The polar period may be as long as 36 days.

◀ FIGURE 9.2 Solar Structure   The main regions of the Sun, not drawn to scale, with some physical dimensions labeled.

Luminosity

The Sun radiates an enormous amount of energy into space. We can measure the Sun's total energy output in two stages, as follows. First, by holding a light-sensitive device—a solar cell, perhaps—above Earth's atmosphere perpendicular to the Sun's rays we can measure how much solar energy is received per square meter of area every second at Earth's distance from the Sun. This quantity, known as the solar constant, is approximately 1400 watts per square meter (W/m^2).[1] About 50–70 percent of this energy reaches Earth's surface; the rest is intercepted by the atmosphere (30 percent) or reflected away by clouds (0–20 percent). Thus, for example, a sunbather's body having a total surface area of about 0.5 m^2 receives solar energy at a rate of roughly 500 watts, approximately equivalent to the output of a small electric room heater or five 100-W lightbulbs.

Second, knowing the solar constant, we can calculate the total amount of energy radiated in all directions from the Sun. Imagine a three-dimensional sphere that is centered on the Sun and is just large enough that its surface intersects Earth's center (Figure 9.3). The sphere's radius is one AU, and its surface area is therefore $4\pi \times (1\ AU)^2$, or approximately $2.8 \times 10^{23}\ m^2$. Assuming that the Sun radiates energy equally in all directions, we can determine the total rate at which energy leaves the Sun's surface simply by multiplying the solar constant by the total surface area of our imaginary sphere. The result is just under 4×10^{26} W. This quantity is called the luminosity of the Sun. As we will see, it is quite typical of the luminosities of many of the stars in the sky.

Take a moment to consider the magnitude of the solar luminosity. The Sun is an enormously powerful source of energy. *Every second* it produces an amount of energy equivalent to the detonation of about 100 billion 1-megaton nuclear bombs. There is simply nothing on Earth comparable to a star!

TABLE 9.1	**Some Solar Properties**
Radius	696,000 km
Mass	1.99×10^{30} kg
Average density	1410 kg/m³
Rotation period	25.1 days (equator)
	30.8 days (60° latitude)
	36 days (poles)
	26.9 days (interior)
Surface temperature	5780 K
Luminosity	3.86×10^{26} W

CONCEPT CHECK ▶ Why must we assume that the Sun radiates equally in all directions when computing the solar luminosity from the solar constant?

9.2 The Solar Interior

How have we learned about the inside of the Sun? After all, the sunlight we see comes from the photosphere; none of the radiation from the interior reaches us directly. In this section we discuss in a little more detail how we know what we know about conditions below the solar surface.

Lacking any direct measurements of the Sun's interior, researchers must use indirect means to probe in more detail the inner workings of our parent star. To accomplish this, they construct *mathematical models* of the Sun, combining all available observations with theoretical insight into solar physics ∞ *(Sec. 0.5)*. The resulting **standard solar model** has gained widespread acceptance among astronomers. We can construct similar models of other stars, but because the Sun is so close and well studied, the standard solar model is by far the best tested.

Modeling the Structure of the Sun

The Sun's bulk properties—mass, radius, temperature, and luminosity—do not vary much from day to day or from year to year. We will see in Chapter 12 that stars like the Sun do change significantly over periods of *billions* of years, but for our purposes here we can ignore this slow evolution. On "human" time scales, the Sun may reasonably be thought of as unchanging.

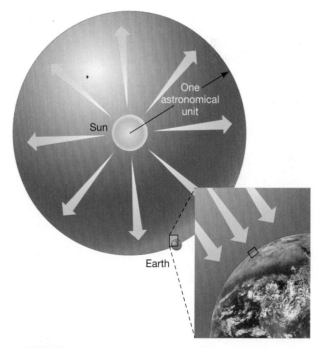

▲ **FIGURE 9.3 Solar Luminosity** If we draw an imaginary sphere around the Sun so that the sphere's surface passes through Earth's center, then the radius of this imaginary sphere is 1 AU. The "solar constant" is the amount of power striking a 1-m^2 detector at Earth's distance, as suggested in the inset. The Sun's luminosity is then determined by multiplying the sphere's surface area by the solar constant. *(NASA)*

---

[1]The SI unit of energy is the *joule* (J) Readers are probably more familiar with the closely related unit, the *watt* (W), which measures *power* and is defined as the rate at which energy is emitted or expended by an object. One watt is equal to 1 joule per second. For example, a lightbulb with a power rating of 100 W radiates 100 J of energy each second.

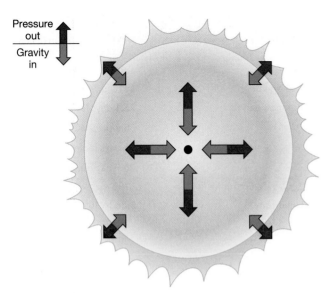

Pressure
out

Gravity
in

▲ FIGURE 9.4 Stellar Balance   In the interior of a star such as the Sun, the outward pressure of hot gas balances the inward pull of gravity. This is true at every point within the star, guaranteeing its stability.

Accordingly, as illustrated in Figure 9.4, theoretical models generally begin by assuming that the Sun is in a state of **hydrostatic equilibrium**, in which pressure's outward push exactly counteracts gravity's inward pull. This stable balance between opposing forces is the basic reason that the Sun neither collapses under its own weight nor explodes into interstellar space. ∞ *(More Precisely 5-1)* It also has an important consequence for the solar interior. Because the Sun is very massive, its gravitational pull is very strong, so very high internal pressure is needed to maintain hydrostatic equilibrium. This high pressure in turn requires a very high central temperature, a fact crucial to our understanding of solar energy generation (Section 9.5). In this way the assumption of hydrostatic equilibrium lets us determine the density and temperature inside the Sun. This information, in turn, allows the model to make predictions about other observable solar properties—luminosity, radius, spectrum, and so on. Scientists fine-tune the internal details of the model until the predictions agree with observations. This is the scientific method at work; the standard solar model is the result. ∞ *(Sec. 0.5)*

To test and refine the standard solar model, astronomers are eager to obtain information about the solar interior. In the 1960s, measurements of the Doppler shifts of solar spectral lines revealed that the surface of the Sun oscillates, or vibrates, like a complex set of bells. ∞ *(Sec. 2.7)* These vibrations, illustrated in Figure 9.5, are the result of internal pressure ("sound") waves that reflect off the photosphere and repeatedly cross the solar interior. These waves penetrate deep inside the Sun, and analysis of their surface patterns allows scientists to study conditions far below the Sun's surface. This process is similar to the way in which seismologists study Earth's interior by observing seismic waves produced by earthquakes. ∞ *(Sec. 5.4)* For this reason, study of solar surface patterns is usually called **helioseismology**, even though solar pressure waves have nothing whatever to do with solar seismic activity (which doesn't exist, as the Sun is entirely gaseous).

The most extensive study of solar vibrations is the ongoing GONG (short for Global Oscillations Network Group) project. By making continuous observations of the Sun from many clear sites around Earth, solar astronomers can obtain uninterrupted high-quality solar data spanning weeks at a time. The space-based *Solar and Heliospheric Observatory* (*SOHO;* see *Discovery 9-1*) has provided continuous monitoring of the Sun's surface and atmosphere since 1995. Analysis of these datasets provides detailed information about the temperature, density, rotation, and convective state of the solar interior, permitting direct comparison with theory over a large portion of the Sun's volume. The agreement between the standard solar model and observations is spectacular—the frequencies and wavelengths of observed solar oscillations are within 0.1 percent of the model predictions.

The data also allow scientists to monitor global circulation patterns in the solar interior, including two gigantic "conveyor belts" that transport subsurface material from the equator to the poles, then return it to the equator at a depth of some

▶ FIGURE 9.5 Solar Oscillations
(a) By observing the motion of the solar surface, scientists can determine the wavelengths and frequencies of the individual waves and deduce information about the Sun's complex vibrations. (b) Waves contributing to the observed oscillations can travel deep inside the Sun, providing vital information about the solar interior. *(National Solar Observatory)*

These colored patches depict gas moving down (red) and up (blue).

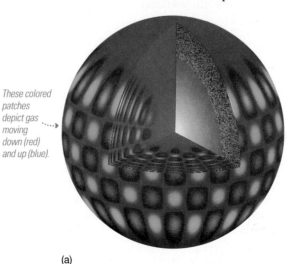

(a)

These are oscillatory patterns in the Sun's convection zone.

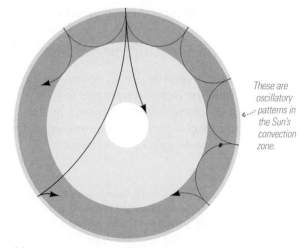

(b)

300,000 km, far below the convection zone. These circulation patterns, which move at 10–15 m/s and take roughly 40 years to complete a single loop, are thought to play crucial roles in regulating the Sunspot cycle (see Section 9.4).

Figure 9.6 shows the solar density and temperature according to the standard solar model, plotted as functions of distance from the Sun's center. Notice how the density drops sharply at first, then more slowly near the photosphere. The variation in density is large, ranging from a core value of about 150,000 kg/m³, 20 times the density of iron, to an extremely small photospheric value of 2×10^{-4} kg/m³, about 10,000 times less dense than air. The *average* density of the Sun (Table 9.1) is 1400 kg/m³, about the same as the density of Jupiter. As shown in Figure 9.6(c), the solar temperature also decreases with increasing radius, but not as rapidly as the density. The temperature is about 15 million K at the center and decreases steadily, reaching the observed value of 5800 K at the photosphere.

Energy Transport

The very hot solar interior ensures violent and frequent collisions among gas particles. In and near the core, the extremely high temperatures guarantee that the gas is completely ionized. ∞ *(Sec. 2.6)* With no electrons left on atoms to capture photons and move into more excited states, the deep solar interior is quite transparent to radiation. Only occasionally does a photon interact with a free electron. The energy produced by nuclear reactions in the core travels outward toward the surface with relative ease in the form of radiation.

As we move outward from the core, the temperature falls, and eventually some electrons can remain bound to nuclei. With more and more atoms retaining electrons that can absorb the outgoing radiation, the gas in the interior changes from being relatively transparent to almost totally opaque. By the outer edge of the radiation zone, 500,000 km from the center, *all* of the photons produced in the Sun's core have been absorbed. Not one of them reaches the surface. But what happens to the energy they carry?

The photon's energy must travel beyond the solar interior—the fact that we see sunlight proves that the energy escapes. The energy is carried to the solar surface by *convection*—the same basic physical process we saw in our study of Earth's atmosphere, although operating in a very different environment in the Sun. ∞ *(Sec. 5.3)* Convection can occur whenever cooler material overlies warmer material, and this is just what happens in the outer part of the Sun's interior. Hot solar gas moves outward, while cooler gas above it sinks, creating a characteristic pattern of convection cells. All through the convection zone, energy is transported to the surface by physical motion of the solar gas. (Note that this actually represents a departure from hydrostatic equilibrium, as defined above, but it can still be handled within the standard solar model.) Remember that there is no physical movement of material when radiation is the energy-transport mechanism; convection and radiation are *fundamentally different* ways in which energy can be transported from one place to another.

Figure 9.7 is a schematic diagram of the solar convection zone. There is a hierarchy of convection cells, organized in tiers at different depths. The deepest tier, about

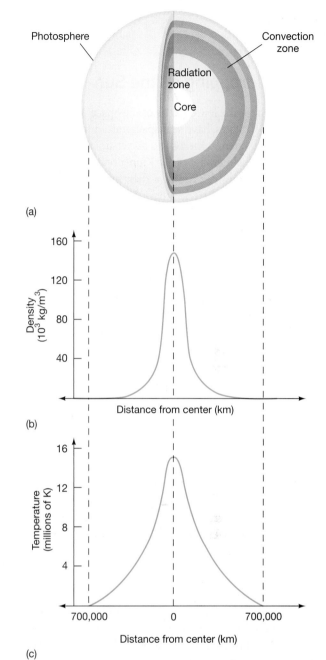

(a)

(b)

(c)

▲ FIGURE 9.6 **Solar Interior** Density and temperature distributions in the interior of the Sun. Parts (b) and (c) show the large variations in solar density and temperature, relative to a cutaway diagram of the Sun's interior (a).

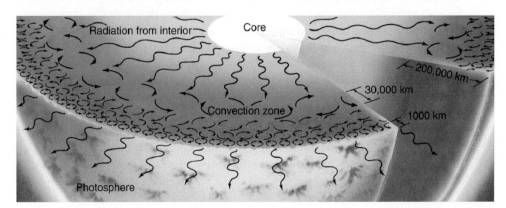

◄ FIGURE 9.7 **Solar Convection** Energy is physically transported in the Sun's convection zone, which is here visualized as a boiling, seething sea of gas. As drawn, the convective cell sizes become progressively larger at greater depths.

9-1 DISCOVERY

Eavesdropping on the Sun

Throughout the few decades of the Space Age, various nations, led by the United States, have sent spacecraft to most of the major bodies in the solar system. One of the as-yet-unexplored bodies is Pluto, the most notable member of the Kuiper belt, which has never been visited by a robot orbiter or even a flyby craft—although this may change soon. ∞ *(Sec. 8.5)* The other unexplored body is the Sun. Currently, the next best thing to a dedicated reconnoitering spacecraft is the *Solar and Heliospheric Observatory (SOHO)*, launched in 1995 and (as of early 2012) still operating, and the *Solar Dynamics Observatory (SDO)*, launched in 2010. Both spacecraft have radioed back to Earth volumes of new data—and a few new puzzles—about our parent star.

SOHO is a billion-dollar mission operated primarily by the European Space Agency. The 2-ton robot is now on-station about 1.5 million kilometers sunward of Earth—about 1 percent of the distance from Earth to the Sun. This is the so-called L₁ Lagrangian point, where the gravitational pull of the Sun and Earth are precisely equal—a good place to park a monitoring platform. By contrast, the American *SDO* spacecraft travels around Earth in a high orbit. Both automated vehicles stare at the Sun 24 hours a day, and both carry several instruments capable of measuring almost everything from the Sun's corona and magnetic field to its solar wind and internal vibrations. The accompanying figure shows a false-color image of the Sun's lower corona, obtained recently by the *SDO* craft in the ultraviolet part of the spectrum.

Both spacecraft are positioned just beyond Earth's magnetosphere, so its instruments can study cleanly the high-speed charged particles of the solar wind. Coordinating these on-site measurements with *SOHO* and *SDO* images of the Sun itself, astronomers now think they can follow solar magnetic field loops expanding and breaking as the Sun prepares itself for mass ejections several days before they actually occur. Given that such coronal storms can endanger pilots and astronauts and play havoc with communications, power grids, satellite electronics, and other human activity, the prospect of having accurate forecasts of disruptive solar events is a welcome development.

In fact, there do seem to be some correlations between the 22-year solar cycle (two sunspot cycles, with oppositely directed magnetic fields) and periods of climatic dryness here on Earth. For example, near the start of the past eight solar cycles, there have been droughts in North America—at least within the middle and western plains from South Dakota to New Mexico. The most recent of these droughts, which typically last 3 to 6 years, came in the late 1950s. The one expected in the early 1980s, however, did not occur as clearly as anticipated.

Other possible Sun–Earth connections include a link between solar activity and increased atmospheric circulation on our planet. As circulation increases, terrestrial storm systems deepen, extend over wider ranges of latitude, and carry more moisture. The relationship is complex, and the subject controversial, because no one has yet shown any physical mechanism (other than the Sun's heat, which does not vary much during the solar cycle) that would allow solar activity to stir our terrestrial atmosphere. Without a better understanding of the physical mechanism involved, none of these effects can be incorporated into our weather-forecasting models.

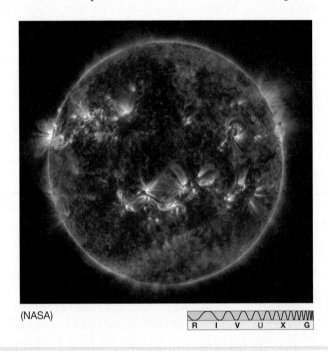

(NASA)   R I V U X G

CONCEPT CHECK ▶ Describe the two distinct ways in which energy moves outward from the solar core to the photosphere.

200,000 km below the photosphere, is thought to contain cells tens of thousands of kilometers in diameter. Energy is carried upward through a series of progressively smaller cells, stacked one upon another until, at a depth of about 1000 km below the photosphere, the individual cells are about 1000 km across. The top of this uppermost tier of convection lies just below the solar photosphere.

Convection does not proceed into the solar atmosphere. In and above the photosphere the density is so low that the gas is transparent and radiation once again becomes the mechanism of energy transport. Photons reaching the photosphere escape freely into space. The solar density falls off so rapidly in this region of the Sun that the gas makes the transition from completely opaque to completely transparent over a very small distance—just a few hundred kilometers—accounting for the thinness of the photosphere and the observed sharp "edge" of the Sun.

Evidence for Solar Convection

In part, our knowledge of solar convection is derived indirectly, from computer models of the solar interior. However, astronomers also have some direct evidence of conditions in the convection zone. Figure 9.8 is a high-resolution photograph of the solar surface. The visible surface is highly mottled with regions of bright and dark gas known as *granules*. This **granulation** of the solar surface is a direct reflection of motion in the convection zone. Each bright granule measures about 1000 km across—comparable to one of the larger states in the United States—and has a lifetime of between 5 and 10 minutes.

Each granule is the topmost part of a solar convection cell. Spectral lines in the bright granules are slightly blueshifted, indicating Doppler-shifted matter approaching us—that is, moving upward—at about 1 km/s. ∞ *(Sec. 2.7)* Similarly, material in the dark granules is sinking into the solar interior, exactly as we would expect for the topmost tier of convection in Figure 9.7. The variation in brightness is due to differences in temperature. The upwelling gas is hotter and therefore (by Stefan's law) emits more radiation than the cooler downwelling gas. ∞ *(Sec. 2.4)* The bright and dark regions appear to contrast considerably, but in reality their temperature difference is less than about 500 K.

Careful measurements also reveal a much larger-scale flow beneath the solar surface. **Supergranulation** is a flow pattern quite similar to granulation except that supergranulation cells measure some 30,000 km across. As with granulation, material upwells at the center of the cells, flows across the surface, then sinks down again at the edges. Scientists think that supergranules are the imprint on the photosphere of the deepest tier of large convective cells depicted in Figure 9.7.

9.3 The Solar Atmosphere

Astronomers can obtain an enormous amount of information about the Sun (or any star) by analyzing the absorption lines in its spectrum. ∞ *(Sec. 2.5)* Figure 9.9 (see also Figure 2.14) is a detailed visible spectrum of the Sun, spanning the range of wavelengths from 360 to 690 nm. Table 9.2 lists the 10 most common elements in the Sun based on spectral analysis. This distribution is very similar to what we saw on the jovian planets, and it is what we will find for the universe as a whole. Hydrogen is by far the most abundant element, followed by helium.

Strictly speaking, analysis of spectral lines allows us to draw conclusions only about the part of the Sun where the lines form—the photosphere and chromosphere. However, the data in Table 9.2 are thought to be representative of the entire Sun (with the exception of the solar core, where nuclear reactions are steadily changing the composition—see Section 9.5).

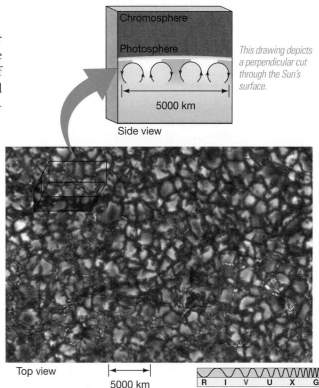

This drawing depicts a perpendicular cut through the Sun's surface.

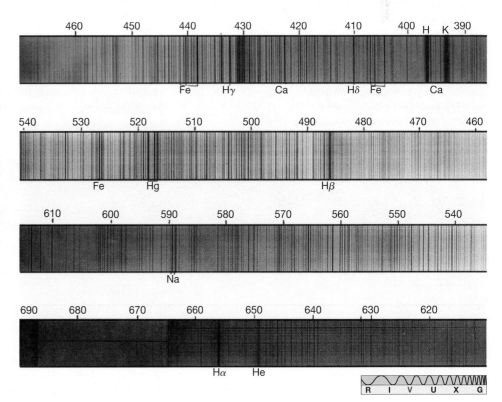

▲ **FIGURE 9.8 Solar Granulation** This photograph of the granulated solar photosphere, taken with the 1-m Swedish Solar Telescope looking directly down on the Sun's surface, shows typical solar granules comparable in size to Earth's continents. The bright portions of the image are regions where hot material is upwelling from below, as illustrated in Figure 9.7. The dark (redder) regions correspond to cooler gas that is sinking back down into the interior. *(SST/Royal Swedish Academy of Sciences)*

▶ **FIGURE 9.9 Solar Spectrum** A detailed visible spectrum of our Sun shows thousands of dark absorption lines, indicating the presence of 67 different elements in various stages of excitation and ionization in the lower solar atmosphere. The numbers give wavelengths, in nanometers. *(Palomar Observatory/Caltech)*

TABLE 9.2	The Composition of the Sun	
Element	Percentage of Total Number of Atoms	Percentage of Total Mass
Hydrogen	91.2	71.0
Helium	8.7	27.1
Oxygen	0.078	0.97
Carbon	0.043	0.40
Nitrogen	0.0088	0.096
Silicon	0.0045	0.099
Magnesium	0.0038	0.076
Neon	0.0035	0.058
Iron	0.0030	0.14
Sulfur	0.0015	0.040

▶ FIGURE 9.10 Solar Chromosphere This photograph of a total solar eclipse shows the solar chromosphere a few thousand kilometers above the Sun's surface. Note the prominence at left. (G. Schneider)

ANIMATION/VIDEO Solar Granulation

ANIMATION/VIDEO Solar Chromosphere

▼ FIGURE 9.11 Solar Spicules Short-lived, narrow jets of gas that typically last mere minutes can be seen sprouting up from the chromosphere in this ultraviolet image of the Sun. These so-called spicules are the thin spikelike regions whose gas escapes from the Sun at speeds of about 100 km/s. (NASA)

The Chromosphere

Density continues to decrease rapidly as we move outward through the solar atmosphere. The low density of the chromosphere means that it emits very little light of its own and cannot be observed visually under normal conditions. The photosphere is just too bright, dominating the chromosphere's radiation. Although we don't normally see the chromosphere, astronomers have long been aware of its existence. Figure 9.10 shows the Sun during an eclipse in which the photosphere was obscured by the Moon, but the chromosphere was not. The chromosphere's characteristic pinkish hue, the result of the red H emission line of hydrogen, is plainly visible. ⚭ (Sec. 2.6)

The chromosphere is far from tranquil. Every few minutes, small solar storms erupt there, expelling jets of hot matter known as *spicules* into the Sun's upper atmosphere (Figure 9.11). These long, thin spikes of matter leave the Sun's surface at typical speeds of about 100 km/s, reaching several thousand kilometers above the photosphere. Spicules tend to accumulate around the edges of supergranules. The Sun's magnetic field is also known to be somewhat stronger than average in those

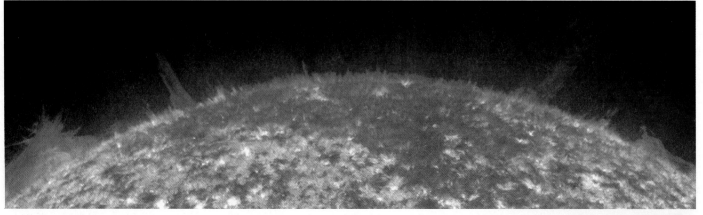

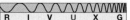

regions. Scientists speculate that the downwelling material there tends to strengthen the solar magnetic field and that spicules are the result of magnetic disturbances in the Sun's churning outer layers.

The Transition Zone and Corona

During the brief moments of a solar eclipse, if the Moon's angular size is large enough that both the photosphere and the chromosphere are blocked, the ghostly solar corona can be seen, as in Figure 9.12. With the photospheric light removed, the pattern of spectral lines changes dramatically. The spectrum shifts from absorption to emission, and an entirely new set of spectral lines suddenly appears.

The shift from absorption to emission is entirely in accordance with Kirchhoff's laws, because we see the corona against the blackness of space, not against the bright continuous spectrum from the photosphere below. ∞ *(Sec. 2.5)* The new lines arise because atoms in the corona are much more highly ionized than atoms in the photosphere or chromosphere. ∞ *(Sec. 2.6)* Their internal electronic structure and hence their spectra are therefore quite different from those of atoms at lower atmospheric levels. The cause of this extensive electron stripping is the high coronal temperature. The degree of ionization inferred from spectra observed during solar eclipses tells us that the temperature of the upper chromosphere exceeds that of the photosphere. Furthermore, the temperature of the solar corona, where even more ionization is seen, is higher still.

Figure 9.13 shows how the temperature varies with altitude above the photosphere. The temperature decreases to a minimum of about 4500 K some 500 km above the photosphere, after which it rises steadily. About 1500 km above the photosphere, in the transition zone, the gas temperature increases sharply, reaching more than 1 million K at an altitude of 10,000 km. Thereafter, in the corona, the temperature remains roughly constant at around 3 million K, although *SOHO* and other orbiting instruments have detected coronal "hot spots" having temperatures many times higher than this average value. The rapid temperature rise in the transition zone is still not fully understood. The temperature profile runs contrary to intuition: Moving away from a heat source, we would normally expect the heat to diminish, but this is not the case in the lower atmosphere of the Sun. The corona must have another energy source. Astronomers think that magnetic disturbances in the solar photosphere (see Section 9.4) are ultimately responsible for heating the corona.

The Solar Wind

Electromagnetic radiation and fast-moving particles—mostly protons and electrons—escape from the Sun all the time. The radiation moves away from the photosphere at the speed of light, taking 8 minutes to reach Earth. The particles travel much more slowly, although at the still considerable speed of about 500 km/s, reaching Earth in a few days. This constant stream of escaping solar particles is the solar wind.

The solar wind is a direct consequence of the high coronal temperature. About 10 million km above the photosphere, the coronal gas is hot enough to escape the Sun's gravity, and it begins to flow outward into space. At the same time, the solar atmosphere is continuously replenished from below. If that were not the case, the corona would disappear in about a day. The Sun is, in effect, "evaporating"—constantly shedding mass through the solar wind. The wind is an extremely thin medium, however. Although it carries away about a million tons of solar matter each second, less than 0.1 percent of the Sun's mass has been lost since the solar system formed 4.6 billion years ago.

CONCEPT CHECK ▶ Describe two ways in which the spectrum of the solar corona differs from that of the photosphere.

R I V U X G

▲ FIGURE 9.12 **Solar Corona** When both the photosphere and the chromosphere are obscured by the Moon during a solar eclipse, the faint solar corona becomes visible. This photograph shows clearly the emission of radiation from a relatively active corona. *(NCAR High Altitude Observatory)*

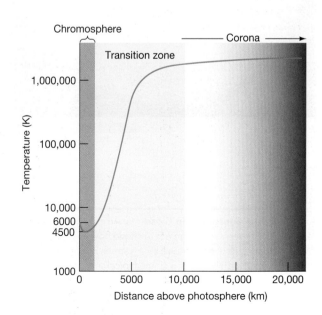

▶ FIGURE 9.13 **Solar Atmospheric Temperature** The change of gas temperature in the lower solar atmosphere is dramatic. The temperature, indicated by the blue line, reaches a minimum of 4500 K in the chromosphere and then rises sharply in the transition zone, finally leveling off at around 3 million K in the corona.

9.4 The Active Sun

Most of the Sun's luminosity results from continuous emission from the photosphere. However, superimposed on this steady, predictable aspect of our star's energy output is a much more irregular component, characterized by explosive and unpredictable surface activity. This **solar activity** contributes little to the Sun's total luminosity and has little effect on the evolution of the Sun as a star, but it does affect us here on Earth. The strength of the solar wind is strongly influenced by the level of solar activity, and this, in turn, directly affects Earth's magnetosphere.

Sunspots

Figure 9.14 is an optical photograph of the entire Sun showing numerous dark **sunspots** on the surface. ∞ *(Sec. 1.2)* They typically measure about 10,000 km across—about the size of Earth. At any given instant, the Sun may have hundreds of sunspots, or it may have none at all.

Detailed observations of sunspots show an *umbra*, or dark center, surrounded by a grayish *penumbra*. The close-up view of a pair of sunspots in Figure 9.15(a) shows both of these dark areas against the bright background of the undisturbed photosphere. This gradation in darkness indicates a gradual change in photospheric temperature. Sunspots are simply cooler regions of the photospheric gas. The temperature of the umbra is about 4500 K, that of the penumbra 5500 K. The spots, then, are certainly composed of hot gases. They seem dark only because they appear against an even brighter background (the 5800 K photosphere).

▼ **FIGURE 9.14 Sunspots**   This photograph of the entire Sun, taken during a period of maximum solar activity, shows several groups of sunspots. The largest spots in this image are more than 20,000 km across, nearly twice the diameter of Earth. *(Palomar Observatory/Caltech)*

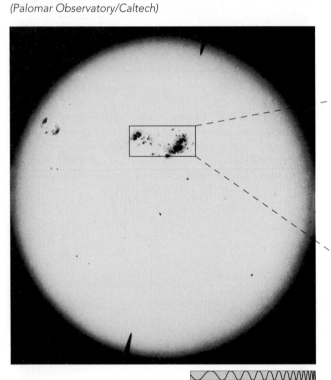

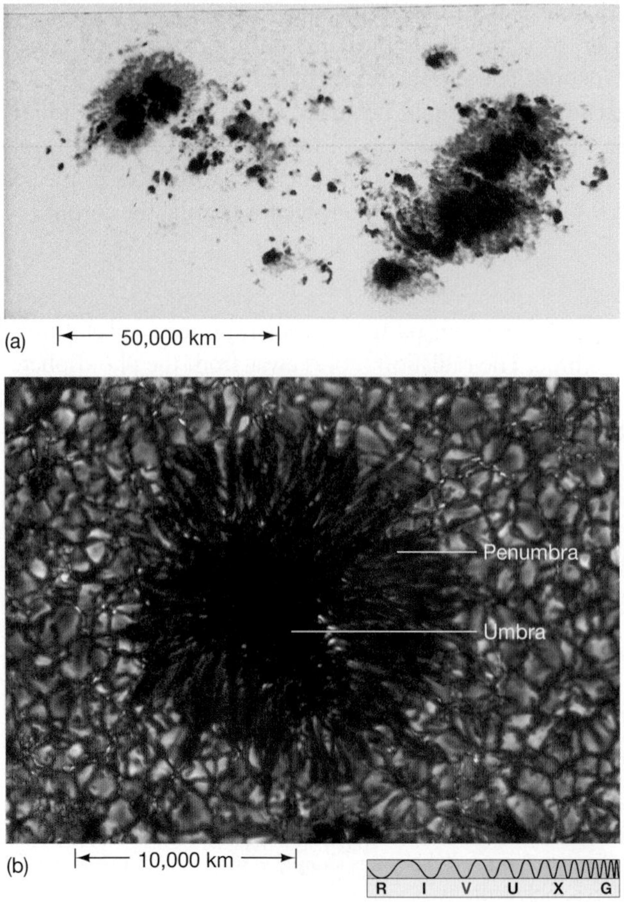

Sunspots appear dark because they are slightly cooler than the surrounding gas.

(a) |← 50,000 km →|

R I V U X G

This spot is about the size of Earth.

Penumbra

Umbra

(b) |← 10,000 km →|

R I V U X G

▶ **FIGURE 9.15 Sunspots, Up Close**   (a) An enlarged photo of the largest pair of sunspots in Figure 9.14 shows how each spot consists of a cool, dark umbra surrounded by a warmer, brighter penumbra. (b) A high-resolution image of a single typical sunspot shows details of its structure as well as the granules surrounding it. *(Palomar Observatory/Caltech; SST/Royal Swedish Academy of Sciences)*

As mentioned in Sec. 9.1 studies of sunspots indicate that the Sun does not rotate as a solid body (see Table 9.1). Instead, it spins *differentially*—faster at the equator and more slowly at the poles, like Jupiter and Saturn. The combination of differential rotation and convection radically affects the character of the Sun's magnetic field, which in turn plays a major role in determining the numbers and location of sunspots.

Solar Magnetism

Spectroscopic studies indicate that the magnetic field in a typical sunspot is about 1000 times greater than the field in the surrounding, undisturbed photosphere (which is itself several times stronger than Earth's magnetic field). Furthermore, the field lines are not randomly oriented, but instead are directed roughly perpendicular to (out of or into) the Sun's surface. Scientists think that sunspots are cooler than their surroundings because these abnormally strong fields interfere with the normal convective flow of hot gas toward the solar surface.

The **polarity** of a sunspot indicates which way its magnetic field is directed. We conventionally label spots as S where the field lines emerge from the interior and as N where the lines dive below the photosphere (so that field lines above the surface run from S to N, as on Earth). Sunspots almost always come in pairs whose members lie at roughly the same latitude and have opposite magnetic polarities. As illustrated in Figure 9.16(a), magnetic field lines emerge from the solar interior through one member (S) of a sunspot pair, loop through the solar atmosphere, then reenter the photosphere through the other member (N). As in Earth's magnetosphere, charged particles tend to follow the solar magnetic field lines. ⚭ *(Sec. 5.7)* Figure 9.16(b) shows an actual image of solar magnetic loops; the light we see here is emitted by charged particles trapped by the solar field.

Despite the irregular appearance of the sunspots themselves, there is a great deal of order in the underlying solar field. *All* the sunspot pairs in the same solar hemisphere (north or south) at any instant have the *same* magnetic configuration. That is, if the leading spot (measured in

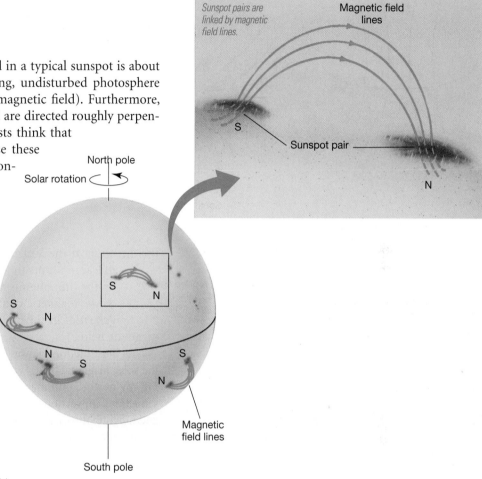

▶ FIGURE 9.16 Sunspot Magnetism (a) The Sun's magnetic field lines emerge from the surface through one member of a sunspot pair and reenter the Sun through the other member. If the magnetic field lines are directed into the Sun in one leading spot, they are inwardly directed in all other leading spots in that hemisphere. The opposite is the case in the southern hemisphere, where the polarities are always opposite those in the north. The entire magnetic field pattern reverses itself roughly every 11 years. (b) A far-ultraviolet image taken by NASA's *Transition Region and Coronal Explorer (TRACE)* satellite, showing magnetic field lines arching between two sunspot groups. *(NASA)*

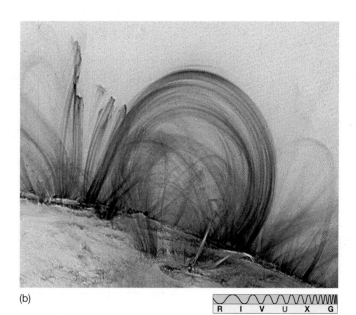

R I V U X G

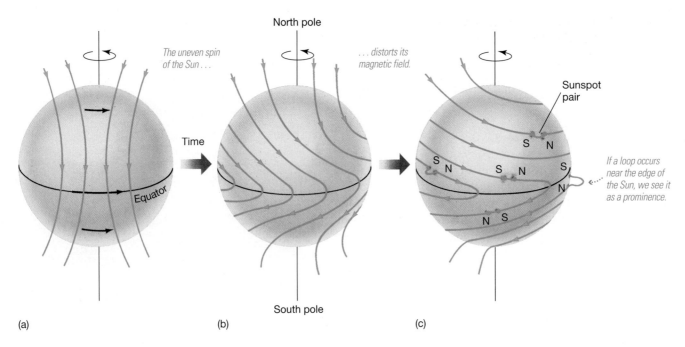

▲ FIGURE 9.17 **Solar Rotation**   (a, b) The Sun's differential rotation wraps and distorts the solar magnetic field. (c) Occasionally, the field lines burst out of the surface and loop through the lower atmosphere, thereby creating a sunspot pair. The underlying pattern of the solar field lines explains the observed pattern of sunspot polarities.

ANALOGY: A good analogy to a tangled solar magnetic field is a garden hose with loops and kinks.

the direction of the Sun's rotation) has N polarity, as shown in the figure, then all leading spots in that hemisphere have the same polarity. What's more, in the other hemisphere at the same time, all sunspot pairs have the *opposite* magnetic configuration (S polarity leading). To understand these regularities in sunspot polarities, we must look in more detail at the Sun's magnetic field.

As illustrated in Figures 9.17 (a–c), the Sun's differential rotation greatly distorts the solar magnetic field, "wrapping" it around the solar equator and eventually causing any north–south magnetic field to reorient itself in an east–west direction. At the same time, convection causes the magnetized gas to upwell toward the surface, twisting and tangling the magnetic field pattern. In some places, the field lines become kinked like a knot in a garden hose, causing the field strength to increase. Occasionally, the field becomes so strong that it overwhelms the Sun's gravity, and a "tube" of field lines bursts out of the surface and loops through the lower atmosphere, forming a sunspot pair (Figure 9.17c). The general east–west organization of the underlying solar field accounts for the observed polarities of the resulting sunspot pairs in each hemisphere.

The Solar Cycle

Sunspots are not steady. Most change their size and shape, and all come and go. Individual spots may last anywhere from 1 to 100 days. A large group of spots typically lasts 50 days. However, centuries of observations have established a clear **sunspot cycle.** Figure 9.18(a) shows the number of sunspots observed monthly during the 20th century. The average number of spots reaches a maximum every 11 or so years, then falls off almost to zero before the cycle begins afresh. The latitudes at which sunspots appear vary as the sunspot cycle progresses. Individual sunspots do not move up or down in latitude, but new spots appear closer to the equator as older ones at higher latitudes fade away. Figure 9.18(b) is a plot of observed sunspot latitudes as a function of time.

In fact, the 11-year sunspot cycle is only half of a 22-year **solar cycle**—the period needed for both the average number of spots *and* the Sun's overall magnetic polarity to repeat themselves. For the first 11 years of the solar cycle, the leading spots of all sunspot pairs in the same solar hemisphere have one polarity, and spots in the other hemisphere have the opposite polarity (Figure 9.16). These polarities then reverse for the next 11 years.

▶ FIGURE 9.18 Sunspot Cycle (a) Monthly number of sunspots during the 20th century clearly displays the (roughly) 11-year solar cycle. At the time of minimum solar activity, hardly any sunspots are seen. About 4 years later, at maximum solar activity, about 100 spots are observed per month. (b) Sunspots cluster at high latitudes when solar activity is at a minimum. They appear at lower and lower latitudes as the number of sunspots peaks. They are again prominent near the Sun's equator as solar minimum is again approached. The blue lines in the upper plot indicate how the "average" sunspot latitude varies over the course of a cycle.

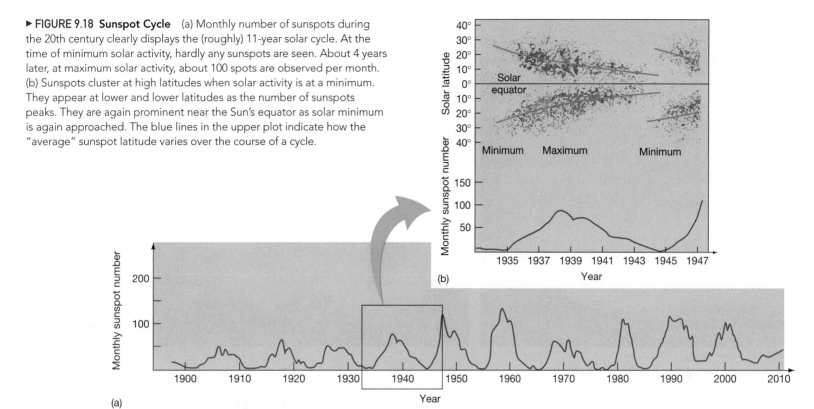

(a)

Astronomers think that the Sun's differential rotation and convection cause constant stretching, twisting, and folding of magnetic field lines. This ongoing process both generates and amplifies the solar magnetic field. The theory is similar to the dynamo theory that accounts for the magnetic fields of Earth and the jovian planets, except that the solar dynamo operates much faster and on a much larger scale. ∞ (Sec. 5.7) One important prediction of this theory is that the Sun's magnetic field should rise to a maximum, then fall to zero and reverse itself in a roughly periodic way, just as observed. Solar surface activity, such as the sunspot cycle, simply follows the changes in the underlying field.

Figure 9.19 plots all sunspot data recorded since the invention of the telescope. As can be seen, the 11-year "periodicity" of the solar sunspot cycle is far from exact. Not only does the period vary from 7 to 15 years, but the sunspot cycle has disappeared entirely in the relatively recent past. The lengthy period of solar inactivity from 1645 to 1715 is called the *Maunder minimum*, after the British astronomer who drew attention to these historical records. (Interestingly, the period also seems to correspond fairly well to the coldest years of the so-called Little Ice Age that chilled northern Europe during the late 17th century, suggesting a link between solar activity and climate on Earth.) Absent a complete understanding of the solar cycle, the cause of the Sun's century-long variations remains a mystery.

In fact, the most recent sunspot minimum (in 2008 and 2009) has resulted in the least active Sun in almost a century—no sunspots at all were seen on almost 80 percent of the days during those years, and the solar wind was anomalously weak. The new cycle, due to peak in mid-2013, is also predicted to be much less active than normal. Scientists attribute this reduced activity to changes in the subsurface "conveyor belt" flow described in Section 9.2, which is thought to directly affect sunspot behavior. For unknown reasons, the flow sped up during the 1990s by a few meters per second, greatly suppressing the numbers of spots during the next cycle (and quite possibly the one after that, now underway). It has since slowed, but for now no one is sure how long the effects will last.

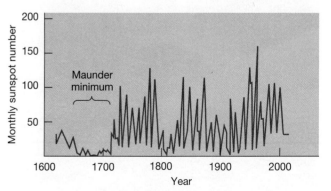

▲ FIGURE 9.19 Maunder Minimum Average number of sunspots occurring each month over the past four centuries. Note the absence of spots during the late 17th century.

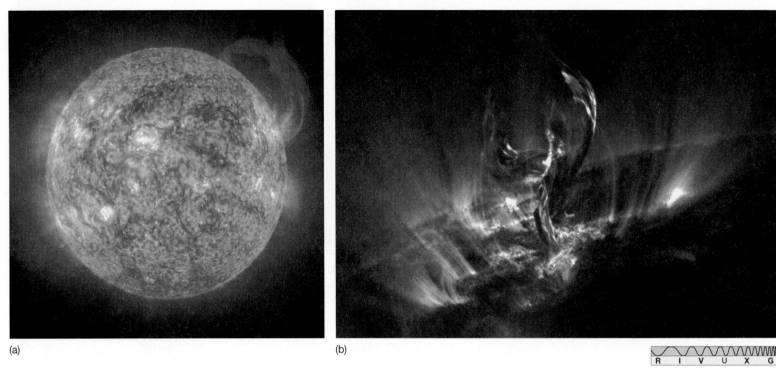

(a)   (b)

▲ **FIGURE 9.20 Solar Prominences**   (a) This particularly large solar prominence was observed by ultraviolet detectors aboard the *SOHO* spacecraft in 2002. (b) Like a phoenix rising from the solar surface, this filament of hot gas measures more than 100,000 km in length. Dark regions in this *TRACE* image have temperatures less than 20,000 K; the brightest regions are about 1 million K. Most of the gas will subsequently cool and fall back into the photosphere. *(NASA)*

Active Regions

Sunspots are relatively gentle aspects of solar activity. However, the photosphere surrounding them occasionally erupts violently, spewing forth into the corona large quantities of energetic particles. The sites of these explosive events are known simply as **active regions.** Most pairs or groups of sunspots have active regions associated with them. Like all other aspects of solar activity, these phenomena tend to follow the solar cycle and are most frequent and violent around the time of sunspot maximum.

Figure 9.20 shows two solar **prominences**—loops or sheets of glowing gas ejected from an active region on the solar surface. Prominences move through the inner parts of the corona under the influence of the Sun's magnetic field. Magnetic instabilities in the strong fields found in and near sunspot groups may cause the prominences, although the details are still not completely understood. A typical solar prominence measures some 100,000 km in extent, nearly 10 times the diameter of planet Earth. Some may persist for days or even weeks. Prominences as large as the one shown in Figure 9.20(b), which spanned almost half a million kilometers of the solar surface (half the Sun's diameter), are less common and usually appear only at times of greatest solar activity.

Flares are another type of solar activity observed near active regions (Figure 9.21). Also the result of magnetic instabilities, flares are much more violent (and even less well understood) than prominences. They flash across a region of the Sun in minutes, releasing enormous amounts of energy as they go. Temperatures in the extremely compact hearts of flares can reach 100 million K (roughly six times hotter than the solar core). So energetic are these cataclysmic explosions that some researchers have likened them to bombs exploding in the lower regions of the Sun's atmosphere. Unlike the trapped gas that makes up the characteristic loop of a prominence, the particles produced by a flare

ANIMATION/VIDEO Solar Flare **(MA)**

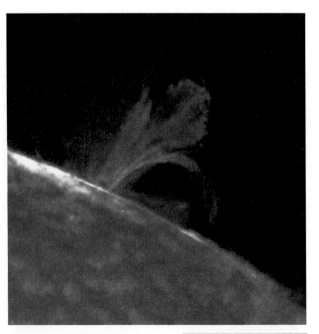

◄ **FIGURE 9.21 Solar Flare**   Much more violent than a prominence, a solar flare is an explosion on the Sun's surface that sweeps across an active region in a matter of minutes, accelerating solar material to high speeds and blasting it into space. *(USAF)*

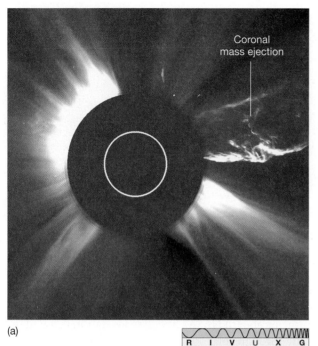

(a)

R I V U X G

◄ **FIGURE 9.22 Coronal Mass Ejection** (a) A few times per week, on average, a giant magnetized "bubble" of solar material detaches itself from the Sun and rapidly escapes into space, as shown in this *SOHO* image taken in 2002. The circles are artifacts of an imaging system designed to block out the light from the Sun itself and exaggerate faint features at larger radii. (b) Should such a coronal mass ejection encounter Earth with its magnetic field oriented opposite to our own, as illustrated, the field lines can join together as in part (c), allowing high-energy particles to enter and possibly severely disrupt our planet's magnetosphere. By contrast, if the fields are in the same direction, the coronal mass ejection can slide by Earth with little effect. *(NASA/ESA)*

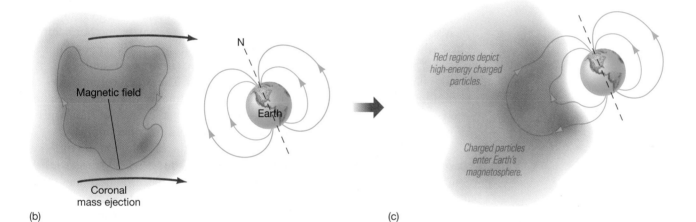

(b)

(c)

are so energetic that the Sun's magnetic field is unable to hold them and shepherd them back to the surface. Instead, the particles are simply blasted into space by the violence of the explosion. Flares are thought to be responsible for the internal pressure waves that give rise to solar surface oscillations.

Sometimes (but not always) a flare or a prominence is associated with a **coronal mass ejection** (Figure 9.22a). This giant magnetic "bubble" of ionized gas separates from the rest of the solar atmosphere and escapes into interplanetary space. Such ejections occur about once per week at times of sunspot minimum, but up to two or three times per day at solar maximum. Carrying an enormous amount of energy, they can—if their fields are properly oriented—connect with Earth's magnetic field via a process known as *reconnection*, dumping some of their energy into the magnetosphere and potentially causing widespread communications and power disruptions on our planet (Figures 9.22b and c).

The Sun in X-Rays

Unlike the 5800 K photosphere, which emits most strongly in the visible part of the electromagnetic spectrum, the million-kelvin coronal gas radiates at much higher frequencies—primarily in X-rays. ⚬ *(Sec. 2.4)* For this reason, X-ray telescopes have

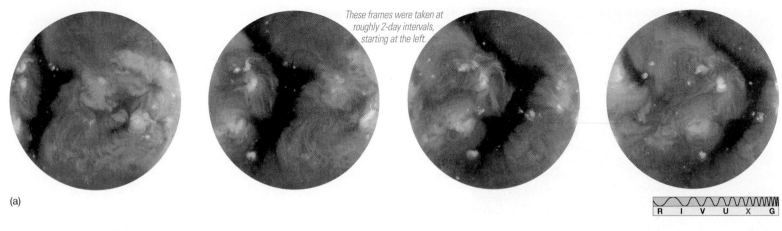

These frames were taken at roughly 2-day intervals, starting at the left.

(a)

R I V U X G

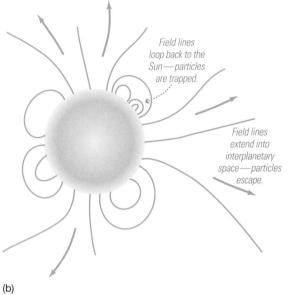

Field lines loop back to the Sun—particles are trapped.

Field lines extend into interplanetary space—particles escape.

(b)

▲ **FIGURE 9.23 Coronal Hole** (a) Images of X-ray emission from the Sun observed by the *Yohkoh* satellite. Note the dark, V-shaped coronal hole traveling from left to right, where the X-ray observations outline in dramatic detail the abnormally thin regions through which the high-speed solar wind streams forth. (b) Charged particles follow magnetic field lines that compete with gravity. When the field is trapped and loops back toward the photosphere, the particles are also trapped; otherwise, they can escape as part of the solar wind. (ISAS/Lockheed Martin)

ANIMATION/VIDEO Coronal Mass Ejections (**MA**)

become important tools in studying the solar corona. Figure 9.23 shows a sequence of X-ray images of the Sun. The full corona extends well beyond the regions shown, but the density of coronal particles emitting the radiation diminishes rapidly with distance from the Sun. The intensity of X-ray radiation farther out is too dim to be seen here.

In the mid-1970s instruments aboard NASA's *Skylab* space station revealed that the solar wind escapes mostly through solar "windows" called **coronal holes.** The dark area moving from left to right in Figure 9.23(a), which shows more recent data from the Japanese *Yohkoh* X-ray solar observatory, is an example. Not really holes, such structures are simply deficient in matter—vast regions of the Sun's atmosphere where the density is about 10 times lower than the already tenuous, normal corona. The underlying solar photosphere looks black in these images because it is far too cool to emit X-rays in any significant quantity. The largest coronal holes can be hundreds of thousands of kilometers across. Structures of this size are seen only a few times each decade. Smaller holes—perhaps only a few tens of thousands of kilometers in size—are much more common, appearing every few hours.

Coronal holes are lacking in matter because the gas there is able to stream freely into space at high speeds, driven by disturbances in the Sun's atmosphere and magnetic field. Figure 9.23(b) illustrates how, in coronal holes, the solar magnetic field lines extend from the surface far out into interplanetary space. Because charged particles tend to follow the field lines, they can escape, particularly from the Sun's polar regions, according to findings from *SOHO* and NASA's *Ulysses* spacecraft, which flew high above the ecliptic plane to explore the Sun's polar regions. (Note, however, that the coronal hole shown in Figure 9.23a spans the Sun's equator.) The speed of the solar wind can reach 800 km/s over some coronal holes. In other regions of the corona, the solar magnetic field lines stay close to the Sun, keeping charged particles near the surface and inhibiting the outward flow of the solar wind (just as Earth's magnetic field tends to prevent the incoming solar wind from striking Earth), so the density remains relatively high. ∞ *(Sec. 5.7)*

The Changing Solar Corona

The solar corona also varies in step with the sunspot cycle. The photograph of the corona in Figure 9.10 shows the Sun at sunspot minimum. The corona is fairly regular in appearance and surrounds the Sun more or less uniformly. Compare this image with Figure 9.12, which was taken in 1994, close to a peak of the sunspot cycle. The active corona here is much more irregular than the inactive one in Figure 9.10 and extends farther from the solar surface. The streamers of coronal material pointing away from the Sun are characteristic of this phase.

Astronomers think that the corona is heated primarily by solar surface activity, which can inject large amounts of energy into the upper solar atmosphere. *SOHO* observations have implicated a "magnetic carpet" as the source of much of the heating, where small magnetic loops continually sprout up, then disappear, dumping vast amounts of energy into the lower solar atmosphere. These small-scale magnetic loops probably provide most of the energy needed to heat the corona. In addition, more extensive disturbances often move through the corona above an active site in the photosphere, distributing the energy throughout the coronal gas. Given this connection, it is hardly surprising that both the appearance of the corona and the strength of the solar wind are closely correlated with the solar cycle.

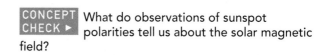

CONCEPT CHECK ► What do observations of sunspot polarities tell us about the solar magnetic field?

9.5 The Heart of the Sun

How does the Sun shine, day after day, year after year, eon after eon? The answer to this question is central to all of astronomy. Without it we can understand neither the evolution of stars and galaxies in the universe nor the existence of life on Earth.

Nuclear Fusion

Only one known energy-generation mechanism can conceivably account for the Sun's enormous energy output. That process is **nuclear fusion**—the combining of light nuclei into heavier ones.[2] We can represent a typical fusion reaction as

$$\text{nucleus } 1 + \text{nucleus } 2 \rightarrow \text{nucleus } 3 + \text{energy}.$$

For powering the Sun and other stars, the most important piece of this equation is the energy produced.

The crucial point here is that, during a fusion reaction, the total mass *decreases*—the mass of nucleus 3 is less than the combined masses of nuclei 1 and 2. Where does the mass go? It is converted to energy according to Einstein's famous equation

$$E = mc^2,$$

or

$$\text{energy} = \text{mass} \times (\text{speed of light})^2.$$

This equation says that to determine the amount of energy corresponding to a given mass, simply multiply the mass by the square of the speed of light (c in the equation). The speed of light is so large that even a small amount of mass translates into an enormous amount of energy.

The production of energy by a nuclear fusion reaction is an example of the **law of conservation of mass and energy,** which states that the sum of mass and energy (properly converted to the same units, using Einstein's equation) must always remain constant in any physical process. There are no known exceptions. The fact that we see light coming from the Sun means that the Sun's mass must be slowly, but steadily, decreasing with time.

The Proton–Proton Chain

The lightest and most common element in the universe is hydrogen, and it is the fusion of hydrogen nuclei (protons) to form nuclei of helium, the next lightest element, that powers the Sun. But all atomic nuclei are positively charged, so they repel one another. Furthermore, by the inverse-square law, the closer two nuclei come to one another, the greater is the repulsive force between them (Figure 9.24a). ⬭ *(Sec. 2.2)* How then do nuclei—two protons, say—ever manage to fuse into anything heavier? The answer is that if they collide at high enough speeds one proton can momentarily plow deep into the other, eventually coming within the exceedingly short range of the **strong nuclear force,** which binds nuclei together (see *More Precisely 9-1*). At distances less than about 10^{-15} m, the attraction of the nuclear force overwhelms the

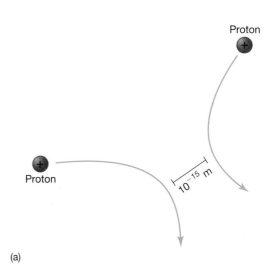

(a)

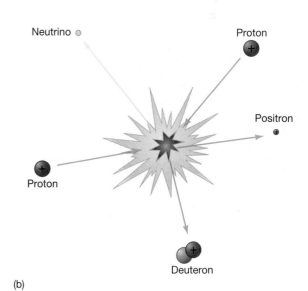

(b)

▲ FIGURE 9.24 Proton Interactions (a) Since like charges repel, two low-speed protons veer away from one another, never coming close enough for fusion to occur. (b) Higher-speed protons may succeed in overcoming their mutual repulsion, approaching close enough for the strong force to bind them together—in which case they collide violently, triggering nuclear fusion that ultimately powers the Sun.

[2]Contrast *fusion* with nuclear *fission*, which powers nuclear reactors on Earth. In this process, a heavy nucleus (such as uranium or plutonium) splits into lighter nuclei, releasing energy in the process.

9-1 MORE PRECISELY

The Strong and Weak Nuclear Forces

As far as we can tell, the behavior of all matter in the universe, from elementary particles to clusters of galaxies, is ruled by just four (or fewer) basic forces, which are fundamental to everything in the universe. We have already encountered two of them in this text—the *gravitational force*, which binds planets, stars, and solar systems together, and the much stronger *electromagnetic force*, which governs the properties of atoms and molecules, as well as magnetic phenomena on planets and stars. ∞ *(Secs. 1.4, 2.2)*

Both of these forces are governed by inverse-square laws and are long range. In contrast, the other two fundamental forces, known as the *weak nuclear force* and the *strong nuclear force* are of extremely short range—about 10^{-18} m and 10^{-15} m, respectively, less than the size of an atomic nucleus.

The weak force governs the emission of radiation during some radioactive decays, as well as interactions between neutrinos and other particles. In fact, according to current theory, the electromagnetic force and the weak force are not really distinct entities, but are actually just different aspects of a single "electroweak" force.

The strong force holds atomic nuclei together. It binds particles with enormous strength, but only when they approach one another very closely. Two protons must come within about 10^{-15} m of one another before the attractive strong force can overcome their electromagnetic repulsion.

Very loosely speaking, we can say that the strong force is about 100 times stronger than electromagnetism, 100,000 times stronger than the weak force, and 10^{39} times stronger than gravity. But not all particles are subject to all types of force. All particles interact through gravity because all have mass (or equivalently, energy). However, only charged particles interact electromagnetically. Protons and neutrons are affected by the strong force, but electrons are not. Under the right circumstances, the weak force can affect any type of subatomic particle, regardless of its charge.

---

electromagnetic repulsion, and fusion can occur (Figure 9.24b). Speeds in excess of a few hundred kilometers per second are needed to slam protons together fast enough to initiate fusion. Such high speeds are associated with extremely high temperatures—at least 10 million K. ∞ *(More Precisely 5-1)* Such conditions are found in the core of the Sun and at the centers of all stars. The radius at which the temperature drops below this value defines the outer edge of the solar core (Figure 9.6).

In the reaction shown in Figure 9.24(b), two protons combine to form a **deuteron**—a nucleus of *deuterium*, or "heavy hydrogen," containing a proton and an extra neutron. Energy is released in the form of two new particles: a *positron* and a *neutrino*. The **positron** is the positively charged *antiparticle* of an electron. Its properties are identical to those of a normal, negatively charged electron, except for the positive charge. Particles and antiparticles annihilate (destroy) one another when they meet, producing pure energy in the form of gamma-ray photons. The **neutrino** is a chargeless and virtually massless elementary particle. (The name derives from the Italian for "little neutral one.") Neutrinos move at nearly the speed of light and interact with hardly anything. They can penetrate, without stopping, several light-years of lead. Their interactions with matter are governed by the **weak nuclear force**, described in *More Precisely 9-1*. Yet despite their elusiveness, neutrinos can be detected with carefully constructed instruments. In the final section we discuss some rudimentary neutrino "telescopes" and the important contribution they have made to solar astronomy.

The basic set of nuclear reactions powering the Sun (and the vast majority of all stars) is sketched in Figure 9.25. In fact, it is not a single reaction, but rather a sequence of reactions called the **proton–proton chain:**

(I). Two protons combine, as shown in Figure 9.24(b), forming a deuteron and releasing a positron and a neutrino.

(II). The deuteron then combines with another proton to create a type of helium called helium-3, which contains only one neutron, while the positron annihilates with an electron, producing gamma rays. The neutrino escapes into space. Figure 9.25 shows two of each of these sets of reactions.

(III). Finally, two helium-3 nuclei combine to form helium-4 and two protons.

Setting aside the temporary intermediate nuclei produced, the net effect is that four hydrogen nuclei (protons) combine to create one nucleus of the next lightest

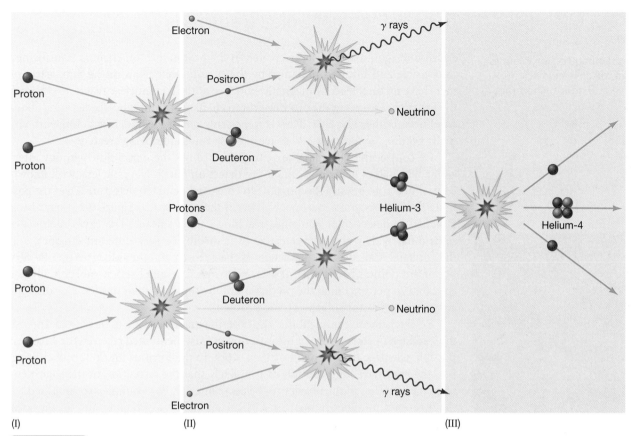

INTERACTIVE ▲ **FIGURE 9.25 Solar Fusion**   In the proton–proton chain, a total of six protons (and two electrons) are converted to two protons, one helium-4 nucleus, and two neutrinos. The two leftover protons are available as fuel for new proton–proton reactions, so the net effect is that four protons are fused to form one helium-4 nucleus. Energy, in the form of gamma rays, is produced at each stage. The three stages indicated here correspond to reactions (I), (II), and (III) described in the text.

element, helium-4 (containing 2 protons and 2 neutrons, for a total mass of 4), creating two neutrinos and releasing energy in the form of gamma-ray radiation: ∞ *(Sec. 2.6)*

$$4 \text{ protons} \rightarrow \text{helium-4} + 2 \text{ neutrinos} + \text{energy}.$$

Other reaction sequences, involving the more massive nuclei of carbon, nitrogen, and oxygen, can produce the same end result—four protons fuse into helium-4 and energy is produced in the form of gamma rays. However, the sequence shown in Figure 9.25 is the simplest and is responsible for almost 90 percent of the Sun's luminosity. Notice how, at each stage of the chain, more complex nuclei are created from simpler ones. We will see later that not all stars are powered by hydrogen fusion; nevertheless, this basic idea of simpler elements combining to form more complex ones, releasing energy in the process, continues to apply.

Gargantuan quantities of protons are fused into helium by the proton–proton chain in the core of the Sun every second. As discussed in more detail in *More Precisely 9-2*, to fuel the Sun's present energy output, hydrogen must be fused into helium in the core at a rate of 600 million tons per second—a lot of mass, but only a tiny fraction of the total amount available. As we will see in Chapter 12, the Sun will be able to sustain this rate of core burning for about another 5 billion years.

The Sun's nuclear energy is produced in the core in the form of gamma rays. However, as the energy passes through the cooler overlying layers of the solar interior and photons are repeatedly absorbed and reemitted, the radiation's blackbody spectrum steadily shifts toward lower and lower temperatures and, by Wien's law, the characteristic wavelength of the radiation increases. ∞ *(Sec. 2.4)* The energy eventually leaves the solar photosphere mainly in the form of visible and infrared radiation. A comparable amount of energy is carried off by the neutrinos, which escape unhindered into space at almost the speed of light.

CONCEPT CHECK ▶ Why does the fact that we see sunlight imply that the Sun's mass is slowly decreasing?

Observations of Solar Neutrinos

Using the solar neutrino problem as an example, discuss how scientific theory and observation evolve and adapt when they come into conflict.

Because the gamma-ray energy created in the proton–proton chain is transformed into visible and infrared radiation by the time it emerges from the Sun, astronomers have no direct electromagnetic evidence of the core nuclear reactions. Instead, the *neutrinos* created in the proton–proton chain are our best bet for learning about conditions in the solar core. They travel cleanly out of the Sun, interacting with virtually nothing, and escape into space a few seconds after being created.

Of course, the fact that they can pass through the entire Sun without interacting also makes neutrinos difficult to detect on Earth! Nevertheless, with knowledge of neutrino physics it is possible to construct neutrino detectors. Over the past three decades, several experiments (two of them shown in Figure 9.26) have been designed to detect solar neutrinos reaching Earth's surface. The experiments' designs differ widely, they are sensitive to neutrinos of very different energies, and they disagree somewhat in the details of their results, but they all agree on one very important point: The number of solar neutrinos reaching Earth is substantially less (by 50 to 70 percent) than the prediction of the standard solar model. This discrepancy is known as the **solar neutrino problem.**

Researchers are confident that the experimental results are to be trusted. Indeed, the lead investigators of two of the experiments just mentioned received the strongest possible scientific endorsement of their work, in the form of the 2002 Nobel Prize! But, theorists have long considered it unlikely that the resolution of the solar neutrino problem lies in the physics of the Sun's interior. For example, we could reduce the theoretical number of neutrinos by imagining a lower temperature in the solar core, but the nuclear reactions just described are too well known and the agreement between the standard solar model and helioseismological observations (Section 9.2) is far too close for conditions in the core to deviate so much from the model.

How, then, do we explain the clear disagreement between theory and observation? The answer involves the properties of the neutrinos themselves and has caused scientists to rethink some very fundamental concepts in particle physics. If neutrinos have even a minute amount of mass, theory indicates that it should be possible for them to change their properties—even to transform into other particles—during their 8-minute flight from the solar core to Earth, through a process known as **neutrino oscillations**. In this picture, neutrinos are produced in the Sun at the rate required by the standard solar model, but some turn into something else—actually, other types of neutrinos—on their way to Earth and hence go undetected in the experiments just described. (In the jargon of the field, the neutrinos are said to "oscillate" into other particles.)

In 1998 the Japanese group operating the Super Kamiokande detector shown in Figure 9.26(a) reported the first experimental evidence of neutrino oscillations (and hence of nonzero neutrino masses), although the observed oscillations did not involve neutrinos of the type produced in the Sun. In 2001, measurements made at the Sudbury Neutrino Observatory (SNO) in Ontario, Canada (Figure 9.26b), revealed strong evidence for the "other" neutrinos into which the Sun's neutrinos have been transformed. Subsequent SNO observations confirmed the result. The total numbers of neutrinos observed were completely consistent with the standard solar model. The solar neutrino problem was solved—and neutrino astronomy claimed its first major triumph!

(a)

(b)

◀ **FIGURE 9.26 Neutrino Telescope** (a) This swimming pool–sized "neutrino telescope" is buried beneath a mountain near Tokyo, Japan. Called Super Kamiokande, it is filled with 50,000 tons of purified water and contains 13,000 individual light detectors (some shown here being inspected by technicians) to sense the telltale signature—a brief burst of light—of a neutrino passing through the apparatus. (b) The Sudbury Neutrino Observatory (SNO), situated 2 km underground in Ontario, Canada, is similar in design to the Kamiokande device, but, by using "heavy" water (with hydrogen replaced by deuterium) instead of ordinary water (H_2O), and adding 2 tons of salt, it also becomes sensitive to other neutrino types. The device contains 10,000 light-sensitive detectors arranged on the inside of the large sphere shown here. (*ICRR; SNO*)

9-2 MORE PRECISELY

Energy Generation in the Proton–Proton Chain

Let's look in a little more detail at the energy produced by fusion in the solar core and compare it with the energy needed to account for the Sun's luminosity. As discussed in the text and illustrated below, the net effect of the fusion process is that four protons combine to produce a nucleus of helium-4, in the process creating two neutrinos and two positrons (which are quickly converted into energy by annihilation with electrons).

We can calculate the total amount of energy released by accounting carefully for the total masses of the nuclei involved and applying Einstein's famous formula $E = mc^2$. This allows us to relate the Sun's total luminosity to the consumption of hydrogen fuel in the core. Careful laboratory experiments have determined the masses of all the particles involved in the above reaction: The total mass of the protons is 6.6943×10^{-27} kg, the mass of the helium-4 nucleus is 6.6466×10^{-27} kg, and the neutrinos are virtually massless. We omit the positrons here—their masses will end up being counted as part of the total energy released. The difference between the total mass of the four protons and that of the final helium-4 nucleus, 0.0477×10^{-27} kg, is not great—only about 0.71 percent of the original mass—but it is easily measurable.

Multiplying the vanished mass by the square of the speed of light yields 0.0477×10^{-27} kg $\times (3.00 \times 10^8$ m/s$)^2 = 4.28 \times 10^{-12}$ J. This is the energy produced in the form of radiation when 6.69×10^{-27} kg (the rounded-off mass of the four protons) of hydrogen

fuses to helium. It follows that fusion of 1 kg of hydrogen generates $4.28 \times 10^{-12}/6.69 \times 10^{-27} = 6.40 \times 10^{14}$ J.

Thus we have established a direct connection between the Sun's energy output and the consumption of hydrogen in the core. The Sun's luminosity of 3.86×10^{26} W (see Table 9.1), or 3.86×10^{26} J/s (joules per second), implies a mass consumption rate of 3.86×10^{26} J/s $\div 6.40 \times 10^{14}$ J/kg $= 6.03 \times 10^{11}$ kg/s—roughly 600 million tons of hydrogen every second. That sounds like a lot—the mass of a small mountain and 600 times the mass loss rate in the solar wind—but it represents only a few million million millionths of the total mass of the Sun. Our parent star will be able to sustain this loss for a very long time—this is the subject of Chapter 12.

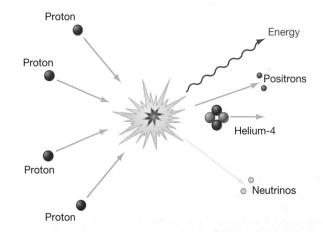

CHAPTER REVIEW

SUMMARY

LO1 Our Sun is a **star** (p. 246), a glowing ball of gas held together by its own gravity and powered by nuclear fusion at its center. The **photosphere** (p. 246) is the region at the Sun's surface from which virtually all the visible light is emitted. The main interior regions of the Sun are the **core** (p. 246), where nuclear reactions generate energy; the **radiation zone** (p. 246), where the energy travels outward in the form of electromagnetic radiation; and the **convection zone** (p. 246), where the Sun's matter is in constant convective motion.

LO2 The amount of solar energy reaching a square meter area at the top of Earth's atmosphere each second is the **solar constant** (p. 247). The Sun's **luminosity** (p. 247) is the total amount of energy radiated from the solar surface per second.

LO3 Much of our knowledge of the solar interior comes from mathematical models. The model that best fits the observed properties of the Sun is the **standard solar model** (p. 247). **Helioseismology** (p. 248)—the study of vibrations of the solar surface caused by pressure waves in the interior—provides further insight into the Sun's structure. The effect of the solar convection zone can be seen on the surface in the form of **granulation** (p. 251) of the photosphere. Lower levels in the convection zone also leave their mark on the photosphere in the form of larger transient patterns called **supergranulation** (p. 251).

LO4 Above the photosphere lies the **chromosphere** (p. 246), the Sun's lower atmosphere. In the **transition zone** (p. 246) above the chromosphere, the temperature increases from a few thousand kelvins to around a million kelvins.

Above the transition zone is the Sun's thin, hot upper atmosphere, the solar **corona** (p. 246). At a distance of about 15 solar radii, the gas in the corona is hot enough to escape the Sun's gravity, and the corona begins to flow outward as the **solar wind** (p. 246).

L05 **Sunspots** (p. 254) are Earth-sized regions of intense magnetism on the solar surface that are a little cooler than the surrounding photosphere. The numbers and locations of sunspots vary in a roughly 11-year **sunspot cycle** (p. 256) as the Sun's magnetic field rises and falls. The overall direction of the field reverses from one sunspot cycle to the next. The 22-year cycle that results when the direction of the field is taken into account is called the **solar cycle** (p. 256). Solar activity tends to be concentrated in **active regions** (p. 258) associated with sunspot groups; this activity heats the corona and drives the solar wind, which flows along open field lines from low-density regions called **coronal holes** (p. 260).

L06 The Sun generates energy by converting hydrogen to helium in its core by the process of **nuclear fusion** (p. 261). When four protons are converted to a helium nucleus in the **proton–proton chain** (p. 262), some mass is lost. The **law of conservation of mass and energy** (p. 261) requires that this mass appear as energy, eventually resulting in the light we see.

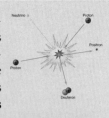

L07 **Neutrinos** (p. 262) are nearly massless particles that are produced in the proton–proton chain and escape from the Sun. It is possible to detect a small fraction of the neutrinos streaming from the Sun. Observations over several decades led to the **solar neutrino problem** (p. 264)—substantially fewer neutrinos are observed than were predicted by theory. The explanation, supported by observations, is that **neutrino oscillations** (p. 264) convert some neutrinos to other (undetected) particles en route from the Sun to Earth.

MasteringAstronomy® For instructor-assigned homework go to www.masteringastronomy.com

Problems labeled **POS** explore the process of science. **VIS** problems focus on reading and interpreting visual information. **LO** connects to the introduction's numbered Learning Outcomes.

REVIEW AND DISCUSSION

1. **L01** Name and briefly describe the main regions of the Sun.
2. How massive is the Sun, compared with Earth?
3. How hot is the solar surface? The solar core?
4. **L02** How is the Sun's luminosity measured?
5. **L03 POS** How is helioseismology used to model the Sun?
6. Describe how energy from the solar core eventually reaches Earth.
7. Why does the Sun appear to have a sharp edge?
8. What evidence do we have for solar convection?
9. **L04** What is the solar wind, and how is it related to the Sun's outer atmosphere?
10. **L05** What is the cause of sunspots, flares, and prominences?
11. What fuels the Sun's enormous energy output?
12. **POS** What is the law of conservation of mass and energy? How is it relevant to nuclear fusion in the Sun?
13. **L06** What are the ingredients and the end result of the proton–proton chain? Why is energy released in the process?
14. **L07 POS** Why are observations of solar neutrinos important to both astronomers and physicists?
15. What would we observe on Earth if the Sun's internal energy suddenly shut off? How long do you think it would take for the Sun's light to begin to fade? Answer the same question for solar neutrinos.

CONCEPTUAL SELF-TEST: TRUE OR FALSE?/MULTIPLE CHOICE

1. The most abundant element in the Sun is hydrogen. (T/F)
2. There are as many absorption lines in the solar spectrum as there are elements present in the Sun. (T/F)
3. The density and temperature in the solar corona are much higher than in the photosphere. (T/F)
4. Sunspots appear dark because they are hotter than the surrounding gas of the photosphere. (T/F)
5. Flares are caused by magnetic disturbances in the lower atmosphere of the Sun. (T/F)
6. The proton–proton chain releases energy because mass is created in the process. (T/F)
7. Neutrinos have never been detected experimentally. (T/F)
8. The Sun is roughly (a) the same size; (b) 10 times larger; (c) 100 times larger; (d) 1 million times larger than Earth.
9. The Sun spins on its axis roughly once each (a) hour; (b) day; (c) month; (d) year.
10. Astronomers on Venus would measure a solar constant (a) larger than; (b) smaller than; (c) the same as on Earth.
11. A typical solar granule is about the size of (a) a U.S. city; (b) a large U.S. state; (c) the Moon; (d) Earth.
12. The time between successive sunspot maxima is about (a) a month; (b) a year; (c) a decade; (d) a century.

13. Solar energy is produced by (a) fusion of light nuclei into heavier ones; (b) fission of heavy nuclei into lighter ones; (c) the release of heat left over from the Sun's formation; (d) solar magnetism.

14. VIS In the standard solar model (Figure 9.6, Solar Interior), as the distance from the center increases, (a) the density and temperature decrease at about the same rate; (b) the density decreases faster than the temperature; (c) the density decreases more slowly than the temperature; (d) the density decreases but the temperature increases.

15. VIS Comparing Figures 9.6 (Solar Interior) and 9.13 (Solar Atmospheric Temperature), the temperature in the solar corona is about the same as in (a) the core; (b) the bottom of the radiation zone; (c) the bottom of the convection zone; (d) the photosphere.

PROBLEMS

The number of squares preceding each problem indicates its approximate level of difficulty.

1. ■ Use the reasoning presented in Section 9.1 to calculate the value of the "solar constant" (a) on Mercury, at perihelion, and (b) on Jupiter.

2. ■■ The largest-amplitude solar pressure waves have periods of about 5 minutes and move at about 10 km/s. (a) How far does the wave move during one wave period? (b) Roughly how many wavelengths are needed to circle the Sun? (c) Compare the wave period with the orbital period in the solar photosphere.

3. ■ Given that the solar spectrum corresponds to a temperature of 5800 K and peaks at a wavelength of 500 nm, use Wien's law to determine the wavelength corresponding to the peak of the blackbody curve (a) in the core of the Sun, where the temperature is 10^7 K, (b) in the solar convection zone (10^5 K), and (c) just below the solar photosphere (10^4 K). ∞ (More Precisely 2-2) What form (visible, infrared, X-ray, etc.) does the radiation take in each case?

4. ■ If convected solar material moves at 1 km/s, how long does it take to cross the 1000-km expanse of a typical granule? Compare this time with the 10-minute lifetimes observed for most solar granules.

5. ■■ Use Stefan's law to calculate how much less energy is emitted per unit area of a 4500-K sunspot than is emitted per unit area of the surrounding 5800-K photosphere. ∞ (More Precisely 2-2)

6. ■■ The solar wind carries mass away from the Sun at a rate of about 2 million ton/s (where 1 ton = 1000 kg). (a) Compare this rate with the rate at which the Sun loses mass in the form of radiation. (b) At this rate, how long would it take for all of the Sun's mass to escape?

7. ■■ The Sun's differential rotation is responsible for wrapping the solar magnetic field around the Sun (Figure 9.17). Using the figures presented in Table 9.1, estimate how long it takes for material at the solar equator to "lap" the material near the poles—that is, to complete one extra trip around the Sun's rotation axis.

8. ■■ How long does it take the Sun to convert one Earth mass of hydrogen to helium?

9. ■■ Assuming that (1) the solar luminosity has been constant since the Sun formed, and (2) the Sun was initially of uniform composition throughout, as described by Table 9.2, estimate how long it would take the Sun to convert all of its original hydrogen into helium.

10. ■■■ The entire reaction sequence shown in Figure 9.25 generates 4.3×10^{-12} joules of electromagnetic energy and releases two neutrinos. If all of the Sun's energy comes from this sequence and neutrino oscillations transform half of the neutrinos produced into other particles by the time they reach Earth, estimate the total number of solar neutrinos passing through Earth each second.

ACTIVITIES

NEVER LOOK DIRECTLY AT THE SUN WITHOUT A FILTER!

Collaborative

The safest way to observe the Sun is to project its image onto a screen. Here are two ways to do it.

1. To build a "pinhole camera," you'll need two sheets of stiff white paper and a pin. Use the pin to punch a hole in the center of one of your pieces of paper. Go outside, hold the paper up and aim the hole at the Sun. **Don't look directly at the Sun!** Now find the image of the Sun coming through the hole. Move the other piece of paper back and forth until the image looks best. What happens to the image as you vary the size of the pinhole?

2. Alternatively, you can project an image of the Sun using a pair of binoculars or a small telescope. You'll need a sheet of stiff white paper and a tripod. Mount the binoculars or telescope firmly on the tripod and point it at the Sun. **Don't attempt to view the Sun directly!** Hold the paper about 10 inches behind the eyepiece to act as a screen. You should see a bright blurred circle of light. Focus the instrument until the circle is sharp. This is the disk of the Sun. Experiment with moving the card closer and farther away from the telescope. What effect does this have on the image?

Use one of these projectors to study some sunspots, as in Project 1 below. Have half your group construct a pinhole camera and the other half build a telescopic projector, and compare results. What are the pros and cons of each?

Individual

1. A filtered telescope will easily show you sunspots. Count the number of sunspots you see on the Sun's surface. Notice that sunspots often come in pairs or groups. Look again a few days later and you'll see that the Sun's rotation has caused the spots to move, and the spots themselves have changed. If a sufficiently large sunspot (or sunspot group) is seen, continue to watch it as the Sun rotates. It will be out of sight for about 2 weeks.

2. View some solar prominences and flares. Hydrogen-alpha (Hα) filters are available for small telescopes. You can often see prominences and flares even during times of sunspot minimum. You are actually viewing the chromosphere rather than the photosphere, so the Sun looks quite different from normal.

10

Measuring the Stars

Giants, Dwarfs, and the Main Sequence

We have now studied Earth, the Moon, the solar system, and the Sun. To continue our inventory of the contents of the universe, we must move away from our local environment into the depths of space. In this chapter we take a great leap in distance and consider stars in general. Our primary goal is to comprehend the nature of the stars that make up the constellations, as well as the myriad more distant stars we cannot perceive with unaided eyes. Rather than studying their individual peculiarities, however, we will concentrate on determining the physical and chemical properties they share. There is order in the legions of stars scattered across the sky. Like comparative planetology in the solar system, comparing and cataloging the stars play vital roles in furthering our understanding of the galaxy and the universe we inhabit.

MasteringAstronomy® Visit www.masteringastronomy.com for quizzes, animations, videos, interactive figures, and self-guided tutorials.

LEARNING OUTCOMES

Studying this chapter will enable you to:

LO1 Explain how stellar distances are determined.

LO2 Distinguish between luminosity and apparent brightness, and explain how stellar luminosity is determined.

LO3 Describe how stars are classified according to their colors, surface temperatures, and spectral characteristics.

LO4 Explain how physical laws can be used to estimate stellar sizes.

LO5 Describe how a Hertzsprung–Russell diagram is constructed and used to identify stellar properties.

LO6 Explain how knowledge of a star's spectroscopic properties can lead to an estimate of its distance.

LO7 Explain how stellar masses are measured and how mass is related to other stellar properties.

◄ The total number of stars, even in our local cosmic neighborhood, is virtually beyond our ability to count. Yet astronomers have learned a great deal about stars in general, as well as their ranges of properties—their masses, their luminosities, even their ages, and (as we shall see in later chapters) their destinies. Here, the *Hubble Space Telescope* has imaged the magnificent star cluster NGC 265, a rich group of more than a thousand stars spread across about 65 light-years. This star field is part of the Small Magellanic Cloud, which itself is located nearly 200,000 light-years away in the southern hemisphere. *(ESA/NASA)*

10.1 The Solar Neighborhood

Looking deep into space, astronomers can observe and study millions of individual stars. By observing distant galaxies, we can infer statistically the properties of trillions more. All told, the observable universe probably contains several tens of sextillions (1 sextillion = 10^{21}) stars. Yet, despite their incredible numbers, the essential properties of stars—their appearance in the sky, their births, lives, deaths, even their interactions with their environment—can be understood in terms of just a few basic physical stellar quantities: luminosity (brightness), temperature (color), chemical composition, size, and mass. In this chapter we examine these basic stellar properties, describe how we measure them, and show how they lead to a simple but powerful means of classifying stars of all kinds. With this foundation we can then discuss the life cycles of stars in the next three chapters.

As with the planets, knowing the distances to the stars is essential to understanding many of their other properties. We therefore begin our study of stellar astronomy by reviewing the use of simple geometry in determining the distances to our neighbors.

Stellar Parallax

The distances to the nearest stars can be measured using *parallax*—the apparent shift of an object relative to some distant background as the observer's point of view changes. ∞ *(Sec. 0.4)* However, even the closest stars are so far away that no baseline on Earth is adequate to measure their parallaxes. But, by comparing observations made of a star *at different times of the year*, as shown in Figure 10.1 (compare Figure 0.20), we can effectively extend the baseline to the diameter of Earth's orbit around the Sun, 2 AU. The parallax is determined by comparing photographs made from the two ends of the baseline. As indicated on the figure, a star's parallactic angle—or, more commonly, just its "parallax"—is conventionally defined to be *half* of its apparent shift relative to the background as we move from one side of Earth's orbit to the other.

Because the stars are so far away, stellar parallaxes are always very small, and astronomers find it convenient to measure them in arc seconds rather than in degrees. ∞ *(More Precisely 0-1)* If we ask at what distance from the Sun a star must lie in order for its observed parallax to be exactly 1″, we get an answer of 206,265 AU, or 3.1×10^{16} m. Astronomers call this distance one **parsec** (1 pc), from "*par*allax in arc *sec*onds." Because parallax decreases as distance increases, we can relate a star's parallax to its distance from the Sun by the following simple formula:

$$\text{distance (in parsecs)} = \frac{1}{\text{parallax (in arc seconds)}}.$$

Thus, a star with a measured parallax of 1″ lies at a distance of 1 pc from the Sun. The parsec is defined so as to make the conversion between parallactic angle and distance easy. An object with a parallax of 0.5″ lies at a distance of 1/0.5 = 2 pc, an object with a parallax of 0.1″ lies at 1/0.1 = 10 pc, and so on. One parsec is approximately equal to 3.3 light-years.

(a)

This is the view as seen in January . . .

. . . and in July, when the star shifts.

(b)

INTERACTIVE ◄ FIGURE 10.1 **Stellar Parallax** (a) For observations made 6 months apart, the baseline is twice the Earth–Sun distance, or 2 AU. (b) The parallactic shift (exaggerated here, and indicated by the white arrow) is usually measured photographically, as illustrated by the red star. Images of the same region of the sky made at different times of the year **(MA)** determine a star's apparent movement relative to background stars.

Our Nearest Neighbors

The closest star to the Sun is called Proxima Centauri. It is part of a triple-star system (three stars orbiting one another, bound by gravity) known as the Alpha Centauri complex. Proxima Centauri has the largest known stellar parallax, 0.77″, meaning that it is about 1/0.77 = 1.3 pc away—about 270,000 AU, or 4.3 light-years. That's the *nearest* star to the Sun—at almost 300,000 times the Earth–Sun distance! This is a fairly typical interstellar distance in the Milky Way Galaxy.

Vast distances can sometimes be grasped by means of analogies. Imagine Earth as a grain of sand orbiting a golfball-sized Sun at a distance of 1 m. The nearest star, also a golfball-sized object, is then 270 *kilo*meters away. Except for the planets in our solar system, ranging in size from grains of sand to small marbles within 50 m of the "Sun," and the Kuiper belt and Oort cloud, trillions of tiny dust grains scattered over a volume 100 km across, nothing of consequence exists in the 270 km separating the Sun and the other star.

The next nearest neighbor to the Sun beyond the Alpha Centauri system is called Barnard's Star. Its parallax is 0.55″, so it lies at a distance of 1.8 pc, or 6.0 light-years—370 km in our model. All told, fewer than 100 stars lie within 5 pc (1000 km in our model) of the Sun. Such is the void of interstellar space. Figure 10.2 is a map of our nearest galactic neighbors—the 30 or so stars lying within 4 pc of Earth. *Discovery 10-1* (p. 273) describes how some of the stars in this figure got their peculiar names.

Ground-based images of stars are generally smeared out into a seeing disk of radius 1″ or so by turbulence in Earth's atmosphere. ∞ *(Sec. 3.3)* However, astronomers use special techniques to routinely measure stellar parallaxes of 0.03″ or less, corresponding to stars within about 30 pc (100 light-years) of Earth. Several thousand stars lie within this range, most of them of much lower luminosity than the Sun and invisible to the naked eye. Adaptive optics systems allow more accurate measurements of stellar positions, extending the parallax range to over 100 pc. ∞ *(Sec. 3.3)* In the 1990s, the European *Hipparcos* satellite expanded this range to well over 200 pc, encompassing nearly a million stars. Even so, almost all of the stars in our Galaxy are far more distant. Upcoming next-generation space missions will extend out to 25,000 pc, spanning our entire Galaxy and likely revolutionizing our knowledge of its structure.

Stellar Motion

In addition to the apparent motion caused by parallax, stars have real spatial motion, too. A star's *radial* velocity—along the line of sight—can be measured using the Doppler effect. ∞ *(Sec. 2.7)* For many nearby stars the *transverse* velocity—perpendicular to our line of sight—can also be determined by careful monitoring of the star's position on the sky.

Figure 10.3(a) shows two photographs of the sky around Barnard's star, taken on the same day of the year, but 22 years apart. Note that the star (marked by the arrow) has moved during the 22-year interval shown. Because Earth was at the same point in its orbit when the photographs were taken, the observed displacement is *not* the result of parallax. Instead, it indicates *real* space motion of Barnard's star relative to the Sun. This annual movement of a star across the

CONCEPT CHECK ▶ Why can't astronomers use simultaneous observations from different parts of Earth's surface to determine stellar distances?

▼ **FIGURE 10.2 Sun's Neighborhood** A plot of the 30 closest stars to the Sun, projected so as to reveal their three-dimensional relationships. All lie within 4 pc (about 13 light-years) of Earth. The gridlines represent distances in the Galactic plane, and the vertical lines denote distances perpendicular to that plane.

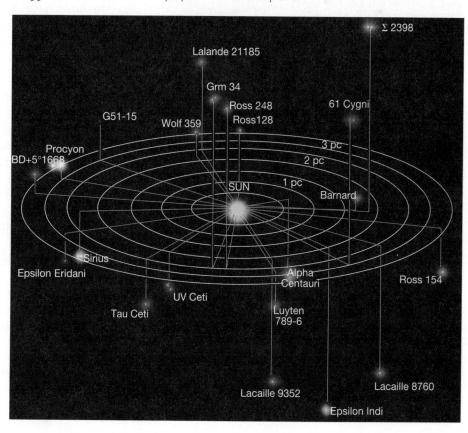

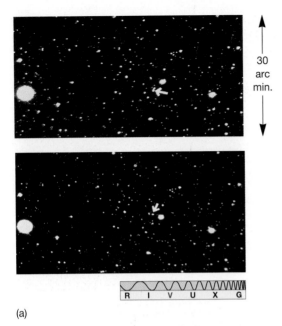

(a)

(b)

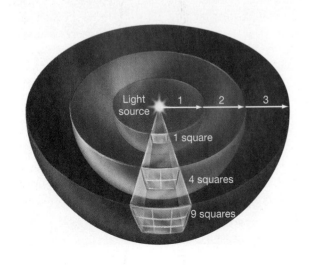

◀ FIGURE 10.3 **Real Space Motion**   (a) Comparison of two photographs taken 22 years apart shows real space motion for Barnard's star (denoted by an arrow). (b) The motion of the Alpha Centauri star system relative to our solar system. The radial component of the velocity is measured using the Doppler shift of lines in Alpha Centauri's spectrum. The transverse component is derived from the system's proper motion. The true spatial velocity, indicated by the red arrow, results from the combination of the two. (*Harvard College Observatory*)

sky, as seen from Earth and corrected for parallax, is called **proper motion.** Barnard's star moved 228″ in 22 years. Its proper motion is therefore 228″/22 years, or 10.3″/yr.

A star's transverse velocity is easily calculated once its proper motion and its distance are known. At the distance of Barnard's star (1.8 pc), an angle of 10.3″ corresponds to a distance of 0.00009 pc, or about 2.8 billion km. Barnard's star takes a year to travel this distance, so its transverse velocity is 2.8 billion km/3.2×10^7 s, or 88 km/s. Even though stars' transverse velocities are often quite large—tens or even hundreds of kilometers per second—their great distances from the Sun mean that it usually takes many years for us to discern their movement across the sky. In fact, Barnard's star has the largest-known proper motion of any star. Only a few hundred stars have proper motions greater than 1″/yr.

A star's total space velocity is the combination of its radial and transverse components. Figure 10.3(b) illustrates the relationship among these quantities for another near neighbor, Alpha Centauri. Spectral lines from Alpha Centauri are blueshifted by a tiny amount—about 0.0067 percent—allowing astronomers to measure the star system's radial velocity (relative to the Sun) as 300,000 km/s $\times$ 6.7×10^{-5} = 20 km/s toward us. ∞ *(Sec. 2.7)* Alpha Centauri's proper motion has been measured to be 3.7″/yr. At Alpha Centauri's distance of 1.35 pc (corresponding to a parallax of 0.74″), that measurement implies a transverse velocity of 24 km/s. We can combine the transverse (24 km/s) and radial (20 km/s) velocities according to the Pythagorean theorem, as indicated in the figure. The total velocity is $\sqrt{24^2 + 20^2}$ or about 31 km/s, in the direction shown by the horizontal red arrow.

10.2 Luminosity and Apparent Brightness

Luminosity—the amount of radiation leaving a star per unit time—is an *intrinsic* stellar property. ∞ *(Sec 9.1)* It does not depend in any way on the location or motion of the observer. It is sometimes referred to as the star's *absolute brightness*. However, when we observe a star, we see not its luminosity but rather its **apparent brightness**—the amount of energy striking unit area of some light-sensitive surface or device, such as a human eye or a CCD chip, per unit time. (Notice that this is precisely the same definition given in Chapter 9 for the solar constant—in our new terminology, the solar constant is just the apparent brightness of the Sun.) In this section we discuss how these important quantities are related to one another.

Another Inverse-Square Law

Figure 10.4 shows light leaving a star and traveling through space. Moving outward, the radiation passes through imaginary spheres of increasing radius surrounding the source. The farther the light travels from the source, the less energy passes through

◀ FIGURE 10.4 **Inverse-Square Law**   As light moves away from a source such as a star, it steadily dilutes while spreading over progressively larger surface areas (depicted here as sections of spherical shells). Thus, the amount of radiation received by a detector (the source's apparent brightness) varies inversely as the square of its distance from the source.

each unit of area. Think of the energy as being spread out over an ever larger area and therefore spread more thinly, or "diluted," as it expands into space. Because the area of a sphere grows as the square of the radius, the energy per unit area—the star's apparent brightness—is inversely proportional to the square of the distance from the star. Doubling the distance from a star makes it appear 2^2, or 4, times fainter. Tripling the distance reduces the apparent brightness by a factor of 3^2, or 9, and so on.

Of course, the star's luminosity also affects its apparent brightness. Doubling the luminosity doubles the energy crossing any spherical shell surrounding the star and hence doubles the apparent brightness. The apparent brightness of a star is therefore directly proportional to the star's luminosity and inversely proportional to the square of its distance:

$$\text{apparent brightness} \propto \frac{\text{luminosity}}{\text{distance}^2}.$$

(Reminder: $\propto$ is the symbol representing proportionality.) Thus, two identical stars can have the same apparent brightness if (and only if) they lie at the same distance from Earth. However, as illustrated in Figure 10.5, two nonidentical stars can also have the same apparent brightness if the more luminous one lies farther away. A bright star, having large apparent brightness, is a powerful emitter of radiation (high luminosity), is near Earth, or both. A faint star (small apparent brightness) is a weak emitter (low luminosity), is far from Earth, or both.

This is the view seen by the observer looking out into space.

▲ FIGURE 10.5 Luminosity Two stars A and B of different luminosities can appear equally bright to an observer on Earth if the brighter star B is more distant than the fainter star A.

10-1 DISCOVERY

Naming the Stars

Looking at Figure 10.2, you might well wonder how those stars got their names. In fact, there is no coherent convention for naming stars—or rather, there are many, often leading to many names for any given object! Here is a brief (and roughly chronological) description of some common naming schemes:

1. The brightest stars in the sky usually have ancient names. Many of the names are of Arabic origin, dating back to around the 10th century A.D., when Muslim astronomy flourished, preserving the scientific knowledge of Greek and other early cultures while European science floundered in the Dark Ages. Examples are Betelgeuse (*Yad al Jauzah,* Hand of the Central One, later corrupted to Bed Elgueze), Rigel (*Rijl Jauzah al Yusra,* Left Foot of the Central One), Aldebaran (*Al Dabaran,* the Follower), and Deneb (*Al Dhanab al Dulfin,* the Dolphin's Tail). The names Procyon and Sirius (see Figure 10.2) are Greek.

2. A more systematic scheme was introduced in 1603 by German lawyer Johann Bayer. He ranked the stars in a given constellation (Orion, say) by brightness, labeling them with Greek letters: Alpha Orionis (Betelgeuse), Beta Orionis (Rigel), or Epsilon Eridani (Figure 10.2). ∞ *(Sec. 0.1)*

3. There are only 24 Greek letters, so eventually the Bayer scheme broke down. In the early 18th century, British Astronomer Royal John Flamsteed suggested simply numbering the stars

from west to east within a constellation. In Flamsteeds scheme, Betelgeuse is 58 Ori, Rigel is 19 Ori, and so on (see also 61 Cygni in Figure 10.2). Sometimes the Bayer and Flamsteed designations are combined, as in "58 α Ori."

4. As telescopes improved and more and more stars were discovered, astronomers started compiling their own catalogs. Some catalogs listed stars by celestial coordinates—for example, star BD + 5°1668 in Figure 10.2, which appears in the 19th-century German Bonner Durchmusterung (BD) catalog and lies just over 5° above the celestial equator. ∞ *(Sec. 0.1)* Other catalogs simply named stars by the order in which they were added to the list—the star Lalande 21185 in Figure 10.2 is the 21,185th entry in a catalog compiled by 18th-century French astronomer Joseph de Lalande. Unfortunately, the catalogs often overlap, so stars can appear in many different catalogs with very different names. For example, Lalande 21185 is also known as HD 95735 (in the Henry Draper catalog), while Betelgeuse is also known as HD 39801 and SAO 113271 (in the Smithsonian Astrophysical Observatory).

5. Today, newly discovered stars are defined simply by their coordinates in the sky and not given any special name. If you like, you can (for a fee, of course) register your own star name, but don't expect any astronomer ever to cite that name, or to take the registry seriously—only "official" names approved by *International Astronomical Union* are used in astronomy.

Determining a star's luminosity is a twofold task. First, the astronomer must determine the star's apparent brightness by measuring the amount of energy detected through a telescope in a given amount of time. Second, the star's distance must be measured—by parallax for nearby stars and by other means (to be discussed later) for more distant stars. The luminosity can then be found using the inverse-square law. This is the same reasoning used earlier in our discussion of how astronomers measure the luminosity of the Sun. ∞ *(Sec. 9.1)*

The Magnitude Scale

Instead of measuring apparent brightness in SI units (for example, watts per square meter, W/m^2, the unit in which we expressed the solar constant in Section 9.1), optical astronomers find it more convenient to work in terms of a construct called the **magnitude scale**—a system of ranking stars by their apparent brightness. This scale dates back to the second century B.C., when the Greek astronomer Hipparchus ranked the naked-eye stars into six groups. He categorized the brightest stars as first magnitude, labeled the next brightest stars as second magnitude, and so on, down to the faintest stars visible to the naked eye, which he classified as sixth magnitude. The range 1 (brightest) through 6 (faintest) spanned all the stars known to the ancients. Notice that a *large* magnitude means a *faint* star.

When astronomers began using telescopes with sophisticated detectors to measure the light received from stars, they quickly discovered two important facts about the magnitude scale. First, the 1 through 6 magnitude range defined by Hipparchus spans about a factor of 100 in apparent brightness—a first-magnitude star is approximately 100 times brighter than a sixth-magnitude star. Second, the characteristics of the human eye are such that a change of one magnitude corresponds to a factor of about 2.5 in apparent brightness. In other words, to the human eye a first-magnitude star is roughly 2.5 times brighter than a second-magnitude star, which in turn is roughly 2.5 times brighter than a third-magnitude star, and so on. By combining factors of 2.5, we confirm that a first-magnitude star is $(2.5)^5$, or roughly 100 times brighter than a sixth-magnitude star.

Because we are really talking about apparent (rather than absolute) brightnesses, astronomers refer to the numbers in Hipparchus's ranking system as **apparent magnitudes.** In the modern version of the magnitude scale, we *define* a change of five in the magnitude of an object to correspond to *exactly* a factor of 100 in apparent brightness. In addition, the scale is no longer limited to whole numbers and has expanded to include magnitudes outside the original range of 1 to 6. Very bright objects can have apparent magnitudes less than 1, and very faint objects can have apparent magnitudes much greater than 6. Figure 10.6 illustrates the apparent magnitudes of some astronomical objects, ranging from the Sun at −26.8 to the faintest object detectable by the *Hubble* or *Keck* telescopes, at an apparent magnitude of +30—about as faint as a firefly seen from a distance equal to Earth's diameter.

Apparent magnitude measures a star's apparent brightness when seen at the star's actual distance from us. To compare intrinsic, or absolute, properties of stars, however, astronomers imagine looking at all stars from a standard distance of 10 pc. (There is no particular reason to use 10 pc—it is simply convenient.) A star's **absolute magnitude** is its apparent magnitude when viewed from a distance of 10 pc. Because distance is fixed in this definition, absolute magnitude is a measure of a star's absolute brightness, or luminosity. Despite the Sun's large negative (that is, very bright) apparent magnitude, our star's absolute magnitude is 4.8. In other words, if the Sun were moved to a distance of 10 pc from Earth, it would be only a little brighter than the faintest stars visible in the night sky. As discussed further in *More Precisely 10-1*, the numerical difference between a star's absolute and apparent magnitudes is a measure of the distance to the star.

CONCEPT CHECK ▶ Two stars are observed to have the same apparent magnitude. Based on this information, what, if anything, can be said about their luminosities?

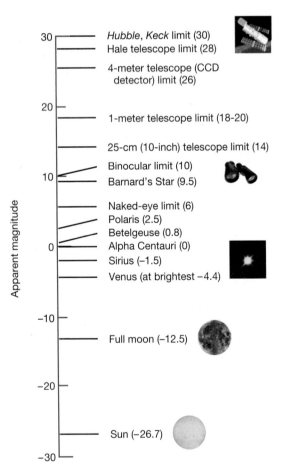

▲ FIGURE 10.6 Apparent Magnitude This graph illustrates the apparent magnitudes of some astronomical objects and the limiting magnitudes (that is, the faintest magnitudes attainable) of some telescopes used to observe them.

The following labels appear along the Apparent magnitude axis:

- *Hubble, Keck* limit (30)
- Hale telescope limit (28)
- 4-meter telescope (CCD detector) limit (26)
- 1-meter telescope limit (18–20)
- 25-cm (10-inch) telescope limit (14)
- Binocular limit (10)
- Barnard's Star (9.5)
- Naked-eye limit (6)
- Polaris (2.5)
- Betelgeuse (0.8)
- Alpha Centauri (0)
- Sirius (−1.5)
- Venus (at brightest −4.4)
- Full moon (−12.5)
- Sun (−26.7)

10-1 MORE PRECISELY

More on the Magnitude Scale

Let's restate our discussion of two important topics—stellar luminosity and the inverse-square law—in terms of magnitudes.

Absolute magnitude is equivalent to luminosity—an intrinsic property of a star. Given that the Sun's absolute magnitude is 4.8, we can construct a conversion chart (shown at right) relating these two quantities. Since an increase in brightness by a factor of 100 corresponds to a decrease in magnitude by 5 units, it follows that a star with luminosity 100 times that of the Sun has absolute magnitude 4.8 − 5 = −0.2, while a 0.01-solar luminosity star has absolute magnitude 4.8 + 5 = 9.8. We can fill in the gaps by noting that 1 magnitude corresponds to a factor of $100^{1/5} \approx 2.512$, 2 magnitudes to $100^{2/5} \approx 6.310$, and so on. A factor of 10 in brightness corresponds to 2.5 magnitudes. You can use this chart to convert between solar luminosities and absolute magnitudes in many of the figures in this and later chapters.

To cast the *inverse-square law* in these terms, recall that increasing the distance to a star by a factor of 10 decreases its apparent brightness by a factor of 100 (by the inverse-square law) and hence increases its apparent magnitude by 5 units. Increasing the distance by a factor of 100 increases the apparent magnitude by 10, and so on. Every increase in distance by a factor of 10 increases the apparent

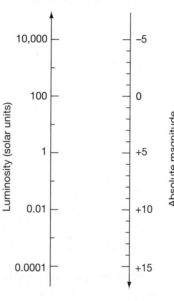

magnitude by 5. Since absolute magnitude is simply apparent magnitude at a distance of 10 pc, we can write:

$$\text{apparent magnitude} - \text{absolute magnitude}$$
$$= 5 \log_{10}\left(\frac{\text{distance}}{10 \text{ pc}}\right).$$

(The *logarithm* function—the LOG key on your calculator—is defined by the property that if $a = \log_{10}(b)$, then $b = 10a$.) Even though it doesn't look much like it, this equation is precisely equivalent to the inverse-square law presented in the text!

For stars more than 10 pc from Earth, the apparent magnitude is greater than the absolute magnitude, while the reverse is true for stars closer than 10 pc. Thus, for example, the Sun, with an absolute magnitude of 4.8, seen from a distance of 100 pc would have an apparent magnitude of $4.8 + 5 \log_{10}(100) = 14.8$, since $\log_{10}(100) = 2$. This is well below the threshold of visibility for binoculars or even a large amateur telescope (see Figure 10.6). We can also turn this around to illustrate how knowledge of a star's absolute and apparent magnitudes tells us its distance. The star Rigel Kentaurus (also known as Alpha Centauri) has absolute magnitude +4.34 and is observed to have apparent magnitude −0.01. Its magnitude difference is −4.35, and its distance must therefore be $10 \text{ pc} \times 10^{-4.34/5} = 1.35$ pc, in agreement with the result (obtained by parallax) given in the text.

10.3 Stellar Temperatures

Color and the Blackbody Curve

Looking at the night sky, you can tell at a glance which stars are hot and which are cool. In Figure 10.7(a), which shows the constellation Orion as it appears through a small telescope, the colors of the cool red star Betelgeuse (α) and the hot blue star Rigel (β) are clearly evident. ∞ *(Sec. 0.2)* Astronomers can determine a star's surface temperature by measuring its apparent brightness (radiation intensity) at several frequencies, then matching the observations to the appropriate blackbody curve. ∞ *(Sec. 2.4)* In the case of the Sun, the theoretical curve that best fits the emission describes a 5800-K emitter. The same technique works for any star, regardless of its distance from Earth.

Because the basic shape of the blackbody curve is so well understood, astronomers can estimate a star's temperature using as few as *two* measurements at selected wavelengths. This is accomplished by using telescope filters that block out all radiation except that within specific wavelength ranges. For example, a B (blue) filter rejects all radiation except for a certain range of violet

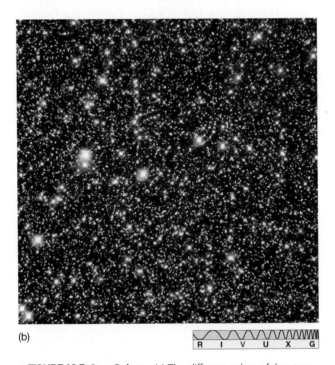

(a)

(b)

R   I   V   U   X   G

▲ **FIGURE 10.7 Star Colors** (a) The different colors of the stars composing the constellation Orion are easily distinguished in this photograph. The bright red star at the upper left (α) is Betelgeuse; the bright blue-white star at the lower right (β) is Rigel. The scale of this photograph is about 20° across. (b) An incredibly rich field of colorful stars, this time in the direction of the center of the Milky Way. Here, the field of view is just 2 arc minutes across—much smaller than in (a). (*P. Sanz/Alamy; NASA*)

to blue light. Similarly, a V (visual) filter passes only radiation in the green to yellow range (the part of the spectrum to which human eyes happen to be particularly sensitive).

Figure 10.8 shows how these filters admit different amounts of light for objects of different temperatures. In curve (a), corresponding to a very hot 30,000-K emitter, considerably more radiation is received through the B filter than through the V filter. The temperature of the emitter in curve (b) is 10,000 K, and the B and V intensities are about the same. The cool 3000-K emitter represented in curve (c) produces far more energy in the V range than in the B range. In each case it is possible to reconstruct the entire blackbody curve on the basis of only those two measurements because no other blackbody curve can be drawn through both measured points.

To the extent that a star's spectrum is well approximated as a blackbody, measurements of the B and V intensities are enough to specify the star's blackbody curve and thus yield its surface temperature. Astronomers often refer to the ratio of a star's B to V intensities (or, equivalently, the difference between the star's apparent magnitudes measured through the B and V filters) as the star's *color index* (or just its "color"). Table 10.1 lists surface temperatures and colors for a few well-known stars.

Stellar Spectra

Color is a useful way to describe a star, but astronomers often use a more detailed scheme to classify stellar properties, incorporating additional knowledge of stellar physics obtained through spectroscopy. ∞ *(Sec. 2.5)* Figure 10.9 compares the spectra of seven stars, arranged in order of decreasing surface temperature. All the spectra extend from 400 to 650 nm. Like the solar spectrum, each shows a series of dark absorption lines superimposed on a background of continuous color. ∞ *(Sec. 9.3)* However, the line patterns differ greatly from one star to another. Some stars display strong (prominent) lines in the long-wavelength part of the spectrum, while others have their strongest lines at short wavelengths. Still others show strong absorption lines spread across the whole visible spectrum. What do these differences tell us?

Although spectral lines of many elements are present at widely varying strengths, the differences among the spectra in Figure 10.9 are not due to differences in chemical composition—all seven stars are, in fact, more or less solar in their makeup. Instead, the differences are due almost entirely to the stars' temperatures. The spectrum at the top of Figure 10.9 is exactly what we expect from a star having solar composition and a surface temperature of about 30,000 K, the second from a 20,000-K star, and so on, down to the 3000-K star at the bottom.

The main differences among the spectra in Figure 10.9 are as follows:

- Spectra of stars having surface temperatures exceeding 25,000 K usually show *strong* absorption lines of singly ionized helium (i.e., helium atoms that have lost one orbiting electron) and multiply ionized heavier elements, such as oxygen, nitrogen, and silicon (the latter lines are not shown in the figure). These strong lines are not seen in the spectra of cooler stars because only very hot stars can excite and ionize such tightly bound atoms.

- In contrast, the hydrogen absorption lines in the spectra of very hot stars are relatively *weak*. The reason is not a lack of hydrogen, which is by far the most abundant element in all stars. Rather, at these high temperatures, much of the hydrogen is ionized, so there are few intact hydrogen atoms to produce strong spectral lines.

TABLE 10.1 Stellar Colors and Temperatures		
Surface Temperature (K)	Color	Familiar Examples
30,000	electric blue	Mintaka (δ Orionis)
20,000	blue	Rigel
10,000	white	Vega, Sirius
7,000	yellow-white	Canopus
6,000	yellow	Sun, Alpha Centauri
4,000	orange	Arcturus, Aldebaran
3,000	red	Betelgeuse, Barnard's Star

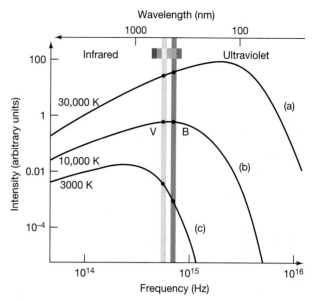

▲ FIGURE 10.8 Blackbody Curves Star (a) is very hot—30,000 K—so its B (blue) intensity is greater than its V (visual) intensity. Star (b) has roughly equal B and V readings and so appears white. Its temperature is about 10,000 K. Star (c) is red; its V intensity greatly exceeds the B value, and its temperature is 3000 K.

- Hydrogen lines are strongest in stars having intermediate surface temperatures of around 10,000 K. This temperature is just right for electrons to move frequently between hydrogen's second and higher orbitals, producing the characteristic visible hydrogen spectrum. ∞ *(Sec. 2.6)*
- Hydrogen lines are again weak in stars with surface temperatures below about 4000 K because the temperature is too low to boost many electrons out of the ground state. ∞ *(Sec. 2.6)* The most intense spectral lines in these stars are due to weakly excited heavy atoms and from some molecules, as noted at the bottom of Figure 10.9.

Stellar spectra are the source of all the detailed information we have on stellar composition, and they do in fact reveal significant differences in composition among stars, particularly in the abundances of carbon, nitrogen, oxygen, and heavier elements. However, as we have just seen, these differences are *not* the primary reason for the different spectra that are observed. Instead, the main determinant of a star's spectral appearance is its temperature, and stellar spectroscopy is a powerful and precise tool for measuring this important stellar property.

Spectral Classification

Astronomers obtained stellar spectra like those shown in Figure 10.9 for numerous stars around the start of the 20th century, as observatories around the world amassed stellar data from both hemispheres of the sky. Between 1880 and 1920, researchers correctly identified many of the observed spectral lines by comparing them with lines obtained in the laboratory. However, modern atomic theory had not yet been developed, so the correct interpretation of the line strengths, as just described, was impossible at the time. Lacking full understanding of how atoms produce spectra, astronomers organized their data by classifying stars according to their most prominent spectral features—especially their hydrogen-line intensities. They adopted an alphabetic scheme in which A stars, with the strongest hydrogen lines, were thought to have more hydrogen than did B stars, and so on. The classification extended as far as the letter P.

In the 1920s scientists began to understand the intricacies of atomic structure and the causes of spectral lines. Astronomers realized that stars could be more meaningfully classified according to surface temperature. But instead of adopting an entirely new scheme, they chose to shuffle the existing alphabetical categories—based on the strength of hydrogen lines—into a new sequence based on temperature. In order of decreasing temperature, the letters now run O, B, A, F, G, K, M. (The other letter classes have been dropped.) These stellar designations are called

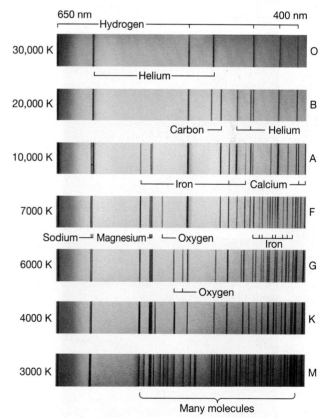

▲ FIGURE 10.9 Stellar Spectra Comparison of spectra observed for seven different stars having a range of surface temperatures. The spectra of the hottest stars, at the top, show lines of helium and multiply ionized heavy elements. In the coolest stars, at the bottom, helium lines are absent, but lines of neutral atoms and molecules are plentiful. At intermediate temperatures, hydrogen lines are strongest.

TABLE 10.2 Spectral Classes

Spectral Class	Temperature (K)	Prominent Absorption Lines	Familiar Examples
O	30,000	Ionized helium strong; multiply ionized heavy elements; hydrogen faint	Mintaka (O9)
B	20,000	Neutral helium moderate; singly ionized heavy elements; hydrogen moderate	Rigel (B8)
A	10,000	Neutral helium very faint; singly ionized heavy elements; hydrogen strong	Vega (A0), Sirius (A1)
F	7000	Singly ionized heavy elements; neutral metals; hydrogen moderate	Canopus (F0)
G	6000	Singly ionized heavy elements; neutral metals; hydrogen relatively faint	Sun (G2), Alpha Centauri (G2)
K	4000	Singly ionized heavy elements; neutral metals strong; hydrogen faint	Arcturus (K2), Aldebaran (K5)
M	3000	Neutral atoms strong; molecules moderate; hydrogen very faint	Betelgeuse (M2)
			Barnard's Star (M5)

PROCESS OF ▶ SCIENCE CHECK The spectral classification scheme developed by astronomers around the start of the 20th century was based on assumptions now known to be incorrect. Why then did it prove to be so useful?

CONCEPT CHECK ▶ Why does a star's spectrum depend on its temperature?

(a)

R I V U X G

(b)

Size of Jupiter's orbit

R I V U X G

spectral classes (or *spectral types*). Use the mnemonic "**O**h, **B**e **A** **F**ine **G**uy/**G**irl, **K**iss **M**e" to remember them in the correct order.[1]

Astronomers further divide each lettered spectral class into 10 subdivisions, numbered 0 to 9. The lower the number, the hotter the star. For example, our Sun is classified as a G2 star (a little cooler than G1 and a little hotter than G3), Vega is A0, Barnard's star is M5, Betelgeuse is M2, and so on. Table 10.2 lists the main properties of each stellar spectral class for the stars presented in Table 10.1.

We should not underestimate the importance of the early work in classifying stellar spectra. Even though the original classification was based on erroneous assumptions, the painstaking accumulation of large quantities of accurate and well-categorized data paved the way for rapid improvements in understanding once a theory came along that explained the observations.

10.4 Stellar Sizes

Direct and Indirect Measurements

Virtually all stars are unresolvable points of light in the sky, even when viewed through the largest telescopes. Still, a few are big enough, bright enough, and close enough to allow us to measure their sizes directly. In some cases, the results are even detailed enough to allow us to distinguish a few surface features (Figure 10.10). By measuring a star's angular size and knowing its distance from the Sun, astronomers can determine its radius using simple geometry. ∞ *(More Precisely 4-1)* The sizes of a few dozen stars have been measured in this way.

Most stars are too distant or too small for such direct measurements to be possible. Their sizes must be inferred by more indirect means, using the radiation laws. ∞ *(Sec. 2.4)* According to the Stefan-Boltzmann law, the rate at which a star emits energy into space—the star's luminosity—is proportional to the fourth power of

[1]Astronomers have since added two new spectral classes—L and T—for low-mass, low-temperature stars whose odd spectra distinguish them from the M-class stars in the current scheme. These new objects are not "true" stars, fusing hydrogen into helium in their cores—rather, they are *brown dwarfs* (see Section 11.4) that never achieved high enough central temperatures for fusion to begin.

◀ **FIGURE 10.10 Betelgeuse** (a) The swollen star Betelgeuse (shown here in false color) is close enough to resolve its size directly, along with some surface features thought to be storms similar to those that occur on the Sun. Betelgeuse is such a huge star (roughly 600 times the size of the Sun) that its photosphere is comparable to Jupiter's orbit. (b) An ultraviolet view of Betelgeuse, in false color, as seen by a European camera onboard the *Hubble Space Telescope*, shows more clearly just how large this huge star is. *(NOAO; NASA/ESA)*

the star's surface temperature. ⚏ *(More Precisely 2-2)* However, the luminosity also depends on the star's surface area, because large bodies radiate more energy than do small bodies having the same surface temperature. Because the surface area of a star is proportional to the square of its radius, we can say

$$\text{luminosity} \propto \text{radius}^2 \times \text{temperature}^4.$$

This **radius–luminosity–temperature relationship** is important because it demonstrates that knowledge of a star's luminosity and temperature can yield an estimate of its radius—an *indirect* determination of stellar size.

Giants and Dwarfs

Let's consider some examples to clarify these ideas. *More Precisely 10-2* (p. 286) presents a little more detail on the calculations involved. The star known as Aldebaran (the orange-red "eye of the bull" in the constellation Taurus) has a surface temperature of about 4000 K and a luminosity of 1.3×10^{29} W. Thus, its surface temperature is 0.7 times, and its luminosity about 330 times, the corresponding quantities for our Sun. The radius–luminosity–temperature relationship *(More Precisely 10-2)* then implies that the star's radius is almost 40 times the solar value. If our Sun were that large, its photosphere would extend halfway to the orbit of Mercury and, seen from Earth, would cover more than 20° on the sky.

A star as large as Aldebaran is known as a *giant*. More precisely, **giants** are stars having radii between 10 and 100 times that of the Sun. Since any 4000-K object is reddish in color, Aldebaran is known as a **red giant.** Even larger stars, ranging up to 1000 solar radii in size, are known as **supergiants.** Betelgeuse is a prime example of a **red supergiant.**

Now consider Sirius B—a faint companion to Sirius A, the brightest star in the night sky. Sirius B's surface temperature is roughly 24,000 K, four times that of the Sun. Its luminosity is 10^{25} W, about 0.025 times the solar value. Substituting these quantities into our radius–luminosity–temperature relationship, we obtain a radius of 0.01 times the solar radius—roughly the size of Earth. Sirius B is much hotter but smaller and fainter than our Sun. Such a star is known as a *dwarf*. In astronomical parlance, the term **dwarf** refers to any star of radius comparable to or smaller than the radius of the Sun (including the Sun itself). Because any 24,000 K object glows bluish-white, Sirius B is an example of a **white dwarf.**

Stellar radii range from less than 0.01 to more than 100 times the radius of the Sun. Figure 10.11 illustrates the sizes of a few well-known stars.

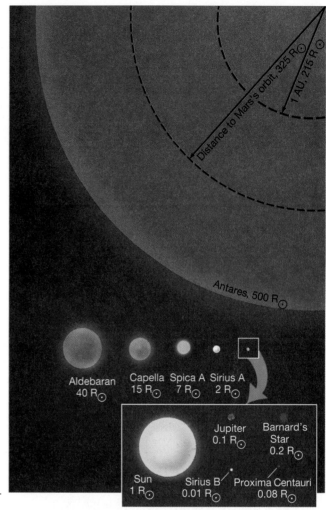

▲ **FIGURE 10.11 Stellar Sizes** Shown here are the greatly varying sizes of several well-known stars. Only part of the red-giant star Antares can be shown on this scale, and the supergiant Betelgeuse would fill the entire page. (Here and in other figures, the symbol "⊙" stands for the Sun, so the symbol "R⊙" here means "solar radius.")

CONCEPT CHECK ▶ Can we measure the radius of a star without knowing its distance?

10.5 The Hertzsprung–Russell Diagram

Figure 10.12 plots luminosity versus temperature for a few well-known stars. A figure of this sort is called a *Hertzsprung–Russell diagram,* or **H–R diagram,** after Danish astronomer Ejnar Hertzsprung and U.S. astronomer Henry Norris Russell, who independently pioneered the use of such plots in the 1920s. The vertical luminosity scale, expressed in units of solar luminosity (3.9×10^{26} W), extends over a large range, from 10^{-4} to 10^4. The Sun appears right in the middle of the luminosity range, at a luminosity of 1. Surface temperature is plotted on the horizontal axis, although in the unconventional sense of temperature increasing to the *left*, so that the spectral sequence O, B, A, . . . reads left to right. (Consult also Overlay 1 of the acetate insert near the middle of this book to see how sizes and colors of the data points can be used to represent stellar radii and temperatures, respectively.)

SELF-GUIDED TUTORIAL Hertzsprung–Russell Diagram

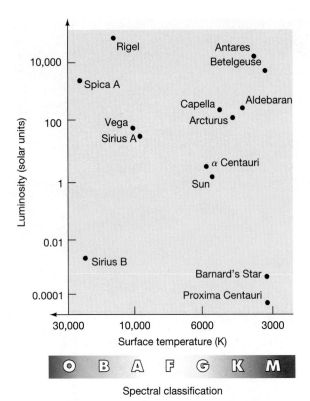

A plot of luminosity against surface temperature (or spectral class) is a useful way to compare stars. Plotted here are the data for some stars mentioned earlier in the text, including the Sun, which has a luminosity of 1 solar unit and a temperature of 5800 K—a G-type star. The B-type star Rigel, at top left, has a temperature of about 11,000 K and a luminosity more than 10,000 times that of the Sun. The M-type star Proxima Centauri, at bottom right, has a temperature of about 3000 K and a luminosity less than 1/10,000 that of the Sun. (See also Overlay 1 of the acetate insert.)

The Main Sequence

The few stars plotted in Figure 10.12 give little indication of any particular connection between stellar properties. However, as Hertzsprung and Russell plotted more and more stellar temperatures and luminosities, they found that a relationship does in fact exist. Stars are *not* uniformly scattered across the H–R diagram. Instead, most are confined to a fairly well-defined band stretching diagonally from top left (high-temperature, high-luminosity) to bottom right (low-temperature, low-luminosity). In other words, cool stars tend to be faint (less luminous) and hot stars tend to be bright (more luminous). This band of stars spanning the H–R diagram is known as the **main sequence.** Figure 10.13 is an H–R diagram for stars lying within 5 pc of the Sun. Note that most stars in the solar neighborhood lie on the main sequence.

A useful analogy involves human beings. Our height and weight are also correlated along a "main sequence." That is, as shown near Figure 10.13, the majority of the human population falls within a diagonal area (bounded by the shaded area). This is true since tall people usually weigh more than shorter ones. Some exceptions—including dwarfs and basketball players—lie outside this well-defined range.

The surface temperatures of main-sequence stars range from about 3000 K (spectral class M) to more than 30,000 K (spectral class O). This temperature range is relatively small—only a factor of 10. In contrast, the observed range in luminosities is very large, covering eight orders of magnitude (that is, a factor of 100 million), from 10^{-4} to 10^4 the luminosity of the Sun.

Using the radius–luminosity–temperature relationship (Section 10.4), astronomers find that stellar radii also vary along the main sequence. The faint, red M-type stars in the bottom right of the H–R diagram are only about 1/10 the size of the Sun, whereas the bright, blue O-type stars in the upper left are about 10 times larger than the Sun. The dashed lines in Figure 10.13 represent constant stellar radius, meaning that any star lying on a given line has the same radius, regardless of its luminosity or temperature. Along a constant-radius line, the radius–luminosity–temperature relationship implies

$$\text{luminosity} \propto \text{temperature}^4.$$

By including such lines on our H–R diagrams, we can indicate stellar temperatures, luminosities, and radii on a single plot.

At the top end of the main sequence, the stars are large, hot, and very luminous. Because of their size and color, they are referred to as **blue giants.** The very largest are called **blue supergiants.** At the other end, stars are small, cool, and faint. They are known as **red dwarfs.** Our Sun lies right in the middle.

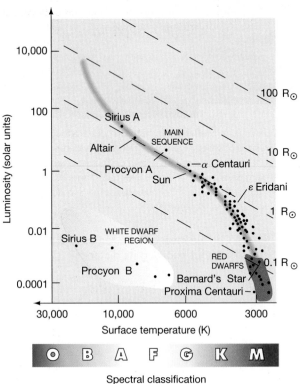

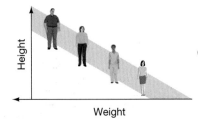

ANALOGY: Analogous to an H–R diagram, people's height and weight fall along a "main sequence."

INTERACTIVE ◀ FIGURE 10.13 H–R Diagram of Nearby Stars Most stars have properties within the long, thin shaded region of the H–R diagram known as the main sequence. The points plotted here are for stars lying within about 5 pc of the Sun. Each dashed diagonal line corresponds to a constant stellar radius. (Recall that the symbol "$R_\odot$" means "solar radius.")

Figure 10.14 shows an H–R diagram for a different group of stars—the 100 stars of known distance having the greatest apparent brightness, as seen from Earth. Notice that here there are many more stars at the upper end of the main sequence than at the lower end. The reason for this excess of blue giants is simple—we can see very luminous stars a long way off. The stars shown in Figure 10.14 are scattered through a much greater volume of space than those shown in Figure 10.13, and the Figure 10.14 sample is heavily biased toward the most luminous objects. In fact, of the 20 brightest stars in the sky, only 6 lie within 10 pc of us. The rest are visible, despite their great distances, because of their high luminosities.

If very luminous blue giants are overrepresented in Figure 10.14, low-luminosity red dwarfs are surely underrepresented. In fact, no dwarfs appear on this diagram. This absence is not surprising, because low-luminosity stars are difficult to observe from Earth. In the 1970s, astronomers began to realize that they had greatly underestimated the number of red dwarfs in our Galaxy. As hinted at by the H–R diagram in Figure 10.13, which shows an unbiased sample of stars in the solar neighborhood, red dwarfs are actually the most common type of star in the sky. They probably account for upward of 80 percent of all stars in the universe. In contrast, O- and B-type supergiants are extremely rare—only about one star in 10,000 falls into this category.

The White-Dwarf and Red-Giant Regions

Some of the points plotted in Figures 10.12–10.14 clearly do not lie on the main sequence. One such point in Figure 10.12 represents Sirius B, the white dwarf we met in Section 10.4 (see *More Precisely 10-2*). Its surface temperature (24,000 K) is about four times greater than that of the Sun, but its luminosity is hundreds of times less than the solar value. A few more such faint bluish-white stars can be seen in the bottom left-hand corner of Figure 10.13, in the **white-dwarf region.**

Also shown in Figure 10.12 is Aldebaran, with a surface temperature (4000 K) about two-thirds that of the Sun, but with a luminosity more than 300 times greater than the Sun's. Another point represents Betelgeuse, the ninth brightest star in the sky, a little cooler than Aldebaran, but nearly 100 times more luminous. The upper right-hand corner of the H–R diagram (marked on Figure 10.14), where these stars are found, is called the **red-giant region.** No red giants are found within 5 pc of the Sun (Figure 10.13), but many of the brightest stars seen in the sky are in fact red giants (Figure 10.14). Red giants are relatively rare, but they are so bright that they are visible at very great distances. They form a third distinct class of stars in the H–R diagram, very different in their properties from both main-sequence stars and white dwarfs.

The *Hipparcos* mission (Section 10.1), in addition to determining hundreds of thousands of stellar parallaxes with unprecedented accuracy, also measured the colors and luminosities of more than 2 million stars. Figure 10.15 shows an H–R diagram based on a tiny portion of the enormous *Hipparcos* dataset. The main sequence and red giant regions are clearly evident. Few white dwarfs appear, however, because the telescope was limited to observations of relatively bright objects—brighter than apparent magnitude 12. Almost no white dwarfs lie close enough to Earth that their magnitudes fall below this limit.

About 90 percent of all stars in our solar neighborhood, and presumably a similar percentage elsewhere in the universe, are main-sequence stars. About 9 percent of stars are white dwarfs, and 1 percent are red giants.

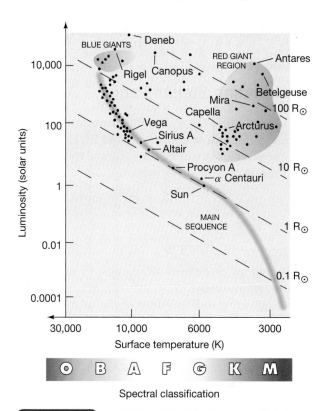

INTERACTIVE ▲ **FIGURE 10.14 H–R Diagram of Brightest Stars** An H–R diagram for the 100 brightest stars in the sky is biased in favor of the most luminous stars—which appear toward the upper left—because we can see them more easily than we can the faintest stars. (Compare with Figure 10.13, which shows only the closest stars.) **(MA)**

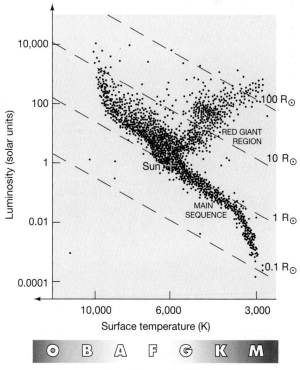

▶ **FIGURE 10.15** *Hipparcos* **H–R Diagram** This simplified version of the most complete H–R diagram ever compiled represents more than 20,000 data points, as measured by the European *Hipparcos* spacecraft for stars within a few hundred parsecs of the Sun.

10-2 MORE PRECISELY

Estimating Stellar Radii

We can combine the Stefan-Boltzmann law $F = \sigma T^4$ with the formula for the area of a sphere, $A = 4\pi R^2$, to obtain the relationship between radius (R), luminosity (L), and temperature (T) described in the text ∞ *(More Precisely 2-2)*:

$$L = FA = 4\pi\sigma R^2 T^4$$

or

$$\text{luminosity} \propto \text{radius}^2 \times \text{temperature}^4.$$

If we adopt convenient solar units, in which L is measured in solar luminosities (3.9×10^{26} W), R is measured in solar radii (696,000 km), and T in units of the solar temperature (5800 K), we can remove the constant $4\pi\sigma$ and write this equation as

L(in solar luminosities)

$$= R^2 \text{ (in solar radii)} \times T^4 \text{ (in units of 5800 } K\text{)}.$$

As illustrated in the accompanying figure, both the radius and the temperature are important in determining the star's luminosity.

To compute the radius of a star from its luminosity and temperature, we rearrange this equation to read (in the same units)

$$R = \frac{\sqrt{L}}{T^2}.$$

This simple application of the radiation laws is the basis for almost every estimate of stellar size made in this book.

Let's take as examples the two stars discussed in the text. The red giant star Aldebaran has luminosity $L = 330$ units and temperature $T = 4000/5800 = 0.69$ unit. According to the foregoing equation for R, its radius is therefore $\sqrt{330}/0.69^2 = 18/0.48 = 39$ times the radius of the Sun. At the opposite extreme, the white dwarf Sirius B has $L = 0.025$ and $T = 4.7$, so its radius is $\sqrt{0.025}/4.7^2 = 0.16/22 \approx 0.007$ solar radii.

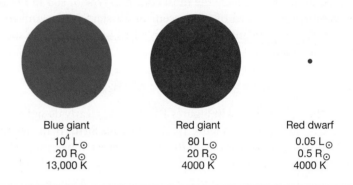

Blue giant	Red giant	Red dwarf
$10^4 \, L_\odot$	$80 \, L_\odot$	$0.05 \, L_\odot$
$20 \, R_\odot$	$20 \, R_\odot$	$0.5 \, R_\odot$
13,000 K	4000 K	4000 K

CONCEPT CHECK ▶ Only a tiny fraction of all stars are giants. Why, then, do giants account for so many of the stars visible to the naked eye in the night sky?

ANIMATION/VIDEO White Dwarfs in Globular Cluster **(MA)**

10.6 Extending the Cosmic Distance Scale

We have already discussed the connections between luminosity, apparent brightness, and distance. Knowledge of a star's apparent brightness and its distance allows us to determine its luminosity using the inverse-square law. But we can also turn the problem around. If we somehow knew a star's luminosity and then measured its apparent brightness, the inverse-square law would tell us its distance from the Sun.

Spectroscopic Parallax

Most of us have a rough idea of the approximate intrinsic brightness (that is, the luminosity) of a typical traffic signal. Suppose you are driving down an unfamiliar street and see a red traffic light in the distance. Your knowledge of the light's luminosity enables you immediately to make a mental estimate of its distance. A light that appears relatively dim (low apparent brightness) must be quite distant (assuming it's not just dirty). A bright one must be relatively close. Thus, *measurement of the apparent brightness of a light source, combined with some knowledge of its luminosity, can yield an estimate of its distance.*

For a star the trick is to find an independent measure of luminosity without knowing the distance. The H–R diagram can provide just that. For example, suppose we observe a star and determine its apparent magnitude to be 10. By itself, that doesn't tell us much—the star could either be faint and close, or bright and distant (Figure 10.5). But suppose we have some additional information: The star lies on the main sequence and has spectral type A0. Then we can read the star's luminosity off a graph such as Figure 10.13 or Figure 10.14. A main-sequence A0 star has a luminosity of approximately 100 solar units. According to *More Precisely 10-1*, this corresponds to an absolute magnitude of 0 and hence to a distance of 1000 pc.

This process of using stellar spectra to infer distances is called **spectroscopic parallax**.[2] The key steps are as follows:

1. we measure the star's apparent brightness and spectral type *without* knowing how far away it is;
2. we use the spectral type to estimate the star's luminosity, assuming that it lies on the main sequence; and
3. finally, we apply the inverse-square law to determine the distance to the star.

The main sequence allows us to make a connection between an easily measured quantity (spectral type) and the star's luminosity, which would otherwise be unknown. As we will see in upcoming chapters, this essential logic (with a variety of different techniques replacing step 2) is used again and again as a means of distance measurement in astronomy. In practice, the "fuzziness" of the main sequence translates into a small (10–20 percent) uncertainty in the distance, but the basic idea remains valid.

In Chapter 1 we introduced the first rung on a ladder of distance-measurement techniques that will ultimately carry us to the edge of the observable universe. That rung is radar ranging on the inner planets. ∞ *(Sec.1.3)* It establishes the scale of the solar system and defines the astronomical unit. At the beginning of this chapter we discussed a second rung on the cosmic distance ladder—stellar parallax—which is based on the first, since Earth's orbit is the baseline. Now, having used the first two rungs to determine the distances and other physical properties of many nearby stars, we use that knowledge to construct the third rung—spectroscopic parallax. Figure 10.16 illustrates schematically how this third rung builds on the first two and expands our cosmic field of view still deeper into space.

Spectroscopic parallax can be used to determine stellar distances out to several thousand parsecs. Beyond that, spectra and colors of individual stars are difficult to obtain. Note that, in using this method, we are assuming (without proof) that distant stars are basically similar to those nearby—in particular, that main-sequence stars *fall on the same main sequence*. Only by making this assumption can we use spectroscopic parallax to expand the boundaries of our distance-measurement techniques.

Each rung in the distance ladder is calibrated using data from the lower rungs, so changes made at any level will affect measurements made on *all* larger scales. As a result, the impact of the enormous dataset of high-quality observations returned by the *Hipparcos* mission (Section 10.1) extends far beyond the volume of space actually surveyed by the satellite. By recalibrating the local foundations of the cosmic distance scale, *Hipparcos* has caused astronomers to revise their estimates of distances on *all* scales—up to and including the scale of the universe itself. All distances quoted throughout this text reflect updated values based on *Hipparcos* data.

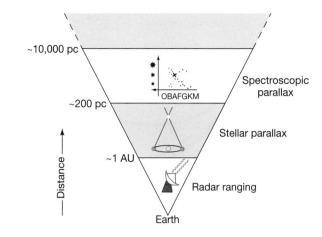

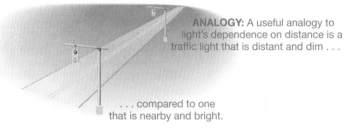

ANALOGY: A useful analogy to light's dependence on distance is a traffic light that is distant and dim . . .

. . . compared to one that is nearby and bright.

INTERACTIVE ▲ **FIGURE 10.16 Stellar Distances** Knowledge of a star's luminosity and apparent brightness can yield an estimate of its distance. Astronomers use this third rung on the distance ladder, called spectroscopic parallax, to measure distances as far out as individual stars can be clearly discerned—several thousand parsecs.

CONCEPT CHECK ▶ Suppose astronomers discover that due to a calibration error all distances measured by geometric parallax are 10 percent larger than currently thought. What effect would this have on the "standard" main sequence used in spectroscopic parallax?

[2]Despite its name, the method has nothing in common with stellar (geometric) parallax other than its use as a means of determining stellar distances.

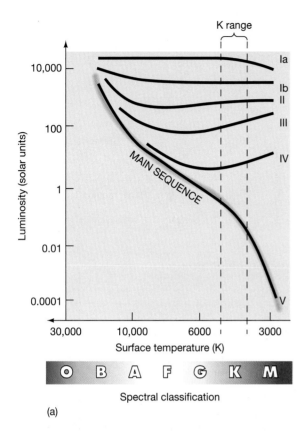

(a)

(a) Approximate locations of the standard stellar luminosity classes in the H–R diagram. The widths of absorption lines also provide information on the density of a star's atmosphere. The denser atmosphere of a main-sequence K-type star has broader lines (c) than a giant star of the same spectral class (b).

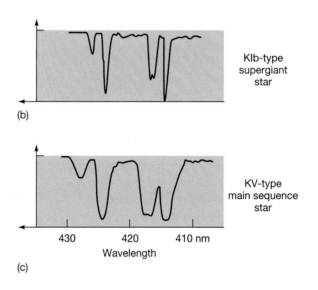

TABLE 10.3   Stellar Luminosity Classes

Class	Description
Ia	bright supergiants
Ib	supergiants
II	bright giants
III	giants
IV	subgiants
V	main-sequence stars/dwarfs

Luminosity Class

What if the star in question happens to be a red giant or a white dwarf and does not lie on the main sequence? As mentioned in Chapter 2, detailed analysis of spectral *line widths* can provide information on the pressure, and hence the density, of the gas where the line formed. ∞ *(Sec. 2.6)* The atmosphere of a red giant is much less dense than that of a main-sequence star, and this in turn is much less dense than the atmosphere of a white dwarf. Figures 10.17(a) and (b) illustrate the difference between the spectra of a main-sequence star and a red giant of the same spectral type.

Over the years, astronomers have developed a system for classifying stars according to the widths of their spectral lines. Because line width depends on pressure in the stellar photosphere, and because this pressure in turn is well correlated with luminosity, this stellar property is known as **luminosity class.** The standard luminosity classes are listed in Table 10.3 and shown on the H–R diagram in Figure 10.17(c). By determining a star's luminosity class, astronomers can usually tell with a high degree of confidence what sort of object it is and the full specification of a star's spectral properties includes its luminosity class. For example, the Sun, a G2 main-sequence star, is of class G2V, the B8 blue supergiant Rigel is B8Ia, the red dwarf Barnard's star is M5V, the red supergiant Betelgeuse is M2Ia, and so on.

Consider, for example, a K2-type star (Table 10.4) with a surface temperature of approximately 4500 K. If the widths of the star's spectral lines tell us that it lies

PROCESS OF ►
SCIENCE CHECK What assumptions are we making when we apply spectroscopic parallax to measure the distance to a star?

TABLE 10.4   Variation in Stellar Properties within a Spectral Class

Surface Temperature (K)	Luminosity (solar luminosities)	Radius (solar radii)	Object	Example
4900	0.3	0.8	K2V main-sequence star	ε Eridani
4500	110	21	K2III red giant	Arcturus
4300	4000	140	K2Ib red supergiant	ε Pegasi

on the main sequence (i.e., it is a K2V star), then its luminosity is about 0.3 times the solar value. If the star's spectral lines are observed to be narrower than lines normally found in main-sequence stars, the star may be recognized as a K2III giant, with a luminosity 100 times that of the Sun (Figure 10.17c). If the lines are very narrow, the star might instead be classified as a K2Ib supergiant, brighter by a further factor of 40, at 4000 solar luminosities. In each case, knowledge of luminosity classes allows astronomers to identify the object and make an appropriate estimate of its luminosity and hence its distance.

10.7 Stellar Masses

As with all other astronomical objects, we measure a star's mass by observing its gravitational influence on some nearby body—another star, perhaps, or a planet. If we know the distance between the two bodies, we can use Newton's laws to calculate their masses. ∞ (Sec. 1.4)

Binary Stars

Most stars are members of *multiple-star systems*—groups of two or more stars in orbit around one another. The majority form **binary-star systems** (or simply *binaries*), which consist of two stars in orbit about their common center of mass, held together by their mutual gravitational attraction. (The Sun is not part of a multiple star system; if it has anything at all uncommon about it, it is this lack of stellar companions.)

Astronomers classify binaries according to their appearance from Earth and the ease with which they can be observed. **Visual binaries** have widely separated components bright enough to be observed and monitored separately (Figure 10.18a). The more common **spectroscopic binaries** are too distant from us to be resolved into two distinct stars, but they can be indirectly perceived by monitoring the back-and-forth Doppler shifts of their spectral lines as the stars orbit one another and their line-of-sight velocities vary periodically. ∞ (Sec. 2.7)

In a *double-line* spectroscopic binary, two distinct sets of spectral lines—one for each component star—shift back and forth as the stars move. The periodic shifts of the lines tell us that the objects emitting them are in orbit. In the more common *single-line* systems (Figure 10.18b), one star is too faint for its spectrum to be distinguished, so we see only one set of lines shifting back and forth. This shifting means that the detected star must be in orbit around another

MA SELF-GUIDED TUTORIAL Binary Stars–Radial Velocity Curves

MA SELF-GUIDED TUTORIAL Eclipsing Binary Stars–Light Curves

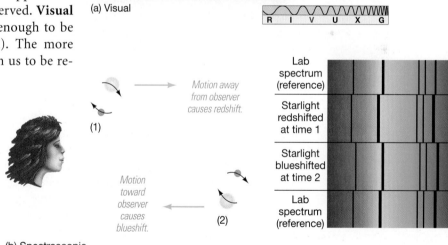

(a) Visual

These are real photos of the binary star Kurger 60.

R I V U X G

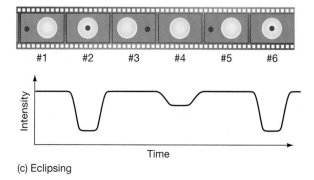

(1) *Motion away from observer causes redshift.*

Motion toward observer causes blueshift. (2)

(b) Spectroscopic

Lab spectrum (reference)

Starlight redshifted at time 1

Starlight blueshifted at time 2

Lab spectrum (reference)

#1 #2 #3 #4 #5 #6

Intensity

Time

(c) Eclipsing

INTERACTIVE ► **FIGURE 10.18 Binary Stars** (a) The period and separation of a double-star system can be observed directly if each star is clearly seen. (b) Binary properties can be determined indirectly by measuring the periodic Doppler shift of one star relative to the other while moving in their orbits. This is a single-line system, in which only one spectrum (from the brighter component) is visible. (c) If two stars eclipse one another, additional information can be obtained by observing the periodic decrease in starlight as one star passes in front of the other. *(Harvard College Observatory)* **MA**

INTERACTIVE Exploring the Light Curve of an Eclipsing Binary Star System

star, even though the companion cannot be observed directly. (If this idea sounds familiar, it should—all the extrasolar planetary systems discovered to date are extreme examples of single-line spectroscopic binaries.) ∞ *(Sec. 4.4)*

In the much rarer **eclipsing binaries,** the orbital plane of the pair of stars is almost edge-on to our line of sight. In this situation, we observe a periodic decrease of starlight intensity as one member of the binary passes in front of the other (Figure 10.18c). By studying the variation of the light from an eclipsing binary system—called the binary's **light curve**—astronomers can derive detailed information not only about the stars' orbits and masses, but also about their radii.

Note that these categories are not mutually exclusive. For example, an eclipsing binary may also be (and in fact often is) a spectroscopic binary system.

Mass Determination

The period of a binary can be measured by observing the orbits of the component stars, the back-and-forth motion of the stellar spectral lines, or the dips in the light curve—whatever information is available. Observed binary periods span a broad range—from hours to centuries. How much additional information can be extracted depends on the type of binary involved.

If the distance to a *visual binary* is known, then the orbits of each component can be individually tracked and the masses of the components can be determined. ∞ *(Sec.1.4)*

For *spectroscopic binaries,* Doppler-shift measurements give us information only on the radial velocities of the component stars, and this limits the information we can obtain. Simply put, we cannot distinguish between a slow-moving binary seen edge-on and a fast-moving binary seen almost face-on (so that only a small component of the orbital motion is along the line of sight). For a double-line system, only lower limits on the individual masses can be obtained. For single-line systems, even less information is available, and only a rather complicated relation between the component masses (known as the mass function) can be derived.

If a spectroscopic binary happens also to be an eclipsing system, then the uncertainty in the orbital inclination is removed, as the binary is known to be edge-on (or very nearly so). In that case, both masses can be determined for a double-line binary. For a single-line system, the mass function is simplified to the point where the mass of the unseen component is known if the mass of the brighter component can be obtained by other means (for example, if it is recognized as a main-sequence star of a certain spectral class—see Figure 10.19).

Despite all these qualifications, individual component masses have been obtained for many nearby binary systems. A simple example is presented in *More Precisely 10-3.* Virtually all we know about the masses of stars is based on such observations.

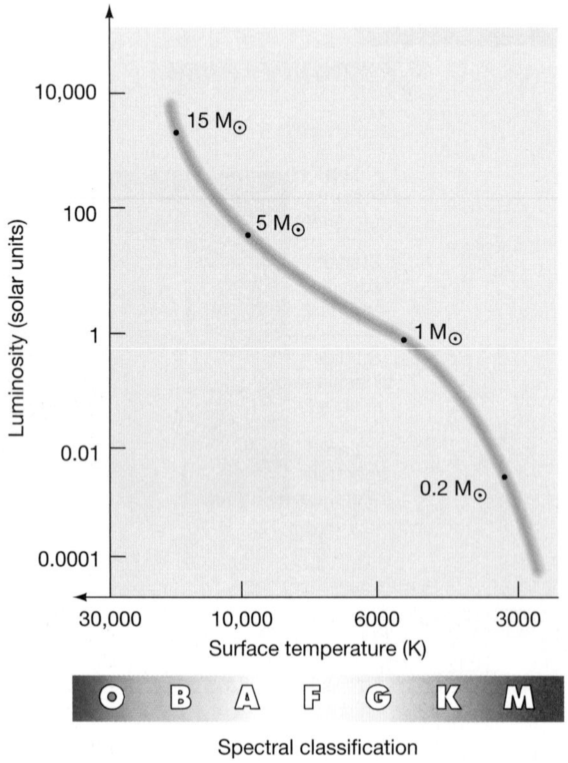

Mass and Other Stellar Properties

Table 10.5 summarizes the various observational and theoretical techniques used to measure the stars, listing the quantities that are assumed known (usually as a result of the application of other techniques listed in the table), those that are measured, and the theory that is applied to turn the observations into the desired result. We end our introduction to the stars with a brief look at how mass is correlated with the other stellar properties discussed in this chapter.

Figure 10.19 is a schematic H–R diagram showing how stellar mass varies along the main sequence. There is a clear progression from low-mass red dwarfs

▲ **FIGURE 10.19 Stellar Masse** More than any other stellar property, mass determines a star's position on the main sequence. Low-mass stars are cool and faint; they lie at the bottom of the main sequence. Very massive stars are hot and bright; they lie at the top of the main sequence. (The symbol "$M_\odot$" means "solar mass.")

10-3 MORE PRECISELY

Measuring Stellar Masses in Binary Stars

As discussed in the text, most stars are members of binary systems—where two stars orbit one another, bound together by gravity. Here we describe how—in an idealized case where the relevant orbital parameters are known—we can use the observed orbital data, together with our knowledge of basic physics, to determine the masses of the component stars.

Consider the nearby visual binary system made up of the bright star Sirius A and its faint companion Sirius B, sketched in the accompanying figure. The binary's orbital period can be measured simply by watching the stars orbit one another, or alternatively by following the back-and-forth velocity wobbles of Sirius A due to its faint companion. It is almost exactly 50 years. The orbital semimajor axis can also be obtained by direct observation of the orbit, although in this case we must use some additional knowledge of Kepler's laws to correct for the binary's 46° inclination to the

line of sight. ∞ *(Sec. 1.3)* It is 20 AU—an angular size of 7.5″ at a distance of 2.7 pc. ∞ *(Sec. 1.3)* Once we know these two key orbital parameters, we can use the modified version of Kepler's third law to calculate the sum of the masses of the two stars. The result is $20^3/50^2 = 3.2$ times the mass of the Sun. ∞ *(Sec. 1.4)*

Further study of the orbit allows us to determine the individual stellar masses. Doppler observations show that Sirius A moves at approximately half the speed of its companion relative to their center of mass. ∞ *(Secs. 1.4, 2.7)* This implies that Sirius A must have twice the mass of Sirius B. It then follows that the masses of Sirius A and Sirius B are 2.1 and 1.1 solar masses, respectively.

Often the calculation of the masses of binary components is complicated by the fact that only partial information is available—we might only be able to see one star, or perhaps only spectroscopic velocity information is available (see Section 10.7). Nevertheless, this technique of combining elementary physical principles with detailed observations is how virtually every stellar mass quoted in this text has been determined.

to high-mass blue giants. With few exceptions, main-sequence stars range in mass from about 0.1 to 20 times the mass of the Sun. The hot O- and B-type stars are generally about 10 to 20 times more massive than our Sun. The coolest K- and M-type stars contain only a few tenths of a solar mass. The mass of a star at the time of its formation determines its location on the main sequence. Based on observations of stars within a few hundred light-years of the Sun, Figure 10.20

TABLE 10.5 Measuring the Stars

Stellar Property	Measurement Technique	"Known" Quantity	Measured Quantity	Theory Applied	Section
Distance	stellar parallax spectroscopic parallax	astronomical unit main sequence	parallactic angle spectral type apparent magnitude	elementary geometry inverse-square law	10.1 10.6
Radial velocity		speed of light atomic spectra	spectral lines	Doppler effect	10.1
Transverse velocity	astrometry	distance	proper motion	elementary geometry	10.1
Luminosity		distance main sequence	apparent magnitude spectral type color	inverse-square law	10.2 10.6
Temperature	photometry spectroscopy		spectral type	blackbody law atomic physics	10.3 10.3
Radius	direct indirect	distance	angular size luminosity temperature	elementary geometry radius–luminosity–temperature relationship	10.4 10.4
Composition	spectroscopy		spectrum	atomic physics	10.3
Mass	observations of binary stars	distance	binary period binary orbit orbital velocity	Newtonian gravity and dynamics	10.7

Star	Spectral Type/ Luminosity Class[1]	Mass (solar masses)	Central Temperature (10^6 K)	Luminosity (solar luminosities)	Estimated Lifetime (millions of years)
Spica B*	B2V	6.8	25	800	90
Vega	A0V	2.6	21	50	500
Sirius	A1V	2.1	20	22	1000
Alpha Centauri	G2V	1.1	17	1.6	7000
Sun	G2V	1.0	15	1.0	10,000
Proxima Centauri	M5V	0.1	0.6	0.00006	16,000,000

TABLE 10.6 Key Properties of Some Well-Known Main-Sequence Stars

*The "star" Spica is, in fact, a binary system comprising a B1III giant primary (Spica A) and a B2V main-sequence secondary (Spica B).

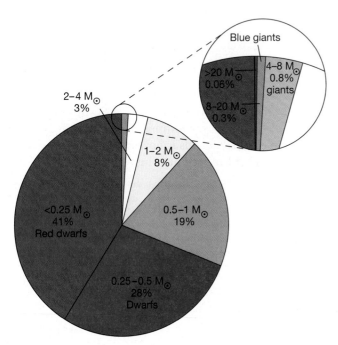

▲ FIGURE 10.20 **Stellar Mass Distribution**   The distribution of masses of main-sequence stars, as determined from careful measurement of stars in the solar neighborhood.

illustrates how the masses of main-sequence stars are distributed. Notice the huge fraction of low-mass stars, as well as the tiny fraction contributed by stars of more than a few solar masses.

Figure 10.21 illustrates how the radii and luminosities of main-sequence stars depend on mass. The two plots are called the *mass–radius* (part a) and *mass–luminosity* (part b) relationships. Along the main sequence, both radius and luminosity increase with mass. The radius rises slightly more slowly than the mass, whereas luminosity increases much faster—more like the *fourth power* of the mass (indicated by the straight line in Figure 10.21b). For example, a 2-solar mass main-sequence star has a radius roughly twice that of the Sun and a luminosity of 16 (2^4) solar luminosities. A 0.2-solar mass main-sequence star has a radius of about 0.2 solar radii and a luminosity of 0.0016 (0.2^4) solar luminosities.

Table 10.6 compares key properties of some well-known main-sequence stars, arranged in order of decreasing mass. Notice that the central temperature (obtained from mathematical models similar to those discussed in Chapter 9) differs relatively little from star to star, compared to the large spread in stellar luminosities. The final column in the table presents an estimate of each star's *lifetime*,

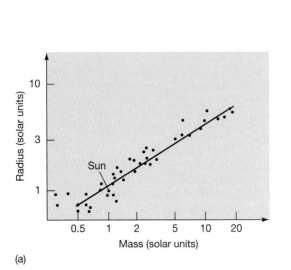

(a)

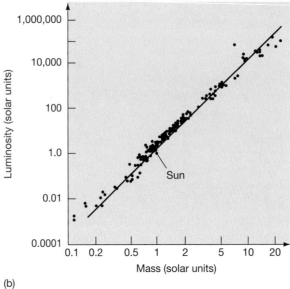

(b)

◀ FIGURE 10.21 **Stellar Radii and Luminosities**   (a) Actual measurements of main-sequence stars show that radius increases almost in proportion to mass over much of the range (as indicated by the straight line drawn through the data). (b) Stellar luminosity increases roughly as the fourth power of the mass (indicated again by the straight line).

obtained by dividing the amount of fuel available (that is, the star's mass) by the rate at which the fuel is consumed (the luminosity),

$$\text{stellar lifetime} \propto \frac{\text{stellar mass}}{\text{stellar luminosity}},$$

and noting that the lifetime of the Sun (see Chapter 12) is about 10 billion years.

Because luminosity increases so rapidly with mass, the most massive stars are by far the shortest lived. For example, according to the mass–luminosity relationship, the lifetime of a 10-solar mass O-type star is roughly $10/10^4 = 1/1000$ that of the Sun, or about 10 million years. We can be sure that all the O- and B-type stars we now observe are quite young—less than a few tens of millions of years old. Their nuclear reactions proceed so rapidly that their fuel is quickly depleted despite their large masses. At the opposite end of the main sequence, the low core density and temperature of an 0.1-solar mass M-type star mean that its proton–proton reactions churn away much more sluggishly than in the Sun's core, leading to a very low luminosity and a correspondingly long lifetime. ∞ *(Sec. 9.5)* Many of the K- and M-type stars now visible in the sky will shine on for at least another trillion years. The evolution of stars—large and small—is the subject of the next two chapters.

CONCEPT CHECK ▶ How do we know about the masses of stars that aren't members of binaries?

CHAPTER REVIEW

SUMMARY

L01 The distances to the nearest stars can be measured by stellar parallax. A star with a parallax of 1 arc second is 1 **parsec** (p. 270)—about 3.3 light-years—away from the Sun. A star's **proper motion** (p. 272), which is its true motion across the sky, is a measure of the star's velocity perpendicular to our line of sight.

L02 The **apparent brightness** (p. 272) of a star is the rate at which energy from the star reaches a detector. Apparent brightness is proportional to luminosity and falls off as the inverse square of the distance. Optical astronomers use the **magnitude scale** (p. 274) to express and compare stellar brightnesses. The greater the magnitude, the fainter the star; a difference of five magnitudes corresponds to a factor of 100 in brightness. **Apparent magnitude** (p. 274) is a measure of apparent brightness. The **absolute magnitude** (p. 274) of a star is the apparent magnitude it would have if placed at a standard distance of 10 pc from the viewer. It is a measure of the star's luminosity.

L03 Astronomers often determine the temperatures of stars by measuring their brightnesses through two or more optical filters, then fitting a blackbody curve to the results. Spectroscopic observations of stars provide an accurate means of determining both stellar temperatures and stellar composition. Astron-

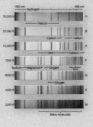

omers classify stars according to the absorption lines in their spectra. The standard stellar **spectral classes** (p. 276), in order of decreasing temperature, are O B A F G K M.

L04 Only a few stars are large enough and close enough to Earth that their radii can be measured directly. The sizes of most stars are estimated indirectly through the **radius–luminosity–temperature relationship** (p. 279). Stars categorized as **dwarfs** (p. 279) are comparable in size to, or smaller than, the Sun; **giants** (p. 279) are up to 100 times larger than the Sun; and **supergiants** (p. 279) are more than 100 times larger than the Sun. In addition to "normal" stars such as the Sun, two other important classes of star are **red giants** (p. 279), which are large, cool, and luminous, and **white dwarfs** (p. 279), which are small, hot, and faint.

L05 A plot of stellar luminosity versus stellar spectral class (or temperature) is called an **H–R diagram** (p. 279). About 90 percent of all stars plotted on an H–R diagram lie on the **main sequence** (p. 280), which stretches from hot, bright **blue supergiants** (p. 280) and **blue giants** (p. 280), through intermediate stars such as the Sun, to cool, faint **red dwarfs** (p. 280).

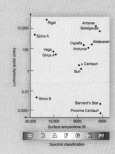

Most main-sequence stars are red dwarfs; blue giants are quite rare. About 9 percent of stars lie in the **white-dwarf region** (p. 281) and the remaining 1 percent lie in the **red-giant region** (p. 281).

L06 If a star is known to lie on the main sequence, measurement of its spectral type allows its luminosity to be estimated and its distance to be measured. This method of distance determination, which is valid for stars up to several thousand parsecs from Earth, is called **spectroscopic parallax** (p. 283). A star's **luminosity class** (p. 284) allows astronomers to distinguish main-sequence stars from giants and supergiants of the same spectral type, making distance estimates much more reliable.

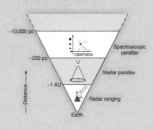

L07 Most stars are not isolated in space but instead orbit other stars in **binary-star systems** (p. 285). In a **visual binary** (p. 285), both stars can be seen and their orbits charted. In a **spectroscopic binary** (p. 285), the stars cannot be resolved, but their orbital motion can be detected spectroscopically. In an **eclipsing binary** (p. 286), the orbits are oriented in such a way that one star periodically passes in front of the other as seen from Earth and dims the light we receive. Studies of binary stars often allow stellar masses to be measured. The mass of a star determines its size, temperature, and brightness. Hot blue giants are much more massive than the Sun; cool red dwarfs are much less massive. High-mass stars burn their fuel rapidly and have much shorter lifetimes than the Sun. Low-mass stars consume their fuel slowly and may remain on the main sequence for trillions of years.

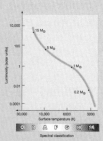

MasteringAstronomy® For instructor-assigned homework go to www.masteringastronomy.com

Problems labeled **POS** explore the process of science. **VIS** problems focus on reading and interpreting visual information. **LO** connects to the introduction's numbered Learning Outcomes.

REVIEW AND DISCUSSION

1. **L01 POS** How is stellar parallax used to measure the distances to stars?

2. What is a parsec? Compare it to the astronomical unit.

3. Explain how a star's real motion through space translates into motion observable from Earth.

4. **L04** Describe some characteristics of red giants and white dwarfs.

5. **L02** What is the difference between absolute and apparent brightness?

6. **POS** How do astronomers measure star temperatures?

7. **L03** Briefly describe how stars are classified according to their spectral characteristics.

8. **L05** What information is needed to plot a star on the H–R diagram?

9. What is the main sequence? What basic property of a star determines where it lies on the main sequence?

10. **L06** How are distances determined using spectroscopic parallax?

11. Which stars are most common in the Galaxy? Why don't we see many of them in H–R diagrams?

12. **L07 POS** How can stellar masses be determined by observing binary-star systems?

13. High-mass stars start off with much more fuel than low-mass stars. Why don't high-mass stars live longer?

14. In general, is it possible to determine the life span of an individual star simply by noting its position on an H–R diagram?

15. Why are visual binaries and eclipsing binaries relatively rare compared to spectroscopic binaries?

CONCEPTUAL SELF-TEST: TRUE OR FALSE?/MULTIPLE CHOICE

1. Star A appears brighter than star B, as seen from Earth. Therefore, star A must be closer to Earth than star B. (T/F)

2. A star of apparent magnitude +5 looks brighter than one of apparent magnitude +2. (T/F)

3. Red giants are very bright because they are extremely hot. (T/F)

4. The radius of a star can be determined if the star's distance and luminosity are known. (T/F)

5. Astronomers can distinguish between main-sequence and giant stars by purely spectroscopic means. (T/F)

6. In a spectroscopic binary, the orbital motion of the component stars appears as variations in the overall apparent brightness of the system. (T/F)

7. There are no billion-year-old main-sequence O- or B-type stars. (T/F)

8. From a distance of 1 parsec, the angular size of Earth's orbit would be (a) 1 degree; (b) 2 degrees; (c) 1 arc minute; (d) 2 arc seconds.

9. Compared with a star of absolute magnitude −2 at a distance of 100 pc, a star of absolute magnitude 5 at a distance of 10 pc will appear (a) brighter; (b) fainter; (c) to have the same brightness; (d) bluer.

10. Stars of spectral class M do not show strong lines of hydrogen in their spectra because (a) they contain very little hydrogen; (b) their surfaces are so cool that most hydrogen is in the ground state; (c) their surfaces are so hot that most hydrogen is ionized; (d) the hydrogen lines are swamped by even stronger lines of other elements.

11. Cool stars can be very luminous if they are very (a) small; (b) hot; (c) large; (d) close to our solar system.

12. The mass of a star may be determined (a) by measuring its luminosity; (b) by determining its composition; (c) by measuring its Doppler shift; (d) by studying its orbit around a binary companion.

13. VIS Pluto's apparent magnitude is approximately 14. According to Figure 10.6 (Apparent Magnitude), Pluto can be seen (a) with the naked eye on a dark night; (b) using binoculars; (c) using a 1-meter telescope; (d) only with the *Hubble Space Telescope*.

14. VIS According to Figure 10.12 (H–R Diagram of Well-Known Stars), Barnard's star must be (a) hotter; (b) larger; (c) closer to us; (d) bluer than Proxima Centauri.

15. VIS According to Figures 10.19 (Stellar Masses) and 10.20 (Stellar Mass Distribution), compared to the Sun, most stars are (a) much hotter; (b) much cooler; (c) much larger; (d) much brighter.

PROBLEMS

The number of squares preceding each problem indicates its approximate level of difficulty.

1. ■ How far away is the star Spica, whose parallax is 0.013″? What would Spica's parallax be if it were measured from an observatory on Neptune's moon Triton as Neptune orbited the Sun?

2. ■■ A star lying 20 pc from the Sun has proper motion of 0.5″/yr. What is its transverse velocity? If the star's spectral lines are observed to be redshifted by 0.01 percent, calculate the magnitude of its three-dimensional velocity relative to the Sun.

3. ■■ (a) What is the luminosity of a star having three times the radius of the Sun and a surface temperature of 10,000 K? (b) A certain star has a temperature twice that of the Sun and a luminosity 64 times greater than the solar value. What is its radius, in solar units?

4. ■■ Two stars—A and B, of luminosities 0.5 and 4.5 times the luminosity of the Sun, respectively—are observed to have the same apparent brightness. Which star is more distant, and how much farther away is it than the other?

5. ■■ Two stars—A and B with absolute magnitudes 3 and 8, respectively—are observed to have the same apparent magnitude.

Which star is more distant, and how much farther away is it than the other?

6. ■■ Calculate the amount of energy received per unit area per unit time from the sun, as seen from a distance of 10 pc. Compare this with the solar constant at Earth. ⌒ *(Sec. 9.1)*

7. ■ Astronomical objects visible to the naked eye range in apparent brightness from faint sixth-magnitude stars to the Sun, with magnitude −27. What is the range in energy flux corresponding to this magnitude range?

8. ■ A star has apparent magnitude 10.0 and absolute magnitude 2.5. How far away is it?

9. ■■■ Two stars in an eclipsing spectroscopic binary system are observed to have an orbital period of 25 days. Further observations reveal that the orbit is circular, with a separation of 0.3 AU, and that one star is 1.5 times the mass of the other. What are the masses of the stars?

10. ■■ Given that the Sun's lifetime is about 10 billion years, estimate the life expectancy of (a) a 0.2-solar mass, 0.01-solar luminosity red dwarf; (b) a 3.0-solar mass, 30-solar luminosity star; (c) a 10-solar mass, 1000-solar luminosity blue giant.

ACTIVITIES

Collaborative

1. Estimate the total number of stars visible in the night sky. Each member of your group should be equipped with identical cardboard tubes—the tube at the center of a roll of paper towels or toilet paper is perfect for the task. On a clear, moonless night, hold your tube up to your eye and count the total number of stars you can see. Do this several times, randomly choosing different areas of the sky and avoiding obvious obstacles such as clouds and trees. Try to sample all directions roughly equally. Be sure to allow time for your eyes to adapt to the dark—10 to 15 minutes at least—before taking any measurements. Add up all your measurements and divide by the total number of observations to calculate the average number of stars observed— call it n. You can convert this number into an estimate of the total number N of visible stars by multiplying by the square of the ratio of the length L of the tube to its diameter D, that is: $N = (L/D)^2 \times n$. (Can you figure out where this formula came from?) Repeat your measurement at a variety of sites—a city, the suburbs, and a dark rural location. Can you understand why astronomers are so concerned about light pollution?

Individual

1. The Winter Circle is an asterism—or pattern of stars—made up of six bright stars in five different constellations: Sirius, Rigel, Betelgeuse, Aldebaran, Capella, and Procyon. These stars span nearly the entire range of colors (and therefore temperatures) of normal stars. Rigel is of type B; Sirius, A; Procyon, F; Capella, G; Aldebaran, K; and Betelgeuse, M. The color differences are easy to see. Why do you suppose there is no O-type star in the Winter Circle?

2. In the winter sky, find the red supergiant Betelgeuse in the constellation Orion. It's easy to see because it's one of the brightest stars in the entire night sky. Betelgeuse is a variable star with a period of about 6.5 years. Its brightness changes as it expands and contracts. At maximum size, it fills a volume of space that would extend from the Sun to beyond the orbit of Jupiter. Betelgeuse is thought to be about 10 to 15 times more massive than our Sun and probably between 4 and 10 million years old. A similar star can be found shining prominently in midsummer. This is the red supergiant Antares in the constellation Scorpius.

11

The Interstellar Medium

Star Formation in the Milky Way

Stars and planets are not the only inhabitants of our Galaxy. The space around us harbors invisible matter throughout the dark voids between the stars. The density of this interstellar matter is extremely low—approximately a trillion trillion times less dense than matter in either stars or planets, far thinner than the best vacuum attainable on Earth. Only because the volume of interstellar space is so vast does the mass of interstellar matter amount to anything at all. Why bother to study this near-perfect vacuum? We do so for three important reasons. First, there is as much mass in the "voids" between the stars as there is in the stars themselves. Second, interstellar space is the region out of which new stars are born. Third, it is the region into which old stars expel their matter when they die. It is one of the most significant crossroads through which matter passes in the history of our universe.

MasteringAstronomy® Visit www.masteringastronomy.com for quizzes, animations, videos, interactive figures, and self-guided tutorials.

LEARNING OUTCOMES

Studying this chapter will enable you to:

LO1 Summarize the composition and properties of interstellar matter.

LO2 Describe the characteristics of emission nebulae, and explain their connection with star formation.

LO3 Outline the properties of dark interstellar dust clouds and molecular clouds.

LO4 Specify some of the radio techniques used to probe the nature of interstellar gas.

LO5 Summarize the sequence of events leading from a cold, dense interstellar cloud to the formation of a star like our Sun.

LO6 Describe some of the observational evidence supporting the modern theory of star formation.

LO7 Explain how the formation of a star depends on its mass.

LO8 Describe the properties of star clusters, and discuss how they form.

◀ Interstellar space is the place both where stars are "born" and to which they return at "death." Rich in gas and dust, yet spread extraordinarily thinly throughout the vast, dark regions among the stars, interstellar matter occasionally glows as nebulae or contracts to form new stars (and sometimes planets). Here, in a visible, true-color image captured by the *Hubble Space Telescope*, pillars of gas and dust within the Carina Nebula stand out against the background sky. These flimsy structures, about 7500 light-years away and extending a few light-years across, will not survive long; radiation from hidden stars is slowly destroying them. In about 100,000 years, a cluster of stars will form here. (*STScI*)

11.1 Interstellar Matter

Figure 11.1 is a mosaic of photographs covering a much greater expanse of universal "real estate" than anything we have studied thus far. From our vantage point on Earth, the panoramic view shown here stretches all the way across the sky. On a clear night it is visible to the naked eye as the *Milky Way*. In Chapter 14 we will come to recognize this band as the flattened disk, or *plane*, of our own Galaxy.

The bright regions in this image are made up of innumerable unresolved distant stars, merging together into a continuous blur at the resolution of the telescope. However, the dark areas are not simply "holes" in the stellar distribution. They are regions of space where *interstellar matter* obscures the light from stars beyond, blocking from our view what would otherwise be a smooth distribution of bright starlight. In this chapter we focus not on the stars themselves but on the vast reaches of interstellar space in which they form.

Gas and Dust

The matter between the stars is collectively termed the **interstellar medium.** It is made up of two components—*gas* and *dust*—intermixed throughout all of space. The gas is made up mainly of individual atoms and small molecules. The dust consists of clumps of atoms and molecules—not unlike the microscopic particles that make up smoke or soot.

Apart from numerous narrow atomic and molecular absorption lines, the gas alone does not block electromagnetic radiation to any great extent. The patchy obscuration evident in Figure 11.1 is caused by dust. Light from distant stars cannot penetrate the densest accumulations of interstellar dust any more than a car's headlights can illuminate the road ahead in a thick fog. As a rule of thumb, a beam of light can be absorbed or scattered only by particles having diameters comparable to or larger than the wavelength of the radiation involved. The amount of obscuration (absorption or scattering) produced by particles of a given size increases with decreasing wavelength. The typical diameter of an interstellar dust particle—or **dust grain**—is about 10^{-7} m, comparable in size to the wavelength of visible light. Consequently,

▼**FIGURE 11.1 Milky Way Mosaic**   The Milky Way photographed panoramically, across 360° of the entire southern and northern celestial sphere. This band, which constitutes the central plane of our Galaxy, contains high concentrations of stars, as well as interstellar gas and dust. The white box outlines the field of view of Figure 11.4. *(ESO/S. Brunier)*

► **FIGURE 11.2 Reddening** (a) Starlight passing through a dusty region of space is both dimmed and reddened, but spectral lines are still recognizable in the light that reaches Earth. (b) This dusty interstellar cloud, called Barnard 68, is opaque to visible light, except near the edges. The cloud spans about 0.2 pc and lies about 160 pc away. Frame (c) illustrates (in false color) how infrared radiation can penetrate Barnard 68. *(ESO)*

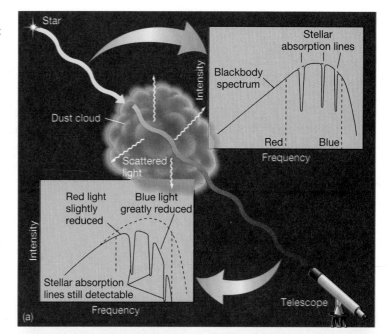

dusty regions of interstellar space are transparent to long-wavelength radio and infrared radiation but opaque to shorter-wavelength optical, ultraviolet, and X-ray radiation.

Because the opacity of the interstellar medium increases with decreasing wavelength, light from distant stars is preferentially robbed of its higher-frequency ("blue") components. Hence, in addition to being generally diminished in overall brightness, stars also appear redder than they really are. Illustrated in Figure 11.2, this effect, known as **reddening**, is similar to the process that produces spectacular red sunsets here on Earth (see Figure 11.3). Reddening can be seen very clearly in Figure 11.2(b), which shows a type of compact, dusty interstellar cloud called a *globule*. (We will discuss such dark interstellar clouds in more detail in Section 11.3.) The center of this cloud, called Barnard 68, is opaque to all optical wavelengths, so starlight cannot pass through it. However, near the edges, where there is less intervening cloud matter, some light does make it through; in the infrared image of Figure 11.2(c), even more radiation gets through. Notice how stars seen through the cloud are both dimmed and reddened relative to those seen directly.

As indicated in Figure 11.2(a), absorption lines in a star's spectrum are still recognizable in the radiation reaching Earth, allowing the star's spectral class, and hence its true luminosity and color, to be determined. ∞ *(Sec. 10.6)* Astronomers can then measure the degree to which the star's light has been dimmed and reddened en route to Earth, allowing them to estimate the amount of dust along the line of sight.

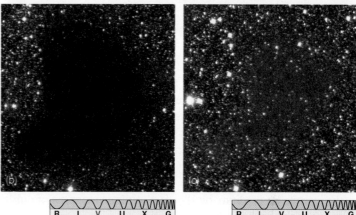

▼ **FIGURE 11.3 Reddening in Earth's Atmosphere**
Sunlight is reddened as it passes through Earth's atmosphere at the end of a hot summer day. Airborne dust particles and water molecules scatter the Sun's blue light, leaving only its dimmed, red light—a spectacular sunset. *(Joyce Photographics)*

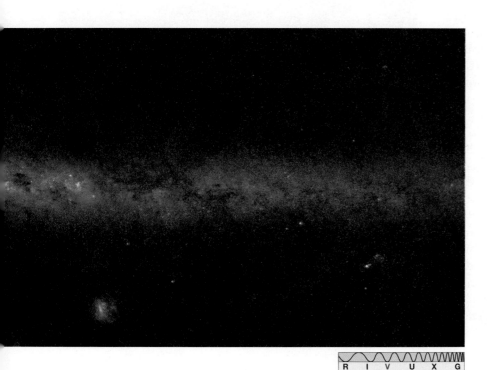

Density and Composition of the Interstellar Medium

By measuring its effect on the light from many different stars, astronomers have built up a picture of the distribution and chemical properties of interstellar matter in the solar neighborhood.

Gas and dust are found everywhere in interstellar space. No part of our Galaxy is truly devoid of matter, although the density of the interstellar medium is extremely low. Overall, the gas averages roughly 10^6 atoms per cubic meter—just one atom per cubic centimeter. (For comparison, the best vacuum presently attainable in laboratories on Earth contains about 10^9 atoms per cubic meter.) Interstellar dust is even rarer—about one dust particle for every trillion or so atoms. The space between the stars is populated with matter so thin that harvesting all the gas and dust in an interstellar region the size of Earth would yield barely enough matter to make a pair of dice.

11-1 DISCOVERY

Ultraviolet Astronomy and the "Local Bubble"

The ultraviolet (UV) region is that part of the spectrum where we expect to witness events and see objects involving temperatures in the hundreds of thousands, even millions, of kelvins—hot regions like the seething atmospheres or eruptive flares of stars; violent events, such as massive stars caught in the act of exploding; and active galaxies whose spinning hearts may harbor black holes. Yet ultraviolet astronomy has also contributed greatly to the study of the interstellar medium, by allowing a unique mapping of our local cosmic neighborhood.

Only 30 years ago astronomers had presumed that the gases filling the spaces among the stars would absorb virtually all short-wavelength UV radiation before it had a chance to reach Earth. But in 1975, during a historic linkup of the *Apollo* (U.S.) and *Soyuz* (USSR) space capsules, the onboard astronauts and cosmonauts performed a key experiment—they used a small telescope to detect extreme ultraviolet radiation from a few nearby, very hot stars. Shortly thereafter, theoretical ideas changed as astronomers began to realize that interstellar gas is distributed very unevenly in space, in cool dense clumps interspersed by irregularly shaped regions of hot, low-density gas.

Since then, space-based ultraviolet observations have shown that some regions of interstellar space are much thinner (5000 atoms/m^3) and hotter (500,000 K) than previously expected. Part of the space among the dust clouds and the emission nebulae seems to contain extremely dilute yet seething plasma, probably the result of the concussion and expanding debris from stars that exploded long ago. These superheated interstellar "bubbles," or *intercloud medium,* may extend far into interstellar space beyond our local neighborhood and conceivably into the even vaster spaces among the galaxies.

The Sun seems to reside in one such low-density region—a huge cavity called the *Local Bubble,* sketched in the accompanying figure. Only because we live within this vast, low-density bubble can we detect so many stars in the far UV; the hot, tenuous interstellar gas is virtually transparent to this radiation. The Local Bubble contains about 200,000 stars and extends for hundreds of trillions of kilometers, or nearly 100 pc. It was probably carved out by a supernova explosion that occurred in these parts roughly 300,000 years ago. ∞ *(Sec. 12.4)* Our distant ancestors must have seen it—a stellar catastrophe as bright as the full Moon.

(Rice University)

Despite such low densities, over sufficiently large distances interstellar matter accumulates slowly but surely to the point where it can block visible light and other short-wavelength radiation from distant sources. All told, space in the vicinity of the Sun contains about as much mass in the form of interstellar gas and dust as exists there in the form of stars.

Interstellar matter is distributed very unevenly (see *Discovery 11-1*). In some directions it is largely absent, allowing astronomers to study objects literally billions of parsecs from the Sun. In other directions there are small amounts of interstellar matter, so the obscuration is moderate, preventing us from seeing objects more than a few thousand parsecs away, but allowing us to study nearby stars. Still other regions are so heavily obscured that starlight from even relatively nearby stars is completely absorbed before reaching Earth. Even in regions of dense obscuration, however, there are often "windows" through which we can see to great distances—several can be seen in Figure 11.1.

The composition of interstellar gas is reasonably well known from spectroscopic studies of interstellar absorption lines. ∞ *(Sec. 2.5)* It mirrors the composition of other astronomical objects, such as the Sun, the stars, and the Jovian planets: Most of the gas—about 90 percent—is atomic or molecular hydrogen, 9 percent is helium, and the remaining 1 percent consists of heavier elements. The gas is deficient in some heavy elements, such as carbon, oxygen, silicon, magnesium, and iron, most likely because these elements have gone to form the interstellar dust.

In contrast, the composition of the dust is not well known, although there is some infrared evidence for silicates, carbon, and iron, supporting the theory that interstellar dust forms out of interstellar gas. The dust probably also contains some "dirty ice," a frozen mixture of water ice contaminated with trace amounts of ammonia, methane, and other compounds, much like cometary nuclei in our own solar system. ∞ *(Sec. 4.2)*

CONCEPT CHECK ▶ If space is a near-perfect vacuum, how can there be enough dust in it to block starlight?

11.2 Star-Forming Regions

Figure 11.4 shows a magnified view of the central part of Figure 11.1 (the region indicated by the rectangle in the earlier figure), in the general direction of the constellation Sagittarius. The field of view is mottled with stars and dark interstellar matter. In addition, several fuzzy patches of light are clearly visible. Labeled M8, M16, M17, and M20, they correspond to the 8th, 16th, 17th, and 20th entries in a catalog compiled by Charles Messier, an 18th-century French astronomer. Today these four objects are known as examples of **emission nebulae**—glowing clouds of hot interstellar gas. Figure 11.5 enlarges the left side of Figure 11.4, showing the nebulae more clearly.

Historically, astronomers have used the term **nebula** to refer to any "fuzzy" patch (bright or dark) on the sky—any region of space that is clearly distinguishable through a telescope, but not sharply defined, unlike a star or a planet. We now know that many (although not all) nebulae are clouds of interstellar dust and gas. If a cloud happens to obscure stars lying behind it, we see it as a dark

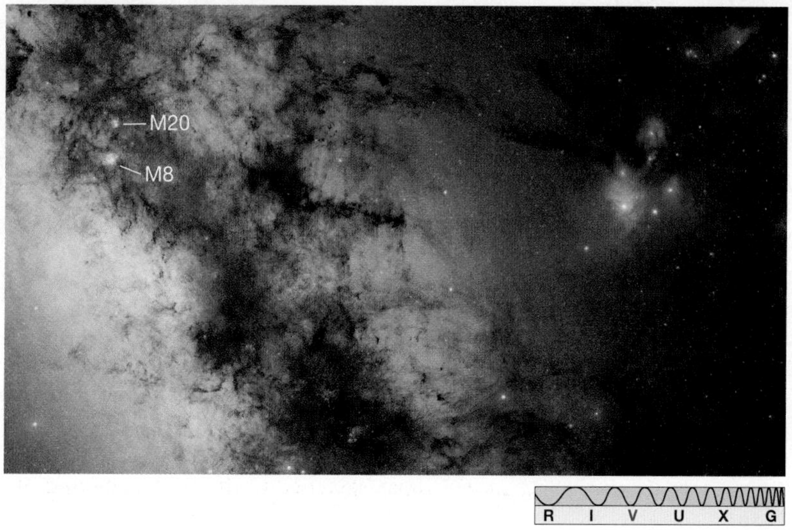

▶ FIGURE 11.4 **Milky Way in Sagittarius** Enlargement of the central portion of Figure 11.1, showing regions of brightness (vast fields of stars) as well as regions of darkness (where interstellar matter obscures the light from more distant stars). The field of view is about 35° across. Two of the emission nebulae discussed in the text are labeled. *(ESO/S. Guisard)*

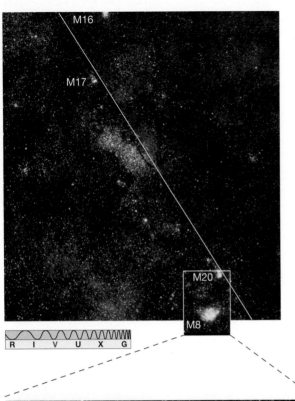

◄ FIGURE 11.5 **Galactic Plane** A black-and-white photograph of part of the region of the sky shown in Figure 11.4, showing stars, gas, and dust, as well as several distinct fuzzy patches of light known as emission nebulae. The plane of the Milky Way is marked with a white diagonal line. *(Harvard College Observatory)*

patch on a bright background, as in Figure 11.2(b)—a *dark nebula*. But if something within the cloud—a group of hot young stars, for example—causes it to glow, then we see a bright emission nebula instead. The method of spectroscopic parallax applied to stars visible within the emission nebulae shown in Figure 11.5 indicates that their distances from Earth range from 1200 pc (M8) to 1800 pc (M16). ∞ *(Sec. 10.6)* M16, at the top left, is approximately 1000 pc from M20, near the bottom.

We can gain a better appreciation of these nebulae by examining progressively smaller fields of view. Figure 11.6 is an enlargement of the region near the bottom of Figure 11.5, showing M20 at the top and M8 at the bottom. Figure 11.7 further enlarges the top of Figure 11.6, presenting a close-up of M20 and its immediate environment in both the visible and infrared parts of the spectrum. The total area displayed measures some 12 pc across. Emission nebulae are among the most spectacular objects in the universe, yet they appear only as small, undistinguished patches of light when viewed in the larger context of the Milky Way. Perspective is crucial in astronomy.

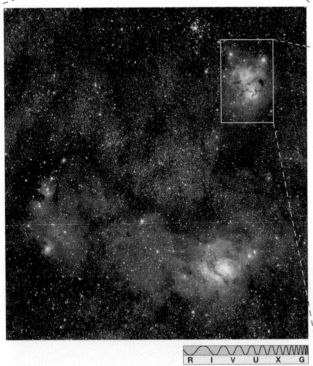

▲ FIGURE 11.6 **M20–M8 Region** A true-color enlargement of the bottom of Figure 11.5, showing M20 (top) and M8 (bottom) more clearly. *(R. Gendler)*

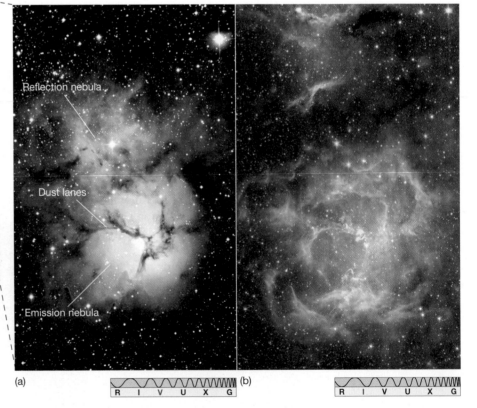

▲ FIGURE 11.7 **Trifid Nebula** (a) Further enlargement of the top of Figure 11.6, showing only M20 and its interstellar environment. Called the Trifid Nebula because of the dust lanes (in black) that trisect its midsection, the nebula itself (in red) is about 20 light-years across. (b) A false-color infrared image taken by the *Spitzer Space Telescope* reveals bright regions of star-forming activity mostly in those lanes of dust. *(AURA; NASA)*

The nebulae shown in Figures 11.4–11.7 are regions of glowing, ionized gas. At or near the center of each is at least one newly formed hot O- or B-type star producing copious amounts of ultraviolet light. As ultraviolet photons travel outward from the star, they heat and ionize the surrounding gas. As electrons recombine with nuclei, they emit visible radiation, causing the gas to glow. The predominant red coloration is the result of hydrogen atoms emitting Hα light in the red part of the visible spectrum. ∞ *(Sec. 2.6)*

Woven through the glowing nebular gas, and plainly visible in Figures 11.5–11.7, are dark **dust lanes** obscuring the nebular light. These dust lanes are part of the nebulae and are not just unrelated dust clouds that happen to lie along our line of sight. The bluish region visible in Figure 11.7 immediately above M20 is another type of nebula unrelated to the red emission nebula itself. Called a *reflection nebula,* it is caused by starlight reflecting off intervening dust particles along the line of sight. The blue coloration occurs because short-wavelength blue light is more easily scattered by interstellar matter back toward Earth and into our detectors. Figure 11.8 sketches some of the key features of emission nebulae, illustrating the connection between the central stars, the nebula itself, and the surrounding interstellar medium.

Figure 11.9 shows enlargements of two of the other nebulae visible in Figure 11.5. Notice again the hot bright stars embedded within the glowing nebular gas and the overall red hue of the reemitted radiation in parts (a) and (c). (The colors in the *Hubble* images parts [b] and [d] accentuate observations made at different wavelengths.) The relationship between the nebula and the dust lanes is again evident in Figure 11.9 (b) and (d), where regions of gas and dust are simultaneously silhouetted against background nebular emission and illuminated by foreground nebular stars. The interaction between stars and gas is particularly striking in Figure 11.9(b). The three dark "pillars" visible in this spectacular image are part of the interstellar cloud from which the stars formed. The rest of the cloud in the vicinity of the new stars has already been dispersed by their radiation. The fuzz around the edges of the pillars, especially at top right and center, is the result of this ongoing process, where intense stellar radiation continues to "eat away" the cloud, heating and dispersing the dense molecular gas. The process removes the less dense material first, leaving behind delicate sculptures composed of the denser parts of the original cloud, much as wind and water create spectacular structures in Earth's deserts and shores by eroding away the softest rock. The pillars will eventually be destroyed, but probably not for another hundred thousand or so years.

INTERACTIVE Multiwavelength Cloud with **MA**™
Embedded Stars

▼ **FIGURE 11.8 Nebular Structure** An emission nebula results when ultraviolet radiation from one or more hot stars ionizes part of an interstellar cloud. If starlight happens to encounter another dusty cloud, some of the radiation, particularly at the shorter wavelength blue end of the spectrum, may be scattered back toward Earth, forming a reflection nebula.

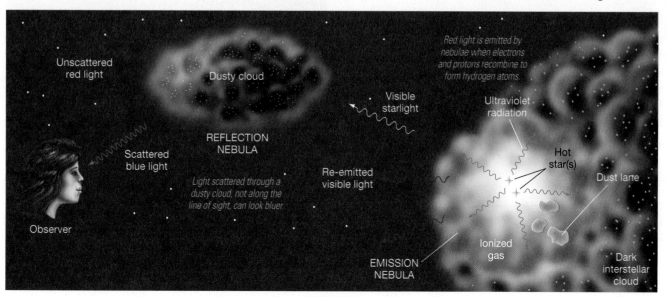

Unscattered red light

Dusty cloud

Visible starlight

Red light is emitted by nebulae when electrons and protons recombine to form hydrogen atoms.

Ultraviolet radiation

REFLECTION NEBULA

Scattered blue light

Hot star(s)

Dust lane

Re-emitted visible light

Light scattered through a dusty cloud, not along the line of sight, can look bluer.

Observer

Ionized gas

EMISSION NEBULA

Dark interstellar cloud

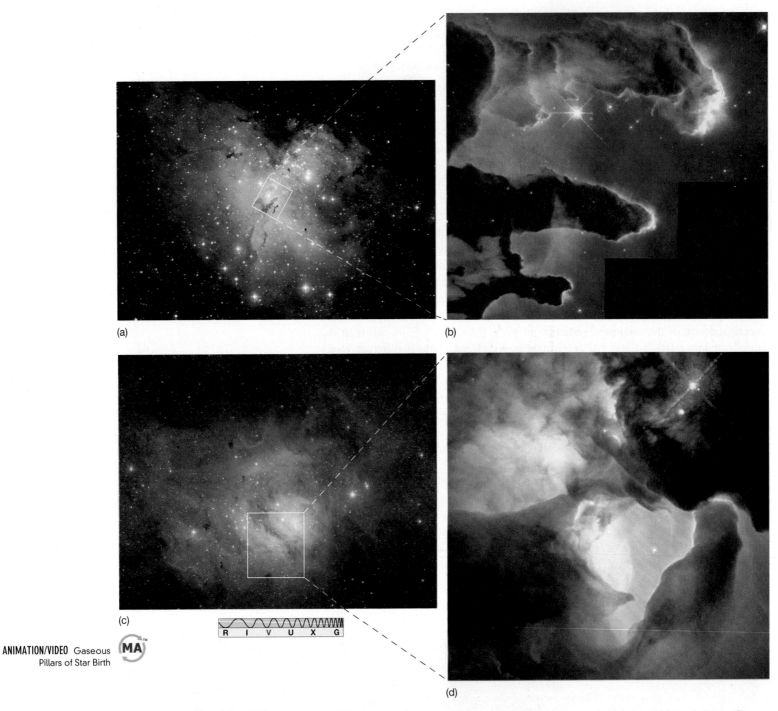

(a)

(b)

(c)

ANIMATION/VIDEO Gaseous
Pillars of Star Birth

R I V U X G

(d)

▲ **FIGURE 11.9 Emission Nebulae** Enlargements of selected portions of Figure 11.5. (a) M16, the Eagle Nebula and (b) a close-up of its huge pillars of cold gas and dust within, showing delicate sculptures created by the action of stellar ultraviolet radiation on the original cloud. (c) M8, the Lagoon Nebula and (d) a high-resolution view its core, a region known as the Hourglass. The varied colors of the insets result from observations at different wavelengths. *(NASA/AURA/ESO)*

CONCEPT CHECK If emission nebulae are powered by ultraviolet light from hot stars, why do they appear red?

Table 11.1 lists some vital statistics for the nebulae shown in Figure 11.5. Unlike stars, nebulae are large enough for their sizes to be measurable by simple geometry. (see *More Precisely 4-1*) Coupling this size information with estimates of the amount of matter along our line of sight (as revealed by a nebula's total light emission), we can find the nebula's density. Generally, emission nebulae have densities of a few hundred particles, mostly protons and electrons, per cubic centimeter (10^8 per cubic meter). Figure 11.10 shows a typical emission nebula, along with its spectrum, spanning part of the visible

TABLE 11.1 Nebular Properties

Object	Approx Distance (pc)	Average Diameter (pc)	Density (10^6 particles/m^3)	Mass (Solar Masses)	Temperature (K)
M8	1200	14	80	2600	7500
M16	1800	8	90	600	8000
M17	1500	7	120	500	8700
M20	1600	6	100	250	8200

and near-ultraviolet wavelength interval. ⇨ *(Sec. 2.5)* Numerous emission lines can be seen, and information on the nebula can be extracted from all of them. Analyses of such spectra reveal compositions similar to those found in the Sun and other stars and elsewhere in the interstellar medium. Spectral-line widths imply that the gas atoms and ions in this nebula have temperatures around 8000 K. ⇨ *(Sec. 2.8)*

Spectroscopists describe the *ionization state* of an atom by attaching a roman numeral to the chemical symbol for the atom—I for the neutral (that is, not ionized) atom, II for a singly ionized atom (an atom missing one electron), and so on. ⇨ *(Sec. 2.6)* Because emission nebulae are composed mainly of ionized hydrogen, astronomers often refer to them as *HII regions*. Regions of space containing primarily neutral (atomic) hydrogen are known as *HI regions*.

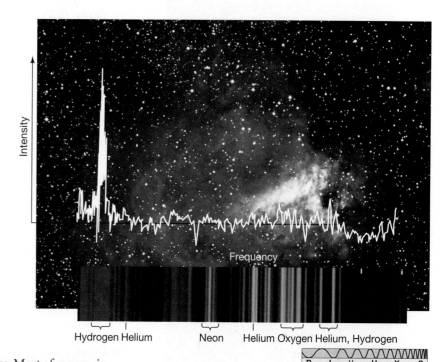

▲ FIGURE 11.10 **Nebular Spectrum** The visible spectrum of the hot gases in a nearby star-forming region known as the Omega Nebula (M17). Shining by the light of several very hot stars, the nebula produces a complex spectrum of bright and dark lines (bottom), also shown here as an intensity trace from red to blue (center). *(ESO)*

11.3 Dark Dust Clouds

Emission nebulae are only one small component of interstellar space. Most of space—in fact, more than 99 percent of it—is devoid of such regions and contains no stars. It is simply dark. The average temperature of a typical dark region of interstellar space is about 100 K. Compare this with 273 K, at which water freezes, and 0 K, at which atomic and molecular motions cease. ⇨ *(More Precisely 2-1)*

Within the dark voids among the nebulae and the stars lurks another distinct type of astronomical object, the **dark dust cloud.** These clouds are cooler than their surroundings, with temperatures as low as a few tens of kelvins, and thousands or even millions of times denser. In some regions, densities exceeding 10^9 atoms/m^3 (1000 atoms/cm^3) are found. Researchers often refer to dark dust clouds as *dense* interstellar clouds, but recognize that even these densest interstellar regions are barely denser than the best vacuum achievable in terrestrial laboratories. Still, it is because their density is much larger than the average value of 10^6 atoms/m^3 in interstellar space that we can distinguish these clouds from the surrounding expanse of the interstellar medium.

Obscuration of Visible Light

Dark dust clouds bear little resemblance to terrestrial clouds. Most are bigger than our solar system, and some are many parsecs across, although they still make up no more than a few percent of the entire volume of interstellar space. Despite their name, these clouds are composed primarily of gas, just like the rest of the interstellar medium. However, their absorption of starlight is due almost entirely to the dust they contain.

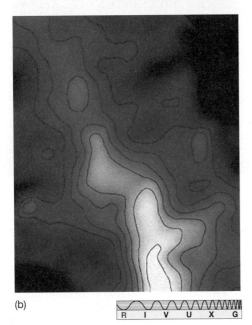

(a)

(b)

▲ FIGURE 11.11 **Obscuration and Emission**   (a) At optical wavelengths, this dark dust cloud (known as L977) can be seen only by its obscuration of background stars. (b) At radio wavelengths, it emits strongly in the CO molecular line, with the most intense radiation coming from the densest part of the cloud. *(C. and E. Lada)*

▶ FIGURE 11.12 **Dark Dust Cloud**   The Ophiuchus dark dust cloud resides only 170 pc away, surrounded by colorful stars and nebulae that are actually small illuminated parts of a much bigger and invisible molecular cloud engulfing much of the region shown. *(R. Gendler/J. Misti/S. Mazlin)*

Figure 11.2(b) is a good example of a dark dust cloud. Figure 11.11(a), a region called L997 in the constellation Cygnus, is another. Some early (18th-century) observers thought that these dark patches on the sky were simply empty regions of space that happened to contain no bright stars. However, by the late 19th century, astronomers had discounted this idea. They realized that seeing clear spaces among the stars would be like seeing clear tunnels between the trees in a forest, and it was statistically impossible that so many tunnels would lead directly away from Earth.

Before the advent of radio astronomy, astronomers had no direct means of studying clouds like L977. Emitting no visible light, they are generally undetectable to the eye except by the degree to which they dim starlight. However, as shown in Figure 11.11(b), the cloud's radio emission—in this case from carbon monoxide (CO) molecules contained within its volume (see p. 000)—outlines it clearly at radio wavelengths, providing an indispensable tool for the study of such objects.

Figure 11.12 is a spectacular wide-field image of another dark dust cloud. Taking its name from a neighboring star system, Rho Ophiuchus, this dust cloud resides relatively nearby, making it one of the most intensely studied regions of star formation in the Milky Way. Pockets of heavy blackness mark regions where the dust and gas are especially concentrated and the light from the background stars is completely obscured. Measuring several parsecs across, the Ophiuchus cloud is only a tiny part of the grand mosaic shown in Figure 11.1. Note that this cloud, like most interstellar clouds, is very irregularly shaped. Some of the features labeled in the image are part of the cloud itself, whereas others are newly formed stars near the edge of the cloud. Still other objects have no connection to the cloud and just happen to lie along the line of sight.

Figure 11.13 shows a particularly striking and well-known example of a dark dust cloud—the Horsehead Nebula in Orion. This curiously shaped finger of gas and dust projects out from the much larger dark cloud in the bottom half of the image and stands out clearly against the red glow of a background emission nebula.

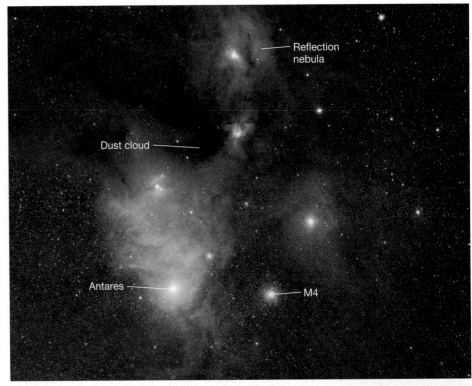

Reflection nebula

Dust cloud

Antares

M4

21-Centimeter Radiation

Often, dark dust clouds absorb so much light from background stars that they cannot easily be studied by the optical techniques described in Section 11.1. Fortunately, astronomers have an important alternative means of probing their structure that relies on low-energy radio emission from the interstellar gas itself.

Much of the gas in interstellar space is atomic hydrogen. Recall that a hydrogen atom consists of one electron orbiting a single-proton nucleus. ∞ *(Sec. 2.6)* Besides its orbital motion around the central proton, the electron also has some rotational motion—that is, *spin*—about its own axis. The proton also spins. This model parallels a planetary system, in which, in addition to the orbital motion of a planet about a central star, both the planet (electron) and the star (proton) rotate about their axes.

The laws of physics dictate that there are just two possible spin configurations for a hydrogen atom in its ground state. As illustrated in Figure 11.14, the electron and proton can either rotate in the *same* direction, with their spin axes parallel, or they can rotate in *opposite* directions, with their axes parallel but oppositely oriented. The former configuration has slightly higher energy than the latter. When a slightly excited hydrogen atom having the electron and proton spinning in the same direction drops down to the less-energetic, opposite-spin state, the transition releases a photon with energy equal to the energy difference between the two states.

Because the energy difference between the two states is very small, the energy of the emitted photon is very low. ∞ *(Sec. 2.6)* Consequently, the wavelength of the radiation is long—about 21 cm. That wavelength lies in the radio portion of the electromagnetic spectrum. Researchers refer to the spectral emission line that results from this hydrogen–spin-flip process as the **21-centimeter line.** It provides a vital probe into any region of the universe containing atomic hydrogen gas. Needing no visible starlight to help calibrate their signals, radio astronomers can observe any interstellar region that contains enough hydrogen to produce a detectable signal. Even the low-density regions between dark dust clouds can be studied.

Of great importance is the fact that the wavelength of 21-cm radiation is much larger than the typical size of interstellar dust particles. Accordingly, this radio radiation reaches Earth completely unaffected by interstellar debris. The opportunity to observe interstellar space well beyond a few thousand parsecs, and in directions lacking detectable background stars, makes 21-cm observations among the most important and useful in all of astronomy. As we will see in Chapters 14 and 15, it has been indispensable in allowing astronomers to map out the large-scale structure of our own galaxy and many others.

Molecular Gas

In certain interstellar regions of cold (typically 10–20 K) neutral gas, densities can reach as high as 10^{12} particles/m^3. The gas particles in these regions are not atoms but *molecules.* Because of the predominance of molecules in these dense interstellar regions, they are known as **molecular clouds.**

Only long-wavelength radio radiation can escape from these dense, dusty parts of interstellar space. Molecular hydrogen (H_2) is by far the most common constituent

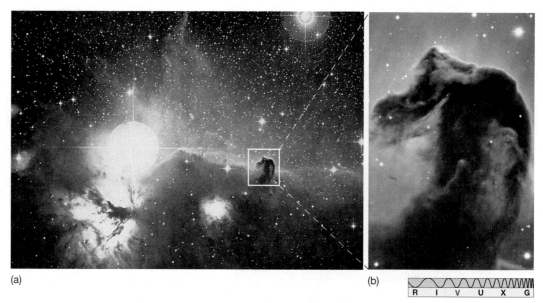

(a)

(b)

R I V U X G

▲ FIGURE 11.13 **Horsehead Nebula**   (a) The Horsehead Nebula is a striking example of a dark dust cloud, silhouetted against the bright background of an emission nebula. (b) A stunning image of the Horsehead, taken at highest resolution by the Very Large Telescope (VLT) in Chile. ∞ *(Sec. 3.2)* The "neck" of the horse is about 0.25 pc across. This nebular region lies roughly 1500 pc from Earth, in the constellation Orion. *(Royal Observatory of Belgium; ESO)*

MA™   ANIMATION/VIDEO Pillars Behind the Dust

MA™   ANIMATION/VIDEO The Tarantula Nebula

MA™   ANIMATION/VIDEO Horsehead Nebula

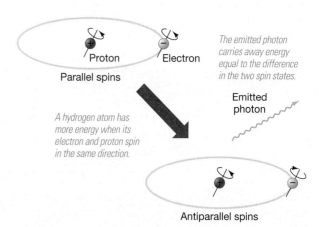

The emitted photon carries away energy equal to the difference in the two spin states.

Proton Electron

Parallel spins

A hydrogen atom has more energy when its electron and proton spin in the same direction.

Emitted photon

Antiparallel spins

▲ FIGURE 11.14 **Hydrogen 21-cm Emission**   A ground-level hydrogen atom changing from a higher-energy state (top) to a lower-energy state (bottom).

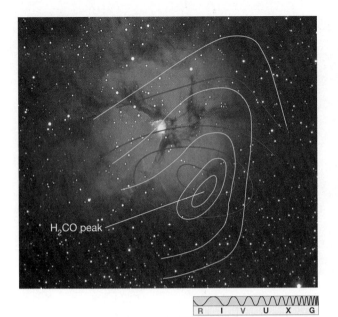

R I V U X G

▲ **FIGURE 11.15 Molecules Near M20** Contour map of formaldehyde near the M20 nebula, showing how the molecule is more abundant in the darkest interstellar regions. The maximum density of formaldehyde lies just to the bottom right of the visible nebula. The colors show the intensity of formaldehyde spectral lines at two different frequencies. *(AURA)*

PROCESS OF ▶
SCIENCE CHECK In mapping molecular clouds, why do astronomers use observations of "minority" molecules such as carbon monoxide and formaldehyde when these molecules constitute only a tiny fraction of the total in interstellar space?

▼ **FIGURE 11.16 Molecular Cloud Complexes** This false-color radio map shows the outer portion of the Milky Way as it appears in CO emission. The bright regions are molecular cloud complexes, dense regions of interstellar space where molecules abound and, apparently, stars are forming. The map is huge, extending over about a quarter of the sky. It was made from 1,696,800 observations of CO spectra. *(Five College Radio Astronomy Observatory)*

of molecular clouds, but unfortunately this molecule does not emit or absorb radio radiation, so it cannot easily be used as a probe of cloud structure. Nor are 21-cm observations helpful—they are sensitive only to *atomic* hydrogen, not to the molecular form of the gas. Instead, astronomers use radio observations of other molecules, such as carbon monoxide (CO), hydrogen cyanide (HCN), ammonia (NH_2), water (H_2O), and formaldehyde (H_2CO), to study the dark interiors of these dusty regions. These molecules are produced by chemical reactions within molecular clouds and have emission lines in the radio part of the spectrum. They are found only in very small quantities—perhaps one part per billion—but when we observe them we know that the regions under study must also contain high densities of molecular hydrogen, dust, and other important constituents. Figure 11.15 shows a contour map of the distribution of formaldehyde molecules around M20. Notice that the amount of formaldehyde (and, we assume, the amount molecular hydrogen) peaks well away from the visible nebula.

Molecular observations such as this reveal that the spectacular star-forming regions we see are actually parts of much larger molecular clouds. The bright emission nebula "bursts out" of its dark parent cloud when the radiation from a group of hot, newborn stars heats and ionizes the cold molecular gas. Radio maps of interstellar gas and infrared maps of interstellar dust reveal that molecular clouds do not exist as distinct and separate objects in space. Rather, they make up huge **molecular cloud complexes,** some spanning as much as 50 pc across and containing enough gas to make millions of stars like our Sun. About 1000 such complexes are known in our Galaxy. Figure 11.16 is a radio map of a large (330 square degree) region of the sky, made using certain spectral lines of the CO molecule, showing numerous molecular cloud complexes stretching across the entire field of view.

Why are molecules found only in the densest and darkest of interstellar clouds? One possible reason is that the dust protects the fragile molecules from the harsh interstellar environment. The same absorption that prevents high-frequency radiation from getting out to our detectors also prevents it from getting in to destroy the molecules. Another possibility is that the dust acts as a catalyst that helps form the molecules. The dust grains provide both a place where atoms can stick and react and a means of dissipating any heat associated with the reaction, which might other- wise destroy the newly formed molecules. Probably the dust plays both roles, although the details are still debated. In recent years, astronomers have come to realize that the interstellar medium is a dynamic, ever-changing environment, in which energy released by newborn stars (Section 11.5) and supernovae (Chapter 12) drives large-scale, turbulent motion in the interstellar gas. In this view, the cold molecular clouds we see are simply regions of dense gas temporarily compressed by the overall flow—transient islands in a sea of surrounding chaos.

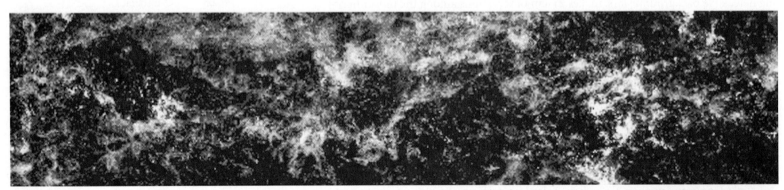

R I V U X G

11.4 The Formation of Stars Like the Sun

Let us now turn our attention to the connection between the interstellar medium and the stars in our Galaxy. How do stars form? What factors determine their masses, luminosities, and spatial distribution? In short, what basic processes are responsible for the appearance of our night sky?

Gravity and Heat

Simply stated, star formation begins when part of the interstellar medium—one of the cold, dark clouds discussed in the previous section—collapses under its own weight. An interstellar cloud is maintained in hydrostatic equilibrium by a balance between two basic opposing influences: gravity (which is always directed inward) and heat (in the form of outwardly directed pressure). ∞ *(Sec. 9.2)* If gravity somehow begins to dominate over heat, then the cloud can lose its equilibrium and start to contract.

Consider a small portion of a large interstellar cloud. Concentrate first on just a few atoms, as shown in Figure 11.17(a). Even though the cloud's temperature is very low, each atom has some random motion. ∞ *(More Precisely 2-1)* Each atom is also influenced by the gravitational attraction of all its neighbors. The gravitational force is not large, however, because the mass of each atom is so small. When a few atoms accidentally cluster for an instant, as sketched in Figure 11.17(b), their combined gravity is insufficient to bind them into a lasting, distinct clump of matter. This accidental cluster disperses as quickly as it formed (Figure 11.17c). The effect of heat—the random motion of the atoms—is much stronger than the effect of gravity.

As the number of atoms increases, their gravitational attraction increases too, and eventually the collective gravity of the clump is strong enough to prevent it from dispersing back into interstellar space. How many atoms are required for this to be the case? The answer, for a typical cool (100 K) cloud, is about 10^{57}—in other words, the clump must have mass comparable to that of the Sun. If our hypothetical clump is more massive than that, it will not disperse. Instead, its gravity will cause it to contract, ultimately to form a star.

Of course, 10^{57} atoms don't just clump together by random chance. Rather, star formation is *triggered* when a sufficiently massive pocket of gas is squeezed by some external event. Perhaps it is compressed by the shock wave produced when a nearby group of O- or B-type stars form and heat their surroundings, creating an emission nebula, or when a neighboring star explodes as a supernova (see Section 12.4). Or possibly part of an interstellar cloud simply becomes too cold for its internal pressure to support it against its own gravity. Whatever the cause, theory suggests that once the collapse begins, star formation inevitably follows.

Table 11.2 lists seven evolutionary stages that an interstellar cloud goes through before becoming a main-sequence star like our Sun. These stages are characterized

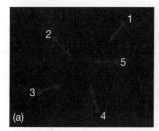
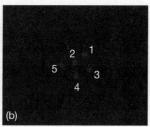
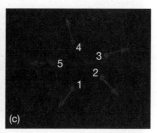

(a)   (b)   (c)

If more than a few atoms interact, the group would come together, start to slide by, but then pause, and finally contract into a clump.

◀ **FIGURE 11.17 Atomic Motions** The motions of a few atoms within an interstellar cloud are influenced by gravity so slightly that the atoms' paths are hardly changed (a) before, (b) during, and (c) after an accidental, random encounter.

TABLE 11.2 Prestellar Evolution of a Sun-like Star

Stage	Approximate Time to Next Stage (yr)	Central Temperature (K)	Surface Temperature (K)	Central Density (particles/m^3)	Diameter[1] (km)	Object
1	2×10^6	10	10	10^9	10^{14}	Interstellar cloud
2	3×10^4	100	10	10^{12}	10^{12}	Cloud fragment
3	10^5	10,000	100	10^{18}	10^{10}	Cloud fragment/protostar
4	10^6	1,000,000	3000	10^{24}	10^8	Protostar
5	10^7	5,000,000	4000	10^{28}	10^7	Protostar
6	3×10^7	10,000,000	4500	10^{31}	2×10^6	Star
7	10^{10}	15,000,000	6000	10^{32}	1.5×10^6	Main-sequence star

[1]For comparison, recall that the diameter of the Sun is 1.4×10^6 km and that of the solar system roughly 1.5×10^{10} km.

by different central temperatures, surface temperatures, central densities, and radii of the prestellar object. They trace its progress from a quiescent interstellar cloud to a genuine star. The numbers given in Table 11.2 and the following discussion are valid only for stars of approximately the same mass as the Sun. In the next section we will relax this restriction and consider the formation of stars of other masses.

Stage 1—An Interstellar Cloud

The first stage in the star-formation process is a dense interstellar cloud—the core of a dark dust cloud or perhaps a molecular cloud. These clouds are truly vast, sometimes spanning tens of parsecs ($10^{14} - 10^{15}$ km) across. Typical temperatures are about 10 K throughout, with a density of perhaps 10^9 particles/m^3. Stage-1 clouds contain thousands of times the mass of the Sun, mainly in the form of cold atomic and molecular gas. (The dust they contain is important for cooling the cloud as it contracts and also plays a crucial role in planet formation, but it constitutes a negligible fraction of the total mass.) ⚭ *(Sec. 4.3)*

The dark region outlined by the red and green radio contours in Figure 11.15 (*not* the emission nebula itself, where stars have already formed) probably represents a stage-1 cloud just starting to contract. Doppler shifts of the observed formaldehyde lines indicate that it is probably infalling. Less than a light year across, this region has a total mass over 1000 times the mass of the Sun—considerably greater than the mass of M20 itself.

Once the collapse begins, theory indicates that fragmentation into smaller and smaller clumps of matter naturally follows due to gravitational instabilities in the gas. As illustrated in Figure 11.18, a typical cloud can break up into tens, hundreds, even thousands of fragments, each imitating the shrinking behavior of the parent cloud and contracting ever faster. The whole process, from a single quiescent cloud

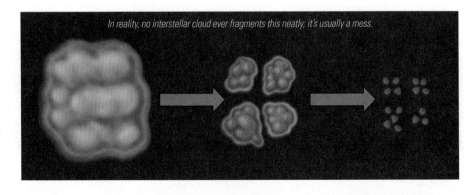

▶ **FIGURE 11.18 Cloud Fragmentation** As an interstellar cloud contracts, gravitational instabilities cause it to fragment into smaller pieces. The pieces themselves continue to fall inward and fragment, eventually forming many tens or hundreds of individual stars.

to many collapsing fragments, takes a few million years. Depending on the precise conditions under which fragmentation takes place, an interstellar cloud can produce either a few dozen stars, each much larger than our Sun, or a collection of hundreds of stars comparable to or smaller than our Sun. There is little evidence for stars born in isolation, one star from one cloud. Most stars—perhaps all—appear to originate as members of multiple systems.

The process of continuing fragmentation is eventually stopped by the increasing density within the shrinking cloud. As fragments continue to contract, they eventually become so dense that radiation cannot get out easily. The trapped radiation causes the temperature to rise, the pressure to increase, and the fragmentation to stop. However, the contraction continues.

Stages 2 and 3—A Contracting Cloud Fragment

As it enters stage 2, a fragment destined to form a star like the Sun—the end product of the process sketched in Figure 11.18—contains between one and two solar masses of material. Estimated to span a few hundredths of a parsec across, this fuzzy, gaseous blob is still about 100 times the size of our solar system. Its central density is about 10^{12} particles/m^3.

Even though the fragment has shrunk substantially, its average temperature is not much different from that of its parent cloud. The reason is that the gas constantly radiates large amounts of energy into space. The material of the fragment is so thin that photons produced anywhere within it easily escape without being reabsorbed, so virtually all the energy released in the contraction is radiated away and does not raise the temperature. Only at the center, where the radiation must traverse the greatest amount of material in order to escape, is there any appreciable temperature increase. The gas there might be as warm as 100 K by this stage. For the most part, however, the fragment stays cold as it shrinks.

Several tens of thousands of years after it first began contracting, the stage-2 fragment has shrunk to a gaseous sphere with a diameter roughly the size of our solar system (still 10,000 times the size of our Sun). Now, at the start of stage 3, the inner regions have become opaque to their own radiation and have started to heat up considerably, as noted in Table 11.2. The central temperature has reached about 10,000 K—hotter than the hottest steel furnace on Earth. However, the gas near the edge is still able to radiate its energy into space and so remains cool. The central density by this time is approximately 10^{18} particles/m^3 (still only 10^{-9} kg/m^3 or so).

For the first time, our fragment is beginning to resemble a star. The dense, opaque region at the center is called a **protostar.** Its mass increases as more and more material rains down on it from outside, although its radius continues to shrink because its pressure is still unable to overcome the relentless pull of gravity. By the end of stage 3, we can distinguish a "surface" on the protostar—its *photosphere.* Inside the photosphere, the protostellar material is opaque to the radiation it emits. (Note that this is the same operational definition of *surface* we used for the Sun.) ∞ *(Sec. 9.1)* From here on, the surface temperatures listed in Table 11.2 refer to the photosphere and not to the edge of the collapsing fragment, where the temperature remains low.

Figure 11.19 shows a star-forming region in Orion. Lit from within by several O-type stars, the bright Orion Nebula is partly surrounded by a vast molecular cloud that extends well beyond the roughly 1-parsec-square region bounded by the photograph in Figure 11.19(c). The Orion complex harbors several smaller sites of intense radio emission from molecules deep within its core. Shown in Figure 11.19(d), they measure about 10^{10} km, about the diameter of our solar system. Their density is about 10^{15} particles/m^3, much higher than the density of the surrounding cloud. Although the temperatures of these regions cannot be estimated reliably, many researchers regard the regions as objects between stages 2 and 3, on the threshold of becoming protostars. Figures 11.24 and 11.30 are visible and infrared images of part of the nebula showing other evidence for protostars.

Stage 4—A Protostar

As the protostar evolves, it shrinks, its density increases, and its temperature rises, both in the core and at the photosphere. Some 100,000 years after the fragment formed, it reaches stage 4, where its center seethes at about 1,000,000 K. The electrons and protons ripped from atoms whiz around at hundreds of kilometers per second, but the temperature is still short of the 10^7 K needed to ignite the proton–proton nuclear reactions that fuse hydrogen into helium. Still much larger than the Sun, our gassy heap is now about the size of Mercury's orbit. Its surface temperature has risen to a few thousand kelvins.

Knowing the protostar's radius and surface temperature, we can calculate its luminosity using the radius–luminosity–temperature relationship. ∞ *(Sec. 10.4)* Remarkably, it turns out to be around 1000 times the luminosity of the Sun. Because nuclear reactions have not yet begun in the protostar's core, this luminosity is due entirely to the release of gravitational energy as the protostar continues to shrink

ANIMATION/VIDEO Orion Nebula Mosaic (MA)

ANIMATION/VIDEO Visit to Orion Nebula (MA)

INTERACTIVE ▼ FIGURE 11.19 Orion Nebula, Up Close (a) The constellation Orion, with the region around its famous emission nebula marked by a rectangle (see Figure 0.2). (b) Enlargement of the framed region in part (a), but here shown in the infrared, revealing how the nebula is partly surrounded by a vast molecular cloud. Various parts of this cloud are probably fragmenting and contracting, with even smaller sites forming protostars. The three frames at the right show some of the evidence for those protostars: (c) real-color visible image of embedded nebular "knots" within the Orion Nebula, (d) false-color radio image of some intensely emitting molecular sites, and (e) high-resolution image of one of several young stars surrounded by disks of gas and dust where planets might ultimately form. (P. Sanz/Alamy; SST; CfA; NASA)

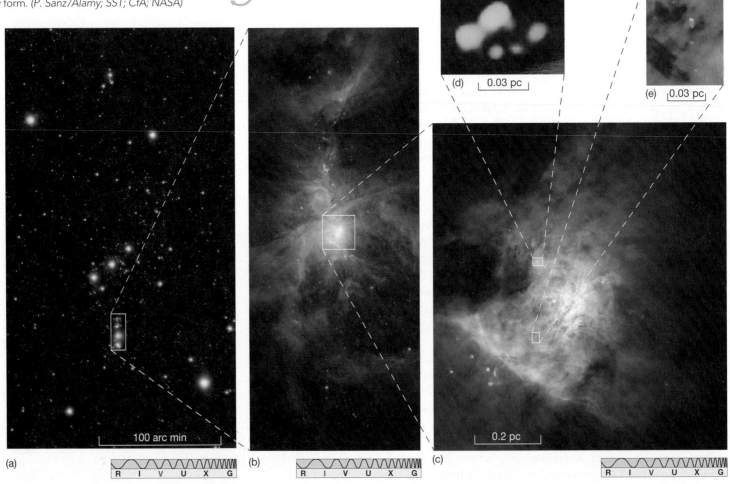

(d) | 0.03 pc |

(e) | 0.03 pc |

(a) R I V U X G

(b) R I V U X G 100 arc min

(c) R I V U X G 0.2 pc

▶ FIGURE 11.20 **Protostar on the H–R Diagram** The red arrow indicates the approximate evolutionary track followed by an interstellar cloud fragment before becoming a stage-4 protostar. The boldface numbers on this and subsequent H–R plots refer to the prestellar evolutionary stages listed in Table 11.2.

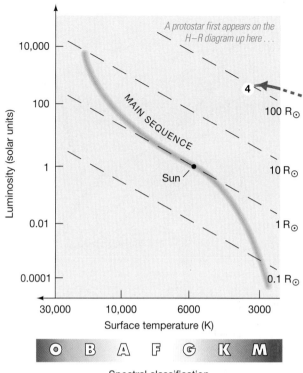

and material from the surrounding fragment (which we called the solar nebula back in Chapter 4) rains down on its surface. ∞ *(Sec. 4.3)*

In the hunt for examples of more advanced stages of star formation, radio techniques become less useful because objects in stages 4, 5, and 6 have increasingly higher temperatures. By Wien's law, their emission shifts toward shorter wavelengths, and so they shine most strongly in the infrared. ∞ *(Sec. 2.4)* One particularly bright infrared emitter, known as the Becklin–Neugebauer object, was detected in the core of the Orion molecular cloud in the 1970s. Its luminosity is around 1000 times the luminosity of the Sun. Most astronomers agree that this warm, dense blob is a high-mass protostar, probably around stage 4.

By the time stage 4 is reached, our protostar's physical properties can be plotted on a Hertzsprung–Russell (H–R) diagram. ∞ *(Sec. 10.5)* At each phase of a star's evolution, its surface temperature and luminosity can be represented by a single point on the diagram. The motion of that point around the diagram as the star evolves is known as the star's **evolutionary track.** It is a graphical representation of a star's life. Note that it has nothing to do with any real spatial motion of the star. The red track on Figure 11.20 depicts the approximate path followed by our interstellar cloud fragment since it became a protostar at the end of stage 3 (which itself lies off the right-hand edge of the figure). Figure 11.21 is an artist's sketch of an interstellar gas cloud proceeding along the evolutionary path outlined so far.

Stage 5—Protostellar Evolution

Our protostar is still not in equilibrium. Even though its temperature is now so high that outward-directed pressure has become a powerful countervailing influence against gravity's inward pull, the balance is not yet perfect. The protostar's internal heat gradually diffuses out from the hot center to the cooler surface, where it is radiated away into space. As a result, the contraction slows, but it does not stop completely.

After stage 4, the protostar moves downward on the H–R diagram (toward lower luminosity) and slightly to the left (toward higher temperature), as shown in Figure 11.22. By stage 5, the protostar has shrunk to about 10 times the size of the Sun, its surface temperature is about 4000 K, and its luminosity has fallen to about 10 times the solar value. The central temperature has risen to about 5,000,000 K. The gas is completely ionized by now, but the protons still do not have enough thermal energy for nuclear fusion to begin.

Protostars often exhibit violent surface activity during this phase of their evolution, resulting in extremely strong protostellar winds, much denser than that of our Sun. This portion of the evolutionary track is often called the **T-Tauri phase,** after T-Tauri, the first "star" (actually protostar) to be observed in this stage of prestellar development. The interaction between the strong winds and the nebular disk out of which the protostar (and possibly planets) is still forming

▼ FIGURE 11.21 **Interstellar Cloud Evolution** Artist's conception of the changes in an interstellar cloud during the early evolutionary stages outlined in Table 11.2. (Not drawn to scale.) The duration of each stage, in years, is indicated.

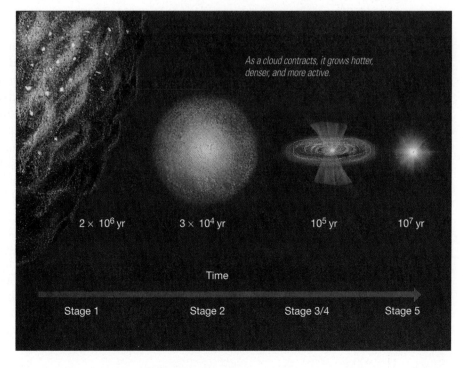

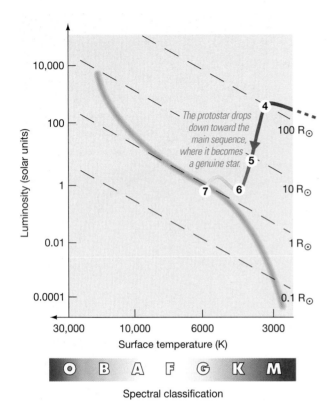

▲ FIGURE 11.22 Newborn Star on the H–R Diagram   The changes in a protostar's observed properties are shown by the path of decreasing luminosity, from stage 4 to stage 6. At stage 7, the newborn star has arrived on the main sequence. (MA)

often results in a *bipolar flow*—expelling two "jets" of matter perpendicular to the disk. Figure 11.23(a) shows a particularly striking example.

These outflows can be very energetic. Figure 11.23(b) shows a portion of the Orion molecular cloud, south of the Orion Nebula, where a protostar is seen still surrounded by a bright nebula, its turbulent wind spreading out into the interstellar medium. Below it (enlarged in the inset) are twin jets known as HH1 and HH2. (HH stands for Herbig–Haro, after the investigators who first cataloged such objects.) Formed in another (unseen) protostellar disk, these jets have traveled outward for almost half a light-year before colliding with interstellar matter. More Herbig–Haro objects can be seen in the upper right portion of the figure.

Events proceed more slowly as the protostar approaches the main sequence. The initial contraction and fragmentation of the interstellar cloud occurred quite rapidly, but by stage 5, as the protostar nears the status of a full-fledged star, its evolution slows. Its contraction is governed largely by the rate at which it can radiate its internal energy into space. As the luminosity decreases, so, too, does the contraction rate.

Until the *Infrared Astronomy Satellite (IRAS)* was launched in the early 1980s, astronomers were aware only of very massive stars forming in clouds far away. But *IRAS* showed that stars are forming much closer to home, and some of these protostars have masses comparable to that of our Sun. Figure 11.24 shows two examples of low-mass protostars—both spotted by the *Hubble Space Telescope* in a rich star-forming region in Orion. Their infrared heat signatures are those expected of a stage-5 object.

Stages 6 and 7—A Newborn Star

Some 10 million years after reaching stage 4, the protostar finally becomes a true star. By stage 6, when our roughly one-solar-mass object has shrunk to a radius of about 1,000,000 km, the contraction has raised the central temperature to 10,000,000 K—enough to ignite nuclear burning. Protons begin fusing into helium nuclei in the core, and a star is born. As shown in Figure 11.22, the star's surface temperature at this point is about 4500 K, still a little cooler than the Sun. Even though the radius of the newly formed star is slightly larger than that of the Sun, its lower temperature means that its luminosity is slightly less than (actually, about two-thirds of) the present solar value.

Indirect evidence for stage-6 stars comes from infrared observations of objects that seem to be luminous hot stars hidden from optical view by surrounding dark clouds. Their radiation is mostly absorbed by a "cocoon" of dust, then reemitted by the dust as infrared radiation. Two considerations support the idea that the hot stars responsible for heating the clouds have only recently ignited: (1) dust cocoons are predicted to disperse quite rapidly once their central stars form, and (2) these objects are invariably found in the dense cores of molecular clouds, consistent with the star-formation sequence just outlined.

Over the next 30 million years or so, the stage-6 star contracts a little more. Its central density rises to about 10^{32} particles/m^3 (more conveniently expressed as 10^5 kg/m^3), the central temperature increases to 15,000,000 K, and the surface temperature reaches 6000 K. By stage 7 the star finally reaches the main sequence just about where our Sun now resides. Pressure and gravity are finally balanced in the stellar interior, and the rate at which nuclear energy is generated in the core exactly matches the rate at which energy is radiated from the surface.

The journey from interstellar cloud to star occurs over the course of 40–50 million years. Although this is a long time by human standards, it is still less than 1 percent of the Sun's lifetime on the main sequence. Once an object begins fusing hydrogen and establishes a "gravity-in/pressure-out" hydrostatic equilibrium, it burns steadily for a very long time. The star's location on the H–R diagram—that is, its surface temperature and luminosity—will remain almost unchanged for the next 10 billion years.

CONCEPT CHECK ▶   What distinguishes a collapsing cloud from a protostar, and a protostar from a star?

▼ **FIGURE 11.23 Protostellar Outflow** (a) This remarkable image (at right) shows two jets emanating from the young star system HH30, the result of matter accreting onto an embryonic star near the center. (b) The view below of the Orion molecular cloud shows the outflow from a newborn star, still surrounded by nebular gas. The inset shows a pair of jets called HH1/HH2, formed when matter falling onto another protostar (still obscured by the dusty cloud fragment from which it formed) creates a pair of high-speed jets of gas perpendicular to the flattened protostellar disk. The jets are nearly 1 light-year across. Several more Herbig–Haro objects can be seen at top right—one of them resembling a "waterfall." (*AURA; NASA*)

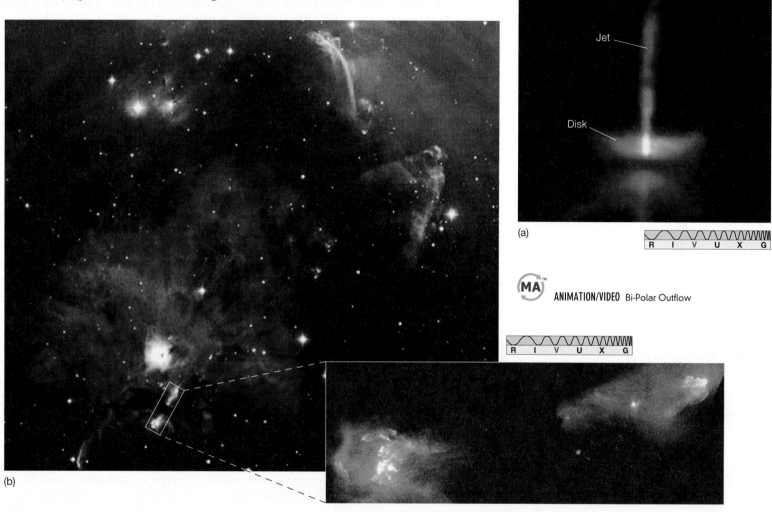

(a)

R I V U X G

(MA) **ANIMATION/VIDEO** Bi-Polar Outflow

R I V U X G

(b)

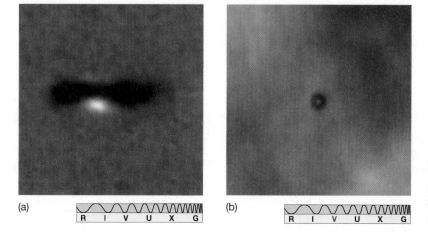

(a) R I V U X G

(b) R I V U X G

◄ **FIGURE 11.24 Protostars** (a) An infrared, edge-on image of a planetary system–size dusty disk in the Orion region, showing heat and light emerging from its center. This unnamed source seems to be a low-mass protostar around stage 5 in the H–R diagram. (b) An optical, face-on image of a slightly more advanced circumstellar disk surrounding an embedded protostar in Orion. (*NASA*)

11.5 Stars of Other Masses

The numerical values and evolutionary track just described are valid only for 1-solar-mass stars. The temperatures, densities, and radii of prestellar objects of other masses exhibit similar trends, but the details differ, in some cases quite considerably. Perhaps not surprisingly, the most massive fragments formed within interstellar clouds tend to produce the most massive protostars and eventually the most massive stars. Similarly, low-mass fragments give rise to low-mass stars.

The Zero-Age Main Sequence

Figure 11.25 compares the theoretical pre-main-sequence evolutionary track taken by our Sun with the corresponding tracks of a 0.3-solar-mass star and a 3.0-solar-mass star. All three tracks traverse the H–R diagram in the same general manner, but cloud fragments that eventually form stars more massive than the Sun approach the main sequence along a higher track on the diagram, while those destined to form less massive stars take a lower track. The *time* required for an interstellar cloud to become a main-sequence star also depends strongly on its mass. ⚬⚬ *(Sec. 10.7)* The most massive fragments contract into O-type stars in a mere million years, roughly 1/50 the time taken by the Sun. The opposite is the case for prestellar objects having masses much less than our Sun. A typical M-type star, for example, requires nearly a billion years to form.

Whatever the mass, the endpoint of the prestellar evolutionary track is the main sequence. A star is considered to have reached the main sequence when hydrogen burning begins in its core and the star's properties settle down to stable values. The main-sequence band predicted by theory is usually called the **zero-age main sequence** (ZAMS). It agrees quite well with the main sequences observed for stars in the vicinity of the Sun and those observed in more distant stellar systems.

It is important to realize that *the main sequence is not an evolutionary track—stars do not evolve along it.* Rather, it is a "waystation" on the H–R diagram where stars stop and spend most of their lives—low-mass stars at the bottom, high-mass stars at the top. Once on the main sequence, a star stays in roughly the same location in the H–R diagram during its whole time as a stage-7 object. In other words, a star that arrives on the main sequence as, say, a G-type star can never "work its way up" to become a B- or an O-type main-sequence star, nor move down to become an M-type red dwarf. As we will see in Chapter 12, the next stage of stellar evolution occurs when a star leaves the main sequence. When this occurs, the star will have pretty much the same surface temperature and luminosity it had when it arrived on the main sequence millions (or billions) of years earlier.

"Failed" Stars

Some cloud fragments are too small ever to become stars. Pressure and gravity come into equilibrium before the central temperature becomes hot enough to fuse hydrogen, so they never evolve beyond the protostar stage. Rather than turning into stars, such low-mass fragments continue to cool, eventually becoming compact, dark "clinkers"—cold fragments of unburned matter—in interstellar space. Small, faint, and cool (and growing ever colder), these objects are known collectively as **brown dwarfs.** On the basis of theoretical modeling, astronomers calculate that the minimum mass of gas needed to generate core temperatures high enough to begin nuclear fusion is about 0.08 solar masses—80 times the mass of Jupiter.

CONCEPT CHECK ▶ Do stars evolve along the main sequence?

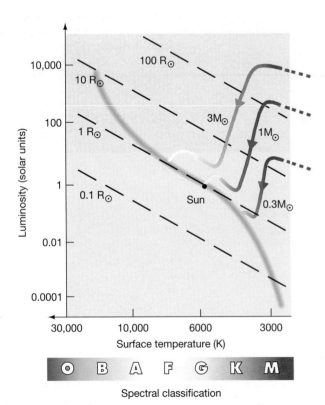

▲ **FIGURE 11.25 Prestellar Evolutionary Tracks** Prestellar evolutionary paths for stars more massive and less massive than our Sun.

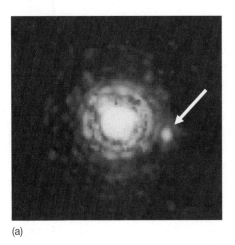

(a)

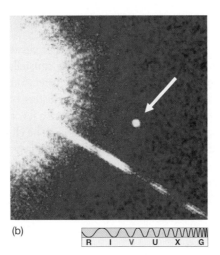

(b)

R I V U X G

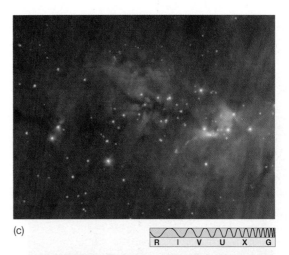

(c)

R I V U X G

(d)

Vast numbers of brown dwarfs may well be scattered throughout the universe—fragments frozen in time somewhere in the cloud-contraction phase. With our current technology, we have great difficulty in detecting them, whether they orbit other stars or float alone in space. We can telescopically detect stars and spectroscopically infer atoms and molecules, but low-mass astronomical objects far beyond our solar system are very hard to see. However, recent advances in observational hardware and image-processing techniques have identified many likely brown dwarf candidates, often using techniques similar to those employed in the search for extrasolar planets. ∞ *(Sec. 4.4)* (Recall from Section 10.7 that most stars are found in binaries; the same may very well be true of brown dwarfs.)

Figure 11.26 shows Gliese 623, a binary system containing a brown dwarf candidate, originally identified by radial velocity measurements and Gliese 229, first identified as a possible brown dwarf by ground-based infrared observations. More recent infrared images from space show many more brown dwarf candidates. Figure 11.26(c) is a *Spitzer* image of a small star cluster showing not only many bright stars, but also numerous much fainter objects thought to be brown dwarfs. Based on current observations, it seems that up to 100 billion cold, dark, brown dwarfs may lurk in the depths of interstellar space—comparable to the total number of "real" stars in our Galaxy. Figure 11.26(d) compares brown dwarfs, notably Gleise 229 of part (b), to some other well-known objects.

▲ **FIGURE 11.26 Brown Dwarfs** (a) This image shows Gliese 623, a binary system that may contain a brown dwarf (marked by an arrow). (b) This image of the binary-star system Gliese 229 shows two objects only 7″ apart; the fainter "star" (arrow) has a luminosity only a few millionths that of the Sun and an estimated mass about 50 times that of Jupiter. (The rings in part a and the spike in part b are instrumental artifacts.) (c) An infrared image of a star cluster just north of the Orion Nebula, showing bright objects that are stars and many faint specks that are brown-dwarf candidates. (d) This rendering compares the sizes of some stars, brown dwarfs, and planets. *(NASA)*

11.6 Star Clusters

The end result of the contraction and fragmentation of an interstellar gas cloud is a group of stars, all formed from the same parent cloud and lying in the same region of space. Such a collection of stars is called a **star cluster.** Figure 11.27 shows a spectacular view of a newborn star cluster and (part of) the interstellar cloud from which it came. Because all the stars formed at the same time out of the same cloud and under the same environmental conditions, clusters in many cases are near-ideal "laboratories" for stellar studies—not in the sense that astronomers can perform experiments on them, but because the properties of the stars are very tightly constrained. The only factor distinguishing one star from another in the same cluster is mass, so theoretical models of star formation and evolution can be compared with reality without the complications introduced by broad spreads in age, chemical composition, and place of origin.

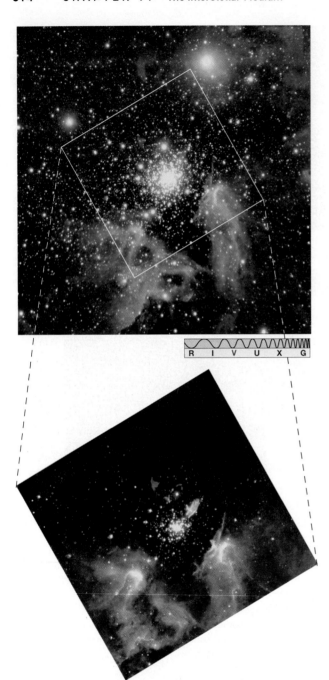

▲ **FIGURE 11.27 Newborn Cluster** The star cluster NGC 3603 and part of the larger molecular cloud in which it formed. The cluster contains about 2000 bright stars and lies some 6000 pc from Earth. The field of view shown here spans about 20 light-years. Radiation from the cluster has cleared a cavity in the cloud several light-years across. The inset shows the central area more clearly, revealing many small stars less massive than the Sun. *(ESO; NASA)*

Clusters and Associations

Figure 11.28(a) shows a small star cluster called the Pleiades, or Seven Sisters, a well-known naked-eye object in the constellation Taurus, lying about 120 pc from Earth. This type of loose, irregular cluster, found mainly in the plane of the Milky Way, is called an **open cluster.** Open clusters typically contain from a few hundred to a few tens of thousands of stars and are a few parsecs across. Figure 11.28(b) shows the H–R diagram for stars in the Pleiades. The cluster contains stars in all parts of the main sequence. The blue stars must be relatively young, for, as we saw in Chapter 10, they burn their fuel very rapidly. ∞ *(Sec. 10.7)* Thus, even though we have no direct evidence of the cluster's birth, we can estimate its age at less than 20 million years, the lifetime of an O-type star. The wisps of leftover gas evident in the photograph are further evidence of the cluster's youth.

Less massive, but more extended, clusters are known as **associations.** These typically contain no more than a few hundred stars, but may span many tens of parsecs. Associations tend to be rich in very young stars and are very loosely bound—if they are bound at all. Many associations appear to be expanding freely into space and dissolving following their formation.

Figure 11.29(a) shows a very different type of star cluster, called a **globular cluster.** All globular clusters are roughly spherical (which accounts for their name), are generally found away from the Milky Way plane, and contain hundreds of thousands, and sometimes millions, of stars spread out over about 50 pc. The H–R diagram for this cluster (called Omega Centauri) is shown in Figure 11.29(b).

The most outstanding feature of globular clusters is their lack of upper main-sequence stars. In fact, globular clusters contain no main-sequence stars with masses greater than about 0.8 times the mass of the Sun. (The A-type stars in this plot are stars at a much later evolutionary stage that happen to be passing through the location of the upper main sequence.) Their more massive O- through F-type stars have long since exhausted their nuclear fuel and disappeared from the main sequence. On the basis of these and other observations, astronomers estimate that all globular clusters are at least 10 billion years old. They contain the oldest known stars in our Galaxy. Astronomers speculate that the 150 or so globular clusters observed today are just the survivors of a much larger population of clusters that formed long ago.

Clusters and Nebulae

How many stars form in a cluster, and of what type are they? What does the collapsed cloud look like once star formation has run its course? At present, although the main stages (3–7) in the formation of individual stars are fairly well established, the answers to these broader questions (involving stages 1 and 2) remain sketchy. Low-mass stars are much more common than high-mass ones, but an explanation of exactly why this is so still awaits a more thorough understanding of the star-formation process. ∞ *(Sec. 10.7)*

At present, researchers are divided on whether more massive stars simply form from denser or more massive clumps of interstellar gas or whether all stars form as relatively lightweight objects (with masses much less than the mass of the Sun). Stars subsequently grow by accretion from their surroundings, in a manner reminiscent of the growth of planetesimals in the early solar system. ∞ *(Sec. 4.3)* In the latter view, the most massive stars are the ones that win the competition for resources.

In either case, computer simulations of star-forming clouds (see Figure 11.30) suggest that the sequence of events leading to main-sequence stars may be strongly influenced by physical interactions—close encounters and even collisions—between protostars within the cluster. The simulations indicate that the strong gravitational fields of the most massive protostars give them a competitive advantage over their smaller rivals in attracting gas from the surrounding nebula, causing giant protostars to grow even

ANIMATION/VIDEO Carina Nebula

faster. Close encounters with massive protostars also tend to disrupt smaller protostellar disks, terminating the growth of their central protostars and ejecting planets and low-mass brown dwarfs from the disk into intracluster space. In dense clusters these interactions may even lead to mergers and further growth of massive objects.

Thus, even though many details remain to be filled in, we can see several ways in which the cluster environment affects the kinds of stars that form there. The most massive stars form fastest and in the process tend to prevent the formation of more high-mass stars, both by stealing their raw material and by helping destroy their

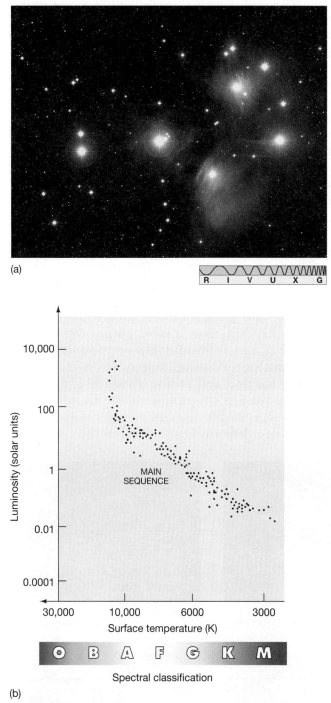

(a)

(b)

▲ FIGURE 11.28 **Open Cluster**   (a) The Pleiades cluster (also known as the Seven Sisters because only six or seven of its stars can be seen with the naked eye) lies about 120 pc from the Sun. (b) An H–R diagram for all the stars of this well-known open cluster. *(NOAO)*

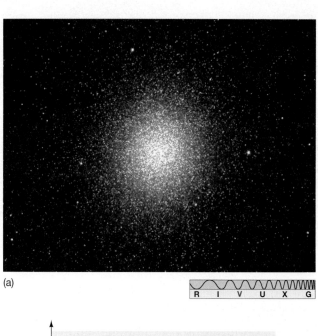

(a)

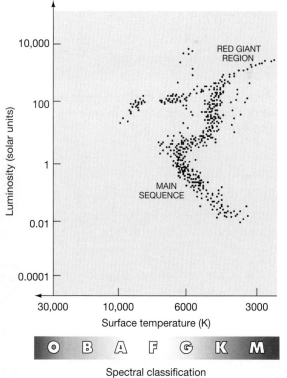

(b)

▲ FIGURE 11.29 **Globular Cluster**   (a) The globular cluster Omega Centauri is approximately 5000 pc from Earth and some 40 pc in diameter. (b) An H–R diagram for some of its stars. *(P. Seitzer)*

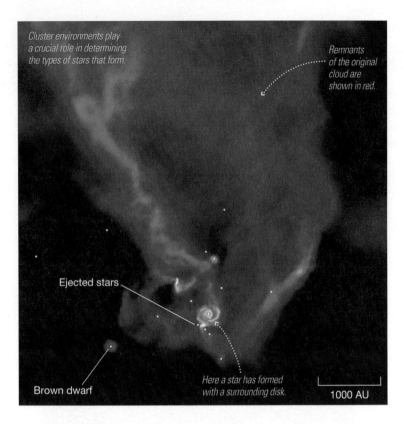

Cluster environments play a crucial role in determining the types of stars that form.

Remnants of the original cloud are shown in red.

Ejected stars

Brown dwarf

Here a star has formed with a surrounding disk.

1000 AU

In the congested environment of a young cluster, star formation is a competitive and violent process. Large protostars may grow by "stealing" gas from smaller ones, and the extended disks surrounding most protostars can lead to collisions and even mergers. This frame from a supercomputer simulation shows a small star cluster emerging from an interstellar cloud that originally contained some 50 solar masses of material spread over a volume 1 light-year across. (M. Bate, I. Bonnell, and V. Bromm)   (MA)

disks. Eventually, the intense radiation from newborn O and B stars disrupts the environment in which lower-mass stars are forming, "freezing in" their masses. This also helps explain the existence of brown dwarfs by providing a mechanism whereby star formation can stop before nuclear fusion begins in a low-mass stellar core.

Young star clusters are often shrouded in gas and dust, making them hard to see in visible light. However, infrared observations clearly demonstrate that clusters really are found within star-forming regions. Figure 11.31 compares optical (a) and infrared (b) views of the central regions of the Orion Nebula. The optical image shows the Trapezium, the group of four bright stars responsible for ionizing the nebula; the infrared image reveals an extensive cluster of stars within and behind the visible nebula and shows many stages of star formation, including nearly 1000 new stars forming and interacting with the surrounding cloud.

Eventually, star clusters dissolve into individual stars. In some cases, the irregular star formation process illustrated in Figure 11.30 simply leaves the newborn cluster gravitationally unbound. In others, the ejection of leftover gas reduces the cluster's mass so much that it becomes unbound and quickly disperses. The tidal gravitational field of the Galaxy slowly destroys the remainder. ∞ (Sec. 5.2) The most massive systems—the globular clusters—may survive for billions of years, but most clusters surrender their stars to the Galactic population in less than a few hundred million years.

Take another look at the sky one clear, dark evening. Ponder all of the cosmic activity you have learned about as you peer upward at the stars. After studying this chapter, you may find that you have to modify your view of the night sky. Even the seemingly quiet nighttime darkness is dominated by continual change.

CONCEPT CHECK ► If stars in a cluster all start to form at the same time, how can some influence the formation of others?

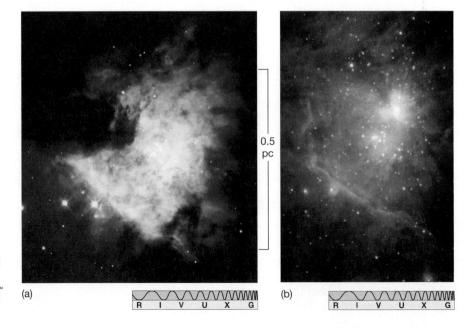

(a) Visible *Hubble* and (b) infrared *Spitzer* views of the central part of the Orion Nebula. The visible image is dominated by the emission nebula and shows few stars. However, the infrared image shows an extensive star cluster containing stars of many masses, possibly including many brown dwarfs (see also Figures 3.24 c and d). (NASA)   (MA)

0.5 pc

(a) R I V U X G

(b) R I V U X G

CHAPTER REVIEW

SUMMARY

L01 The **interstellar medium** (p. 294) occupies the space between stars. It is composed of cold (less than 100 K) gas, mostly atomic or molecular hydrogen and helium, and **dust grains** (p. 294) containing carbon, silicates, and iron. Interstellar dust is very effective at blocking our view of distant stars, even though the density of the interstellar medium is very low. The spatial distribution of interstellar matter is very patchy. The dust preferentially absorbs short-wavelength radiation, leading to **reddening** (p. 295) of light passing through interstellar clouds.

L02 **Emission nebulae** (p. 297) are extended clouds of hot, glowing interstellar matter. Those associated with star formation are caused by hot O- and B-type stars heating and ionizing their surroundings. They are often crossed by dark **dust lanes** (p. 299)—part of the larger molecular cloud from which they formed.

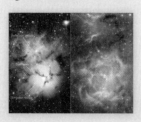

L03 **Dark dust clouds** (p. 301) are cold, irregularly shaped regions in the interstellar medium that diminish or completely obscure the light from background stars. The interstellar medium also contains many cold, dark **molecular clouds** (p. 303). Dust within these clouds probably both protects the molecules and acts as a catalyst to help them form. Molecular clouds are likely sites of future star formation. Often, several molecular clouds are found close to one another, forming a **molecular cloud complex** (p. 304) millions of times more massive than the Sun.

L04 Astronomers can study dark interstellar clouds by observing their effect on the light from more distant stars. Another way to observe these regions of interstellar space is through spectral analysis of the **21-centimeter line** (p. 303) produced whenever the electron in a hydrogen atom reverses its spin, changing its energy very slightly in the process. Astronomers usually study these clouds through observations of other molecules that are less common than hydrogen but much easier to detect.

L05 Stars form when an interstellar cloud collapses under its own gravity and breaks up into smaller pieces. The evolution of the contracting cloud can be represented as an **evolutionary track** (p. 309) on the Hertzsprung–Russell

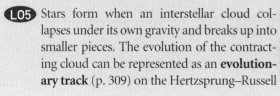

diagram. A cold interstellar cloud (stage 1) containing a few thousand solar masses of gas can fragment into tens or hundreds of smaller clumps of matter, from which stars eventually form. As a collapsing prestellar fragment heats up and becomes denser (stages 2 and 3), it eventually becomes a **protostar** (stage 4; p. 307)—a warm, very luminous object that emits radiation mainly in the infrared portion of the electromagnetic spectrum. The protostar contracts as it radiates its internal energy into space (stage 5). Eventually, the central temperature becomes high enough for hydrogen fusion to begin (stage 6), and the protostar becomes a star (stage 7).

L06 Dark dust clouds and globules are examples of stage-1 clouds and stage-2 collapsing fragments. Examples of stage-3 objects have been observed in some star-forming regions, such as Orion. Stars in the **T-Tauri phase** (p. 309) are examples of stage-4/5 protostars. Low-mass stage-6 stars on their final approach to the main sequence are found in star-forming regions.

L07 Stars of different masses go through similar formation stages, but end up at different locations along the main sequence, high-mass stars at the top, low-mass stars at the bottom. The **zero-age main sequence** (p. 312) is the main sequence predicted by stellar evolutionary theory. It agrees very well with observed main sequences. The most massive stars have the shortest formation times and the shortest main-sequence lifetimes. At the other extreme, some low-mass fragments never reach the point of nuclear ignition. Objects not massive enough to fuse hydrogen to helium are called **brown dwarfs** (p. 312). They may be very common in the universe.

L08 A single contracting and fragmenting cloud can give rise to hundreds or thousands of stars—a **star cluster** (p. 313). **Open clusters** (p. 314), which are loose, irregular clusters that typically contain from a few tens to a few thousands of stars, are found mostly in the Milky Way plane. They typically contain many bright blue stars, indicating that they formed relatively recently. **Globular clusters** (p. 314) are roughly spherical and may contain millions of stars. They include no main-sequence stars more massive than the Sun, indicating that they formed long ago. Infrared observations have revealed young star clusters within several emission nebulae. Eventually, star clusters break up into individual stars, although the process may take billions of years to complete.

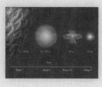

MasteringAstronomy®   For instructor-assigned homework go to www.masteringastronomy.com

Problems labeled POS explore the process of science. VIS problems focus on reading and interpreting visual information. LO connects to the introduction's numbered Learning Outcomes.

REVIEW AND DISCUSSION

1. LO1 What is the composition of interstellar gas? Of interstellar dust?

2. LO3 How is interstellar matter distributed through space?

3. What are some methods that astronomers use to study interstellar dust?

4. LO2 What is an emission nebula?

5. LO4 What is 21-centimeter radiation, and why is it useful to astronomers?

6. If our Sun were surrounded by a cloud of gas, would this cloud be an emission nebula? Why or why not?

7. LO5 Briefly describe the basic chain of events leading to the formation of a star like the Sun.

8. What is an evolutionary track?

9. Why do stars tend to form in groups?

10. What critical event must occur in order for a protostar to become a star?

11. LO7 What are brown dwarfs?

12. POS Stars live much longer than we do, so how do astronomers test the accuracy of theories of star formation?

13. LO6 At what evolutionary stages must astronomers use radio and infrared radiation to study prestellar objects? Why can't they use visible light?

14. POS Explain the usefulness of the H–R diagram in studying the evolution of stars. Why can't evolutionary stages 1–3 be plotted on the diagram?

15. LO8 Compare and contrast the properties of open and globular star clusters.

CONCEPTUAL SELF-TEST: TRUE OR FALSE?/MULTIPLE CHOICE

1. Interstellar matter is quite evenly distributed throughout the Milky Way Galaxy. (T/F)

2. Emission nebulae radiate mainly in the ultraviolet part of the electromagnetic spectrum. (T/F)

3. Twenty-one-centimeter radiation can be used to probe the interiors of molecular clouds. (T/F)

4. More massive stars form more rapidly. (T/F)

5. Brown dwarfs take a long time to form, but will eventually become visible as stars on the lower main sequence. (T/F)

6. The formation of the first high-mass stars in a collapsing cloud tends to inhibit further star formation within that cloud. (T/F)

7. Most stars form as members of groups or clusters. (T/F)

8. VIS The dark interstellar globule shown in Figure 11.2(b) is about the same size as (a) a cloud in Earth's atmosphere; (b) the entire planet Earth; (c) a star like the Sun; (d) the Oort cloud.

9. The red glow of an emission nebula (a) is emitted by warm gas falling onto the stars at the center; (b) is produced by hydrogen gas heated to high temperatures by massive stars within the nebula; (c) is a reflection of light from stars near the nebula; (d) is the unresolved light of many faint red stars within the nebula.

10. VIS The Rho Ophiuchus cloud, shown in Figure 11.12(a) (Dark Dust Cloud), is dark because (a) there are no stars in this region; (b) the stars in this region are young and faint; (c) starlight from behind the cloud does not penetrate the cloud; (d) the region is too cold to sustain stellar fusion.

11. Of the following telescopes, the one best suited to observing dark dust clouds is (a) an X-ray telescope; (b) a large visible-light telescope; (c) an orbiting ultraviolet telescope; (d) a radio telescope.

12. A protostar that will eventually turn into a star like the Sun is significantly (a) smaller; (b) more luminous; (c) fainter; (d) less massive than the Sun.

13. VIS The main difference between the Pleiades cluster (Figure 11.28a) and Omega Centauri (Figure 11.29a) is that the Pleiades cluster is much (a) larger; (b) younger; (c) farther away; (d) denser.

14. VIS In Figure 11.2 (Reddening), the effect of interstellar absorption on spectral lines is (a) to shift them to longer wavelengths; (b) to shift them to shorter wavelengths; (c) to leave their wavelengths unchanged; (d) to erase them.

15. VIS In Figure 11.22 (Newborn Star on the H–R Diagram), as the star moves from stage 6 to stage 7 it becomes (a) cooler and fainter; (b) hotter and smaller; (c) redder and brighter; (d) larger and cooler.

PROBLEMS

The number of squares preceding each problem indicates its approximate level of difficulty.

1. ■ The average density of interstellar gas within the Local Bubble is much lower than the value mentioned in the text—in fact, it is roughly 10^3 hydrogen atoms/m^3. Given that the mass of a hydrogen atom is 1.7×10^{-27} kg, calculate the total mass of interstellar matter contained within a Bubble volume equal in size to planet Earth.

2. ■■ Calculate the frequency of 21-cm radiation. If interstellar clouds along our line of sight have radial velocities in the range 75 km/s (receding) to 50 km/s (approaching), calculate the range of frequencies and wavelengths over which the 21-cm line will be observed. ∞ *(Sec. 2.7)*

3. ■ Calculate the radius of a spherical molecular cloud whose total mass equals the mass of the Sun. Assume a cloud density of 10^{12} hydrogen molecules per cubic meter.

4. ■■ A beam of light shining through a dense molecular cloud is diminished in intensity by a factor of 2 for every 5 pc it travels. By how many magnitudes is the light from a background star dimmed if the total thickness of the cloud is 60 pc?

5. ■■ Estimate the escape speeds near the edges of the four emission nebulae listed in Table 11.1, and compare them with the average speeds of hydrogen nuclei in those nebulae. ∞ *(More*

Precisely 5-1) Is it likely that the nebulae are held together by their own gravity?

6. ■■ For an interstellar gas cloud to contract, the average speed of its constituent particles must be less than half the cloud's escape speed. Will a (spherical) molecular hydrogen cloud of mass of 1000 solar masses, a radius 10 pc, and temperature of 10 K begin to collapse? Why or why not? ∞ *(More Precisely 5-1)*

7. ■ A protostar evolves from a temperature $T = 3500$ K and a luminosity $L = 5000$ times that of the Sun to $T = 5000$ K and $L = 3$ solar units. What is its radius (a) at the start, and (b) at the end of the evolution? ∞ *(Sec. 10.4)*

8. ■ Through roughly how many magnitudes does a three-solar-mass protostar decrease in brightness between stage 4 and stage 6?

9. ■■ What is the luminosity, in solar units, of a brown dwarf whose radius is 0.1 solar radii and whose surface temperature is 600 K (0.1 times that of the Sun)? ∞ *(Sec. 10.4)*

10. ■■ Approximating the gravitational field of our Galaxy as a mass of 10^{11} solar masses at a distance of 8000 pc (see Chapter 14), estimate the "tidal radius" of a 20,000-solar-mass star cluster—that is, the distance from the cluster center outside of which the tidal force due to the Galaxy overwhelms the cluster's gravity. ∞ *(Sec. 5.2)*

ACTIVITIES

Collaborative

1. The Messier catalog is a list of nebulae and other "fuzzy" objects in the night sky, compiled by 18th century astronomer Charles Messier to avoid confusion with his main interest, comets. Today, the catalog is a treasure trove of fascinating deep-sky objects. Many Messier objects are star-forming regions or star clusters. Star-forming regions mentioned in this chapter include M8 (the Lagoon Nebula), M16 (Eagle Nebula), M17 (Omega Nebula), M20 (Trifid Nebula), M42 (Orion Nebula). Interesting star clusters and associations include M6, M7, M11, M35, M37, M44, M45, M52, M67, and M103. Not all of these objects are easily observable on any given night, so consult a sky chart, or do some research online, and make a list of which are visible. A small telescope will give the best results in most cases, and you may want to observe in shifts over the course of the night. For each Messier object on your list, find out what sort of object it is (emission nebula, association, young or old open cluster), carefully follow the finding instructions to locate it, and sketch it (or photograph it, if you have the equipment). Compare your sketch or photograph to images found in this book or online.

Individual

1. Observe the Milky Way on a dark, clear night away from city lights. Is it a continuous band of light across the sky or is it mottled? The parts of the Milky Way that appear missing are actually dark dust clouds lying relatively near the Sun. Identify the constellations in which you see these clouds. Make a sketch and

compare with a star atlas. Find other small clouds in the atlas and try to find them with your eye or with binoculars.

2. The constellation Orion the Hunter is prominent in the winter sky. Its most noticeable feature is a short, straight row of three medium-bright stars: the famous belt of Orion. A line of stars extends from the easternmost star of the belt, toward the south. This is Orion's sword. Toward the bottom of the sword is the sky's most famous emission nebula, the Orion Nebula (M42). Observe the Orion Nebula with your eye, with binoculars, and with a telescope. What is its color? How can you account for this? With the telescope, try to find the Trapezium, a grouping of four stars in the center of the nebula. These are hot, young stars; their energy causes the Orion Nebula to glow.

3. The Trifid Nebula (M20) is another place where new stars are forming. It has been called a "dark night revelation, even in modest apertures." An 8- to 10-inch telescope is needed to see the nebula's triple-lobed structure. Ordinary binoculars reveal the Trifid as a hazy patch located in the constellation Sagittarius, set against the richest part of the Milky Way. What are the dark lanes? Why are other parts of the nebula bright? There have been reports of large-scale changes occurring in this nebula over the last 150 years, based on old drawings showing the nebula looking slightly different from its appearance today. Do you think it possible for an interstellar cloud to undergo a change in appearance on a time scale of years, decades, or centuries?

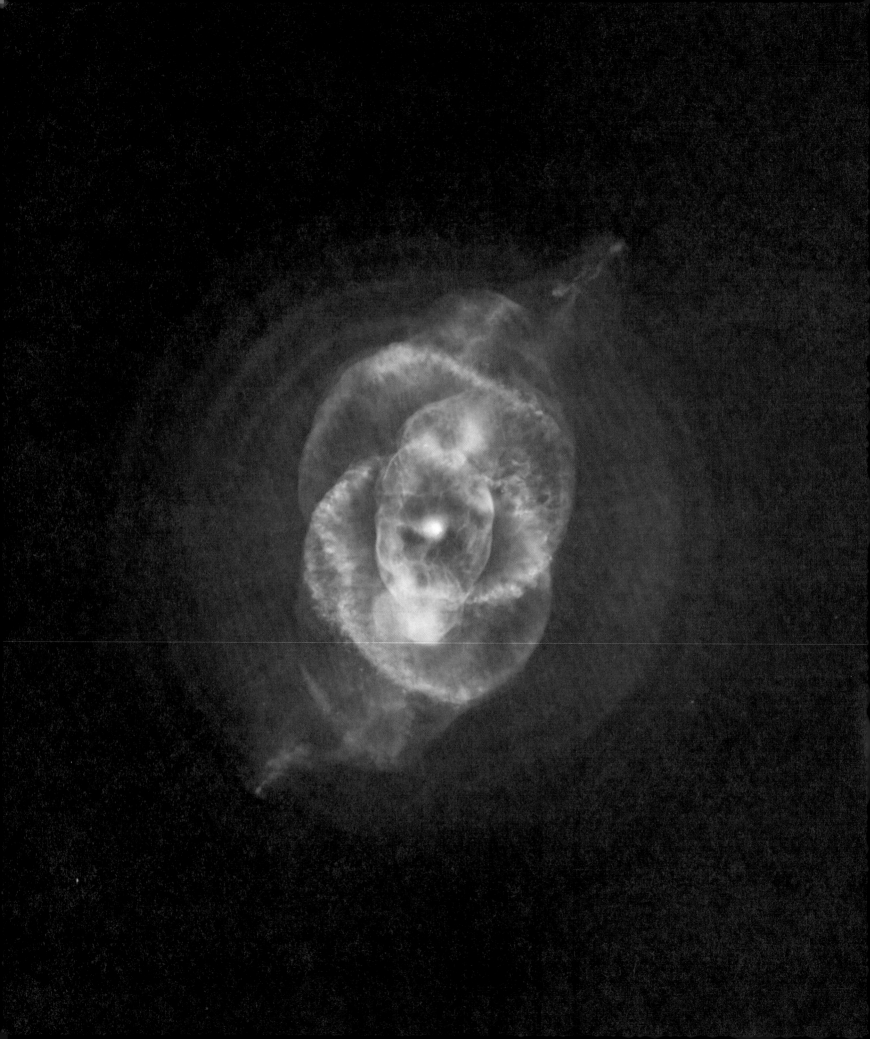

12
Stellar Evolution
The Lives and Deaths of Stars

Once nuclear fusion begins, a newborn star changes little in outward appearance for more than 90 percent of its lifetime. However, at the end of this period, as the star begins to run out of fuel and die, its properties once again change greatly. Aging stars travel along evolutionary tracks that take them far from the main sequence. The fate of a star depends primarily on its mass. Low-mass stars are destined to end their lives quiescently, their outer layers eventually escaping into interstellar space, while very massive stars die violently, in spectacular explosions of almost unimaginable fury. In either case, the deaths of stars enrich the Galaxy with newly created heavy elements. By continually comparing theoretical calculations with detailed observations of stars of all types, astronomers have refined the theory of stellar evolution into a precise and powerful tool for understanding the universe.

MasteringAstronomy® Visit www.masteringastronomy.com for quizzes, animations, videos, interactive figures, and self-guided tutorials.

◄ No one has seen a single star move through all of its evolutionary paces. Like archaeologists who study bones and artifacts of different ages to understand more about human development, astronomers observe stars of different ages and then try to assemble the pieces of data into a consistent model of how stars evolve over billions of years. This striking composite image was taken in visible light (colored red and purple) by the *Hubble* telescope and in X-ray radiation (colored blue) by the *Chandra* telescope. Known technically as NGC 6543, or informally as the Cat's Eye, this complex object is a planetary nebula—an old star (at center) that is shedding its outer layers, which then interact with surrounding gas to create the bizarre-shaped object spread over about a light-year. (*STScI/CXC/NASA*)

12.1 Leaving the Main Sequence

Most stars spend most of their lives on the main sequence of the H–R diagram. A star like the Sun, for example, after spending a few tens of millions of years in formation (stages 1–6 in Chapter 11), will reside on or near the main sequence (stage 7), with roughly constant temperature and luminosity, for some 10 billion years before evolving into something else. ∞ *(Sec. 11.3)* That "something else" is the topic of this chapter.

Stars and the Scientific Method

No one has ever witnessed the complete evolution of any star from birth to death. Stars take a very long time—millions, billions, even trillions of years—to evolve. ∞ *(Sec. 10.7)* Yet in less than a century, astronomers have developed a comprehensive theory of stellar evolution that is one of the best-tested in all of astronomy. How can we can talk so confidently about what took place billions of years in the past, and what will happen billions of years in the future? The answer is that we can observe billions of stars in the universe, enough to see examples of every stage of stellar development, allowing us to test and refine our theoretical ideas. Just as we can piece together a picture of the human life cycle by studying a snapshot of all the residents of a large city, so we can construct a picture of stellar evolution by studying the myriad stars we see in the night sky.

The modern theory of the lives and deaths of stars is another example of the scientific method in action. ∞ *(Sec. 0.5)* Faced with a huge volume of observational data, with little theory to organize or explain it, astronomers in the late 19th and early 20th centuries painstakingly classified and categorized the properties of the stars they observed. ∞ *(Sec. 10.5)* During the first half of the 20th century, as quantum mechanics yielded detailed insight into the behavior of light and matter on subatomic scales, theoretical understanding of key stellar properties fell into place. ∞ *(Sec. 2.6)* Since the 1950s, a truly comprehensive theory has emerged, tying together the basic disciplines of atomic and nuclear physics, electromagnetism, thermodynamics, and gravitation into a coherent whole. Today, theory and observation proceed hand in hand, each refining and validating the details of the other as astronomers continue to hone their understanding of stellar evolution.

Note that astronomers always use the term *evolution* in this context to mean change "during the lifetime of an individual star." Contrast this usage with the meaning of the term in biology, where evolution refers to changes in the characteristics of a *population* of plants or animals over many generations. In fact, populations of stars also evolve in the "biological" sense, as the overall composition of the interstellar medium (and hence of each new stellar generation) changes slowly over time due to nuclear fusion in stars. However, in astronomical parlance, *stellar evolution* refers to changes that occur during a single stellar lifetime.

Structural Change

On the main sequence, a star slowly fuses hydrogen into helium in its core. This process is usually called **core hydrogen burning,** but realize that here is another instance where astronomers use a familiar term in an unfamiliar way: To astronomers, "burning" always means nuclear fusion in a star's core, not a chemical reaction, such as the combustion of wood or gasoline in air, that we would normally mean in everyday speech. Chemical burning does not directly affect atomic nuclei.

As illustrated schematically in Figure 12.1, a main-sequence star is in a state of hydrostatic equilibrium, in which pressure's outward push exactly counteracts

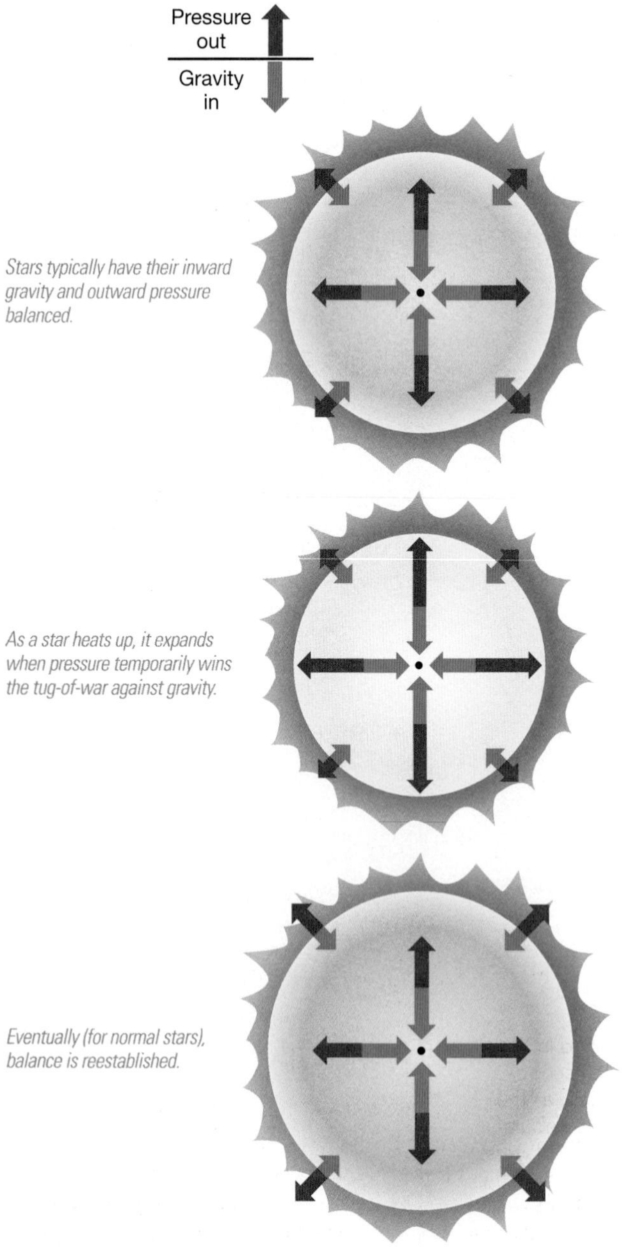

Pressure
out

Gravity
in

Stars typically have their inward gravity and outward pressure balanced.

As a star heats up, it expands when pressure temporarily wins the tug-of-war against gravity.

Eventually (for normal stars), balance is reestablished.

◀ **FIGURE 12.1 Hydrostatic Equilibrium**   In a steadily burning star on the main sequence, the outward pressure exerted by hot gas balances the inward pull of gravity.

gravity's inward pull. ⚭ *(Sec. 9.2)* This is a stable balance between gravity and pressure, in which a small change in one always results in a small compensating change in the other. For example, as shown in the figure, a small increase in the star's central temperature leads to an increase in pressure, causing the star to expand and cool, thus recovering its equilibrium. Conversely, a small temperature decrease leads to a slight pressure decrease, and gravity causes the star to contract and heat up; again, equilibrium is restored. You should keep Figure 12.1 in mind as you study the various stages of stellar evolution described next. Much of a star's complex behavior can be understood in these simple terms.

As the hydrogen in the star's core is consumed, the balance between these opposing forces starts to shift. Both the star's internal structure and its outward appearance begin to change. The star leaves the main sequence. The subsequent stages of stellar evolution—the end of the star's life—depend critically on the star's mass. As a rule of thumb, we can say that low-mass stars die gently, while high-mass stars die catastrophically. The dividing line between these two very different outcomes lies around eight times the mass of the Sun, and in this chapter we will refer to stars of more than 8 solar masses as "high-mass" stars. However, within both the "high-mass" and the "low-mass" (i.e., less than 8 solar masses) categories, there are substantial variations, some of which we will point out as we proceed.

Rather than dwelling on the many details, we will concentrate on a few representative evolutionary sequences. We begin by considering the evolution of a fairly low-mass star like the Sun. Later, we will broaden our discussion to include all stars, large and small.

12.2 Evolution of a Sun-like Star

Figure 12.2 illustrates how the interior composition of a main-sequence star changes as the star ages. The star's helium content increases fastest at the center, where temperatures are highest and the burning (creating helium from hydrogen) is fastest. ⚭ *(Sec. 9.5)* The helium content also increases near the edge of the core, but more slowly because the burning rate is less rapid there. The inner, helium-rich region becomes larger and more hydrogen deficient as the star continues to shine.

Stages 8 and 9—Subgiant to Red Giant

Recall from Chapter 9 that a temperature of 10^7 K is needed to fuse hydrogen into helium. Only above that temperature do colliding hydrogen nuclei (that is, protons) have enough speed to overwhelm their mutual electromagnetic repulsion. ⚭ *(Sec. 9.5)* Because helium nuclei (with two protons each, compared to one for hydrogen) carry a greater positive charge, their electromagnetic repulsion is greater, and so higher temperatures are needed to cause them to fuse. The core temperature at this stage is too low for helium fusion to begin. Eventually, hydrogen becomes depleted at the center, the nuclear fires there subside, and the location of principal burning moves to higher layers in the core. An inner core of nonburning pure helium starts to grow.

Without nuclear burning to maintain it, the outward-pushing gas pressure weakens in the helium inner core; however, the inward pull of gravity does not. Once the pressure is relaxed—even a little—structural changes in the star become inevitable. As soon as hydrogen becomes substantially depleted, about 10 billion years after the star arrived on the main sequence, the helium core begins to contract.

CONCEPT CHECK ▶ In what sense is the Sun stable?

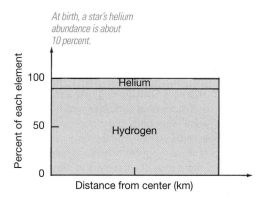

At birth, a star's helium abundance is about 10 percent.

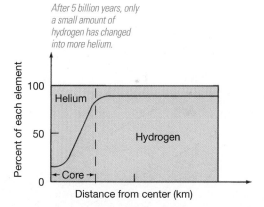

After 5 billion years, only a small amount of hydrogen has changed into more helium.

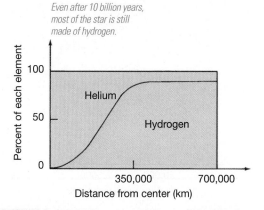

Even after 10 billion years, most of the star is still made of hydrogen.

▲ **FIGURE 12.2 Solar Composition Change** Theoretical estimates of the changes in a Sun-like star's composition show how hydrogen (yellow) and helium (blue) abundance vary within the star from birth to death (top to bottom).

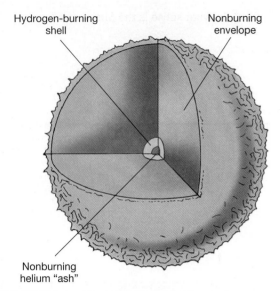

Hydrogen-burning shell

Nonburning envelope

Nonburning helium "ash"

▲ **FIGURE 12.3 Hydrogen Shell Burning** As a star's core converts more and more of its hydrogen into helium, the hydrogen in the shell surrounding the nonburning helium "ash" burns ever more violently. By the time the star has reached the bottom of the giant branch, its core has shrunk to a few tens of thousands of kilometers in diameter, and its surface is 10 times the star's original size.

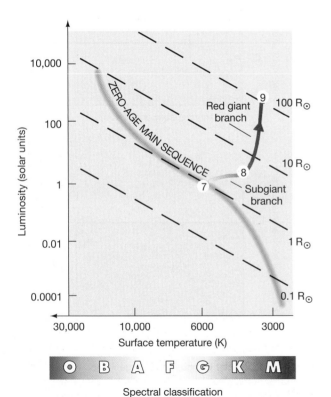

▲ FIGURE 12.4 **Red Giant on the H–R Diagram** As its helium core shrinks and its outer envelope expands, the star leaves the main sequence (stage 7). At stage 8, the star is well on its way to becoming a red giant. The star continues to brighten and grow as it ascends the red giant branch to stage 9.

The shrinkage of the helium core releases gravitational energy, driving up the central temperature, heating the overlying layers of the core, and causing the hydrogen there to fuse even more rapidly than before. Figure 12.3 depicts this **hydrogen shell-burning** stage, in which hydrogen is burning at a furious rate in a relatively thin layer surrounding the nonburning inner core of helium "ash." The hydrogen shell generates energy faster than did the original main-sequence star's hydrogen-burning core, and the shell's energy production continues to increase as the helium core contracts. Strange as it may seem, the star's response to the disappearance of the nuclear fire at its center is to get brighter!

The pressure exerted by this enhanced hydrogen burning causes the star's nonburning outer layers to increase in radius. Not even gravity can stop them. Even as the core shrinks and heats up, the overlying layers expand and cool. The star, aged and unbalanced, is on its way to becoming a red giant. The change from normal main-sequence star to elderly red giant takes about 100 million years.

We can trace these large-scale changes on an H–R diagram. Figure 12.4 shows the star's path away from the main sequence, labeled as stage 7. ∞ *(Sec. 11.3)* The star evolves to the right on the diagram, its surface temperature dropping while its luminosity increases only slightly. The star's roughly horizontal track from the main-sequence (stage 7) to stage 8 on the figure is called the **subgiant branch.** By stage 8, the star's radius has increased to about three times the radius of the Sun. The nearly vertical (constant temperature) path followed by the star between stages 8 and 9 is known as the **red giant branch** of the H–R diagram. (Figure 12.3 corresponds to a point just after stage 8, as the star begins to ascend the giant branch.)

The red giant is huge. By stage 9 its luminosity is many hundreds of times the solar value. Its radius is around 100 solar radii—about the size of Mercury's orbit. In contrast, its helium core is surprisingly small, less than 1/1000 the size of the entire star, or just a few times larger than Earth. Continued shrinkage of the red giant's core has compacted its helium gas to a density of approximately 10^8 kg/m^3—10,000 times denser than any material found on Earth. About 25 percent of the mass of the entire star is packed into its planet-sized core.

A familiar example of a low-mass star in the red giant phase is the KIII giant Arcturus (see Figure 10.14), one of the brightest stars in our night sky. Its mass is about 1.5 times that of the Sun. Currently in the hydrogen shell-burning stage and ascending the red giant branch, its radius is some 21 times the solar value. ∞ *(Table 10.4)* It emits more than 160 times more energy than the Sun, much of it in the infrared part of the spectrum.

Stage 10—Helium Fusion

This simultaneous contraction of the red giant's core and expansion of its envelope—the nonburning layers surrounding the core—does not continue indefinitely. A few hundred million years after a solar-mass star leaves the main sequence, the central temperature reaches the 10^8 K needed for helium to fuse into carbon, and the nuclear fires reignite.

For stars comparable in mass to the Sun, the high densities and pressures found in the stage-9 core mean that the onset of helium fusion is a very violent event. Once the burning starts, the core cannot respond quickly enough to the rapidly changing conditions within it, and its temperature rises sharply in a runaway explosion called the **helium flash.** For a few hours, the helium burns ferociously, like an uncontrolled bomb. Eventually, the star's structure "catches up" with the flood of energy dumped into it by helium burning. The core expands, its density drops, and equilibrium is restored as the inward pull of gravity and the outward push of gas pressure come back into balance. The core, now stable, begins to burn helium into carbon at temperatures well above 10^8 K.

The helium flash terminates the star's ascent of the red giant branch of the H–R diagram. Yet despite the explosive detonation of helium in the core, the flash does *not* increase the star's luminosity. Rather, the helium flash produces an expansion and cooling of the core that ultimately results in a *reduction* in the energy output as the star jumps from stage 9 to stage 10. As indicated in Figure 12.5, the surface temperature is now higher than it was on the red giant branch, although the luminosity is less than at the helium flash. This adjustment in the star's properties occurs quite quickly—in about 100,000 years.

At stage 10 our star is now stably burning helium in its inner core and fusing hydrogen in a shell surrounding that core. It resides in a well-defined region of the H–R diagram known as the **horizontal branch,** where core helium-burning stars remain for a time before resuming their journey around the H–R diagram. The star's specific position within this region is determined mostly by its mass—not its original mass, but whatever mass remains after the red giant phase. The two masses differ because strong stellar winds can eject large amounts of matter—as much as 20–30 percent of the initial mass—from a red giant's surface. The more massive stars have thicker envelopes and lower surface temperatures at this stage, but all stars have roughly the same luminosity after the helium flash. As a result, stage-10 stars tend to lie along a horizontal line on the H–R diagram, with more massive stars to the right and less massive ones to the left.

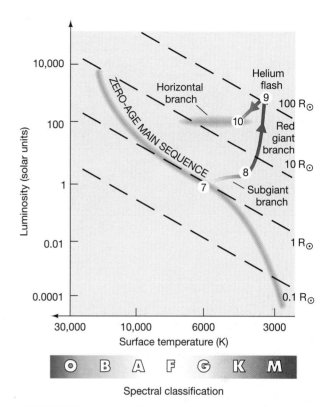

▲ **FIGURE 12.5 Horizontal Branch** A large increase in luminosity occurs as a star ascends the red giant branch, ending in the helium flash. The star then settles down into another equilibrium state at stage 10, on the horizontal branch.

CONCEPT CHECK ► Why does a star get brighter as it runs out of fuel in its core?

Stage 11—A Red Giant Again

Nuclear reactions in stars proceed at rates that increase very rapidly with temperature. At the extremely high temperatures found in the core of a horizontal-branch star, the helium fuel doesn't last long—no more than a few tens of millions of years after the initial flash.

As helium fuses to carbon, a new carbon-rich inner core begins to form and phenomena similar to the earlier buildup of helium occur. Helium becomes depleted at the center, and eventually fusion ceases there. The nonburning carbon core shrinks in size—even as its mass increases due to helium fusion—and heats up as gravity pulls it inward, causing the hydrogen- and helium-burning rates in the overlying layers to increase. The star now contains a contracting carbon-ash inner core surrounded by a helium-burning shell, which is in turn surrounded by a hydrogen-burning shell (Figure 12.6). The outer envelope of the star expands, much as it did earlier in the first red giant stage. By the time it reaches stage 11 in Figure 12.7, the star has become a swollen red giant for a second time.

The burning rates in the shells surrounding the inner core are much fiercer during the star's second trip into the red giant region, and the radius and luminosity increase to values even greater than those reached during the first visit (stage 9). Its carbon core continues to shrink, driving the hydrogen- and helium-burning shells to higher and higher temperatures and luminosities. To distinguish this second visit

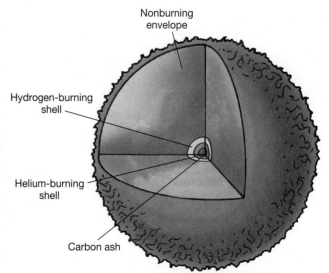

▲ **FIGURE 12.6 Helium Shell Burning** Within a few million years after the onset of helium burning (stage 9), carbon ash accumulates in the star's inner core (not to scale). Above this core, hydrogen and helium are still burning in concentric shells.

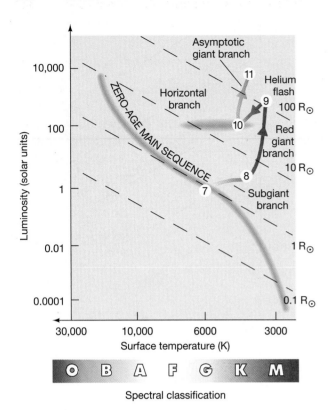

▲ FIGURE 12.7 Reascending the Giant Branch A carbon-core star reenters the giant region of the H–R diagram (stage 11) for the same reason it evolved there the first time around: Lack of nuclear fusion at the center causes the core to contract and the overlying layers to expand.

to the red giant region from the first, the star's evolutionary track during this stage is often referred to as the *asymptotic giant branch*.[1]

Table 12.1 summarizes the key stages through which our solar-mass star evolves. It is a continuation of Table 11.2, except that the density units are now the more convenient kilograms per cubic meter and we now express radii in units of the solar radius. As in Chapter 11, the numbers in the "Stage" column refer to the evolutionary stages noted in the figures and discussed in the text. Figure 12.8 illustrates the stages through which a G-type star like our Sun will evolve over the course of its lifetime.

12.3 The Death of a Low-Mass Star

As our star moves from stage 10 to stage 11, its envelope swells while its inner carbon core, too cool for further nuclear burning, continues to contract. If the central temperature could become high enough for carbon fusion to occur, still heavier elements could be synthesized, and the newly generated energy might again support the star, restoring for a time the equilibrium between gravity and pressure. For solar-mass stars, however, this does not occur—a lot more mass is needed, as we will see in the next section. The temperature never reaches the 600 million K needed for new nuclear reactions to occur. The red giant is now very close to the end of its nuclear-burning lifetime.

Stage 12—A Planetary Nebula

Our aging star is now in quite a predicament. Its carbon core is, for all practical purposes, dead, while the outer hydrogen- and helium-burning shells consume fuel at a furious and increasing rate. As it expands, cools, and reascends the giant branch, the star begins to fall apart. Driven by the intense radiation from within, its outer layers begin to drift away into interstellar space. Slowly at first, then more rapidly as the core luminosity increases, the star loses virtually its entire envelope in less than a million years. The former red giant now consists of two distinct pieces:

[1]The term is borrowed from mathematics. An asymptote to a curve is a second curve that approaches ever closer to the first as the two are extended to infinity. Theoretically, if the star remained intact, the asymptotic-giant branch would approach the red giant branch from the left as the luminosity increased and would effectively merge with the red giant branch near the top of Figure 12.10. However, a Sun-like star will not live long enough for that to occur.

ANIMATION/VIDEO Death of the Sun Part I   (MA)

ANIMATION/VIDEO Death of the Sun Part II   (MA)

INTERACTIVE ▼ **FIGURE 12.8 G-type Star Evolution** Artist's conception of the relative sizes and changing colors of a normal G-type star (such as our Sun) during its formative stages, on the main sequence, and while passing through the red giant and white dwarf stages. At maximum swelling, the red giant is approximately 70 times the size of its main-sequence parent; the core of the giant is about $1/15$ the main-sequence size and would be barely discernible if this figure were drawn to scale. The duration of time spent in the various stages—protostar, main-sequence star, red giant, and white dwarf—is roughly proportional to the lengths shown in this imaginary trek through space. (MA)

Protostar

Main-sequence G-type star

Stage 4

Stage 7

TABLE 12.1 Evolution of a Sun-like Star

Stage	Approx. Time to Next Stage (yr)	Central Temperature (K)	Surface Temperature (K)	Central Density (kg/m^3)	Radius (km)	Radius (solar radii)	Object
7	10^{10}	1.5×10^7	6,000	10^5	7×10^5	1	Main-sequence star
8	10^8	5×10^7	4,000	10^7	2×10^6	3	Subgiant
9	10^5	10^8	4,000	10^8	7×10^7	100	Red giant/helium flash
10	5×10^7	2×10^8	5,000	10^7	7×10^6	10	Horizontal branch
11	10^4	2.5×10^8	4,000	10^8	4×10^8	500	Red giant (AGB)
	10^5	3×10^8	100,000	10^{10}	10^4	0.01	Carbon core
12	—	—	3,000	10^{-17}	7×10^8	1,000	Planetary nebula*
13	—	10^8	50,000	10^{10}	10^4	0.01	White dwarf
14	—	Close to 0	Close to 0	10^{10}	10^4	0.01	Black dwarf

*Values in columns 2–7 refer to the envelope.

the exposed core, very hot and still very luminous, surrounded by a cloud of dust and cool gas—the escaping envelope—expanding away at a typical speed of a few tens of kilometers per second.

As the core exhausts its last remaining fuel, it contracts and heats up, moving to the left in the H–R diagram. Eventually, it becomes so hot that its ultraviolet radiation ionizes the inner parts of the surrounding cloud, producing a spectacular display called a **planetary nebula** (see Figure 12.9). The word *planetary* here is misleading, for these objects have no connection whatever with planets. The name originated in the 18th century when, viewed at poor resolution through small telescopes, these shells of gas looked to some astronomers like the circular disks of the planets in our solar system.

Note that the mechanism by which planetary nebulae shine is very similar to that powering the emission nebulae we studied earlier—ionizing radiation from a hot star embedded in a cool gas cloud. ∞ *(Sec. 11.2)* However, recognize that these two classes of object have very different origins and represent completely separate phases of stellar evolution. The emission nebulae in Chapter 11 were the signposts of recent stellar birth. Planetary nebulae indicate impending stellar death.

Astronomers once thought that the escaping giant envelope would be more or less spherical in shape, completely surrounding the core in three dimensions, just as it had while still part of the star. Figure 12.9(a) shows an example in which this may well, in fact, be the case. The nebula looks brighter near the edges simply because there is more emitting gas along the line of sight there, creating the illusion of a bright ring. However, such cases now seem to be in the minority. There is growing evidence that, for reasons not yet fully understood, the final stages of red giant mass loss are often decidedly *non*spherical. The Ring Nebula shown in Figure 12.9(b) may very well *be* a ring and not just our view of a glowing shell of gas as was once thought. As illustrated in Figure 12.9(c), some planetary nebulae exhibit much more complex

 ANIMATION/VIDEO Helix Nebula

 ANIMATION/VIDEO Helix Nebula Animation

 ANIMATION/VIDEO Helix Nebula White Dwarf

 ANIMATION/VIDEO White Dwarf Cooling Sequence

 ANIMATION/VIDEO Bi-Polar Planetary Nebula

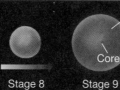

Red giant

White dwarf

Core

Stage 8 Stage 9

(a)

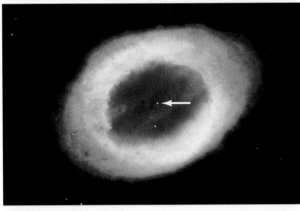

(b)

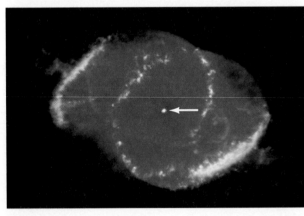

(c)

▲ FIGURE 12.9 Planetary Nebulae   A planetary nebula is an extended region of glowing gas surrounding a small, dense core (visible in each case denoted by an arrow). (a) Abell 39, some 2100 pc away, is a classic planetary nebula shedding a spherical shell of gas about 1.5 pc across; (b) Ring Nebula, perhaps the most famous of all planetaries at 1500 pc distant and 0.5 pc across, is probably a genuine ring, yet too small and dim to be seen with the naked eye; (c) Cat's Eye Nebula (see also page 318) about 1000 pc away and 0.1 pc across, is an example of a much more complex planetary nebula, possibly produced by a pair of binary stars (unresolved at center) that have both shed envelopes. (AURA; NASA)

structures, suggesting that the star's environment—including the existence of a binary companion—can play an important role in determining the nebula's shape and appearance.

The central star fades and eventually cools, and the expanding gas cloud becomes more and more diffuse, gradually dispersing into interstellar space. In doing so it plays a vital role in the evolution of our Galaxy. During the final stages of the red giant's life, nuclear reactions between carbon and unburned helium in the core create oxygen, and in some cases even heavier elements, such as neon and magnesium. Some of these reactions also release neutrons that, carrying no electrical charge, have no electrostatic repulsion to overcome and hence can interact with existing nuclei to form still heavier elements. ∞ (Secs. 2.6, 9.5) All of these elements—helium, carbon, oxygen, and heavier—are "dredged up" from the depths of the core into the envelope by convection during the star's final years and enrich the interstellar medium when the giant envelope escapes. ∞ (Sec. 9.2) The evolution of low-mass stars is the source of virtually all of the carbon-rich dust observed throughout the plane of our Galaxy. ∞ (Sec. 11.1)

Dense Matter

Before the shrinking carbon core can become hot enough for carbon to fuse, the rising pressure stops the contraction and stabilizes the temperature. However, this pressure is not the "normal" thermal pressure of high-speed particles in a very hot gas. ∞ (More Precisely 5-1) Instead, as we now discuss, the core enters a state in which the free electrons in the core—a vast sea of charged particles stripped from their parent nuclei by the ferocious heat in the stellar interior—play a critical role in determining the star's future.

The core density at this stage (stage 12 in Table 12.1) is enormous—about 10^{10} kg/m^3. This density is much higher than anything we have seen so far in our study of the cosmos. A single cubic centimeter of core matter would weigh 1000 kg on Earth—a ton of matter compressed into a volume about the size of a grape! Under these extreme conditions, a law of quantum physics known as the *Pauli Exclusion Principle* comes into play, preventing the electrons in the core from being crushed any closer together. (In fact, the core was in a similar high-density state at the moment of helium ignition, causing the helium flash.)

In essence, the electrons behave like tiny rigid spheres that can be squeezed relatively easily up to the point of contact, but become virtually incompressible thereafter. By stage 12 the pressure in the carbon core resisting the force of gravity is supplied almost entirely by tightly packed electrons, and this pressure brings the core back into equilibrium at a central temperature of "only" about 300 million K—too cool to fuse carbon into any of the heavier elements. This stage represents the maximum compression that the star can achieve—there is simply not enough matter in the overlying layers to bear down any harder. Supported by the resistance of its electrons to further compression, the contraction of the core stops.

Stage 13—A White Dwarf

Formerly concealed by the atmosphere of the red giant star, the carbon core becomes visible as a *white dwarf* as the envelope recedes. ∞ (Sec. 10.4) The core is very small, about the size of Earth, with a mass about half that of the Sun. Shining only by stored heat, not by nuclear reactions, this small star has a white-hot surface when it first becomes visible, although it appears dim because of its small size. This is stage 13 of Table 12.1. The approximate path followed by the star on the H–R diagram as it evolves from a stage 11 red giant to a stage 13 white dwarf is shown in Figure 12.10.

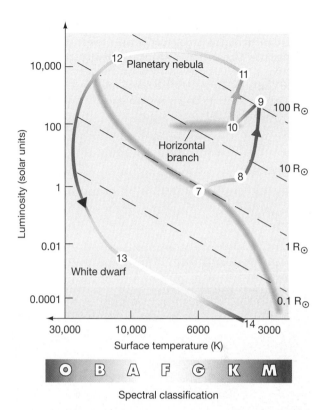

INTERACTIVE ▶ FIGURE 12.10 White Dwarf on H–R Diagram   A star's passage from the horizontal branch (stage 10) to the white dwarf stage (stage 13) creates an evolutionary path that cuts across the H–R diagram.

Not all white dwarfs are seen as the cores of planetary nebulae. Several hundred have been discovered "naked," their envelopes expelled to invisibility long ago. Figure 12.11 shows an example of a white dwarf, Sirius B, that happens to lie particularly close to Earth; it is the faint binary companion of the much brighter Sirius A. ⚬⚬ *(Sec. 10.7)* Some of its properties are listed in Table 12.2. With more than a solar mass of material packed into a volume smaller than Earth, Sirius B's density is about a million times greater than anything familiar to us in the solar system. Sirius B has an unusually high mass for a white dwarf. It is thought to be the evolutionary product of a star roughly four times the mass of the Sun.

Hubble Space Telescope observations of nearby globular clusters have revealed the white dwarf sequences long predicted by theory but previously too faint to detect at such large distances. Figure 12.12(a) shows a ground-based view of the globular cluster M4, lying 2100 pc from Earth. Part (b) of the figure shows a *Hubble* close-up of a small portion of the cluster, revealing dozens of white dwarfs (some marked) among the cluster's much brighter main-sequence, red giant, and horizontal-branch stars. When plotted on an H–R diagram (see Figure 12.26), the white dwarfs fall nicely along the line indicated on Figure 12.10.

The white dwarf continues to cool and dim with time, following the white–yellow–red line near the bottom of Figure 12.10, eventually becoming a **black dwarf**—a cold, dense, burned-out cinder in space. This is stage 14 of Table 12.1, the graveyard of stars. The cooling dwarf does not shrink much as it fades away, however. Even though its heat is leaking away into space, gravity does not compress it further. Its tightly packed electrons will support the star even as its temperature drops (after trillions of years) almost to absolute zero. As the dwarf cools, it remains about the size of Earth.

Most white dwarfs are composed of carbon and oxygen, but theory predicts that very low-mass stars (less than about one-quarter the mass of the Sun) will never reach the point of helium fusion. Instead, their cores will become supported by the resistance of its electrons to further compression before the central temperature reaches the 100 million K needed for helium to fuse. The interiors of such stars are completely convective, ensuring that fresh hydrogen continually mixes from the envelope into the core. As a result, unlike the case of the Sun illustrated in Figure 12.2, a nonburning helium inner core never appears, and eventually *all* of the star's hydrogen is converted to helium, forming a *helium white dwarf.*

The time needed for this kind of transformation to occur is very long—hundreds of billions of years—so no helium white dwarfs have ever actually formed in this way. ⚬⚬ *(Sec. 10.7)* However, if a solar-mass star is a member of a *binary* star system, it is possible for its envelope to be stripped away during the red giant stage by the gravitational pull of its companion, exposing the helium core and terminating

TABLE 12.2	Sirius B—A Nearby White Dwarf
Mass	1.1 Solar masses
Radius	0.008 Solar radii (5500 km)
Luminosity (total)	0.0025 Solar luminosities
Surface temperature	24,000 K
Average density	3×10^9 kg/m³

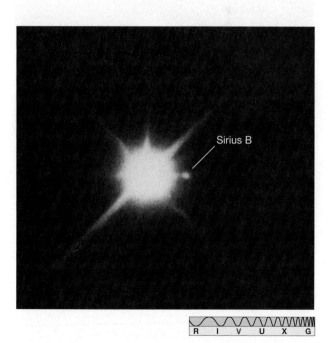

R I V U X G

▲ FIGURE 12.11 Sirius Binary System   Sirius B (the speck of light to the right of the much larger and brighter Sirius A) is a white dwarf star, a companion to Sirius A. The "spikes" on the image of Sirius A are not real; they are caused by the support struts of the telescope. *(Palomar Observatory)*

▶ FIGURE 12.12 **Distant White Dwarfs**   (a) The globular cluster M4, as seen through a large ground-based telescope at Kitt Peak National Observatory in Arizona, is the closest globular cluster to us, at 1700 pc away; it spans some 16 pc. (b) A peek at M4's "suburbs" by the *Hubble Space Telescope* shows nearly a hundred white dwarfs within a small 0.2 square-parsec region. *(AURA; NASA)*

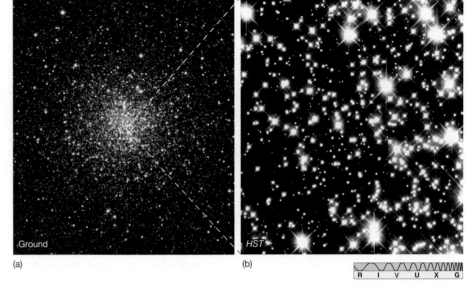

the star's evolution before helium fusion can begin. Several such low-mass helium white dwarfs have in fact been detected in binary systems in our Galaxy.

In stars much more massive than the Sun (close to the 8-solar mass limit on "low-mass" stars at the time the carbon core forms—see Section 12.4), temperatures in the core may become high enough that oxygen can combine with helium to form neon, eventually leading to the formation of a rare *neon–oxygen white dwarf* when fusion finally stops.

Novae

In some cases, the white dwarf stage does not represent the end of the road for a Sun-like star. Given the right circumstances, it is possible for a white dwarf to become explosively active, in the form of a highly luminous **nova** (plural: novae). The word *nova* means "new" in Latin, and to early observers, these stars did indeed seem new as they suddenly appeared in the night sky. Astronomers now recognize that a nova is what we see when a white dwarf undergoes a violent explosion on its surface, resulting in a rapid, temporary increase in luminosity. Figures 12.13(a) and (b) illustrate the brightening of a typical nova over a period of 3 days. Figure 12.13(c) shows the nova's light curve, demonstrating how the luminosity rises dramatically in a matter of days, then fades slowly back to normal over the course of several months. On average, two or three novae are observed each year.

What could cause such an explosion on a faint, dead star? The energy involved is far too great to be explained by flares or other surface activity, and as we have just seen, there is no nuclear activity in the dwarf's interior. The answer to this question lies in the white dwarf's surroundings. If the white dwarf is isolated, then it will indeed cool and ultimately become a black dwarf, as just described. However, should the white dwarf be part of a binary system in which the other star is either a main-sequence star or a giant, an important new possibility exists.

If the distance between the dwarf and the other star is small enough, the dwarf's gravitational field can pull matter—primarily hydrogen and helium—away from the surface of the companion. Because of the binary's rotation, material leaving the

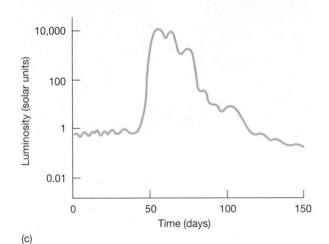

(c)

◀ FIGURE 12.13 **Nova**   Shown here is Nova Herculis 1934 in (a) March 1935 and (b) May 1935, after brightening by a factor of 60,000. (c) The light curve of a typical nova displays a rapid rise followed by a slow decline in the light received from the star, in good agreement with the explanation of the nova as a nuclear flash on a white dwarf's surface. *(UC/Lick Observatory)*

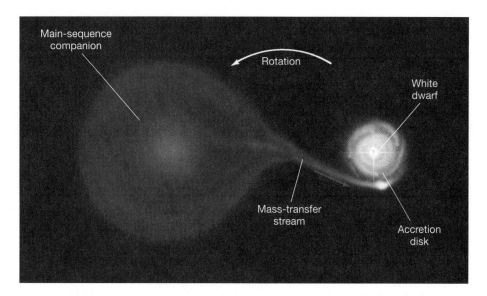

◄ FIGURE 12.14 Close Binary System   If a white dwarf in a binary system is close enough to its companion, its gravitational field can tear matter from the companion's surface. Notice that the matter does not fall directly onto the white dwarf's surface. Instead, it forms an accretion disk of gas spiraling down onto the dwarf.

companion does not fall directly onto the white dwarf. Instead, it "misses" the dwarf, loops around behind it, and goes into orbit around it, forming a swirling, flattened disk of matter called an **accretion disk,** as shown in Figure 12.14. As it builds up on the white dwarf's surface, the stolen gas becomes hotter and denser. Eventually, its temperature exceeds 10^7 K and the hydrogen ignites, fusing into helium at a furious rate. This surface-burning stage is as brief as it is violent. The star suddenly flares up in luminosity, then fades away as some of the fuel is exhausted and the rest is blown off into space (Figure 12.15). If the event happens to be visible from Earth, we see a nova.

Once the nova explosion is over and the binary has returned to normal, the mass-transfer process can begin again. Astronomers know of many *recurrent novae*—stars that have been observed to "go nova" several times over the course of a few decades. Such systems can, in principle, repeat their violent outbursts many dozens, if not hundreds, of times.

CONCEPT CHECK ►   Will the Sun ever become a nova?

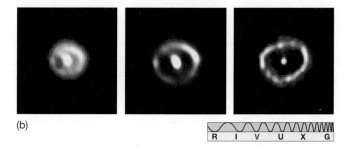

(a)

(b)

MA   ANIMATION/VIDEO Recurrent Nova

◄ FIGURE 12.15 Nova Matter Ejection   (a) The ejection of material from the star's surface can clearly be seen in this image of Nova Persei, taken some 50 years after it suddenly brightened by a factor of 40,000 in 1901. (b) Nova Cygni, imaged here with a European camera on the *Hubble Space Telescope*, erupted in 1992. At left, more than a year after the blast, a rapidly billowing bubble is seen; at right, 7 months after that, the shell continued to expand and distort. (*Palomar Observatory; ESA*)

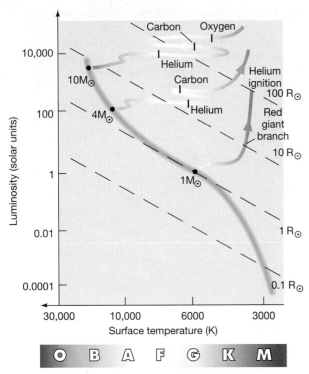

INTERACTIVE ▲ FIGURE 12.16 **High-Mass Evolutionary Tracks** Evolutionary tracks for stars of 1, 4, and 10 solar masses. Stars with masses comparable to the mass of the Sun ascend the giant branch almost vertically, whereas higher-mass stars move roughly horizontally across the H–R diagram from the main sequence into the red giant region. Some points are labeled with the element that has just started to fuse in the inner core.

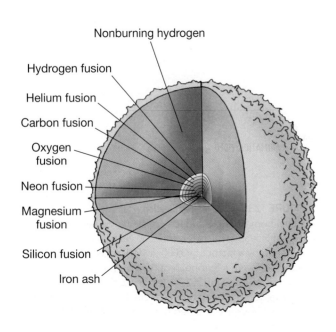

FIGURE 12.17 **Heavy-Element Fusion** Cutaway diagram of the interior of a highly evolved high-mass star shows how the interior resembles the layers of an onion, with shells of progressively heavier elements burning at smaller and smaller radii and at higher and higher temperatures. The core is actually only a few times larger than planet Earth, while the star is hundreds of times larger than the Sun.

12.4 Evolution of Stars More Massive than the Sun

All stars leaving the main sequence on their way toward the red giant region of the H–R diagram have similar internal structure. Thereafter, however, their evolutionary tracks diverge.

Formation of Heavy Elements

Figure 12.16 compares the post-main-sequence evolution of three stars having masses 1, 4, and 10 times the mass of the Sun. Note the qualitative differences between the solar-mass and higher-mass evolutionary tracks. The solar-mass star ascends the red giant branch almost vertically after leaving the main sequence, its luminosity and radius increasing dramatically while its temperature changes only a little. The higher-mass stars tend to do just the opposite. They loop back and forth across the top of the H–R diagram, their luminosities staying roughly constant as their radii alternately increase and decrease and their surface temperatures fall and rise. The 4- and 10-solar-mass stars do not undergo helium flashes when helium fusion begins, and they do reach core temperatures high enough for helium and carbon to fuse into oxygen.

In the 4-solar-mass star, nuclear burning ceases after the carbon-fusion stage, eventually leading to a carbon–oxygen white dwarf, much as described previously. The fate of the 10-solar-mass star is a little less certain. Depending on how much mass it loses as it evolves (see below), it may also end up as a white dwarf, or it might retain enough mass for fusion to proceed to more massive nuclei. Even higher-mass stars can fuse not just hydrogen, helium, and carbon, but also oxygen, neon, magnesium, silicon, and even heavier elements as their inner cores continue to contract and their central temperatures continue to rise.

Evolution proceeds so rapidly for stars bigger than 10 solar masses that they don't even reach the red giant region before helium fusion begins. Such stars achieve central temperatures of 10^8 K while still quite close to the main sequence. As each element is burned to depletion at the center, the core contracts, heats up, and fusion starts again. A new inner core forms, contracts again, heats again, and so on. The star's evolutionary track continues smoothly across the H–R diagram, seemingly unaffected by each new phase of burning. The star's luminosity stays roughly constant as its radius increases and its surface temperature drops. It swells to become a **red supergiant.**

Figure 12.17 is a cutaway diagram of an evolved high-mass star nearing the end of its red supergiant stage. Note the numerous layers where various nuclei burn. As the temperature increases with depth, the product of each burning stage becomes the fuel for the next. At the relatively cool periphery of the core, hydrogen fuses into helium. In the intermediate layers, shells of helium, carbon, and oxygen burn to form heavier nuclei. Deeper down reside neon, magnesium, silicon, and other heavy nuclei, all produced by nuclear fusion in the layers overlying the nonburning inner core. The inner core itself is composed of iron.

Observations of Supergiants

A good example of a post-main-sequence blue supergiant is the bright star Rigel in the constellation Orion. With a radius some 70 times that of the Sun and a luminosity of around 60,000 solar luminosities, Rigel is thought to have had an original mass about 17 times that of the Sun. A strong stellar wind has probably carried away a significant fraction of the star's mass since it formed. Although still near the main sequence, Rigel is probably already fusing helium into carbon in its core.

Astronomers have discovered that, while stars of all spectral types have stellar winds, the highly luminous supergiants have by far the strongest. ∞ *(Sec. 9.3)* Satellite and rocket observations have shown that the winds of some blue supergiants may reach speeds of 3000 km/s and carry off more than 10^{-6} solar masses of material per year. Over the relatively short span of a few million years, these stars can blow a tenth or more of their total mass—a Sun's worth of matter—off into space. During violent outbursts near the end of the star's life, the mass loss rate may exceed 10^{-3} solar masses per year. Figure 12.18 shows a luminous blue variable star called Eta Carinae that is currently displaying this type of behavior. The glowing nebula surrounding the star itself probably contains some 2–3 solar masses of gas shed during this and previous outbursts over the past few hundred years.

Perhaps the best-known red supergiant is Betelgeuse (shown in Figures 10.7 and 10.10), also in Orion and Rigel's rival for the title of brightest star in the constellation. Its luminosity is 10,000 times that of the Sun in visible light and perhaps four times that in the infrared. Astronomers think that Betelgeuse is currently fusing helium into carbon and oxygen in its core, but its eventual fate is uncertain. As best we can tell, the star's mass at formation was between 12 and 17 times the mass of the Sun. However, like Rigel and many other supergiants, Betelgeuse has a strong stellar wind and is known to be surrounded by a huge shell of dust of its own making. This suggests that Betelgeuse has lost a lot of mass since it formed, but just how much remains uncertain.

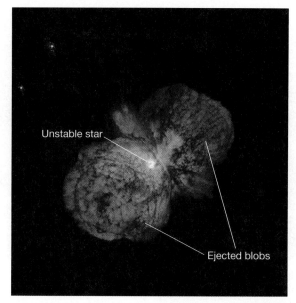

Unstable star

Ejected blobs

▲ **FIGURE 12.18 Mass Loss from Supergiants**   With an estimated mass of around 100 times the mass of the Sun and a luminosity of 5 million times the solar value, Eta Carinae is one of the most massive and most luminous stars known. This *HST* image shows blobs of ejected material racing away from the star at hundreds of kilometers per second, following a series of explosive outbursts that ejected 2 or 3 solar masses of material into space. *(NASA)*

(MA) **ANIMATION/VIDEO** Light Echo

The End of the Road

Through each period of stability and instability and each new burning stage, the star's central temperature increases, the nuclear reactions speed up, and the newly released energy supports the star for ever shorter periods of time. For example, in round numbers, a star 20 times more massive than the Sun burns hydrogen for 10 million years, helium for 1 million years, carbon for 1000 years, oxygen for 1 year, and silicon for a week. Its iron core grows for less than a day.

Once the inner core begins to change to iron, our high-mass star is in trouble. Nuclear fusion involving iron does not produce energy, because iron nuclei are so tightly bound that energy cannot be extracted by combining them into heavier elements. In effect, iron plays the role of a fire extinguisher, damping the inferno in the stellar core. With the appearance of substantial quantities of iron, the central fires cease for the last time, and the star's internal support begins to dwindle. The star's foundation is destroyed, and its equilibrium is gone forever. Even though the temperature in the iron core has reached several billion kelvins by this stage, the enormous inward gravitational pull of matter ensures catastrophe in the very near future. Gravity overwhelms the pressure of the hot gas, and the star implodes, falling in on itself.

The core temperature rises to nearly 10 billion K. At these temperatures individual photons are energetic enough to split iron into lighter nuclei and then

INTERACTIVE Life and Death of a High-Mass Star

R   I   V   U   X   G

▲ **FIGURE 12.19 Supernova 1987A**   A supernova called SN1987A (arrow) was exploding near this nebula (called 30 Doradus) at the moment the photograph at right was taken. The photograph at left is the normal appearance of the star field. (See *Discovery 12-1.) (AURA)*

CONCEPT CHECK ▶   Why does the iron core of a high-mass star collapse?

TABLE 12.3   End Points of Evolution for Stars of Different Masses

Initial Mass (Solar Masses)	Final State
Less than 0.08	(Hydrogen) brown dwarf
0.08–0.25	Helium white dwarf
0.25–8	Carbon–oxygen white dwarf
8–12 (approx.)*	Neon–oxygen white dwarf
Greater than 12*	Supernova

*Precise numbers depend on the (poorly known) amount of mass lost while the star is on, and after it leaves, the main sequence.

break those lighter nuclei apart until only protons and neutrons remain. This process is known as *photodisintegration.* In less than a second, the collapsing core undoes all the effects of nuclear fusion that occurred during the previous 10 million years. But to split iron and lighter nuclei into smaller pieces requires a lot of energy. After all, this splitting is the opposite of the fusion reactions that generated the star's energy during earlier times. The process thus *absorbs* some of the core's heat energy, reducing the pressure and accelerating the collapse.

Now the core consists entirely of electrons, protons, neutrons, and photons at enormously high densities, and it is still shrinking. As the density continues to rise, the protons and electrons are crushed together, combining to form more neutrons and releasing neutrinos. Even though the central density by this time may exceed 10^{12} kg/m³, most of the neutrinos pass through the core as if it weren't there. ∞ *(Sec. 9.5)* They escape into space, carrying away still more energy as they go.

There is nothing to prevent the collapse from continuing all the way to the point at which the neutrons themselves come into contact with each other, at the incredible density of about 10^{15} kg/m³. At this density, the neutrons in the shrinking core play a role similar in many ways to that of the electrons in a white dwarf. When far apart, they offer little resistance to compression, but when brought into contact, they produce enormous pressures that strongly oppose further contraction. The collapse finally begins to slow. By the time it is actually halted, however, the core has overshot its point of equilibrium and may reach a density as high as 10^{18} kg/m³ before beginning to reexpand. Like a fast-moving ball hitting a brick wall, the core becomes compressed, stops, then rebounds—with a vengeance!

The events just described do not take long. Only about a second elapses from the start of the collapse to the "bounce" at nuclear densities. Driven by the rebounding core, an enormously energetic shock wave then sweeps outward through the star at high speed, blasting all the overlying layers—including the heavy elements outside the iron inner core—into space. Although the details of how the shock reaches the surface and destroys the star are still uncertain, the end result is not: The star explodes in one of the most energetic events known in the universe (Figure 12.19). This spectacular death rattle of a high-mass star is known as a **core-collapse supernova.**

Our earlier dividing line of 8 solar masses between "low mass" and "high mass" really refers to the mass at the time the carbon core forms. Since very luminous stars often have strong stellar winds, it is possible that main-sequence stars as massive as 10 to 12 times the mass of the Sun may still manage to avoid going supernova. Unfortunately, we do not know exactly how much mass either Rigel or Betelgeuse has lost, so we cannot yet tell whether they are above or below the supernova threshold. Either might explode or instead become a (neon–oxygen) white dwarf, but for now we can't say which. We may just have to wait and see—in a few million years we will know for sure!

Table 12.3 summarizes the possible outcomes of stellar evolution for stars of different masses. For completeness, brown dwarfs—the end product of low-mass protostars unable even to fuse hydrogen in their cores—are included in the list. ∞ *(Sec. 11.4)*

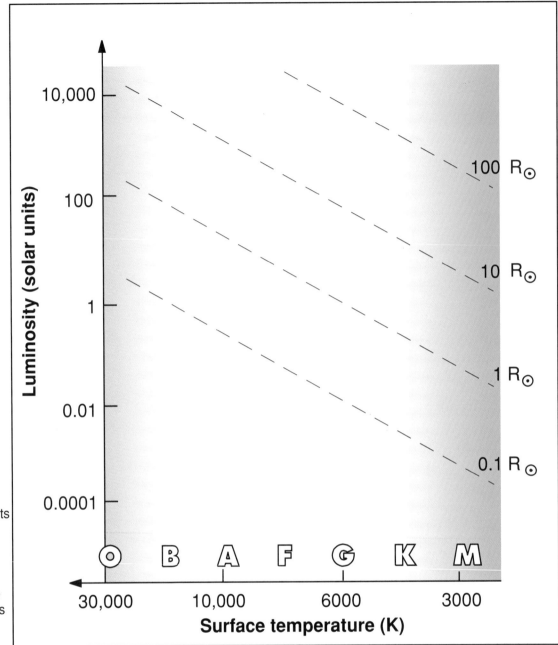

The H–R diagram plots stars by luminosity (vertical axis) and temperature or spectral class (horizontal axis). The dashed diagonal lines are lines of constant radius.

R I V U X G

12.5 Supernova Explosions

Novae and Supernovae

Observationally, a *supernova,* like a nova, is a "star" that suddenly increases dramatically in brightness, then slowly dims again, eventually fading from view. However, despite some similarities in their light curves, novae and supernovae are very different phenomena. As we saw in Section 12.3, a nova is a violent explosion on the surface of a white dwarf in a binary system.[2] Supernovae are much more energetic—about a million times brighter than novae—and are driven by very different underlying physical processes. A supernova produces a burst of light billions of times brighter than the Sun, reaching that brightness within just a few hours of the start of the outburst. The total amount of electromagnetic energy radiated by a supernova during the few months it takes to brighten and fade away is roughly the same as the Sun will radiate during its *entire* 10^{10}-year lifetime!

Astronomers divide supernovae into two classes. **Type I supernovae** contain very little hydrogen, according to their spectra, and have light curves (see Figure 12.20) somewhat similar in shape to those of typical novae—a sharp rise in intensity followed by steady, gradual decline. **Type II supernovae** are hydrogen-rich and usually have a characteristic "plateau" in the light curve a few months after the maximum. Observed supernovae are divided roughly evenly between these two categories.

Type I and Type II Supernovae Explained

How do we explain the two types of supernova just described? In fact, we already have half of the answer—the characteristics of Type II supernovae are exactly consistent with the core-collapse supernovae discussed in the previous section. The Type II light curve is in good agreement with computer simulations of a stellar envelope expanding and cooling as it is blown into space by a shock wave sweeping up from below. And since the expanding material consists mainly of unburned hydrogen and helium, it is not surprising that those elements dominate the supernova's spectrum.

What of Type I supernovae? Is there more than one way for a supernova explosion to occur? The answer is yes. To understand the alternative supernova mechanism, we must reconsider the long-term consequences of the accretion–explosion cycle that causes a nova. A nova explosion ejects matter from a white dwarf's surface, but it does not necessarily expel or burn all the material that has accumulated since the last outburst. In other words, there is a tendency for the dwarf's mass to increase slowly with each new nova cycle. As its mass grows and the internal pressure required to support its weight rises, the white dwarf can enter into a new period of instability—with disastrous consequences.

Recall that a white dwarf is held up by the pressure of electrons that have been squeezed so close together that they have effectively come into contact with one another. However, there is a limit to the mass of a white dwarf, beyond which the electrons cannot support the star against its own gravity. Detailed calculations show that the maximum mass of a white dwarf is about 1.4 solar masses, a mass often called the *Chandrasekhar mass,* after the Indian-American astronomer Subramanyan Chandrasekhar, whose work in theoretical astrophysics earned him a Nobel Prize in Physics in 1983. More recently, in 1999, NASA's latest X-ray telescope was

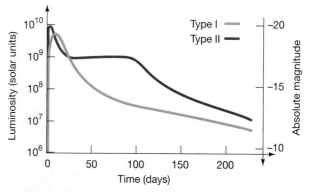

▲ **FIGURE 12.20 Supernova Light Curves** The light curves of typical Type I and Type II supernovae both show that their maximum luminosities can sometimes reach that of a billion suns, but there are characteristic differences in the decline of luminosity after the initial peak. Type I light curves resemble those of novae (see Figure 12.13c), but the total energy released is much larger. Type II curves have a characteristic plateau during the declining phase.

[2]When discussing novae and supernovae, astronomers tend to blur the distinction between the observed event (the sudden appearance and brightening of an object in the sky) and the process responsible for it (a violent explosion in or on a star). The terms can have either meaning, depending on context.

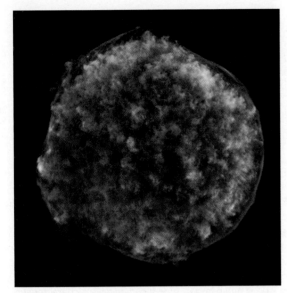

▲ **FIGURE 12.21 Tycho's Supernova**   This X-ray image, captured by the *Chandra* telescope in orbit in 2009, shows the aftermath of a titanic stellar explosion that was observed by many people on Earth in 1572. It is named after the famous Danish astronomer who studied it closely, even before the invention of the telescope ∞ *(Sec. 1.2)*. *(CXC)*

named *Chandra* in his honor. ∞ *(Sec. 3.5)* Figure 12.21 shows a *Chandra* image of a (relatively) recent Type I supernova.

If an accreting white dwarf exceeds the Chandrasekhar mass, it starts to collapse. Its internal temperature rapidly rises to the point at which carbon can (at last) fuse into heavier elements. Carbon fusion begins everywhere throughout the white dwarf almost simultaneously, and the entire star explodes in a **carbon-detonation supernova**—an event comparable in violence to the core-collapse supernova associated with the death of a high-mass star. In an alternative and (many astronomers think) possibly more common scenario, two white dwarfs in a binary system may collide and merge to form a massive, unstable star. The end result is the same.

The detonation of a carbon white dwarf, the descendant of a *low-mass* star, is a Type I supernova.[3] Because the explosion occurs in a system containing virtually no hydrogen, we can readily see why the supernova spectrum shows little evidence of that element. The light curve is almost entirely due to the radioactive decay of unstable heavy elements produced during the explosion. These explosions have one other important property: Because they all originate in similar systems—white dwarfs whose masses have just exceeded the Chandrasekhar limit—Type I supernovae show remarkably little variation in their properties from one supernova to the next, no matter where (or when) they occur. As we will see in Chapters 15 and 17, this fact, coupled with the huge luminosities involved, make Type I supernovae invaluable tools for mapping out the structure and evolution of the universe on the largest scales.

Figure 12.22 summarizes the processes responsible for the two types of supernovae. Recognize that, despite the similarity in their peak luminosities, Type I and

---

[3]In the jargon of the field, these explosions are technically known as Type Ia supernovae, to distinguish them from the rarer Type Ib and Ic explosions, which are core-collapse events in stars that have lost their hydrogen envelopes.

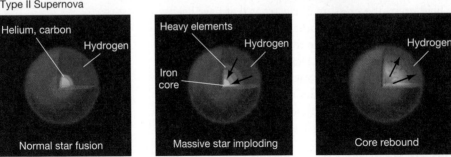

▲ **FIGURE 12.22 Two Types of Supernova**   Type I and Type II supernovae have different causes. These sequences depict the evolutionary history of each type. (a) A Type I supernova usually results when a carbon-rich white dwarf pulls matter onto itself from a nearby red giant or main-sequence companion. (b) A Type II supernova occurs when the core of a high-mass star collapses and then rebounds in a catastrophic explosion.

Type II supernovae are unrelated to one another. They occur in stars of very different types, under very different circumstances. All high-mass stars become Type II (core-collapse) supernovae, but only a tiny fraction of low-mass stars evolve into white dwarfs that ultimately explode as Type I (carbon-detonation) supernovae. However, there are far more low-mass stars than high-mass stars, resulting in the remarkable coincidence that the two types of supernova occur at roughly the same rate.

Supernova Remnants

We have plenty of evidence that supernovae have occurred in our Galaxy. Occasionally, the supernova explosions are visible from Earth. In other cases, we can detect their glowing remains, or **supernova remnants.** One of the best-studied supernova remnants is the Crab Nebula (Figure 12.23a). Its brightness has greatly

MA ANIMATION/VIDEO Supernova Explosions

INTERACTIVE Supernova Remnant Cassiopeia A **MA**

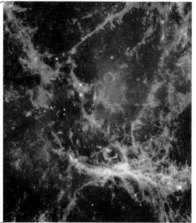

(a)

(b)

INTERACTIVE ▲ FIGURE 12.23 **Supernova Remnants** (a) This remnant of an ancient Type II supernova is called the Crab Nebula. It resides about 1800 pc from Earth, and its debris is scattered over a region about 2 pc across. The main image was taken with the Very Large Telescope of the European Southern Observatory in Chile, the inset by the *Hubble* telescope in orbit. (b) The glowing gases of the Vela supernova remnant are spread across 6° of the sky. *(ESO; NASA; D. Malin/Anglo-Australian Observatory)* **MA**

R I V U X G

dimmed now, but the original explosion in the year A.D. 1054 was so brilliant that it is prominently recorded in the manuscripts of Chinese and Middle Eastern astronomers. For nearly a month, this exploded star reportedly could be seen in broad daylight. Even today, the knots and filaments give a strong indication of past violence. The nebula—the envelope of the high-mass star that exploded to create this Type II supernova—is still expanding into space at a speed of several thousand kilometers per second.

Figure 12.23b shows another example (also of Type II). The expansion velocity of the Vela supernova remnant implies that its central star exploded around 9000 B.C. This remnant lies only 500 pc from Earth. Given its proximity, it may have been as bright as the Moon for several months. See also Figure 3.29 for an X-ray image of a nearby supernova remnant.

Although hundreds of supernovae have been observed in other galaxies during the 20th century, no one using modern equipment has ever observed one in our own Galaxy. (See *Discovery 12-1* for a discussion of a recent supernova in a galaxy very close to our own.) A viewable Milky Way star has not exploded since Galileo first turned his telescope to the heavens almost four centuries ago. Based on stellar evolutionary theory, astronomers estimate that an observable supernova ought to occur in our Galaxy every 100 years or so. Our local neighborhood seems long overdue for one. Unless stars explode much less frequently than predicted by theory, we should be treated to a (relatively) nearby version of nature's most spectacular cosmic event any day now.

Formation of the Heavy Elements

From the point of view of life on Earth, probably the most important aspect of stellar evolution is its role in creating, and then dispersing, the heavy elements out of which both our planet and our bodies are made. Since the 1950s, astronomers have come to realize that all of the hydrogen and most of the helium in the universe are *primordial*—that is, they date back to the very earliest times, long before the first stars formed (see Section 17.6). All other elements (and, in particular, virtually everything we see around us on Earth) formed later, in stars.

We have already seen how heavy elements are created from light ones by nuclear fusion. This is the basic process that powers all stars. Hydrogen fuses to helium, then helium to carbon and oxygen. Subsequently, in high-mass stars, carbon and oxygen can fuse to form still heavier elements. Neon, magnesium, sulfur, silicon—in fact, elements up to and including iron—are created in turn by fusion reactions in the cores of the most massive stars.

However, fusion stops at iron. The fact that iron nuclei will not fuse to release energy and create more massive nuclei is the basic underlying cause of Type II supernovae. How then were even heavier elements, such as copper, lead, gold, and uranium, formed? As mentioned in Section 12.3, some of these elements were formed during the late red giant stages of low-mass stars via reactions involving neutrons. Similar reactions occur in supernovae too, where neutrons and protons, produced when some nuclei are ripped apart by the almost unimaginable violence of the blast, are crammed into other nuclei, creating heavy elements that cannot have formed by any other means. Many of the heaviest elements were formed *after* their parent stars had already died, even as the debris from the explosion signaling the stars' deaths was hurled into interstellar space. In this way stellar evolution steadily enriches the composition of each new generation of stars.

Although no one has ever observed directly the formation of heavy nuclei in stars, astronomers are confident that the chain of events just described really occurs. When

CONCEPT CHECK ▶ How did astronomers know, even before the mechanisms were understood, that there were at least two distinct physical processes at work in creating supernovae?

nuclear reaction rates (determined from laboratory experiments) are incorporated into detailed computer models of stars and supernovae, the results agree very well, point by point, with the composition of the universe inferred from analysis of meteorites and spectroscopic studies of planets, stars, and the interstellar medium. The reasoning is indirect, but the agreement between theory and observation is so striking that most astronomers regard it as strong evidence supporting the entire theory of stellar evolution.

12.6 Observing Stellar Evolution in Star Clusters

Star clusters provide excellent test sites for the theory of stellar evolution. Every star in a given cluster formed at nearly the same time, from the same interstellar cloud, with virtually the same composition. ∞ *(Sec. 11.5)* Only the mass varies from one star to another. This uniformity allows us to check the accuracy of our theoretical models in a very straightforward way. Having studied in some detail the evolutionary tracks of individual stars, let us now consider how their collective appearance changes in time.

We begin our study shortly after the cluster's formation, with the high-mass stars already fully formed and burning steadily on the upper main sequence and lower-mass stars just beginning to arrive on the main sequence (Figure 12.24a). The appearance of the cluster at this early stage is dominated by its most massive stars—the bright blue supergiants.

Figure 12.24(b) shows our cluster's H–R diagram after 10 million years. The most massive O-type stars have evolved off the main sequence. Most have already exploded and vanished, but one or two may still be visible as supergiants traversing the top of the diagram. The remaining cluster stars are largely unchanged in appearance. The cluster's H–R diagram has a slightly shortened main sequence. Figure 12.25 shows the twin open clusters h and χ (the Greek letter chi) Persei, along with their observed H–R diagram. Comparing Figure 12.25(b) with Figure 12.24(b), astronomers estimate the age of this double cluster to be about 10 million years.

After 100 million years (Figure 12.24c), stars brighter than type B5 or so (about 4 to 5 solar masses) have left the main sequence, and a few more supergiants are visible. By this time most of the cluster's low-mass stars have finally arrived on the main sequence, although the dimmest M-type stars may still be in their contraction phase. The

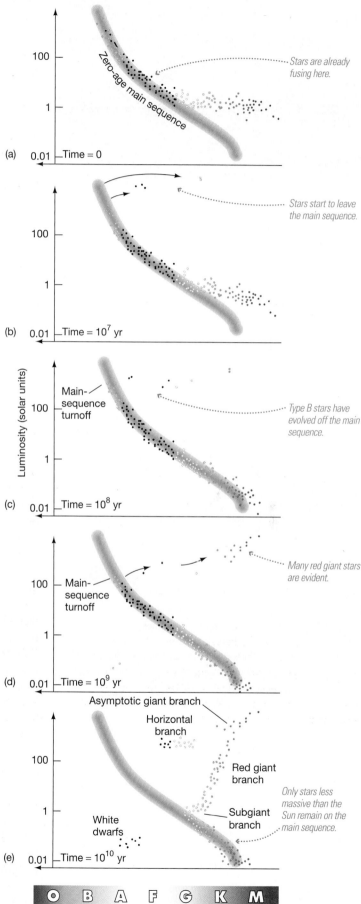

► **FIGURE 12.24 Cluster Evolution on the H–R Diagram** (a) Initially, stars on the upper main sequence are already burning steadily while the lower main sequence is still forming. (b) At 10^7 years, O-type stars have already left the main sequence, and a few red giants are visible. (c) By 10^8 years, more red giants are visible, and the lower main sequence is almost fully formed. (d) At 10^9 years, the subgiant and red giant branches are just becoming evident, and the formation of the lower main sequence is complete. (e) At 10^{10} years, the cluster's subgiant, red giant, horizontal, and asymptotic giant branches are all discernible, and many white dwarfs have now formed.

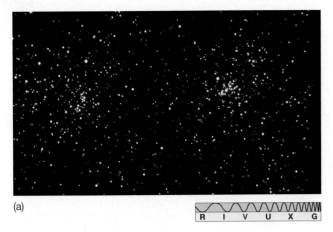

(a)

| R | I | V | U | X | G |

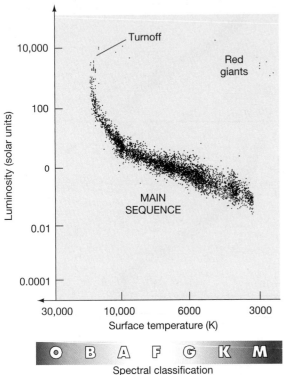

(b)

▲ FIGURE 12.25 Newborn Cluster H–R Diagram   (a) The "double cluster" h and χ Persei, two open clusters that apparently formed at the same time. (b) The H–R diagram of the pair indicates that the stars are very young—probably only 10–15 million years old. Even so, the most massive stars have already left the main sequence. (NOAO; data from T. Currie)

appearance of the cluster is now dominated by bright B-type stars and brighter supergiants.

At any time during the evolution, the cluster's original main sequence is intact up to some well-defined stellar mass, corresponding to the stars that are just leaving the main sequence at that instant. We can imagine the main sequence being "peeled away" from the top down, with fainter and fainter stars turning off and heading for the giant branch as time goes on. Astronomers refer to the high-luminosity end of the observed main sequence as the **main-sequence turnoff.** The mass of the star that is just evolving off the main sequence at any moment is known as the *turnoff mass.* Knowing the turnoff mass is equivalent to knowing the age of the cluster—stars less massive than the turnoff are still on the main sequence, while more massive stars have already evolved into something else.

At 1 billion years (Figure 12.24d) the main-sequence turnoff mass is around 2 solar masses, corresponding roughly to spectral type A2. The subgiant and giant branches associated with the evolution of low-mass stars are just becoming apparent, and the formation of the lower main sequence is now complete. In addition, the first white dwarfs have just appeared, although they are often too faint to be observed at the distances of most clusters. Figure 12.26 shows the Hyades open cluster and its H–R diagram. The H–R diagram appears to lie between Figures 12.24 (c) and (d). More careful measurements yield a cluster age of about 600 million years.

At 10 billion years, the turnoff point has reached solar-mass stars, of spectral type G2. The subgiant and giant branches are now clearly discernible in the H–R diagram (Figure 12.24e), and the horizontal branch appears as a distinct region. Many white dwarfs are also present in the cluster. Figure 12.27 shows a composite H–R diagram for several globular clusters in our Galaxy. The various evolutionary stages predicted by theory are all clearly visible in this figure, although the individual points are shifted somewhat to the left relative to our previous H–R diagrams because of differences in composition between stars like the Sun and stars in globular clusters. (Globular cluster stars tend to be hotter than solar-type stars of the same mass.) The clusters represented in Figure 12.27(b) are quite deficient in heavy elements compared with the Sun. Figure 12.27(a) shows a *Hubble* image of the globular cluster M80, which has a similarly low concentration of heavy elements—only about 2 percent the solar value—and whose H–R diagram looks qualitatively very similar to Figure 12.27.

By carefully adjusting their theoretical models until the main sequence, subgiant, red giant, and horizontal branches are all well matched, astronomers find that this diagram corresponds to a cluster age of roughly 12 billion years, a little older than the hypothetical cluster in Figure 12.24(e). The deficiency of heavy elements is consistent with these clusters being among the first objects to form in our Galaxy (see Section 12.7). In fact, globular cluster ages determined this way show a remarkably small spread. Most of the globular clusters in our Galaxy formed between about 10 and 12 billion years ago.

The objects labeled as *blue stragglers* in Figure 12.27(b) appear at first sight to contradict the theory just described. They lie on the main sequence, but above the turnoff, suggesting that they should have evolved into white dwarfs long ago, given the cluster's age of 12 billion years. Blue stragglers are observed in many star clusters. They are main-sequence stars, but they did not form when the cluster did. Instead, they formed much more recently, through *mergers* of lower-mass stars—so recently, in fact, that they have not yet had time to evolve into giants. In some cases, the mergers were probably the result of stellar evolution in binary systems, as the component stars evolved, grew, and came into contact. In others, the mergers are thought to be the result of actual *collisions*

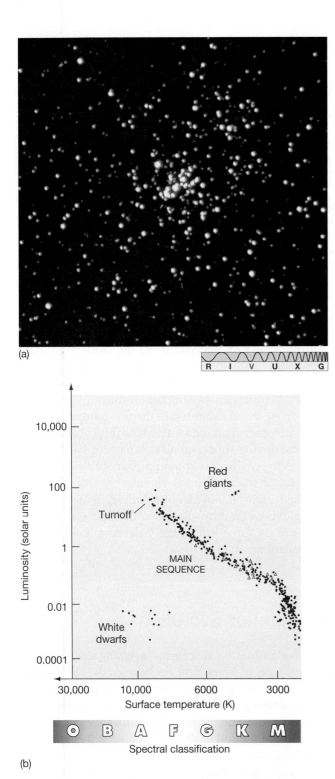

(a)

R I V U X G

(b)

▲ FIGURE 12.26 Young Cluster H–R Diagram (a) The Hyades cluster, a relatively young group of stars visible to the naked eye, is found 46 pc away in the constellation Taurus. (b) The H–R diagram for this cluster is cut off at about spectral type A, implying an age of about 600 million years. A few massive stars have already become white dwarfs. (*NOAO; ESA*)

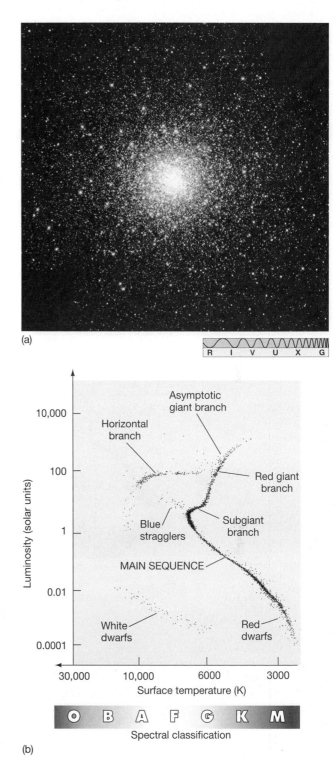

(a)

R I V U X G

(b)

▲ FIGURE 12.27 Old Cluster H–R Diagram (a) The globular cluster M80, which lies some 8 kpc from Earth. (b) Combined H–R diagram, based on ground- and space-based observations, for several globular clusters similar in overall composition to M80. Fitting the main-sequence turnoff and the giant and horizontal branches to theoretical models implies an age of about 12 billion years, making these clusters among the oldest known objects in the Milky Way Galaxy. (*NASA; data courtesy of William E. Harris*)

(a)

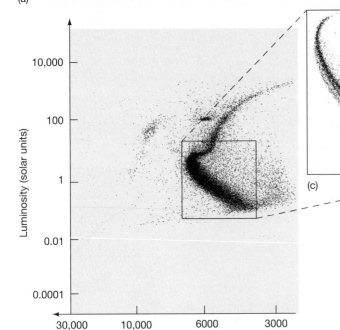

(b)

◄ **FIGURE 12.28 Multiple Stellar Generations**   The ground-based H–R diagram (b) of the globular cluster NGC 2808 (a) shows an apparently normal main sequence, but more precise observations from *HST* (c) reveal that the main sequence is actually made up of three distinct sequences (as noted), increasing in helium content from right to left. These observations imply multiple generations of star formation shortly after the cluster formed, but no theory can yet explain just how this occurred. (*NASA*)

Three separate main sequences are evident here.

(c)

between stars. The dense central cores of globular clusters—with a million or more times as many stars per unit volume as are found in the neighborhood of the Sun—are among the few places in the entire universe where stellar collisions are likely to occur.

High-precision observations from *HST* have revealed a new, and as yet unresolved, mystery about globular clusters that may yet force astronomers to change significantly their views on how massive star clusters form. Figure 12.28 presents the H–R diagram for the cluster NGC 2808, showing (in the inset) what appear to be *three* distinct main sequences, undetected in earlier ground-based observations. The stars in the three sequences contain different amounts of helium, carbon, and nitrogen and are thought to be the result of multiple episodes of star formation occurring over the course of about 100 million years. Models suggest that the two more helium-rich generations formed from gas enriched by stellar evolution in the first-generation stars, but astronomers do not know how this could have occurred in the time available. Whatever happened, it seems to have been a common phenomenon, as high-resolution studies of many globular clusters now reveal similar multiple populations. Indeed, some observers would go so far as to claim that multiple stellar populations are the norm in the Galactic globular cluster population.

12.7 The Cycle of Stellar Evolution

The theory of stellar evolution outlined in this chapter can naturally account for the observed differences in the abundances of heavy elements between the old globular-cluster stars and stars now forming in our Galaxy. Even though an evolved star continuously creates new heavy elements in its interior, changes in the star's composition are confined largely to the core, and the star's spectrum may give little indication of the events within. Only at the end of the star's life are its newly created elements released and scattered into space.

Thus the spectra of the *youngest* stars show the *most* heavy elements, because each new generation of stars increases the concentration of these elements in the interstellar clouds from which the next generation forms. Accordingly, the photosphere of a recently formed star contains a much greater abundance of heavy elements than that of a star formed long ago. Knowledge of stellar evolution allows astronomers to estimate the ages of stars from purely spectroscopic studies, even when the stars are isolated and are not members of any cluster.

We have now seen all the ingredients that make up the complete cycle of star formation and evolution. Let us now briefly summarize that process, which is illustrated in Figure 12.29.

PROCESS OF ▶
SCIENCE CHECK Why are observations of star clusters so important to the theory of stellar evolution?

INTERACTIVE ◄ **FIGURE 12.29 Stellar Recycling** The cycle of star formation and evolution continuously replenishes the Galaxy with new heavy elements and provides the driving force for the creation of new generations of stars. Clockwise from the top are an interstellar cloud (Barnard 68), a star-forming region in our Galaxy (RCW 38), a massive star ejecting a "bubble" and about to explode (NGC 7635), and a supernova remnant and its heavy-element debris (N49). *(ESO; NASA)*

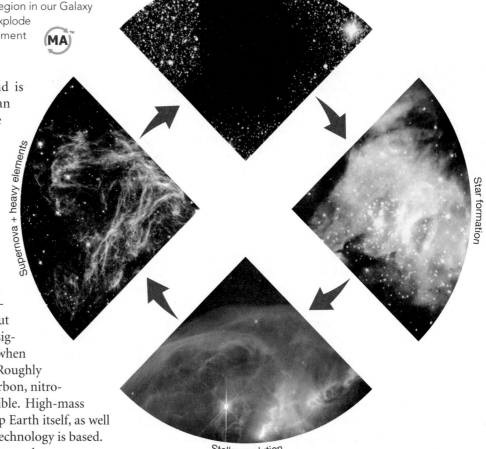

Interstellar medium

Star formation

Stellar evolution

Supernova + heavy elements

1. Stars form when part of an interstellar cloud is compressed beyond the point at which it can support itself against its own gravity. The cloud collapses and fragments, forming a cluster of stars. The hottest stars heat and ionize the surrounding gas, sending shock waves through the surrounding cloud, modifying the formation of lower-mass stars, and possibly triggering new rounds of star formation. ∞ *(Secs. 11.2, 11.3)*

2. Within the cluster, stars evolve. The most massive stars evolve fastest, creating heavy elements in their cores and spewing them forth into the interstellar medium in supernova explosions. Low-mass stars evolve more slowly, but they too create heavy elements and contribute significantly to the "seeding" of interstellar space when they shed their envelopes as planetary nebulae. Roughly speaking, low-mass stars created most of the carbon, nitrogen, and oxygen that make life on Earth possible. High-mass stars produced the iron and silicon that make up Earth itself, as well as the heavier elements on which much of our technology is based.

3. The creation and explosive dispersal of new heavy elements are accompanied by further shock waves. The passage of these shock waves through the interstellar medium simultaneously enriches the medium and compresses it into further star formation. Each generation of stars increases the concentration of heavy elements in the interstellar clouds from which the next generation forms. As a result, recently formed stars contain a much greater abundance of heavy elements than stars that formed long ago.

In this way, although some material is used up in each cycle—turned into energy or locked up in stars less massive than the Sun (most of which have not yet left the main sequence)—the Galaxy continuously recycles its matter. Each new round of formation creates stars containing more heavy elements than the preceding generation had. From the old globular clusters, which are observed to be deficient in heavy elements relative to the Sun, to the young open clusters, which contain much larger amounts of these elements, we observe this enrichment process in action. Our Sun is the product of many such cycles. We ourselves are another. Without the heavy elements synthesized in the hearts of stars, life on Earth would not exist.

Stellar evolution is one of the great success stories of astrophysics. Like all good scientific theories, it makes definite, testable predictions about the universe, at the same time remaining flexible enough to incorporate new discoveries as they occur. Theory and observation have advanced hand in hand. At the start of the 20th century, many scientists despaired of ever knowing the compositions of the stars, let alone why they shine and how they change. Today, the theory of stellar evolution is a cornerstone of modern astronomy.

CONCEPT CHECK ► Why is stellar evolution important to life on Earth?

12-1 DISCOVERY

Supernova 1987A

Although there have been no observable supernovae in our own Galaxy since the invention of the telescope, astronomers were treated to the next best thing in 1987 when a 15-solar mass B-type supergiant exploded in the Large Magellanic Cloud (LMC), a small satellite galaxy orbiting our own (see Section 15.1). For a few weeks the Type II supernova, designated SN1987A and shown in Figure 12.19, outshone all the other stars in the LMC combined. Because the LMC is relatively close to Earth and the explosion was detected soon after it occurred, SN1987A has provided astronomers with a wealth of detailed information on supernovae, allowing them to make key comparisons between theoretical models and observational reality. By and large, the theory of stellar evolution has held up very well. Still, SN1987A did hold a few surprises.

The light curve of SN1987A, shown below, differed somewhat from the "standard" Type II shape (Figure 12.20). The peak brightness was only about 1/10 the standard value and occurred much later than expected. These differences were mainly the result of the (relatively) small size of SN1987A's parent star, which was a blue supergiant at the time of the explosion and not a red supergiant, as suggested by Figure 12.16. The reason was that the parent star's envelope was deficient in heavy elements, significantly altering the star's evolutionary track. The star was located at the top right of the H–R diagram following the ignition of carbon, with a surface temperature of around 20,000 K, when the rapid chain of events leading to the supernova occurred.

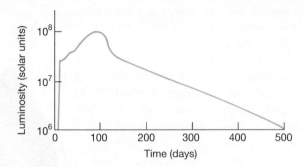

Because the parent star was small and quite tightly bound by gravity, a lot of the energy produced in the explosion was used in expanding SN1987A's stellar envelope, so far less was left over to be radiated into space. Thus the luminosity during the first few months was lower than expected, and the early peak evident in Figure 12.20 did not occur. The peak in the SN1987A light curve at about 80 days actually corresponds to the "plateau" in the Type II light curve in Figure 12.20.

About 20 hours before the supernova was detected optically, underground detectors in Japan and the United States recorded a brief (13-second) burst of neutrinos. The neutrinos preceded the light because they escaped during the collapse, whereas the first light of the explosion was emitted only after the supernova shock had plowed through the body of the star to the surface. Theoretical models, consistent with these observations, suggest that many tens of thousands of times more energy was emitted in the form of neutrinos than in any other form. Detection of this neutrino pulse was a brilliant confirmation of theory and may well herald a new age of astronomy. For the first time, astronomers received information from beyond the solar system by radiation outside the electromagnetic spectrum.

The accompanying photographs show the barely resolved remnant of SN1987A (at the center of each image) surrounded by a much larger shell of glowing gas (in yellow). Scientists reason that the progenitor star expelled this shell during its red giant phase, some 40,000 years before the explosion. The image we see results from the initial flash of ultraviolet light from the supernova hitting the ring and causing it to glow brightly. The insets at the bottom show supernova debris moving outward toward the ring at nearly 3000 km/s. The fastest-moving ejecta have already reached the ring, forming the small (1000-AU diameter) glowing regions that now surround the supernova remnant. The main image also revealed two additional faint rings that might be caused by radiation sweeping across an hourglass-shaped bubble of gas, itself perhaps the result of a nonspherical "bipolar" stellar wind from the progenitor star before the supernova occurred.

Buoyed by the success of stellar-evolution theory and armed with firm theoretical predictions of what should happen next, astronomers eagerly await further developments in the story of this remarkable object.

(NASA)

1994   1998   2001   2003   2004   2007

R I V U X G

ANIMATION/VIDEO Shockwaves Hit the Ring Around Supernova 1987A

CHAPTER REVIEW

SUMMARY

LO1 Stars spend most of their lives on the main sequence, in the **core hydrogen-burning** (stage 7; p. 322) phase of stellar evolution. Stars leave the main sequence when the hydrogen in their cores is exhausted. With no internal energy source, the star's helium core is unable to support itself against its own gravity and begins to shrink. The star at this stage is in the **hydrogen shell-burning** (p. 324) phase, with nonburning helium at the center surrounded by a layer of burning hydrogen. The energy released by the contracting helium core heats the hydrogen-burning shell, greatly increasing the nuclear reaction rates there. As a result, the star becomes much brighter, while the envelope expands and cools. A solar-mass star moves off the main sequence on the H–R diagram first along the **subgiant branch** (stage 8; p. 324), then almost vertically up the **red giant branch** (stage 9; p. 324).

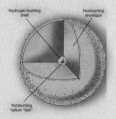

LO2 As the helium core contracts, it heats up. Eventually, helium begins to fuse into carbon. In a star like the Sun, helium burning begins explosively, in the **helium flash** (p. 325). The flash expands the core and reduces the star's luminosity, sending it onto the **horizontal branch** (stage 10; p. 325) of the H–R diagram. The star now has a core of burning helium surrounded by a shell of burning hydrogen. An inner core of nonburning carbon forms, shrinks, and heats the overlying burning layers, and the star once again becomes a red giant, even more luminous than before (stage 11). The core of a solar-mass star never becomes hot enough to fuse carbon. Such a star continues to brighten and expand until its envelope is ejected into space, forming a **planetary nebula** (stage 12; p. 327). At that point the core becomes visible as a hot, faint, and extremely dense white dwarf (stage 13). The white dwarf cools and fades, eventually becoming a cold **black dwarf** (stage 14; p. 329).

LO3 Although most white dwarfs simply cool and dim with time, some may become explosively active in the form of a **nova** (p. 330). A nova is a star that suddenly increases greatly in brightness, then slowly fades back to its normal appearance over a period of months. It is the result of a white dwarf in a binary system drawing hydrogen-rich material from its companion. The gas spirals inward in an **accretion disk** (p. 331) and builds up on the white dwarf's surface, eventually becoming hot and dense enough for the hydrogen to burn explosively, temporarily causing a large increase in the dwarf's luminosity.

LO4 Stars more massive than about 8 solar masses form heavier and heavier elements in their cores, at a more and more rapid pace. As they evolve into **red supergiants** (p. 332), their cores form a layered structure consisting of burning shells of successively heavier elements. The process stops at iron, whose nuclei can neither be fused together nor split to produce energy. As a star's iron core grows in mass, it eventually becomes unable to support itself against gravity and begins to collapse. At the high temperatures produced during the collapse, iron nuclei are broken down into protons and neutrons. The protons combine with electrons to form more neutrons. Eventually, when the core becomes so dense that the neutrons are effectively brought into physical contact with one another, their resistance to further squeezing stops the collapse and the core rebounds, sending a violent shock wave out through the rest of the star. The star explodes in a **core-collapse supernova** (p. 334).

LO5 **Type I supernovae** (p. 335) are hydrogen poor and have light curves similar in shape to those of novae. **Type II supernovae** (p. 335) are hydrogen rich and have a characteristic bump in the light curve a few months after maximum. A Type II supernova is a core-collapse supernova. A Type I supernova is a **carbon-detonation supernova** (p. 336), which occurs when a white dwarf in a binary system collapses and then explodes as its carbon ignites. We can see evidence for a past supernova in the form of a **supernova remnant** (p. 337), a shell of exploded debris surrounding the site of the explosion and expanding into space at a speed of thousands of kilometers per second.

LO6 All elements heavier than helium formed in evolved stars or in supernova explosions. With each new generation of stars, the fraction of heavy elements in the universe increases. Comparisons between theoretical predictions of element production and observations of element abundances in the Galaxy provide strong support for the theory of stellar evolution.

LO7 The theory of stellar evolution can be tested by observing star clusters. At any instant, stars with masses above the cluster's **main-sequence turnoff** (p. 340) have evolved off the main sequence. By comparing a cluster's main-sequence turnoff mass with theoretical predictions, astronomers can determine the cluster's age. Many glubular clusters show evidence of multiple stellar populations, whose origin is not yet understood.

MasteringAstronomy® For instructor-assigned homework go to www.masteringastronomy.com

Problems labeled **POS** explore the process of science. **VIS** problems focus on reading and interpreting visual information. **LO** connects to the introduction's numbered Learning Outcomes.

REVIEW AND DISCUSSION

1. How long can a star like the Sun keep burning hydrogen in its core?

2. **LO1** Why is the depletion of hydrogen in the core of a star such an important event?

3. What makes an ordinary star become a red giant?

4. Roughly how big (in AU) will the Sun become when it enters the red giant phase?

5. **LO2** How long does it take for a star like the Sun to evolve from the main sequence to the top of the red giant branch?

6. What is the helium flash?

7. How do stars of low mass die? How do stars of high mass die?

8. What is a planetary nebula? With what stage of stellar evolution is it associated?

9. What are white dwarfs? What is their ultimate fate?

10. **LO3** Under what circumstances will a binary star produce a nova?

11. **LO4** What occurs in a massive star to cause it to explode?

12. What are the observational differences between Type I and Type II supernovae?

13. **LO5** How do the mechanisms that cause Type I and Type II supernovae explain their observed differences?

14. **LO6 POS** What evidence do we have that many supernovae have occurred in our Galaxy?

15. **LO7 POS** What do stars in clusters tell us about the various stages of stellar evolution?

CONCEPTUAL SELF-TEST: TRUE OR FALSE?/MULTIPLE CHOICE

1. All of the single red-dwarf stars that ever formed are still on the main sequence today. (T/F)

2. The Sun will get brighter as it begins to run out of fuel in its core. (T/F)

3. A planetary nebula is the disk of matter around a star that will eventually form a planetary system. (T/F)

4. Nuclear fusion in the core of a massive star cannot create elements much heavier than iron. (T/F)

5. A nova is a sudden outburst of light coming from an old main-sequence star. (T/F)

6. It takes less and less time to fuse heavier and heavier elements inside a high-mass star. (T/F)

7. In a core-collapse supernova, the outer part of the core rebounds from the inner, high-density core, destroying the entire outer part of the star. (T/F)

8. Because of stellar nucleosynthesis, the spectra of old stars show more heavy elements than those of young stars. (T/F)

9. A white dwarf is supported by the pressure of tightly packed (a) electrons; (b) protons; (c) neutrons; (d) photons.

10. A star like the Sun will end up as a (a) blue giant; (b) white dwarf; (c) binary star; (d) red dwarf.

11. A white dwarf can dramatically increase in brightness only if it (a) has another star nearby; (b) can avoid nuclear fusion in its core; (c) is spinning very rapidly; (d) is descended from a very massive star.

12. Nuclear fusion in the Sun will (a) never create elements heavier than helium; (b) create elements up to and including oxygen; (c) create all elements up to and including iron; (d) create some elements heavier than iron.

13. Most of the carbon in our bodies originated in (a) the core of the Sun; (b) the core of a red giant star; (c) a supernova; (d) a nearby galaxy.

14. **VIS** If the evolutionary track in *Overlay 3*, showing a Sun-like star, were instead illustrating a significantly more massive star, its starting point (stage 7) would be (a) up and to the right; (b) down and to the left; (c) up and to the left; (d) down and to the right.

15. **VIS** Figure 12.20 (Supernova Light Curves) indicates that a supernova whose luminosity declines steadily in time is most likely associated with a star that is (a) without a binary companion; (b) more than eight times the mass of the Sun; (c) on the main sequence; (d) comparable in mass to the Sun.

PROBLEMS

The number of squares preceding each problem indicates its approximate level of difficulty.

1. ■ Calculate the average density of a red giant core of mass 0.25 solar masses and radius 15,000 km. Compare this with the average density of the giant's envelope, if the mass of the envelope is 0.5 solar masses and its radius is 0.5 AU. Compare each with the central density of the Sun. ∞ *(Sec. 9.2)*

2. ■■■ A main sequence star at a distance of 20 pc is barely visible through a certain telescope. The star subsequently ascends the giant branch, during which time its temperature drops by a factor of three and its radius increases 100-fold. What is the new maximum distance at which the star would still be visible using the same telescope?

3. ■■ The Sun will reside on the main sequence for 10^{10} years. If the luminosity of a main-sequence star is proportional to the fourth power of the star's mass, what is the mass of a star that is just now leaving the main sequence in a cluster that formed (a) 400 million years ago? (b) 2 billion years ago?

4. ■ How long will it take the Sun's planetary nebula, expanding at a speed of 20 km/s, to reach the orbit of Neptune? How long to reach the nearest star?

5. ■ What are the escape speed (in km/s) and surface gravity (relative to Earth's gravity) of Sirius B (Table 12.2)? ∞ *(More Precisely 5-1)*

6. ■■■ A certain telescope could just detect the Sun at a distance of 10,000 pc. What is the apparent magnitude of the Sun at this distance? (For convenience, take the Sun's absolute magnitude to be 5.) What is the maximum distance at which the telescope could detect a nova having a peak luminosity of 10^5 solar luminosities? Repeat the calculation for a supernova having a peak luminosity 10^{10} times that of the Sun.

7. ■■ At what distance would the supernova in the previous question look as bright as the Sun (apparent magnitude −27)? Would you expect a supernova to occur that close to us?

8. ■■ A (hypothetical) supernova at a distance of 150 pc has an absolute magnitude of −20. Compare its apparent magnitude with that of (a) the full Moon; (b) Venus at its brightest (see Figure 10.6). Would you expect a supernova to occur so close to us?

9. ■ The Crab Nebula is now about 1 pc in radius. If it was observed to explode in A.D. 1054, roughly how fast is it expanding? (Assume constant expansion velocity. Is that a reasonable assumption?)

10. ■■ A supernova's energy is often compared to the total energy output of the Sun over its lifetime. Using the Sun's current luminosity, calculate the total solar energy output, assuming a 10^{10} year main-sequence lifetime. Using Einstein's formula $E = mc^2$ calculate the equivalent amount of mass, in Earth masses. ∞ *(Sec. 9.5)*

ACTIVITIES

Collaborative

1. Open clusters are generally found in the plane of the Galaxy. If you can see the hazy band of the Milky Way arcing across your night sky—in other words, if you are far from city lights and looking at an appropriate time of night and year—you can simply sweep with your binoculars along the Milky Way. Numerous "clumps" of stars will pop into view. Many will turn out to be open star clusters. The most easily visible clusters are those in the Messier catalog, but there are many others besides those. How many clusters can you find that are *not* on Messier's list?

2. Globular star clusters are harder to find. They are intrinsically larger, but they are also much farther away and therefore appear smaller in the sky. The most famous globular cluster visible from the Northern Hemisphere is M13 in the constellation Hercules, visible on spring and summer evenings. It contains half million or so of the Galaxy's most ancient stars. It may be glimpsed through binoculars as a little ball of light, located about one-third of the way from the star Eta to the star Zeta in the Keystone asterism of the constellation Hercules. Telescopes reveal this cluster as a magnificent, symmetrical grouping of stars. Can you find the following well-known globular clusters: M3, M4, M5, M13, M15? Look online or in a star chart for details on how to locate them.

Individual

1. Can you find the Hyades cluster? It lies about 46 pc away in the constellation Taurus, making up the "face" of the bull. It appears to surround the very bright star Aldebaran, the bull's eye, which makes it easy to locate in the sky. Aldebaran is a low-mass red giant, about twice the mass of the Sun, probably on the asymptotic giant branch of its evolution. Despite appearances, it is not part of the Hyades cluster. In fact, it lies only about half as far away—about 20 pc from Earth

2. In 1758, Charles Messier discovered the sky's most legendary supernova remnant, now called M1, or the Crab Nebula. It is located northwest of Zeta Tauri, the star that marks the southern tip of the horns of Taurus the Bull. Try to find it. An 8-inch telescope reveals the Crab's oval shape, but it will appear faint; a 10-inch or larger telescope reveals some of its famous filamentary structure.

3. The Ring Nebula (M57) is perhaps the most famous planetary nebula. At magnitude 9, it is faint, but a 6-inch or larger telescope should show its structure. To locate it, find Beta and Gamma Lyrae, the second and third brightest stars in the constellation Lyra. The ring lies between them, about one-third the way from Beta to Gamma. Don't expect the Ring to look as colorful as the *Hubble* images you may have seen! Can you see any color in the ring?

13
Neutron Stars and Black Holes
Strange States of Matter

Our study of stellar evolution has led us to some very unusual and unexpected objects. Red giants, white dwarfs, and supernova explosions surely represent extreme states of matter completely unfamiliar to us here on Earth. Yet, stellar evolution can have even more bizarre consequences. The strangest states of all result from the catastrophic implosion–explosion of stars much more massive than our Sun. The almost unimaginable violence of a supernova explosion may bring into being objects so extreme in their behavior that they require us to reconsider some of our most hallowed laws of physics. They open up a science fiction writer's dream of fantastic phenomena. They may even one day force scientists to construct a whole new theory of the universe. Yet, incredibly, as far as we can tell, they, may actually exist!

MasteringAstronomy® Visit www.masteringastronomy.com for quizzes, animations, videos, interactive figures, and self-guided tutorials.

LEARNING OUTCOMES

Studying this chapter will enable you to:

LO1 List the key properties of neutron stars, and outline how these strange objects are formed.

LO2 Describe the nature and origin of pulsars, and explain how pulsars provide evidence for the existence of neutron stars.

LO3 List and explain some of the observable properties of neutron star binary systems.

LO4 Outline the basic characteristics of gamma-ray bursts and the leading theoretical attempts to explain them.

LO5 Describe Einstein's theories of relativity, and explain how they relate to neutron stars and black holes.

LO6 Explain how black holes are formed, and describe their effects on matter and radiation in their vicinity.

LO7 Relate the phenomena that occur near black holes to the warping of space around them.

LO8 Describe the difficulties in observing black holes, and list some ways in which black holes can be detected.

◀ Neutron stars and black holes are among the most exotic members of the vast population of stars throughout the universe. These objects represent the end states of stellar systems, yet despite their bizarre nature, they do seem to agree quite well with models of stellar evolution. This stunning image is actually a composite of three images taken by telescopes in orbit: optical light (in yellow) observed with *Hubble*, X-ray radiation (blue and green) with *Chandra*, and infrared radiation (red) with *Spitzer*. This debris field spread across 10 light-years is known as Cassiopeia A, the remnant of a supernova whose radiation first reached Earth about 300 years ago. The turquoise dot at the center may be a neutron star created at the precise center of the blast. *(NASA)*

13.1 Neutron Stars

What remains after a supernova explosion? Is the original star blown to bits and dispersed into interstellar space, or does some portion of it survive? For a Type I (carbon-detonation) supernova, most astronomers regard it as unlikely that anything is left behind after the explosion. ⚯ *(Sec. 12.5)* The entire star is shattered by the blast. However, for a Type II (core-collapse) supernova, theoretical calculations indicate that part of the star survives the explosion.

Recall that in a Type II supernova, the iron core of a massive star collapses until its neutrons effectively come into contact with one another. At that point the central portion of the core rebounds, creating a powerful shock wave that races outward through the star, violently expelling matter into space. ⚯ *(Sec. 12.4)* The key point here is that the shock wave does not start at the very center of the collapsing core. The innermost part of the core—the region that rebounds—remains intact as the shock wave it produces destroys the rest of the star. After the violence of the supernova has subsided, this ultracompressed ball of neutrons is all that is left. Researchers call this core remnant a **neutron star.**

Neutron stars are extremely small and very massive. Composed purely of neutrons packed together in a tight ball about 20 km across, a typical neutron star is not much bigger than a small asteroid or a terrestrial city (Figure 13.1), yet its mass is greater than that of the Sun. With so much mass squeezed into such a small volume, neutron stars are incredibly dense. Their average density can reach 10^{17} or even 10^{18} kg/m^3, nearly a billion times denser than a white dwarf. ⚯ *(Sec. 12.3)* A single thimbleful of neutron star material would weigh 100 million tons—about as much as a good-sized terrestrial mountain.

Neutron stars are solid objects. Provided that a sufficiently cool one could be found, you might even imagine standing on it. However, this would not be easy, as its gravity is extremely powerful. A 70-kg human would weigh the Earth equivalent of about 1 billion kg (1 million tons). The severe pull of a neutron star's gravity would flatten you much thinner than this piece of paper.

In addition to large mass and small size, newly formed neutron stars have two other very important properties. First, they *rotate* extremely rapidly, with periods measured in fractions of a second. This is a direct result of the law of conservation of angular momentum (Chapter 4), which tells us that any rotating body must spin faster as it shrinks, and the core of the parent star almost certainly had some rotation before it began to collapse. ⚯ *(More Precisely 4-1)* Second, they have very strong *magnetic fields*. The original field of the parent star is amplified as the collapsing core squeezes the magnetic field lines closer together, creating a magnetic field trillions of times stronger than Earth's.

In time, theory indicates, a neutron star will spin more and more slowly as it radiates its energy into space, and its magnetic field will diminish. However, for a few million years after its birth, these two properties combine to provide the primary means by which this strange object can be detected and studied, as we now discuss.

13.2 Pulsars

The first observation of a neutron star occurred in 1967 when Jocelyn Bell, a graduate student at Cambridge University, observed an astronomical object emitting radio radiation in the form of rapid *pulses* (Figure 13.2). Each pulse consisted of a roughly 0.01-s burst of radiation, separated by precisely 1.34 s from the next.

More than 1500 of these pulsating objects are now known. They are called **pulsars.** Each has its own characteristic pulse shape and period. Observed pulsar periods are generally quite short, ranging from about a few milliseconds to about a

CONCEPT CHECK ▶ Are all supernovae expected to lead to neutron stars?

▲ **FIGURE 13.1 Neutron Star** Neutron stars are not much larger than many of Earth's major cities. In this fanciful comparison, a typical neutron star sits alongside Manhattan Island. *(NASA)*

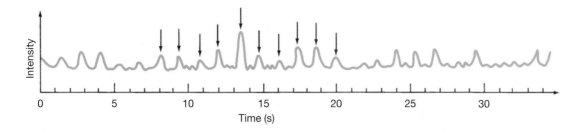

second, corresponding to flashing rates of between one and several hundred times per second. In some cases, the periods are extremely stable—in fact, constant to within a few seconds in a million years—making pulsars the most accurate natural clocks known in the universe, more accurate even than the best atomic clocks on Earth.

A Pulsar Model

When Bell made her discovery in 1967, she did not know what she was looking at. Indeed, no one at the time knew what a pulsar was. The explanation won Bell's thesis advisor, Anthony Hewish, the 1974 Nobel Prize in physics. Hewish reasoned that the only physical mechanism consistent with such precisely timed pulses is a small, rotating source of radiation. Only rotation can cause the high degree of regularity of the observed pulses, and only a small object can account for the sharpness of each pulse (see Section 13.4). The best current model describes a pulsar as a compact, spinning neutron star that periodically flashes radiation toward Earth.

Figure 13.3 outlines the main features of this pulsar model. Two "hot spots," either on the surface of a neutron star or in the magnetosphere just above it, continuously emit radiation in a narrow beam. These spots are most likely localized regions near the neutron star's magnetic poles, where charged particles, accelerated to extremely high energies by the star's rotating magnetic field, emit radiation along the star's magnetic axis. The hot spots radiate more or less

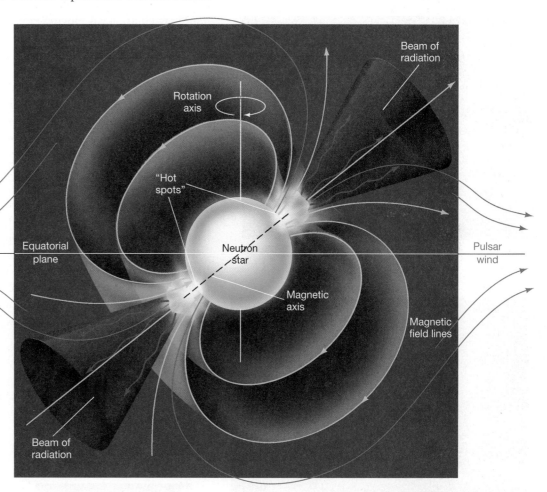

► FIGURE 13.3 Pulsar Model
The "lighthouse model" of neutron star emission explains many of the observed properties of pulsars. Charged particles, accelerated by the magnetism of the neutron star, flow along the magnetic field lines, producing radio radiation that beams outward. At greater distances from the star, the field lines channel these particles into a high-speed outflow in the star's equatorial plane, forming a pulsar wind. The beam sweeps across the sky as the neutron star rotates. If it happens to intersect Earth, we see a pulsar.

ANALOGY:
A lighthouse beacon is a good analogy for a rotating pulsar.

ANIMATION/VIDEO Pulsar in Crab Nebula

steadily, and the resulting beams sweep through space like a revolving lighthouse beacon as the neutron star rotates. Indeed, this pulsar model is often known as the **lighthouse model.** If the neutron star happens to be oriented such that one of the rotating beams sweeps across Earth, then we see the pulses. The period of the pulses is the star's rotation period.

A few pulsars are definitely associated with supernova remnants, clearly establishing those pulsars' explosive origin, although not all such remnants have a detectable pulsar within them. Figure 13.4(c) shows optical images of the Crab pulsar at the center of the Crab supernova remnant (Figure 13.4a). ⚬⚬ *(Sec. 12.5)* By observing the speed and direction of the Crab's ejected matter, astronomers can work backward to pinpoint the location in space at which the explosion must have occurred. It corresponds to the location of the pulsar. This is all that remains of the once massive star whose supernova was observed in 1054.

As indicated in Figure 13.3, the neutron star's strong magnetic field and rapid rotation channel high-energy particles from near the star's surface into the surrounding nebula. The result is an energetic *pulsar wind* that flows outward at almost the speed of light, primarily in the star's equatorial plane. Figure 13.4(b) shows this process in action in the Crab—this combined *Hubble/Chandra* image reveals rings of hot X-ray-emitting gas moving rapidly away from the pulsar. Eventually, the energy from the pulsar wind is deposited in the surrounding Crab Nebula, heating the gas to very high temperatures and powering the spectacular display we see from Earth.

Most pulsars emit pulses in the form of radio radiation. However, some also pulse in the visible, X-ray, and gamma-ray parts of the spectrum. Figure 13.5 shows the Crab and nearby Geminga pulsars in gamma rays. Geminga is unusual in that, while it pulsates strongly in gamma rays, it is barely detectable in visible light and not at all at radio wavelengths. Whatever types of radiation are produced, pulsar flashes at different frequencies all occur at regular, repeated time intervals—as we would expect since they arise from the same object—although the pulses may not all occur at the same instant.

Most known pulsars are observed (usually by Doppler measurements) to have high speeds—much greater than the typical speeds of stars in the Galaxy. The most likely explanation for these anomalously high speeds is that a neutron star can receive a substantial "kick" due to asymmetries in the supernova in which it forms. If the supernova's enormous energy is channeled even slightly in one direction, the newborn neutron star/pulsar can recoil in the opposite direction with a speed of many tens or even hundreds of kilometers per second. Thus, observations of pulsar velocities give theorists additional insight into the detailed physics of supernovae.

CONCEPT CHECK ▶ Why don't we see pulsars at the centers of all supernova remnants?

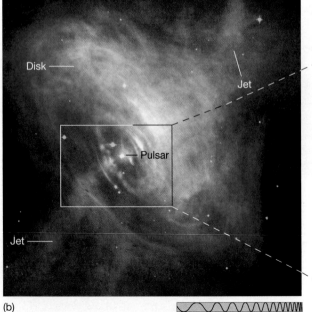

(a)

R I V U X G

(b)

Disk

Jet

Pulsar

Jet

R I V U X G

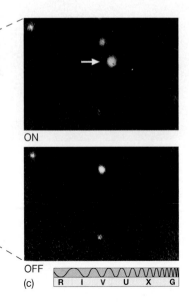

ON

OFF

(c)

R I V U X G

◀ **FIGURE 13.4 Crab Pulsar** In the core of the Crab Nebula (a), the Crab pulsar (c) blinks on and off about 30 times each second. In the top frame, the pulsar (arrow) is on; in the bottom frame, it is off. (b) This more recent *Chandra* X-ray image of the Crab, superimposed on a *Hubble* optical image, shows the central pulsar, as well as rings of hot X-ray-emitting gas in the equatorial plane, driven outward by the pulsar wind. Also visible in the image is a jet of hot gas escaping perpendicular to the equatorial plane. *(ESO; NASA; UC/Lick Observatory)*

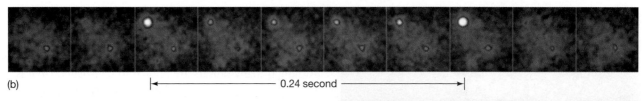

(b) |← 0.24 second →|

Neutron Stars and Pulsars

All pulsars are neutron stars, but not all neutron stars are pulsars, for two reasons. First, the two ingredients that make the neutron star pulse—rapid rotation and strong magnetic field—both diminish with time, so the pulses gradually weaken and become less frequent. Theory indicates that within a few tens of millions of years, the pulses have all but stopped. Second, even a young, bright pulsar is not necessarily visible from Earth. The pulsar beam depicted in Figure 13.3 is relatively narrow—perhaps as little as a few degrees across. Only if the neutron star happens to be oriented in just the right way do we see pulses. When we see the pulses from Earth, we call the object a pulsar. Note that we are using the term "pulsar" here to mean the pulsing object we observe if the beam crosses Earth. However, many astronomers use the word more generically to mean *any* young neutron star producing beams of radiation as in Figure 13.3. Such an object will be a pulsar as seen from some directions—just not necessarily ours!

Figure 13.6 shows *Hubble* images of a neutron star that is not a pulsar. First identified by X-ray observations, this object is just 30 km across and has a surface temperature of about 700,000 K. Despite this enormous temperature, its small size means that it is very faint—just 25th magnitude in visible light. ∞ *(Sec. 10.4)* The star is thought to be about 1 million years old, lies about 60 pc away, and is moving at a speed of 110 km/s across our line of sight. Such faint objects are very difficult to detect, however, and a detection of a "bare" neutron star like this is a rare event. Most cataloged neutron stars have been detected either as pulsars or by their interaction with a "normal" stellar companion in a binary system (see Section 13.3).

Given our current knowledge of star formation, stellar evolution, and neutron stars, observations of pulsars are consistent with the ideas that (1) *every* high-mass star dies in a supernova explosion, (2) most supernovae leave a neutron star behind (a few result in black holes, as discussed in a moment), and (3) *all* young neutron stars emit beams of radiation, just like the pulsars we actually detect. On the basis of estimates of the rate at which massive stars have formed over the lifetime of the Milky Way, astronomers reason that, for every pulsar we know of, there must be several hundred thousand more neutron stars moving unseen somewhere in the Galaxy. Some formed relatively recently—less than a few million years ago—and simply happen not to be beaming their energy toward Earth. However, the vast majority are old, their youthful pulsar phase long past.

Neutron stars (and black holes too) were predicted by theory long before they were actually observed, although their extreme properties made many scientists doubt that they would ever be found in nature. The fact that we now have strong observational evidence, not just for their existence, but also for the vitally important roles they play in many areas of high-energy astrophysics, is yet another testament to the fundamental soundness of the theory of stellar evolution.

(a)

R I V U X G

▲ **FIGURE 13.5 Gamma-Ray Pulsars** (a) The Crab and Geminga pulsars lie fairly close to one another in the sky. Unlike the Crab, Geminga is barely visible at optical wavelengths, and undetectable in the radio region of the spectrum. (b) Sequence of *Compton Gamma-Ray Observatory* images showing Geminga's 0.24-s pulse period. *(NASA)*

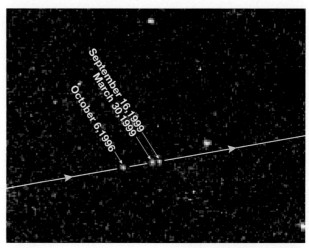

R I V U X G

▶ **FIGURE 13.6 Isolated Neutron Star** This lone neutron star was first detected by its X-ray emission and subsequently imaged by *Hubble*. This 700,000-K object lies about 60 pc from Earth and is thought to be about 1 million years old. This triple exposure shows the star streaking across the sky at more than 100 km/s. *(NASA)*

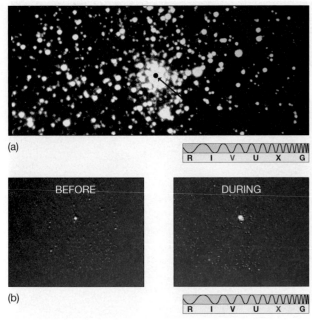

(a)

R I V U X G

BEFORE

DURING

(b)

R I V U X G

▲ **FIGURE 13.7 X-Ray Burster** An X-ray burster produces a sudden, intense flash of X-rays, followed by a period of relative inactivity lasting as long as several hours. Then another burst occurs. (a) An optical photograph of the globular star cluster Terzan 2, showing a 2″ dot at the center where the X-ray bursts originate. (b) X-ray images taken before and during the outburst. The most intense X-rays correspond to the position of the black dot shown in (a). *(SAO)*

13.3 Neutron Star Binaries

We noted in Chapter 10 that most stars are members of binary systems. ∞ *(Sec. 10.7)* Although many pulsars are known to be isolated (that is, not part of any binary), at least some do have binary companions, and the same is true of neutron stars in general (that is, even the ones not seen as pulsars). One important consequence of this is that some neutron star masses have been determined very accurately. Most of the measured masses are fairly close to 1.4 times the mass of the Sun—the Chandrasekhar mass of the stellar core that collapsed to form the neutron star remnant—although a neutron star with a mass twice that of the Sun has recently been reported.

X-Ray Sources

With the launch of the first orbiting X-ray telescopes in the 1970s, numerous X-ray sources were found near the central regions of our Galaxy and also near the centers of a few globular star clusters. ∞ *(Sec. 11.6)* Some of these sources, known as **X-ray bursters,** emit much of their energy in violent eruptions, each thousands of times more luminous than our Sun but lasting only seconds. A typical burst is shown in Figure 13.7.

The X-ray emission is thought to arise on or near neutron stars that are members of binary systems. Matter torn from the surface of the (main-sequence or giant) companion by the neutron star's strong gravitational pull accumulates on the neutron star's surface. (Figure 12.14 shows the white dwarf equivalent.) ∞ *(Sec. 12.3)* The gas forms an accretion disk around the neutron star, then slowly spirals inward. The inner portions of the disk become extremely hot, releasing a steady stream of X-rays.

As gas builds up on the neutron star's surface, its temperature rises due to the pressure of overlying material. Eventually, it becomes hot enough to fuse hydrogen. The result is a sudden period of rapid nuclear burning that releases a huge amount of energy in a brief but intense X-ray burst. After several hours of renewed accumulation, a fresh layer of matter produces the next burst. Thus, the mechanism is very similar to that responsible for a nova explosion on a white dwarf, except that it occurs on a far more violent scale because of the neutron star's much stronger surface gravity.

Millisecond Pulsars

In the mid-1980s astronomers discovered an important new category of pulsars—a class of very rapidly rotating objects called **millisecond pulsars.** These objects spin hundreds of times per second (that is, their pulse periods are a few milliseconds, 0.001 s). This speed is about as fast as a typical neutron star can spin without flying apart. In some cases the star's equator is moving at more than 20 percent of the speed of light. This suggests a phenomenon bordering on the incredible—a cosmic object just 20 km across, more massive than our Sun, spinning almost at breakup speed, making nearly 1000 complete revolutions *every second*. Yet the observations and their interpretation leave little room for doubt. More than 200 millisecond pulsars are now known.

The story of these remarkable objects is further complicated by the fact that many of them—about two-thirds—are found in globular clusters. This is odd because globular clusters are at least 10 billion years old, yet Type II supernovae (the kind that create neutron stars) are associated with massive stars that explode within a few tens of *millions* of years after their formation, and no stars have formed in any globular cluster since the cluster came into being. ∞ *(Secs.11.6, 12.4)* Thus, no new neutron star has been produced in a globular cluster in a very long time. Furthermore, as mentioned earlier, the pulsar produced in a supernova explosion is expected to slow down and fade away in only a few tens of millions of years—after 10 billion years, its rotation rate should be very slow. The rapid rotation of the pulsars found in globular clusters therefore cannot be a relic of their birth. These pulsars must have been "spun up"—that is, had their rotation rates increased—by some other, much more recent, mechanism.

The most likely explanation for the high rotation rate of millisecond pulsars is that they have been spun up by drawing in matter from a companion star. As matter spirals down onto the neutron star's surface in an accretion disk, it provides the "push" needed to make the neutron star spin faster (Figure 13.8). Theoretical calculations indicate that this process can spin the star up to breakup speed in about 100 million years. This picture is supported by the finding that, of the 140 or so millisecond pulsars observed in globular clusters, roughly half are known currently to be members of binary systems. The remaining solo millisecond pulsars were probably formed when an encounter with another star ejected the pulsar from the binary or when the pulsar's own intense radiation destroyed its companion.

Notice that the scenario of accretion onto a neutron star from a binary companion is the same scenario that we just used to explain the existence of X-ray binaries. In fact, the two phenomena are very closely linked. Many X-ray binaries may be on their way to becoming millisecond pulsars, and many millisecond pulsars are X-ray sources, powered by the trickle of material still falling onto them from their binary companions. Figure 13.9 shows the globular cluster 47 Tucanae, together with a remarkable *Chandra* image of its core showing no fewer than 108 X-ray sources—about 10 times the number that had been known in the cluster prior to *Chandra's* launch. Roughly half of these sources are millisecond pulsars; the cluster also contains two or three "conventional" neutron star binaries. Most of the remaining sources are white dwarf binaries, similar to those discussed in Chapter 12. ∞ *(Sec. 12.3)*

Pulsar Planets

Radio astronomers can capitalize on the precision with which pulsar signals repeat themselves to make extremely accurate measurements of pulsar motion. In 1992 radio astronomers at the Arecibo Observatory found that the pulse period of a millisecond pulsar lying some 500 pc from Earth exhibits tiny but regular fluctuations (at the level of about one part in 10^7) on two distinct time scales—67 and 98 days.

These fluctuations are caused by the Doppler effect as the pulsar wobbles back and forth in space under the combined gravitational pulls of two small bodies, each about three times the mass of Earth ∞ *(Sec. 4.4)*. One orbits the pulsar at a distance of 0.4 AU, the other at a distance of 0.5 AU, with orbital periods of 67 and 98 days, respectively. Further observations not only confirmed these findings, but also revealed the presence of a third body, with a mass comparable to that of Earth's Moon, orbiting just 0.2 AU from the pulsar.

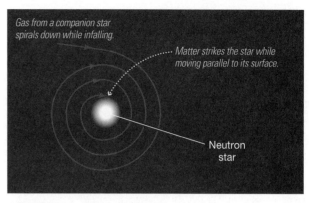

▲ FIGURE 13.8 Millisecond Pulsar As infalling matter strikes the star, it moves almost parallel to the surface, so it tends to make the star spin faster. Eventually, this process can result in a millisecond pulsar—a neutron star spinning at the incredible rate of hundreds of revolutions per second.

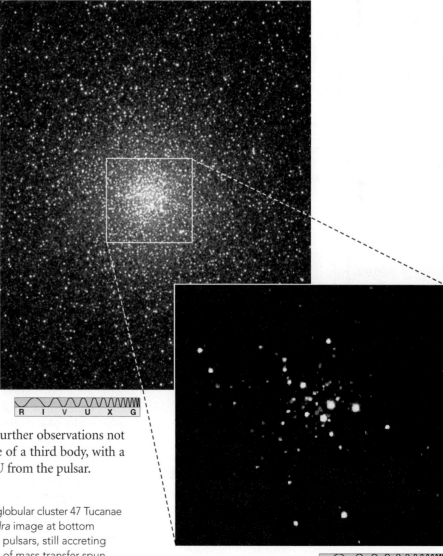

▶ FIGURE 13.9 Cluster X-Ray Binaries The dense core of the old globular cluster 47 Tucanae harbors more than 100 separate X-ray sources (shown in the *Chandra* image at bottom right). More than half of these are thought to be binary millisecond pulsars, still accreting small amounts of gas from their companions after an earlier period of mass transfer spun them up to millisecond speeds. *(ESO; NASA)*

CONCEPT
CHECK ▶   What is the connection between X-ray
sources and millisecond pulsars?

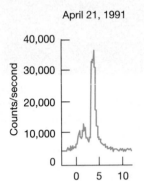

April 21, 1991

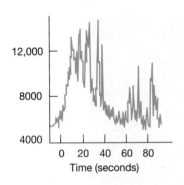

April 25, 1991

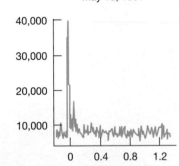

May 18, 1991

(a)

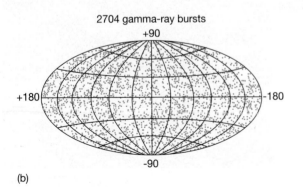

2704 gamma-ray bursts

(b)

These remarkable results constituted the first definite evidence of planet-sized bodies outside our solar system. A few other millisecond pulsars have since been found with similar behavior. However, it is unlikely that any of these planets formed in the same way as our own. Any planetary system orbiting the pulsar's parent star was almost certainly destroyed in the supernova explosion that created the pulsar. As a result, scientists are uncertain about how these planets came into being. One possibility involves the binary companion that provided the matter necessary to spin the pulsar up to millisecond speeds. Possibly, the pulsar's intense radiation and strong gravity destroyed the companion, then spread its matter out into a disk (a little like the solar nebula) in whose cool outer regions the planets might have condensed.

Astronomers searched for decades for planets orbiting main-sequence stars like our Sun on the assumption that planets are a natural by-product of star formation. ∞ *(Sec. 4.3)* As we have seen, these searches have now identified many extrasolar planets, although only a few planets comparable in mass to Earth have so far been detected. ∞ *(Sec. 4.4)* It is ironic that the first Earth-sized planets found outside the solar system orbit a dead star and have little or nothing in common with our own world.

13.4 Gamma-Ray Bursts

Discovered serendipitously in the late 1960s by military satellites looking for violations of the Nuclear Test Ban Treaty, and first made public in the 1970s, **gamma-ray bursts** consist of bright, irregular flashes of gamma rays typically lasting only a few seconds (Figure 13.10a). Until the 1990s it was thought that gamma-ray bursts were basically "scaled-up" versions of X-ray bursters, in which even more violent nuclear burning resulted in the release of more energetic gamma rays. However, it now appears that this is not the case.

Distances and Luminosities

Figure 13.10(b) shows an all-sky plot of the positions of 2704 bursts detected by the *Compton Gamma-Ray Observatory (CGRO)* during its 9-year operational lifetime. ∞ *(Sec. 3.5)* On average, *CGRO* detected gamma-ray bursts at the rate of about one per day. Note that the bursts are distributed uniformly across the sky, rather than being confined to the relatively narrow band of the Milky Way (compare Figure 3.30). This widespread distribution convinced most astronomers that the bursts do not originate within our own Galaxy, as had previously been assumed, but instead are produced at much greater distances.

Actually measuring the distance to a gamma-ray burst is no easy task. The gamma-ray observations do not provide enough information in and of themselves to tell us how far away the burst is, so astronomers must instead associate the burst with some other object in the sky—called the burst *counterpart*—whose distance can be measured by other means. The techniques for studying counterparts generally involve observations in the optical or X-ray parts of the electromagnetic spectrum. The problem is that the resolution of a gamma-ray telescope is quite poor, so

◀ **FIGURE 13.10 Gamma-Ray Bursts**   (a) Plots of intensity versus time for three gamma-ray bursts show that some of them are irregular and spiky, whereas others are much more smoothly varying. (b) Positions on the sky of all the gamma-ray bursts detected by the *Compton Gamma-Ray Observatory* during its 9 years of operation in orbit. The bursts appear to be distributed uniformly across the entire sky, which is mapped here with the plane of the Milky Way running horizontally across its center. *(NASA)*

the burst positions are uncertain by about a degree, and a relatively large region of the sky must be scanned in search of a counterpart. ∞ *(Sec. 3.5)* In addition, the "afterglow" of a burst at X-ray or optical wavelengths fades quite rapidly, severely limiting the time available to complete the search.

The most successful searches for burst counterparts have been carried out by satellites combining gamma-ray detectors with X-ray and/or optical telescopes. For example, NASA's *Swift* mission, launched in 2004 and still operational, combines a wide-angle gamma-ray detector (to monitor as much of the sky as possible) with two telescopes: one X-ray and one optical/ultraviolet instrument. The gamma-ray detector system pinpoints the direction of the burst to an accuracy of about 4 arc minutes, and within seconds the onboard computer automatically repositions the satellite to point the X-ray and optical telescopes in that direction. At the same time, the craft relays the burst position to other instruments in space and on the ground. *Swift* detects burst counterparts at the rate of about one per week and has played a pivotal role in advancing our understanding of these violent phenomena. Figures 13.11(c) and (d) show *Swift* X-ray and optical images of GRB 080319B, one of the brightest bursts to date. Automated observations at many wavelengths began within seconds of *Swift's* detection, making this burst one of the most intensively studied on record.

The first direct measurement of the distance to a gamma-ray burst was made in 1997 when astronomers detected the visible afterglow created as the burst faded and cooled. Using spectroscopic distance-measurement techniques (to be discussed in more detail in Section 16.1), they found that the event occurred more than 2 *billion* parsecs from Earth. Figure 13.11 shows an optical image of another gamma-ray burst, this one almost 5 billion parsecs distant, along with what may be its host galaxy—this is one of the first images to associate a burst with a possible galactic host. To date, the distances to hundreds of gamma-ray bursts have been measured from their afterglows. All are very large, implying that the bursts must be extremely energetic, otherwise they wouldn't be detectable by our equipment.

If we simply assumed that the gamma rays are emitted equally in all directions (a big assumption!), we could estimate the total energy emitted using the inverse-square law. ∞ *(Sec. 10.2)* We would find that each burst generates more energy—in some cases hundreds of times more energy—than a typical supernova explosion, all in a matter of seconds!

However, if the energy is emitted as a *jet*, as most experts now think, then the overall energy of the burst is reduced to more "manageable" levels, as the luminosity we see is representative of only a small fraction of the sky, making the huge luminosities of the most energetic bursts easier to explain. As an analogy, consider a handheld laser pointer of the sort commonly used in talks and lectures. It radiates only a few milliwatts of power, much less than a household lightbulb, but it appears enormously bright if you happen to look directly into the beam. (*Don't* do this, by the way!) The laser beam is bright because all of its energy is concentrated in almost a single direction instead of being radiated in all directions into space.

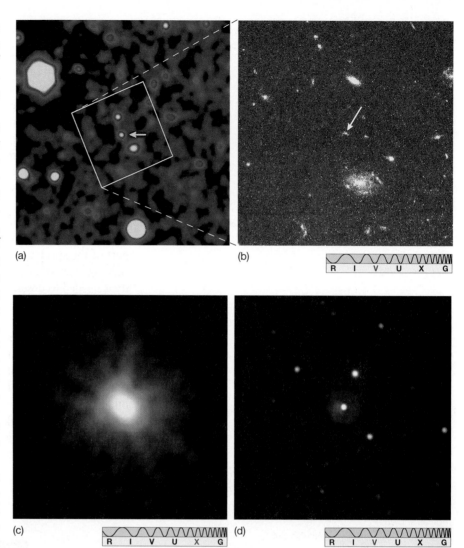

(a)

(b)

R I V U X G

(c)

R I V U X G

(d)

R I V U X G

▲ FIGURE 13.11 **Gamma-Ray Burst Counterpart** (a) Optical image of the gamma-ray burst GRB 971214 taken by the Keck telescope, shows the visible afterglow of the gamma-ray source (arrow) to be quite bright, (b) By the time this *Hubble* image was taken 2 months later, the afterglow had faded, but a faint image of a host galaxy remains. (c and d) Another long-duration gamma-ray burst, GRB 080319B, was one of the brightest yet observed. Its light that reached Earth on March 19, 2008, was emitted 7.5 billion years ago, yet for a few seconds the flash would have been visible to the unaided eye—if anyone had been looking in just the right spot! Only moments after detonation, it was observed in X-rays (c) and in visible light (d). (*Keck; NASA; ESO*)

What Causes the Bursts?

Not only are gamma-ray burst sources extremely energetic, they are also very *small*. The millisecond flickering in the bursting gamma rays detected by *GRO* implies that whatever their origin, all of their energy must come from a volume no larger than a few hundred kilometers across. The reasoning is as follows: If the emitting region were, say, 300,000 km—1 light-second—across, even an instantaneous change in intensity at the source would be smeared out over a time interval of 1 s as seen from Earth, because light from the far side of the object would take 1 s longer to reach us than light from the near side. For the gamma-ray variation not to be blurred by the light-travel time, the source cannot be more than 1 light-millisecond, or 300 km, in diameter.

Theoretical models of gamma-ray bursts describe the burst as a *relativistic fireball*—an expanding jet of superhot gas radiating furiously in the gamma-ray part of the spectrum. (The term "relativistic" here means that the gas particles are moving at speeds comparable to the speed of light, and Einstein's theory of relativity is needed to describe them—see Section 13.6.) The complex burst structure and afterglows we see are produced as the fireball expands, cools, and interacts with its surroundings.

Two leading models for the energy source have emerged. The first (Figure 13.12a) is the "true" end point of a binary star system. Suppose that both members of the binary become neutron stars. As the system continues to evolve, gravitational radiation (see *Discovery 13-1*) is released, and the two ultradense stars spiral in toward each other. Once they are within a few kilometers of one another, coalescence is inevitable. Such a merger will likely produce a violent explosion comparable in energy to a supernova, and energetic enough to explain the flashes of gamma rays we observe. The overall rotation of the binary system may channel the energy into a high-speed, high-temperature jet.

ANIMATION/VIDEO Colliding Binary Neutron Stars (MA)

▼ FIGURE 13.12 **Gamma-Ray Burst Models** Two models have been proposed to explain gamma-ray bursts. Part (a) depicts the merger of two neutron stars; part (b) shows the collapse of a single star. Both models predict (at right) a relativistic fireball, perhaps releasing energy in the form of jets, as shown.

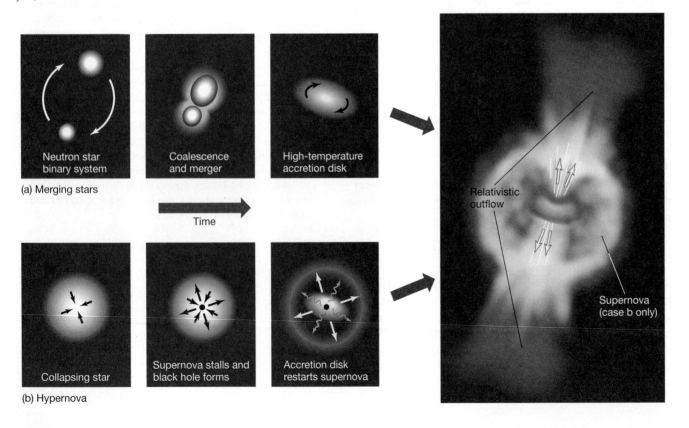

(a) Merging stars

Neutron star binary system   Coalescence and merger   High-temperature accretion disk

Time

(b) Hypernova

Collapsing star   Supernova stalls and black hole forms   Accretion disk restarts supernova

Relativistic outflow

Supernova (case b only)

13-1 DISCOVERY

Gravity Waves—A New Window on the Universe

Electromagnetic waves are common everyday phenomena. They involve periodic changes in the strengths of electric and magnetic fields. ∞ *(Fig. 2.6)* They move through space and transport energy. Any accelerating charged particle, such as an electron in a broadcasting antenna or on the surface of a star, generates electromagnetic waves.

The modern theory of gravity—Einstein's theory of relativity—also predicts waves that move through space. A *gravity wave* is the gravitational counterpart of an electromagnetic wave. *Gravitational radiation* results from changes in the strength of a gravitational field. In theory, any time an object having any mass accelerates, a gravity wave is emitted at the speed of light. The passage of a gravity wave should produce small distortions in the space through which it passes. Gravity is an exceedingly weak force compared with electromagnetism, so these distortions are expected to be very small—in fact, much smaller than the diameter of an atomic nucleus for waves that might be produced by any known astrophysical source. Yet many researchers think that these tiny distortions should be measurable. No experiment has yet succeeded in detecting gravity waves, but their detection would provide such strong support for the theory of relativity that scientists are eager to find them.

The leading candidates for systems likely to produce gravity waves detectable on Earth include (1) close binary systems containing black holes, neutron stars, or white dwarfs and (2) the collapse of a star into a black hole. Each of these possibilities involves the acceleration of huge masses, resulting in rapidly changing gravitational fields. The first probably presents the best chance to detect gravity waves. Binary star systems should emit gravitational radiation as the component stars orbit one another. As energy escapes in the form of gravity waves, the two stars spiral toward one another, orbiting more rapidly and emitting even more gravitational radiation. For sufficiently close systems, the stars merge in a few tens or hundreds of millions of years (although most of the radiation is emitted during the last few seconds). As we saw in Section 13.4, neutron star mergers may well also be the origin of some gamma-ray bursts, so gravitational radiation might provide an alternative means of studying these violent and mysterious phenomena.

Such a slow but steady decay in the orbit of a binary system has in fact been detected. In 1974, radio astronomer Joseph Taylor and graduate student Russell Hulse at the University of Massachusetts discovered a very unusual binary system. Both components are neutron stars, and one is observable from Earth as a pulsar. This system has become known as the *binary pulsar*. Measurements of the periodic Doppler shift of the pulsar's radiation show that its orbit is slowly shrinking at exactly the rate predicted by relativity theory if the energy were being carried off by gravity waves. Even though the gravity waves themselves have not been detected, the binary pulsar is regarded by most astronomers as very strong evidence supporting general relativity. Taylor and Hulse received the 1993 Nobel Prize in Physics for their discovery.

In January 2004, radio astronomers announced the discovery of a *double-pulsar* binary system with an even shorter period than the binary pulsar, implying stronger relativistic effects and a shorter merger time—less than 100 million years. Because both components are pulsars and the system, by pure luck, also happens to be seen almost exactly edge on by observers on Earth, leading to eclipses, this system has provided a wealth of information on both neutron stars and gravitational physics.

The figure below shows part of an ambitious gravity-wave observatory called LIGO—short for Laser Interferometric Gravity-Wave Observatory—which became operational in 2003. Twin detectors, one (shown here) in Hanford, Washington, the other in Livingston, Louisiana, use laser beams to measure the extremely small (less than one-thousandth the diameter of an atomic nucleus) distortions of space produced in the lengths of the 4-km-long arms should a gravity wave pass by. The instruments are in theory capable of detecting waves from many Galactic and extragalactic sources, although, so far, no gravity waves have actually been detected, despite a 2007 upgrade that greatly increased the system's sensitivity. Additional upgrades boosting the detector's sensitivity by a further factor of 10 are planned to come on line around 2014.

If these experiments are successful, the discovery of gravity waves could herald a new age in astronomy, in much the same way that invisible electromagnetic waves, virtually unexplored a century ago, revolutionized classical astronomy and led to the field of modern astrophysics.

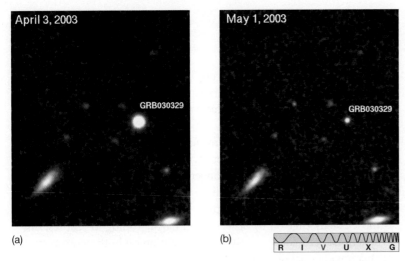

(a)   (b)

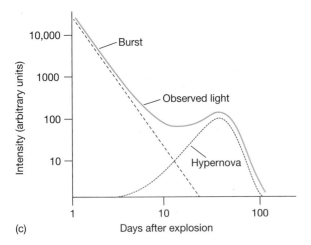

(c)

◄ FIGURE 13.13 **Hypernova?** The gamma-ray burst GRB 030329 may prove crucial to theorists' understanding of the physical processes underlying these violent phenomena. First detected by the *High Energy Transient Explorer 2* satellite and subsequently observed at radio, optical, and X-ray wavelengths, the burst counterpart has all the hallmarks of a high-mass supernova, lending strong support to the hypernova model. Here, the counterpart is shown (a) near the moment of the burst and (b) fading a month later. (c) This graph shows details of the emitted radiation from another, but similar, gamma-ray burst. *(ESO)*

The second model (Figure 13.12b), sometimes called a *hypernova*, is a "failed" supernova—but what a failure! In this picture a very massive star undergoes core collapse just as described earlier for a Type II supernova, but instead of forming a neutron star, the core collapses in on itself, forming a black hole (see Section 13.5). ∞ *(Sec. 12.4)* Instead of being blown to pieces, the star implodes onto the black hole, forming an accretion disk and again generating a relativistic jet. The jet punches its way out of the star, producing a gamma-ray burst as it slams into the surrounding shells of gas expelled from the star during the final stages of its nuclear-burning lifetime. At the same time, intense radiation from the accretion disk may restart the stalled supernova, blasting what remains of the star into space.

The idea of a relativistic fireball has become widely accepted among workers in this branch of astrophysics. Most researchers regard it as likely that the energy is released as a jet and that the most intense gamma-ray bursts occur when the jet is directed toward us ("looking into the laser"). Which of the two models just described is correct? Experts in the field would say that the answer is probably *both*. The hypernova model predicts bursts of relatively long duration and is the leading explanation of the "long" gamma-ray bursts lasting more than about 2 s. Detailed calculations indicate that the neutron star merger model can naturally account for the shorter bursts. Thus, both of the scenarios depicted in Figure 13.12 contribute to the total.

Thanks to the "rapid response network" provided by *Swift* and other instruments, astronomers now have detailed observations of many afterglows of both types of burst. Figures 13.13(a) and (b) show the afterglow of the long burst GRB 030329, as seen by the 8.2-m VLT in Chile. ∞ *(Sec. 3.2)* Both the spectrum and the light curve were consistent with what astronomers expected from the supernova of a very massive (roughly 25-solar mass) star. ∞ *(Sec. 12.4)* Figure 13.13(c) shows a simplified light curve of another burst, illustrating how the "burst" and "hypernova" components can be distinguished.

The rapidly fading X-ray afterglows from the short bursts are consistent with the predictions of the neutron star merger scenario. Very recent observations also reveal that a few of these bursts may actually involve the theoretically predicted, but rare, merger between a neutron star and a black hole, which should have its own characteristic light signature.

CONCEPT CHECK ► What are gamma-ray bursts, and why did they pose such a challenge to theory?

13.5 Black Holes

Neutron stars are supported by the resistance of tightly packed neutrons to further compression. Squeezed together, these elementary particles form a hard ball of ultradense matter that not even gravity can compress further. Or do they? Is it possible that, given enough matter packed into a small enough volume, the collective pull of gravity can eventually crush even a neutron star? Can gravity continue to compress a massive star into an object the size of a planet, a city, a pinhead—even smaller? The answer, apparently, is yes.

The Final Stage of Stellar Evolution

Although the precise figure is uncertain, mainly because the behavior of matter at very high densities is not well understood, most researchers concur that the mass of a neutron star cannot exceed about three times the mass of the Sun. This is the neutron star equivalent of the white dwarf mass limit discussed in the previous chapter. ∞ *(Sec. 12.5)* Above this mass, not even tightly packed neutrons can withstand the star's gravitational pull. In fact, we know of *no* force that can counteract gravity beyond this point. Thus, if enough material is left behind after a supernova explosion that the central core exceeds this limit, or if enough matter falls back onto the neutron star after the supernova has occurred, gravity wins the battle with pressure once and for all, and the central core collapses forever. Stellar evolution theory indicates that this is the fate of any star whose main-sequence mass exceeds about 25 times the mass of the Sun.

As the core shrinks, the gravitational pull in its vicinity eventually becomes so great that nothing—not even light—can escape. The resultant object therefore emits no light, no other form of radiation, no information whatsoever. Astronomers call this bizarre end point of stellar evolution, in which the core of a very-high-mass star collapses in on itself and vanishes forever, a **black hole.**

Escape Speed

Newtonian mechanics—up to now our indispensable tool in understanding the universe—cannot adequately describe conditions in or near black holes. ∞ *(Sec. 1.4)* To comprehend these collapsed objects, we must turn instead to the modern theory of gravity, Einstein's **general theory of relativity,** discussed in Section 13.6. Still, we can usefully describe some aspects of these strange bodies in more or less Newtonian terms. Let's reconsider the familiar Newtonian concept of escape speed—the speed needed for one object to escape from the gravitational pull of another—supplemented by two key facts from relativity: (1) nothing can travel faster than the speed of light and (2) all things, *including light,* are attracted by gravity.

A body's escape speed is proportional to the square root of the body's mass divided by the square root of its radius. ∞ *(More Precisely 5-1)* Earth's radius is 6400 km and the escape speed from Earth's surface is just over 11 km/s. Now consider a hypothetical experiment in which Earth is squeezed on all sides by a gigantic vise. As our planet shrinks under the pressure, the escape speed increases because the radius is decreasing. Suppose Earth were compressed to one-fourth its present size. The escape speed would double (because $1/\sqrt{1/4} = 2$). To escape from this compressed Earth, an object would need a speed of at least 22 km/s.

Imagine compressing Earth by an additional factor of 1000, making its radius hardly more than a kilometer. Now a speed of about 700 km/s—more than the escape speed of the Sun—would be needed to escape. Compress Earth still further, and the escape speed continues to rise. If our hypothetical vise were to squeeze Earth hard enough to crush its radius to about a centimeter, the speed needed to escape its surface would reach 300,000 km/s. But this is no ordinary speed. It is the speed of light, *c,* the fastest speed allowed by the laws of physics as we currently know them.

Thus, if by some fantastic means the entire planet Earth could be compressed to less than the size of a grape, the escape speed would exceed the speed of light. And because nothing can exceed that speed, the compelling conclusion is that nothing—absolutely nothing—could escape from the surface of such a compressed body. Even radiation—radio waves, visible light, X-rays, photons of all wavelengths—would be unable to escape its intense gravity. Our planet would be invisible and uncommunicative—no signal of any sort could be sent to the universe outside. The origin of the term *black hole* becomes clear. For all practical purposes, Earth could be said to have disappeared from the universe. Only its gravitational field would remain behind, betraying the presence of its mass, now shrunk to a point.[1]

The Event Horizon

Astronomers have a special name for the critical radius at which the escape speed from an object would equal the speed of light and within which the object could no longer be seen. It is the **Schwarzschild radius,** after Karl Schwarzschild, the German scientist who first studied its properties. The Schwarzschild radius of any object is proportional to its mass. For Earth, it is 1 cm. For Jupiter, at about 300 Earth masses, it is about 3 m. For the Sun, at 300,000 Earth masses, it is 3 km. For a 3-solar-mass stellar core, the Schwarzschild radius is about 9 km. As a convenient rule of thumb, the Schwarzschild radius of an object is simply 3 km multiplied by the object's mass measured in solar masses.

Every object has a Schwarzschild radius—it is the radius to which the object would have to be compressed for it to become a black hole. Put another way, a black hole is an object that happens to lie within its own Schwarzschild radius. The surface of an imaginary sphere having a radius equal to the Schwarzschild radius and centered on a collapsing star is called the **event horizon.** No event occurring within this region can ever be seen, heard, or known by anyone outside. Even though there is no matter of any sort associated with it, we can think of the event horizon as the "surface" of a black hole.

Theory indicates that, if more than 3 solar masses of material remain behind after a supernova explosion, the remnant core will collapse catastrophically, diving below the event horizon in less than a second. The event horizon is not a physical boundary of any kind—just a communications barrier. The core would shrink right past the Schwarzschild radius to ever-diminishing size on its way to being crushed to a point. It simply "winks out," disappearing and becoming a small dark region from which nothing can escape—a literal black hole in space. This is the predicted fate of stars having more than about 20–25 times the mass of the Sun.

SELF-GUIDED TUTORIAL Escape Speed and Black Hole Event Horizons

CONCEPT CHECK ▶ What happens if you compress an object within its own Schwarzschild radius?

13.6 Einstein's Theories of Relativity

By the latter part of the 19th century, physicists were well aware of the special status of the speed of light, *c*. It was, they knew, the speed at which all electromagnetic waves traveled, and, as best they could tell, it represented an upper limit on the speeds of *all* known particles. Scientists struggled without success to construct a theory of mechanics and radiation in which *c* was a natural speed limit.

Special Relativity

In 1887, a fundamental experiment carried out by American physicists A. A. Michelson and E. W. Morley compounded theorists' problems further by demonstrating another important aspect of light: The measured speed of a beam of light is *independent* of the motion of either the observer or the source.

---

[1]In fact, regardless of the composition or condition of the object that formed the black hole, only three physical properties can be measured from the outside—the hole's mass, charge, and angular momentum. All other information is lost once matter falls into the black hole. Only these three numbers are required to describe completely a black hole's outward appearance. In this chapter we consider only holes that formed from nonrotating, electrically neutral matter. Such objects are completely specified once their masses are known.

Michelson and Morley attempted to determine Earth's motion relative to the "absolute" space in which light supposedly moved by measuring the speed of light at different times of the day (so that the orientation of their equipment would change as Earth rotated) and on different days of the year (so that Earth's velocity would vary as our planet orbited the Sun). As illustrated in Figure 13.14, they expected to measure a faster-moving beam of light when it was traveling opposite to Earth's motion (to the left in the figure) and a slower beam when Earth was "catching up" on it (to the right). But in fact they measured precisely the *same* velocity for any orientation of their apparatus. Far from establishing the properties of absolute space, the Michelson–Morley experiment ultimately demolished the entire concept—and with it, the 19th-century view of the universe.

All subsequent experiments confirm that no matter what our motion may be relative to the source of the radiation, we always measure precisely the same value for *c*: 299,792.458 km/s. A moment's thought tells us that this is a decidedly nonintuitive statement. For example, if we were traveling in a car moving at 100 km/h and we fired a bullet forward with a speed of 1000 km/h relative to the car, an observer standing at the side of the road would see the bullet pass by at 100 + 1000 = 1100 km/h, as illustrated in Figure 13.15(a). However, the Michelson–Morley experiment tells us that if we were traveling in a rocket ship at one-tenth the speed of light, 0.1*c*, and we shone a searchlight beam ahead of us (Figure 13.15b), an outside observer would measure the speed of the beam not as 1.1*c*, as the example of the bullet would suggest, but as *c*. The rules that apply to particles moving at or near the speed of light are different from those we are used to in everyday life.

The *special theory of relativity*, or just *special relativity*, was proposed by Einstein in 1905 to deal with the preferred status of the speed of light. The theory is the mathematical framework that allows us to extend the familiar laws of physics from low speeds (i.e., speeds much less than *c*, which are often referred to as *nonrelativistic*) to very high (or *relativistic*) speeds, comparable to *c*.

The essential features of special relativity are as follows:

1. The speed of light, *c*, is the maximum possible speed in the universe, and all observers measure the *same* value for *c*, regardless of their motion. Einstein broadened this statement into the *principle of relativity*: The basic laws of physics are *the same* to all observers.

2. There is *no* absolute frame of reference in the universe; that is, there is no "preferred" observer relative to whom all other velocities can be measured. Instead, only relative velocities between observers matter (hence the term "relativity").

3. Neither space nor time can be considered independently of one another. Rather, they are each components of a single entity: *spacetime*. There is no absolute, universal time—observers' clocks tick at different rates, depending on the observers' motions relative to one another.

Special relativity is equivalent to Newtonian mechanics in describing objects that move much more slowly than the speed of light, but it differs greatly in its predictions at relativistic velocities (see *More Precisely 13-1*). Yet, despite their often nonintuitive nature, all of the theory's predictions have been repeatedly verified to a high degree of accuracy. Today, special relativity lies at the heart of modern science. No scientist seriously doubts its validity.

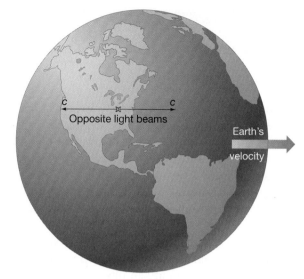

▲ FIGURE 13.14 **Michelson–Morley Experiment**   If light propagates at a fixed speed through space, then common sense suggests that the speed of a beam of light should depend on Earth's velocity. The Michelson–Morley experiment tried but failed to measure this dependence and in the process ushered in the era of modern physics.

INTERACTIVE Exploring the Effects of Relative Motion

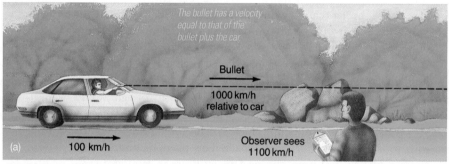

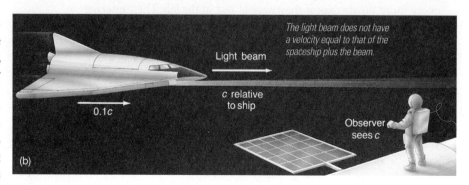

▲ FIGURE 13.15 **Speed of Light**   (a) A bullet fired from a speeding car is measured by an outside observer to have a speed equal to the sum of the speeds of the car and of the bullet. (b) A beam of light shining forward from a high-speed spacecraft is still observed to have speed *c*, regardless of the speed of the spacecraft. The speed of light is thus independent of the speed of the source or of the observer.

13-1 MORE PRECISELY

Special Relativity

Einstein's theory of special relativity was constructed in large part to solve the puzzle of the 1887 Michelson–Morley experiment (see Section 13.6), which had demonstrated that the observed speed of light is *independent* of the observer's motion through space. Einstein elevated the speed of light to the status of a constant of nature, rewrote the laws of mechanics to reflect that new fact, and opened the door to a flood of new physics and a much deeper understanding of the universe. But many commonsense ideas had to be abandoned in the process and replaced with some decidedly less-intuitive concepts. We describe here some of those odd consequences of Einstein's theory.

Imagine that you are an observer watching a rocket ship fly past at relative velocity v and that the craft is close enough for you to make detailed observations of the inside of its cabin. If v is much less than the speed of light, c, you would see nothing out of the ordinary—special relativity is consistent with familiar Newtonian mechanics at low velocities. However, if v is comparable to c, unexpected things start to happen.

As the ship's velocity increases, you begin to notice that the ship appears to *contract* in the direction in which it is moving. A meterstick on board, identical at launch to the one you keep in your laboratory, is now shorter than its twin. This is called *Lorentz contraction* (or sometimes Lorentz–Fitzgerald contraction). The accompanying figure shows the stick's measured length aboard the moving ship: At low velocities (bottom) the meterstick measures 1 m, but at high velocities (top), the stick is shortened considerably. Because of Lorentz contraction, a meterstick moving at 90 percent of the velocity of light would shrink to a little less than half a meter. (This is not an optical illusion.)

At the same time, the ship's clock, synchronized prior to launch with your own, now ticks *more slowly*. This phenomenon, known as *time dilation*, has been observed many times in laboratory experiments in which fast-moving radioactive particles are observed to decay more slowly than if they were at rest in the lab. Their internal clocks are slowed by their rapid motion. Although no material particle can actually reach the speed of light, Einstein's theory implies that, as v approaches c, the measured length of the meterstick will shrink to nearly zero and the clock will slow to a virtual stop.

The effects of Lorentz contraction and time dilation are described by the same quantity, usually called the *Lorentz factor* and denoted by the Greek letter γ (gamma). For a particle moving with velocity v, this factor is given by

$$\gamma = \frac{1}{\sqrt{1 - v^2/c^2}}.$$

As v increases, lengths are reduced and clocks slow down by this factor (which is always greater than or equal to 1). For Earth, orbiting the Sun at 30 km/s, $v = 0.0001c$ and $\gamma = 1/\sqrt{0.99999999} = 1.0000000005$, so the effects of relativity are negligibly small. However, for a charged particle in a pulsar wind (Section 13.2)

traveling at 90 percent of the speed of light, $v/c = 0.9$, and now $\gamma = 1/\sqrt{0.19} = 2.3$. Speeds very close to that of light are needed before the effect becomes large. Even at one-tenth the speed of light, 30,000 km/s, γ is still only 1.005, so relativistic effects are barely noticeable.

Of course, from the point of view of an astronaut on board the spaceship, *you* are the one moving rapidly. Hence, as seen from the ship, *you* appear to be compressed in the direction of motion, and *your* clock runs slowly! How can both observations be correct? After all, you both know that nothing has happened to your metersticks or clocks. You feel no force squeezing your bodies. What is going on?

When measuring the length of the moving meterstick, you do so by noting the positions of the two ends *at the same time, according to your clock*. But those two events—the two measurements you make—do *not* occur at the same time as seen by the astronaut on the spaceship. In relativity, time is relative, and simultaneity (the idea that two events happen "at the same time") is no longer a well-defined concept. From the astronaut's viewpoint, your measurement of the leading end of the meterstick occurs *before* your measurement of the trailing end. This time discrepancy ultimately results in the Lorentz contraction you observe. A similar argument applies to measurements of time, such as the period between two clock ticks. Time dilation occurs because the measurements occur at the same location and different times in one frame, but at *different places and times* in the other.

Einstein's ideas were revolutionary at the time, requiring physicists to abandon some long-held, cherished, and "obvious" facts about the universe. Reading the above discussion, you can perhaps begin to understand why special relativity initially encountered such resistance among some scientists (and why it still confuses nonscientists today!). Nevertheless, the gain in scientific understanding ultimately overcame the price in unfamiliarity. Within just a few years, special relativity had become almost universally accepted (and Einstein was well on his way to becoming the best-known scientist on the planet).

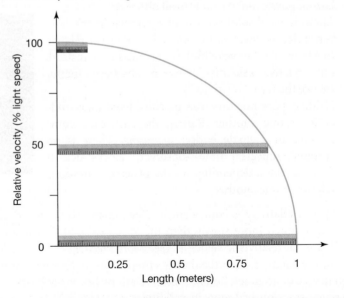

General Relativity

In constructing his special theory of relativity, Einstein rewrote the laws of motion expounded by Newton more than two centuries previously. ∞ *(Sec. 1.4)* Fitting Newton's other great legacy—the theory of gravitation—into the framework of relativity was a much more complex mathematical problem, which took Einstein another decade to complete. The result once again overturned scientists' conception of the universe.

In 1915, Einstein illustrated the connection between special relativity and gravity with the following famous "thought experiment." Imagine that you are enclosed in an elevator with no windows, so that you cannot directly observe the outside world, and the elevator is floating in space. You are weightless. Now suppose that you begin to feel the floor press up against your feet. Weight has apparently returned. There are two possible explanations for this, as shown in Figure 13.16. A large mass could have come nearby, and you are feeling its downward gravitational attraction (Figure 13.16a), *or* the elevator has begun to accelerate upward, and the force you feel is that exerted by the elevator as it accelerates you at the same rate (Figure 13.16b). The crux of Einstein's argument is this: There is *no* experiment that you can perform within the elevator (without looking outside) that will let you distinguish between these two possibilities.

Thus, Einstein reasoned, there is no way to tell the difference between a gravitational field and an accelerated frame of reference (such as the rising elevator in the thought experiment). This statement is known more formally as the *equivalence principle.* Using it, Einstein succeeded in incorporating gravity into special relativity as a general acceleration of all particles. The resulting theory is called *general relativity.* However, Einstein found that another huge conceptual leap was needed. To include the effects of gravity, the mathematics forced Einstein to the unavoidable conclusion that spacetime—the single entity combining space and time, central to special relativity—has to be *curved.*

The central concept of general relativity is this: Matter—all matter—tends to "warp" or curve space in its vicinity. Objects such as planets and stars react to this warping by changing their paths. In the Newtonian view of gravity, particles move on curved trajectories because they are acted upon by a gravitational force. ∞ *(Sec. 1.4)* In Einsteinian relativity, those same particles move on curved trajectories because they are falling freely through space, following the curvature of spacetime produced by some nearby massive object. The more the mass, the greater is the warping. Thus, in general relativity, there is no such thing as a "gravitational force" in the Newtonian sense. Objects move as they do because they follow the curvature of spacetime, which is determined by the amount of matter present. Stated more loosely, "Spacetime tells matter how to move, and matter tells spacetime how to curve."[2]

Some props may help you visualize these ideas. Bear in mind, however, that these props are not real, but only tools to help you grasp some exceedingly strange concepts. Imagine a pool table with the tabletop made of a thin rubber sheet rather than the usual hard felt. As Figure 13.17 suggests, such a rubber sheet becomes distorted when a heavy weight (such as a rock) is placed on it. The heavier the rock (Figure 13.17b), the larger is the distortion.

Trying to play pool on this table, you would quickly find that balls passing near the rock are deflected by the curvature of the tabletop (Figure 13.17b). The pool balls are not attracted to the rock in any way; rather, they respond to the curvature of the sheet produced by the rock's presence. In much the same way, anything that moves through space—matter *or radiation*—is deflected by the

A person inside a windowless elevator could not distinguish between these two cases.

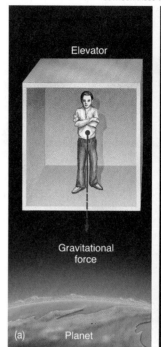

▲ **FIGURE 13.16 Einstein's Elevator** Einstein reasoned that no experiment conducted entirely within an elevator floating in space can tell a passenger whether the force he feels is (a) due to the gravity of a nearby massive object or (b) caused by the acceleration of the elevator itself.

[2]This characterization of gravity is due to the renowned relativist John Archibald Wheeler.

The amount of mass determines the amount of curvature and hence the amount of deflection.

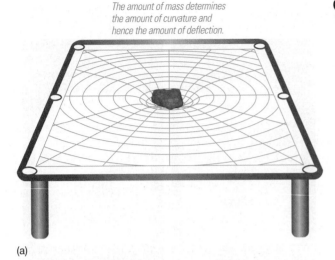

(a)

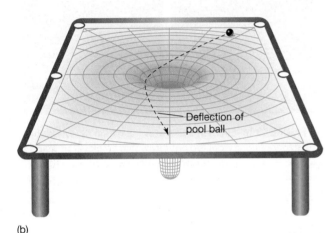

Deflection of pool ball

(b)

INTERACTIVE ◄ FIGURE 13.17 **Curved Space**   (a) A pool table made of a thin rubber sheet sags when a weight is placed on it. Likewise, space is bent, or warped, in the vicinity of any massive object. (b) A ball shown rolling across the table is deflected by the curvature of the surface, in much the same way that a planet's curved orbit is determined by the curvature of spacetime produced by the Sun.

curvature of spacetime near a star. (See *More Precisely 13-2* for a solar system measurement, and Sections 14.6 and 16.3 for examples on much larger scales.) Earth's orbital path is the trajectory that results as our planet falls freely in the relatively gentle curvature of space created by our Sun. When the curvature is small (i.e., gravity is weak), both Einstein and Newton predict the same orbit—the one we observe. However, as the gravitating mass increases, the two theories begin to diverge.

Curved Space and Black Holes

Modern notions about black holes rest squarely on the general theory of relativity. Although white dwarfs and (to a lesser extent) neutron stars can be adequately described by the classical Newtonian theory of gravity, only the modern Einsteinian theory of relativity can properly account for the bizarre physical properties of black holes.

Let's consider another analogy. Imagine a large extended family of people living on a huge rubber sheet—a sort of gigantic trampoline. Deciding to hold a reunion, they converge on a given place at a given time. As shown in Figure 13.18, one person remains behind, not wishing to attend. She keeps in touch with her relatives by means of "message balls" rolled out to her (and back from her) along the surface of the sheet. These message balls are the analog of radiation carrying information through space.

As the people converge, the rubber sheet sags more and more. Their accumulating mass creates an increasing amount of space curvature. The message balls can still reach the lone person far away in nearly flat space, but they arrive less frequently as the sheet becomes more and more warped and stretched—as shown in Figures 13.15(b) and (c)—and the balls have to climb out of a deeper and deeper well. Finally, when enough people have arrived at the appointed spot, the mass becomes too great for the rubber to support them. As illustrated in Figure 13.18(d), the sheet pinches off into a "bubble," compressing the people into oblivion and severing their communications with the lone survivor outside. This final stage represents the formation of an event horizon around the reunion party.

This analogy again shows how more mass...

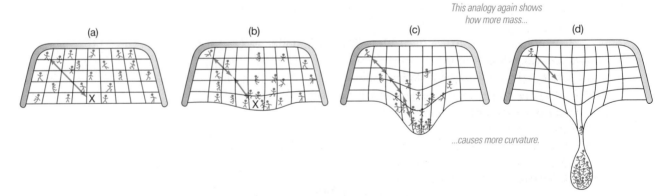

(a)   (b)   (c)   (d)

...causes more curvature.

▲ FIGURE 13.18 **Space Warping**   Mass causes a rubber sheet (or space) to be curved. As people assemble at a fixed spot on the sheet (marked by an "×"), the curvature grows larger, as shown in (a)–(c). The blue arrows represent some directions in which information can be sent from place to place. (d) The people are eventually sealed inside the bubble, forever trapped and cut off from the outside world.

13-2 MORE PRECISELY

Tests of General Relativity

Special relativity is the most thoroughly tested and most accurately verified theory in the history of science. General relativity, however, is on somewhat less firm experimental ground. The problem with verifying general relativity is that its effects on Earth and in the solar system—the places where we can most easily perform tests—are very small. Just as special relativity produces major departures from Newtonian mechanics only when speeds approach the speed of light, general relativity predicts large departures from Newtonian gravity only when extremely strong gravitational fields are involved.

Here we consider just two "classical" tests of the theory—solar system experiments that helped ensure acceptance of Einstein's theory. Bear in mind, however, that there are no known tests of general relativity in the "strong-field" regime—that part of the theory that predicts black holes, for example—so the full theory has never been experimentally tested.

At the heart of general relativity is the premise that everything, including light, is affected by gravity because of the curvature of space-time. Shortly after he published his theory in 1915, Einstein noted that light from a star should be deflected by a measurable amount as it passes the Sun. The closer to the Sun the light comes, the more it is deflected. Thus, the maximum deflection should occur for a ray that just grazes the solar surface. Einstein calculated that the deflection angle should be 1.75″, a small but detectable amount.

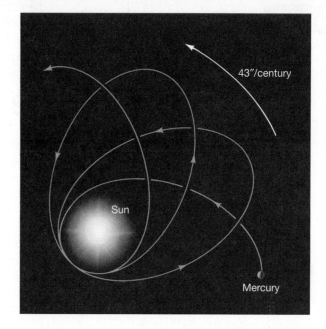

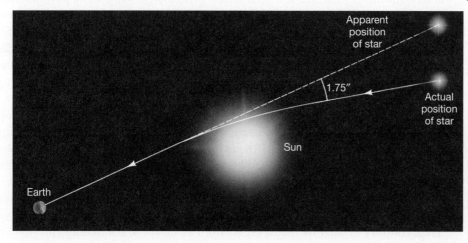

Of course, it is normally impossible to see stars so close to the Sun. During a solar eclipse, however, when the Moon blocks the Sun's light, the observation does become possible, as illustrated in the figure at lower left. In 1919 a team of observers led by British astronomer Sir Arthur Eddington succeeded in measuring the deflection of starlight during an eclipse. The results were in excellent agreement with the prediction of general relativity.

Another prediction of general relativity is that planetary orbits should deviate slightly from the perfect ellipses of Kepler's laws. Again, the effect is greatest where gravity is strongest—that is, closest to the Sun. Thus, the largest relativistic effects are found in the orbit of Mercury. Relativity predicts that Mercury's orbit is not exactly a closed ellipse. Instead, its orbit should rotate slowly, as shown in the second (highly exaggerated) diagram above. The amount of rotation is very small—only 43″ per century—but Mercury's orbit is so well charted that even this tiny effect is measurable.

Right up to the end—the pinching off of the bubble—two-way communication is possible. Message balls can reach the outside from within (but at a slower and slower rate as the rubber stretches), and messages from outside can get in without difficulty. However, once the event horizon (the bubble) forms, balls from the outside can still fall in, but they can no longer be sent back out to the person left behind, no matter how fast they are rolled. They cannot make it past the "lip" of the bubble in Figure 13.18(d). This analogy (very) roughly depicts how a black hole warps space completely around on itself, isolating its interior from the rest of the universe.

PROCESS OF ▶
SCIENCE CHECK How do Newton's and Einstein's theories differ in their descriptions of gravity? Why were scientists initially resistant to Einstein's theories of relativity?

ANIMATION/VIDEO Energy Released from a Black Hole? **(MA)**™

The essential ideas—the slowing down and eventual cessation of outward-going signals and the one-way nature of the event horizon once it forms—all have parallels in the case of stellar black holes. In Einsteinian terms, a *black hole* is a region of space where the gravitational field becomes overwhelming and the curvature of space extreme. At the event horizon itself, the curvature is so great that space "folds over" on itself, causing objects within to become trapped and disappear.

13.7 Space Travel Near Black Holes

Black holes are *not* cosmic vacuum cleaners. They don't cruise around interstellar space, sucking up everything in sight. The orbit of an object near a black hole is essentially the same as its orbit near a "normal" star of the same mass as the black hole. Only if the object happened to pass within a few Schwarzschild radii (perhaps 50–100 km for a typical 5- to 10-solar-mass black hole formed in a supernova explosion) of the event horizon would there be any significant difference between its actual orbit and the one predicted by Newtonian gravity and described by Kepler's laws. Of course, if some matter did fall into the black hole— that is, if its orbit happened to take it too close to the event horizon—it would be unable to get out. Black holes are like turnstiles, permitting matter to flow in only one direction—inward.

Tidal Forces

Matter flowing into a black hole is subject to enormous gravitational stress. An unfortunate explorer falling feet first into a solar mass black hole would find that the pull of gravity is much stronger at her feet (which are closer to the hole) than at her head. This force difference is a *tidal force*—precisely the same basic phenomenon that is responsible for ocean tides on Earth and the spectacular volcanoes on Io— except that the tidal forces at work in and near a black hole are far more powerful than any we are ever likely to find in the solar system. ∞ *(Secs. 5.2, 8.1)* The tidal force due to the black hole is so great that our explorer would be stretched enormously in height and squeezed unmercifully laterally. She would be torn apart long before she reached the event horizon.

The net result of all this stretching and squeezing is numerous and violent collisions among the resulting debris, causing a great deal of frictional heating of any infalling matter. As illustrated (with some artistic license) in Figure 13.19, material is simultaneously torn apart and heated to high temperatures as it plunges into the hole. So efficient is this heating that prior to reaching the hole's event horizon, matter falling into the hole emits radiation of its own accord. For a black hole of roughly solar mass, the energy is expected to be emitted in the form of X-rays. Thus, contrary to what we might expect from an object whose defining property is that nothing can escape from it, the region surrounding a black hole is expected to be a *source* of energy. Of course, once the hot matter falls below the event horizon, its radiation is no longer detectable—it can never leave the hole.

Approaching the Event Horizon

One safe way to study a black hole would be to go into orbit around it, safely beyond the disruptive influence of the hole's strong tidal forces. After all, Earth and the other planets of our solar system all orbit the Sun without falling into it and being torn apart. The gravity field around a black hole is basically no different. Let's imagine sending a less fragile (and more expendable) observer—a robot probe—toward the center of the hole, as illustrated in Figure 13.20. If our robot is small enough (to

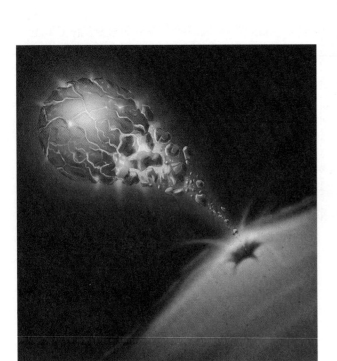

▲ FIGURE 13.19 **Black Hole Heating** Any matter falling into the clutches of a black hole will become severely distorted and heated. This sketch shows an imaginary planet being pulled apart by a black hole's gravitational tides.

minimize the tidal force) and of sufficiently strong construction, it might conceivably survive to reach the event horizon, even if a human could not. Watching from a safe distance in our orbiting spacecraft, we can then examine the nature of space and time, at least down to the event horizon. After that boundary is crossed, there is no way for the probe to return any information about its findings.

Suppose our robot has an accurate clock and a light source of known frequency mounted on it. From our safe vantage point far outside the event horizon, we could use telescopes to read the clock and measure the frequency of the light we receive. What might we discover? We would find that the light from the robot would become more and more redshifted—shifted toward longer wavelengths—as the robot neared the event horizon (Figure 13.21). Even if the robot used rocket engines to remain motionless, the redshift would still be detected. The redshift is not caused by motion of the light source—it is not the result of the Doppler effect. ∞ *(Sec. 2.7)* Rather, it is a result of the black hole's gravitational field, clearly predicted by Einstein's general theory of relativity and known as **gravitational redshift.**

We can explain the gravitational redshift as follows. Photons are attracted by gravity. As a result, in order to escape from a source of gravity, photons must expend some energy—they have to work to get out of the gravitational field. They don't slow down (photons always move at the speed of light); they just lose energy. Because a photon's energy is proportional to the frequency of its radiation, light that loses energy must have its frequency reduced (or, conversely, its wavelength increased).

As photons traveled from the robot's light source to our orbiting spacecraft, they would become gravitationally redshifted. From our standpoint, a yellow light would appear orange, then red as the robot astronaut approached the black hole. As the robot neared the event horizon, the light would become undetectable with an optical telescope. First infrared and then radio telescopes would be needed to detect it. Theoretically, light emitted *from the event horizon itself* would be gravitationally redshifted to infinitely long wavelengths.

What about the robot's clock? What time does it tell? Is there any observable change in the rate at which the clock ticks while moving deeper into the hole's gravitational field? We would find, from our orbiting spacecraft, that any clock close to the hole would appear to tick more *slowly* than an equivalent clock on board our ship. The closer the clock came to the hole, the slower it would appear to run. Upon reaching the event horizon, the clock would seem to stop altogether. All action would become frozen in time. Consequently, we would never actually witness the robot cross the event horizon. The process would appear to take forever.

This apparent slowing down of the robot's clock is known as **time dilation** and is closely related to gravitational redshift. To see why this is so, think of the robot's light source as a clock, with the passage of a wave crest (say) constituting a "tick," so the clock ticks at the frequency of the radiation. As the wave is redshifted, the frequency drops, and fewer wave crests pass us each second—the clock appears to slow down. This thought experiment demonstrates that the redshift of the radiation and the slowing of the clock are one and the same thing.

From the point of view of the robot, relativity theory predicts no strange effects at all. The light source hasn't reddened and the clock keeps perfect time. In the robot's frame of reference, everything is normal. Nothing prohibits it from

Attempt this experiment only if you are in a perfectly circular orbit!

▲ FIGURE 13.20 **Robot Astronaut**   A hypothetical spacecraft launches a robotic probe that travels toward a black hole to perform experiments that humans, farther away in the craft, could monitor to learn about the nature of space near the event horizon.

INTERACTIVE Gravitational Time Dilation and Redshift

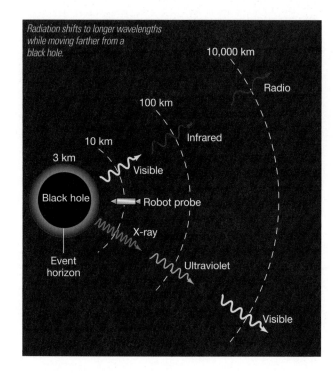

Radiation shifts to longer wavelengths while moving farther from a black hole.

10,000 km

Radio

100 km

Infrared

10 km

Visible

3 km

Black hole

Robot probe

Event horizon

X-ray

Ultraviolet

Visible

▶ FIGURE 13.21 **Gravitational Redshift**   Photons escaping from the strong gravitational field close to a black hole must expend energy to overcome the hole's gravity. As a result, the photons change wavelength; their color changes, and their frequency lessens. This figure shows the effect on two beams of radiation, one of visible light and one of X-rays, emitted from a space probe as it nears the event horizon of a 1-solar-mass black hole.

approaching within the Schwarzschild radius of the hole (so long as the probe is strong enough to avoid being torn apart by tidal forces)—no law of physics constrains an object from passing through an event horizon. There is no barrier at the event horizon and no sudden lurch as it is crossed. It is just a boundary in space, but one that can be crossed in only one direction.

Perhaps counterintuitively, the tidal force at the event horizon is *smaller* for more massive black holes (see problems 8 and 9 at the end of this chapter). Although, as we saw, an intrepid explorer would be shredded long before reaching the event horizon of a 1-solar-mass black hole, travelers passing through the event horizon of a multimillion solar mass hole (such as might lurk in the heart of our own Galaxy, as we will see in the next chapter) might not feel any particular discomfort. They would remain able to contemplate their surroundings—and the all-important fact that they couldn't reverse course—for a few more seconds before the rapidly rising tidal forces in the black hole's interior finally finished them off.

Deep Down Inside

No doubt, you are wondering what lies within the event horizon of a black hole. The answer is simple: No one knows. However, the question raises some very fundamental issues that lie at the forefront of modern physics.

General relativity predicts that without some agent to compete with gravity, the core remnant of a high-mass star will collapse all the way to a point at which both its density and its gravitational field become infinite—a so-called **singularity.** The same applies to any infalling matter. However, we should not take this prediction of infinite density too literally. Singularities always signal the breakdown of the theory producing them. The present laws of physics are simply inadequate to describe the final moments of a star's collapse.

As it stands today, the theory of gravity is incomplete because it does not incorporate a proper description of matter on very small scales. As our collapsing stellar core shrinks to smaller and smaller radii, we eventually lose our ability even to describe, let alone predict, its behavior. Perhaps matter trapped in a black hole never actually reaches a singularity. Perhaps it just approaches this bizarre state in a manner that we will someday understand as the subject of *quantum gravity*—the merger of general relativity (the theory of the universe on the largest scales) with quantum mechanics (the theory of matter on atomic and subatomic scales)—develops.

Having said that, we can at least estimate how small the core can get before current theory fails. It turns out that by the time that stage is reached, the core is already much smaller than any elementary particle. Thus, although a complete description of the end point of stellar collapse may well require a major overhaul of the laws of physics, for all practical purposes the prediction of collapse to a point is valid. Even if a new theory somehow succeeds in doing away with the central singularity, it is very unlikely that the external appearance of the hole, or the existence of its event horizon, will change. Any modifications to general relativity are expected to occur only on submicroscopic scales, not on the macroscopic (kilometer-sized) scale of the Schwarzschild radius.

Singularities are places where the rules break down, and some very strange things may occur near them. Many possibilities have been suggested—gateways to other universes, time travel, or the creation of new states of matter, for instance—but none has been proved, and certainly none has been observed. Because these regions are places where science fails, their presence causes serious problems for many of our cherished laws of physics, including causality (the idea that cause should precede effect, which runs into immediate problems if time travel is possible) and energy conservation (which is violated if material can hop from one universe to another through a black hole). It is unclear whether the removal of the singularity by some future, all-encompassing theory would necessarily also eliminate all of these problematic side effects.

CONCEPT CHECK ▶ Why would you never actually witness an infalling object crossing the event horizon of a black hole?

ANIMATION/VIDEO X-Ray Binary Star

13.8 Observational Evidence for Black Holes

Theoretical ideas aside, is there any observational evidence for black holes? What proof do we have that these strange invisible objects really do exist?

Black Holes in Binary Systems

Perhaps the most promising way to find black holes is to look for their effects on other astronomical objects. Our Galaxy harbors many binary star systems in which only one object can be seen, but we need observe the motion of only one star to infer the existence of an unseen companion and measure some of its properties. ∞ *(Sec. 10.7)* In the majority of cases, the invisible companion is simply small and dim, nothing more than an M-type star hidden in the glare of an O- or B-type partner, or perhaps shrouded by dust or other debris, making it invisible to even the best available equipment. In either case, the invisible object is not a black hole.

A few close binary systems, however, have peculiarities suggesting that one of their members may be a black hole. Figure 13.22(a) shows the area of the sky in the constellation Cygnus, where the evidence is particularly strong. The rectangle outlines the celestial system of interest, some 2000 pc from Earth. The black hole candidate is an X-ray source called Cygnus X-1, discovered by the *Uhuru* satellite in the early 1970s. Its visible companion is a blue B-type giant. Assuming that it lies on the main sequence, its mass must be around 25 times the mass of the Sun. Spectroscopic observations indicate that the binary system has an orbital period of 5.6 days. Combining this information with spectroscopic measurements of the visible component's orbital speed, astronomers estimate the total mass of the binary system to be around 35 solar masses, implying that Cygnus X-1 has a mass of about 10 times that of the Sun.

X-ray radiation emitted from the immediate neighborhood of Cygnus X-1 indicates the presence of high-temperature gas, perhaps as hot as several million kelvins. Furthermore, the X-ray emission has been observed to vary in intensity on time scales as short as a millisecond. As discussed earlier in the context of gamma-ray bursts, for this variation not to be blurred by the travel time of light across the source, Cygnus X-1 cannot be more than 1 light-millisecond, or 300 km, in diameter.

These properties suggest that Cygnus X-1 could be a black hole. Figure 13.23 is an artist's conception of this intriguing object. The X-ray-emitting region is likely an accretion disk formed as matter drawn from the visible star spirals down onto the unseen component. The rapid variability of the X-ray emission indicates that the unseen component must be very compact—a neutron star or a black hole. Its large mass, well above the neutron star mass limit discussed earlier, implies the latter.

A few other black hole candidates are known. For example, LMC X-3—the third X-ray source ever discovered in the Large Magellanic Cloud, a small galaxy orbiting our own (see Section 15.2)—is an invisible object that, like Cygnus X-1, orbits a bright companion star. LMC X-3's visible companion seems to be distorted into the shape of an egg by the unseen object's intense gravitational pull. Reasoning similar to that applied to Cygnus X-1 leads to the conclusion that LMC X-3 has a mass nearly 10 times that of the Sun, making it too massive to be anything but a black hole. The X-ray binary system A0620-00 has been found to contain an invisible compact object of mass 3.8 times the mass of the Sun. In total, about 20 known objects may turn out to be black holes; Cygnus X-1, LMC X-3, and A0620-00 probably have the strongest claims.

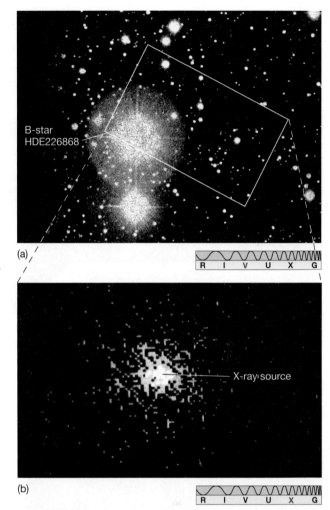

(a)

(b)

▲ **FIGURE 13.22 Cygnus X-1** (a) The brightest star in this photograph (marked with its catalog number) is a member of a binary system whose unseen companion, called Cygnus X-1, is a leading black hole candidate. (b) An X-ray image of the field of view outlined by the rectangle in part (a). *(Harvard-Smithsonian Center for Astrophysics)*

MA ANIMATION/VIDEO Black Hole and Companion Star

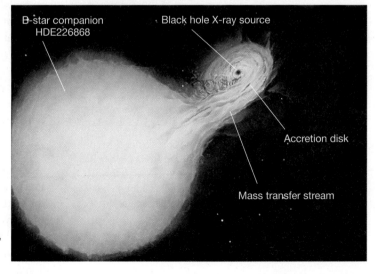

INTERACTIVE ► **FIGURE 13.23 Black Hole** Artist's conception of a binary system containing a large, bright, visible star and an invisible, X-ray-emitting black hole. (Compare Figure 12.14) This painting is based on data obtained from many observations of Cygnus X-1. *(L. Chaisson)* **MA**

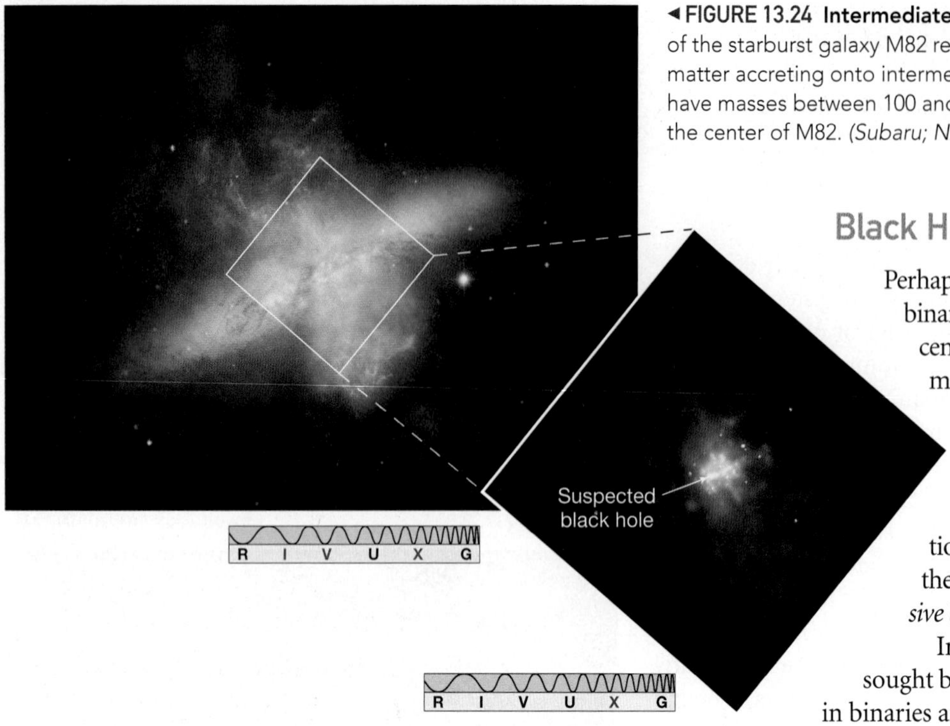

◀ FIGURE 13.24 Intermediate-Mass Black Holes? X-ray observations (inset) of the center of the starburst galaxy M82 reveal a collection of bright sources that may be the result of matter accreting onto intermediate-mass black holes. The black holes are probably young, have masses between 100 and 1000 times the mass of the Sun, and lie relatively far from the center of M82. *(Subaru; NASA)*

Black Holes in Galaxies

Perhaps the strongest evidence for black holes comes not from binary systems in our own Galaxy but from observations of the centers of many galaxies (including our own), where astronomers have found that stars and gas are moving extremely rapidly, orbiting some very massive, unseen object. Masses inferred from Newton's laws range from millions to billions of times the mass of the Sun. ∞ *(More Precisely 1-1)* The leading (and at present, the only) explanation is that these objects are black holes. We will return to these observations, and the question of how such *supermassive black holes* might have formed, in the next three chapters.

In 2000, X-ray astronomers reported evidence for a long-sought but elusive missing link between "stellar-mass" black holes in binaries and the supermassive black holes in the hearts of galaxies. Figure 13.24 shows an unusual looking galaxy called M82, currently the site of an intense and widespread burst of star formation (see Chapter 15). The inset shows a *Chandra* image of the innermost few thousand parsecs of M82, revealing a number of bright X-ray sources close to—but *not* at—the center of the galaxy. Their spectra and X-ray luminosities suggest that they may be accreting compact objects with masses ranging from 100 to almost 1000 times the mass of the Sun. If confirmed, they will be the first *intermediate-mass black holes* ever observed.

Too large to be remnants of normal stars and too small to warrant the "supermassive" label, these objects present a puzzle to astronomers. One possible origin is suggested by an apparent association between some of the X-ray emitters in M82 and elsewhere and dense, young star clusters. Theorists speculate that collisions between high-mass stars in the congested cores of these clusters might lead to the runaway growth of an extremely massive and very unstable star, which then collapses to form an intermediate-mass black hole. However, both the observations and this explanation remain controversial. Figure 13.25 shows the current best candidate for a "nearby" star cluster harboring an intermediate-mass black hole—the globular cluster G1 in the Andromeda Galaxy. ∞ *(Sec. 2.1)* The peculiar orbits of stars near the center of this cluster suggest a black hole of mass 20,000 times that of the Sun, and observations of the cluster in both radio and X-rays are consistent with theoretical expectations of the emission from such a massive object in the cluster's core.

Do Black Holes Exist?

So have black holes really been discovered? The answer is probably yes. Skepticism is healthy in science, but only the most stubborn astronomers (and some do exist!) would take strong issue with the reasoning that supports the case for black holes. Can we guarantee that future modifications to the theory of compact objects will not invalidate our arguments? No, but similar statements could be made in many other areas of astronomy—indeed, about almost any theory in any area of science. We conclude that, strange as they are, black holes have been detected in our Galaxy and beyond. Perhaps someday future generations of space travelers will visit Cygnus X-1 or the center of our Galaxy and (carefully!) test these conclusions firsthand. Until then, we will have to continue to rely on improving theoretical models and observational techniques to guide our discussions of these mysterious and intriguing objects.

ANIMATION/VIDEO Black Hole in Galaxy M87 **MA**

ANIMATION/VIDEO Supermassive Black Hole **MA**

▲ FIGURE 13.25 Black Hole Host? The star cluster G1 is the most massive globular cluster in the Andromeda Galaxy, but the stars near its center do not move as expected if the cluster's mass is as smoothly distributed as its light. Instead, the observations suggest that an intermediate-mass black hole resides at the cluster's center. *(NASA)*

PROCESS OF ▶ SCIENCE CHECK How do astronomers "see" black holes?

CHAPTER REVIEW

SUMMARY

LO1 A core-collapse supernova may leave behind an ultracompressed ball of material called a **neutron star** (p. 350). This is the remnant of the inner core that rebounded and blew the rest of the star apart. Neutron stars are extremely dense and, at formation, are predicted to be extremely hot, strongly magnetized, and rapidly rotating. They cool down, lose much of their magnetism, and slow down as they age.

LO2 According to the **lighthouse model** (p. 352), neutron stars, because they are magnetized and rotating, send regular bursts of electromagnetic energy into space. The beams are produced by charged particles confined by the strong magnetic fields. When we can see the beams from Earth, we call the source neutron star a **pulsar** (p. 350). The pulse period is the rotation period of the neutron star.

LO3 A neutron star that is a member of a binary system can draw matter from its companion, forming an accretion disk, which is usually a strong source of X-rays. As gas builds up on the star's surface, it eventually becomes hot enough to fuse hydrogen. When hydrogen burning starts on a neutron star, it does so explosively, and an **X-ray burster** (p. 354) results. The rapid rotation of the inner part of the accretion disk causes the neutron star to spin faster as new gas arrives on its surface. The eventual result is a very rapidly rotating neutron star—a **millisecond pulsar** (p. 354). Many millisecond pulsars are found in the hearts of old globular clusters. They cannot have formed recently, and must have been spun up by interactions with other stars. Careful analysis of the radiation received has shown that some millisecond pulsars are orbited by planet-sized objects.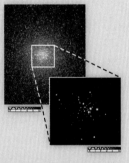

LO4 **Gamma-ray bursts** (p. 356) are very energetic flashes of gamma rays that occur about once a day and are distributed uniformly over the entire sky. In some cases, their distances have been measured, placing them at very large distances and implying that they are extremely luminous. The leading theoretical models for these explosions involve the violent merger of neutron stars in a distant binary system, or the recollapse and subsequent violent explosion following a "failed" supernova in a very massive star.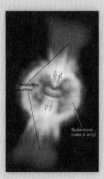

LO5 Einstein's special theory of relativity deals with the behavior of particles moving at speeds comparable to the speed of light. It agrees with Newton's theory at low velocities, but makes many very different predictions for high-speed motion. All of its predictions have been repeatedly verified by experiment. The modern replacement for Newtonian gravity is Einstein's **general theory of relativity** (p. 361), which describes gravity in terms of the warping, or bending, of space by the presence of mass. The more mass, the greater the warping. All particles—including photons—respond to that warping by moving along curved paths.

LO6 The upper limit on the mass of a neutron star is about three solar masses. Beyond that mass, the star can no longer support itself against its own gravity, and it collapses to form a **black hole** (p. 361), a region of space from which nothing can escape. The most massive stars, after exploding in a supernova, form black holes rather than neutron stars. Conditions in and near black holes can only be described by general relativity. The radius at which the escape speed from a collapsing star equals the speed of light is called the **Schwarzschild radius** (p. 362). The surface of an imaginary sphere centered on the collapsing star and having a radius equal to the star's Schwarzschild radius is called the **event horizon** (p. 362).

LO7 To a distant observer, light leaving a spaceship that is falling into a black hole would be subject to **gravitational redshift** (p. 369) as the light climbed out of the hole's intense gravitational field. At the same time, a clock on the spaceship would show **time dilation** (p. 369)—the clock would appear to slow down as the ship approached the event horizon. The observer would never see the ship reach the surface of the hole. Once within the event horizon, no known force can prevent a collapsing star from contracting all the way to a pointlike **singularity** (p. 370), at which point both the density and the gravitational field of the star become infinite. This prediction of relativity theory has yet to be proved. Singularities are places where the known laws of physics break down.

LO8 Once matter falls into a black hole, it can no longer communicate with the outside. However, on its way in, it can form an accretion disk and emit X-rays. The best place to look for a black hole is in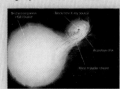

a binary system in which one component is a compact X-ray source. Cygnus X-1, a well-studied X-ray source in the constellation Cygnus, is a long-standing black-hole candidate. Studies of orbital motions imply that some binaries contain compact objects too massive to be neutron stars, leaving black holes as the only alternative. There is also substantial evidence for more massive black holes residing in or near the centers of many galaxies, including our own.

MasteringAstronomy®   For instructor-assigned homework go to www.masteringastronomy.com

Problems labeled POS explore the process of science. VIS problems focus on reading and interpreting visual information. LO connects to the introduction's numbered Learning Outcomes.

REVIEW AND DISCUSSION

1. **LO1** How does the way in which a neutron star forms determine some of its most basic properties?

2. What would happen to a person standing on the surface of a neutron star?

3. **LO2** Why aren't all neutron stars seen as pulsars?

4. What are X-ray bursters?

5. **LO3** What is the favored explanation for the rapid spin rates of millisecond pulsars?

6. Why do you think astronomers were surprised to find a pulsar with a planetary system?

7. Why do astronomers think there are two basically different types of gamma-ray burst?

8. What does it mean to say that the measured speed of a light beam is independent of the motion of the observer?

9. **LO6** Use your knowledge of escape speed to explain why black holes are said to be "black."

10. What is an event horizon?

11. **LO5** **POS** Why is it so difficult to test the predictions of the theory of general relativity? Describe two tests of the theory.

12. **LO7** What would happen to someone falling into a black hole?

13. **LO8** **POS** What makes Cygnus X-1 a good black hole candidate?

14. **POS** Imagine that you had the ability to travel at will through the Milky Way Galaxy. Explain why you would discover many more neutron stars than those known to observers on Earth. Where would you be most likely to find these objects?

15. **POS** Do you think that planet-sized objects discovered in orbit around a pulsar should be called planets? Why or why not?

CONCEPTUAL SELF-TEST: TRUE OR FALSE?/MULTIPLE CHOICE

1. Newly formed neutron stars have extremely strong magnetic fields. (T/F)

2. All millisecond pulsars are now, or once were, members of binary star systems. (T/F)

3. The fact that gamma-ray bursts are so distant means that they must be very energetic events. (T/F)

4. All things, except light, are attracted by gravity. (T/F)

5. According to general relativity, space is warped, or curved, by matter. (T/F)

6. Although visible light cannot escape from a black hole, high-energy radiation, like gamma rays, can. (T/F)

7. Thousands of black holes have now been identified in our Galaxy. (T/F)

8. A neutron star is about the same size as (a) a school bus; (b) a U.S. city; (c) the Moon; (d) Earth.

9. The most rapidly "blinking" pulsars are those that (a) spin fastest; (b) are oldest; (c) are most massive; (d) are hottest.

10. The X-ray emission from a neutron star in a binary system comes mainly from (a) the hot surface of the neutron star itself; (b) heated material in an accretion disk around the neutron star; (c) the neutron star's magnetic field; (d) the surface of the companion star.

11. Gamma-ray bursts are observed to occur (a) mainly near the Sun; (b) throughout the Milky Way Galaxy; (c) approximately uniformly over the entire sky; (d) near pulsars.

12. If the Sun were magically to turn into a black hole of the same mass, (a) Earth would start to spiral inward; (b) Earth's orbit would remain unchanged; (c) Earth would fly off into space; (d) Earth would be torn apart by the black hole's gravity.

13. The best place to search for black holes is in a region of space that (a) is dark and empty; (b) has recently lost some stars; (c) has strong X-ray emission; (d) is cooler than its surroundings.

14. **VIS** According to Figure 13.3 (Pulsar Model), the beam of a pulsar is emitted (a) along the rotation axis; (b) in the equatorial plane; (c) from the magnetic poles; (d) in a single direction in space.

15. **VIS** Figure 13.10 (Gamma-Ray Bursts) shows that gamma-ray bursts (a) are all very similar; (b) sometimes last for less than a second; (c) are confined to the Galactic plane; (d) come from great distances.

PROBLEMS

The number of squares preceding each problem indicates its approximate level of difficulty.

1. ■ The angular momentum of a spherical body is proportional to the body's angular speed times the square of its radius. ⚭ *(More Precisely 4-1)* Using the law of conservation of angular momentum, estimate how fast a collapsed stellar core would spin if its initial spin rate was one revolution per day and its radius decreased from 10,000 km to 10 km.

2. ■ What would your mass be if you were composed entirely of neutron star material of density 3×10^{17} kg/m^3 (Assume that your average density is 1000 kg/m^3.) Compare this with the mass of (a) the Moon; (b) a typical 1-km diameter asteroid.

3. ■ Calculate the surface gravity (relative to Earth's gravity of 9.8 m/s^2) and the escape speed of a 1.4-solar-mass neutron star with a radius of 10 km. What would be the escape speed of a 1-solar-mass object with a radius of 3 km? ⚭ *(More Precisely 5-1)*

4. ■■ Use the radius–luminosity–temperature relation to calculate the luminosity of a 10-km-radius neutron star for temperatures of 10^5 K, 10^7 K, and 10^9 K. What do you conclude about the visibility of neutron stars? Could the coolest of them be plotted on our H–R diagram?

5. ■■ A gamma-ray detector of area 0.5 m^2 observing a gamma-ray burst records photons having total energy 10^{-8} joules. If the burst occurred 1000 Mpc away, calculate the total amount of energy it released (assuming that the energy was emitted equally in all directions). How would this figure change if the burst occurred 10,000 pc away instead, in the halo of our Galaxy? What if it occurred within the Oort cloud of our own solar system, at a distance of 50,000 AU?

6. ■■■ An unstable elementary particle is known to decay into other particles in 2 μs, as measured in a laboratory where the particle are at rest. A beam of such particles is accelerated to a speed of 99.99 percent of the speed of light. How long do the particles in the beam take to decay, in the laboratory frame of reference?

7. ■ Supermassive black holes are thought to exist in the centers of some galaxies. What would be the Schwarzschild radii of black holes of 1 million and 1 billion solar masses? How does the first black hole compare in size with the Sun? How does the second compare in size with the solar system?

8. ■■ Calculate the tidal acceleration on a 2-m-tall human falling feet first into a 1-solar-mass black hole—that is, compute the difference in the accelerations (forces per unit mass) on his head and his feet just before his feet cross the event horizon. Repeat the calculation for a 1-million-solar-mass black hole and for a 1-billion-solar-mass black hole (see the previous question). Compare these accelerations with the acceleration due to gravity on Earth (9.8 m/s^2).

9. ■■■ Endurance tests suggest that the human body cannot withstand stress greater than about 10 times the acceleration due to gravity on Earth's surface. At what distance from a 1-solar-mass black hole would the human in the previous question be torn apart? Calculate the minimum mass of a black hole for which an infalling human could just reach the event horizon intact.

10. ■■ Using the data given in the text (assume the upper limit on the stated range for the black hole mass), calculate the orbital separation of Cygnus X-1 and its B-type stellar companion. ⚭ *(Sec. 1.4)*

ACTIVITIES

Collaborative

1. We can't easily observe a black hole, or make one to study, but theorists have lots to say about their properties! The text focuses on the simplest possible type of black hole—the uncharged, nonrotating *Schwarzchild* black hole—but there is a large body of literature on charged and spinning black holes, too. Charged black holes don't figure much in astronomy—the universe is electrically neutral on macroscopic scales—but rotating *Kerr* black holes are in fact very important. Divide your group in two and research online the properties of Schwarzchild and Kerr black holes. You'll probably find more information on the simpler Schwarzchild case, but some persistence will yield a lot on the rotating Kerr case, too. Combine your research to make a joint presentation on the similarities and differences between the two types. Focus on properties such as the event horizon, the singularity, and the orbits of light and matter near the hole. How fast can a black hole rotate? Which kind of black hole is thought to be most common in nature? A note on researching on the Web: try to stick to "authoritative" sites, such as NASA, ESO, university, and journal pages. Wikipedia is probably trustworthy, but always check what you find. And stay away from blogs and other nonauthoritative sites!

Individual

1. Find the ninth-magnitude companion to Cygnus X-1, the sky's most famous black hole candidate. Because none of us can see X-rays, no sign of anything unusual can be seen. Still, it's fun to gaze toward this region of the heavens and contemplate Cygnus X-1's powerful energy emission and strange properties. Even without a telescope, it is easy to locate the region of the heavens where Cygnus X-1 resides. The constellation Cygnus contains a recognizable star pattern, or asterism, in the shape of a large cross. This asterism is called the Northern Cross. The star in the center of the crossbar is called Sadr. The star at the bottom of the cross is called Albireo. Approximately midway between Sadr and Albireo lies the star Eta Cygni. Cygnus X-1 is located slightly less than 0.5° from this star. Whether or not you are using a telescope, sketch what you see. Create a demonstration of the densities of various astronomical objects—an interstellar cloud, a star, a terrestrial planet, a white dwarf, and a neutron star. Select a common object that is easily held in your hand—an apple, for example. For the lowest densities, calculate how large a volume would contain the object's equivalent mass. For high densities, calculate how many of the objects would have to fit into a standard volume, such as 1 cm^3. (This volume is better for this project than 1 m^3 because most people don't appreciate just how large a volume 1 m^3 is.) Present your demonstration to your class or to some other group of students. Tell them about each astronomical object and how it comes by its density.

PART **4**

Galaxies and the Universe

Our home Galaxy—the Milky Way—is a thousand times bigger than anything we have considered thus far in this book. Yet this vast system, more than 100,000 light-years across and home to 100 billion stars, is but one of billions of galaxies strewn across the vast expanse of deep space.

Part 4 explores the properties of galaxies, the glorious "building blocks" of the universe. Collectively they trace

out truly gargantuan patterns that top the hierarchy of material structures in the cosmos. On these huge scales we enter the realm of fundamental inquiry that has preoccupied philosophers since the dawn of recorded time: How did our universe begin? And how will it end?

The images here illustrate cosmic objects in Part 4, all of them at least a million million million times beyond any familiar terrestrial scale.

Active Galaxy ~ 10^{22} m

Normal Galaxy ~ 10^{21} m

Dwarf Galaxy ~ 10^{20} m

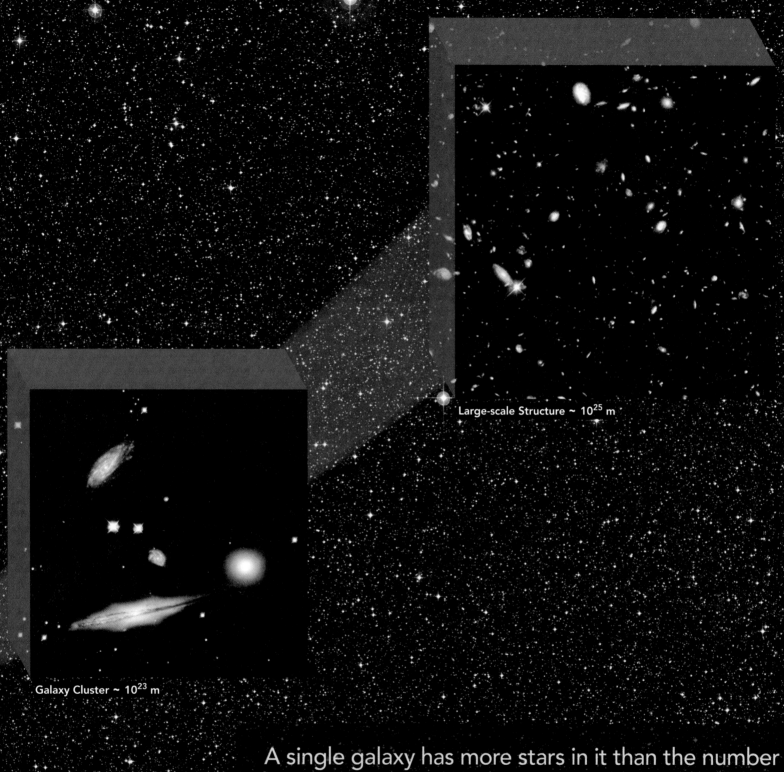

Large-scale Structure ~ 10^{25} m

Galaxy Cluster ~ 10^{23} m

A single galaxy has more stars in it than the number of people who have ever lived on Earth.

14

The Milky Way Galaxy

A Spiral in Space

Looking up on a dark, clear night, we are struck by two aspects of the night sky. The first is that the individual stars we see are roughly uniformly distributed in all directions. They all lie relatively close to us, mapping out the local galactic neighborhood within a few hundred parsecs of the Sun. But this is only a local impression. Ours is a rather provincial view. Beyond those nearby stars, the second thing we notice is a fuzzy band of light—the Milky Way—stretching across the heavens. From the Northern Hemisphere, this band is most easily visible in the summertime, arcing high above the horizon. Its full extent forms a great circle that encompasses the entire celestial sphere. This is the insider's view of the galaxy in which we live, the blended light of countless distant stars. As we consider much larger volumes of space, on scales far, far greater than the distances between neighboring stars, a new level of organization becomes apparent as the large-scale structure of the Milky Way Galaxy is revealed.

MasteringAstronomy® Visit www.masteringastronomy.com for quizzes, animations, videos, interactive figures, and self-guided tutorials.

◄ Stars cluster into gigantic assemblages called galaxies, of which our Milky Way Galaxy is just one among roughly a hundred billion others. Galaxies such as this one, known as NGC 1232 and shown here in true color, in turn contain roughly a hundred billion stars. As we now enter the realm of very big dimensions, the graceful winding arms of this majestic spiral galaxy resemble a curving staircase sweeping across some 100,000 light-years of space. Its size, shape, and mass approximate those of our Galaxy, which has never been photographed in its full grandeur because we live inside it. If this were our Galaxy, the Sun would reside in one of its spiral arms, about two-thirds of the way out from the center. *(ESO)*

14.1 Our Parent Galaxy

A **galaxy** is a gargantuan collection of stellar and interstellar matter—stars, gas, dust, neutron stars, black holes—isolated in space and held together by its own gravity. Astronomers are aware of literally billions of galaxies beyond our own. The particular galaxy we happen to inhabit is known as the **Milky Way Galaxy,** or just "the Galaxy," with a capital "G."

Our Sun lies in a part of the Galaxy known as the **Galactic disk,** an immense, circular, flattened region containing most of our Galaxy's luminous stars and interstellar matter (and almost everything we have studied so far in this book). Figure 14.1 illustrates how, viewed from within, the Galactic disk appears as a band of light—the *Milky Way*—stretching across our nighttime sky. Seen from Earth, the center of the disk (indeed, of the entire Galaxy) lies in the direction of the constellation Sagittarius. As indicated in the figure, if we look in a direction away from the Galactic disk (red arrows), relatively few stars lie in our field of view. However, if our line of sight happens to lie within the disk (white and blue arrows), we see so many stars that their light merges into a continuous blur.

Paradoxically, although we can study individual stars and interstellar clouds near the Sun in great detail, our location within the Galactic disk makes deciphering our Galaxy's large-scale structure from Earth a very difficult task—a little like trying to unravel the layout of paths, bushes, and trees in a city park without being able to leave one particular park bench. In many directions, the interpretation of what we see is inconclusive, with foreground objects obscuring our view of what lies beyond. As a result, astronomers who study the Milky Way Galaxy are often guided in their efforts by comparisons with more distant, but much more easily observable, galaxies. There is every reason to suppose that *all* the features of our Galaxy described in this chapter are shared by billions of other galaxies across the universe.

Figure 14.2 shows three galaxies thought to resemble our own in overall structure. Figure 14.2(a) is the Andromeda Galaxy, the nearest major galaxy to the Milky Way Galaxy, lying about 800 kpc (about 2.5 million light-years) away. Andromeda's apparent elongated shape is a result of the angle at which we happen to view it. In fact, this galaxy, like our own, consists of a thin circular *galactic disk* of matter that fattens to a **galactic bulge** at the center. The disk and bulge are embedded in a roughly spherical ball of faint old stars known as the **galactic halo.** These three basic galactic regions are indicated on the figure (the halo stars are too faint to see here; see Figure 14.9 for a clearer view). Figures 14.2(b) and (c) show images of two other galaxies—one seen face-on, the other edge-on—that illustrate these points more clearly.

CONCEPT CHECK ▶ Why do we see the Milky Way as a band of light across the sky?

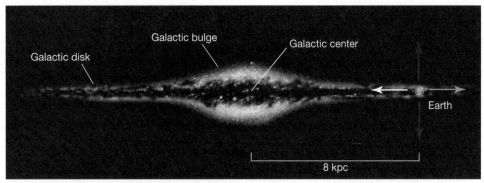

(a) Artist's view of Milky Way from afar

(b) Real image of Milky Way from inside

R I V U X G

◀ **FIGURE 14.1 Galactic Plane**   (a) Gazing from Earth toward the Galactic center (white arrow) in this artist's conception, we see myriad stars stacked up within the thin band of light known as the Milky Way. In the opposite direction (blue arrow), we see little of our Galaxy, much as when looking perpendicular to the disk (red arrows) where far fewer stars can be seen. (b) A real optical view of the sky in the direction of the white arrow shows the fuzzy (mostly white and "milky") band or disk of our Milky Way Galaxy (see Fig 11.1). (*Axel Mellinger*)

(a)

(b)

(c)

▲ FIGURE 14.2 Disk Galaxies (a) The Andromeda Galaxy closely resembles the overall layout of our Milky Way Galaxy. Its disk and bulge are clearly visible in this image, which is about 30,000 pc across. (b) This galaxy, named M101 and seen nearly face-on, is similar in its overall structure to our own Milky Way Galaxy and Andromeda. (c) The galaxy NGC 4565 is oriented edge-on, allowing us to see clearly its disk and central bulge. (R. Gendler; NASA)

14.2 Measuring the Milky Way

Before the 20th century, astronomers' conception of the cosmos differed markedly from the modern view. The growth in our knowledge of our Galaxy, as well as the realization that there are many other distant galaxies similar to our own, has gone hand in hand with the development of the cosmic distance scale.

Star Counts

In the late 18th century, long before the distances to any stars were known, English astronomer William Herschel tried to estimate the size and shape of our Galaxy simply by counting how many stars he could see in different directions in the sky. Assuming that all stars were of about equal brightness, he concluded that the Galaxy was a somewhat flattened, roughly disk-shaped collection of stars lying in the plane of the Milky Way, with the Sun at its *center* (Figure 14.3). Subsequent

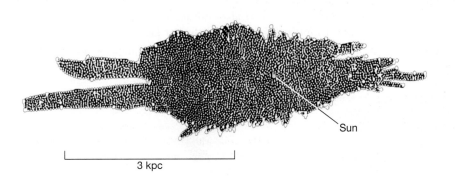

Sun

3 kpc

◄ FIGURE 14.3 Herschel's Galaxy Model Eighteenth-century English astronomer William Herschel constructed this "map" of the Galaxy by counting the number of stars he saw in different directions in the sky. Our Sun (marked by the yellow dot) appears to lie near the center of the distribution. The long axis of the diagram roughly parallels the plane of the disk.

refinements to this approach led to essentially the same picture. Early in the 20th century some researchers went so far as to estimate the dimensions of this "Galaxy" as about 10 kpc in diameter by 2 kpc thick.

In fact, the Milky Way Galaxy is several tens of kiloparsecs across, and the Sun lies far from the center. The flaw in the above reasoning is that the observations were made in the visible part of the electromagnetic spectrum, and astronomers failed to take into account the absorption of visible light by the then unknown interstellar gas and dust. ∞ *(Sec. 11.1)* Only in the 1930s did astronomers begin to realize the true extent and importance of the interstellar medium.

The apparent falloff in the density of stars with distance in the plane of the Milky Way is not a real thinning of their numbers in space, but simply a consequence of the murky environment in the Galactic disk. Objects in the disk lying more than a few kiloparsecs away are hidden from our view by the effects of interstellar absorption. The long "fingers" in Herschel's map are directions where the obscuration happens to be a little less severe than in others.

The falloff in density perpendicular to the disk is real. Radiation coming to us from above or below the plane of the Galaxy, where there is less gas and dust along the line of sight, arrives on Earth relatively unscathed. There is still some patchy obscuration, but the Sun happens to lie in a location where the view out of the disk is largely unimpeded by nearby interstellar clouds.

Observations of Variable Stars

An important by-product of the laborious effort to catalog stars around the turn of the 20th century was the systematic study of **variable stars.** These are stars whose luminosities change significantly over relatively short periods of time—some quite erratically, others more regularly. Only a small fraction of stars fall into this category, but those that do are of great astronomical significance, as we will see in a moment.

We have encountered several examples of variable stars in earlier chapters. Often, the variability is the result of membership in a binary system. Eclipsing binaries and novae are cases in point. However, sometimes the variability is an intrinsic property of the star itself and is not dependent on its being a part of a binary. Particularly important to Galactic astronomy are the **pulsating variable stars,**[1] which vary cyclically in luminosity in very characteristic ways. Two types of pulsating variable stars that have played central roles in revealing both the true extent of our Galaxy and the distances to our galactic neighbors are the **RR Lyrae** and **Cepheid** variables. (Following long-standing astronomical practice, the names come from the first star of each class to be discovered—in this case the variable star labeled RR in the constellation Lyra and the variable star Delta Cephei, the fourth brightest star in the constellation Cepheus.)

RR Lyrae and Cepheid variable stars are recognizable by the characteristic shapes of their light curves. RR Lyrae stars all pulsate in essentially similar ways (Figure 14.4a), with only small differences in period (the time from peak to peak) between one RR Lyrae variable and another. Observed periods range from about 0.5 to 1 day. Cepheid variables also pulsate in distinctive ways (the regular "sawtooth" pattern in Figure 14.4b), but different Cepheids can have very different pulsation periods, ranging from about 1 to 100 days. In either case, the stars can be recognized and identified *just by observing the variations in the light they emit.*

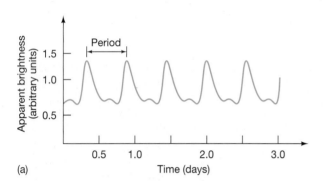

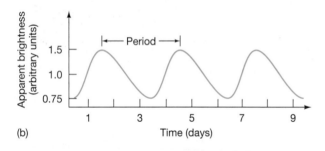

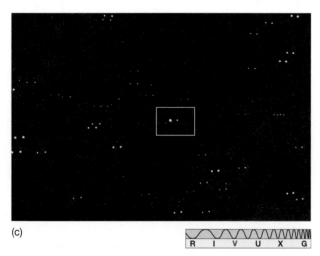

▲ **FIGURE 14.4 Variable Stars**   Light curves of (a) the pulsating variable star RR Lyrae, which has a period of less than a day, and (b) of the Cepheid variable star WW Cygni, having a period of several days. (c) This same Cepheid is shown here (boxed) on successive nights, near its maximum and minimum brightness; two photos, one from each night, were superposed and then slightly displaced. *(Harvard College Observatory)*

---

[1]Note that these stars have *nothing* whatsoever to do with pulsars. ∞ *(Sec. 13.2)* Pulsars are rapidly rotating ultracompact neutron stars, while pulsating variable stars vary in luminosity due to changes in their internal structure.

Pulsating variable stars are normal stars experiencing a brief (perhaps a few million years) period of instability as a natural part of stellar evolution. The conditions necessary to cause pulsations are not found in main-sequence stars. However, they do occur in post-main-sequence stars as they expand and cool on their way to becoming red giants. On the H–R diagram, these unstable stars are found in a region called the *instability strip* (Figure 14.5). When a star's temperature and luminosity place it in this strip, the star becomes internally unstable, and both its temperature and its radius vary in a regular way, causing the pulsations we observe. Cepheid variables are high-mass, high-luminosity stars evolving across the upper part of the H–R diagram. ∞ *(Sec. 12.4)* The less-luminous RR Lyrae variables are lower-mass horizontal-branch (core helium fusion) stars that happen to lie within the lower portion of the instability strip. ∞ *(Sec. 12.2)*

A New Yardstick

The importance of pulsating variable stars to Galactic astronomy lies in the fact that once we recognize a star as being of the RR Lyrae or Cepheid type, we can infer its luminosity, and that in turn allows us to measure its distance. The distance calculation is the same as that presented in Chapter 10 during our discussion of spectroscopic parallax. ∞ *(Sec. 10.6)* Comparing the star's (known) luminosity with its (observed) apparent brightness yields an estimate of its distance, by the inverse-square law. ∞ *(Sec. 10.2)* In this way, astronomers can use pulsating variables as a means of determining distances, both within our own Galaxy and far beyond.

How do we infer a variable star's luminosity? For RR Lyrae variables, this is simple. All such stars have basically the same luminosity (averaged over a complete pulsation cycle)—about 100 times that of the Sun. For Cepheids, we use a correlation between average luminosity and pulsation period, known as the **period–luminosity relationship.** In 1908, Henrietta Leavitt of Harvard College Observatory (see *Discovery 14-1*) discovered that Cepheids that vary slowly—that is, have long periods—have high luminosities, while short-period Cepheids have low luminosities. This new probe of stellar properties had immediate consequences for Galactic astronomy.

Figure 14.6 illustrates the period–luminosity relationship for Cepheids found within a thousand parsecs or so of Earth. Astronomers can plot such a diagram for relatively nearby Cepheid variables because they can measure the stars' distances, and hence their luminosities, using stellar or spectroscopic parallax. Thus, a simple measurement of a Cepheid variable's pulsation period immediately tells us its luminosity—we just read it off the graph in Figure 14.6. (The roughly constant luminosities of the RR Lyrae variables are also indicated in the figure.)

This distance-measurement technique works well provided the variable star can be clearly identified and its pulsation period measured. With Cepheids, this method allows astronomers to estimate distances out to about 25 million parsecs, more than enough to take us to the nearest galaxies. Indeed, the existence of galaxies beyond our own was first established in the late 1920s, when American astronomer Edwin Hubble observed Cepheids in the Andromeda Galaxy and thereby succeeded in measuring its distance. The less luminous RR Lyrae stars are not as easily detectable as Cepheids, so their usable range is not as great. However, they are much more common, so within this limited range, they are actually more useful than Cepheids.

We began our cosmic distance ladder in Chapter 1 with radar ranging in the solar system. ∞ *(Sec. 1.3)* In Chapter 10, we extended it to include stellar and spectroscopic parallax. ∞ *(Secs. 10.1, 10.6)* Figure 14.7 lengthens the distance ladder further by adding variable stars as a fourth method of determining distance.

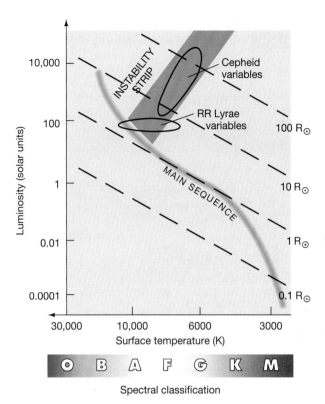

▲ **FIGURE 14.5 Variable Stars on the H–R Diagram** Pulsating variable stars are found in the instability strip of the H–R diagram. As a high-mass star evolves through the strip, it becomes a Cepheid variable. Low-mass horizontal-branch stars in the instability strip are RR Lyrae variables.

ANIMATION/VIDEO Cepheid Variable Star in Distant Galaxy

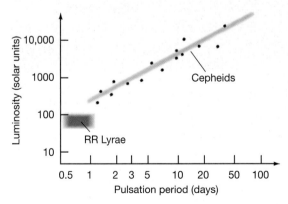

▲ **FIGURE 14.6 Period–Luminosity Relationship** A plot of pulsation period versus average absolute brightness (that is, luminosity) for a group of Cepheid variable stars. The two properties are tightly correlated. The pulsation periods of some RR Lyrae variables are also shown.

CONCEPT CHECK ▶ Can variable stars be used to map out the structure of the Galactic disk?

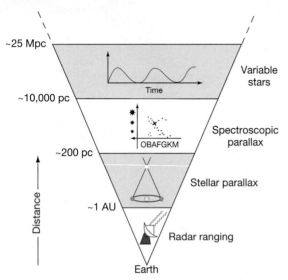

INTERACTIVE ▲ **FIGURE 14.7 Variable Stars on Distance Ladder**   Application of the period–luminosity relationship for Cepheid variable stars allows estimates of distances out to about 25 Mpc with reasonable accuracy. **MA**

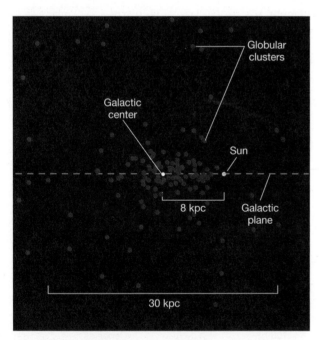

▲ **FIGURE 14.8 Globular Cluster Distribution**   Our Sun does not coincide with the center of the very large collection of globular clusters (indicated by the pink dots). Instead, more globular clusters are found in one direction than in any other. The Sun resides closer to the edge of the collection, which measures roughly 30 kpc across. The globular clusters outline the true distribution of stars in the Galactic halo.

PROCESS OF ▶ SCIENCE CHECK   Go to a library or the Web and research the Great Debate in 1920 between Harlow Shapley and Heber Curtis on the size of our Galaxy and scale of the universe. Which viewpoint was ultimately proved right, and what were the consequences for our understanding of the cosmos?

The Size and Shape of Our Galaxy

Many RR Lyrae variables are found in globular clusters, those tightly bound swarms of old, reddish stars we first met in Chapter 11. ∞ *(Sec. 11.6)* Early in the 20th century, the American astronomer Harlow Shapley used observations of RR Lyrae stars to make two very important discoveries about the Galactic globular cluster system. First, he showed that most globular clusters reside at great distances—many thousands of parsecs—from the Sun. Second, by measuring the direction and distance of each cluster, he determined their three-dimensional distribution in space (Figures 14.8 and 14.9). In this way, Shapley demonstrated that the globular clusters map out a truly gigantic, and roughly *spherical,* volume of space, now known to be about 30 kpc across.[2] However, the center of the distribution lies nowhere near our Sun. It is located nearly 8 kpc away from us, in the constellation Sagittarius.

In a brilliant intellectual leap, Shapley realized that the distribution of globular clusters maps out the true extent of stars in the Milky Way Galaxy—the region that we now call the Galactic halo. The hub of this vast collection of matter, 8 kpc from the Sun, is the **Galactic center.** As illustrated in Figure 14.9, we live in the suburbs of this huge ensemble, in the Galactic disk—the thin sheet of young stars, gas, and dust that cuts through the center of the halo. Since Shapley's time, astronomers have identified many individual stars—that is, stars not belonging to any globular cluster—within the Galactic halo.

Shapley's bold interpretation of the globular clusters as defining the overall structure of our Galaxy was an enormous step forward in human understanding of our place in the universe. Five hundred years ago Earth was considered the center of all things. Copernicus argued otherwise, demoting our planet to an undistinguished place far from the center of the solar system. In Shapley's time the prevailing view was that our Sun was the center not only of the Galaxy but also of the universe. Shapley showed otherwise. With his observations of globular clusters, he simultaneously increased the size of our Galaxy by almost a factor of 10 over earlier estimates and banished our parent Sun to its periphery, virtually overnight!

14.3 Galactic Structure

Based on optical, infrared, and radio studies of our Galaxy, Figure 14.9 illustrates the very different spatial distributions of the disk, bulge, and halo components of the Milky Way Galaxy.

Mapping the Galaxy

The extent of the halo in Figure 14.9 is based largely on optical observations of globular clusters and other halo stars. However, as we have seen, optical techniques can cover only a small portion of the dusty Galactic disk. Much of our knowledge of the structure of the disk on larger scales is based on radio observations, particularly of the 21-cm radio emission line produced by atomic hydrogen. ∞ *(Sec. 11.3)* Because long-wavelength radio waves are largely unaffected by interstellar dust, and hydrogen is by far the most abundant element in interstellar space, the 21-cm signals are strong enough that virtually the entire disk can be observed in this way.

According to the radio studies, the center of the gas distribution coincides roughly with the center of the globular cluster system, lying about 8 kpc from the Sun. In fact, this figure is derived most accurately from radio observations of

---

[2]The Galactic globular cluster system and the Galactic halo of which it is a part are somewhat flattened in the direction perpendicular to the disk, but the degree of flattening is uncertain. The halo is certainly much less flattened than the disk, however.

14-1 DISCOVERY

Early Computers

A large portion of the early research in observational astronomy focused on monitoring stellar luminosities and analyzing stellar spectra. Much of this pioneering work was done using photographic methods. What is not so well known is that most of the labor was accomplished by women. Around the turn of the 20th century, a few dozen dedicated women—assistants at the Harvard College Observatory—created an enormous database by observing, sorting, measuring, and cataloging photographic information that helped form the foundation of modern astronomy. Some of them went far beyond their duties in the lab to make many of the basic astronomical discoveries often taken for granted today.

This 1910 photograph shows several of those women carefully examining star images and measuring variations in luminosity or wavelengths of spectral lines. In the cramped quarters of the Harvard Observatory, these women inspected image after image to collect a vast body of data on hundreds of thousands of stars. Note the plot of stellar luminosity changes pasted on the wall at the left. The cyclical pattern is so regular that it likely belongs to a Cepheid variable. Known as "computers" (there were no electronic devices then), these women were paid 25 cents an hour.

In 1880 these workers began a survey of the skies that would continue for half a century. Their first major accomplishment was a catalog of the brightnesses and spectra of tens of thousands of stars, published in 1890 under the direction of Williamina Fleming. On the

(Harvard College Observatory)

basis of this compilation, several of these women made fundamental contributions to astronomy. In 1897 Antonia Maury undertook the most detailed study of stellar spectra to that time, enabling Hertzsprung and Russell independently to develop what is now called the H–R diagram. In 1898 Annie Cannon proposed the spectral classification system (described in Chapter 10) that is now the international standard for categorizing stars. ⬚ (Sec. 10.5) In 1908 Henrietta Leavitt discovered the period–luminosity relationship for Cepheid variable stars, which later allowed astronomers to recognize our Sun's true position in our Galaxy, as well as our Galaxy's true place in the universe.

Galactic gas. The densities of both stars and gas in the disk decline quite rapidly beyond about 15 kpc from the Galactic center (although some radio-emitting gas has been observed out to at least 50 kpc).

Perpendicular to the Galactic plane, the disk in the vicinity of the Sun is "only" about 300 pc thick, or about 1/100 of the 30-kpc Galactic diameter. Don't be fooled, though. Even if you could travel at the speed of light, it would take you 1000 years to traverse the thickness of the Galactic disk. The disk may be thin compared with the Galactic diameter, but it is huge by human standards.

Also shown in Figure 14.9 is our Galaxy's central bulge, measuring roughly 6 kpc across in the plane of the Galactic disk by 4 kpc perpendicular to that plane. Obscuration by interstellar dust makes it difficult to study

INTERACTIVE ▶ FIGURE 14.9 Stellar Populations in Our Galaxy   Artist's conception of a (nearly) edge-on view of the Milky Way Galaxy, showing schematically the distributions of young blue stars, open clusters, old red stars, and globular clusters. (The brightness and size of our Sun are greatly exaggerated for clarity.)

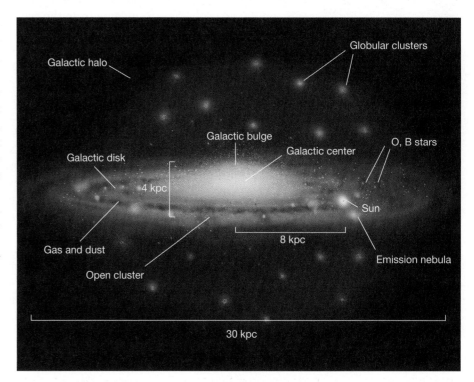

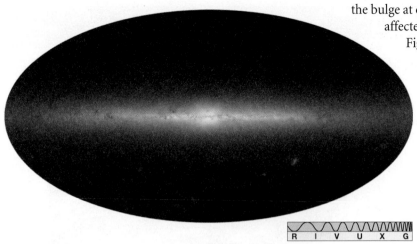

R I V U X G

▲ FIGURE 14.10 **Infrared View of the Milky Way Galaxy** A wide-angle infrared image of the disk and bulge of the Milky Way Galaxy, as observed by the Two Micron All Sky Survey. Compare Figure 14.2(c). *(UMass/Caltech)*

Stellar Populations in a Star Cluster

the bulge at optical wavelengths. However, at longer wavelengths, which are less affected by interstellar matter, a much clearer picture emerges (compare Figure 14.10 with Figures 14.1b and 14.2c). Detailed measurements of the motion of gas and stars in and near the bulge imply that it is actually football shaped, about half as wide as it is long, with the long axis of the football lying in the Galactic plane. (In this case, our Galaxy is probably of the "barred-spiral" type—see Section 15.1.)

Stellar Populations

Aside from their shapes, the three components of the Galaxy—disk, bulge, and halo—have other properties that distinguish them from one another. First, the halo contains essentially *no* gas or dust—just the opposite of the disk and bulge, in which interstellar matter is common. Second, there are clear differences in the *colors* of the stars in the disk, bulge, and halo. Stars in the Galactic bulge and halo are distinctly redder than those found in the disk. Observations of other spiral galaxies also show this trend—the blue-white tint of the disk and the yellowish coloration of the bulge are evident in Figures 14.2(a) and (c).

All the bright, blue stars visible in our night sky are part of the Galactic disk, as are the young open star clusters and star-forming regions. In contrast, the cooler, redder stars—including those found in the old globular clusters—are more uniformly distributed throughout the disk, bulge, and halo. Galactic disks appear bluish because main-sequence O- and B-type blue supergiants are very much brighter than G-, K-, and M-type dwarfs, even though the dwarfs are present in far greater numbers.

The explanation for the marked difference in stellar content between disk and halo is that, whereas the gas-rich Galactic disk is the site of ongoing star formation and thus contains stars of all ages, all the stars in the Galactic halo are *old*. The absence of dust and gas in the halo means that no new stars are forming there, and star formation apparently ceased long ago—at least 10 billion years in the past, judging from the ages of the globular clusters. ⊂⊃ *(Sec. 12.6)* The gas density is very high in the inner part of the Galactic bulge, making this region the site of vigorous ongoing star formation, and both very old and very young stars mingle there. The bulge's gas-poor outer regions have properties more similar to those of the halo. As we will see in Chapter 15, all these statements are equally true for other spiral galaxies—young, bright stars are always found in the disk and inner bulge because the interstellar medium is densest there.

Spectroscopic studies indicate that halo stars are far less abundant in heavy elements (that is, elements heavier than helium) than are stars in the disk. Each successive cycle of star formation and evolution enriches the interstellar medium with the products of stellar evolution, leading to a steady increase in heavy elements with time. ⊂⊃ *(Sec. 12.7)* Thus, the scarcity of these elements in halo stars is consistent with the view that the halo formed long ago.

Astronomers often refer to young disk stars as *Population I* stars and old halo stars as *Population II* stars. The idea of two stellar "populations" dates back to the 1930s, when the differences between disk and halo stars first became clear. It represents something of an oversimplification, as there is actually a continuous variation in stellar ages throughout the Milky Way Galaxy and not a simple division of stars into two distinct "young" and "old" categories. Nevertheless, the terminology is still widely used.

▶ FIGURE 14.11 **Orbital Motion in the Galactic Disk** Stars and interstellar clouds in the neighborhood of the Sun show systematic Doppler motions, implying that the disk of the Galaxy spins in a well-ordered way. These four Galactic quadrants are drawn to intersect not at the Galactic center, but at the Sun, the location from which observations are made. Because the Sun orbits faster than stars and gas at larger radii, it moves away from material at top left and gains on that at top right, resulting in the Doppler shifts indicated. Likewise, stars and gas in the bottom left quadrant are gaining on us, while material at bottom right is pulling away.

Orbital Motion

Are the internal motions of our Galaxy's members chaotic and random, or are they part of some gigantic "traffic pattern"? The answer depends on our perspective. The motion of stars and clouds we see on small scales (within a few tens of parsecs of the Sun) seems random, but on larger scales (hundreds or thousands of parsecs) the motion is much more orderly.

As we look around the Galactic disk in different directions, a clear pattern of motion emerges (see Figure 14.11). Radiation received from stars and interstellar gas clouds in the upper-right and the lower-left quadrants of Figure 14.11 is generally *blueshifted*. At the same time, radiation from stars and gas sampled in the upper-left and lower-right quadrants tends to be *redshifted*. In other words, some regions of the Galaxy (in the blueshifted directions) are approaching the Sun, while others (the redshifted ones) are receding from us. Careful study of the positions and velocities of stars and gas clouds near the Sun leads us to the conclusion that the entire Galactic disk is *rotating* about the Galactic center. In the vicinity of the Sun, 8 kpc from the center, the orbital speed is about 220 km/s, so material takes about 225 million years to complete one circuit. At other distances from the center, the rotation period is different—shorter closer to the center, longer at greater distances—that is, the Galactic disk rotates not as a solid object but *differentially.*

This picture of orderly circular orbital motion about the Galactic center applies only to the Galactic disk. The old globular clusters in the halo and the faint, reddish individual stars in both the halo and the bulge do *not* share the disk's well-defined rotation. Instead, as shown in Figure 14.12, their orbits are oriented randomly.[3] Although they do orbit the Galactic center, these stars move in all directions, their paths filling an entire three-dimensional volume rather than a nearly two-dimensional disk.

At any given distance from the Galactic center, bulge or halo stars move in their orbits at speeds comparable to the disk's rotational speed at that radius. However, unlike stars in the disk, bulge and halo stars orbit in *all* directions, not just in a roughly circular path confined to a narrowly defined plane. Their orbits carry these stars repeatedly through the disk and out the other side. (They don't collide with stars in the disk because interstellar distances are huge compared with the diameters of individual stars—a star or even an entire star cluster passes through the disk almost as though it wasn't there—see *Discovery 15-1.*)

[3]Halo stars may in fact have some small net rotation about the Galactic center, but the rotational component of their motion is overwhelmed by the much larger random component. The motion of bulge stars also has a rotational component, larger than that of the halo but still smaller than the random component of stellar motion within the bulge.

The curved arrows denote the speed of the disk material, which is greater closer to the center.

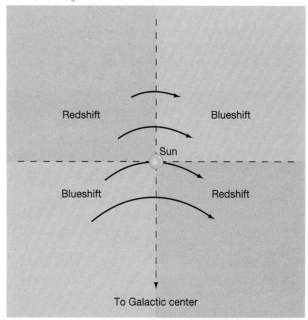

Redshift   Blueshift

Sun

Blueshift   Redshift

To Galactic center

Halo

Bulge   Disk

▲ FIGURE 14.12 **Stellar Orbits in Our Galaxy** Stars in the Galactic disk (blue curves) move in orderly, circular orbits about the Galactic center. In contrast, halo stars (orange curves) move randomly around the center. The orbit of a typical halo star takes it high above the Galactic disk, then down through the disk plane, then out the other side and far below the disk. Orbital properties of bulge stars are intermediate between those of the disk and halo stars.

CONCEPT CHECK ▶ Why do astronomers regard the disk and halo as distinctly different components of our Galaxy?

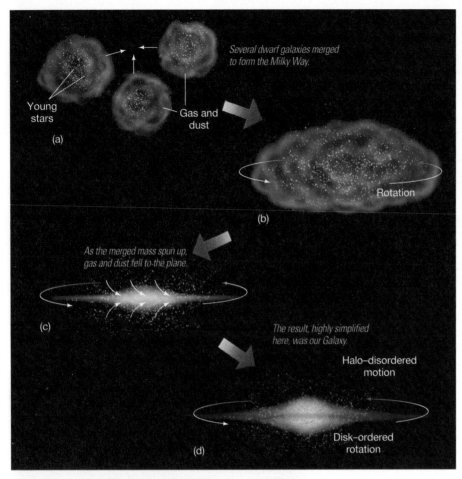

Several dwarf galaxies merged to form the Milky Way.

Young stars

Gas and dust

(a)

Rotation

(b)

As the merged mass spun up, gas and dust fell to the plane.

(c)

The result, highly simplified here, was our Galaxy.

Halo–disordered motion

Disk–ordered rotation

(d)

▲ FIGURE 14.13 Milky Way Galaxy Formation (a) The Milky Way Galaxy likely formed by a merger of several smaller systems. (b) Early on, our Galaxy was irregularly shaped, with gas distributed throughout its volume. When stars formed during this stage, their orbits carried them throughout an extended three-dimensional volume surrounding the newborn Galaxy. (c) In time, the gas and dust fell to the Galactic plane and formed a spinning disk. The stars that had already formed were left behind in the halo. (d) New stars forming in the disk inherit its overall rotation and so orbit the Galactic center on ordered, circular orbits.

14.4 Formation of the Milky Way

Table 14.1 compares some key properties of the three basic components of the Galaxy. Is there some evolutionary scenario that can naturally account for the structure we see today? The answer is yes, and it takes us all the way back to the birth of our Galaxy, more than 10 billion years ago. Not all the details are agreed upon by all astronomers, but the overall picture is now fairly widely accepted. For simplicity we confine our discussion here to the Galactic disk and halo. In many ways the bulge is intermediate in its properties between these two extremes.

Figure 14.13 illustrates the current view of our Galaxy's evolution. It explains, in broad terms at least, the Galactic structure we see today. Not unlike the star-formation scenario outlined in Chapter 11, it starts from a contracting cloud of pregalactic gas. ∞ (Sec. 11.4) When the first Galactic stars and globular clusters formed, the gas in our Galaxy had not yet accumulated into a thin disk. Instead, it was spread out over an irregular, and quite extended, region of space, spanning many tens of kiloparsecs in all directions. When the first stars formed, they were distributed throughout this volume (Figure 14.13b). Their distribution today (the Galactic halo) reflects that fact—it is an imprint of their birth. Many astronomers think that the very first stars formed even earlier, in smaller systems that later merged to create our Galaxy (Figure 14.13b; see Section 16.4). Probably many more stars were born during the mergers themselves, as interstellar gas clouds collided and began to collapse. The present-day halo would look much the same in either case.

Since those early times, rotation has flattened the gas in our Galaxy into a relatively thin disk (Figure 14.13c). Physically, this process is similar to the flattening of the solar nebula during the formation of the solar system, except on a vastly larger scale. ∞ (Sec. 4.3) Star formation in the halo ceased billions of years ago when the raw materials—interstellar gas and dust—fell to the Galactic plane. Ongoing star formation in the disk gives it its bluish tint, but the halo's short-lived blue stars have

TABLE 14.1 Overall Properties of the Galactic Disk, Halo, and Bulge

Galactic Disk	Galactic Halo	Galactic Bulge
Highly flattened	roughly spherical—mildly flattened	somewhat flattened—elongated in the plane of the disk (football-shaped)
Contains both young and old stars	contains old stars only	contains both young and old stars; more old stars at greater distances from the center
Contains gas and dust	contains no gas and dust	contains gas and dust, especially in the inner regions
Site of ongoing star formation	no star formation during the last 10 billion years	ongoing star formation in the inner regions
Gas and stars move in circular orbits in the Galactic plane	stars have random orbits in three dimensions	stars have largely random orbits, but with some net rotation about the Galactic center
Spiral arms (Sec. 14.5)	little discernible substructure; globular clusters, tidal streams (Sec. 14.3)	ring of gas and dust near center; central Galactic nucleus (Sec. 14.7)
Overall white coloration, with blue spiral arms	reddish in color	yellow-white

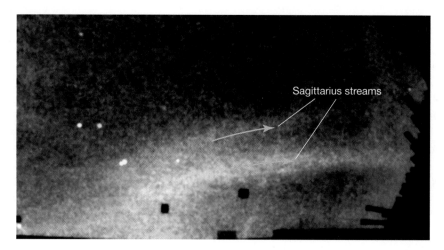

Sagittarius streams

◀ FIGURE 14.14 Galactic Streams   This is a map of the outer regions of the Milky Way showing innumerable stars that have been torn from our Galaxy's disrupted satellite galaxies (colors indicate distance, with blue being the closest). Several streams are evident, the biggest one at center showing two orbits of the enormous, arching death spiral of the Sagittarius dwarf galaxy. The yellow arrows indicate the direction of motion of stars in the two streams. *(SDSS)*

long since burned out, leaving only the long-lived red stars that give it its characteristic pinkish glow (Figure 14.13d). The Galactic halo is ancient, whereas the disk is full of youthful activity.

Recent studies of stars in the Galactic disk suggest that the infall of halo gas is still going on today. The best available models of star formation and stellar evolution predict that the fraction of heavy elements in disk stars should be significantly *greater* than is actually observed, unless the gas in the disk is steadily being "diluted" by fresh gas arriving from the halo at a rate of perhaps 5–10 solar masses per year. ∞ *(Sec. 12.5)* This may not sound like much mass, but accumulated over billions of years it amounts to a significant fraction of the total mass of the disk (see Section 14.6).

This theory also explains the chaotic orbits of the halo stars. When the halo developed, the irregularly shaped Galaxy was rotating very slowly, so there was no strongly preferred direction in which matter tended to move. As a result, halo stars were free to travel along nearly any path once they formed (or when their parent systems merged). As the Galactic disk formed, however, conservation of angular momentum caused it to spin more rapidly. Stars forming from the gas and dust of the disk inherit its rotational motion and so move on well-defined circular orbits.

None of the processes just described are nearly as clean as depicted in Figure 14.13. Both galaxy mergers and gas infall are still occurring today. Astronomers have detected numerous *tidal streams* in the Galactic halo—groups of stars thought to be the remnants of globular clusters and satellite galaxies (see *Secs. 15.1, 16.3*) torn apart relatively recently by our Galaxy's tidal field. Figure 14.14 is a wide-angle mosaic of roughly half of the northern sky, showing numerous streams of stars crossing the field of view. The stars are following the orbits of their defunct parent galaxies, much as micrometeoroid swarms in our solar system follow the orbit of their parent comet long after the comet itself is gone. ∞ *(Sec. 4.2)*

Studies of the composition of stars in the Galactic disk support the idea that the infall of halo gas is ongoing. Models of star formation and stellar evolution predict that the fraction of heavy elements in disk stars should actually be considerably *greater* than observed, unless the gas in the disk is being "diluted" by new gas arriving from the halo at a rate of perhaps 5–10 solar masses per year. ∞ *(Sec. 12.7)* This may not sound like much mass, but accumulated over billions of years it amounts to a substantial fraction of the total mass of the disk (see Sec. 14.6).

In principle, the structure of our Galaxy bears witness to the conditions that created it. In practice, however, the interpretation of the observations is made difficult by the sheer complexity of the system we inhabit and by the many competing physical processes that have modified its appearance since it formed. As a result, the early stages of the Milky Way are still quite poorly understood. We will return to the subject of galaxy formation in Chapters 15 and 16.

CONCEPT CHECK ▶   Why are there no young halo stars?

14.5 Galactic Spiral Arms

Radio studies provide perhaps the best direct evidence that we live in a spiral Galaxy. As illustrated in Figure 14.15, radio observations of interstellar gas, repeated in all directions in the Galactic plane, allow astronomers to map out its distribution and paint a detailed picture of the Galactic disk.

The Galaxy's differential rotation means that gas clouds at different distances are moving at different speeds with respect to Earth, so the radiation received from them is Doppler shifted by different amounts. Astronomers use all available data, coupled with knowledge of Newtonian mechanics, to construct a mathematical model of the rotation of stars and gas throughout the Galactic disk. ∞ *(Sec. 2.7)* The model allows us to turn a measured velocity into a distance along the line of sight. As in so many areas of astronomy, theory and observations complement one another—the data refine the theoretical model, while the model in turn provides the framework needed to understand and interpret further observations. ∞ *(Sec. 0.5)*

Figure 14.16 is an artist's conception (based on observational data) of the appearance of our Galaxy as seen from far above the disk. The figure clearly shows our Galaxy's **spiral arms,** pinwheel-like structures originating close to the Galactic bulge and extending outward throughout much of the Galactic disk. One of these arms wraps around a large part of the disk and contains our Sun. Notice, incidentally, the scale markers on Figures 14.8, 14.9, and 14.16. The Galactic globular-cluster distribution (shown in Figure 14.8), the luminous stellar component of the disk (Figure 14.9), and the known spiral structure (Figure 14.16) all have roughly the same diameter—about 30 kpc. This diameter is fairly typical of spiral galaxies observed elsewhere in the universe.

The spiral arms in our Galaxy are made up of much more than just interstellar gas and dust. Studies of the Galactic disk within a kiloparsec or so of the Sun indicate that young stellar and prestellar objects—emission nebulae, O- and B-type stars, and recently formed open clusters—are also distributed in a spiral pattern that closely follows the distribution of interstellar clouds. The obvious conclusion is that the spiral arms are the part of the Galactic disk where star formation takes place. The brightness of the young stellar objects just listed is the main reason that the spiral arms of other galaxies are easily seen from afar (Figure 14.2b).

▼ **FIGURE 14.15 Gas in the Galactic Disk** Because the disk of our Galaxy is rotating differentially (inside faster than the outside), signals from different clumps of hydrogen matter along any line of sight are Doppler shifted by different amounts. Repeated observations in many different directions allow astronomers to map out the distribution of gas in our Galaxy.

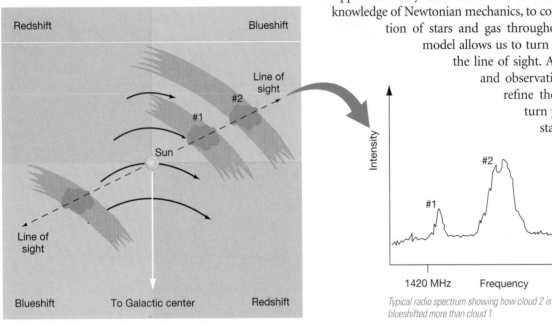

Typical radio spectrum showing how cloud 2 is blueshifted more than cloud 1.

◄ **FIGURE 14.16 Milky Way Spiral Structure** An artist's conception of our Milky Way Galaxy seen face-on, based on radio and infrared maps of stars, gas, and dust in the Galactic disk. The illustration is painted from the perspective of an observer 100 kpc above the Galactic plane, showing the spiral arms emanating from a bar whose length is twice its width. Everything is drawn to scale (except for the oversized yellow dot near the top, which represents our Sun). The two small blotches to the left are dwarf galaxies, called the Magellanic Clouds, which are studied in Chapter 15. *(Adapted from JPL)*

30 kpc

14-2 DISCOVERY

Density Waves

In the late 1960s, American astrophysicists C. C. Lin and Frank Shu proposed a way in which spiral arms in the Galaxy could persist for many Galactic rotations. They argued that the arms themselves contain no "permanent" matter. They should not be viewed as assemblages of stars, gas, and dust moving intact through the disk because such structures would quickly be destroyed by differential rotation. Instead, as described in the text, a spiral arm should be envisaged as a *density wave*—a wave of alternating compression and expansion sweeping through the Galaxy.

A wave in water builds up material temporarily in some places (crests) and lets it down in others (troughs). Similarly, as Galactic matter encounters a spiral density wave, the matter is compressed to form a region of higher than normal density. The matter enters the wave, is temporarily slowed down (by gravity) and compressed as it passes through, then continues on its way. This compression triggers the formation of new stars and nebulae. In this way, the spiral arms are formed and reformed repeatedly, without wrapping up.

The accompanying figure illustrates the formation of a density wave in a more familiar context—a traffic jam triggered by the presence of a repair crew moving slowly down the road. Cars slow down temporarily as they approach the crew, then speed up again as they pass the worksite and continue on their way. The result observed by a traffic helicopter flying overhead is a region of high traffic density concentrated around the work crew and moving with it. An observer on the side of the road, however, sees that the jam never contains the same cars for very long. Cars constantly catch up to the bottleneck, move slowly through it, then speed up again, only to be replaced by more cars arriving from behind.

The traffic jam is analogous to the region of high stellar density in a Galactic spiral arm. Just as the traffic density wave is not tied to any particular group of cars, the spiral arms are not attached to any particular piece of disk material. Stars and gas enter a spiral arm, slow down for a while, then exit the arm and continue on their way around the Galactic center. The result is a moving region of high stellar and gas density, involving different parts of the disk at different times. Notice also that, just as in our Galaxy, the traffic jam wave moves more slowly than, and independently of, the overall traffic flow.

We can extend our traffic analogy a little further. Most drivers are well aware that the effects of such a tie-up can persist long after the road crew has stopped work and gone home for the night. Similarly, spiral density waves can continue to move through the disk even after the disturbance that originally produced them has long since subsided. According to spiral density wave theory, this is precisely what has happened in the Milky Way Galaxy. Some disturbance in the past—an encounter with a satellite galaxy, perhaps, or the effect of the central bar—produced a wave that has been moving through the Galactic disk ever since.

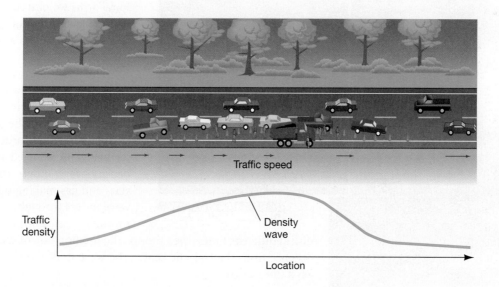

A central problem in understanding spiral structure is explaining how that structure persists over long periods of time. The basic issue is simple: differential rotation makes it impossible for any large-scale structure "tied" to the disk material to survive. Figure 14.17 shows how a spiral pattern always consisting of the same group of stars and gas clouds would necessarily "wind up" and disappear within a few hundred million years. How then do the Galaxy's spiral arms retain their structure

INTERACTIVE Spiral Arms and Star Formations

CONCEPT CHECK ▶ Why can't spiral arms simply be clouds of gas and young stars orbiting the Galactic center?

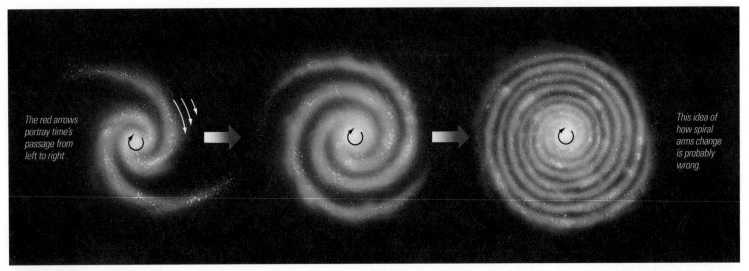

The red arrows portray time's passage from left to right.

This idea of how spiral arms change is probably wrong.

INTERACTIVE ▲ **FIGURE 14.17 Differential Galactic Rotation**   The disk of our Galaxy rotates differentially, as shown by the small white arrows that represent the angular speed of the disk. If spiral arms were somehow tied to the material of the Galactic disk, such uneven rotation would cause the spiral pattern to wind up and disappear in a few hundred million years. Spiral arms would be too short-lived to be consistent with the numbers of spiral galaxies we observe today.

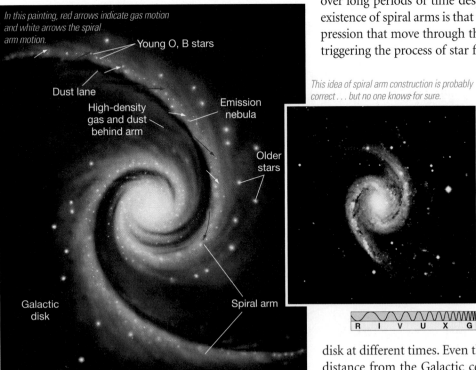

In this painting, red arrows indicate gas motion and white arrows the spiral arm motion.

Young O, B stars

Dust lane

High-density gas and dust behind arm

Emission nebula

Older stars

Galactic disk

Spiral arm

This idea of spiral arm construction is probably correct . . . but no one knows for sure.

R I V U X G

▲ **FIGURE 14.18 Spiral Density Waves**   Density-wave theory holds that the spiral arms seen in our own and many other galaxies are waves of gas compression and star formation moving through the material of the galactic disk. Gas enters an arm from behind, is compressed, and forms stars. The spiral pattern is delineated by dust lanes, regions of high gas density, and newly formed bright stars. The inset at right shows the spiral galaxy NGC 1566, which displays many of the features just described. (*AURA*)

over long periods of time despite differential rotation? A leading explanation for the existence of spiral arms is that they are **spiral density waves**—coiled waves of gas compression that move through the Galactic disk, squeezing clouds of interstellar gas and triggering the process of star formation as they go. ∞ (*Sec. 11.4*) Like traffic slowing as it passes an obstacle on the highway (see *Discovery 14-2*), Galactic gas slows down and becomes denser as it passes through the wave. The spiral arms we observe are defined by the denser-than-normal clouds of gas thus created and by the new stars formed as a result of the spiral waves' passage.

This explanation of spiral structure avoids the problem of differential rotation because the wave pattern is not tied to any particular piece of the Galactic disk. The spirals we see are merely patterns moving through the disk, not matter being transported from place to place. The density wave moves through the collection of stars and gas making up the disk just as a sound wave moves through air or an ocean wave passes through water, compressing different parts of the disk at different times. Even though the rotation rate of the disk material varies with distance from the Galactic center, the wave itself remains intact, defining the Galaxy's spiral arms.

Over much of the visible portion of the Galactic disk (within about 15 kpc of the center), the spiral wave pattern is predicted to rotate *more slowly* than the stars and gas. Thus, as shown in Figure 14.18, Galactic material catches up with the wave, is temporarily slowed down and compressed by the wave's gravitational pull as it passes through, then continues on its way.

As the density wave enters the arm from behind, the gas is compressed and forms stars. Dust lanes mark the regions of highest-density gas. The most prominent stars—the bright O- and B-type blue giants—live for only a short time, so emission nebulae and young star clusters are found only within the arms, near their birth sites, just ahead of the dust lanes. Their brightness emphasizes the spiral structure. Further downstream, ahead of the spiral arms, we see mostly older stars and star clusters, which have had enough time since their formation to pull ahead of the wave. Over millions of years their random individual motions, superimposed on the overall rotation around the Galactic center, distort and eventually destroy their original spiral configuration, and they become part of the general disk population.

Note, incidentally, that although the spirals shown in Figure 14.18 have two arms each, astronomers are not completely certain how many arms make up the spiral structure in our Galaxy. The theory makes no strong predictions on this point. As illustrated in Figure 14.16, the best available data suggest that our Galaxy has two major arms.

An alternative possibility is that the formation of stars drives the waves, instead of the other way around. Imagine a row of newly formed massive stars somewhere in the disk. The emission nebula created when these stars form, and the supernovae when they die, send shock waves through the surrounding gas, possibly triggering new star formation. ∞ (Sec. 12.7) Thus, as illustrated in Figure 14.19(a), the formation of one group of stars provides the mechanism for the creation of more stars. Computer simulations suggest that it is possible for the "wave" of star formation thus created to take on the form of a partial spiral and for this pattern to persist for some time. However, this process, known as **self-propagating star formation,** can produce only pieces of spirals, as are seen in some galaxies (Figure 14.19b). It apparently cannot produce the galaxywide spiral arms seen in other galaxies and present in our own. It may well be that there is more than one process at work in the spectacular spirals we see.

An important question, but one not answered by either of the two theories just described, is: Where do these spirals come from? What was responsible for generating the density wave in the first place or for creating the line of newborn stars whose evolution drives the advancing spiral arm? Scientists speculate that (1) instabilities in the gas near the galactic bulge, (2) the gravitational effects of nearby galaxies, or (3) the elongated shape of the bulge itself may have had a big enough influence on the disk to get the process going. The fact is that we still don't know for sure how galaxies—including our own—acquire such beautiful spiral arms.

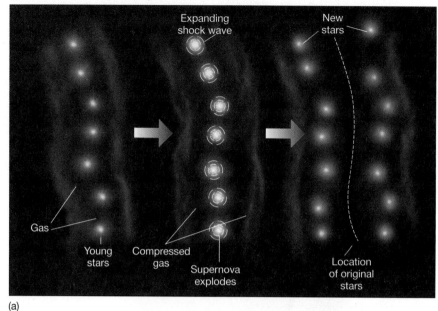

Again, time unfolds left to right, creating new stars over and over . . .

(a)

▲ FIGURE 14.19 Self-Propagating Star Formation (a) In this theory of the formation of spiral arms, the shock waves produced by the formation and later evolution of a group of stars provide the trigger for new rounds of star formation. (b) This process may well be responsible for the partial spiral arms seen in some galaxies, such as NGC 4314, shown here in true color. Its distinct blue appearance derives from the vast numbers of young stars that pepper its ill-defined spiral arms. (R. Gendler)

14.6 The Mass of the Milky Way Galaxy

We can measure our Galaxy's mass by studying the motions of gas clouds and stars in the Galactic disk. Recall from Chapter 1 that Kepler's third law (as modified by Newton) connects the orbital period, orbit size, and masses of any two objects in orbit around one another:

$$\text{total mass (solar masses)} = \frac{\text{orbit size (AU)}^3}{\text{orbit period (years)}^2}.$$

The distance from the Sun to the Galactic center is about 8 kpc, and the Sun's orbital period is 225 million years. Plugging these numbers into the equation, we find a mass of almost 10^{11} solar masses—100 *billion* times the mass of our Sun.

But what mass have we just measured? When we performed the analogous calculation in the case of a planet orbiting the Sun, there was no ambiguity: neglecting the planet's mass, the result of our calculation was the mass of the Sun. ∞ (Sec. 1.4)

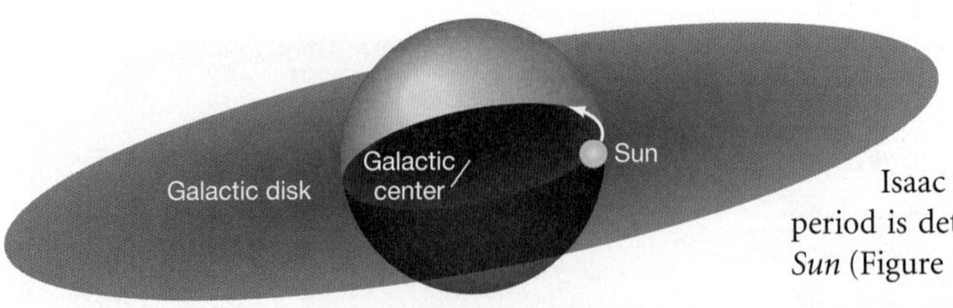

▲ FIGURE 14.20 **Weighing the Galaxy**   The orbital speed of a star or gas cloud moving around the Galactic center is determined only by the mass of the Galaxy lying inside the orbit (within the gray-shaded sphere). Thus, to measure the Galaxy's total mass, we must observe objects orbiting at large distances from the center.

However, the Galaxy's matter is not concentrated at the Galactic center (as the Sun's mass is concentrated at the center of the solar system). Instead, Galactic matter is distributed over a large volume of space. What portion of the Galaxy's mass controls the Sun's orbit? Isaac Newton answered this question three centuries ago: the Sun's orbital period is determined by the portion of the Galaxy that lies *within the orbit of the Sun* (Figure 14.20). This is the mass computed from the above equation.

Dark Matter

To determine the mass of the Galaxy on larger scales, we must measure the orbital motion of stars and gas at greater distances from the Galactic center. Astronomers have found that the most effective way to do this is to make radio observations of gas in the Galactic disk, because radio waves are relatively unaffected by interstellar absorption and allow us to probe to great distances. On the basis of these studies, radio astronomers have determined our Galaxy's rotation rate at various distances from the Galactic center. The resultant plot of rotation speed versus distance from the center (Figure 14.21) is called the Galactic **rotation curve.**

Using the Galactic rotation curve, we can now repeat our earlier calculation to compute the total mass that lies within any distance from the Galactic center. We find, for example, that the mass within about 15 kpc from the center—the volume defined by the globular clusters and the known spiral structure—is roughly 2×10^{11} solar masses, about twice the mass contained within the Sun's orbit. Does the distribution of matter in the Galaxy end at 15 kpc, where the luminosity drops off sharply? Surprisingly, the answer is no.

If all of the mass of the Galaxy were contained within the edge of the visible structure, Newton's laws of motion predict that the orbital speed of stars and gas beyond 15 kpc would decrease with increasing distance from the Galactic center, just as the orbital speeds of the planets diminish as we move outward from the Sun. The dashed line in Figure 14.21 indicates how the rotation curve should look in that case. However, the actual rotation curve is quite different. Far from declining at larger distances, it *rises* slightly out to the limits of our measurement capabilities. This implies that the amount of mass contained within successively larger radii continues to grow beyond the orbit of the Sun, apparently out to a distance of at least 50 kpc.

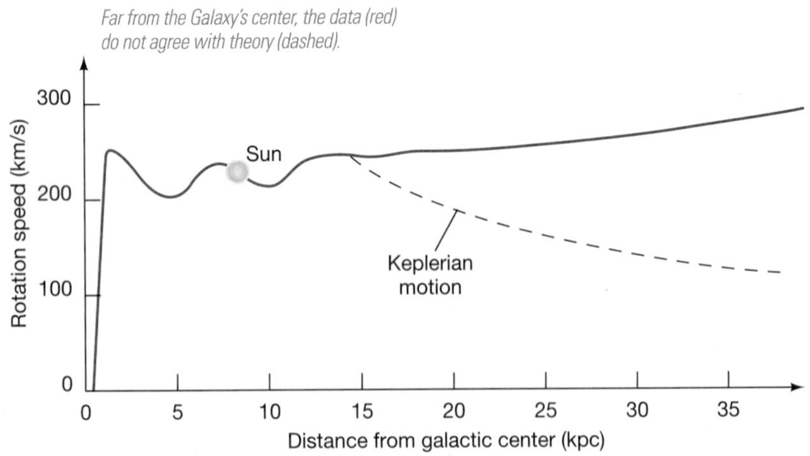

According to the equation above, the amount of mass within 50 kpc is approximately 6×10^{11} solar masses. Since 2×10^{11} solar masses lies within 15 kpc of the Galactic center, we have to conclude that at least twice as much mass lies *outside* the luminous part of our Galaxy—the part made up of stars, star clusters, and spiral arms—as lies inside!

Based on these observations, astronomers conclude that the luminous portion of the Milky Way Galaxy—the region outlined by the globular clusters and by the spiral arms—is merely the "tip of the Galactic iceberg." Our Galaxy is in reality very much larger. The luminous region is surrounded by an extensive, invisible **dark halo,** which dwarfs the inner halo of stars and globular clusters and extends well beyond the 15-kpc radius once thought to represent the limit of our Galaxy. But what is this dark halo made of? We do not detect enough stars or interstellar matter to account for the mass that our computations tell us must be there. We are inescapably drawn to the conclusion that most of the mass in our Galaxy (and, as we will see, of all galaxies) exists in the form of invisible **dark matter,** whose gravitational effects we can measure and quantify, but whose precise nature we do not understand.

▲ FIGURE 14.21 **Galaxy Rotation Curve**   The rotation curve for the Milky Way Galaxy plots rotation speed against distance from the Galactic center. The dashed curve is the rotation curve expected if the Galaxy "ended" abruptly at a radius of 15 kpc, the limit of most of the known spiral structure. The fact that the red curve does not follow this dashed line, but instead stays well above it, indicates that there must be additional unseen matter beyond that radius.

The term *dark* here does not refer just to matter undetectable in visible light. The material has (so far) escaped detection at *all* electromagnetic wavelengths, from radio to gamma rays. Only by its gravitational pull do we know of its existence. Dark matter is not hydrogen gas (atomic or molecular), nor is it made up of ordinary stars. Given the amount of matter that must be accounted for, we would have been able to detect it with present-day equipment if it were in either of those forms. Its composition and its consequences for the evolution of galaxies and the universe are among the most important questions in astronomy today.

Many candidates have been suggested for this dark matter, although none is proven. Among the strongest "stellar" contenders are brown dwarfs and white dwarfs. ∞ *(Secs. 11.5, 12.3)* These objects could in principle exist in great numbers throughout the Galaxy, yet would be exceedingly hard to see. Note that the ultimate "dark" candidates—black holes—are not thought to contribute much to the dark matter total, simply because the massive stars that produce them are too rare. Only about 1 star in 10,000 ends up as a black hole.

A radically different alternative is that the dark matter is made up of exotic *subatomic particles* that pervade the entire universe. In order to account for the properties of dark matter, these particles must have mass (to produce the observed gravitational effects) but otherwise interact hardly at all with "normal" matter (because otherwise we would be able to see them).[4] Many astrophysicists think that such particles could have been produced in abundance during the very earliest moments of our universe. If they survived to the present day, there might be enough of them to account for all the dark matter we think must be out there. These ideas are hard to test, however, because such particles would be very difficult to detect. Several detection experiments have been attempted, so far without success.

A few astronomers have proposed a very different explanation for the "dark matter problem," suggesting that its resolution may lie not in the nature of dark matter, but rather in a modification to Newton's law of gravity that increases the gravitational force on very large (Galactic) scales, doing away with the need for dark matter in the first place. We emphasize that the vast majority of scientists do *not* accept this view, but the very fact that it has been proposed—and is being seriously discussed in (some) scientific circles—underscores our current level of uncertainty. Dark matter is one of the great unsolved mysteries in astronomy today.

PROCESS OF ▶ SCIENCE CHECK The nature of subatomic dark matter particles is completely unknown, yet most scientists regard these particles as the best solution to the dark matter problem. How do you think these statements square with the experimental scientific method presented in Chapter 0?

The Search for Stellar Dark Matter

Recently, researchers have obtained insight into the distribution of stellar dark matter by using a key element of Albert Einstein's theory of general relativity—the prediction that a beam of light can be deflected by a gravitational field, which has already been verified in the case of starlight passing close to the Sun. ∞ *(More Precisely 13-2)* Although this effect is small, it has the potential for making otherwise invisible stellar objects observable from Earth. Here's how.

Imagine looking at a distant star as a faint foreground object (such as a brown dwarf or a white dwarf) happens to cross your line of sight. As illustrated in Figure 14.22, the intervening object deflects a little more starlight than usual toward you, resulting in a temporary, but quite substantial, *brightening* of the distant star. Because the effect is in some ways like the focusing of light by a lens, this process is known as **gravitational lensing.** The foreground object is referred to as a *gravitational lens.* The amount of brightening and the duration of the effect depend on the mass, distance, and speed of the lensing object. Typically, the apparent brightness of the background star increases by a factor of 2 to 5 for a period of several weeks.

 SELF-GUIDED TUTORIAL Gravitational Lensing

[4]One class of candidate particles satisfying these requirements has been dubbed *Weakly Interacting Massive Particles,* or WIMPs. Not to be outdone, astronomers searching for the more mundane stellar dark matter have labeled them *MAssive Compact Halo Objects,* or MACHOs. Who says astronomers don't have a sense of humor?

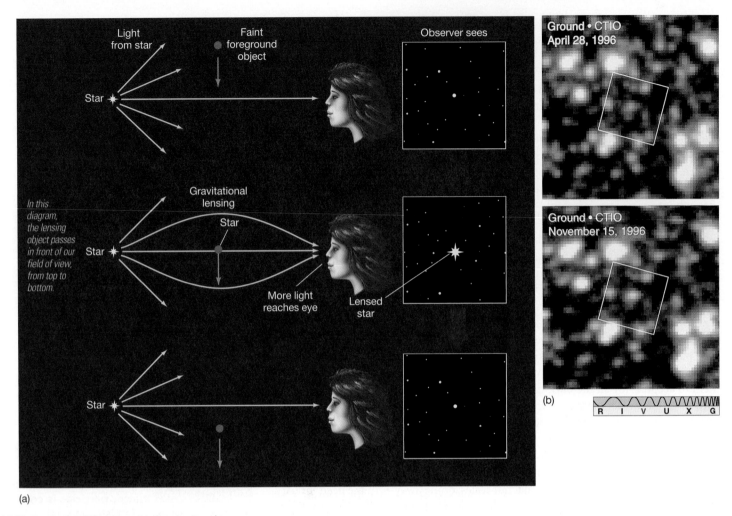

(a)

▲ **FIGURE 14.22 Gravitational Lensing**   (a) Gravitational lensing by a faint foreground object (such as a brown dwarf) can temporarily cause a background star to brighten significantly, providing a means of detecting otherwise invisible stellar dark matter. (b) These two images show the brightening of a star during a lensing event, this one implying that a massive, but unseen, object passed in front of the unnamed star at the center of the two boxes imaged 6 months apart. *(AURA)*

CONCEPT
CHECK ▶   In what sense is dark matter "dark"?

Thus, even though the foreground object cannot be seen directly, its effect on the light of the background star makes it detectable.

Of course, the probability of one star passing almost directly in front of another, as seen from Earth, is extremely small. However, by observing millions of stars every few days over a period of years (using automated telescopes and high-speed computers to reduce the burden of coping with so much data), astronomers have so far seen thousands of events. The observations are consistent with lensing by low-mass white dwarfs and suggest that such stars could account for at least half—but apparently not all—of the Galactic dark matter inferred from dynamical studies.

Bear in mind, though, that the identity of the dark matter is not necessarily an "all-or-nothing" proposition. It is quite possible that more than one type of dark matter exists. For example, most of the dark matter in the inner (visible) parts of galaxies could be in the form of low-mass stars, while the dark matter farther out might be primarily in the form of exotic particles. We will return to this perplexing problem in later chapters.

14.7 The Galactic Center

Theory predicts that the Galactic bulge, and especially the region close to the Galactic center, should be densely populated with billions of stars. However, we are unable to see this region of our Galaxy—the interstellar medium in the Galactic disk shrouds what otherwise would be a stunning view. Figure 14.23 shows the (optical) view we do have of the region of the Milky Way toward the Galactic center, in the general direction of the constellation Sagittarius.

Galactic Activity

With the help of infrared and radio techniques we can peer more deeply into the central regions of our Galaxy than we can by optical means. Infrared observations (Figure 14.23 inset) show that the innermost parsec of our Galaxy harbors a dense cluster containing roughly 1 million stars. That's a stellar density about 100 *million* times greater than in our solar neighborhood, high enough that stars must experience frequent close encounters and even collisions.

Figure 14.24(a) is an infrared view of part of Figure 14.23, with the Galactic plane now horizontal. On this scale, infrared radiation has been detected from what appear to be huge clouds rich in dust. Radio observations indicate a ring of molecular gas nearly 400 pc across, containing hundreds of thousands of solar masses of material and rotating around the Galactic center at about 100 km/s. Its origin is unclear, although researchers think that the gravity of the Galaxy's central rotating bar may deflect gas from farther out into the dense central regions.

Higher-resolution radio studies show much more structure. Figure 14.24(b) shows a region called Sagittarius A. (The name just means that it is the brightest radio source in Sagittarius.) It lies at the center of the boxed region in Figures 14.23 and 14.24(a) and, we think, at the center of our Galaxy. On a scale of about 25 pc, extended filaments can be seen. Their presence suggests to many astronomers that strong magnetic fields operate near the center, creating structures similar in appearance to (but much larger than) those observed on the active Sun.

INTERACTIVE Spiral Galaxy M81

R I V U X G

0.3 pc

R I V U X G

INTERACTIVE ◄ FIGURE 14.23 Galactic Center Photograph of stellar and interstellar matter in the direction of the Galactic center. The field is roughly 20°, top to bottom, and is a continuation of the bottom part of Figure 11.5. The overlaid box indicates the location of the center of our Galaxy. The inset shows an adaptive-optics infrared view of the dense stellar cluster surrounding the Galactic center, whose very core is indicated by the twin arrows. (AURA; ESO)

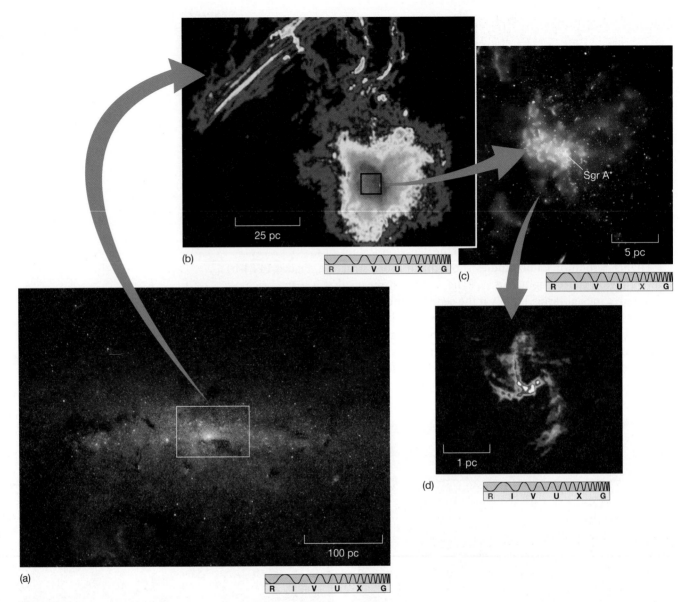

INTERACTIVE ▲ **FIGURE 14.24 Galactic Center Close-up**   (a) An infrared image of the region around the center of our Galaxy shows many bright stars packed into a relatively small volume. (b) The central portion of our Galaxy, as observed in the radio part of the spectrum, shows a region about 100 pc across surrounding the Galactic center (which lies within the orange-yellow region at bottom right). The long-wavelength radio emission cuts through the Galaxy's dust, providing a view of matter in the immediate vicinity of the Galaxy's center. (c) This *Chandra* X-ray image shows the relation of a hot supernova remnant (red) and Sgr A*, the suspected black hole at the very center of our Galaxy. (d) The spiral pattern of radio emission arising from Sagittarius A itself suggests a rotating ring of matter only a few parsecs across. All images are false color, since they lie outside the visible spectrum. (*SST; NRAO; NASA*)

On even smaller scales (Figure 14.24c), *Chandra* observations indicate an extended region of hot X-ray-emitting gas, apparently associated with a supernova remnant, in addition to many other individual bright X-ray sources. Within that region lies a rotating ring of star-forming molecular gas only a few parsecs across, with streams of matter spiraling inward toward the center (Figure 14.24d).

What could cause all this activity? An important clue comes from the Doppler broadening of infrared spectral lines emitted from the central swirling whirlpool of gas. The extent of the broadening indicates that the gas is moving very rapidly. In order to keep this gas in orbit, whatever is at the center must be extremely massive—several million solar masses. Given the twin requirements of large mass and small size, a leading contender is a *supermassive black hole.* The strong magnetic fields are thought to be generated within the accretion disk around the black hole as matter spirals inward, and may act as "particle accelerators," creating the extremely high-energy particles detected on Earth as *cosmic rays*. ∞ *(Sec. 13.3)* In the late 1990s, the *Compton Gamma Ray Observatory* found indirect evidence for a fountain of high-energy particles, possibly produced by violent processes close to the event horizon, gushing into the halo more than a thousand parsecs beyond the Galactic center. Astronomers have reason to suspect that similar events are occurring at the centers of many other galaxies.

The Central Black Hole

At the very center of our Galaxy, at the heart of Sagittarius A, lies a remarkable object with the odd-sounding name **Sgr A*** (pronounced "saj A star"). By the standards of the active galaxies to be studied in Chapter 15, this compact **Galactic nucleus** is not particularly energetic. Still, radio observations made during the past two decades, along with more recent X- and gamma-ray observations, suggest that it is nevertheless a pretty violent place. Its total energy output (at all wavelengths) is estimated to be 10^{33} W, which is more than a million times that of the Sun. Very-long-baseline interferometry (VLBI) observations using radio telescopes arrayed from Hawaii to Massachusetts imply that Sgr A* cannot be much larger than 10 AU, and it is probably a good deal smaller than that. ∞ *(Sec. 3.4)* This size is consistent with the view that the energy source is a massive black hole. Figure 14.25 is perhaps the strongest evidence to date supporting the black hole picture. It shows a high-resolution infrared image of an 0.04-pc (8000 AU) field near the Galactic center, centered on Sgr A*. Using advanced adaptive-optics techniques on the Keck telescope and the VLT, two teams of researchers from the United States and Europe have created the first-ever diffraction-limited (0.05″ resolution) images of the region. ∞ *(Sec. 3.3)*

Remarkably, the image quality is good enough that the *proper motions* of several of the stars—their orbits around the Galactic center—can clearly be seen. The inset shows a series of observations of one of the brightest stars—called S2—over a 10-year period. The motion is consistent with an orbit around a massive object at the location of Sgr A*, in accordance with Newton's laws of motion. ∞ *(Sec. 1.4)* The solid curve on the figure shows the elliptical orbit that best fits the observations. It corresponds to a 15-year orbit with a semimajor axis of 950 AU, corresponding (from Kepler's third law, as modified by Newton) to a central mass of 4 million solar masses. The small size of the central object is clearly demonstrated by the orbit of another star (S16), whose extremely eccentric orbit brings it within 45 AU of the center. Note that, even with this large mass, if Sgr A* is indeed a genuine black hole, the size of its event horizon is still only 0.08 AU. ∞ *(Sec. 13.5)* Such a small region, 8 kpc away, is currently unresolvable, although radio astronomers are hopeful that improving VLBI techniques will allow them to "see" the event horion and study the surrounding accretion disk within the next decade.

Figure 14.26 places these findings into a simplified perspective—all frames are artist's conceptions, but are based on real data. Each frame is centered on the Galaxy's core, with an increase in resolution of a factor of 10 from one frame to the next. Frame (a) renders the Galaxy's overall shape, as shown in Figure 14.16.

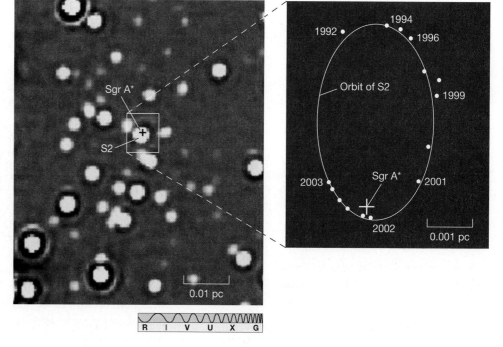

▲ FIGURE 14.25 Orbits Near the Galactic Center This extremely close-up map of the Galactic center (left) was obtained by infrared adaptive optics, resulting in an ultra-high-resolution image of the innermost 0.1 pc of the Milky Way. The inset shows the orbit of the innermost star in the frame, labeled S2, between 1992 and 2003. The solid line shows the best-fitting orbit for S2 around a black hole of 4 million solar masses, located at Sgr A* (marked with a cross). *(ESO)*

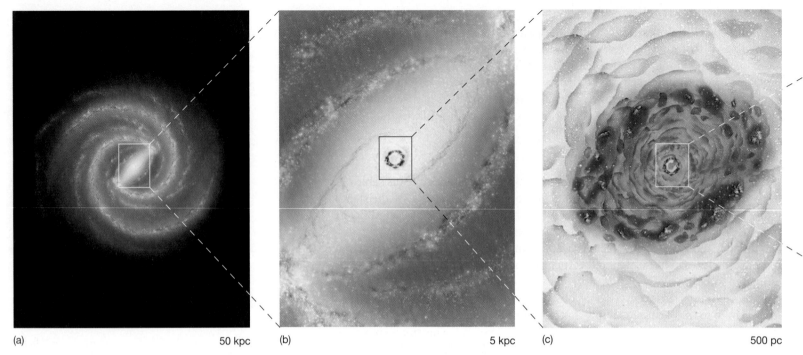

(a)   50 kpc   (b)   5 kpc   (c)   500 pc

▲ **FIGURE 14.26 Galactic Center Zoom** This series of artist's conceptions of the Galactic center depicts each frame increasing in resolution by a factor of 10. Frame (a) is the same scene as Figure 14.16. Frame (f) is a rendition of a vast whirlpool within the innermost 0.5 parsec of our Galaxy. The data imaged in Figure 14.24 do not closely match these artistic renderings because the Figure 14.24 view is parallel to the Galactic disk—along the line of sight from the Sun to the Galactic center—whereas these six paintings portray a simplified view perpendicular to the disk, while progressively zooming down onto that disk.

ANIMATION/VIDEO Black Hole in the Center of the Milky Way?

CONCEPT CHECK ▶ What is the most likely explanation of the energetic events observed at the Galactic center?

This frame measures about 50 kpc across. Frame (b) spans a distance of 5 kpc and is nearly filled by the Galactic bar and the great circular sweep of the innermost spiral arm. Moving in to a 500-pc span, frame (c) depicts the 400-pc ring of gas mentioned earlier and some young dense star clusters, evidence of recent star formation near the Galactic center. The dark blobs represent giant molecular clouds, the pink patches emission nebulae associated with star formation within those clouds.

In frame (d), at 50 pc, a pinkish (thin, warm) region of ionized gas surrounds the reddish (thicker, warmer) heart of the Galaxy. The energy responsible for this vast ionized region comes from frequent supernovae and other violent activity in the Galactic center. The central star cluster (diluted in the painting for clarity) and the surrounding star-forming ring can also be seen. Frame (e), spanning 5 pc, depicts the ring in more detail as well as the central cluster (again diluted lest we not be able to portray anything in this congested region), along with the tilted, spinning whirlpool of hot (10^4 K) gas surrounding the center of our Galaxy. The innermost part of this gigantic whirlpool is shown in frame (f), in which a swiftly spinning, white-hot disk of gas with temperatures in the millions of kelvins nearly engulfs the central black hole (represented here as a black dot). Two rings of stars, possibly the remains of disrupted star clusters, have also been detected, as roughly sketched in frame (f). The black hole itself, and the stellar orbits shown in Figure 14.25, are far too small to be pictured on this scale.

The last decade has seen an explosion in our knowledge of the innermost few parsecs of our Galaxy, and astronomers are working hard to decipher the clues hidden within its invisible radiation. Still, we are only now beginning to appreciate the full complexity of this strange realm deep in the heart of the Milky Way.

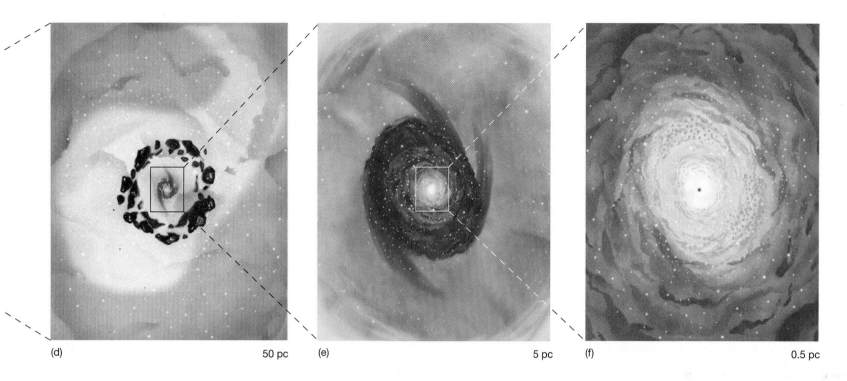

(d)   50 pc   (e)   5 pc   (f)   0.5 pc

CHAPTER REVIEW

SUMMARY

L01 A **galaxy** (p. 380) is a huge collection of stellar and interstellar matter isolated in space and bound together by its own gravity. Because we live within it, the **Galactic disk** (p. 380) of our **Milky Way Galaxy** (p. 380) appears as a broad band of light across the sky—the Milky Way. Near the center, the Galactic disk thickens into the **Galactic bulge** (p. 380). The disk is surrounded by a roughly spherical **Galactic halo** (p. 380) of old stars and star clusters. Our Galaxy, like many others visible in the sky, is a spiral galaxy. Disk and halo stars differ in their spatial distributions, ages, colors, and orbital motion. The luminous portion of our Galaxy has a diameter of about 30 kpc. In the vicinity of the Sun, the Galactic disk is about 300 pc thick.

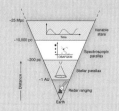

L02 The halo can be studied using **variable stars** (p. 382), whose luminosity changes with time. Two types of pulsating variable stars of particular importance to astronomers are **RR Lyrae variables** (p. 382) and **Cepheid variables** (p. 382). All RR Lyrae stars have roughly the same luminosity. For Cepheids, the luminosity can be determined using the **period–luminosity relationship** (p. 383). Knowing the luminosity, astronomers apply the inverse-square law to find

the distance. The brightest Cepheids can be seen at distances of millions of parsecs, extending the cosmic distance ladder well beyond our own Galaxy. In the early 20th century, Harlow Shapley used RR Lyrae stars to determine the distances to many of the Galaxy's globular clusters and found that they have a roughly spherical distribution in space, but the center of the sphere lies far from the Sun—it is close to the **Galactic center** (p. 384), about 8 kpc away.

L03 Stars and gas within the Galactic disk move on roughly circular orbits around the Galactic center. Stars in the halo and bulge move on largely random three-dimensional orbits that pass repeatedly through the disk plane but have no preferred orientation.

L04 The halo lacks gas and dust, so no stars are forming there. All halo stars are old. The gas-rich disk is the site of current star formation and contains many young stars. Halo stars appeared early on, before the Galactic disk took shape, when there was still no preferred orientation for their orbits. As the gas and dust formed a rotating disk, stars that formed in the disk inherited its overall spin and so moved on circular orbits in the Galactic plane, as they do today.

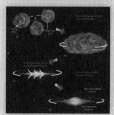

L05 Radio observations clearly reveal the extent of our Galaxy's **spiral arms** (p. 390), regions of the densest interstellar gas where star formation is taking place. The spirals cannot be "tied" to the disk material, as differential rotation would have wound them up long ago. Instead, they may be **spiral density waves** (p. 392) that move through the disk, triggering star formation as they pass by. Alternatively, the spirals may arise from **self-propagating star formation** (p. 393), when shock waves produced by the formation and evolution of one generation of stars trigger the formation of the next.

L06 The Galactic **rotation curve** (p. 394) plots the orbital speed of matter in the disk against distance from the Galactic center. By applying Newton's laws of motion, astronomers can determine the mass of the Galaxy. The Galactic mass continues to increase beyond the radius defined by the globular clusters and the spiral structure we observe,

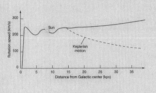

indicating that our Galaxy has an invisible **dark halo** (p. 394). The **dark matter** (p. 394) making up this dark halo is of unknown composition. Candidates include low-mass stars and exotic subatomic particles. Recent attempts to detect stellar dark matter have used the fact that a faint foreground object can occasionally pass in front of a more distant star, deflecting the star's light and causing its apparent brightness to increase temporarily. This deflection is called **gravitational lensing** (p. 395).

L07 Astronomers working at infrared and radio wavelengths have uncovered evidence for energetic activity within a few parsecs of the Galactic center. The leading explanation is that a black hole roughly 4 million times more massive than the Sun resides there. The hole lies at the center of a dense star cluster containing millions of stars, which is in turn surrounded by a star-forming disk of molecular gas. The observed activity is thought to be powered by accretion onto the black hole, as well as by supernova explosions in the cluster surrounding it.

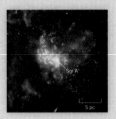

MasteringAstronomy®   For instructor-assigned homework go to www.masteringastronomy.com

Problems labeled **POS** explore the process of science. **VIS** problems focus on reading and interpreting visual information. **LO** connects to the introduction's numbered Learning Outcomes.

REVIEW AND DISCUSSION

1. **L01 POS** What do globular clusters tell us about our Galaxy and our place within it?

2. How are Cepheid variables used in determining distances?

3. How far away can we use Cepheids to measure distance?

4. **L02** What important discoveries were made early in this century using RR Lyrae variables?

5. Why can't optical astronomers easily study the center of our Galaxy?

6. Why are radio studies often more useful in the study of Galactic structure than observations made in visible light?

7. **L03** Contrast the motions of disk and halo stars.

8. **L05 POS** Explain why Galactic spiral arms are thought to be regions of recent and ongoing star formation.

9. Describe what happens to interstellar gas as it passes through a spiral density wave.

10. What is self-propagating star formation?

11. **L04** What do halo stars tell us about the history of our Galaxy?

12. What does a galaxy's rotation curve tell us about its total mass?

13. **L06 POS** What evidence is there for dark matter in the Galaxy?

14. Describe some candidates for Galactic dark matter.

15. **L07 POS** Why do astronomers think that a supermassive black hole lies at the center of the Milky Way Galaxy?

CONCEPTUAL SELF-TEST: TRUE OR FALSE?/MULTIPLE CHOICE

1. Herschel's attempt to map the Milky Way by counting stars led to an inaccurate estimate of the Galaxy's size because he was unaware of absorption by interstellar dust. (T/F)

2. Cepheid variables can be used to determine the distances to the nearest galaxies. (T/F)

3. Globular clusters trace out the structure of the Galactic disk. (T/F)

4. The Galactic halo contains as much gas and dust as the disk. (T/F)

5. The Galactic disk contains only old stars. (T/F)

6. Stars and gas in the Galactic disk move in roughly circular orbits around the Galactic center. (T/F)

7. We can use 21-cm radiation to study molecular clouds. (T/F)

8. The most likely explanation of the high-speed motion of stars and gas near the Galactic center is that they are orbiting a supermassive black hole. (T/F)

9. In the Milky Way Galaxy, our Sun is located (a) near the Galactic center; (b) about halfway out from the center; (c) at the outer edge; (d) in the halo.

10. A telescope searching for newly formed stars would make the most discoveries if it were pointed (a) directly away from the Galactic center; (b) perpendicular to the Galactic disk; (c) within a spiral arm; (d) between spiral arms.

11. The first stars that formed in the Milky Way now (a) have chaotic orbits; (b) orbit in the Galactic plane; (c) orbit near the Galactic center; (d) orbit in the same direction as the Milky Way spins.

12. Stars in the outermost regions of the Milky Way Galaxy (a) are youngest; (b) orbit faster than astronomers would expect based on the Galactic mass we can see; (c) are more likely to explode as supernovae; (d) are more luminous than other stars.

13. Most of the mass of the Milky Way exists in the form of (a) stars; (b) gas; (c) dust; (d) dark matter.

14. **VIS** In Figure 14.6 (Period–Luminosity Relationship), a Cepheid variable star with luminosity 1000 times that of the Sun has a pulsation period of roughly (a) 1 day; (b) 3 days; (c) 10 days; (d) 50 days.

15. **VIS** Figure 14.21 (Galaxy Rotation Curve) tells us that (a) the Galaxy rotates like a solid body; (b) far from the center, the Galaxy rotates more slowly than we would expect based on the light we see; (c) far from the center, the Galaxy rotates more rapidly than we would expect based on the light we see; (d) there is no matter beyond about 15 kpc from the Galactic center.

PROBLEMS

The number of squares preceding each problem indicates its approximate level of difficulty.

1. ■ At one time some astronomers claimed that the Andromeda "nebula" (Figure 14.2a) was a star-forming region lying within our Galaxy. Calculate the angular diameter of a prestellar nebula of radius 100 AU, lying 100 pc from Earth. ∞ *(More Precisely 1-1)* How close to Earth would the nebula have to lie for it to have the same angular diameter (roughly 6°) as Andromeda?

2. ■ What is the greatest distance at which an RR Lyrae star of absolute magnitude 0 could be seen by a telescope capable of detecting objects as faint as 20th magnitude? ∞ *(Sec. 10.2)*

3. ■ A typical Cepheid variable is 100 times brighter than a typical RR Lyrae star. On average, how much farther away than RR Lyrae stars can Cepheids be used as distance-measuring tools?

4. ■■ An astronomer using *HST* can just see a star with solar luminosity at a distance of 100,000 pc. The brightest Cepheids are 30,000 times more luminous than the Sun. If the Sun's absolute magnitude is 5, calculate the absolute magnitudes of these bright Cepheids. Neglecting interstellar absorption, how far away can *HST* see them?

5. ■■ Calculate the proper motion (in arc seconds per year) of a globular cluster with a transverse velocity (relative to the Sun) of

200 km/s and a distance of 3 kpc. Might this motion be measurable? ∞ *(Sec.10.1)*

6. ■■ Calculate the total mass of the Galaxy lying within 20 kpc of the Galactic center if the rotation speed at that radius is 240 km/s.

7. ■■■ Using the data presented in Figure 14.21, calculate how long it takes the Sun to "lap" stars orbiting 15 kpc from the Galactic center.

8. ■ A two-arm spiral density wave is moving through the Galactic disk. At the 8-kpc radius of the Sun's orbit around the Galactic center, the wave's speed is 120 km/s, and the Galactic rotation speed is 220 km/s. How many times has the Sun passed through a spiral arm since the Sun formed 4.6 billion years ago?

9. ■■ Given the data in the previous question and the fact that O-type stars live at most 10 million years before exploding as supernovae, calculate the maximum distance at which an O-type star (orbiting at the Sun's distance from the Galactic center) can be found from the density wave in which it formed.

10. ■■■ Material at an angular distance of 0.2″ from the Galactic center is observed to have an orbital speed of 1200 km/s. If the Sun's distance to the Galactic center is 8 kpc and the material's orbit is circular and is seen edge-on, calculate the radius of the orbit and the mass of the object around which the material is orbiting.

ACTIVITIES

Collaborative

1. Construct your own electronic version of the Messier catalog, listing the names, types, and coordinates of all 110 Messier objects. (You should be able to find all the needed information on the Web, although you may end up combining information from more than one source.) Use a spreadsheet to plot the celestial coordinates—right ascension and declination—of all objects. ∞ *(Sec. 0.1)* Color-code the objects to distinguish among emission nebulae, young star clusters, older open star clusters, globular clusters, and galaxies. What do you notice about the distributions of these objects on the sky? It might help to sketch the approximate location of the Milky Way (more research!) on the plot. Why are the young objects mainly found in the plane of the Milky Way. What about the globular clusters? Why do you think the galaxies apparently avoid the Galactic plane? If you have time, add deep-sky objects from other catalogs to your plot.

Individual

1. If you are far from city lights, look for a hazy band of light arching across the sky. This is our edgewise view of the Milky Way Galaxy. The

Galactic center is located in the direction of the constellation Sagittarius, highest in the sky in summer, but visible from spring through fall. Look at the band making up the Milky Way and sketch what you see. Look for dark regions and faint fuzzy spots in the Milky Way; note their positions on your sketch. Add the major constellations for reference. Compare your sketch with a map of the Milky Way in a star atlas. Can you identify the dark regions and faint fuzzy spots?

2. Observe the Andromeda galaxy, M31. It's the most distant object visible to the naked eye, but don't expect to see anything like Figure 14.2(a)! To the unaided eye, from all but the darkest sites, only its nucleus will be visible, looking like a slightly fuzzy star. To locate M31, find Polaris, the pole star, and the constellations Cassiopeia and Andromeda. Follow a line from Polaris through the second "V" in the "W" of Cassiopeia, and continue south. That line will pass through M31 before you reach the northern arc of stars in Andromeda. Use binoculars or a wide-angle eyepiece to view the galaxy and its disk. Switch to higher magnification to view the nucleus and the small satellite galaxies M32, just to its south, and M110, to the northwest.

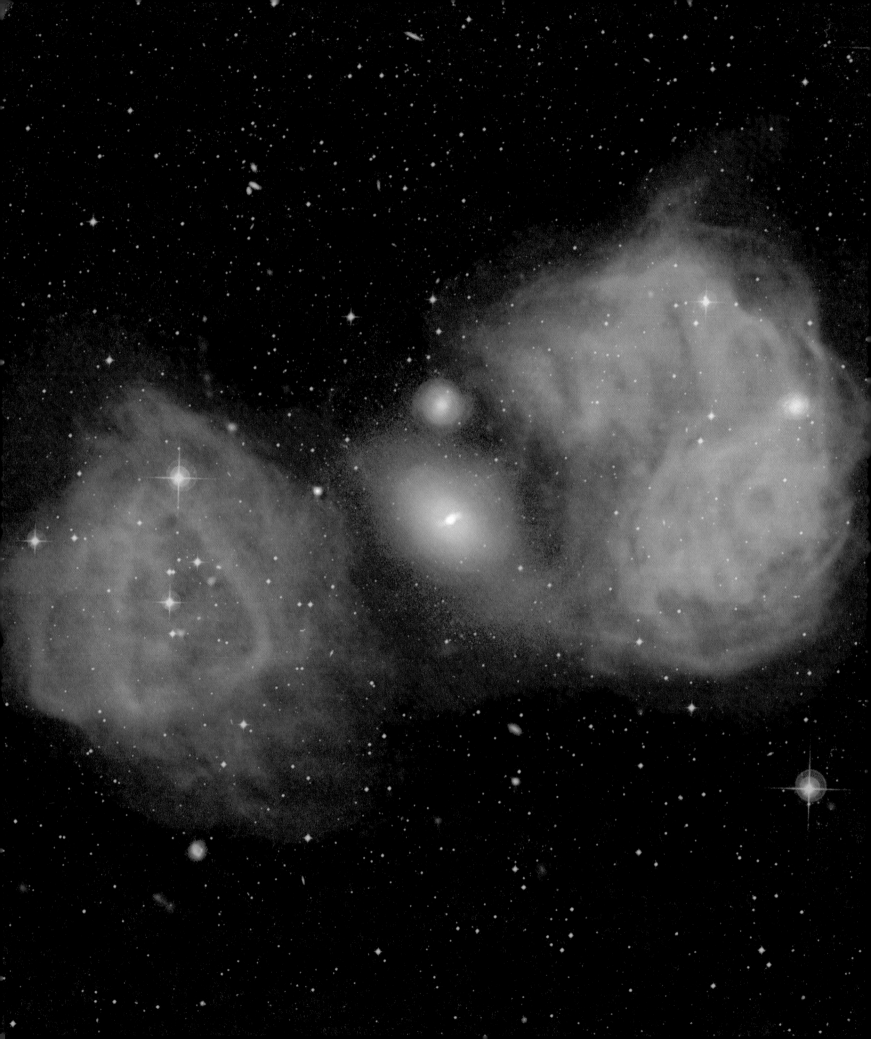

15

Normal and Active Galaxies

Building Blocks of the Universe

As our field of view expands to truly cosmic scales, the focus of our studies shifts dramatically. Planets become inconsequential, stars themselves mere points of hydrogen consumption. Now entire galaxies become the "atoms" from which the universe is built—distant realms completely unknown to scientists just a century ago. We know of literally millions of galaxies beyond our own. Most are smaller than the Milky Way, some comparable in size, a few much larger. Many are sites of explosive events far more energetic than anything ever witnessed in our own Galaxy. All are vast, gravitationally bound assemblages of stars, gas, dust, dark matter, and radiation separated from us by almost incomprehensibly large distances. The light we receive tonight from the most distant galaxies was emitted long before Earth existed. By studying the properties of galaxies and the violence that ensues when they collide, we gain insight into the history of our Galaxy and the universe in which we live.

MasteringAstronomy® Visit www.masteringastronomy.com for quizzes, animations, videos, interactive figures, and self-guided tutorials.

◀ Active galaxies are much more energetic than normal galaxies such as the Milky Way in which we live. The "central engines" powering active galaxies are thought to be supermassive black holes. This is a double image, mixing optical light (acquired by the *Hubble Space Telescope* in Earth orbit) with radio emission (captured by the Very Large Array in New Mexico). At center (in white) is a giant, visible elliptical galaxy catalogued as NGC 1316, which is probably devouring its small northern neighbor. At its center is the likely black hole whose complex radio emission (in orange), called Fornax A, extends more than a million light-years across. *(NRAO/STScI)*

15.1 Hubble's Galaxy Classification

Figure 15.1 shows a vast expanse of space lying about 100 million pc from Earth. Almost every patch or point of light in this figure is a galaxy—several hundred can be seen in just this one photograph. Images of galaxies look distinctly nonstellar. They have fuzzy edges, and many are quite elongated—not at all like the sharp, point-like images normally associated with stars. Although it is difficult to tell from the photograph, some of the blobs of light in Figure 15.1 are spiral galaxies like the Milky Way Galaxy and Andromeda. Others, however, are definitely not spirals—no disks or spiral arms can be seen. Even when we take into account their different orientations in space, galaxies do *not* all look the same.

American astronomer Edwin Hubble was the first to categorize galaxies in a comprehensive way. Working with the then recently completed 2.5-m optical telescope on Mount Wilson in California in 1924, he classified the galaxies he saw into four basic types—*spirals, barred spirals, ellipticals,* and *irregulars*—solely on the basis of their visual appearance. Many modifications and refinements have been incorporated over the years, but the basic **Hubble classification scheme** is still widely used today.

Spirals

We saw several examples of **spiral galaxies** in Chapter 14—for example, our own Milky Way Galaxy and our neighbor Andromeda. All galaxies of this type contain a flattened galactic disk in which spiral arms are found, a central galactic bulge with a dense nucleus, and an extended halo of faint, old stars. ∞ *(Sec. 14.3)* The stellar density (that is, the number of stars per unit volume) is greatest in the galactic nucleus, at the center of the bulge. ∞ *(Sec. 14.7)*

(a)

R I V U X G

(b)

▲ **FIGURE 15.1 Coma Cluster** (a) A collection of many galaxies, each consisting of hundreds of billions of stars. Called the Coma Cluster, this group of galaxies lies more than 100 million pc from Earth. (b) A recent *Hubble Space Telescope* image of part of the cluster. *(AURA; NASA)*

(a) M81   Type Sa

(b) M51   Type Sb

(c) NGC 2997   Type Sc

In Hubble's scheme, a spiral galaxy is denoted by the letter S and classified as type a, b, or c according to the size of its central bulge: Type Sa galaxies have the largest bulges, type Sc the smallest (Figure 15.2). The tightness of the spiral pattern is quite well correlated with the size of the bulge (although the correspondence is not perfect). Type Sa spiral galaxies tend to have tightly wrapped, almost circular, spiral arms. Type Sb galaxies typically have more open spiral arms, while type Sc spirals often have loose, poorly defined spiral structure. The arms also tend to become more "knotty," or clumped, in appearance as the spiral pattern becomes more open. Based on all available evidence, the Milky Way is a barred spiral galaxy, most likely of type SBb. ⚭ *(Sec. 14.5)*

The bulges and halos of spiral galaxies contain large numbers of reddish old stars and globular clusters, similar to those observed in our own Galaxy and in Andromeda. ⚭ *(Sec. 14.1)* Most of the light from spirals, however, comes from younger A- through G-type stars in the disk, giving these galaxies an overall whitish glow. As in the Milky Way, the disks of typical spiral galaxies are rich in gas and dust. The 21-cm radio radiation emitted by spirals betrays the presence of interstellar gas, and obscuring dust lanes are clearly visible in many systems (see Figures 15.2b and c). Stars are forming within the spiral arms, which contain numerous emission nebulae and newly formed O- and B-type stars. ⚭ *(Sec. 14.5)* The arms appear bluish because of the bright blue O- and B-type stars there. Type Sc galaxies contain the most interstellar gas and dust, Sa galaxies the least.

Most spirals are not seen face-on as in Figure 15.2. Many are tilted with respect to our line of sight, making their spiral structure hard to discern. However, we do not need to see spiral arms to classify a galaxy as a spiral. The presence of the disk, with its gas, dust, and newborn stars, is sufficient. For example, the galaxy shown in Figure 15.3 is classified as a spiral because of the clear line of obscuring dust seen

▲ **FIGURE 15.2 Spiral Galaxy Shapes** Variation in shape among spiral galaxies. As we progress from type Sa to Sb to Sc, the bulges become smaller, while the spiral arms tend to become less tightly wound. *(R. Gendler; NASA; D. Malin/AAT)*

▲ **FIGURE 15.3 Sombrero Galaxy** The Sombrero Galaxy (M104), a spiral system seen edge-on, has a dark band composed of interstellar gas and dust. The large size of this galaxy's central bulge marks it as type Sa, even though its spiral arms cannot be seen from our perspective. The inset shows this galaxy in the infrared part of the spectrum, highlighting its dust content in false-colored pink. *(NASA)*

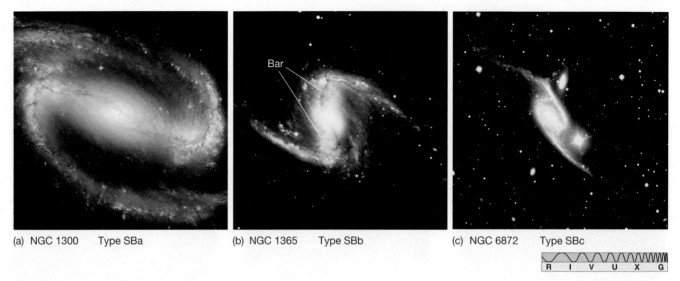

(a) NGC 1300   Type SBa   (b) NGC 1365   Type SBb   (c) NGC 6872   Type SBc

▲ **FIGURE 15.4 Barred-Spiral Galaxy Shapes**   Variation in shape among barred-spiral galaxies from SBa to SBc is similar to that for the spirals in Figure 15.2, except that here the spiral arms begin at either end of a bar through the galactic center. In frame (c), the bright star is a foreground object in our own Galaxy. *(NASA; D. Malin/AAT; ESO)*

along its midplane. That dust is evident most clearly in the inset infrared image, which was acquired by the *Spitzer Space Telescope.*

A variation of the spiral category in Hubble's classification scheme is the **barred-spiral galaxy**. Barred spirals differ from ordinary spirals mainly by the presence of an elongated "bar" of stellar and interstellar matter passing through the center and extending beyond the bulge, into the disk. The spiral arms project from near the ends of the bar rather than from the bulge (as they do in normal spirals). Barred spirals are designated by the letters SB and are subdivided, like the ordinary spirals, into categories SBa, SBb, and SBc, depending on the size of the bulge. Again like ordinary spirals, the tightness of the spiral pattern is correlated with bulge size. Figure 15.4 shows the variation among barred-spiral galaxies. In the case of the SBc category, it is often hard to tell where the bar ends and the spiral arms begin.

Ellipticals

Unlike the spirals, **elliptical galaxies** have no spiral arms and, in most cases, no obvious galactic disk—in fact, other than a dense central nucleus, they often exhibit little internal structure of any kind. As with spirals, the stellar density increases sharply in the nucleus. Denoted by the letter E, these systems are subdivided according to how elliptical they appear on the sky. The most circular are designated E0, slightly flattened systems are labeled E1, and so on, all the way to the most elongated ellipticals, of type E7 (Figure 15.5).

Note that an elliptical galaxy's Hubble type depends both on its intrinsic three-dimensional shape *and* on its orientation relative to the line of sight. A spherical galaxy, a cigar-shaped galaxy seen end-on, and a disk-shaped galaxy seen face-on, would all appear to be circular on the sky and be classified as E0. It can be difficult to decipher a galaxy's true shape solely from its visual appearance.

There is a large range in both the size and the number of stars contained in elliptical galaxies. The largest elliptical galaxies are much larger than our Milky Way Galaxy. These *giant ellipticals* can range up to a few megaparsecs across and contain trillions of stars. At the other extreme, *dwarf ellipticals* may be as small as 1 kpc in diameter and contain fewer than a million stars. Their many differences suggest to

► FIGURE 15.5 **Elliptical Galaxy Shapes**   (a) The E2 elliptical galaxy M49 is nearly circular in appearance. (b) M84 is slightly more elongated and classified as E3. Both galaxies lack spiral structure, and neither shows evidence of interstellar dust or gas, although each has an extensive X-ray halo of hot gas that extends far beyond the visible portion of the galaxy. (c) M110 is a dwarf elliptical companion to the much larger Andromeda Galaxy. (*AURA; SAO; R. Gendler*)

astronomers that giant and dwarf ellipticals represent distinct galaxy classes, with quite dissimilar formation histories and stellar content. The dwarfs are by far the most common type of ellipticals, outnumbering their brighter counterparts by about 10 to 1. However, most of the *mass* that exists in the form of elliptical galaxies is contained in the larger systems.

Lack of spiral arms is not the only difference between spirals and ellipticals. Most ellipticals also contain little or no cool gas and dust and display no evidence of young stars or ongoing star formation. Like the halo of our Galaxy, ellipticals are composed mostly of old, reddish, low-mass stars. Again like the halo of our Galaxy, the orbits of stars in ellipticals are disordered, exhibiting little or no overall rotation; objects move in all directions, not in regular, circular paths as in our Galaxy's disk. Ellipticals differ from our Galaxy's halo in at least one important respect, however: X-ray observations reveal large amounts of very *hot* (several million kelvin) interstellar gas distributed throughout their interiors, often extending well beyond the visible portions of the galaxies.

Some giant ellipticals are exceptions to many of these general statements, as they have been found to contain disks of gas and dust in which stars are forming. Astronomers think that these systems may be the results of mergers between gas-rich galaxies (see Chapter 16). Indeed, galactic collisions may have played an important role in determining the appearance of many of the systems we observe today.

Intermediate between the E7 ellipticals and the Sa spirals in the Hubble classification is a class of galaxies that show evidence of a thin disk and a flattened bulge yet contain neither loose gas nor spiral arms. Two such objects are shown in Figure 15.6. These galaxies are called **S0 galaxies** (S-zero) if no bar is evident and **SB0 galaxies** if a bar is present. They are also known as *lenticular* galaxies, because of their lens-shaped appearance. They look a little like spirals whose dust and gas have been stripped away, leaving behind just a stellar disk. Observations in recent years have shown that many normal elliptical galaxies have faint disks within them, like the S0 galaxies. As with the S0s, the origin of these disks is uncertain, but some researchers suspect that S0s and ellipticals may be closely related.

(a) M49   Type E2

(b) M84   Type E3

(c) M110   Type E5

R I V U X G

(a) NGC 1201   Type S0

(b) NGC 2859   Type SB0

R I V U X G

◄ FIGURE 15.6 **S0 Galaxies**   (a) S0 (or lenticular) galaxies contain a disk and a bulge, but no interstellar gas and no spiral arms. Their properties are intermediate between E7 ellipticals and Sa spirals. (b) SB0 galaxies are similar to S0 galaxies, except for a bar of stellar material extending beyond the central bulge. (*Palomar/Caltech*)

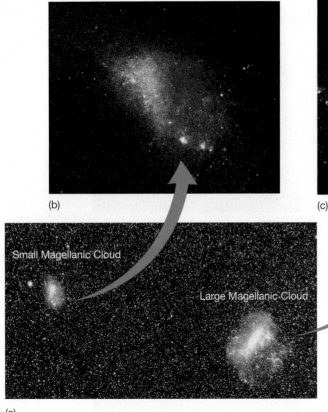

(a)

Small Magellanic Cloud

Large Magellanic Cloud

(b)

(c)

R I V U X G

▲ FIGURE 15.7 Magellanic Clouds The Magellanic Clouds are prominent features of the night sky in the Southern Hemisphere. Named for the 16th-century Portuguese explorer Ferdinand Magellan, whose around-the-world expedition first brought word of these fuzzy patches of light to Europe, these dwarf irregular galaxies orbit our Galaxy and accompany it on its trek through the cosmos. (a) The Clouds' relationship to one another in the southern sky reveals both the Small (b) and the Large (c) Clouds to have distorted, irregular shapes. *(Mount Stromlo & Siding Spring Observatories)*

Irregulars

The final galaxy class identified by Hubble is a catch-all category—**irregular galaxies**—so named because their visual appearance excludes them from the other categories just discussed. Irregulars tend to be rich in interstellar matter and young, blue stars, but they lack any regular structure, such as well-defined spiral arms or central bulges. Irregular galaxies tend to be smaller than spirals but somewhat larger than dwarf ellipticals. They typically contain between 10^8 and 10^{10} stars. The smallest are called *dwarf irregulars*. As with elliptical galaxies, the dwarf type is the most common. Dwarf ellipticals and dwarf irregulars occur in approximately equal numbers and together make up the vast majority of galaxies in the universe. They are often found close to a larger "parent" galaxy.

Irregular galaxies are divided into two subclasses—Irr I and Irr II. The Irr I galaxies often look like misshapen spirals. Figure 15.7 shows the **Magellanic Clouds**, a famous pair of Irr I galaxies that orbit the Milky Way Galaxy. They are shown to proper scale in Figure 14.16. Studies of Cepheid variables within the Clouds show them to be approximately 50 kpc from the center of our Galaxy. ∞ *(Sec. 14.2)* The Large Cloud contains about 6 billion solar masses of material and is a few kiloparsecs across. Both Magellanic Clouds contain lots of gas, dust, and blue stars (as well as the well-documented supernova discussed in *Discovery 12-1*), indicating ongoing star formation. Both also contain many old stars and several old globular clusters, so we know that star formation has been taking place there for a very long time.

The much rarer Irr II galaxies (Figure 15.8), in addition to their irregular shapes, have other peculiarities, often exhibiting a distinctly explosive or filamentary appearance. Their appearance once led astronomers to suspect that violent events had occurred within them. However, it now seems more likely that in some (but probably not all) cases, we are seeing the result of a close encounter or collision between two previously "normal" systems.

The Hubble Sequence

Table 15.1 summarizes the basic characteristics of the various galaxy types. When he first developed his classification scheme, Hubble arranged the galaxy types into the "tuning fork" diagram shown in Figure 15.9. The variation in galaxy types across the diagram, from ellipticals to spirals to irregulars, is often referred to as the **Hubble sequence.**

Hubble's primary aim in creating this diagram was to indicate similarities in appearance among galaxies, but he also regarded the tuning fork as an evolutionary sequence from left to right, with E0 ellipticals evolving into flatter ellipticals and S0 systems that ultimately formed disks and spiral arms. Indeed, Hubble's terminology referring to ellipticals as "early-type" galaxies and spirals as "late-type" is still widely used today. But as far as modern astronomers can tell, there is *no* evolutionary connection of this sort along the Hubble sequence.

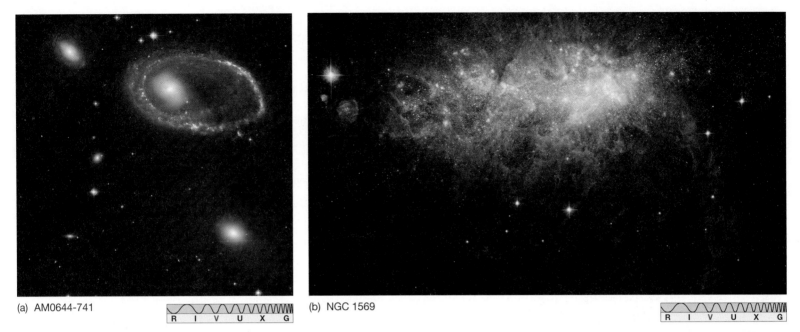

(a) AM0644-741

(b) NGC 1569

▲ FIGURE 15.8 **Irregular Galaxy Shapes**   (a) The strangely shaped galaxy AM0644-741 is probably plunging headlong into a group of several other galaxies (two of them shown here), causing huge rearrangements of its stars, gas, and dust. (b) The galaxy NGC 1569 seems to show an explosive appearance, probably the result of a recent galaxywide burst of star formation. *(NASA)*

Isolated normal galaxies do *not* evolve from one type to another. Spirals are not ellipticals with arms, nor are ellipticals spirals that have somehow shed their star-forming disks. In short, astronomers know of no simple parent–child relationship among Hubble types.

However, the key word here is *isolated*. As we will see, there is now strong observational evidence indicating that collisions and tidal interactions *between* galaxies are commonplace, and these may be the main physical processes driving galaxy evolution. We will return to this important subject in Chapter 16.

ANIMATION/VIDEO The Evolution of Galaxies

CONCEPT CHECK ▶ In what ways are large spirals such as the Milky Way and Andromeda not representative of galaxies as a whole?

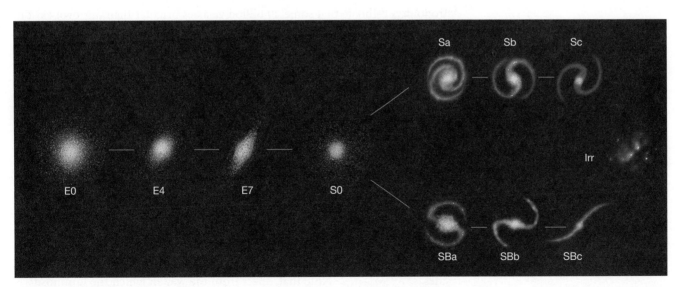

▲ FIGURE 15.9 **Galactic "Tuning Fork"**   The placement of the four basic types of galaxies—ellipticals, spirals, barred spirals, and irregulars—in Hubble's "tuning-fork" diagram is suggestive, but this galaxy classification scheme has no known physical meaning.

TABLE 15.1   Basic Galaxy Properties by Type

	Spiral/Barred Spiral (S/SB)	Elliptical (E)[1]	Irregular (Irr)
Shape and structural properties	Highly flattened disk of stars and gas containing spiral arms and thickening to central bulge. SB galaxies have an elongated central "bar" of stars and gas.	No disk. Stars smoothly distributed through an ellipsoidal volume. No obvious substructure other than a dense central nucleus.	No obvious structure. Irr II galaxies often have "explosive" appearance.
Stellar content	Disks contain both young and old stars; halos consist of old stars only.	Contain old stars only.	Contain both young and old stars.
Gas and dust	Disks contain substantial amounts of gas and dust; halos contain little of either.	Contain hot X-ray-emitting gas, little or no cool gas and dust.	Very abundant in gas and dust.
Star formation	Ongoing star formation in spiral arms.	No significant star formation during the last 10 billion years.	Vigorous ongoing star formation.
Stellar motion	Gas and stars in disk move in circular orbits around the galactic center; halo stars have random orbits in three dimensions.	Stars have random orbits in three dimensions.	Stars and gas have very irregular orbits.

[1] As noted in the text, some giant ellipticals appear to be the result of mergers between gas-rich galaxies and are exceptions to many of the statements listed here.

15.2 The Distribution of Galaxies in Space

Now that we have seen some of their basic properties, let us ask how galaxies are spread throughout the universe beyond the Milky Way. Galaxies are not distributed uniformly in space. Rather, they tend to clump into still larger agglomerations of matter. As always in astronomy, our understanding hinges on knowing how far away an object lies. We therefore begin by looking more closely at the means used by astronomers to measure distances to galaxies.

Extending the Distance Scale

Astronomers estimate that some 40 billion galaxies brighter than our own exist in the observable universe. Some reside close enough for the Cepheid variable technique to work—astronomers have detected and measured the periods of Cepheids in galaxies as far away as 25 Mpc. ⚬⚬ *(Sec. 14.2)* However, some galaxies contain no Cepheid stars (can you think of some reasons why this might be?), and in any case most known galaxies lie much farther away than 25 Mpc. Cepheid variables in very distant galaxies simply cannot be observed well enough, even through the world's most sensitive telescopes, to allow us to measure their apparent brightness and periods. To extend our distance-measurement ladder, we must find some new object to study.

One way in which researchers have tackled this problem is by using **standard candles**—bright, easily recognizable astronomical objects whose luminosities are confidently known. The basic idea is very simple. Once an object is identified as a standard candle—by its appearance or by the shape of its light curve, say—its luminosity can be estimated. Comparison of the luminosity with the apparent brightness then gives the object's distance and hence the distance to the galaxy in which it resides. ⚬⚬ *(Sec. 10.2)* Note that, apart from the way in which the luminosity is determined, the Cepheid variable technique relies on identical reasoning—the measured period tells us the luminosity and the inverse-square law then tells us the distance.

PROCESS OF ▶ SCIENCE CHECK What are some of the problems in measuring accurately the distances to faraway galaxies?

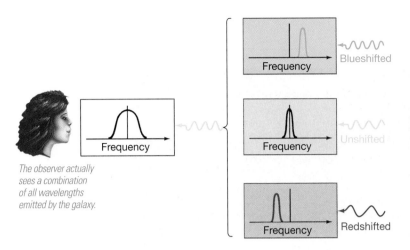

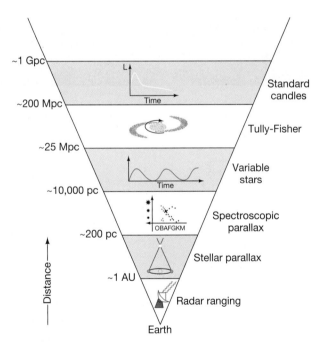

▲ FIGURE 15.10 **Galaxy Rotation** A galaxy's rotation causes some of the radiation it emits to be blueshifted and some redshifted. From a distance, when all the radiation from the galaxy is combined into a single beam and analyzed spectroscopically, the redshifted and blueshifted components produce a broadening of the galaxy's spectral lines. The amount of broadening is a direct measure of the rotation speed of the galaxy, such as NGC 4603 shown here. *(NASA)*

To be most useful, a standard candle must (1) have a well-defined luminosity so that the uncertainty in estimating its brightness is small and (2) be bright enough to be seen at great distances. Currently, Cepheid variable stars are usable for distance measurement "only" out to about 25 Mpc. In recent years, planetary nebulae and Type I supernovae have proved particularly reliable as standard candles. ⚭ *(Secs. 12.3, 12.5)* The latter have remarkably consistent peak luminosities and are bright enough to be identified and measured out to distances of more than a gigaparsec (1000 Mpc).

An important alternative to standard candles was discovered in the 1970s, when astronomers found a close correlation between the rotational speeds and the luminosities of spiral galaxies within a few tens of megaparsecs of the Milky Way Galaxy. Rotation speed is a measure of a spiral galaxy's total mass, so it is perhaps not surprising that this property should be related to luminosity. ⚭ *(Sec. 14.5)* What *is* surprising, though, is how tight the correlation is. The **Tully–Fisher relation**, as it is now known (after its discoverers), allows us to obtain a remarkably accurate estimate of a spiral galaxy's luminosity simply by observing how fast it rotates (Figure 15.10). As usual, comparing the galaxy's (true) luminosity with its (observed) apparent brightness yields its distance.

The Tully–Fisher relation can be used to measure distances to spiral galaxies out to about 200 Mpc, beyond which the line broadening in Figure 15.10 becomes increasingly difficult to measure accurately. A somewhat similar connection, relating line broadening to a galaxy's *diameter*, exists for elliptical galaxies. Once the galaxy's diameter and angular are known, its distance can be computed from elementary geometry. ⚭ *(More Precisely 4-1)* These methods bypass many of the standard candles often used by astronomers and so provide independent means of determining distances to faraway objects.

As indicated in Figure 15.11, standard candles and the Tully–Fisher relation form the fifth and sixth rungs of our cosmic distance ladder, introduced in Chapter 1 and expanded in Chapters 10 and 14. ⚭ *(Secs. 1.4, 10.6, 14.2)* In fact, they stand for perhaps a dozen or so related but separate techniques that astronomers have employed in their quest to map out the universe on large scales. Just as with the lower rungs, we calibrate the properties of these new techniques using distances measured by more local means. In this way, the distance-measurement process "bootstraps" itself to greater and greater distances. However, at the same time, the errors and uncertainties in each step accumulate, so the distances to the farthest objects are the least well known.

INTERACTIVE ▲ FIGURE 15.11 **Extragalactic Distance Ladder** An inverted pyramid summarizes the distance techniques used to study different realms of the universe. The techniques shown in the bottom four layers—radar ranging, stellar parallax, spectroscopic parallax, and variable stars—take us as far as the nearest galaxies. To go farther, we must use other techniques—the Tully–Fisher relation and the use of standard candles, among other methods—based on distances determined by the four lowest techniques.

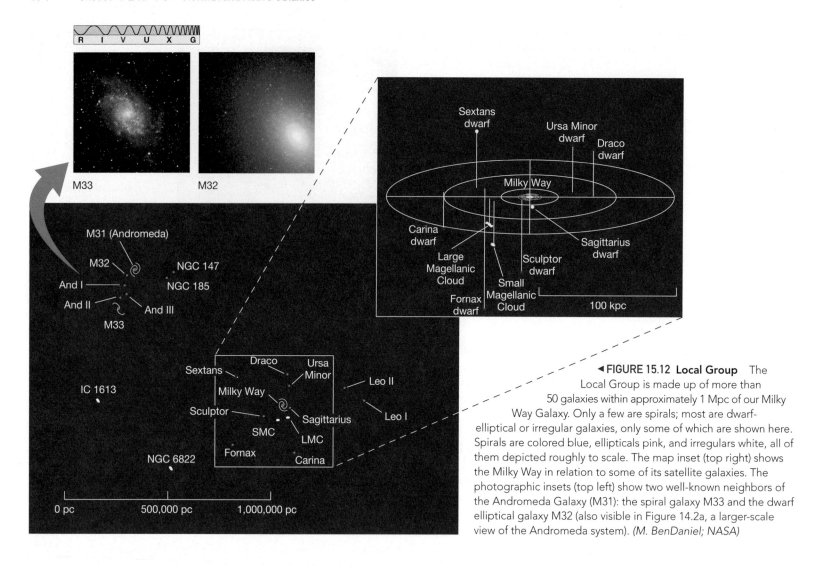

◀**FIGURE 15.12 Local Group** The Local Group is made up of more than 50 galaxies within approximately 1 Mpc of our Milky Way Galaxy. Only a few are spirals; most are dwarf-elliptical or irregular galaxies, only some of which are shown here. Spirals are colored blue, ellipticals pink, and irregulars white, all of them depicted roughly to scale. The map inset (top right) shows the Milky Way in relation to some of its satellite galaxies. The photographic insets (top left) show two well-known neighbors of the Andromeda Galaxy (M31): the spiral galaxy M33 and the dwarf elliptical galaxy M32 (also visible in Figure 14.2a, a larger-scale view of the Andromeda system). *(M. BenDaniel; NASA)*

Clusters of Galaxies

Figure 15.12 sketches the locations of all the known major astronomical objects within about 1 Mpc of the Milky Way. Our Galaxy appears with its dozen or so satellite galaxies—including the two Magellanic Clouds discussed earlier and a small companion (labeled "Sagittarius" in the figure) lying almost within our own Galactic plane. The Andromeda Galaxy, lying 800 kpc from us, is also shown, surrounded by satellites of its own. Two of Andromeda's galactic neighbors are shown in the inset to the figure. M33 is a spiral, while M32 is a dwarf elliptical, easily seen in Figure 14.2(a) to the bottom right of Andromeda's central bulge.

All told, some 55 galaxies are known to populate our Galaxy's neighborhood. Three of them (the Milky Way, Andromeda, and M33) are spirals; the remainder are dwarf irregulars and dwarf ellipticals. Together, these galaxies form the **Local Group**—a new level of structure in the universe above the scale of our Galaxy. As indicated in Figure 15.12, the Local Group's diameter is a little over 1 Mpc. The Milky Way Galaxy and Andromeda are by far its largest members, and most of the smaller galaxies are gravitationally bound to one or other of them. The combined gravity of the galaxies in the Local Group binds them together,

▶ FIGURE 15.13 **Virgo Cluster** In the central region of the Virgo Cluster of galaxies, about 17 Mpc from Earth, many large spiral and elliptical galaxies can be seen. The inset shows several galaxies surrounding the giant elliptical M86. An even bigger elliptical galaxy, M87 noted at bottom, will be discussed later in this chapter. *(M. BenDaniel; AURA)*

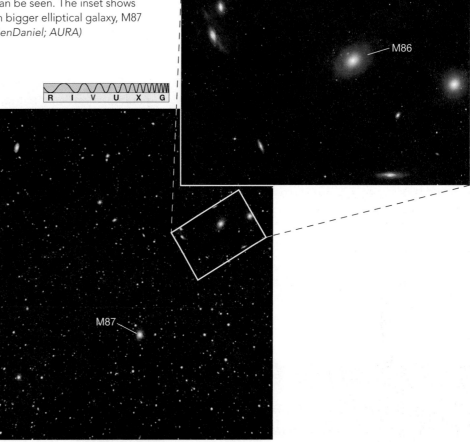

like stars in a star cluster, but on a scale a millionfold larger. More generally, a group of galaxies held together by their mutual gravitational attraction is called a **galaxy cluster.**

Moving beyond the Local Group, the next large concentration of galaxies we come to is the *Virgo Cluster* (Figure 15.13), named for the constellation in which it is found. It lies some 18 Mpc from the Milky Way. The Virgo Cluster does not contain a mere 45 galaxies, however. Rather, it houses more than 2500 galaxies, bound by gravity into a tightly knit group about 3 Mpc across.

Wherever we look in the universe we find galaxies, and most galaxies are members of galaxy clusters. Small clusters, such as the Local Group, contain only a few galaxies and are quite irregular in shape. Large, "rich" clusters such as Virgo contain thousands of individual galaxies distributed fairly smoothly in space. The Coma Cluster, shown in Figure 15.1, is another example of a rich cluster. It lies approximately 100 Mpc away. Figure 15.14 is a long-exposure photograph of a much more distant rich cluster, residing roughly 2 billion parsecs from Earth. A sizeable minority of galaxies (possibly as many as 40 percent) are not members of any cluster. They are apparently isolated systems, moving alone through intercluster space.

ANALOGY: The dynamics of galaxy clusters resemble a bunch of buzzing fireflies trapped within a jar moving away from us.

▶ FIGURE 15.14 **Distant Galaxy Cluster** The galaxy cluster Abell 1689 contains huge numbers of galaxies and resides roughly 2 billion parsecs from Earth. Virtually every patch of light in this photograph is a separate galaxy. With the most powerful telescopes, astronomers can now discern, even at this great distance, spiral structure in some of the galaxies, as well as many galaxies colliding—some tearing matter from one another, others merging into single systems. *(NASA)*

15.3 Hubble's Law

Now that we have seen some basic properties of galaxies across the universe, let's turn our attention to the large-scale *motions* of galaxies and galaxy clusters. Within a galaxy cluster, individual galaxies move more or less randomly. You might expect that, on even larger scales, the clusters themselves would also have random, disordered motion—some clusters moving this way, some that, but this is not the case. On the largest scales, galaxies and galaxy clusters alike move in a very *ordered* way.

Universal Recession

In 1912 American astronomer Vesto M. Slipher, working under the direction of Percival Lowell, discovered that virtually every spiral galaxy he observed had a redshifted spectrum—they were *receding* from our Galaxy. ∞ *(Sec. 2.7)* In fact, with the exception of a few nearby systems, *every* known galaxy is part of a general motion away from us in all directions. Individual galaxies that are not part of galaxy clusters are steadily receding. Galaxy clusters, too, have an overall recessional motion, although their individual member galaxies move randomly with respect to one another. (Consider a jar full of fireflies that has been thrown into the air. The fireflies within the jar, like the galaxies within the cluster, have random motions due to their individual whims, but the jar as a whole, like the galaxy cluster, has some directed motion as well.)

Figure 15.15 shows the optical spectra of several galaxies, arranged in order of increasing distance from the Milky Way Galaxy. These spectra are redshifted, indicating that the galaxies are receding, and the extent of the redshift increases from top to bottom in the figure. There is a connection between Doppler shift and distance: the greater the distance, the greater the redshift. This trend holds for nearly all galaxies in the universe. (Two galaxies within our Local Group, including Andromeda, and a few galaxies in the Virgo Cluster display blueshifts and so are moving toward us, but this results from their local motions within their parent clusters. Recall the fireflies in the jar.)

Figure 15.16(a) plots recessional velocity against distance for the galaxies shown in Figure 15.15. A similar plot for more galaxies within about 1 billion parsecs of Earth is shown in Figure 15.16(b). Plots like these were first made by Edwin Hubble in the 1920s and now bear his name—*Hubble diagrams*. The data points fall close to a straight line, indicating that the rate at which a galaxy recedes is *directly proportional* to its distance from us. This rule is called **Hubble's law**. We could construct such a diagram for any group of galaxies, provided we could determine their distances and velocities. The universal recession described by the Hubble diagram is called the *Hubble flow*.

The recessional motions of the galaxies prove that the cosmos is not steady and unchanging on the largest scales. The universe (actually, *space itself*—see Sec. 17.2) is expanding. But let's be clear on just *what* is expanding and what is not. Hubble's law does not mean that humans, Earth, the solar system, or even individual galaxies and galaxy clusters are physically increasing in size. These groups of atoms, rocks, planets, stars, and galaxies are held together by their own internal forces and are not themselves

▼ **FIGURE 15.15 Galaxy Spectra**   Optical spectra (left) of several galaxies shown on the right. Both the extent of the redshift (denoted by the horizontal red arrows) and the distance from the Milky Way Galaxy to each galaxy (numbers in center column) increase from top to bottom. The vertical yellow arrows denote a pair of dark absorption lines in the observed spectra. The many white vertical lines at the top and bottom of each spectrum are laboratory references. *(Adapted from Palomar/Caltech)*

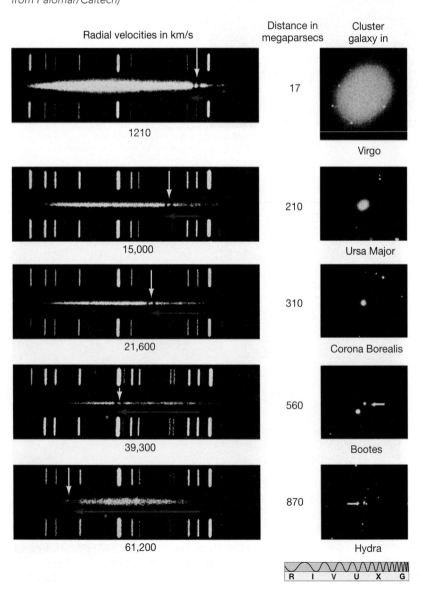

Radial velocities in km/s	Distance in megaparsecs	Cluster galaxy in
1210	17	Virgo
15,000	210	Ursa Major
21,600	310	Corona Borealis
39,300	560	Bootes
61,200	870	Hydra

R I V U X G

getting bigger. Only the largest scales in the universe—the distances separating the galaxy clusters—are expanding.

To distinguish recessional redshift from redshifts caused by motion *within* an object—for example, galaxy orbits within a cluster or explosive events in a galactic nucleus—astronomers refer to the redshift resulting from the Hubble flow as the **cosmological redshift**. Objects that lie so far away that they exhibit a large cosmological redshift are said to be at *cosmological distances*—distances comparable to the scale of the universe itself.

Hubble's law has some fairly dramatic implications. If nearly all galaxies show recessional velocities according to Hubble's law, then doesn't that mean that they all started their journey from a single point? If we could run time backward, wouldn't all the galaxies fly back to this one point, perhaps the site of some explosion in the remote past? The answer is yes—but not in the way you might expect! In Chapter 17 we will explore the ramifications of the Hubble flow for the past and future evolution of our universe. For now, however, we set aside its cosmic implications and use Hubble's law simply as a convenient distance-measuring tool.

Hubble's Constant

The constant of proportionality between recessional velocity and distance in Hubble's law is known as **Hubble's constant**, denoted by the symbol H_0. The data shown in Figure 15.16 then obey the equation:

$$\text{recessional velocity} = H_0 \times \text{distance}.$$

The value of Hubble's constant is the slope of the straight line—recessional velocity divided by distance—in Figure 15.16(b). Reading the numbers off the graph, this comes to roughly 70,000 km/s divided by 1000 Mpc, or 70 km/s/Mpc (kilometers per second per megaparsec, the most commonly used unit for H_0). Astronomers continually strive to refine the accuracy of the Hubble diagram and the resulting estimate of H_0 because Hubble's constant is one of the most fundamental quantities of nature—it specifies the rate of expansion of the entire cosmos.

Hubble's original value for H_0 was about 500 km/s/Mpc, far higher than the currently accepted value. This overestimate was due almost entirely to errors in the cosmic distance scale at the time, particularly the calibrations of Cepheid variables and standard candles. The measured value dropped rapidly as various observational errors were recognized and resolved and distance measurements became more reliable. Published estimates of H_0 entered the "modern" range (within, say, 20 percent of the current value) in roughly the mid-1960s.

As measurement techniques have continued to improve, the uncertainty in the result has steadily decreased to the point that now, early in the 21st century, all leading measurements of H_0, determined by a variety of different techniques—Tully–Fisher measurements, studies of Cepheid variables in the Virgo cluster, and observations of standard candles, such as Type I supernovae—are all remarkably consistent with one another. We will adopt a rounded-off value of $H_0 = 70$ km/s/Mpc (a choice roughly in the middle of all recent results and also in line with some precise cosmological measurements to be discussed in Chapter 17) as the best current estimate of Hubble's constant for the remainder of the text.

The Top of the Distance Ladder

Using Hubble's law, we can derive the distance to a remote object simply by measuring the object's recessional velocity and dividing by Hubble's constant. Hubble's law thus tops our inverted pyramid of distance-measurement techniques (Figure 15.17). This seventh method simply assumes that Hubble's law holds. If this assumption

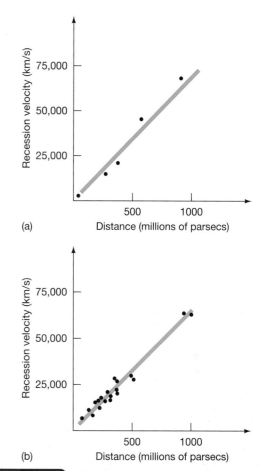

INTERACTIVE ▲ **FIGURE 15.16 Hubble's Law** Plots of recessional velocity against distance (a) for the galaxies shown in Figure 15.15 and (b) for numerous other galaxies within about 1 billion pc of Earth.

CONCEPT CHECK ► How does the use of Hubble's law differ from the other extragalactic distance-measurement techniques we have seen in this text?

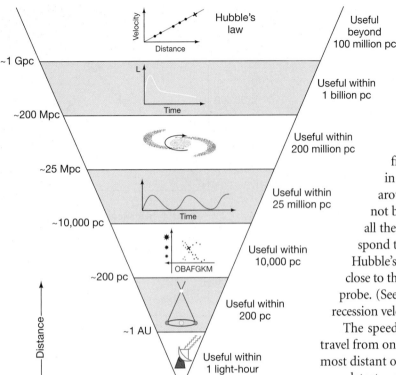

INTERACTIVE ▲ FIGURE 15.17 **Cosmic Distance Ladder** Hubble's law tops the hierarchy of distance-measurement techniques. It is used to find the distances of astronomical objects all the way out to the limits of the observable universe.   (MA)

is correct, then Hubble's law enables us to measure great distances in the universe—so long as we can obtain an object's spectrum, we can determine how far away it is.

The precise astronomical definition of *redshift* is the fractional increase in the wavelength of a beam of radiation due to the source's motion away from us. For many nearby objects, recession velocities are small compared to the speed of light, and the redshifts are correspondingly low—just a few percent. ∞ *(Sec. 2.7)* However, more distant objects can have recessional motions that are substantial fractions of the speed of light. The most distant objects thus far observed in the universe—some young galaxies and quasars—have redshifts of around 8, meaning that their radiation has been stretched in wavelength not by a few percent, but *ninefold.* Their ultraviolet spectral lines are shifted all the way into the infrared part of the spectrum! Such huge redshifts correspond to recessional velocities of more almost 98 percent of the speed of light. Hubble's law implies that such objects lie more than 9000 Mpc away from us, as close to the limits of the observable universe as astronomers have yet been able to probe. (See *More Precisely 15-1* on p. 422, for a fuller discussion of redshifts and recession velocities comparable to the speed of light.)

The speed of light is finite. It takes time for light or any kind of radiation to travel from one point in space to another. The radiation that we now see from these most distant objects originated long ago. Incredibly, the radiation that astronomers now detect was emitted roughly billion years ago, well before our planet, our Sun, and perhaps even our Galaxy came into being.

15.4 Active Galactic Nuclei

The galaxies described in Section 15.1—those falling into the various Hubble classes—are generally referred to as **normal galaxies.** As we have seen, their luminosities range from a million or so times that of the Sun for dwarf ellipticals and irregulars to more than a trillion solar luminosities for the largest giant ellipticals. For comparison, in round numbers, the luminosity of the Milky Way Galaxy is 2×10^{10} solar luminosities, or roughly 10^{37} W.

In these last two sections we focus our attention on "bright" galaxies, conventionally taken to mean galaxies with luminosities more than about 10^{10} times the solar value. By this measure, our Galaxy is bright, but not abnormally so.

Galactic Radiation

A substantial fraction of bright galaxies—perhaps as many as 40 percent—don't fit well into the "normal" category. Their *spectra* differ significantly from those of their normal cousins, and their *luminosities* can be extremely large. Known collectively as **active galaxies,** they are of great interest to astronomers. The brightest among them are the most energetic objects known in the universe, and all may represent an important, if intermittent, phase of galactic evolution (see Section 16.4). At optical wavelengths, active galaxies often *look* like normal galaxies—familiar components such as disks, bulges, stars, and dark dust lanes can be identified. At other wavelengths, however, their unusual properties are much more apparent.

As illustrated schematically in Figure 15.18, most of a normal galaxy's energy is emitted in or near the visible portion of the electromagnetic spectrum, much like the radiation from stars. Indeed, to a large extent, the light we see from a normal galaxy *is* just the accumulated light of its many component stars (once the effects of interstellar dust are taken into account), each described approximately

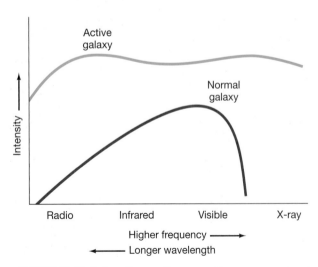

▲ FIGURE 15.18 **Galaxy Energy Spectra** The energy emitted by a normal galaxy differs significantly from that emitted by an active galaxy. This plot illustrates the general spread of radiation intensity for all galaxies of a particular type and does not represent any one individual galaxy.

by a blackbody curve. ∞ *(Sec. 2.4)* By contrast the radiation from active galaxies does *not* peak in the visible portion of the spectrum. Most active galaxies do emit substantial amounts of visible radiation, but far more of their energy is emitted at invisible wavelengths, both longer and shorter than those in the visible range. Put another way, the radiation from active galaxies is *inconsistent* with what we would expect if it were the combined radiation of myriad stars. Their radiation is said to be *nonstellar*.

Many luminous galaxies with nonstellar emission are known to be **starburst galaxies**—previously normal systems currently characterized by widespread episodes of star formation, most likely as a result of interactions with a neighbor. The irregular galaxy NGC 1569 shown in Figure 15.8(b) is a prime example. We will study these important systems and their role in galaxy evolution in Chapter 16. For purposes of this text, however, we will use the term "active galaxy" to mean a system whose abnormal activity is related to violent events occurring in or near the galactic nucleus. The nuclei of such systems are called **active galactic nuclei**.

Astronomers have identified and cataloged a bewildering array of systems falling into the "active" category. For example, Figure 15.19 shows an active galaxy exhibiting both nuclear activity and widespread star formation, with a blue-tinted ring of newborn stars surrounding an extended 1-kpc-wide core of intense emission. Rather than attempting to describe the entire "zoo" of active galaxies, we will confine ourselves to three basic types: the energetic *Seyfert galaxies* and *radio galaxies* and the even more luminous *quasars*. Their properties will allow us to identify and discuss features common to active galaxies in general.

The association of galactic activity with the galactic nucleus is reminiscent of the discussion of the center of the Milky Way in Chapter 14. ∞ *(Sec. 14.7)* In our Galaxy, it seems that the activity in the nucleus is associated with the central *supermassive black hole,* whose presence is inferred from observations of stellar orbits in the innermost fraction of a parsec. As we will see, most astronomers think that basically the same thing occurs in the nuclei of active galaxies and that "normal" and "active" galaxies may differ principally in the degree to which the nonstellar nuclear component of the radiation outshines the light from the rest of the galaxy. This is a powerful unifying theme for understanding the evolution of galaxies, and we will return to it in Chapter 16. For the remainder of this chapter, we concentrate on describing the properties of active galaxies and the black holes that power them.

Seyfert Galaxies

In 1943 Carl Seyfert, an American optical astronomer studying spiral galaxies from Mount Wilson Observatory, discovered the type of active galaxy that now bears his name. **Seyfert galaxies** are a class of astronomical objects whose properties lie between normal galaxies and the most energetic active galaxies known.

Superficially, Seyferts resemble normal spiral galaxies (Figure 15.20). Indeed, the stars in a Seyfert's galactic disk and spiral arms produce about the same amount of visible radiation as the stars in a normal spiral galaxy. However, most of a Seyfert's energy is emitted from the galactic nucleus the center of the overexposed white patch in Figure 15.20. The nucleus of a Seyfert galaxy is some 10,000 times brighter than the center of our own Galaxy. In fact, the brightest Seyfert nuclei are 10 times more energetic than the *entire* Milky Way.

Some Seyferts produce radiation spanning a broad range in wavelengths, from the infrared all the way through ultraviolet and even X-rays. However, the majority (about 75 percent) emit most of their energy in the infrared. Scientists think that much of the high-energy radiation in these Seyferts is absorbed by dust in or near the nucleus, then reemitted as infrared radiation.

▲ FIGURE 15.19 **Active Galaxy** This image of the galaxy NGC 7742 resembles a fried egg, with a ring of blue star-forming regions surrounding a very bright yellow core that spans about 1 kpc. This active galaxy combines star formation with intense emission from its central nucleus and lies roughly 24 Mpc away. *(NASA)*

CONCEPT CHECK ▶ The energy emission from an active galactic nucleus does not resemble a blackbody curve. Why is this important?

▲ FIGURE 15.20 **Seyfert Galaxy** The Circinus Galaxy, a Seyfert with a bright compact core, lies some 4 Mpc away. It is one of the closest active galaxies. *(NASA)*

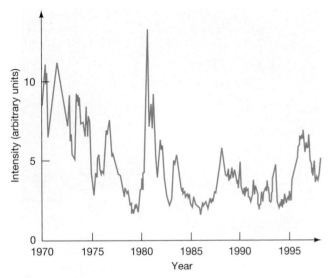

▲ FIGURE 15.21 Seyfert Time Variability This graph illustrates the irregular variations of the luminosity of Seyfert galaxy 3C 84 over three decades. These observations were made in the radio part of the electromagnetic spectrum; the optical and X-ray luminosities vary as well. *(NRAO)*

(a)

▲ FIGURE 15.22 Centaurus A Radio Lobes Radio galaxies, such as Centaurus A (a), often have giant radio-emitting lobes (b) extending a million light-years or more beyond the central galaxy. The lobes cannot be imaged in visible light and must be observed with radio telescopes; they are shown here in false color, with decreasing intensity from red to yellow to green to blue. The inset is a *Chandra* X-ray image of one of the lobes, showing that the jets in the inner parts of the lobes also emit higher-energy radiation. *(ESO; NRAO; SAO)*

Seyfert spectral lines have many similarities to those observed toward the center of our own Galaxy. ∞ *(Sec. 14.7)* The lines are very broad, most likely indicating rapid (5000 km/s or more) internal motion within the nuclei. ∞ *(Sec. 2.8)* In addition, the energy emission often varies in time (Figure 15.21). A Seyfert's luminosity can double or halve within a fraction of a year. These rapid fluctuations in luminosity lead us to conclude that the source of energy emissions must be quite compact—simply put, an object cannot "flicker" in less time than radiation takes to cross it. ∞ *(Sec. 13.4)* The emitting region must therefore be less than 1 light-year across—an extraordinarily small region, considering the amount of energy emanating from it.

Taken together, the rapid time variability and large radio and infrared luminosities observed in Seyfert galaxies imply violent nonstellar activity in their nuclei. As just mentioned, this activity is probably similar in *nature* to processes occurring at the center of our own Galaxy, but its *magnitude* is thousands of times greater than the comparatively mild events within our own Galaxy's heart. ∞ *(Sec. 14.7)*

Radio Galaxies

As the name suggests, **radio galaxies** are active galaxies that emit large amounts of energy in the radio portion of the electromagnetic spectrum. They differ from Seyferts not only in the wavelengths at which they radiate, but also in both the appearance and the extent of their emitting regions.

Figure 15.22(a) shows the radio galaxy Centaurus A, which lies about 4 Mpc from Earth. Almost none of this galaxy's radio emission comes from a compact nucleus. Instead, the energy is released from two huge extended regions called **radio lobes**—roundish clouds of gas spanning about half a megaparsec, lying well beyond the visible galaxy.[1] Undetectable in visible light, the radio lobes of radio galaxies are truly enormous. From end to end, they typically span more than 10 times the size of the Milky Way Galaxy, comparable in scale to the entire Local Group.

---

[1]The term *visible galaxy* is commonly used to refer to those components of an active galaxy that emit visible "stellar" radiation, as opposed to the nonstellar and invisible "active" component of the galaxy's emission.

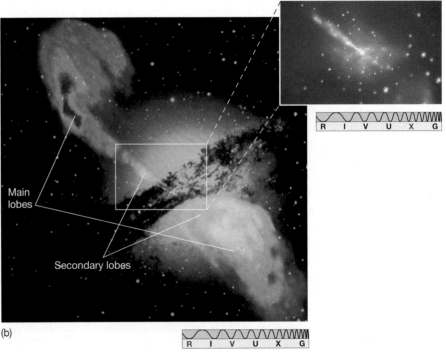

Main lobes

Secondary lobes

(b)

▶ FIGURE 15.23 **Cygnus A** (a) A visible-light image of Cygnus A appears to show two galaxies in collision. (b) On a much larger scale, it displays radio-emitting lobes (mapped in blue) on both sides of the visible image. The galaxy in (a) is about the size of the small dot at the center of (b). The distance from one lobe to the other is approximately a million light-years. (NOAO; NRAO)

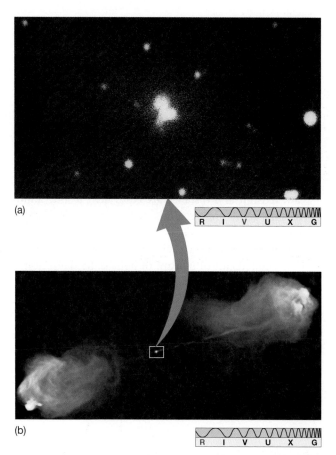

(a)

(b)

Figure 15.22(b) shows the relationship between the galaxy's visible, radio, and X-ray emission. In visible light, Centaurus A is apparently a large E2 galaxy some 500 kpc in diameter, bisected by an irregular band of dust. Centaurus A is a member of a small cluster of galaxies, and numerical simulations suggest that this peculiar galaxy is probably the result of a collision between an elliptical galaxy and a smaller spiral galaxy about 500 million years ago (see Chapter 16). The radio lobes are roughly symmetrically placed, jutting out from the center of the visible galaxy, roughly perpendicular to the dust lane, suggesting that they consist of material ejected in opposite directions from the galactic nucleus. This conclusion is strengthened by the presence of a pair of smaller secondary lobes closer to the visible galaxy and by the presence of a roughly 1-kpc-long jet of matter in the galactic center, all aligned with the main lobes (and marked in the figure).

If the material was ejected from the nucleus at close to the speed of light, then Centaurus A's outer lobes were created a few hundred million years ago, quite possibly around the time of the elliptical/spiral merger thought to be responsible for the galaxy's odd optical appearance. The secondary lobes were expelled more recently. Apparently, some violent process at the center of Centaurus A—most probably triggered by the collision—started up around that time and has been intermittently firing jets of matter out into intergalactic space ever since.

Centaurus A is a relatively low-luminosity source that happens to lie very close to us, astronomically speaking, making it particularly easy to study. Figure 15.23 shows a much more powerful emitter, called Cygnus A, lying roughly 250 Mpc from Earth. The high-resolution radio map in Figure 15.23(b) clearly shows two narrow, high-speed jets joining the radio lobes to the center of the visible galaxy (the dot at the center of the radio image). Notice also that, as with Centaurus A, Cygnus A is a member of a small group of galaxies, and the optical image (Figure 15.23a) appears to show a galaxy collision in progress.

The radio lobes of the brightest radio galaxies (such as Cygnus A) emit roughly 10 times more energy than the Milky Way Galaxy at all wavelengths, coincidentally about the same as the most luminous Seyfert nuclei. However, despite their names, radio galaxies actually radiate far more energy at shorter wavelengths. Their total energy output can be 100 times or more greater than their radio emission. Most of this energy comes from the nucleus of the visible galaxy. Having total luminosities up to 1000 times that of the Milky Way Galaxy, bright radio galaxies are among the most energetic objects known in the universe. Their radio emission is important to astronomers because it lets them study in detail the connection between the small-scale nucleus and the large-scale radio lobes.

Not all radio galaxies have obvious radio lobes. Figure 15.24 shows a *core-dominated* radio galaxy, most of whose energy is emitted from a small central nucleus (which radio astronomers refer to as the *core*) less than 1 pc across. Weaker radio emission comes from an extended region surrounding the nucleus. Probably all radio galaxies have jets and lobes, but what we observe depends on our perspective. As illustrated in Figure 15.25, when we view the radio galaxy from the side we see the jets and lobes. However, if we view the jet almost head-on—in other words, looking *through* the lobe—we see a core-dominated system.

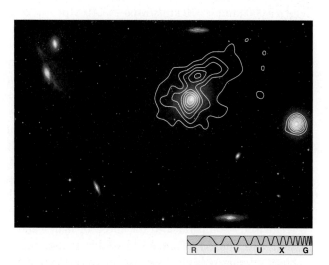

▲ FIGURE 15.24 **Core-Dominated Radio Galaxy** As shown by this radio contour map of the galaxy M86, the radio emission comes from a bright central nucleus, which is surrounded by an extended region of less intense emission. The radio map is superimposed on an optical image of the galaxy and some of its neighbors, a wider-field version of which was shown previously in Figure 15.13. (Harvard-Smithsonian Center for Astrophysics)

15-1 MORE PRECISELY

Relativistic Redshifts and Look-Back Time

When discussing very distant objects, astronomers usually talk about their redshifts rather than their distances. Indeed, it is very common for researchers to speak of an event occurring "at" a certain redshift—meaning that the light received today from that event is redshifted by the specified amount. Of course, because of Hubble's law, redshift and distance are equivalent to one another. However, redshift is the preferred quantity because it is a directly observable property of an object, whereas distance is derived from redshift using Hubble's constant, whose value is not accurately known.

The redshift of a beam of light is, by definition, the *fractional* increase in its wavelength resulting from the recessional motion of the source. ⚬ *(Sec. 2.7)* A redshift of 1 corresponds to a *doubling* of the wavelength. Using the formula for the Doppler shift given previously, the redshift of radiation received from a source moving away from us with speed v is

$$\text{redshift} = \frac{\text{observed wavelength} - \text{true wavelength}}{\text{true wavelengh}}$$

$$= \frac{\text{recessional velocity, } v}{\text{speed of light, } c}.$$

Let's illustrate this with two examples, rounding the speed of light, c, to 300,000 km/s. A galaxy at a distance of 100 Mpc has a recessional speed (by Hubble's law) of 70 km/s/Mpc × 100 Mpc = 7000 km/s. Its redshift therefore is 7000 km/s ÷ 300,000 km/s = 0.023. Conversely, an object that has a redshift of 0.05 has a recessional velocity of 0.05 × 300,000 km/s = 15,000 km/s and hence a distance of 15,000 km/s ÷ 70 km/s/Mpc = 210 Mpc.

Unfortunately, while the foregoing equation is correct for low speeds, it does not take into account the effects of relativity. As we saw in Chapter 13, the rules of everyday physics have to be modified when speeds begin to approach the speed of light, and the formula for the Doppler shift is no exception. ⚬ *(More Precisely 13-1)* Our formula is valid for speeds much less than the speed of light, but when $v = c$, the redshift is not 1, as the equation suggests, but is in fact *infinite*. Radiation received from an object moving away from us at nearly the speed of light is redshifted to almost infinite wavelength.

Thus, do not be alarmed to find that many galaxies and quasars have redshifts greater than 1. This does not mean that they are receding faster than light! It simply means that their recessional speeds are relativistic—comparable to the speed of light—and the preceding simple formula is not applicable. Table 15.2 presents a conversion chart relating redshift, recession speed, and present distance. The column headed "v/c" gives recession velocities based on the Doppler effect, taking relativity properly into account (but see Sec. 17.2 for a more correct interpretation of the redshift). All values are based on reasonable assumptions and are usable even for $v \approx c$. We take Hubble's constant to be 70 km/s/Mpc and assume a flat universe (see

Sec. 17.3 for details) in which matter (mostly dark) contributes just over one-quarter of the total density. The conversions in the table are used consistently throughout this text.

Because the universe is expanding, the "distance" to a galaxy is not very well defined—do we mean the distance when the galaxy emitted the light we see today or the present distance (as listed in the table), even though we do not see the galaxy as it is today, or is some other measure more appropriate? Largely because of this ambiguity, astronomers prefer to work in terms of a quantity known as the *look-back time* (last column of Table 15.2), which is simply how long ago an object emitted the radiation we see today. While astronomers talk frequently about redshifts and sometimes about look-back times, they hardly ever talk of distances to high-redshift objects.

For nearby sources, the look-back time is numerically equal to the distance in light-years—the light we receive tonight from a galaxy at a distance of 100 million light-years was emitted 100 million years ago. However, for more distant objects, the look-back time and the present distance in light-years differ because of the expansion of the universe, and the divergence increases dramatically with increasing redshift (see Sec. 17.2). For example, a galaxy now located 15 billion light-years from Earth was much closer to us when it emitted the light we now see. Consequently, its light has taken considerably less than 15 billion years—in fact, a little less than 10 billion years—to reach us.

TABLE 15.2 Redshift, Distance, and Look-Back Time

Redshift	v/c	Present Distance (Mpc)	Present Distance (10^6 light-years)	Look-Back Time (millions of years)
0.000	0.000	0	0	0
0.010	0.010	42	137	137
0.025	0.025	105	343	338
0.050	0.049	209	682	665
0.100	0.095	413	1,350	1,290
0.250	0.220	999	3,260	2,920
0.500	0.385	1,880	6,140	5,020
1.000	0.600	3,320	10,800	7,730
2.000	0.800	5,250	17,100	10,300
3.000	0.882	6,460	21,100	11,500
4.000	0.923	7,310	23,800	12,100
5.000	0.946	7,940	25,900	12,500
6.000	0.960	8,420	27,500	12,700
8.000	0.976	9,150	29,800	13,000
10.000	0.984	9,660	31,500	13,200
20.000	0.995	11,000	35,900	13,500
100.000	1.000	12,900	42,200	13,700
∞	1.000	14,600	47,500	13,700

Our precise location with respect to the jet can also radically affect the *type* of radiation we see. The theory of relativity tells us that radiation emitted by particles moving close to the speed of light is strongly concentrated, or beamed, in the direction of motion. ⚭ *(More Precisely 13-1)* As a result, if the observer in Figure 15.25 happens to be directly in line with the beam, the radiation received will be both very intense and Doppler shifted toward short wavelengths. ⚭ *(Sec. 2.7)* The resulting object is called a **blazar.** Much of the luminosity of the several hundred known blazars is received in the form of X-rays or gamma rays.

Jets are common in active galaxies of all types. Figure 15.26 presents several images of the giant elliptical galaxy M87, a prominent member of the Virgo Cluster (see Figure 15.13). A long time exposure (Figure 15.26a) shows a large, fuzzy ball of light—a fairly normal-looking E1 galaxy about 100 kpc across. A shorter exposure of M87 (Figure 15.26b), capturing only the galaxy's bright inner regions, reveals a long (2 kpc) thin jet of matter ejected from the galactic center at nearly the speed of light. Computer enhancement shows that it is made up of a series of distinct "blobs" more or less evenly spaced along its length, suggesting that the material was ejected during bursts of activity. The jet has also been imaged in the radio, infrared (Figure 15.26c), and X-ray regions of the spectrum.

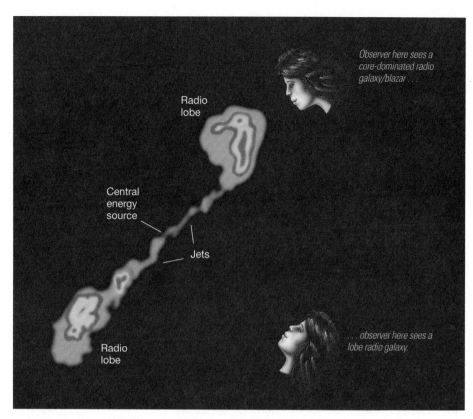

▲ **FIGURE 15.25 Radio Galaxy** A central energy source produces high-speed jets of matter that interact with intergalactic gas to form radio lobes. The system may appear to us as either radio lobes or a core-dominated radio galaxy, depending on our location with respect to the jets and lobes.

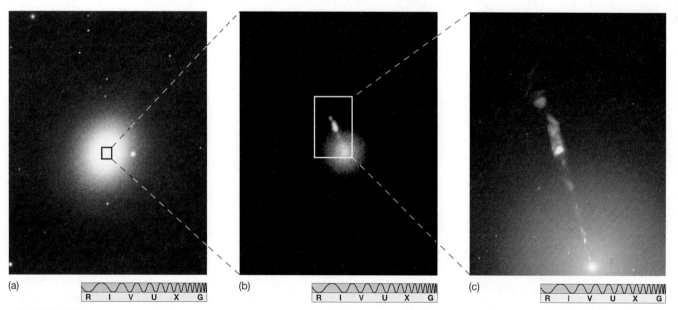

INTERACTIVE ▲ **FIGURE 15.26 M87 Jet** The giant elliptical galaxy M87 (also called Virgo A) is displayed here in a series of zooms. (a) A long optical exposure of the galaxy's halo and embedded central region. (b) A short optical exposure of its core and an intriguing jet of matter, on a smaller scale. (c) An infrared image of M87's jet, examined more closely than in (b). The bright point at bottom in (c) marks the nucleus of the galaxy. *(NOAO; AURA; NASA)*

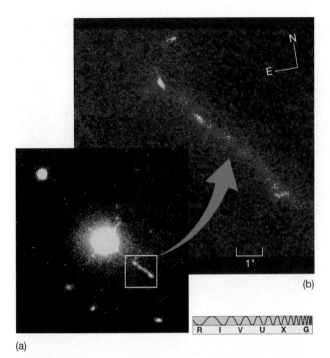

▲ FIGURE 15.27 Quasar 3C 273   (a) The bright quasar 3C 273 displays a luminous jet of matter, but the main body of the quasar is starlike in appearance. (b) The jet extends for about 30 kpc and can be seen better in this high-resolution image. *(AURA)*

Quasars

In the early days of radio astronomy, many radio sources were detected for which no corresponding visible object was known. By 1960 several hundred such sources were listed in the *Third Cambridge Catalog* and astronomers were scanning the skies in search of visible counterparts to these radio sources. Their job was made difficult both by the low resolution of the radio observations (which meant that the observers did not know exactly where to look) and by the faintness of these objects at visible wavelengths.

In 1960 astronomers detected what appeared to be a faint blue star at the location of the radio source 3C 48 (the 48th object on the third *Cambridge* list) and obtained its spectrum. Containing many unknown broad emission lines, the unusual spectrum defied interpretation. 3C 48 remained a unique curiosity until 1962, when another similar-looking, and similarly mysterious, faint blue object with "odd" spectral lines was discovered and identified with the radio source 3C 273 (Figure 15.27).

The following year saw a breakthrough when astronomers realized that the strongest unknown lines in 3C 273's spectrum were simply familiar spectral lines of hydrogen redshifted by a very unfamiliar amount—about 16 percent, corresponding to a recession velocity of 48,000 km/s. Figure 15.28 shows the spectrum of 3C 273, with prominent emission lines and their redshifts marked. Once the nature of the strange spectral lines was known, astronomers quickly found a similar explanation for the spectrum of 3C 48. Its 37 percent redshift implied that it is receding from Earth at almost one-third the speed of light.

These huge speeds mean that neither of these two objects can possibly be members of our Galaxy. In fact, their large redshifts mean that they are very far away indeed. Applying Hubble's law (with our adopted value of the Hubble constant $H_0 = 70$ km/s/Mpc), we obtain distances of 650 Mpc for 3C 273 and 1400 Mpc for 3C 48. (See *More Precisely 15-1* and Chapter 17 for an account of how these distances are determined and an explanation of what these large redshifts really mean.)

However, this explanation of the unusual spectra created a greater mystery. A simple calculation using the inverse-square law reveals that despite their unimpressive optical appearance these faint "stars" are in fact the brightest known objects in the universe! ⚬ *(Sec. 10.2)* 3C 273, for example, has a luminosity of about 10^{40} W, comparable to 20 trillion Suns or 1000 Milky Way Galaxies. More generally, quasars range in luminosity from around 10^{38} W—about the same as the brightest Seyferts—up to nearly 10^{42} W. A value of 10^{40} W (comparable to the luminosity of a bright radio galaxy) is fairly typical.

Clearly not stars (with such enormous luminosities), these objects became known as *quasi-stellar radio sources* (quasi-stellar means "starlike"), or **quasars.** Because we now know that not all such highly redshifted, starlike objects are strong radio sources, the term **quasi-stellar object** (or QSO) is more common today. However, the name quasar persists, and we will continue to use it here. More than 200,000 quasars are now known, and the numbers are increasing rapidly as large-scale surveys probe deeper and deeper into space. The distance to the *closest* quasar is 240 Mpc; the farthest lies some 9000 Mpc away. Most quasars lie well over 1000 Mpc

This is the observed spectrum of 3C 273.

Hβ   Hγ   Hδ

Redshift

Red

Blue

This is a lab comparison spectrum.

Hβ   Hγ   Hδ

600 nm   500 nm   400 nm

◄ FIGURE 15.28 Quasar Spectrum   Optical spectrum of the distant quasar 3C 273. (This is a negative, as the lines are actually in emission.) Notice the redshift of the three hydrogen spectral lines marked as Hβ, Hγ, and Hδ. ⚬ *(Sec. 2.6)* The widths of these lines imply rapid internal motion within the quasar. *(Adapted from Palomar/Caltech)* from Earth. Since light travels at a finite speed, these faraway objects represent the universe as it was in the distant past. Most quasars date back to much earlier periods of galaxy formation and evolution, rather than more recent times. The prevalence of these energetic objects at great distances tells us that the universe was once a much more violent place than it is today.

Quasars share many properties with Seyferts and radio galaxies. Their radiation is nonstellar and may vary irregularly in brightness over periods of months, weeks, days, or (in some cases) even hours. Some show evidence of jets and extended emission features. Note the luminous jet in 3C 273 (Figure 15.27), reminiscent of the jet in M87 and extending nearly 30 kpc from the center of the quasar. Figure 15.29 shows a quasar with radio lobes similar to those in Cygnus A (Figure 15.23b). Quasars have been observed in all parts of the electromagnetic spectrum, but many emit most of their energy in the infrared. About 10–15 percent of quasars (called "radio-loud" quasars) also emit substantial amounts of energy at radio wavelengths, presumably as a result of unresolved jets.

Astronomers once distinguished between active galaxies and quasars on the basis of their appearance, spectra, and distance from us, but today most astronomers think that quasars are in fact just the intensely bright nuclei of distant active galaxies lying too far away for the galaxies themselves to be seen. (Figure 16.17 presents *Hubble Space Telescope* observations of two relatively nearby quasars in which the surrounding galaxies are clearly visible.)

15.5 The Central Engine of an Active Galaxy

The present consensus among astronomers is that, despite their differences in appearance and luminosity, Seyferts, radio galaxies, and quasars share a common energy-generation mechanism. As a class, active galactic nuclei have some or all of the following properties:

1. They have *high luminosities,* generally greater than the 10^{37} W characteristic of a bright normal galaxy.
2. Their energy emission is mostly *nonstellar*—it cannot be explained as the combined radiation of even trillions of stars.
3. Their energy output can be highly *variable,* implying that it is emitted from a small central nucleus much less than a parsec across.
4. They often exhibit *jets* and other signs of explosive activity.
5. Their optical spectra may show broad emission lines, indicating *rapid internal motion* within the energy-producing region.
6. Often the activity appears to be associated with *interactions* between galaxies.

The principal questions then are: How can such vast quantities of energy arise from these relatively small regions of space? Why is the radiation nonstellar? And what is the origin of the jets and extended radio-emitting lobes? We first consider how the energy is *produced,* then turn to the question of how it is actually *emitted* into intergalactic space.

Energy Production

As illustrated in Figure 15.30, the leading model for the central engines of active galaxies is a scaled-up version of the process powering X-ray binaries in our Galaxy and the activity in the Galactic center—accretion of gas onto a supermassive black hole, releasing huge amounts of energy as the matter sinks onto the central object. ∞ *(Secs. 13.3, 13.8, 14.7)* In order to power the brightest active galaxies, theory suggests that the black holes involved must be *billions* of times more massive than the Sun.

PROCESS OF ▶
SCIENCE CHECK How did the determination of quasar distances change astronomers' understanding of these objects?

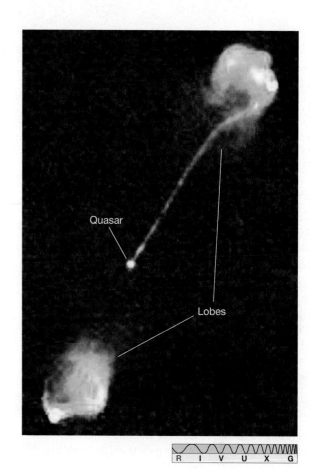

R I V U X G

▲ **FIGURE 15.29 Quasar Jets** This radio image of the quasar 3C 175, which is some 3000 Mpc away, shows radio jets feeding faint radio lobes. The lobes span approximately a million light-years—comparable in size to the radio galaxies discussed earlier (see also the chapter opener on page 404). *(NRAO)*

Quasar

Lobes

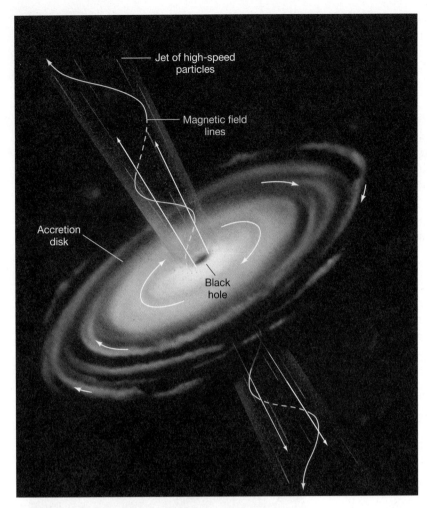

Jet of high-speed
particles

Magnetic field
lines

Accretion
disk

Black
hole

▲ FIGURE 15.30 **Active Galactic Nucleus**   The leading theory for the energy source in active galactic nuclei holds that these objects are powered by material accreting onto a supermassive black hole. As matter spirals toward the hole, it heats up, producing large amounts of energy. High-speed jets of gas may be ejected perpendicular to the accretion disk, forming the jets and lobes seen in many active objects. Magnetic fields generated in the disk by charged matter in motion are carried by the jets out to the radio lobes, where they play a crucial role in producing the observed radiation.

ANIMATION/VIDEO Cosmic Jets

As with this model's smaller-scale counterparts, infalling gas forms an accretion disk and spirals down toward the black hole. It is heated to high temperatures by friction within the disk and emits large amounts of radiation as a result. In this case, however, the origin of the accreted gas is not a binary companion, as in stellar X-ray sources, but rather entire stars and clouds of interstellar gas, most likely diverted into the galactic center by an encounter with another galaxy, that come too close to the hole and are torn apart by its strong gravity.

Accretion is extremely efficient at converting infalling mass (in the form of gas) into energy (in the form of electromagnetic radiation). Detailed calculations indicate that as much as 10 or 20 percent of the total mass–energy of the infalling matter can be radiated away before it crosses the hole's event horizon and is lost forever. ∞ *(Sec. 13.5)* Since the total mass–energy of a star like the Sun—the mass times the speed of light squared—is about 2×10^{47} J, it follows that the 10^{38} W luminosity of a bright active galaxy can be accounted for by the consumption of "only" 1 solar mass of gas per decade by a billion-solar-mass black hole. More or less luminous active galaxies would require correspondingly more or less fuel. Thus, to power a 10^{40}-W quasar, which is 100 times brighter, the black hole simply consumes 100 times more fuel, or 10 stars per year. The central black hole of a 10^{36}-W Seyfert galaxy would devour only one Sun's worth of material every thousand years.

The small size of the emitting region is a direct consequence of the compact central black hole. Even a billion-solar-mass black hole has a radius of only 3×10^9 km, or 10^{-4} pc—about 20 AU. Theory suggests that the part of the accretion disk responsible for most of the emission would be much less than 1 pc across. ∞ *(Sec. 13.5)* Instabilities in the accretion disk can cause fluctuations in the energy released, leading to the variability observed in many objects. The broadening of the spectral lines observed in the nuclei of many active galaxies results from the rapid orbital motion of the gas in the black hole's intense gravity.

Jets appear to be a common feature of accretion flows, large and small, and consist of material (mainly protons and electrons) blasted out into space—and completely out of the visible portion of the galaxy—from the inner regions of the disk. They are most likely formed by strong magnetic fields produced within the accretion disk itself, which accelerate charged particles to nearly the speed of light and eject them parallel to the disk's rotation axis. Figure 15.31 shows a *Hubble Space Telescope* image of a disk of gas and dust at the core of the radio galaxy NGC 4261 in the Virgo Cluster. Consistent with the theory just described, the disk is perpendicular to the huge jets emanating from the galaxy's center.

Figure 15.32 shows further evidence for this model in the form of imaging and spectroscopic data from the center of M87, suggesting a rapidly rotating disk of matter orbiting the galaxy's center, again perpendicular to the jet. Measurements of the gas velocity on opposite sides of the disk indicate that the mass within a few parsecs of the center is approximately 3×10^9 solar masses—we assume that this is the mass of the central black hole. At M87's distance, *HST*'s resolution of 0.05 arc second corresponds to a scale of about 5 pc, so we are still far from seeing the (solar system–sized) central black hole itself, but the improved "circumstantial" evidence has convinced many astronomers of the basic correctness of the theory.

▶ **FIGURE 15.31 Giant Elliptical Galaxy**   (a) A combined optical/radio image of the giant elliptical galaxy NGC 4261, in the Virgo Cluster, shows a white visible galaxy at center, from which blue-orange (false color) radio lobes extend for about 60 kpc. (b) A close-up of the galaxy's nucleus reveals a 100-pc-diameter disk surrounding a bright hub thought to harbor a black hole. *(NRAO; NASA)*

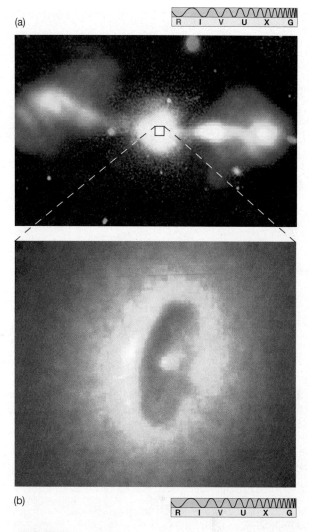

Energy Emission

Theory suggests that the radiation emitted by the hot accretion disk surrounding a supermassive black hole would span a broad range of wavelengths, from infrared through X-rays, corresponding to the broad range of temperatures in the disk as the gas heats up. This accounts for the spectra of some active galactic nuclei, but as mentioned earlier, it appears that in many cases the high-energy radiation emitted from the accretion disk itself is "reprocessed"—that is, absorbed and reemitted at longer, particularly infrared, wavelengths—by dust surrounding the nucleus before eventually reaching our detectors.

Researchers think that the most likely site of this reprocessing is in a rather fat, donut-shaped ring of gas and dust surrounding the inner accretion disk where the energy is actually produced. As illustrated in Figure 15.33, if our line of sight to the black hole does not intersect the dusty donut, we see the "bare" energy source, emitting large amounts of high-energy radiation. If the donut intervenes, we see instead large amounts of infrared radiation reradiated from the dust. The structure of the donut itself is quite uncertain and may in reality bear little resemblance to the rather regular-looking ring shown in the figure. Many astronomers suspect that the absorbing region may actually be a dense *outflow* of gas driven from the accretion disk's outer edge by the intense radiation within.

A different reprocessing mechanism operates in many jets and radio lobes. It involves the *magnetic fields* possibly produced within the accretion disk and transported by the jets into intergalactic space (Figure 15.30). As sketched in Figure 15.34(a), whenever a charged particle (here an electron) encounters a magnetic field, the particle tends to spiral around the magnetic field lines. We have already encountered this idea on smaller scales in Earth's magnetosphere ∞ *(Sec. 5.7)*, in solar activity ∞ *(Sec. 9.4)*, near neutron stars ∞ *(Sec. 13.2)*, and at the center of our own Galaxy. ∞ *(Sec. 14.7)*

As the particles whirl around, they emit electromagnetic radiation. ∞ *(Sec. 2.2)* The radiation produced in this way—called **synchrotron radiation**, after the type of

CONCEPT CHECK ▶ How does accretion onto a supermassive black hole power the energy emission from the extended radio lobes of a radio galaxy?

▶ **FIGURE 15.32 M87 Disk**   Both images and spectra of M87 support the idea of a rapidly whirling accretion disk at this galaxy's heart. (a) The central region of M87, similar to that shown in Figure 15.26(c), shows its bright nucleus and jet. (b) A magnified view of the nucleus implies a spiral swarm of stars, gas, and dust. (c) Spectral-line features observed on opposite sides of the nucleus show contrasting red and blue Doppler shifts, implying that matter on one side is coming toward us while matter on the other side is moving away. Apparently, an accretion disk spins perpendicular to the jet, and at its center is a black hole having some three billion times the mass of the Sun. *(NASA)*

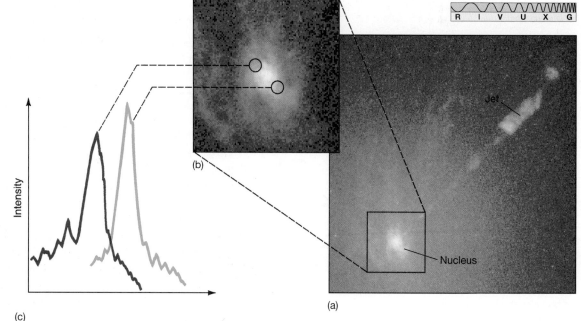

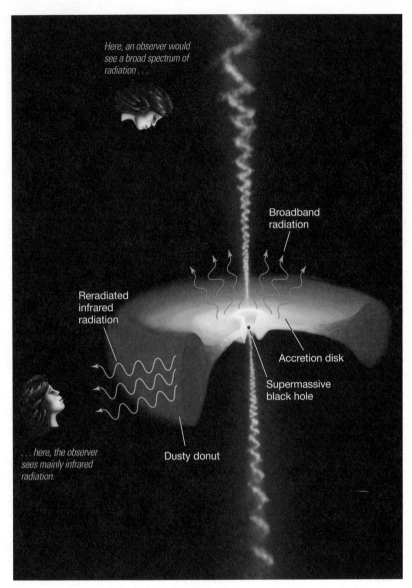

Here, an observer would see a broad spectrum of radiation . . .

Broadband radiation

Reradiated infrared radiation

Accretion disk

Supermassive black hole

Dusty donut

. . . here, the observer sees mainly infrared radiation.

◀ **FIGURE 15.33 Dusty Donut** The accretion disk surrounding a massive black hole, drawn here with some artistic license, consists of hot gas at many different temperatures (hottest nearest the center). When viewed from above or below, the disk radiates a broad spectrum of electromagnetic energy extending into the X-ray band. However, the dusty infalling gas that ultimately powers the system is thought to form a fat, donut-shaped region outside the accretion disk (shown here in red), which effectively absorbs much of the high-energy radiation reaching it, reemitting it mainly in the form of cooler, infrared radiation. When the accretion disk is viewed from the side, strong infrared emission is observed. (Compare with Figure 15.25).

particle accelerator in which it was first observed—is *nonthermal* in nature. This means that there is no link between the emission and the temperature of the radiating object, so the radiation is not described by a blackbody curve. Instead, its intensity decreases with increasing frequency, as shown in Figure 15.34(b). This is just what is needed to explain the overall spectrum of radiation from active galaxies [compare Figure 15.34(b) with Figure 15.18]. Observations of the radiation received from the jets and radio lobes of active galaxies are completely consistent with this process.

Eventually, the jet is slowed and stopped by the intergalactic medium, the flow becomes turbulent, and the magnetic field grows tangled. The result is a gigantic radio lobe emitting virtually all of its energy in the form of synchrotron radiation. Thus, even though the radio *emission* comes from an enormously extended volume of space that dwarfs the visible galaxy, the *source* of the energy is still the accretion disk—a billion billion times smaller in volume than the radio lobe—lying at the galactic center. The jets transport energy from the nucleus, where it is generated, into the lobes, where it is finally radiated into space.

The existence of the inner lobes of Centaurus A and the blobs in M87's jet imply that the formation of a jet may be an intermittent process (or, as in the case of the Seyferts discussed earlier, may not occur at all). Many nearby active galaxies (like Centaurus A) appear to have been "caught in the act" of interacting with another galaxy, suggesting that the fuel supply can be "turned on" by a companion. The tidal forces involved divert gas and stars into the galactic nucleus, triggering an outburst that may last for many millions of years.

What do active galaxies look like between active outbursts? What is their connection with the normal galaxies we see? These important questions will be answered in Chapter 16, where we explore the subject of *galaxy evolution*.

▶ **FIGURE 15.34 Nonthermal Radiation** (a) Charged particles, especially fast-moving electrons (red), emit synchrotron radiation (blue) while spiraling in a magnetic field (black). Thermal and synchrotron (nonthermal) radiation vary differently with frequency. Thermal radiation, described by a blackbody curve, peaks at a frequency that depends on the temperature of the source. By contrast, nonthermal synchrotron radiation is more intense at low frequencies and is independent of the temperature of the emitting object. (Compare with Figure 15.18.)

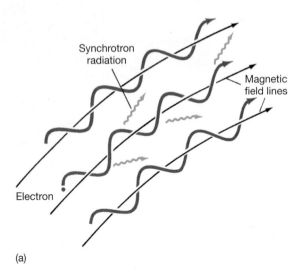

Synchrotron radiation

Magnetic field lines

Electron

(a)

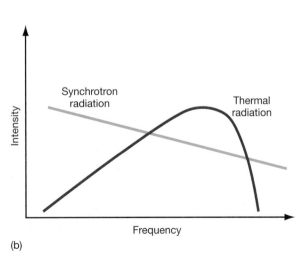

Synchrotron radiation

Thermal radiation

Intensity

Frequency

(b)

CHAPTER REVIEW

SUMMARY

L01 The **Hubble classification scheme** (p. 406) divides galaxies into several classes, depending on their appearance. **Spiral galaxies** (p. 406) have flattened disks, central bulges, and spiral arms. Their halos consist of old stars, whereas  the gas-rich disks are the sites of ongoing star formation. **Barred-spiral galaxies** (p. 408) contain an extended "bar" of material projecting beyond the central bulge. **Elliptical galaxies** (p. 408) have no disk and contain little or no cool gas or dust, although very hot interstellar gas is observed. In most cases, they consist entirely of old stars. They range in size from dwarf ellipticals, which are much less massive than the Milky Way Galaxy, to giant ellipticals, which may contain trillions of stars. **S0 and SB0 galaxies** (p. 409) are intermediate in their properties between ellipticals and spirals. **Irregular galaxies** (p. 410) are galaxies that do not fit into either of the other categories. Some may be the result of galaxy collisions or close encounters. Many irregulars are rich in gas and dust and are the sites of vigorous star formation.

L02 Astronomers often use **standard candles** (p. 412) as distance-measuring tools. These are objects that are easily identifiable and whose luminosities lie in some reasonably  well-defined range. Comparing luminosity and apparent brightness, astronomers determine the distance using the inverse-square law. An alternative approach is the **Tully–Fisher relation** (p. 413), an empirical correlation between rotational velocity and luminosity in spiral galaxies.

L03 The Milky Way, Andromeda, and several other smaller galaxies form the **Local Group** (p. 414), a small **galaxy cluster** (p. 415). Galaxy clusters consist of a collection 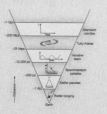 of galaxies orbiting one another, bound together by their own gravity. The nearest large galaxy cluster to the Local Group is the Virgo Cluster.

L04 Distant galaxies are observed to be receding from the Milky Way at speeds proportional to their distances from us. This relationship is called **Hubble's law** (p. 416). The constant 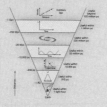 of proportionality in the law is **Hubble's constant** (p. 417). Its value is approximately 70 km/s/Mpc. Astronomers use Hubble's law to determine distances to the most remote objects in the universe simply by measuring their redshifts and converting the corresponding speed directly to a distance. The redshift associated with the Hubble expansion is called the **cosmological redshift** (p. 417).

L05 **Active galaxies** (p. 418) are much more luminous than normal galaxies and have nonstellar spectra, emitting most of their energy outside the visible part of the elec- 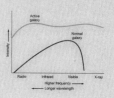 tromagnetic spectrum. Often the nonstellar activity suggests rapid internal motion and is associated with a bright central **active galactic nucleus** (p. 419). Many active galaxies have high-speed, narrow jets of matter shooting out from their central nuclei. Often the jets transport energy from the nucleus, where it is generated, to enormous **radio lobes** (p. 420) lying far beyond the visible portion of the galaxy, where it is radiated into space. The jets often appear to be made up of distinct "blobs" of gas, suggesting that the process generating the energy is intermittent.

L06 A **Seyfert galaxy** (p. 419) looks like a normal spiral, except that the Seyfert has an extremely bright central **galactic nucleus** (p. 419). Spectral lines from Seyfert nuclei are very broad,  indicating rapid internal motion, and rapid variability implies that the source of the radiation is much less than 1 light-year across. **Radio galaxies** (p. 420) emit large amounts of energy in the radio part of the spectrum. The corresponding visible galaxy is usually elliptical. **Quasars** (p. 424), or **quasistellar objects** (p. 424), are the most luminous objects known. In visible light they appear starlike, and their spectra are usually substantially redshifted. All quasars are very distant, indicating that we see them as they were in the distant past.

L07 The generally accepted explanation for the observed properties of all active galaxies is that the energy is generated by accretion of galactic gas onto a supermassive (million- to  billion-solar-mass) black hole lying in the galactic center. The small size of the accretion disk explains the compact extent of the emitting region, and the high-speed orbit of gas in the black hole's intense gravity accounts for the rapid motion observed. Much of the high-energy radiation emitted from the disk is absorbed by a fat, donut-shaped region outside the disk and reemitted in the form of infrared radiation. Typical active galaxy luminosities require the consumption of about 1 solar mass of material every few years. Some of the infalling matter is blasted out into space, producing magnetized jets that create and feed the radio lobes. Charged particles spiraling around the magnetic field lines produce **synchrotron radiation** (p. 427), whose spectrum is consistent with the nonstellar radiation observed in radio galaxies and jets.

MasteringAstronomy® For instructor-assigned homework go to www.masteringastronomy.com

Problems labeled POS explore the process of science. VIS problems focus on reading and interpreting visual information. LO connects to the introduction's numbered Learning Outcomes.

REVIEW AND DISCUSSION

1. **LO1** Distinguish among the various types of spiral galaxies.

2. Compare an elliptical galaxy with the halo of the Milky Way.

3. **LO2** Describe the four rungs in the distance-measurement ladder involved in determining the distance to a galaxy lying 5 Mpc away.

4. How does the Tully–Fisher relation allow astronomers to measure the distances to galaxies?

5. **LO3** What is the Virgo Cluster?

6. **LO4** What is Hubble's law, and how is it used by astronomers to measure distances to galaxies?

7. What is the most likely range of values for Hubble's constant?

8. Why do astronomers prefer to speak in terms of redshifts rather than distances to faraway objects?

9. **LO5** Name two differences between normal and active galaxies.

10. **LO6 POS** What is the evidence that the radio lobes of some active galaxies consist of material ejected from the galaxy's nucleus?

11. **POS** How do we know that the energy-emitting regions of many active galaxies must be very small?

12. **LO7** Briefly describe the leading model for the central engine of an active galaxy.

13. **POS** How do astronomers account for the differences in the spectra observed from active galaxies?

14. **POS** How is the process of synchrotron emission related to observations of active galaxies?

15. How do we know that quasars are extremely luminous?

CONCEPTUAL SELF-TEST: TRUE OR FALSE?/MULTIPLE CHOICE

1. Most galaxies are spirals like the Milky Way. (T/F)

2. Most elliptical galaxies contain only old stars. (T/F)

3. Irregular galaxies, although small, often have lots of star formation taking place in them. (T/F)

4. Type I supernovae can be used to determine distances to galaxies. (T/F)

5. Most galaxies are receding from the Milky Way Galaxy. (T/F)

6. Hubble's law can be used to determine distances to the farthest objects in the universe. (T/F)

7. Active galaxies can emit thousands of times more energy than our own Galaxy. (T/F)

8. Using the method of standard candles, we could, in principle, estimate the distance to a campfire if we knew (a) the number of logs used; (b) the fire's temperature; (c) the length of time the fire has been burning; (d) the type of wood used in the fire.

9. Within 30 Mpc of the Sun, there are about (a) 3 galaxies; (b) 30 galaxies; (c) a few thousand galaxies; (d) a few million galaxies.

10. Hubble's law states that (a) more distant galaxies are younger; (b) the greater the distance to a galaxy, the greater is the galaxy's redshift; (c) most galaxies are found in clusters; (d) the greater the distance to a galaxy, the fainter the galaxy appears.

11. If the light from a galaxy fluctuates in brightness very rapidly, the region producing the radiation must be (a) very large; (b) very small; (c) very hot; (d) rotating very rapidly.

12. Quasar spectra (a) are strongly redshifted; (b) show no spectral lines; (c) look like the spectra of stars; (d) contain emission lines from unknown elements.

13. Active galaxies are very luminous because they (a) are hot; (b) contain black holes in their cores; (c) are surrounded by hot gas; (d) emit jets.

14. **VIS** If the galaxy in Figure 15.10 (Galaxy Rotation) were smaller and spinning more slowly, then the figure should be redrawn to show (a) a greater blueshift; (b) a greater redshift; (c) a narrower combined line; (d) a larger combined amplitude.

15. **VIS** According to Figure 15.18 (Galaxy Energy Spectra), active galaxies (a) emit most of their energy at long wavelengths; (b) emit very little energy at high frequencies; (c) emit large amounts of energy at all wavelengths; (d) emit most of their energy in the visible part of the spectrum.

PROBLEMS

The number of squares preceding each problem indicates its approximate level of difficulty.

1. ■ The Andromeda Galaxy lies 800 kpc from our Galaxy and is approaching it with a radial velocity of 266 km/s. Neglecting both the transverse component of the velocity and the effect of gravity in accelerating the motion, estimate when the two galaxies will collide.

2. ■■ A Cepheid variable star in the Virgo Cluster has an apparent magnitude of 26 and is observed to have a pulsation period of 20 days. Use these numbers and Figure 14.6 to estimate the distance to the Virgo Cluster. The absolute magnitude of the Sun is +4.85.

3. ■ A supernova 1 billion times more luminous than the Sun is used as a standard candle to measure the distance to a faraway galaxy. From Earth, the supernova appears as bright as the Sun would appear from a distance of 10 kpc. What is the distance to the galaxy?

4. ■ According to Hubble's law, with $H_0 = 70$ km/s/Mpc, what is the recessional speed of a galaxy at a distance of 200 Mpc? How far away is a galaxy whose recessional speed is 4000 km/s? How would these answers change if (a) $H_0 = 50$ km/s/Mpc, (b) $H_0 = 80$ km/s/Mpc?

5. ■■ If $H_0 = 70$ km/s/Mpc, how long will it take for the distance from the Milky Way Galaxy to the Virgo Cluster to double? How long for the distance to the Coma Cluster (100 Mpc) to double?

6. ■■ A quasar is observed to have luminosity fluctuations and broadened emission lines indicating a speed of 5000 km/s within 0.1 light-years of its center. Assuming circular orbits, use Kepler's laws to estimate the mass within this radius. ∞ *(Sec. 14.6)*

7. ■ Assuming a jet speed of 75 percent of the speed of light, calculate the time taken for material in Cygnus A's jet to cover the 250 kpc between the galaxy's nucleus and its radio-emitting lobes.

8. ■■ Assuming the upper end of the efficiency range indicated in the text, calculate how much energy an active galaxy would generate if it consumed one Earth mass of material every day. Compare this value with the luminosity of the Sun.

9. ■ Using the data presented in the text, calculate the orbital speed of material orbiting at a distance of 0.25 pc from the center of M87.

10. ■■■ Calculate the amount of energy received per unit area per unit time that would be observed at Earth from a 10^{37}-W Seyfert nucleus located at the Galactic center, 8 kpc away, neglecting the effects of absorption by interstellar dust. Using the data presented in Appendix 3, Table 3, compare this with the energy received from Sirius A, the brightest star in the night sky. From what you know about active galaxy energy emission, is it reasonable to ignore interstellar absorption?

ACTIVITIES

Collaborative

1. Observe the Virgo Cluster of galaxies. An 8-inch telescope is the optimal size for this project. The constellation Virgo is visible from the United States during much of fall through spring. To locate the cluster, first find the constellation Leo. The eastern part of Leo is composed of a distinct triangle of stars, Denebola (β), Chort (θ), and Zosma (δ). Go from Chort to Denebola in a straight line east; continue on the same distance as between the two stars, and you will be approximately at the center of the Virgo Cluster. Look for the following Messier objects, the brightest galaxies in the cluster: M49, M58, M59, M60, M84, M86, M87, M89, and M90. Examine each galaxy for unusual features; some have very bright nuclei. Sketch or photograph what you see, and construct your own pictorial catalog of the brightest galaxies in Virgo.

Individual

1. In the previous exercise, you were given directions for finding the Virgo Cluster of galaxies. M87, in the central part of this cluster, is the nearest core–halo radio galaxy. M87 has coordinates RA = $12^h 30.8^m$, dec = $+12° 24'$. At magnitude 8.6 it should not be difficult to find with an 8-inch telescope. Its distance is roughly 18 Mpc. Describe its nucleus; compare what you see with other nearby ellipticals in the Virgo Cluster.

2. 3C 273 is the nearest and brightest quasar, but that does not mean it is easy to find! Its coordinates are RA = $12^h 29.2^m$, dec = $+2° 03'$. It is located in the southern part of the Virgo Cluster, but it is not associated with it. At magnitude 12–13 (again, it is variable), it may require a 10- or 12-inch telescope to see, but try it first with an 8-inch. It should appear as a very faint star. The significance of seeing this object is that it is 640 Mpc distant. The light you are seeing left this object over 2 billion years ago! It is the most distant object observable with a small telescope.

16

Galaxies and Dark Matter

The Large-Scale Structure of the Cosmos

On scales much larger than even the largest galaxy clusters, new levels of structure are revealed, and a humbling new reality emerges. We may be star stuff, the product of countless cycles of stellar evolution, but we are not the stuff of the cosmos. The universe in the large is composed of matter fundamentally different from the familiar atoms and molecules that compose our bodies, our planet, our star and galaxy, and all the luminous matter we observe in the heavens. Only its gravity announces the presence of this strange kind of matter. By comparing and classifying the properties of galaxies near and far, astronomers have begun to understand their evolution. By mapping out their distribution in space, astronomers trace out the immense realms of the universe. Points of light in the uncharted darkness, the galaxies remind us that our position in the universe is no more special than that of a boat adrift at sea.

MasteringAstronomy® Visit www.masteringastronomy.com for quizzes, animations, videos, interactive figures, and self-guided tutorials.

LEARNING OUTCOMES

Studying this chapter will enable you to:

LO1 Describe some methods used to determine the masses of galaxies and galaxy clusters.

LO2 Explain why astronomers think that most of the matter in the universe is dark.

LO3 Describe how galaxies form and evolve, and outline the role of collisions in the process.

LO4 Present the evidence for supermassive black holes in the centers of galaxies, and explain how active galaxies fit into current theories of galactic evolution.

LO5 Describe how galaxies are distributed on large scales in the universe.

LO6 Outline some techniques used by astronomers to probe the universe on very large scales.

◄ Galaxies are among the grandest, most beautiful objects in the universe—each one a colossal collection of hundreds of billions of stars held together loosely by gravity. Some are bright and splendid, like the two big ones in this image. Others are dim and distant, like several that appear smaller in the background. This pair of galaxies, nearly 300 million light-years away and known collectively as Arp 273, is in the process of colliding over millions of years. Notice the rose-like shape of the top galaxy, caused by the gravitational pull of the bottom one, and the swath of clusters of young blue stars glowing like jewels. Mergers and acquisitions are apparently common among galaxies, but astronomers still don't understand well how the galaxies originated long ago. *(STScI)*

16.1 Dark Matter in the Universe

In Chapter 14 we saw how measurements of the orbital velocities of stars and gas in our Galaxy reveal the presence of an extensive *dark-matter halo* surrounding the galaxy we see. ∞ *(Sec. 14.6)* Do other galaxies have similar dark halos? And what evidence do we have for dark matter on larger scales? To answer these questions, we need to know the masses of galaxies and galaxy clusters.

How can we measure the masses of such large systems? Surely, we can neither count all their stars nor estimate their interstellar content very well. Galaxies are just too complex to take direct inventory of their material makeup. Instead, we must rely on indirect techniques. Despite their enormous sizes, galaxies and galaxy clusters obey the same physical laws as do the planets in our own solar system. To determine galaxy masses, we turn as usual to Newton's law of gravity.

Galaxy Masses

Astronomers can calculate the masses of some spiral galaxies by determining their *rotation curves*, which plot rotation speed versus distance from the galactic center (Figure 16.1a). ∞ *(Sec. 14.6)* The mass within any given radius then follows directly from Newton's laws. Rotation curves for a few nearby spirals are shown in Figure 16.1(b). They imply masses ranging from about 10^{11} to 5×10^{11} solar masses within about 25 kpc of the center—comparable to the results obtained for our own Galaxy using the same technique. For galaxies too distant for detailed rotation curves to be drawn, the overall rotation speed can still be inferred from the broadening of spectral lines, as described in the previous chapter. ∞ *(Sec. 15.2)*

This approach is useful for measuring the mass lying within about 50 kpc of a galaxy's center (the extent of the electromagnetic emission from stellar and interstellar material). To probe farther from the centers of galaxies, astronomers turn to *binary* galaxy systems (Figure 16.2a), whose components may lie hundreds of kiloparsecs apart. The orbital period of such a system is typically billions of years, far too long for the orbit to be accurately measured. However, by estimating the period and semimajor axis using the information available—the line-of-sight velocities and angular separation of the components—an approximate total mass can be derived. ∞ *(Sec. 1.4)*

Galaxy masses obtained in this way are fairly uncertain, but by combining many such measurements, astronomers can obtain quite reliable *statistical* information about galaxy masses. Most normal spirals (the Milky Way Galaxy included) and large ellipticals contain between 10^{11} and 10^{12} solar masses of material. Irregular galaxies often contain less mass, about 10^8 to 10^{10} times that of the Sun. Dwarf ellipticals and dwarf irregulars can contain as little as 10^6 or 10^7 solar masses of material.

We can use another statistical technique to derive the combined mass of all the galaxies within a galaxy cluster. As depicted in Figure 16.2(b), each galaxy within a cluster moves relative to all other cluster members, and we can estimate the cluster's mass simply by asking how massive it must be in order to bind its galaxies gravitationally. Typical cluster masses obtained in this way lie in the range of 10^{13}–10^{14} solar masses. Notice that this calculation gives us no information whatsoever about the masses of individual galaxies. It tells us only about the *total* mass of the entire cluster.

INTERACTIVE Rotation Curve for a Merry-Go-Round (MA)

ANIMATION/VIDEO Dark Matter (MA)

Observer

(a)

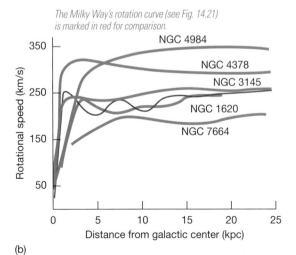

The Milky Way's rotation curve (see Fig. 14.21) is marked in red for comparison.

NGC 4984

NGC 4378

NGC 3145

NGC 1620

NGC 7664

Rotational speed (km/s)

350

250

150

50

0   5   10   15   20   25

Distance from galactic center (kpc)

(b)

INTERACTIVE ◀ **FIGURE 16.1 Galaxy Rotation Curves** (a) Orbital velocities can be measured at different distances from the center of a disk galaxy, as illustrated here for M64, the "Evil Eye" galaxy, some 5 Mpc distant. (b) Rotation curves for some nearby spiral galaxies indicate masses of a few hundred billion times the mass of the Sun. *(NASA)*

Visible Matter and Dark Halos

The rotation curves of the spiral galaxies shown in Figure 16.1(b) remain flat (that is, do not decline and may even rise slightly) far beyond the visible images of the galaxies themselves, implying that these galaxies—and perhaps all spiral galaxies—have invisible dark halos similar to that surrounding the Milky Way. ∞ *(Sec. 14.5)* Spiral galaxies seem to contain from 3 to 10 times more mass than can be accounted for in the form of visible matter. Studies of elliptical galaxies suggest similarly large dark halos surrounding those galaxies, too.

Astronomers find even greater discrepancies between visible light and total mass when they study galaxy clusters. Calculated cluster masses range from 10 to nearly 100 times the mass suggested by the light emitted by individual cluster galaxies. Put another way, a lot more mass is needed to bind galaxy clusters than we can see. Thus, the problem of dark matter exists not just in our own Galaxy, but also in other galaxies and, to an even greater degree, in galaxy clusters, too. In that case, we are compelled to accept the fact that *upward of 90 percent of the matter in the universe is dark.* As noted in Chapter 14, this matter is not just dark in the visible portion of the spectrum—it is undetected at *any* electromagnetic wavelength. ∞ *(Sec. 14.5)*

The dark matter in clusters is not simply the accumulation of dark matter within individual galaxies. Even including the galaxies' dark halos, we still cannot account for all the dark matter in galaxy clusters. As we look on larger and larger scales, we find that a larger and larger fraction of the matter in the universe is dark.

Intracluster Gas

In addition to the luminous matter observed within the cluster galaxies themselves, astronomers also have evidence for large amounts of *intracluster gas*—superhot (more than 10 million K), diffuse intergalactic matter filling the space among the galaxies. Satellites orbiting above Earth's atmosphere have detected X-rays from hot gas in many clusters. ∞ *(Sec. 3.5)* Figure 16.3 shows false-color X-ray images of two such systems. In each case, the X-ray–emitting region is centered on, and comparable in size to, the visible cluster image.

Further evidence for intracluster gas can be found in the appearance of the radio lobes of some active galaxies. ∞ *(Sec. 15.4)* In some systems, known as *head–tail radio galaxies*, the lobes seem to form a "tail" behind the main part of the galaxy. For example, the lobes of radio galaxy NGC 1265, shown in Figure 16.4, appear to be "swept back" by some onrushing wind, and, indeed, this is the most likely explanation for the galaxy's appearance. If NGC 1265 were at rest, it would be just another double-lobe source, perhaps quite similar to Centaurus A (see Figure 15.22). However, the galaxy is traveling through the intergalactic medium of its parent galaxy cluster (known as the Perseus Cluster), and the outflowing matter forming the lobes tends to be left behind as NGC 1265 moves.

How much gas do these observations reveal? At least as much matter—and in a few cases substantially more—exists within clusters in the form of hot gas as is visible in the form of stars. This is a lot of material, but it still doesn't solve the dark-matter problem. To account for the total masses of galaxy clusters implied by dynamical studies, we would have to find from 10 to 100 times more mass in gas than in stars.

Why is the intracluster gas so hot? Simply because its particles are bound by gravity and hence are moving at speeds comparable to those of the galaxies in the cluster—1000 km/s or so. Since temperature is just a measure of the speed at which the gas particles move, this speed translates (for protons, which make up the bulk of the mass) to a temperature of 40 million K. ∞ *(More Precisely 5-1)*

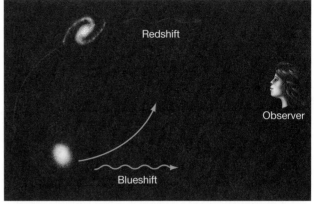

(a)

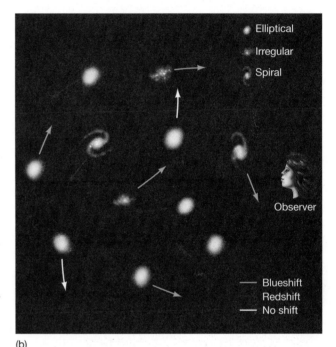

(b)

▲ **FIGURE 16.2 Galaxy Masses** (a) In a binary galaxy system, galaxy masses can be estimated by observing the orbit of one galaxy about the other. (b) The mass of a galaxy cluster can be estimated by observing the motion of many galaxies in the cluster and then estimating how much mass is needed to prevent the cluster from flying apart.

Images like these show that the space between the galaxies within galaxy clusters is filled with superheated gas.

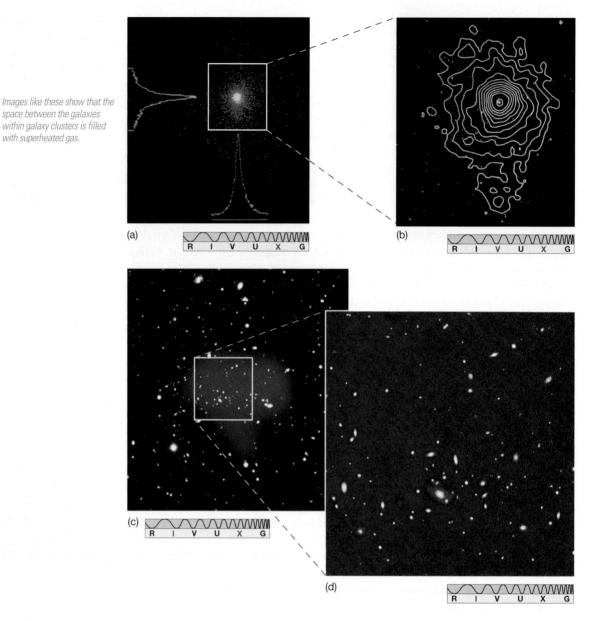

(a)

(b)

(c)

(d)

▲ FIGURE 16.3 **Galaxy Cluster X-Ray Emission** (a) X-ray image of Abell 85, an old, distant cluster of galaxies, taken by the *Einstein X-ray Observatory*. The cluster's X-ray emission is shown in orange, scans of its cross-section in green. (b) A contour map of the X-rays is superimposed on an optical photo, showing that its X-rays are strongest around Abell 85's central supergiant galaxy. (c) Superposition of infrared (red) and X-ray (blue) radiation from another distant galaxy cluster. Part (d) is a longer infrared exposure of the central region, showing the richness of this cluster, which spans about a million parsecs. *(NASA; AURA; ESA)*

CONCEPT CHECK ▶ What assumptions are we making when we infer the mass of a galaxy cluster from observations of the spectra of its constituent galaxies?

Where did the gas come from? There is so much of it that it could not have been expelled from the galaxies themselves. Instead, astronomers think that it is mainly *primordial*—gas that has been around since the universe began and that never became part of a galaxy. However, the intracluster gas does contain some heavy elements—carbon, nitrogen, and so on—implying that at least some of it is material ejected from galaxies after enrichment by stellar evolution. ∞ *(Sec. 12.7)* Just how this occurred remains a mystery.

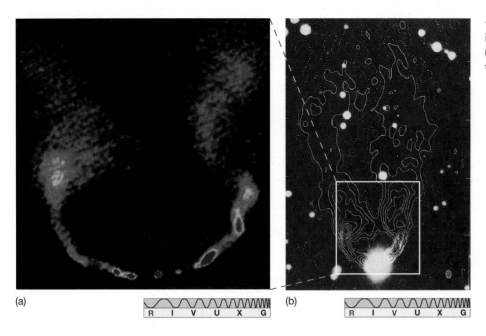

(a)

R I V U X G

(b)

R I V U X G

◄ **FIGURE 16.4 Head–Tail Radio Galaxy**   (a) Radiograph, in false color, of the head–tail radio galaxy NGC 1265. (b) The same radio data, in contour form, superimposed on the optical image of the galaxy. *(NRAO; Palomar/Caltech)*

16.2 Galaxy Collisions

Contemplating the congested confines of a rich galaxy cluster (such as Virgo or Coma) with thousands of member galaxies orbiting within a few megaparsecs, we might expect that collisions among galaxies would be common. ∞ *(Sec. 15.2)* Gas particles collide in our atmosphere, and hockey players collide in the rink. Do galaxies in clusters collide too? The answer is yes.

Figure 16.5 apparently shows the aftermath of a bull's-eye collision between a small galaxy (perhaps one of the two at the right, although that is by no means certain) and the larger galaxy at the left. The result is the Cartwheel Galaxy, about 150 Mpc from Earth, its halo of young stars resembling a vast ripple in a pond. The ripple is most likely a density wave created by the passage of the smaller galaxy through the disk of the larger one. ∞ *(Sec. 14.5)* The disturbance is now spreading outward from the region of impact, creating new stars as it goes.

Figure 16.6 shows an example of a close encounter that hasn't (yet) led to an actual collision. Two spiral galaxies are apparently passing each other like majestic ships in the night. The larger and more massive galaxy at the bottom is called NGC 2207; the smaller one at the top is IC 2163. Analysis of this image suggests that IC 2163 is now swinging past NGC 2207 in a counterclockwise direction, having made a close approach some 40 million years ago. The two galaxies seem destined to undergo further close encounters, as IC 2163 apparently does not have enough energy to escape the gravitational pull of NGC 2207. In roughly a billion years, these two galaxies will probably merge into a single, massive galaxy.

No human will ever witness an entire galaxy collision, for it lasts many millions of years. However, modern computers can follow the event in a matter of hours. Large-scale simulations modeling in detail the gravitational interactions among stars and gas, and incorporating the best available models of gas dynamics, allow astronomers to better understand the effects of a collision on the galaxies involved and even estimate the eventual outcome of the interaction. The computer simulation shown in Figure 16.7(b) began with two colliding spiral galaxies, not so different from those shown in Figure 16.6, but the details of the original structure have been largely obliterated by the collision. Notice the similarity to the real image of NGC 4038/4039 (Figure 16.7a), the so-called Antennae galaxies, which show extended tails, as well as double galactic centers only a few hundred parsecs across.

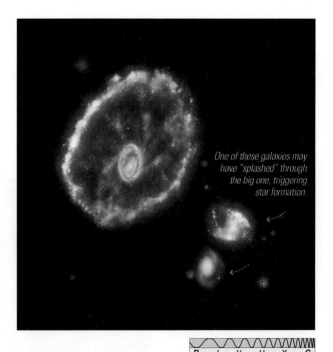

One of these galaxies may have "splashed" through the big one, triggering star formation.

R I V U X G

▲ **FIGURE 16.5 Cosmic Splash**   The Cartwheel Galaxy (left) may have resulted from a collision (possibly with one of the smaller galaxies at right) that has led to an expanding ring of star formation moving outward through the galactic disk. This is a false color composite image combining four spectral bands: infrared in red (from *Spitzer*), optical in green (from *Hubble*), ultraviolet in blue (from *Galex*), and X-ray in purple (from *Chandra*). *(NASA)*

► FIGURE 16.6 **Galaxy Encounter** This encounter between two spiral galaxies, NGC 2207 (left) and IC2163 (right), has already led to bursts of star formation in each. Eventually the two will merge, but not for a billion years or so. *(NASA)*

INTERACTIVE Colliding Galaxies

Star formation induced by the collision is clearly traced by the blue light from thousands of young, hot stars. The simulations indicate that, as with the galaxies in Figure 16.6, ultimately the two galaxies will merge into one.

Galaxies in clusters apparently collide quite often. In the smaller groups, the galaxies' speeds are low enough that interacting galaxies tend to "stick together," and mergers, as shown in the computer simulation, are a very common outcome. In larger groups, galaxies move faster and tend to pass through one another without sticking. Either way, as we will see, the encounters have substantial effects on the galaxies involved. Some researchers go so far as to suggest that most galaxies have been strongly influenced by collisions, in many cases in the relatively recent past. Incidentally, if we wait long enough, we will have an opportunity to see for ourselves what a galaxy collision is really like: Our nearest large neighbor, the Andromeda Galaxy (see Figure 14.2), is currently approaching us at a velocity of 120 km/s. In a few billion years it will collide with the Milky Way, and we will be able to test astronomers' theories firsthand!

The computer simulation at right...   *...tries to match the photos at left.*

► FIGURE 16.7 **Galaxy Collision** (a) The long tidal "tails" (black-and-white image at left) mark the final plunge of the "Antennae" galaxies a few tens of millions of years ago. Strings of young, bright "super star clusters" (magnified color image at center) were caused by violent shock waves produced in the gas disks of the two colliding galaxies. (b) A computer simulation of the encounter shows many of the same features as the real object at left. *(AURA; NASA; J. Barnes)*

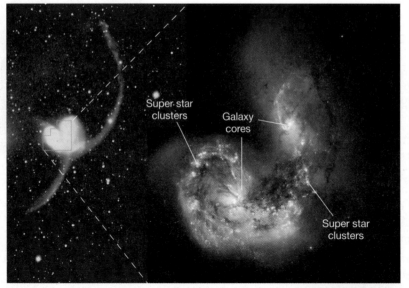

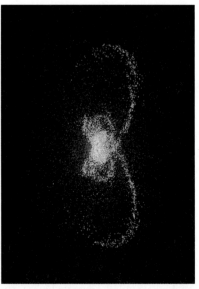

(a)   (b)

Curiously, although a collision may wreak havoc on the large-scale structure of the galaxies involved, it has essentially no effect on the individual stars they contain. The stars within each galaxy just glide past one another. In contrast to galaxies in the cluster, the stars in a galaxy are so small compared with the distances between them, that when two galaxies collide, the star population merely doubles for a time, and the stars continue to have so much space that they do not run into each other. Collisions can rearrange the stellar and interstellar contents of each galaxy, often producing a spectacular burst of star formation that may be visible to enormous distances, but from the point of view of the stars, it's clear sailing.

ANIMATION/VIDEO Galaxy Collision I

16.3 Galaxy Formation and Evolution

With Hubble's law as our indispensable guide to distances in the universe, and armed now with knowledge of the distribution of dark matter on galactic and larger scales, let's finally turn to the question of how galaxies came to be the way they are. Can we explain the different galaxy types we see? Astronomers know of no simple evolutionary connections among the various categories in the Hubble classification scheme. ∞ *(Sec. 15.1)* To answer the question, then, we must understand how galaxies formed.

Unfortunately, compared with the theories of star formation and stellar evolution, the theory of galaxy formation is still very much in its infancy. Galaxies are much more complex than stars, they are harder to observe, and the observations are harder to interpret. In addition, we have only a partial understanding of conditions in the universe immediately preceding galaxy formation, quite unlike the corresponding situation for stars. ∞ *(Sec. 11.2)* Finally, and most important, stars almost never *collide* with one another (apart from situations where stellar evolution causes binary components to merge), so most stars in galaxies evolve in isolation. ∞ *(Sec. 12.3)* Galaxies, however, may suffer numerous collisions during their lives, making it much harder to decipher their pasts. Indeed, collisions like those described in the previous section blur the distinction between formation and evolution to the point where it can be hard to separate one from the other.

Nevertheless, some general ideas have gained widespread acceptance, and we can offer some insights into the processes responsible for the galaxies we see. We first describe a general scenario for how large galaxies form from smaller ones, then discuss how galaxies change in time due to both internal stellar evolution and external influences. Finally, we consider how the galaxy types in Hubble's classification fit in to this broad picture.

Mergers and Acquisitions

The seeds of galaxy formation were sown in the very early universe, when small density fluctuations in the primordial matter began to grow (see Section 17.7). Our discussion here begins with these "pregalactic" blobs of gas already formed. The masses of these fragments were quite small—only a few million solar masses, comparable to the masses of the smallest present-day dwarf galaxies, which may in fact be remnants of this early time. Most astronomers think that galaxies grew by repeated *merging* of smaller objects, as illustrated in Figure 16.8(a). Contrast this

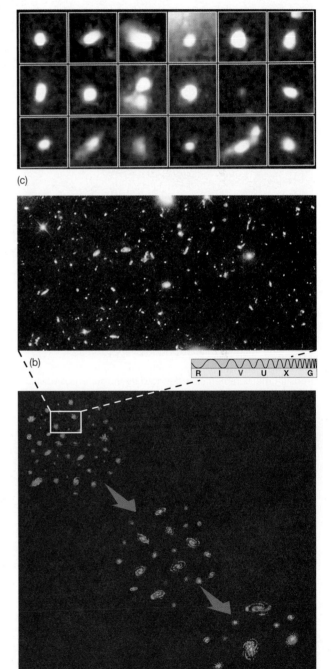

(c)

R I V U X G

(b)

(a)

▶ **FIGURE 16.8 Galaxy Formation** (a) The best current theory of galaxy formation holds that large systems were built up from smaller ones through collisions and mergers, as shown schematically in the drawing at bottom. (b) This photograph, one of the deepest ever taken of the universe, provides "fossil evidence" for hundreds of galaxy shards and fragments, up to 5000 Mpc distant. (c) Enlargements of selected portions of (b) reveal rich (billion-star) "star clusters," all lying within a relatively small volume of space (about 1 Mpc across). Such pregalactic fragments might be about to merge to form a galaxy. The events pictured took place about 10 billion years ago. *(NASA)*

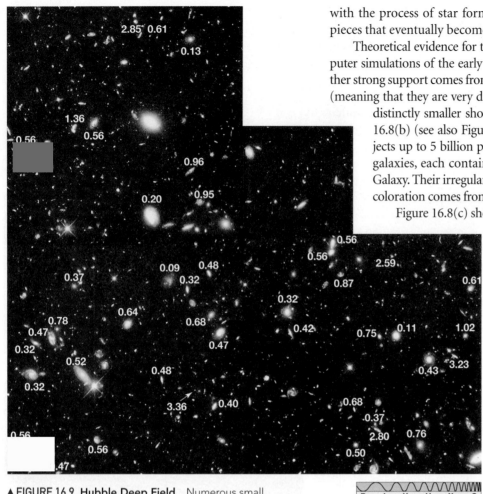

▲ FIGURE 16.9 Hubble Deep Field Numerous small, irregularly shaped young galaxies can be seen in this very deep optical image. Known as the Hubble Deep Field-North, this image, made with an exposure of approximately 100 hours, captured objects as faint as 30th magnitude. Redshift measurements (as denoted by the superposed values observed at the Keck Observatory in Hawaii) indicate that some of these galaxies lie well over 1000 Mpc from Earth. ∞ *(Sec 15.3)* The field of view is about 2 arc minutes across, or less than 1 percent of the area subtended by the full Moon. *(NASA; Keck)*

with the process of star formation, in which a large cloud fragments into smaller pieces that eventually become stars. ∞ *(Sec. 11.3)*

Theoretical evidence for this picture of **hierarchical merging** is provided by computer simulations of the early universe, which clearly show merging taking place. Further strong support comes from observations that indicate that galaxies at large redshifts (meaning that they are very distant and the light we see was emitted long ago) appear distinctly smaller shows and more irregular than those found nearby. Figure 16.8(b) (see also Figure 16.9) shows some of these images, which include objects up to 5 billion parsecs away. The vague bluish patches are separate small galaxies, each containing only a few percent of the mass of the Milky Way Galaxy. Their irregular shape is probably the result of galaxy mergers; the bluish coloration comes from young stars that formed during the merging process.

Figure 16.8(c) shows more detailed views of some of the objects in Figure 16.8(b), all lying in the same region of space, about 1 Mpc across and almost 5000 Mpc from Earth. Each blob seems to contain several billion stars spread throughout a distorted spheroid about a kiloparsec across. Their decidedly bluish tint suggests that active star formation is already under way. We see them as they were nearly 10 billion years ago, a group of young galaxies possibly poised to merge into one or more larger objects.

Hierarchical merging provides the conceptual framework for all modern studies of galaxy evolution. It describes a process that began billions of years ago and continues (albeit at a greatly reduced rate) to the present day, as galaxies continue to collide and merge. By studying how galaxy properties vary with distance, and hence look-back time, astronomers try to piece together the merger history of the universe. ∞ *(More Precisely 15-1)*

Figure 16.9 is a remarkable *Hubble* image showing billions of years of galaxy evolution in a single tiny patch of the sky. The large, bright galaxies with easily discernible Hubble types are mostly (according to their redshifts) relatively nearby objects. They are seen here against a backdrop of small, faint, irregular galaxies lying much farther away. The size and appearance of these distant galaxies compared with those in the foreground strongly support the basic idea that galaxies grew by mergers and were smaller and less regular in the past.

Evolution and Interaction

Left alone, a galaxy will evolve slowly and fairly steadily as interstellar clouds of gas and dust are turned into new generations of stars and main sequence stars evolve into giants and ultimately into compact remnants—white dwarfs, neutron stars, and black holes. ∞ *(Secs. 11.3, 13.5)* The galaxy's overall color, composition, and appearance change in a more or less predictable way as the cycle of stellar evolution recycles and enriches the galaxy's interstellar matter. ∞ *(Sec. 12.7)*

If the galaxy is an elliptical, lacking interstellar gas, it will tend to become fainter and redder in time as its more massive stars burn out and are not replaced. ∞ *(Sec. 12.6)* For a gas-rich galaxy, such as a spiral or irregular, hot, bright stars will lend a bluish coloration to the overall light for as long as gas remains available to form them. As in our own Galaxy, the star-forming lifetime of a spiral disk may be prolonged by infall of fresh gas from the galaxy's surroundings. ∞ *(Sec. 14.4)*

But many galaxies—perhaps most—are not alone. They reside in small groups and clusters, and as we have just seen, their orderly "internal" evolution may be significantly

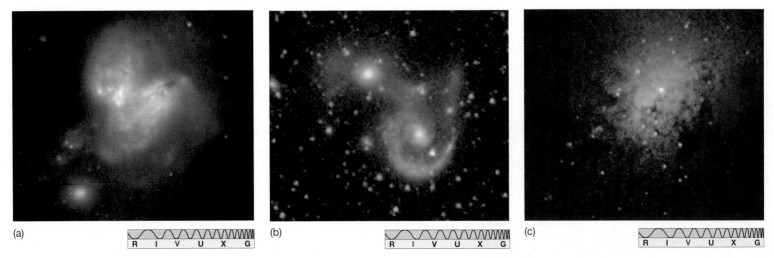

▲ FIGURE 16.10 Starburst Galaxies (a) This interacting galaxy pair (IC 694, at left, and NGC 3690) shows starbursts now under way in both galaxies—hence the bluish tint. (b) This infrared image of a pair of starburst galaxies (called Arp 107) shows numerous young star clusters arrayed as though along a pearl necklace. The colors are false, representing different infrared wavelengths observed by the *Spitzer Space Telescope*. (c) The peculiar (Irr II) galaxy NGC 1275 contains a system of long filaments (in its outskirts) that seem to be exploding outward into space. Its blue blobs are probably young globular clusters formed by the collision of two galaxies. *(W. Keel; NASA)*

complicated by external events—close encounters, mergers, and the accumulation of smaller satellite galaxies over extended periods of time. As described in Section 16.2, these interactions can rearrange a galaxy's internal structure and trigger sudden, intense bursts of star formation. The result is a *starburst galaxy*, some examples of which are shown in Figure 16.10. ⚬ *(Sec. 15.4)* Encounters may also divert fuel to a central black hole, powering violent activity in some galactic nuclei. ⚬ *(Sec. 15.5)* Thus, starbursts and nuclear activity are key indicators of interactions and mergers between galaxies.

Careful studies of starburst galaxies and active galactic nuclei indicate that most such encounters probably took place long ago—in galaxies having redshifts greater than about 1, meaning that we see them as they were roughly 10 billion years ago (see Table 15.2). Nevertheless, the galaxy interactions observed in the local universe are the extension of this same basic process into the present day. We have ample evidence that galaxies have evolved in response to external factors long after the first pregalactic fragments formed and merged.

Computer simulations have shown that the extensive dark-matter halos surrounding most, if not all, galaxies are crucial to galaxy interactions. Consider first two galaxies orbiting one another—a binary galaxy system. As they orbit, the galaxies interact with each other's dark halos, one galaxy stripping halo material from the other by tidal forces. The freed matter is redistributed between the galaxies or is entirely lost from the binary system. In either case, the interaction changes the orbits of the galaxies, causing them to spiral toward one another and eventually to merge.

If one galaxy of the pair happens to have a much lower mass than the other, the process is colloquially termed *galactic cannibalism*. Such cannibalism might explain why supermassive galaxies are often found at the cores of rich galaxy clusters. Having dined on their companions, they now lie at the center of the cluster, waiting for more food to arrive. Figure 16.11 is a remarkable combination of images that has apparently captured this process at work.

▶ FIGURE 16.11 Galactic Cannibalism This computer-enhanced, false-color composite optical photograph of the galaxy cluster Abell 2199 is thought to show an example of galactic cannibalism. The large central galaxy of the cluster is displayed with a superimposed "window." (This results from a shorter time exposure, which shows only the brightest objects that appear within the frame.) Within the core of the large galaxy are several smaller galaxies (the three bright yellow images at center) apparently already "eaten" and now being "digested." *(SAO)*

INTERACTIVE Starburst Galaxy M82 **MA**

MA ANIMATION/VIDEO Hubble Deep Field Zoom I

MA ANIMATION/VIDEO Hubble Deep Field Zoom II

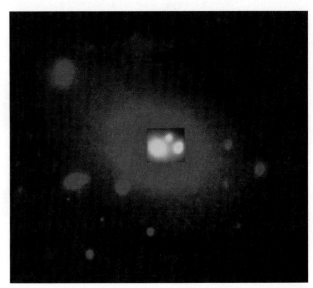

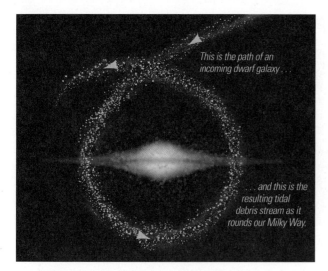

▲ FIGURE 16.12 **Tidal Streams in the Milky Way**   This illustration depicts the breakup and dispersal of the contents of an incoming galaxy companion captured by our Milky Way. Eventually, the smaller galaxy dissolves within the larger one as the smaller one is "digested," much as other dwarf companion galaxies were probably "consumed" by our Galaxy long ago.

CONCEPT CHECK ▶   Why would we expect the evolution of galaxies in voids to differ from that of galaxies in clusters?

Astronomers also have examples of galactic cannibalism closer to home. The small Sagittarius dwarf galaxy (Figure 15.12) is already well on its way to suffering a similar fate at the center of the Milky Way, and theory suggests that the Magellanic Clouds (Figure 15.7) will eventually meet the same fate. As illustrated in Figure 16.12, the "streams" of halo stars, all with similar orbits and composition, in the halo of the Milky Way Galaxy are most likely the stellar remnants of such disruption in the past. ∞ *(Sec. 14.4)*

Now consider two interacting disk galaxies, one a little smaller than the other but each having a mass comparable to the Milky Way Galaxy. As shown in the computer-generated frames of Figure 16.13, the smaller galaxy can substantially distort the larger one, causing spiral arms to appear where none existed before. The entire event requires several hundred million years—a span of evolution that a supercomputer can model in minutes. The final frame of Figure 16.13 looks remarkably similar to the double galaxy shown in Figure 15.2(b), demonstrating how the two galaxies might have interacted millions of years ago and how spiral arms might have been created or enhanced as a result.

What if the colliding galaxies are comparable in size and mass? Computer simulations reveal that such a merger can destroy a spiral galaxy's disk, creating a galaxywide starburst episode. ∞ *(Sec. 15.4)* The violence of the merger and the effects of subsequent supernovae eject much of the remaining gas into intergalactic space, creating the hot intracluster gas noted in Section 16.1. Once the burst of star formation has subsided, the resulting object looks very much like an elliptical galaxy. The elliptical's hot X-ray halo is the last vestige of the original spiral's disk. ∞ *(Sec. 15.1)* The merging galaxies in Figure 16.8 and the irregular galaxies in Figure 16.10 may be examples of this phenomenon in progress. The blue blobs in some of these images are young star clusters formed during the starburst, and the explosive appearance of especially Figure 16.10(c) suggests that we are witnessing gas and dust being ejected.

Making the Hubble Sequence

If galaxies form and evolve by repeated mergers, can we account for the Hubble sequence and, specifically, differences between spirals and ellipticals? ∞ *(Sec. 15.1)* The details are still far from certain, but remarkably, the answer now seems to be a qualified yes. Collisions and close encounters are random events and do not represent a genuine evolutionary sequence linking all galaxies. However, computer simulations suggest a plausible way in which the observed Hubble types might have arisen, starting from a universe populated only by irregular, gas-rich galaxy fragments.

Time ➡

▲ FIGURE 16.13 **Galaxy Interaction**   Galaxies can change their shapes long after their formation. In this computer-generated sequence, two galaxies closely interact over several hundred million years. The smaller galaxy, in red, has gravitationally disrupted the larger galaxy, in blue, changing it into a spiral galaxy. Compare the result of this supercomputer simulation with Figure 15.2(b), a photograph of M51 and its small companion. *(J. Barnes & L. Hernquist)*

As we have just seen, the simulations reveal that "major" mergers—collisions between large galaxies of comparable size—tend to destroy galactic disks, effectively turning spirals into ellipticals. On the other hand, "minor" mergers, in which a small galaxy interacts with, and ultimately is absorbed by, a larger one, generally leave the larger galaxy intact, with more or less the same Hubble type as it had before the merger. This is the most likely way for large spirals to grow—in particular, our own Galaxy probably formed in such a manner.

Supporting evidence for this general picture comes from observations that spiral galaxies are relatively rare in regions of high galaxy density, such as the central regions of rich galaxy clusters. This is consistent with the view that their fragile disks are easily destroyed by collisions, which are more common in dense galactic environments. Spirals also seem to be more common at larger redshifts (that is, in the past), implying that their numbers are decreasing with time, presumably also as the result of collisions. However, nothing in this area of astronomy is clear-cut. Astronomers know of numerous isolated elliptical galaxies in low-density regions of the universe that are hard to explain as the result of mergers.

In principle, the starbursts associated with galaxy mergers leave their imprint on the star-formation history of the universe in a way that can be correlated with the properties of galaxies. As a result, studies of star formation in distant galaxies have become a very important way of testing and quantifying the details of the entire hierarchical merger scenario.

PROCESS OF ▶
SCIENCE CHECK In what ways can astronomers test the predictions of the hierarchical merger scenario?

ANIMATION/VIDEO Galaxy Collision II

16.4 Black Holes in Galaxies

Now let's ask how quasars and active galaxies fit into the framework of galaxy evolution just described. The fact that quasars are more common at great distances from us demonstrates that they were much more prevalent in the past than they are today. ∞ *(Sec. 15.4)* Quasars have been observed with redshifts of up to 7.1 (the current record, as of early 2012), so the process must have started at least 13 billion years ago (see Table 15.2). However, most quasars have redshifts between 2 and 3, corresponding to an epoch some 2 billion years later. Most astronomers agree that quasars represent an early stage of galaxy evolution—an "adolescent" phase of development, prone to frequent flare-ups and outbursts before settling into more steady "adulthood." This view is reinforced by the fact that the same black hole energy-generation mechanism can account for the luminosities of quasars, active galaxies, and the central regions of normal galaxies like our own.

Black Hole Masses

In Chapter 15 we saw the standard model of active galactic nuclei accepted by most astronomers—accretion of gas onto a supermassive black hole. ∞ *(Sec. 15.5)* We also saw that a large fraction of all "bright" galaxies exhibit activity of some sort, even though in many cases it represents only a small fraction of the galaxy's total energy output. This suggests that these galaxies may also harbor central black holes with the potential of far greater activity under the right circumstances. Our own Galaxy is a case in point. ∞ *(Sec. 14.7)* The 3- to 4-million-solar-mass black hole at the center of the Milky Way is not currently active, but if fresh fuel were supplied (say, by a star or molecular cloud coming too close to the hole's intense gravitational field), it might well become a (relatively weak) active galactic nucleus.

In recent years, astronomers have found that many bright normal galaxies have supermassive black holes at their centers. Figure 16.14 presents perhaps the most

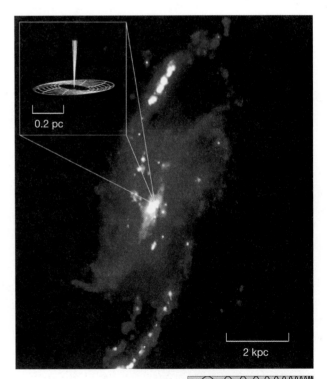

▲ **FIGURE 16.14 Galactic Black Hole** A network of radio telescopes has probed the core of the spiral galaxy NGC 4258, shown here in the light of mostly hydrogen emission. Within the innermost region (inset) a disk of Doppler-shifted molecular clouds (designated by red, green, and blue dots) obey Kepler's third law perfectly, apparently revealing a huge black hole at the center of the disk. *(J. Moran)*

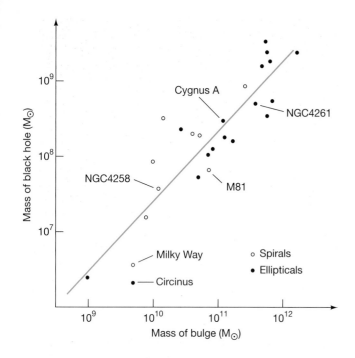

Mass of black hole ($M_\odot$)

Mass of bulge ($M_\odot$)

Cygnus A

NGC4261

NGC4258

M81

Milky Way

Circinus

○ Spirals

● Ellipticals

◄ **FIGURE 16.15 Black Hole Masses**   Observations of nearby normal and active galaxies show that the mass of the central black hole is closely correlated with the mass of the galactic bulge. The straight line is the best fit to the data points for many galaxies, implying a black hole mass of $^1/_{200}$ the mass of the bulge. *(L. Ferrarese)*

compelling evidence for a supermassive black hole at the center of a normal galaxy. Using the Very Long Baseline Array, a continent-wide interferometer comprising 10 radio telescopes, a U.S.-Japanese team has achieved an angular resolution hundreds of times better than that attainable with the *Hubble* telescope. ∞ *(Sec. 3.4)* The observations reveal a group of molecular clouds swirling in an organized fashion about the galaxy's center. Doppler measurements indicate a slightly warped, spinning disk centered precisely on the galaxy's heart. The rotation speeds imply the presence of more than 40 million solar masses packed into a region less than 0.2 pc across.

Similar evidence exists for supermassive black holes in the nuclei of several dozen bright galaxies—some normal, some active—within a few tens of megaparsecs of the Milky Way. Some observers would go so far as to say that in every case where a galaxy has been surveyed and a black hole *could* have been detected, given the resolution and the sensitivity of the observations, a black hole *has* in fact been found. It is a small step to the remarkable conclusion that *every bright galaxy—active or not—contains a central supermassive black hole.* This unifying principle connects our theories of normal and active galaxies in a fundamental way.

Astronomers have also found a correlation between the masses of the central black holes and the properties of the galaxy in which they reside. As illustrated in Figure 16.15, the largest black holes tend to be found in the most massive galaxies (as measured by the mass of the bulge). The reason for this correlation is not fully understood, but most astronomers take it to mean that the evolution of normal and active galaxies must be very closely connected, as we now discuss.

The Quasar Epoch

Where did the supermassive black holes in galaxies come from? To be honest, the processes whereby the first billion-solar-mass black holes formed early in the history of the universe are not fully understood. However, the accretion responsible for the energy emission also naturally accounts for the mass of the black hole—only a few percent of the infalling mass is converted to energy; the rest is trapped forever in the black hole once it crosses the event horizon. ∞ *(Sec. 13.5)*. Simple estimates suggest that the accretion rates needed to power the quasars are generally consistent with black-hole masses inferred by other means.

Since the brightest known quasars devour about a thousand solar masses of material every year, it is unlikely that they could maintain their luminosity for very long—even a million years would require a billion solar masses, enough to account for the most massive black holes known. ∞ *(Sec. 15.4)* This suggests that a typical quasar spends only a fairly short amount of time in its highly luminous phase—perhaps a few million years—before running out of fuel. Thus, most quasars were relatively brief events that occurred long ago.

To make a quasar, we need a black hole and enough fuel to power it. While fuel was abundant at early times in the universe's history in the form of gas and newly formed stars, black holes were not. They had yet to form, most probably by the same basic stellar evolutionary processes we saw in Chapter 12, although the details are still not known. ∞ *(Sec. 12.4)* The building blocks of the supermassive black holes that would ultimately power the quasars may well have been relatively small black holes having masses tens or perhaps a few hundreds of times the mass of the Sun. These small black holes sank to the center of their still-forming–parent galaxy and merged to form a single, more massive black hole.

As galaxies merged, so, too, did their central black holes, and eventually supermassive (1-million- to 1-billion-solar mass) black holes existed in the centers of many young galaxies. Some supermassive black holes may have formed directly by the gravitational collapse of the dense central regions of a protogalactic fragment or perhaps by accretion or a rapid series of mergers in a particularly dense region of the universe. These events resulted in the earliest (redshift 6–7) quasars known, shining brightly 13 billion years ago. However, in most cases, the mergers took longer—roughly another 2 billion years. By then (at redshifts between 2 and 3, roughly 11 billion years ago), many supermassive black holes had formed, and there was still plenty of merger-driven fuel available to power them. This was the height of the "quasar epoch" in the universe.

Until recently, astronomers were confident that black holes would merge when their parent galaxies collided, but they had no direct evidence of the process— no image of two black holes "caught in the act." In 2002, the *Chandra* X-ray observatory discovered a binary black hole—two supermassive objects, each having a mass a few tens of millions of times that of the Sun—in the center of the ultraluminous starburst galaxy NGC 6240, itself the product of a galaxy merger some 30 million years ago. Figure 16.16 shows optical and X-ray views of the system. The black holes are the two blue-white objects near the center of the (false-color) X-ray image. Orbiting just 1000 pc apart, they are losing energy through interactions with stars and gas and are predicted to merge in about 400 million years. Astronomers now know of several binary black holes in relatively nearby galaxies, caught in the act of spiraling together on their way to merging. NGC 6240 lies just 120 Mpc from Earth, so we are far from seeing a quasar merger in the early universe. Nevertheless, astronomers think that events similar to this must have occurred countless times billions of years ago, as galaxies collided and quasars blazed.

Distant galaxies are generally much fainter than their bright quasar cores. As a result, until quite recently, astronomers were hard-pressed to discern any galactic structure in quasar images. Since the mid-1990s, several groups of astronomers have used the *Hubble Space Telescope* to search for the "host" galaxies of moderately distant quasars. After removing the bright quasar core from the *HST* images and carefully analyzing the remnant light, the researchers have reported that in every case studied—several dozen quasars so far—a host galaxy can be seen enveloping the quasar. Figure 16.17 shows some of the longest quasar exposures ever taken. Even without sophisticated computer processing, the hosts are clearly visible.

As we saw in Chapter 15, the connection between active galaxies and galaxy clusters is well established, and many relatively nearby quasars are also known to be members of clusters. ⊂⊃ *(Sec. 15.4)* The link is less clear-cut for the most distant quasars, however, simply because they are so far away that other cluster members are very faint and extremely hard to see. However, as the number of known quasars continues to increase, evidence for quasar clustering (and presumably, therefore, for quasar membership in young galaxy clusters) mounts. Thus, as best we can tell, quasar activity—and, in fact, galaxy activity of *all* sorts—is intimately related to interactions and collisions in galaxy clusters.

This connection also suggests a possible way in which the growth of black holes might be tied to the growth of their parent galaxies. Astronomers speculate that a process called **quasar feedback,** in which some fraction of the quasar's enormous energy output is absorbed by the surrounding galactic gas, might help explain the correlation of black hole and bulge masses shown in Figure 16.15. According to

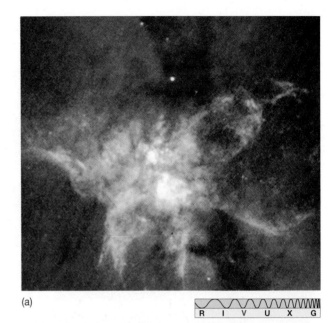

(a)

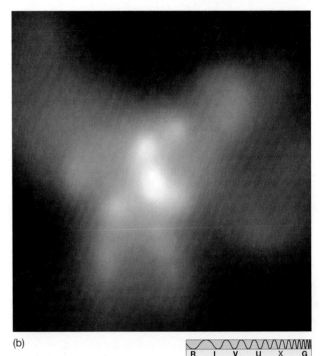

(b)

▲ **FIGURE 16.16 Binary Black Hole** These (a) optical (*Hubble*) and (b) X-ray (*Chandra*) images of the starburst galaxy NGC 6420 show two supermassive black holes (the blue-white objects near the center of the X-ray image) orbiting about 1 kpc apart. Theoretical estimates imply that they will merge in about 400 million years, releasing an intense burst of gravitational radiation. ⊂⊃ *(Discovery 13-1)* *(NASA)*

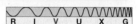

▲ FIGURE 16.17 **Quasar Host Galaxies**   These long-exposure images of distant quasars show the young host galaxies in which the quasars reside, lending support to the idea that quasars represent an early, very luminous phase of galactic evolution. The quasar at the left is the best example, having the catalog name PG0052 + 251 and residing roughly 690 Mpc from Earth. *(NASA)*

this view, which is appealing but by no means certain, the absorbed energy expels the gas from the galaxy, simultaneously shutting down both galactic star formation and the quasar's own fuel supply, thus tying the growth of the central black hole to the formation of new stars in the bulge.

Active and Normal Galaxies

Early on, frequent mergers may have replenished the quasar's fuel supply, extending its luminous lifetime. However, as the merger rate declined, these systems spent less and less of their time in the "bright" phase. The rapid decline in the number of bright quasars roughly 10 billion years ago marks the end of the quasar epoch. Today, the number of quasars has dropped virtually to zero (recall that the nearest lies hundreds of megaparsecs away). ∞ *(Sec. 15.4)*

Large black holes do not simply vanish. If a galaxy contained a bright quasar 10 billion years ago, the black hole responsible for all that youthful activity must still be present in the center of the galaxy today. We see some of these black holes as active galaxies. The remainder reside dormant in normal galaxies all around us. In this view, *the difference between an active galaxy and a normal one is mainly a matter of fuel supply.* When the fuel runs out and a quasar shuts down, its central black hole remains behind, its energy output reduced to a relative trickle. The black holes at the hearts of normal galaxies are simply quiescent, awaiting another interaction to trigger a new active outburst. Occasionally, two nearby galaxies may interact with each other, causing a flood of new fuel to be directed toward the central black hole of one or both. The engine starts up for a while, giving rise to the nearby active galaxies—Seyferts, radio galaxies, and others— we observe.

Should this general picture be correct, it follows that many relatively nearby galaxies (but probably *not* our own Milky Way, whose central black hole is even now only a paltry 3–4 million solar masses) must once have been brilliant quasars. ∞ *(Sec. 14.7)* Perhaps some alien astronomer, thousands of megaparsecs away, is at this very moment observing the progenitor of M87 in the Virgo Cluster—seeing it as it was billions of years ago—and is commenting on its enormous luminosity, nonstellar spectrum, and high-speed jets and wondering what exotic physical process can possibly for its violent activity! ∞ *(Sec. 14.4)*

Finally, Figure 16.18 suggests some possible (but unproven) evolutionary connections among quasars, active galaxies, and normal galaxies. If the largest black holes reside in the most massive galaxies, and also tend to power the brightest active galactic nuclei, then we would expect that the most luminous nuclei should reside in the largest galaxies, which probably came into being via "major" mergers of other large galaxies. Since the products of such mergers are elliptical galaxies, we have a plausible explanation of why the brightest active galaxies— the radio galaxies—should be associated with large ellipticals. ∞ *(Sec. 15.4)* Furthermore, the path to spiral galaxies would necessarily have entailed a series of mergers involving smaller galaxies, resulting in the less violent Seyferts along the way.

PROCESS OF ▶
SCIENCE CHECK   How does the existence of supermassive black holes in normal galaxies fit in with theories of active galaxy evolution?

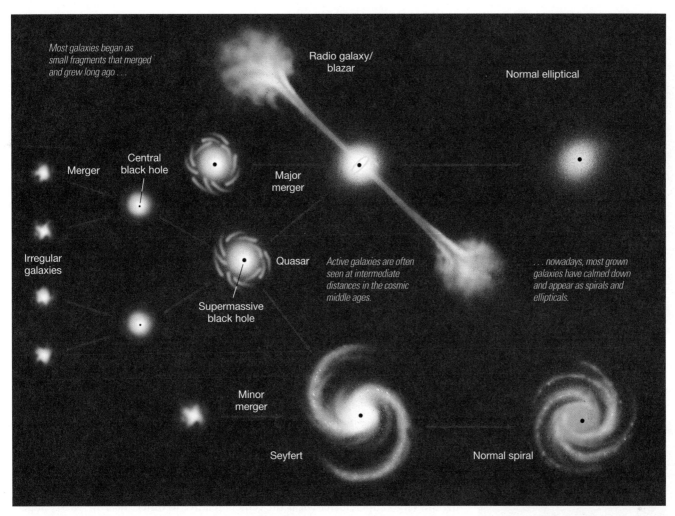

▲ **FIGURE 16.18 Galaxy Evolution**   Most evolutionary sequences for galaxies began with galaxy mergers that led to highly luminous quasars, after which they decreased in violence through the radio and Seyfert galaxies, eventually resulting in normal ellipticals and spirals. The central black holes that powered the early activity remain at later times, but many of them have run out of fuel.

Active Galaxies and the Scientific Method

When active galactic nuclei—and especially quasars—were first discovered, their extreme properties defied conventional explanation. Initially, the idea of supermassive (million- to billion-solar mass) black holes in galaxies was just one of several competing, and very different, hypotheses advanced to account for the enormous luminosities and small sizes of those baffling objects. Some astronomers suggested instead that the energy source might be multiple supernova explosions, others that perhaps some exotic form of matter–antimatter annihilation was taking place. A few went so far as to propose that these inexplicable objects demanded an even more radical explanation: that Hubble's law was incorrect and that the large redshifts of quasars had some other, unknown cause.

However, as observational evidence mounted, the other hypotheses were abandoned one by one, and massive black holes in galactic nuclei became first the leading, and eventually the standard, theory of active galaxies. As often happens in science, a theory once itself considered extreme is now the accepted explanation. Far from threatening the laws of physics, as some astronomers once feared, active galaxies are now an integral part of our understanding of how galaxies form and evolve. The synthesis of studies of normal and active galaxies, galaxy formation, and large-scale structure is one of the great triumphs of extragalactic astronomy.

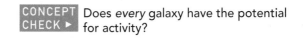

CONCEPT CHECK ▶   Does *every* galaxy have the potential for activity?

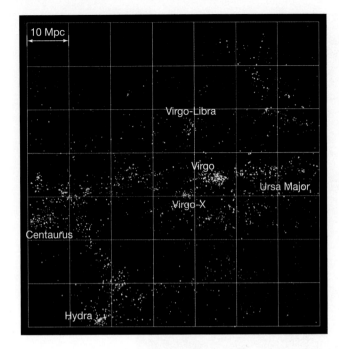

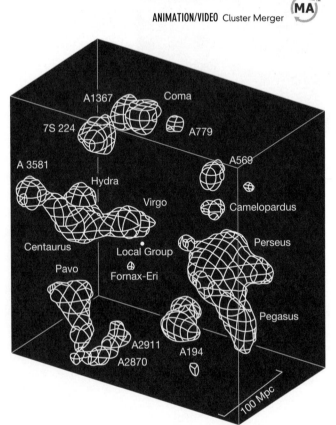

▲ **FIGURE 16.20 Virgo Supercluster in 3-D** The elongated shape of the Virgo Supercluster (left) is mapped relative to other neighboring galaxy superclusters within about 100 Mpc of the Milky Way (within the Local Group near center). Individual galaxies are not shown; rather, smoothed contour plots outline galaxy clusters, each named or numbered by its most prominent member.

◄ **FIGURE 16.19 Local Supercluster** More than 4500 galaxies are plotted here in the vicinity of the Virgo cluster, and several prominent galaxy clusters are labeled. The diagram depicts the Virgo Supercluster roughly as we see it from our own Galaxy, which is located approximately 20 Mpc (two grid squares) above the page. Notice the supercluster's irregular, elongated shape. (B. Tully and S. Levy)

16.5 The Universe on Very Large Scales

Many galaxies, including our own, are members of galaxy clusters—megaparsec-sized structures held together by their own gravity. ∞ *(Sec. 15.2)* Our own small cluster is called the Local Group. Figure 16.19 shows the locations of the Virgo Cluster, the closest "large" cluster, and of several other well-defined clusters in our cosmic neighborhood. The region displayed is about 70 Mpc across. Each point in the figure represents an entire galaxy whose distance has been determined by one of the methods described in Chapter 15.

Clusters of Clusters

Do galaxy clusters top the cosmic hierarchy, or does the universe have even larger groupings of matter? Most astronomers have concluded that the galaxy clusters are themselves clustered, forming titanic agglomerations of matter known as **superclusters.**

Together, the galaxies and clusters shown in Figure 16.19 form the *Local Supercluster*, also known as the Virgo Supercluster. Aside from the Virgo Cluster itself, it contains the Local Group and numerous other clusters lying within about 20–30 Mpc of Virgo. Most of the galaxies depicted in the figure are fairly large spirals and ellipticals; the fainter irregulars and dwarfs are not included in the diagram. Figure 16.20 shows a three-dimensional rendering of an even wider view, illustrating the Virgo Supercluster (near the center) relative to other "nearby" galaxy superclusters within a vast imaginary rectangle roughly 100 Mpc on its short side.

All told, the Local Supercluster is about 40–50 Mpc across, contains some 10^{15} solar masses of material (several tens of thousands of galaxies), and is very irregular in shape. The Local Supercluster is significantly elongated perpendicular to the line joining the Milky Way to Virgo, with its center lying near the Virgo Cluster. By now it should come as no surprise that the Local Group is *not* found at the heart of the Local Supercluster—we live far off in the periphery, about 18 Mpc from the center.

Redshift Surveys

The farther we peer into deep space, the more galaxies, clusters of galaxies, and superclusters we see. Is there structure on even larger scales? To answer this question, astronomers use Hubble's law to map out the distribution of galaxies in the universe and have developed indirect techniques to probe the dark recesses of intergalactic space.

Figure 16.21 shows part of an early survey of the universe performed by astronomers at Harvard University in the 1980s. Using Hubble's law as a distance indicator, the team systematically mapped out the locations of galaxies within about 200 Mpc of the Milky Way in a series of wedge-shaped "slices," each 6° thick, starting in the northern sky. The first slice (Figure 16.21) covered a region of the sky containing the Coma Cluster (see Figure 15.1), which happens to lie in a direction almost perpendicular to our Galaxy's plane. Because redshift is used as the primary distance indicator, these studies are known as *redshift surveys*.

▶ **FIGURE 16.21 Galaxy Survey** A "slice" of the universe, covering 1732 galaxies out to a distance of about 200 Mpc, shows that galaxies and clusters are not randomly distributed on large scales. Instead, they appear to have a filamentary structure, surrounding vast, nearly empty voids. *(Harvard-Smithsonian Center for Astrophysics)*

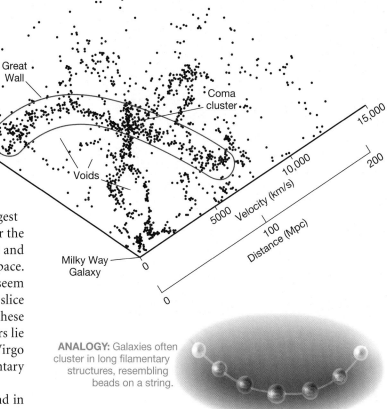

ANALOGY: Galaxies often cluster in long filamentary structures, resembling beads on a string.

The most striking feature of maps such as this is that the distribution of galaxies on very large scales is decidedly nonrandom. The galaxies appear to be arranged in a network of strings, or filaments, surrounding large, relatively empty regions of space known as **voids.** Voids account for some 50 percent of the total volume of the nearby universe, but only 5–10 percent of the mass. The biggest voids measure some 100 Mpc across. The most likely explanation for the voids and filamentary structure in Figure 16.21 is that the galaxies and galaxy clusters are spread across the surfaces of vast "bubbles" in space. The voids are the interiors of these gigantic bubbles. The galaxies seem to be distributed like beads on strings only because of the way our slice of the universe cuts through the bubbles. Like suds on soapy water, these bubbles fill the entire universe. The densest clusters and superclusters lie in regions where several bubbles meet. The elongated shape of the Virgo Supercluster (see Figure 16.19) is a local example of this same filamentary structure.

Most theorists think that this "frothy" distribution of galaxies, and in fact all structure on scales larger than a few megaparsecs, traces its origin directly to conditions in the very earliest stages of the universe (Chapter 17). As such, studies of large-scale structure are vital to our efforts to understand the origin and nature of the cosmos itself. The idea that the filaments are the intersection of the survey slice with much larger structures (the bubble surfaces) was confirmed when the next three slices of the survey, lying above and below the first, were completed. The region of Figure 16.21 indicated by the red outline was found to continue through both the other slices. This extended sheet of galaxies, which has come to be known as the *Great Wall*, measures at least 70 Mpc (out of the plane of the page) by 200 Mpc (across the page). It is one of the largest known structures in the universe.

Figure 16.22 shows a more recent redshift survey, considerably larger than the one presented in Figure 16.21. This survey includes nearly 24,000 galaxies within about 750 Mpc of the Milky Way. Numerous voids and "Great Wall-like" filaments can be seen (some are marked), but apart from the general fall-off in numbers of galaxies at large distances—basically because the more distant galaxies are harder to see due to the inverse-square law—there is no obvious evidence for any structures on scales larger than about 200 Mpc. Careful statistical analysis and even larger surveys (see *Discovery 16-1*) confirm this impression. Apparently, voids and walls represent the largest structures in the universe. We will return to the far-reaching implications of this fact in Chapter 17.

Quasar Absorption Lines

How can we probe the structure of the universe on very large scales? As we have seen, much of the matter is dark, and even the "luminous" component is so faint that it is hard to detect at large distances. One way to study large-scale structure is to take advantage of the great distances, point-like appearance, and large luminosities of quasars, in much the same way as astronomers use bright

CONCEPT CHECK ▶ What do observations of distant quasars tell us about the structure of the universe closer to home?

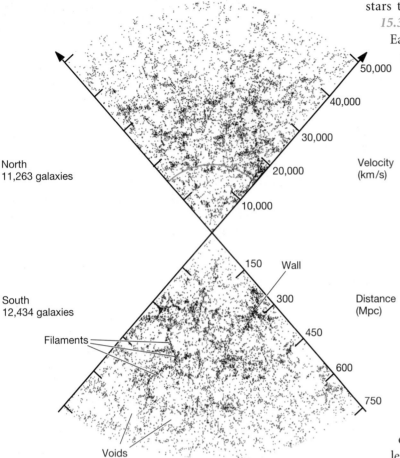

North
11,263 galaxies

South
12,434 galaxies

Velocity (km/s)

Distance (Mpc)

Wall

Filaments

Voids

▲ FIGURE 16.22 **The Universe on Larger Scales**   This large-scale galaxy survey, carried out at the Las Campanas Observatory in Chile, consists of 23,697 galaxies within about 1000 Mpc, in two 80° (wide) × 4.5° (thick) wedges of the sky. Many voids and "walls" on scales of up to 100–200 Mpc are evident, but nothing much larger. For scale, the extent of the survey shown in Figure 16.21 is marked as a thin blue arc in the northern sky.

INTERACTIVE Quasar Spectrum due to Absorption from Distant Clouds

stars to probe the interstellar medium near the Sun. ∞ *(Secs. 11.1, 15.3)* Since quasars are so far away, light traveling from a quasar to Earth has a pretty good chance of passing through or near something "interesting" en route. By analyzing quasar images and spectra, it is possible to piece together a partial picture of the intervening space.

In addition to their own strongly redshifted spectra, many quasars also show additional absorption features that are redshifted by substantially *less* than the lines from the quasar itself. For example, the quasar PHL 938 has an emission-line redshift of 1.954, placing it at a distance of some 5200 Mpc, but it also shows absorption lines having redshifts of just 0.613. These are interpreted as arising from intervening gas that is much closer to us (only about 2200 Mpc away) than the quasar itself. Most probably this gas is part of an otherwise invisible galaxy lying along the line of sight. Quasar spectra, then, afford astronomers a means of probing previously undetected parts of the universe.

The absorption lines of atomic hydrogen are of particular interest, since hydrogen makes up so much of all matter in the cosmos. Specifically, hydrogen's ultraviolet (122 nm) "Lyman-alpha" line, associated with transitions between the ground and first excited states, is often used in this context. ∞ *(Sec. 2.6)* As illustrated in Figure 16.23, when astronomers observe the spectrum of a high-redshift quasar, they typically see a "forest" of absorption lines starting at the (redshifted) wavelength of the quasar's own Lyman-alpha emission line and extending to shorter wavelengths. These lines are interpreted as Lyman-alpha absorption features produced by gas clouds in foreground structures—galaxies, clusters, and so on—giving astronomers crucial information about the distribution of matter along the line of sight.

Quasar light thus explores an otherwise invisible component of cosmic gas. In principle, every intervening cloud of atomic hydrogen leaves its own characteristic imprint on the quasar's spectrum, in a form that lets us explore the distribution of matter in the universe. By comparing these *Lyman-alpha forests* with the results of simulations, astronomers hope to refine many key elements of the theories of galaxy formation and the evolution of large-scale structure.

▶ FIGURE 16.23 **Absorption Line "Forest"**   The huge number of absorption lines in the spectrum of quasar QSO 1422 + 2309 are the ultraviolet Lyman-alpha lines from hundreds of clouds of foreground hydrogen gas, each redshifted by a slightly different amount (but less than the quasar itself). The peak at left marks the Lyman-alpha emission line from the quasar, emitted 122 nm but redshifted here to 564 nm, in the visible range.

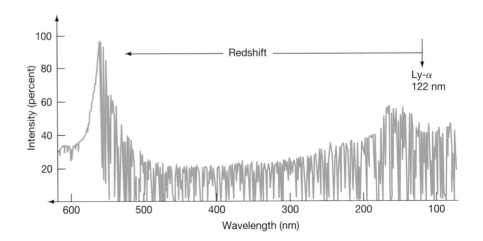

16-1 DISCOVERY

The Sloan Digital Sky Survey

Many of the photographs used in this book—not to mention most of the headline-grabbing imagery found in the popular media—come from large, high-profile, and usually very expensive, instruments such as NASA's *Hubble Space Telescope* and the European Southern Observatory's Very Large Telescope in Chile. ⊂⊃ *(Secs. 3.3, 3.4)* Their spectacular views of deep space have revolutionized our view of the universe, yet a less well-known, considerably cheaper, but no less ambitious, project completed in 2008 may, in the long run, have every bit as great an impact on astronomy and our understanding of the cosmos.

The Sloan Digital Sky Survey (SDSS) was originally a 5-year project, and was subsequently extended for 3 more years. It was designed to systematically map out a quarter of the entire sky on a scale and at a level of precision never before attempted. It cataloged almost 500 million celestial objects, recording their apparent brightnesses at five different colors (wavelength ranges) spread across the optical part of the spectrum. In addition, spectroscopic follow-up observations determined redshifts and hence distances to 1 million galaxies and 120,000 quasars. These data have been used to construct even more detailed redshift surveys than those described in the text and to probe the structure of the universe on very large scales. The sensitivity of the survey was such that it could detect bright galaxies like our own out to distances of more than 1 billion parsecs. Very bright objects, such as quasars and young starburst galaxies, were detectable almost throughout the entire observable universe.

The first figure shows the Sloan Survey telescope, a special-purpose 2.5-m instrument sited in Apache Point Observatory, near Sunspot, New Mexico. The telescope is not space-based, does not employ active or adaptive optics, and cannot probe as deeply (i.e., far)

into space as do larger instruments. How could it possibly compete with these other systems? The answer is that, unlike most other large telescopes in current use, where hundreds or even thousands of observers share the instrument and compete for its time, the SDSS telescope was designed specifically for the purpose of the survey. It has a wide field of view and was dedicated to the task, carrying out observations of the sky on *every* clear night during the duration of the project.

The survey used this single instrument night after night, applying tight quality controls on which nights' data are actually incorporated into the survey. Nights with poor seeing or other problematic conditions were discarded and the observations repeated. The end product was a database of exceptionally high quality and uniformity spanning an enormous volume of space—a monumental achievement and an indispensable tool for cosmology. The survey field of view covers much of the sky away from the Galactic plane in the north, together with three broad "wedges" in the south.

Archiving images and spectra on millions of galaxies produces a lot of data. The total amount of data gathered by the survey's 120-megapixel camera comprises roughly 15 *trillion* bytes of information—comparable to the entire Library of Congress! All of the survey data have now been released to the public. The second figure shows an image of the Perseus galaxy cluster, just one of hundreds of thousands of images that make up the full dataset. Among recent highlights, SDSS has detected the largest known structure in the universe, observed the most distant known galaxies and quasars, and has been instrumental in pinning down the key observational parameters describing our universe (see Chapter 17).

SDSS impacts astronomy in areas as diverse as the large-scale structure of the universe, the origin and evolution of galaxies, the nature of dark matter, the structure of the Milky Way, and the properties and distribution of interstellar matter. Its uniform, accurate, and detailed database is likely to be used by generations of scientists for decades to come, and its success has spawned several even more ambitious follow-up surveys; the first is due to become operational in 2015.

(SDSS; R. Lupton)

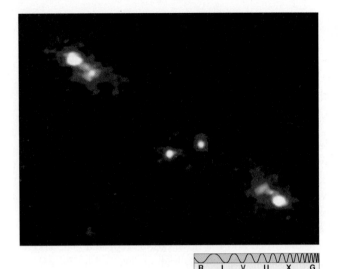

R I V U X G

◀ FIGURE 16.24 Twin Quasar   This twin quasar (designated AC114 and located about 2 billion parsecs away) is not two separate objects at all. Instead, the two large "blobs" (at upper left and lower right) are images of the same object, created by a gravitational lens. The lensing galaxy itself is probably not visible in this image—the two objects near the center of the frame are thought to be unrelated galaxies in a foreground cluster. (NASA)

Quasar "Mirages"

In 1979 astronomers were surprised to discover what appeared to be a binary quasar—two quasars with exactly the same redshift and very similar spectra, separated by only a few arc seconds on the sky. Remarkable as the discovery of such a binary would have been, the truth about this pair of quasars turned out to be even more amazing. Closer study of the quasars' radio emission revealed that they were *not* two distinct objects. Instead, they were two separate images of the *same* quasar! Optical views of such a *twin quasar* are shown in Figure 16.24.

What could produce such a "doubling" of a quasar image? The answer is gravitational lensing—the deflection and focusing of light from a background object by the gravity of some foreground body (Figure 16.25). In Chapter 14 we saw how lensing by compact objects in the halo of the Milky Way Galaxy may amplify the light from a distant star, allowing astronomers to detect otherwise invisible stellar dark matter. ∞ *(Sec. 14.5)* In the case of quasars, the idea is the same, except that the foreground lensing object is an entire galaxy or galaxy cluster and the deflection of the light is so great (a few arc seconds) that several separate images of the quasar may be formed, as shown in Figure 16.26. About two dozen such gravitational lenses are known.

The existence of these multiple images provides astronomers with a number of useful observational tools. First, lensing by a foreground galaxy tends to amplify the light of the quasar, as just mentioned, making it easier to observe. At the same time, *microlensing* by individual stars within the galaxy may cause large fluctuations in the quasar's brightness, allowing astronomers to study the galaxy's stellar content.

Second, because the light rays forming the images usually follow paths of different lengths, there is often a time delay, ranging from days to years, between them. This delay provides advance notice of interesting events, such as sudden changes in the quasar's brightness—if one image flares up, in time the other(s) will too, giving astronomers a second chance to study the event. The time delay also allows astronomers to determine the distance to the lensing galaxy. This method provides an alternative means of measuring Hubble's constant that is independent of any of the techniques discussed previously. The average value of H_0 reported by workers using this approach is consistent with the 70 km/s/Mpc value used throughout this text.

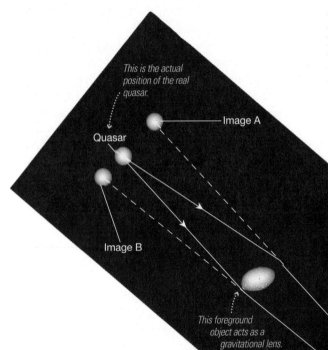

This is the actual position of the real quasar.

Image A

Quasar

Image B

This foreground object acts as a gravitational lens.

An observer at Earth sees the quasar image duplicated at A and B.

INTERACTIVE ▶ FIGURE 16.25 Gravitational Lens
When light from a distant object passes close to a galaxy or cluster of galaxies along the line of sight, the image of the background object (here, the quasar) can sometimes be split into two or more separate images (A and B). The foreground object is a gravitational lens.

MA

▶ FIGURE 16.26 Einstein Cross (a) The Einstein Cross, a multiply-imaged quasar, spans only a couple of arc seconds, showing four separate images of the same quasar produced by the galaxy at the center. (b) A simplified artist's conception of what might be occurring here, with Earth at right and the distant quasar at left. *(NASA; D. Berry)*

Mapping Dark Matter

Astronomers have extended the ideas first learned from studies of quasars to use lensing of any distant object to probe the universe. Distant faint irregular galaxies—the raw material of the universe, if current theories are correct (see Section 16.4)—are of particular interest here, as they are far more common than quasars; thus, they provide much better coverage of the sky. By studying the lensing of background quasars and galaxies by foreground galaxy clusters, astronomers can obtain a better understanding of the distribution of dark matter on large scales.

Figures 16.27(a) and (b) show how the images of faint, background galaxies are bent into arcs by the gravity of a foreground galaxy cluster. The degree of bending allows the total mass of the cluster (including the dark matter) to be measured. The (mostly blue) loop- and arc-shaped features visible in Figure 16.27(b) are multiple images of a single, distant (unseen) spiral or ring-shaped galaxy, lensed by the foreground galaxy cluster (the yellow-red blobs in the image).

It is even possible to reconstruct the foreground dark mass distribution by carefully analyzing the distortions of the background objects, providing a means of tracing out the distribution of matter on scales far larger than have previously been possible Figure 16.28(a) is an optical image of a (hard to see) foreground galaxy cluster set against a background of much fainter distant galaxies. Figure 16.28(b) is the reconstructed dark-matter image, revealing the presence of dark mass many megaparsecs from the cluster center, extending far beyond the visible galaxies. Evidently, the dark matter fraction is even greater on supercluster scales than it is on the scales of individual galaxies or galaxy clusters. Notice also the elongated structure of the dark-matter distribution, reminiscent of the Virgo Supercluster and filamentary structure seen in large-scale galaxy surveys. ∞ *(Sec. 14.5)*

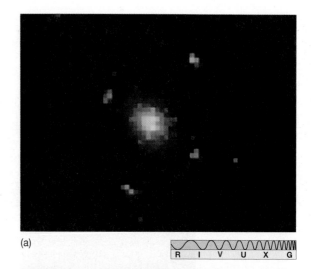

(a)

(b)

MA ANIMATION/VIDEO Simulation of Gravitational Lens in Space

(a)

(b)

INTERACTIVE ▲ FIGURE 16.27 Galaxy Cluster Lensing (a) This spectacular example of gravitational lensing shows more than a hundred faint arcs from very distant galaxies. The wispy pattern spread across the foreground galaxy cluster (A2218, about a billion parsecs distant) resembles a spider's web, but it is really an illusion caused by A2218's gravitational field, which deflects the light from background galaxies and distorts their appearance. (b) Another galaxy cluster, known only by its catalog name 0024 + 1654 and residing some 1.5 billion pc away, shows reddish-yellow blobs that are mostly normal elliptical galaxies and bluish loop-like features **MA** that are images of a single background galaxy. *(NASA)*

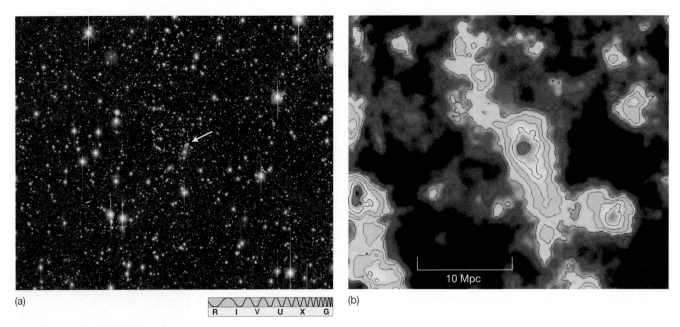

(a)

(b)

10 Mpc

R I V U X G

▲ FIGURE 16.28 **Dark Matter Map**   By measuring distortions in the images of background objects, astronomers can make maps of dark matter in the universe. Analysis of an optical view (a) of a region of the sky containing a small galaxy cluster (the clump of yellowish galaxies near the center of the frame indicated by the arrow) reveals the distribution of dark matter (b) in and near the visible cluster and on the same scale as (a). *(J. A. Tyson, Bell Labs; Alcatel-Lucent; NOAO)*

ANIMATION/VIDEO The Cosmic Web of Dark Matter **(MA)**

In 2006, astronomers used these techniques to obtain what may be the first direct observational evidence for dark matter. Figure 16.29 shows combined optical and X-ray images of a distant galaxy cluster called 1E 0657-56. The fuzzy red region shows the location of the hot X-ray–emitting gas in the system, the dominant *luminous* component of the mass. The blue regions indicate where most of the mass actually lies, as determined from lensing studies of background galaxies. Note that

The arrows indicate the directions in which the two clusters are now moving, subsequent to what might have been the most energetic collision in the universe since the Big Bang.

INTERACTIVE ▶ FIGURE 16.29 **Cluster Collision**   Clusters of galaxies must also occasionally collide, as is apparently the case here for this combined cluster with the innocuous name 1E 0657-56 and the nickname "bullet cluster." This is a composite image of a region about 1 billion parsecs away, showing optical light from the galaxies themselves in white and X-ray emission from the hot intracluster gas in red. By contrast, the blue color represents the inferred dark matter within the two large clusters that is distinctly displaced from their normal matter. *(NOAO/NASA)* **(MA)**

R I V U X G

the bulk of the mass is *not* found in the form of hot gas, implying that the dark matter is distributed differently from the "normal" matter in the cluster.

The explanation for this odd state of affairs is that we are witnessing a collision between two clusters. Each initially contained hot gas and dark matter distributed throughout the cluster, but when the two collided, the pressure of each gas cloud effectively stopped the other, leaving the gas behind in the middle as the galaxies and dark matter moved on. This separation between the gas and the dark matter directly contradicts some alternative theories of gravity that have been invoked to avoid the "dark matter problem" in galaxies and clusters and may prove to be a crucial piece of evidence in our understanding of large-scale structure in the universe.

CHAPTER REVIEW

SUMMARY

L01 The masses of nearby spiral galaxies can be determined by studying their rotation curves. Astronomers also use studies of binary galaxies and distant galaxy clusters to obtain statistical mass estimates of the galaxies involved.

L02 Measurements of galaxy and cluster masses reveal the presence of large amounts of dark matter. The fraction of dark matter grows as the scale under consideration increases. More than 90 percent of the mass in the universe is dark. Substantial quantities of hot X-ray–emitting gas have been detected among the galaxies in many clusters, but not enough to account for the dark matter inferred from dynamical studies.

L03 Researchers know of no simple evolutionary sequence that links spiral, elliptical, and irregular galaxies. Most astronomers think that large galaxies formed by the merger of smaller ones and collisions and mergers of galaxies play very important roles in galactic evolution. A starburst galaxy may result when a galaxy has a close encounter or a collision with a neighbor. The strong tidal distortions caused by the encounter compress galactic gas, resulting in a widespread burst of star formation. Mergers between spirals most likely result in elliptical galaxies.

L04 Quasars, active galaxies, and normal galaxies may represent an evolutionary sequence. When galaxies began to form and merge, conditions may have been suitable for the formation of large black

holes at their centers, and highly luminous quasars could have been the result. The brightest quasars consume so much fuel that their energy-emitting lifetimes must be quite short. As the fuel supply diminished, the quasar dimmed, and the galaxy in which it was embedded became intermittently visible as an active galaxy. At even later times, the nucleus became virtually inactive, and a normal galaxy was all that remained. Many normal galaxies have been found to contain central massive black holes, suggesting that most galaxies in clusters have the capacity for activity if they should interact with a neighbor.

L05 Galaxy clusters themselves tend to clump together into **superclusters** (p. 448). The Virgo Cluster, the Local Group, and several other nearby clusters form the Local Supercluster. On even larger scales, galaxies and galaxy clusters are arranged on the surfaces of enormous "bubbles" of matter surrounding vast low-density regions called **voids** (p. 449). The origin of this structure is thought to be closely related to conditions in the very earliest epochs of the universe.

L06 Quasars can be used as probes of the universe along the line of sight. Some quasars have been observed to have double

or multiple images. These result from gravitational lensing, in which the gravitational field of a foreground galaxy or galaxy cluster bends and focuses the light from the more distant quasar. Analysis of the images of distant galaxies, distorted by the gravitational effect of a foreground cluster, provides a means of determining the masses of galaxy clusters—including the dark matter—far beyond the information that optical images of the galaxies themselves afford.

MasteringAstronomy®   For instructor-assigned homework go to www.masteringastronomy.com

Problems labeled **POS** explore the process of science.   **VIS** problems focus on reading and interpreting visual information.   **LO** connects to the introduction's numbered Learning Outcomes.

REVIEW AND DISCUSSION

1. **LO1** Describe two techniques for measuring the mass of a galaxy.

2. **LO2** Why do astronomers think that galaxy clusters contain more mass than we can see?

3. Why do some clusters of galaxies emit X-rays?

4. What evidence do we have that galaxies collide with one another?

5. **LO3** Describe the role of collisions in the formation and evolution of galaxies.

6. **POS** Do you think that collisions between galaxies constitute "evolution" in the same sense as the evolution of stars? Why or why not?

7. What are starburst galaxies, and what do they have to do with galaxy evolution?

8. Why do astronomers think that quasars represent an early stage of galactic evolution?

9. **LO4 POS** Why does the theory of galaxy evolution suggest that there should be supermassive black holes at the centers of many normal galaxies?

10. What evidence do we have for supermassive black holes in galaxies?

11. How might a normal galaxy become active?

12. What is a redshift survey?

13. **LO5** What are voids? What is the distribution of galactic matter on very large (more than 100 Mpc) scales?

14. **LO6 POS** How can observations of distant quasars be used to probe the space between us and them?

15. **POS** How can astronomers "see" dark matter?

CONCEPTUAL SELF-TEST: TRUE OR FALSE?/MULTIPLE CHOICE

1. Intergalactic gas in galaxy clusters emits large amounts of energy in the form of radio waves. (T/F)

2. Distant galaxies appear to be much larger than those nearby. (T/F)

3. Collisions between galaxies are rare and have little or no effect on the stars and interstellar gas in the galaxies involved. (T/F)

4. The quasar stage of a galaxy ends because the central black hole swallows up all the matter around it. (T/F)

5. Elliptical galaxies may be formed by mergers between spirals. (T/F)

6. The fact that a typical quasar would consume an entire galaxy's worth of mass in 10 billion years suggests that quasar lifetimes are relatively long. (T/F)

7. On the largest scales, galaxies in the universe appear to be arranged on huge sheets surrounding nearly empty voids. (T/F)

8. The image of a distant quasar can be split into several images by the gravitational field of a foreground cluster. (T/F)

9. A galaxy containing substantial amounts of dark matter will (a) appear darker; (b) spin faster; (c) repel other galaxies; (d) have more tightly wound arms.

10. According to X-ray observations, the space between galaxies in a galaxy cluster is (a) completely devoid of matter; (b) very cold; (c) very hot; (d) filled with faint stars.

11. The fraction of mass of the universe made up of dark matter is (a) zero; (b) less than 10 percent; (b) roughly 50 percent; (d) more than 90 percent.

12. In current theories of galaxy evolution, quasars occur (a) early in the evolutionary sequence; (b) near the Milky Way; (c) when elliptical galaxies merge; (d) late in the evolutionary sequence.

13. Many nearby galaxies (a) may be much more active in the future; (b) contain quasars; (c) have radio lobes; (d) may have been much more active in the past.

14. **VIS** If light from a distant quasar did not pass through any intervening atomic hydrogen clouds, then Figure 16.23 (Absorption Line "Forest") would have to be redrawn to show (a) more absorption features; (b) fewer absorption features; (c) a single, large absorption feature; (d) more features at short wavelengths, but fewer at long wavelengths.

15. **VIS** If Figure 16.25 (Gravitational Lens) were to be redrawn using a more massive lensing galaxy, the quasar images would be (a) farther apart; (b) closer together; (c) fainter; (d) redder.

PROBLEMS

The number of squares preceding each problem indicates its approximate level of difficulty.

1. ■ Two galaxies orbit one another at a separation of 500 kpc. Their orbital period is estimated to be 30 billion years. Use Kepler's third law to find the total mass of the pair. ∞ *(Sec. 14.6)*

2. ■ Based on the data in Figure 16.1, estimate the mass of the galaxy NGC 4984 inside 20 kpc.

3. ■■■ A small satellite galaxy is moving in a circular orbit around a much more massive parent and just happens to be moving exactly parallel to the line of sight as seen from Earth. The recession velocities of the satellite and the parent galaxy are measured to be 6450 km/s and 6500 km/s, respectively, and the two galaxies are separated by an angle of 0.1° on the sky. Assuming $H_0 = 70$ km/s/Mpc, calculate the mass of the parent galaxy.

4. ■ Calculate the average speed of hydrogen nuclei (protons) in a gas of temperature 20 million K. ∞ *(More Precisely 5-1)* Compare this with the speed of a galaxy moving in a circular orbit of radius 1 Mpc around the center of a cluster of mass 10^{14} solar masses.

5. ■ In a galaxy collision, two similar-sized galaxies pass through each other with a combined relative velocity of 1500 km/s. If each galaxy is 100 kpc across, how long does the event last?

6. ■■ A certain quasar has a redshift of 0.25 and an apparent magnitude of 13. Using the data from Table 15.2, calculate the quasar's absolute magnitude and hence its luminosity. ∞ *(More Precisely 10-1)* Compare the apparent brightness of the quasar, viewed from a distance of 10 pc, with that of the Sun as seen from Earth.

7. ■ What are the absolute magnitude and luminosity of a quasar with a redshift of 5 and an apparent magnitude of 22?

8. ■■ Assuming an energy-generation efficiency (that is, the ratio of energy released to total mass-energy consumed) of 20 percent, calculate how long a 10^{40}-W quasar can shine if a total of 10^8 solar masses of fuel is available.

9. ■■ The spectrum of a quasar with a redshift of 0.20 contains two sets of absorption lines, redshifted by 0.15 and 0.155, respectively. If $H_0 = 70$ km/s/Mpc, estimate the distance between the intervening galaxies responsible for the two sets of lines.

10. ■■■ Light from a distant quasar is deflected through an angle of 3″ by an intervening lensing galaxy and is subsequently detected on Earth (see Figure 16.25). If Earth, the galaxy, and the quasar are all aligned, the quasar's redshift is 3.0 (see Table 15.2), and the galaxy lies midway between Earth and the quasar, calculate the minimum distance between the light ray and the center of the galaxy.

ACTIVITIES

Collaborative

1. Figure 16.9 is called the Hubble Deep Field. It contains too many galaxies for one person to easily count. Each group member should count the galaxies in a random area 2 cm × 2 cm and then determine a group average. Since the entire image is approximately 500 cm², multiply your group's average number of galaxies in a 2 cm × 2 cm area by 125 to estimate the number of galaxies in the image. How does your value compare to that of another group?

Individual

1. Look for a copy of the *Atlas of Peculiar Galaxies* by Halton Arp. It is available in book form, but it will be more convenient to find a version online. Search for examples of interacting galaxies of various types: (1) tidal interactions, (2) starburst galaxies, (3) collisions between two spirals, and (4) collisions between a spiral and an elliptical. For (1) look for galactic material pulled away from a galaxy by a neighboring galaxy. Is the latter galaxy also tidally distorted? In (2) the surest signs of starburst activity are bright knots of star formation. In what type(s) of galaxies do you find starburst activity? For (3) and (4) how do collisions differ depending on the types of galaxies involved? What typically happens to a spiral galaxy after a near miss or collision? Do ellipticals suffer the same fate?

17

Cosmology

The Big Bang and the Fate of the Universe

Our field of view now extends for billions of parsecs into space and billions of years back in time. We have asked and answered many questions about the structure and evolution of planets, stars, and galaxies. At last we are in a position to address the central issues of the biggest puzzle of all: How big is the universe? How long has it been around, and how long will it last? What was its origin, and what will be its fate? What is it made of? Many cultures have asked these questions, in one form or another, and have developed their own cosmologies—theories about the nature, origin, and destiny of the universe—to answer them. In this chapter, we see how modern scientific cosmology addresses these important issues and what it has to tell us about the universe we inhabit. After more than 10,000 years of civilization, science may be ready to provide some insight into the ultimate origin of all things.

MasteringAstronomy® Visit www.masteringastronomy.com for quizzes, animations, videos, interactive figures, and self-guided tutorials.

LEARNING OUTCOMES

Studying this chapter will enable you to:

LO1 State the cosmological principle, and explain its significance and observational underpinnings.

LO2 Describe the Big Bang theory of the expanding universe, and explain why the night sky is dark.

LO3 Outline the factors that determine whether the universe will expand forever, and explain the connection between the density of the universe and the overall geometry of space.

LO4 Describe the observational evidence that the expansion of the universe is accelerating, and discuss the role of dark energy in that acceleration.

LO5 Identify the cosmic microwave background as a major piece of evidence in favor of the Big Bang theory, and explain how it formed.

LO6 Explain how nuclei and atoms emerged from the primeval fireball and why astronomers think most of the helium observed in the universe was not formed by nuclear fusion in stars.

LO7 Summarize the horizon and flatness problems, and describe how the theory of cosmic inflation solves them.

LO8 Describe the formation of large-scale structure in the cosmos, and present observational evidence for the leading theory of structure formation.

◀ The universe began in a fiery expansion some 14 billion years ago, and out of this maelstrom emerged all the energy that would later form galaxies, stars, and planets. The story of the origin, evolution, and fate of all these systems comprises the subject of cosmology. This image—called the Ultra Deep Field—was taken with the Advanced Camera for Surveys aboard the *Hubble* telescope. More than a thousand galaxies are crowded into this one image, displaying many different types and shapes. In all, astronomers estimate that the observable universe contains about 100 billion such galaxies. *(NASA/ESA)*

17.1 The Universe on the Largest Scales

Galaxy surveys have revealed the existence of structures in the universe up to 200 Mpc across. ∞ *(Sec. 16.5)* Are there even larger structures? Does the hierarchy of clustering of stars into galaxies and galaxies into galaxy clusters, superclusters, voids, and filaments continue forever, on larger and larger scales? Perhaps surprisingly, given the trend we have just described, most astronomers think the answer is *no*.

We saw in Chapter 16 how astronomers use redshift surveys to construct three-dimensional maps of the universe on truly "cosmic" scales. ∞ *(Sec. 16.5)* Figure 17.1 is a map similar to those shown previously, but based on data from the most extensive redshift survey to date—the Sloan Digital Sky Survey. ∞ *(Discovery 16-1)* It extends out to a distance of almost 1000 Mpc, comparable to Figure 16.22, but because it includes much fainter galaxies, the Sloan map contains many more galaxies than the earlier figure, making structure easier to discern, particularly at large distances. Many voids and filaments can be seen on scales of a few tens of megaparsecs, but on the largest scales, apart from the general falloff in numbers of galaxies at large distances—basically because more distant galaxies are harder to see due to the inverse-square law—there is no obvious evidence for any structures on scales larger than about 200 Mpc. ∞ *(Sec. 10.2)* Careful statistical analysis of the galaxy distribution confirms this impression—even the Sloan Great Wall in Figure 17.1 can be explained statistically as a chance superposition of smaller structures.

The results of this and other large-scale surveys strongly suggest that the universe is **homogeneous** (the same everywhere) on scales greater than a few hundred megaparsecs. In other words, if we took a huge cube—300 Mpc on a side, say—and placed it anywhere in the universe, its overall contents would look much the same no matter where it was centered. The universe also appears to be **isotropic** (the same in all directions) on these scales. Excluding directions that are obscured by

CONCEPT CHECK ▶ In what sense is the universe homogeneous and isotropic?

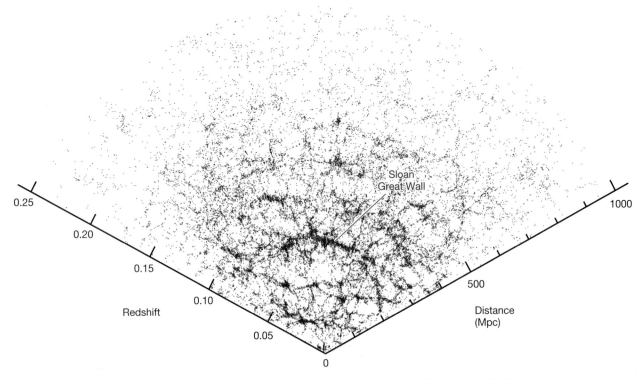

▲ **FIGURE 17.1 Galaxy Survey** This map of the universe is drawn using data from the Sloan Digital Sky Survey. ∞ *(Discovery 16-1)* It shows the locations of 66,976 galaxies lying within 12 degrees of the celestial equator and extending to a distance of almost 1000 Mpc (or a redshift of about 0.25). The largest known structure in the universe, the Sloan Great Wall, stretches nearly 300 Mpc across the center of the frame. *(SDSS Collaboration)*

our Galaxy, we count roughly the same number of galaxies per square degree in any patch of the sky we choose to observe, provided we look deep (far) enough that local inhomogeneities don't distort our sample.

In the science of **cosmology**—the study of the structure and evolution of the entire universe—researchers generally assume that the universe is homogeneous and isotropic on sufficiently large scales. No one knows if these assumptions are precisely correct, but we can at least say that they are consistent with current observations. In this chapter we simply assume that they hold. The twin assumptions of homogeneity and isotropy are known as the **cosmological principle.**

The cosmological principle has some very far-reaching implications. For example, it implies that there can be no *edge* to the universe because that would violate the assumption of homogeneity. And there can be no *center* because then the universe would not look the same in all directions from any noncentral point, violating the assumption of isotropy. This is the familiar Copernican principle expanded to truly cosmic proportions—not only are we not central to the universe, but *no one* can be central, because *the universe has no center*!

17.2 The Expanding Universe

Every time you go outside at night and notice that the sky is dark, you are making a profound cosmological observation. Here's why.

Olbers's Paradox

Let's assume for a moment that, in addition to being homogeneous and isotropic, the universe is also infinite in spatial extent and unchanging in time. On average, then, the universe is uniformly populated with galaxies filled with stars. In that case, as illustrated in Figure 17.2, when you look up at the night sky, your line of sight must *eventually* encounter a star. The star may lie at an enormous distance, in some remote galaxy, but the laws of probability dictate that, sooner or later, any line drawn outward from Earth will run into a bright stellar surface.

Faraway stars appear fainter than those nearby because of the inverse-square law, but they are also much more numerous—the number of stars we see at any given distance in fact increases as the *square* of the distance (just consider the area of a sphere of increasing radius). ∞ *(Sec. 10.2)* Thus the diminishing brightnesses of distant stars are exactly balanced by their increasing numbers, and stars at all distances contribute equally to the total amount of light received on Earth.

This fact has a dramatic implication: No matter where you look, the sky should be as bright as the surface of a star—the entire night sky should be as brilliant as the surface of the Sun! The obvious difference between this prediction and the actual appearance of the night sky is known as **Olbers's paradox,** after the 19th-century German astronomer Heinrich Olbers, who popularized the idea.

So why is the sky dark at night? Given that the universe is homogeneous and isotropic, then one (or both) of the other two assumptions must be false. Either the universe is finite in extent, or it evolves in time. In fact, the answer involves a little of each and is intimately tied to the behavior of the universe on the largest scales.

The Birth of the Universe

We have seen that all the galaxies in the universe are rushing away from us as described by Hubble's law:

$$\text{recession velocity} = H_0 \times \text{distance},$$

where we take Hubble's constant H_0 to be 70 km/s/Mpc. ∞ *(Sec. 15.3)* We have used this relationship as a convenient means of determining the distances to galaxies and quasars, but it is much more than that.

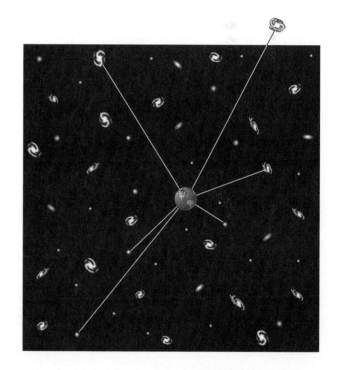

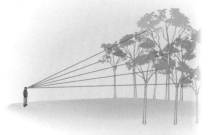

ANALOGY: In a forest, every line of sight eventually intersects a tree, much as Olbers's paradox suggests for stars seen from Earth.

▲ **FIGURE 17.2 Olbers's Paradox** If the universe were homogeneous, isotropic, infinite in extent, and unchanging, then any line of sight from Earth should eventually meet a star and the entire night sky should be bright. This obvious contradiction of the facts is known as Olbers's paradox.

Assuming that all velocities have remained constant in time, we can ask how long it has taken for any given galaxy to reach its present distance from us. The answer follows from Hubble's law. The time taken is simply the distance traveled divided by the velocity:

$$\text{time} = \frac{\text{distance}}{\text{velocity}}$$

$$= \frac{\text{distance}}{H_0 \times \text{distance}} \quad \text{(using Hubble's law for the velocity)}$$

$$= \frac{1}{H_0}.$$

For $H_0 = 70$ km/s/Mpc, this time is about 14 billion years. Notice that it is *independent* of the distance—galaxies twice as far away are moving twice as fast, so the time they took to cross the intervening distance is the same.

Hubble's law therefore implies that, at some time in the past—14 billion years ago, according to the foregoing simple calculation (see also Section 17.4)—*all* the galaxies in the universe lay right on top of one another. In fact, astronomers think that *everything* in the universe—matter and radiation alike—was confined at that instant to a single point of enormously high temperature and density. Then the universe began to expand at a furious rate, its density and temperature falling rapidly as the volume increased. This stupendous, unimaginably violent event, involving literally everything in the cosmos, is known as the **Big Bang.** Astronomers often refer to the superhot, expanding universe just after the Big Bang as the primeval fireball. It marked the beginning of the universe.

The Big Bang provides the resolution of Olbers's paradox. Whether the universe is actually finite or infinite in extent is irrelevant, at least as far as the appearance of the night sky is concerned. The key point is that we see only a *finite* part of it—the region lying within roughly 14 billion light-years of us. What lies beyond is unknown—its light has not had time to reach us.

Where Was the Big Bang?

Now we know *when* the Big Bang occurred. Is there any way of telling *where*? Astronomers think that the universe is the same everywhere, yet we have just seen that the observed recession of the galaxies described by Hubble's law suggests that all the galaxies expanded from a point at some time in the past. Wasn't that point, then, different from the rest of the universe, violating the assumption of homogeneity expressed in the cosmological principle? The answer is a definite *no*!

To understand why there is no center to the expansion, we must make a great leap in our perception of the universe. If the Big Bang was simply an enormous explosion that spewed matter out into space, ultimately to form the galaxies we see, then the foregoing reasoning would be quite correct. The universe would have a center and an edge, and the cosmological principle would not apply. But the Big Bang was *not* an explosion in an otherwise featureless, empty universe. The only way that we can have Hubble's law *and* retain the cosmological principle is to realize that the Big Bang involved the entire universe—not just the matter and radiation within it, but the universe *itself.* In other words, the galaxies are not flying apart into the rest of the universe. The universe itself is expanding. Like raisins in a loaf of raisin bread that move apart as the bread expands in an oven, the galaxies are just along for the ride.

Let's reconsider some of our earlier statements in this new light. We now recognize that Hubble's law describes the expansion of the universe itself. Apart from small-scale individual random motions, galaxies are not moving with respect to the

INTERACTIVE The Expanding Raisin Cake (Universe)

fabric of space. The component of the galaxies' motion that makes up the Hubble flow is really an expansion of space itself. The expanding universe remains homogeneous at all times. There is no "empty space" beyond the galaxies into which they rush. At the time of the Big Bang, the galaxies did not reside at a point located at some well-defined place within the universe. The *entire universe* was a point. The Big Bang happened *everywhere* at once.

To illustrate these ideas, imagine an ordinary balloon with coins taped to its surface, as shown in Figure 17.3. (Better yet, do the experiment yourself.) The coins represent galaxies, and the two-dimensional surface of the balloon represents the "fabric" of our three-dimensional universe. The cosmological principle applies here because every point on the balloon looks pretty much the same as every other. Imagine yourself as a resident of one of the coin "galaxies" on the leftmost balloon, and note your position relative to your neighbors. As the balloon inflates (that is, as the universe expands), the other galaxies recede from you and more distant galaxies recede more rapidly.

Regardless of which galaxy you chose to consider, you would see all the other galaxies receding from you, but nothing is special or peculiar about this fact. There is no center to the expansion and no position that can be identified as the location from which the universal expansion began. Everyone sees an overall expansion described by Hubble's law, with the same value of Hubble's constant in all cases. Now imagine letting the balloon deflate, running the "universe" backward from the present time to the moment of the Big Bang. *All* the galaxies (coins) would arrive at the same place at the same time at the instant the balloon reached zero size, but no single point on the balloon could be said to be *the* place where that occurred.

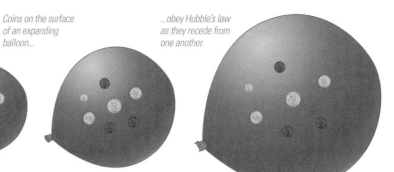

Coins on the surface of an expanding balloon...

...obey Hubble's law as they recede from one another.

INTERACTIVE ▲ FIGURE 17.3 **Receding Galaxies**
Coins taped to the surface of a spherical balloon recede from each other as the balloon inflates (left to right). Similarly, galaxies recede from each other as the universe expands. As the coins separate, the distance between any two of them increases, and the rate of increase of this distance **(MA)** is proportional to the distance between them.

PROCESS OF ►
SCIENCE CHECK Why does Hubble's law imply a Big Bang?

The Cosmological Redshift

Up to now, we have explained the cosmological redshift of galaxies as a Doppler shift, a consequence of their motion relative to us. ∞ *(Sec. 15.3)* However, we have just argued that the galaxies are not moving with respect to the universe, in which case the Doppler interpretation cannot be correct. The true explanation is that, as a photon moves through space, its wavelength is influenced by the expansion of the universe. Think of the photon as being attached to the expanding fabric of space, so its wavelength expands along with the universe, as illustrated in Figure 17.4. Although it is common in astronomy to refer to the cosmological redshift in terms of recessional velocity, bear in mind that, strictly speaking, this is not the right thing to do. The cosmological redshift is a consequence of the changing size of the universe and is *not* related to velocity at all.

The redshift of a photon measures the amount by which the universe has expanded since that photon was emitted. For example, when we measure the light from a quasar to have a redshift of 7, it means that the observed wavelength is 8 times (1 plus the redshift) greater than the wavelength at the time of emission, which means that the light was emitted at a time when the universe was just one-eighth its present size (and we are observing the quasar as it was at that time). ∞ *(More Precisely 15-1)* In general, the larger a photon's redshift, the smaller the universe was at the time the photon was emitted, and so the longer ago that emission occurred. Because the universe expands with time and redshift is related to that expansion, cosmologists routinely use redshift as a convenient means of expressing time.

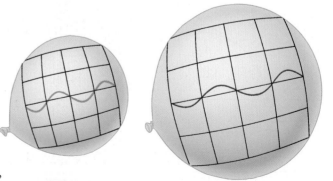

INTERACTIVE ▲ FIGURE 17.4 **Cosmological Redshift**
As the universe expands, photons of radiation are stretched in wavelength, giving rise to the cosmological redshift.

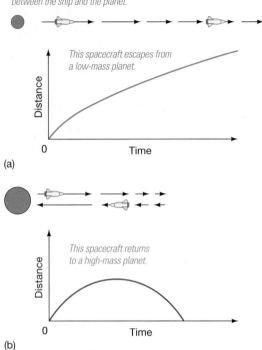

These graphs show the changing distance between the ship and the planet.

This spacecraft escapes from a low-mass planet.

Distance

0   Time

(a)

This spacecraft returns to a high-mass planet.

Distance

0   Time

(b)

▲ **FIGURE 17.5 Escape Speed** (a) A spacecraft (arrow) leaving a planet (blue ball) with a speed greater than the planet's escape speed follows an unbound trajectory and escapes. (b) If the launch speed is less than the escape speed, the ship eventually drops back to the planet. Its distance, as graphed, first rises and then falls.

CONCEPT CHECK ▶ What are the two basic possibilities for the future expansion of the universe?

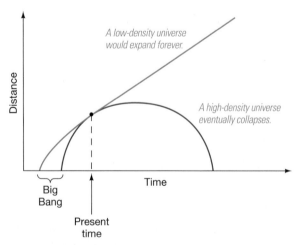

A low-density universe would expand forever.

A high-density universe eventually collapses.

Distance

Time

Big Bang

Present time

▲ **FIGURE 17.6 Model Universes** Distance between two galaxies as a function of time in each of the two basic universes discussed in the text: a low-density universe that expands forever and a high-density cosmos that collapses. The point where the two curves touch represents the present time.

17.3 Cosmic Dynamics and the Geometry of Space

Will the universe expand forever? This fundamental question about the fate of the universe has been at the heart of cosmology since Hubble's law was first discovered. Until the late 1990s, the prevailing view among cosmologists was that the answer would most likely be found by determining the extent to which gravity would slow, and perhaps ultimately reverse, the current expansion. However, it now appears that the answer is rather more subtle—and perhaps a lot more profound in its implications—than was hitherto thought.

Two Futures

Consider a rocket ship launched from the surface of a planet. What are the likely outcomes of the its motion? According to Newtonian mechanics, there are just two basic possibilities, depending on the launch speed of the ship relative to the escape speed of the planet. ∞ *(More Precisely 5-1)* If the launch speed is high enough, it will exceed the planet's escape speed, and the ship will never return to the surface. The speed will diminish because of the planet's gravitational pull, but it will never reach zero. The spacecraft leaves the planet on an unbound trajectory, as illustrated in Figure 17.5(a). Alternatively, if the launch speed is lower than the escape speed, the ship will reach a maximum distance from the planet and then fall back to the surface. Its bound trajectory is shown in Figure 17.5(b).

Similar reasoning applies to the expansion of the universe. Imagine two galaxies at some known distance from one another, their present relative velocity given by Hubble's law. The same two basic possibilities exist for these galaxies as for our rocket ship. The distance between them can increase forever, or it can increase for a while and then start to decrease. What's more, the cosmological principle says that whatever the outcome, it must be the same for *any* two galaxies—in other words, the same statement applies to the universe *as a whole*. Thus, as illustrated in Figure 17.6, the universe has only two options: It can continue to expand forever, or the present expansion will someday stop and turn around into a contraction. The two curves in the figure are drawn so that they pass through the same point at the present time. Both are possible descriptions of the universe, given its current size and expansion rate.

What determines which of these possibilities will actually occur? In the case of a rocket ship of fixed launch speed, the *mass* of the planet determines whether or not escape will occur—a more massive planet has a higher escape speed, making it less likely that the rocket can escape. For the universe, the corresponding quantity is the *density* of the cosmos. A high-density universe contains enough mass to stop the expansion and eventually cause a recollapse. Such a universe is destined ultimately to shrink toward a superdense, superhot singularity much like the one from which it originated—a "Big Crunch." A low-density universe, conversely, will expand forever. In that case, in time, even with the most powerful telescopes, we will see no galaxies in the sky beyond the Local Group (which is not itself expanding). The rest of the observable universe will appear dark, the distant galaxies too faint to be seen.

The dividing line between these outcomes—the density corresponding to a universe in which *gravity acting alone* would be just sufficient to halt the present expansion—is called the universe's **critical density.** For $H_0 = 70$ km/s/Mpc, the critical density is about 9×10^{-27} kg/m³. That's an extraordinarily low density—just five hydrogen atoms per cubic meter, a volume the size of a typical household closet. In more "cosmological" terms, it corresponds to about 0.1 Milky Way Galaxy (including the dark matter) per cubic megaparsec.

We will see in a moment that the separation between never-ending expansion and cosmic collapse is not nearly as straightforward as the foregoing simple reasoning would suggest. Several independent lines of evidence now indicate that gravity is *not* the only influence on the dynamics of the universe on large scales (Section 17.4). As a result, while the "futures" just described are still the only two possibilities for the long-term evolution of the universe, the distinction between them turns out to be more than just a matter of density alone. Nevertheless, the density of the universe—or, more precisely, the *ratio of the total density to the critical value*—is a vitally important quantity in cosmology.

The Shape of the Universe

We have reverted to the familiar notions of Newtonian mechanics and gravity and moved away from general relativity and the more correct concept of warped spacetime, because speaking in Newtonian terms makes our discussion of the evolution of the universe easier to understand. ∞ *(Secs. 1.4, 13.5)* However, general relativity makes some predictions about the universe that do not have a simple description within Newton's theory.

Foremost among these non-Newtonian predictions is the fact that space is curved and that the degree of curvature is determined by the total density of the cosmos. ∞ *(Sec. 13.5)* But general relativity is clear on what "density" means here. Both matter *and* energy must be taken into account, with energy (E) properly "converted" into mass (m) units via Einstein's famous relation $E = mc^2$. ∞ *(More Precisely 13-1)* The density of the universe includes not just the atoms and molecules that make up the familiar normal matter around us, but also the invisible dark matter that dominates the masses of galaxies and galaxy clusters, as well as *everything* that carries energy—photons, relativistic neutrinos, gravity waves, and anything else we can think of.

In a homogeneous universe, the curvature (on sufficiently large scales) must be the same everywhere, so there are really only three distinct possibilities for the large-scale geometry of space. General relativity tells us that the geometry of the universe depends only on the ratio of the density of the universe to the critical density (as defined in the previous section). Cosmologists call the ratio of the universe's actual density to the critical value the *cosmic density parameter* and denote it by the symbol Ω_0 ("omega nought"). In terms of this quantity, then, a universe with density equal to the critical value has $\Omega_0 = 1$, a "low-density" cosmos has Ω_0 less than 1, and a "high-density" universe has Ω_0 greater than 1.

In a high-density universe (Ω_0 greater than 1), space is curved so much that it bends back on itself and closes off, making the universe *finite* in size. Such a universe is known as a **closed universe.** It is difficult to visualize a three-dimensional volume uniformly arching back on itself in this way, but the two-dimensional version is well known: It is the surface of a sphere, like the balloon we discussed earlier. Figure 17.3, then, is the two-dimensional likeness of a three-dimensional closed universe. Like the surface of a sphere, a closed universe has no boundary, yet it is finite in extent.[1] One remarkable property of a closed universe is illustrated in Figure 17.7: A flashlight beam shining in some direction in space might eventually traverse the entire universe and return from the opposite direction.

The surface of a sphere curves, loosely speaking, "in the same direction" no matter which way we move from a given point. A sphere is said to have *positive curvature.* However, if the density of the universe is below the critical value, then the curvature of space is qualitatively quite different from that of a sphere. The two-dimensional surface corresponding to such a space curves like a saddle and is said to have *negative curvature.* Most people have seen a saddle—it curves "up" in one direction and "down" in another, but no one has ever seen a uniformly negatively curved surface, for the good reason that it cannot be constructed in three-dimensional space! It is just "too big" to fit. A low-density (Ω_0 less than 1) saddle-curved universe is infinite in extent and is usually called an **open universe.**

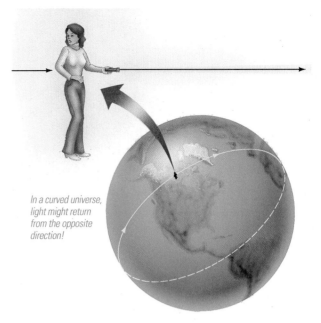

In a curved universe, light might return from the opposite direction!

▲ **FIGURE 17.7 Einstein's Curve Ball** In a closed universe, a beam of light launched in one direction might return someday from the opposite direction after circling the universe, just as motion in a "straight line" on Earth's surface will eventually encircle the globe.

CONCEPT
CHECK ► How is the curvature of space related to the size and density of the universe?

[1]Notice that for the sphere analogy to work, we must imagine ourselves as two-dimensional "flatlanders" who cannot visualize or experience in any way the third dimension perpendicular to the sphere's surface. Flatlanders are confined to the sphere's surface, just as we are confined to the three-dimensional volume of our universe.

The intermediate case, when the density is precisely equal to the critical density (i.e., $\Omega_0 = 1$), is the easiest to visualize. This **critical universe** has no curvature. It is said to be "flat," and it is infinite in extent. In this case, and only in this case, the geometry of space on large scales is precisely the familiar Euclidean geometry taught in high school. Apart from its overall expansion, this is basically the universe that Newton knew.

17.4 The Fate of the Cosmos

Is there any way for us to determine which of the futures just described actually applies to our universe (apart from just waiting to find out)? Will the universe end as a small, dense point much like that from which it began? Or will it expand forever? And can we hope to determine the geometry of the vast cosmos we inhabit? Finding answers to these questions has been the dream of astronomers for decades. We are fortunate to live at a time when we can subject these questions to intensive observational tests and come up with definite answers—even though they aren't what most cosmologists expected! Let's begin by looking at the density of the universe (or, equivalently, the cosmic density parameter Ω_0).

The Density of the Universe

How might we determine the average density of the universe? On the face of it, it would seem simple—just measure the average mass of the galaxies residing within a large parcel of space, calculate the volume of that space, and compute the total mass density. When astronomers do this, they usually find a little less than 10^{-28} kg/m^3 in the form of luminous matter. Largely independent of whether the chosen region contains only a few galaxies or a rich galaxy cluster, the resulting density is about the same, within a factor of 2 or 3. Galaxy counts thus yield a value of Ω_0 of only a few percent. If that measure were correct, and galaxies were all that existed, then we would live in a low-density open universe destined to expand forever.

But there is a catch. Most of the matter in the universe is *dark*—it exists in the form of invisible material that has been detected only through its gravitational effect in galaxies and galaxy clusters. ∞ *(Secs. 14.6, 16.1)* We currently do not know *what* the dark matter is, but we *do* know that it is there. Galaxies may contain as much as 10 times more dark matter than luminous material, and the figure for galaxy clusters is even higher—perhaps as much as 95 percent of the total mass in clusters is invisible. Even though we cannot see it, dark matter contributes to the average density of the universe and plays its part in opposing the cosmic expansion. Including all the dark matter that is known to exist in galaxies and galaxy clusters increases the value of Ω_0 to about 0.25.

Although we can detect and quantify the effects of dark matter in galaxies and galaxy clusters, its distribution on larger scales is harder to measure. Astronomers have developed techniques to study matter on supercluster and larger scales, using gravitational lensing of distant objects and the large-scale motions of galaxies and clusters to probe the gravitational fields of cosmic clumps of invisible matter. ∞ *(Sec. 16.5)* Yet all these studies add little to the overall density. As best we can tell, there doesn't seem to be much additional dark matter "tucked away" on very large scales. Most cosmologists agree that the overall density of matter (luminous plus dark) in the universe is between 25 and 30 percent of the critical value—not enough to halt the universe's current expansion.

Cosmic Acceleration

Determining the density of the universe is an example of a *local* measurement that provides an estimate of Ω_0. But the result we obtain depends on just how local our measurement is, and there are many uncertainties in the result, especially on large scales. In an attempt to get around this problem, astronomers have devised alternative

▶ FIGURE 17.8 Accelerating Universe (a) In a decelerating universe (purple and red curves), redshifts of distant objects are greater than would be predicted from Hubble's law (black curve). The reverse is true for an accelerating universe. The data points show observations of some 50 supernovae that strongly suggest cosmic expansion is accelerating. (b) The bottom frames show three Type I supernovae (marked by arrows) that exploded in distant galaxies when the universe was nearly half its current age. The top frames show the same areas prior to the explosions. (CfA/NASA)

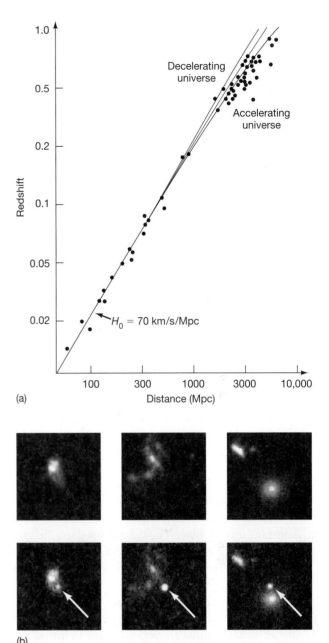

methods that rely instead on *global* measurements, covering much larger regions of the observable universe. In principle, such global tests should indicate the universe's overall density, not just its value in our cosmic neighborhood.

One such global method is based on observations of Type I (carbon detonation) supernovae. ∞ *(Sec. 12.5)* Recall that these objects are both very bright and have a remarkably narrow spread in luminosities, making them particularly useful as standard candles. ∞ *(Sec. 15.2)* They can be used as probes of the universe because, by measuring their distances (*without* using Hubble's law) and their redshifts, we can determine the rate of cosmic expansion in the distant past. Here's how the method works.

Suppose the universe is decelerating, as we would expect if gravity were slowing its expansion. Then, because the expansion rate is decreasing, objects at great distances—that is, objects that emitted their radiation long ago—should appear to be receding *faster* than Hubble's law predicts. Figure 17.8 illustrates this concept. If the universal expansion were constant in time, recessional velocity and distance would be related by the black line in the figure. (The line is not quite straight because it takes the expansion of the universe properly into account in computing the distance.) ∞ *(More Precisely 15-1)* In a decelerating universe, the velocities of distant objects should lie *above* the black curve, and the deviation from that curve is greater for a denser universe, in which gravity has been more effective at slowing the expansion.

How does theory compare with reality? In the late 1990s two groups of astronomers announced the results of independent, systematic surveys of distant supernovae. Some of these supernovae are shown in Figure 17.8(b); the data are marked on Figure 17.8(a). Far from clarifying the picture of cosmic deceleration, however, these findings indicated that the expansion of the universe is not slowing, but actually accelerating! According to the supernova data, galaxies at large distances are receding *less* rapidly than Hubble's law would predict. The deviations from the decelerating curves appear small in the figure, but they are statistically very significant, and both groups report similar findings.

These observations are *inconsistent* with the standard "gravity only" Big Bang theory just described and sparked a major revision of our view of the cosmos. The measurements are difficult, and the results depend quite sensitively on just how "standard" the supernovae luminosities really are. Not surprisingly, since so much hangs on this result, the reliability of the supernova measurement technique has been the subject of intense scrutiny by cosmologists. However, no convincing argument against the method has yet been put forward, so there is no reason to believe that we are somehow being "fooled" by nature. As far as we can tell, the measurements are good, and the cosmic acceleration is real. This unexpected but crucially important result earned the leaders of the two groups who discovered it the 2011 Nobel Prize in Physics.

Dark Energy

What could cause an overall acceleration of the universe? Frankly, cosmologists don't know, although several possibilities have been suggested. Whatever it is, the mysterious cosmic force causing the universe to accelerate is neither matter nor radiation. Although it carries energy, it exerts an overall *repulsive* effect on the universe, causing empty space to expand. It has come to be known simply as **dark energy,** and it is perhaps *the* leading puzzle in astronomy today.

17-1 DISCOVERY

Einstein and the Cosmological Constant

Even the greatest minds are fallible. The first scientist to apply general relativity to the universe was (not surprisingly), the theory's inventor, Albert Einstein. When he derived and solved the equations describing the behavior of the universe, Einstein discovered that they predicted a universe that evolved in time. But in 1917 neither he nor anyone else knew about the expansion of the universe, as described by Hubble's law, which would not be discovered for another 10 years. ∞ *(Sec. 15.3)* At the time Einstein, like most scientists, believed that the universe was static—that is, unchanging and everlasting. The discovery that there was no static solution to his equations seemed to Einstein a near-fatal flaw in his new theory.

To bring his theory into line with his beliefs, Einstein tinkered with his equations, introducing a "fudge factor" now known as the cosmological constant. The accompanying figure shows the effect of introducing a cosmological constant into the equations describing the expansion of a critical-density universe. One possible solution is a "coasting" universe, whose radius does indeed remain constant for an indefinite period of time. Einstein took this to be the static universe he expected. Instead of predicting an evolving cosmos, which would have been one of general relativity's greatest triumphs, Einstein yielded to a preconceived notion of the way the universe "should be," unsupported by observational evidence. Later, when the expansion of the universe was discovered and Einstein's equations—without the cosmological constant—were found to describe it perfectly, he declared that the cosmological constant was the biggest mistake of his scientific career.

For many researchers—Einstein included—the main problem with the cosmological constant was (and still is) the fact that it had no clear physical interpretation. Einstein introduced it to fix what he thought was a problem with his equations, but he discarded it immediately once he realized that no problem actually existed. Scientists are very reluctant to introduce unknown quantities into their equations purely to make the results "come out right." As a result, the cosmological constant fell out of favor among astronomers for many years. However, as described in the text, it has apparently been completely rehabilitated and is now identified as a leading candidate for the "dark energy" whose existence is inferred from studies of the universe on very large scales.

Attempts are underway to construct theories of dark energy that preserve the character of the cosmological constant, yet also account for its value in some natural way, perhaps somehow coupling it to the density of matter. Before we make too many sweeping statements about the role of the cosmological constant in cosmology, we should probably remember the experience of its inventor and bear in mind that—at least for now—its physical meaning remains completely unknown.

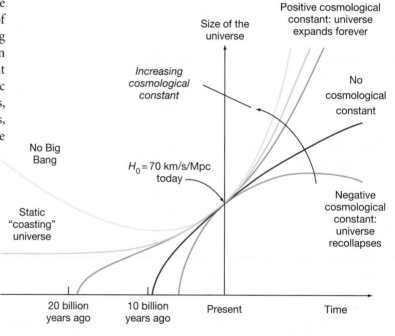

As illustrated in Figure 17.9, the repulsive effect of dark energy is proportional to the size of the universe, so it increases as the universe expands. Thus, it was negligible at early times, but today it is the major factor controlling the cosmic expansion. Furthermore, since the effect of gravity *weakens* as the expansion proceeds, it follows that once dark energy begins to dominate, gravity can never catch up, and the universe will continue to accelerate at an ever-increasing pace. Thus, by opposing the attractive force of gravity, the repulsive effect of dark energy strengthens our earlier conclusion that the universe will expand forever.

A leading candidate for dark energy is an additional "vacuum pressure" force associated with empty space and effective only on very large scales. Known simply as the **cosmological constant,** it has a long and checkered history (see *Discovery 17-1*) and has gone in and out of favor since its invention by Einstein almost a century ago. Note, however, that while models including the cosmological constant can fit

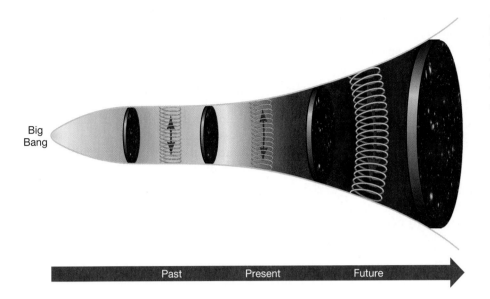

◄ **FIGURE 17.9 Dark Energy**   The expansion of the universe is opposed by the attractive force of gravity and sped up by the repulsion due to dark energy. As the universe expands, the gravitational force weakens, whereas the force due to dark energy increases. A few billion years ago, dark energy began to dominate, and the expansion of the cosmos has been accelerating ever since.

Big Bang

Past   Present   Future

the observational data, astronomers as yet have no clear physical interpretation of what it actually means—it is neither required nor explained by any known law of physics.

The fact that the present value of the repulsive dark energy force is comparable to the attractive force of gravity poses another problem to astronomers. When we calculate the evolution of a universe governed by a cosmological constant, consistent with current observations (see Figure 17.12), we find that this state of affairs was not true in the early universe (when galaxies were forming, say), nor will it be true in 10 or 20 billion years' time. In other words, the observations seem to suggest that we live at a special time in the history of the universe—a conclusion viewed with great suspicion by astronomers schooled in the Copernican principle. ∞ *(Sec. 1.1)* Cosmologists have tried to address this problem by constructing alternative theories in which the behavior of dark energy depends on the densities of matter and radiation in the universe, but for now at least, the cosmological constant seems to give just as good a fit to the observational data.

Cosmic Composition

In addition to measurements of density and acceleration, astronomers have several other means of estimating the "cosmological parameters" that describe the large-scale properties of our universe. Theoretical studies of the very early universe (Section 17.7) strongly suggest that the geometry of the universe should be precisely flat—and hence the density should be exactly critical. These ideas first became widespread in the 1980s and for many years seemed to be at odds with observations indicating a cosmic matter density of just 20–30 percent of the critical value, even taking the dark matter into account. The cosmic acceleration just discussed actually resolves that conflict, albeit at the price of introducing yet another unknown component—dark energy—into the cosmic mix.

Detailed measurements of the radiation field known to fill the entire cosmos (Sections 17.5 and 17.8) support the theoretical prediction of critical density and are also consistent with the dark energy inferred from the supernova studies. Further support comes from careful analysis of galaxy surveys such as those discussed in Section 17.1, which allow astronomers to measure the growth of large-scale structure in the universe, and in turn to constrain the value of Ω_0.

Remarkably, all the approaches just described yield consistent results! The current consensus among cosmologists is that the universe is of precisely critical

 ANIMATION/VIDEO Gravitational Curvature of Space

PROCESS OF ►
SCIENCE CHECK   Why have astronomers concluded that dark energy is the major constituent of the universe?

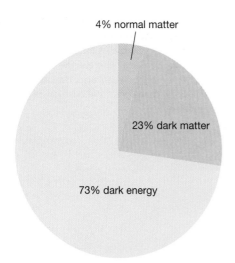

4% normal matter

23% dark matter

73% dark energy

▲ **FIGURE 17.10 Composition of the Universe**   The universe today is made mostly of mysterious dark energy, accounting for nearly three-quarters of the total. Dark matter accounts for nearly a quarter. Normal matter amounts to only a few percent—and of that 3.6% is mostly in galactic and intergalactic gas, and only a miniscule 0.4% comprises stars, planets, and life-forms.

▲ **FIGURE 17.11 Geometry of the Universe**   The universe on the largest scales seems geometrically flat—governed by the same familiar Euclidean geometry taught in high schools.

density, $\Omega_0 = 1$, but this density is made up of both matter (mostly dark) and dark energy (suitably converted into mass units using the relation $E = mc^2$). ∞ *(Sec. 9.5)* Note that, while dark energy *opposes* the force of gravity in its effect on cosmic expansion, it still *adds* to the total density (and Ω_0) in its effect on the curvature of space. Based on all available data, as illustrated in Figure 17.10, the best estimate is that matter accounts for 27 percent of the total and dark energy for the remaining 73 percent. This is the assumption underlying Table 15.1 and used consistently throughout this book. ∞ *(More Precisely 15-1)*

Note that such a universe will expand forever and, the heavy machinery of general relativity and curved spacetime notwithstanding, is perfectly flat (Figure 17.11)—an irony that would no doubt have amused Newton!

The Age of the Universe

In Section 17.2, when we estimated the age of the universe from the accepted value of Hubble's constant, we made the assumption that the expansion speed of the cosmos was constant in the past. However, as we have just seen, this is a considerable oversimplification. Gravity tends to slow the universe's expansion, while dark energy acts to accelerate it. In the absence of a cosmological constant, the universe expanded faster in the past than it does today, so the assumption of a constant expansion rate leads to an overestimate of the universe's age—such a universe is younger than the 14 billion years calculated earlier. Conversely, the repulsive effect of dark energy tends to increase the age of the cosmos.

Figure 17.12 illustrates these points. It is similar to Figure 17.6, except that here we have added three curves, including one (labeled "Empty") corresponding to a constant expansion rate at the present value and another to an accelerating universe with the "best estimate" parameters given above. The age of a critical-density universe with no cosmological constant (whose curve is also added here) is about 9 billion years. A low-density, unbound universe is older than 9 billion years but still less than 14 billion years old. The age corresponding to the accelerating universe is about 14 billion years, coincidentally very close to the value calculated for constant expansion.

How does this compare with an age estimated by other means? On the basis of the theory of stellar evolution, the oldest globular clusters formed about 12 billion years ago, and most are estimated to be between 10 and 12 billion years old. ∞ *(Sec. 12.6)* This range is indicated on Figure 17.12. These ancient star clusters are thought to have formed at around the same time as our Galaxy, so they date the epoch of galaxy formation. More important, they can't be older than the universe! Figure 17.12 shows that globular cluster ages are consistent with a 14 billion-year-old cosmos, and even allow a couple of billion years for galaxies to grow, as discussed in Chapter 16. ∞ *(Sec. 16.3)* Note also that the cluster ages are *not* consistent with a critical-density universe without dark energy. This independent check of a key prediction is an important piece of evidence supporting the modern version of the Big Bang theory.

Thus, for $H_0 = 70$ km/s/Mpc, our current best guess of the history of the universe places the Big Bang at 14 billion

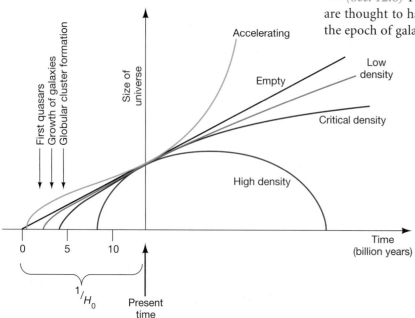

◄ **FIGURE 17.12 Cosmic Age**   The age of a universe without dark energy (represented by all three lower curves colored red, purple, and blue) is always less than $1/H_0$ and decreases for larger values of the present-day density. The existence of a repulsive cosmological constant (green curve) increases the age of the cosmos.

years ago. The first quasars appeared about 13 billion years ago (at a redshift of 7), the peak quasar epoch (redshifts 2–3) occurred during the next 1 billion years, and the oldest known stars in our Galaxy formed during the 2 billion years after that. ∞ *(Sec. 16.3)* Even though astronomers do not presently understand the nature of dark energy, the good agreement between so many separate lines of reasoning has convinced many that the dark-matter/dark-energy Big Bang theory just described is the correct description of the universe.

Bear in mind as we proceed through the remainder of this chapter that the Big Bang *is* a scientific theory and as such is continually being challenged and scrutinized. ∞ *(Sec. 0.5)* It makes testable predictions about the state and history of the cosmos and must change—or fail—if these predictions are found to be at odds with observations. Astronomers aren't ready to relax just yet—the history of this subject suggests that there may be a few more unexpected twists and turns in the road before the details are finally resolved.

17.5 The Early Universe

We now turn from studies of the far future of our universe to the quest to understand its distant past. Just how far back in time can we probe? Is there any way to study the universe beyond the most remote quasar? How close can we come to perceiving directly the edge of time, the very origin of the universe?

The Cosmic Microwave Background

A partial answer to these questions was discovered by accident in 1964, during an experiment designed to improve America's telephone system. As part of a project to identify and eliminate unwanted interference in satellite communications, Arno Penzias and Robert Wilson, scientists at Bell Telephone Laboratories in New Jersey, were studying the Milky Way's emission at microwave (radio) wavelengths, using the horn-shaped antenna shown in Figure 17.13. In their data they noticed a bothersome background "hiss"—a little like the background static on an AM radio station. Regardless of where and when they pointed their antenna, the hiss persisted. Never diminishing or intensifying, the weak signal was detectable at any time of the day and on any day of the year, apparently filling all of space.

Eventually, after discussions with colleagues at Bell Labs and theorists at nearby Princeton University, the two experimentalists realized that the origin of the mysterious static was nothing less than the fiery creation of the universe itself. The radio hiss that Penzias and Wilson detected is now known as the **cosmic microwave background**. Their discovery won them the 1978 Nobel Prize in Physics.

In fact, theorists had predicted the existence and general properties of the microwave background well before its discovery. They reasoned that, in addition to being extremely dense, the early universe must also have been very *hot*, and so, shortly after the Big Bang, the universe must have been filled with extremely high-energy thermal radiation—gamma rays of very short wavelength. As illustrated in Figure 17.14, this intensely hot primordial radiation has been redshifted by the expansion of the universe from gamma rays, through optical and infrared wavelengths, all the way into the radio (microwave) range of the electromagnetic spectrum. ∞ *(Sec. 2.4)* The radiation observed today is the "fossil remnant" of the primeval fireball that existed at those very early times. The discovery of the microwave background was thus a powerful piece of evidence in support of the Big Bang theory of the universe and was instrumental in convincing many astronomers of the basic correctness of the idea.

 CONCEPT CHECK ▶ Why do astronomers think the universe will expand forever?

▲ **FIGURE 17.13 Microwave Background Discoverers** This "sugarscoop" antenna was built to communicate with Earth-orbiting satellites but was used by Robert Wilson (right) and Arno Penzias to discover the 2.7 K cosmic background radiation. *(Alcatel-Lucent)*

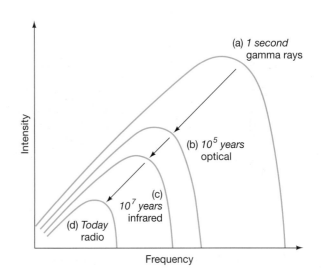

▲ **FIGURE 17.14 Cosmic Blackbody Curves** Theoretically derived blackbody curves for the entire universe (a) 1 second after the Big Bang, (b) 100,000 years later, (c) 10 million years after that, and (d) today, approximately 14 billion years after the Big Bang.

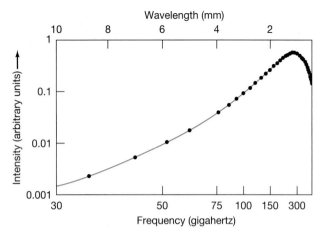

▲ FIGURE 17.15 Microwave Background Spectrum The intensity of the cosmic background radiation, as measured by the *COBE* satellite, agrees well with theory. The curve is the best fit to the data, corresponding to a temperature of 2.725 K. Experimental errors in this remarkably accurate observation are smaller than the dots representing the data points.

CONCEPT CHECK ► When was the cosmic microwave background formed? Why is it still a blackbody curve?

Researchers at Princeton confirmed the existence of the microwave background and estimated its temperature at about 3 K. However, this part of the electromagnetic spectrum happens to be very difficult to observe from the ground, and it was 25 years before astronomers could demonstrate conclusively that the radiation was indeed described by a blackbody curve. ∞ *(Sec. 2.3)* In 1989 the *Cosmic Background Explorer* satellite—*COBE* for short—measured the intensity of the microwave background at wavelengths straddling the peak, from half a millimeter to about 1 cm. The results are shown in Figure 17.15. The solid line is the blackbody curve that best fits the *COBE* data. The near-perfect fit corresponds to a universal temperature of just over 2.7 K.

A striking aspect of the cosmic microwave background is its high degree of *isotropy*. When we correct for Earth's motion through space (which causes the microwave background to appear a little hotter than average in front of us and slightly cooler behind, by the Doppler effect), the intensity of the radiation is virtually constant (to about one part in 10^5) from one direction in the sky to another.

Remarkably, the cosmic microwave background contains more energy than has been emitted by all the stars and galaxies that have ever existed in the history of the universe. Stars and galaxies, though very intense sources of radiation, occupy only a small fraction of space. When their energy is averaged out over the volume of the entire cosmos, it falls short of the energy in the microwave background by at least a factor of 10. For our present purposes, then, the cosmic microwave background is the only significant form of radiation in the universe.

Radiation in the Universe

What role does radiation play in the evolution of the universe on large scales? We can begin to address this question by expressing the energy in the microwave background as an equivalent density, calculating the number of photons in a cubic meter of space, and then converting the total energy of these photons into a mass using the relation $E = mc^2$. ∞ *(Sec. 16.6)* When we do this, we find an equivalent density for the microwave background of about 5×10^{-21} kg/m^3—much less than the current value of the cosmic density, which as we have just seen (Figure 17.10) is roughly 9×10^{-27} kg/m^3, virtually all in the form of dark energy and matter. Thus, *at the present moment,* the densities of both dark energy and matter in the universe far exceed the density of radiation.

Has this always been the case? Have dark energy and matter always dominated the cosmos? To answer this we must understand how the densities of dark energy, matter, and radiation change as the universe expands. According to current theory, dark energy is a property of space itself, and its density remains *constant* as the universe evolves. In contrast, the densities of matter and radiation both *decrease*, as the expansion dilutes the numbers of atoms and photons alike, but the radiation is further diminished in energy by the cosmological redshift, so its equivalent density falls even faster than the density of matter as the universe grows. Thus, as illustrated in Figure 17.16, even though

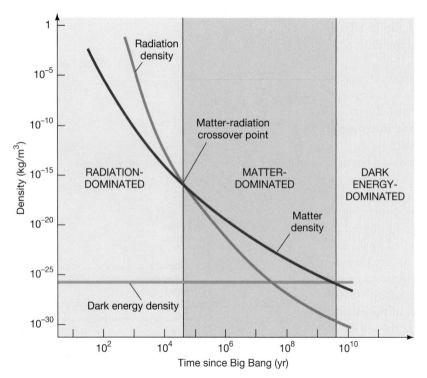

◀ FIGURE 17.16 Changing Dominance As the universe expanded, the number of both matter particles and photons per unit volume decreased. The photons were additionally reduced in energy by the cosmological redshift. As a result, the density of radiation fell faster than the density of matter as the universe grew, and that radiation dominated matter at early times, before the crossover point. Today, dark energy dominates both matter and radiation.

dark energy dominates today, it was *unimportant* in the early universe. Furthermore, even though the present density of radiation is much less than that of matter, there must have been a time in the past when they were equal, and before that time, radiation was the *main* constituent of the cosmos. At early times the universe was **radiation-dominated.**

17.6 Formation of Nuclei and Atoms

According to the Big Bang theory, at the very earliest times the cosmos consisted entirely of radiation. During the first minute or so of the universe's existence, temperatures were high enough that individual photons of radiation had sufficient energy to transform themselves into matter in the form of elementary particles. This period saw the creation of all the basic building blocks of matter we know today—both "normal" matter, such as protons, neutrons, and electrons, and the "exotic" particles that may compose the dark matter in galaxies and galaxy clusters. ⚭ *(Sec. 14.5)* Since then, matter has evolved, clumping together into more and more complex structures forming the nuclei, atoms, planets, stars, galaxies, and large-scale structure we know today, but no *new* matter has been created. *Everything we see around us formed out of radiation as the early universe expanded and cooled.*

Helium Formation

As the temperature of the universe fell below about 900 million K, roughly 2 minutes after the Big Bang, conditions became favorable for protons and neutrons to fuse to form deuterium.[2] Prior to that time, deuterium nuclei were broken apart by high-energy gamma rays as quickly as they were created. Once the deuterium was at last able to survive the radiation background, other fusion reactions quickly converted it into heavier elements, especially helium-4. In just a few minutes, most of the available neutrons were consumed, leaving a universe whose "normal" matter content was primarily hydrogen and helium, with a trace of deuterium. Figure 17.17 illustrates some of the reactions responsible for helium formation. Contrast it with Figure 9.25, which depicts how helium is formed today in the cores of main-sequence stars such as the Sun. ⚭ *(Sec. 9.5)* The production of elements heavier than hydrogen by nuclear fusion shortly after the Big Bang is called **primordial nucleosynthesis.**

This early burst of fusion did not last long. Conditions in the cooling, thinning universe became less and less conducive to further fusion reactions as time went by. For all practical purposes, primordial nucleosynthesis stopped at helium-4. Detailed calculations show that, by about 15 minutes after the Big Bang, the cosmic elemental abundance was set, and helium accounted for approximately one-quarter of the total mass of normal matter in the universe. The remaining 75 percent was hydrogen. It would be almost a billion years before nuclear reactions in stars would change these figures.

Astronomers find that no matter where they look, and no matter how low a star's supply of heavy elements may be, there is a minimum amount of helium—a little less than 25 percent by mass—in all stars. This is the case even in stars containing very few heavy elements, indicating that the helium already existed when those ancient stars formed. The accepted explanation is that this base level of helium is *primordial*—created during the early, hot epochs of the universe, before any stars had formed, as just described. The striking agreement between the theoretical calculation and actual observations of helium in stars is another key point in favor of the Big Bang theory.

---

[2]Recall from Chapter 9 that deuterium is simply a heavy form of hydrogen. Its nucleus contains one proton and one neutron. ⚭ *(Sec. 9.5)*

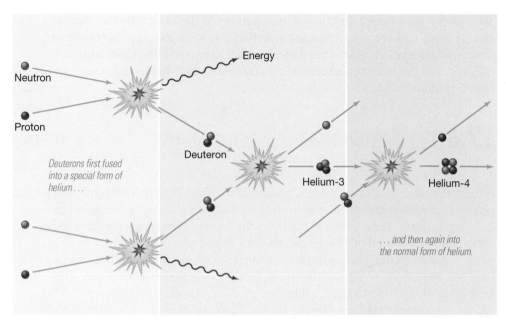

▲ FIGURE 17.17 **Helium Formation**   Some of the reaction sequences that led to the formation of helium in the early universe.

Nucleosynthesis and the Composition of the Universe

Although most deuterium was quickly burned into helium as soon as it formed, a small amount was left over when the primordial nuclear reactions ceased. Nucleosynthesis calculations indicate that the amount of deuterium remaining is a very sensitive indicator of the present-day density of matter in the universe: The *denser* the universe is today, the more matter (protons and neutrons) there was at those early times to react with the deuterium and the less deuterium was left when nucleosynthesis ended. Comparison between theoretical calculations of the production of deuterium in the early universe and the observed abundance of deuterium in stars and the interstellar medium implies a present-day density of at most 3 or 4 percent of the critical value.

But there is an important qualification. Primordial nucleosynthesis depends only on the presence of protons and neutrons in the early universe. Thus, measurements of the abundance of helium and deuterium tell us only about the density of "normal" matter—matter made up of protons and neutrons—in the cosmos. This finding has a momentous implication for the overall composition of the universe. Since the total matter density is about one-third the critical value, if the density of normal matter is only a few percent of this density, then we are forced to conclude that not only is most of the matter in the universe dark, but most of the dark matter is *not* composed of protons and neutrons (see Figure 17.10). The bulk of the matter in the universe apparently exists in the form of elusive subatomic particles whose nature we do not fully understand and whose very existence has yet to be conclusively demonstrated in laboratory experiments. ∞ *(Sec. 14.5)*

For the sake of brevity, we will adopt from here on the convention that the term *dark matter* refers only to these unknown particles and not to "stellar" dark matter, such as brown or white dwarfs, which are made of normal matter.

CONCEPT CHECK ► How do we know that most of the dark matter in the universe is not of "normal" composition?

The Formation of Atoms

When the universe was tens of thousands of years old, matter (consisting of electrons, protons, helium nuclei, and dark matter) began to dominate over radiation.

INTERACTIVE Creation of the Cosmic Microwave Background

The temperature at that time was several tens of thousands of kelvins—too hot for atoms of hydrogen to exist, although some helium ions may already have formed. During the next few hundred thousand years, the universe expanded by another factor of 10, the temperature dropped to a few thousand kelvins, and electrons and nuclei combined to form neutral atoms. By the time the temperature had fallen to 3000 K, the universe consisted of atoms, photons, and dark matter.

The period during which nuclei and electrons combined to form atoms is often called the epoch of **decoupling,** for it was during this period that the radiation background parted company with normal matter. At early times, when matter was ionized, the universe was filled with large numbers of free electrons that interacted frequently with electromagnetic radiation of all wavelengths. A photon could not travel far before encountering an electron and scattering off it. The universe was *opaque* to radiation. However, when the electrons combined with nuclei to form atoms of hydrogen and helium, only certain wavelengths of radiation—those corresponding to the spectral lines of those atoms—could interact with matter. ∞ *(Sec. 2.6)* Light of all other wavelengths could travel virtually forever without being absorbed. The universe became nearly *transparent.* After that time, as the universe expanded, the radiation simply cooled, eventually to become the microwave background we observe today.

The microwave photons now detected on Earth have been traveling through the universe ever since they decoupled. According to the models that best fit the observational data, their last interaction with matter (at the epoch of decoupling) occurred when the universe was about 400,000 years old and roughly 1100 times smaller (and hotter) than it is today—that is, at a redshift of 1100. As illustrated in Figure 17.18, the epoch of atom formation created a kind of "photosphere" in the universe, often referred to as the *surface of last scattering,* completely surrounding Earth at a distance of approximately 14,000 Mpc, corresponding to a redshift of 1100. ∞ *(More Precisely 15-1)* On our side of the photosphere—that is, since decoupling—the universe is transparent. On the far side—before decoupling—it was opaque. Thus, by observing the microwave background, we are probing conditions in the universe almost all the way back in time to the Big Bang itself, in much the same way as studying sunlight tells us about the surface layers of the Sun.

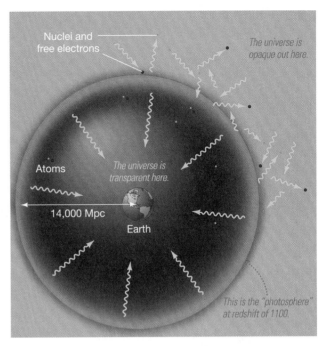

▲ FIGURE 17.18 Radiation–Matter Decoupling When atoms formed, the universe became transparent to radiation. Thus, observations of the cosmic background radiation reveal conditions in the universe around a time when the redshift was 1100 and the temperature was less than about 3000 K. For an explanation of how we can see a region of space 14,000 Mpc (46 billion light-years) away when the universe is just 14 billion years old, see *More Precisely 15-1.*

17.7 Cosmic Inflation

In the late 1970s, cosmologists were confronted with two nagging problems that had no easy explanation within the standard Big Bang model. Their resolution has caused theorists to completely rethink their views of the very early universe.

The Horizon and Flatness Problems

The first problem is known as the **horizon problem.** Imagine observing the microwave background in two opposite directions on the sky, as illustrated in Figure 17.19. As we have just seen, in doing so we are actually observing two distant regions of the universe, marked A and B on the figure, where the radiation background last interacted with matter. The fact that the background is known to be *isotropic* to high accuracy means that regions A and B must have had very similar densities and temperatures at the time the radiation we see left them.

The problem is, within the Big Bang theory as just described, there is no particular reason *why* these regions should have been so similar. At the time in question, a few hundred thousand years after the Big Bang, regions A and B were

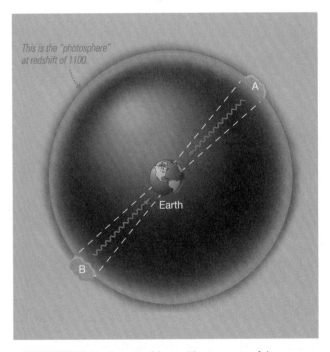

▲ FIGURE 17.19 Horizon Problem The isotropy of the microwave background indicates that regions A and B in the universe were very similar to each other when the radiation we now observe left them, but there has not been enough time since the Big Bang for them ever to have interacted with one another. Why then should they look the same?

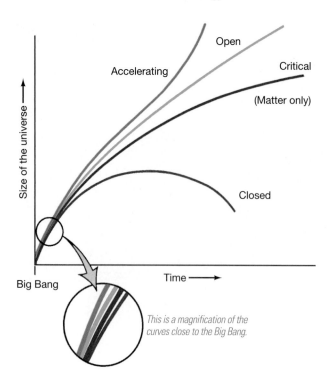

▲ FIGURE 17.20 **Flatness Problem** If the universe deviates even slightly from critical density, that deviation grows rapidly in time. For the universe to be as close to critical as it is today, it must have differed from the critical density in the past by only a tiny amount.

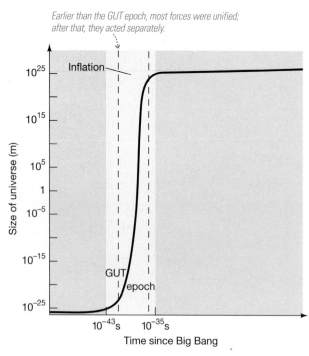

▲ FIGURE 17.21 **Cosmic Inflation** During the period of inflation, the universe expanded enormously in a very short time. Afterward, it resumed its earlier "normal" expansion rate, except that the size of the cosmos had become about 10^{50} times bigger than it was before inflation.

separated by many megaparsecs, and no signal—sound waves, light rays, or anything else—had had time to travel from one to the other. In cosmological terminology, each region lay outside the other's *horizon.* But with no possibility of information or energy exchange between these regions, there was no way for them to "know" that they should look the same. The only alternative is that they simply started off looking alike—an assumption that cosmologists are very unwilling to make.

The second problem with the standard Big Bang model is called the **flatness problem.** Whatever the exact value of the cosmic density, it is quite close to the critical value. In terms of spacetime curvature, the universe is remarkably close to being *flat.* But again, there is no good reason why the universe should have formed with a density very close to critical. Why not a millionth or a million times that value? Furthermore, as can be seen in Figure 17.20, a universe that starts off close to, but not exactly on, the critical curve soon deviates greatly from it, so if the universe is close to critical now, it must have been *extremely* close to critical in the past. For example, even if Ω_0 were as low as 0.1 today, the departure from the critical density at the time of nucleosynthesis would have been only change: 1 part in 10^{15} (a thousand trillion).

These observations constitute "problems" because cosmologists want to be able to explain the present condition of the universe, not just accept it "as is." They would prefer to resolve the horizon and flatness problems in terms of physical processes that could have taken a universe with no special properties and caused it to evolve into the cosmos we now see. The resolution of both problems takes us back in time even earlier than nucleosynthesis or the formation of any of the elementary particles we know today—back, in fact, almost to the instant of the Big Bang itself.

The Epoch of Inflation

In the 1970s and 1980s, theoretical physicists succeeded in combining, or *unifying*, the three nongravitational forces in the universe—electromagnetism and the strong and weak forces—into a single all-encompassing "superforce." ∞ *(More Precisely 9-1)* A general prediction of the **Grand Unified Theories** (GUTs) that describe this superforce is that the three forces are unified and indistinguishable from one another *only* at enormously high energies—corresponding to temperatures in excess of 10^{28} K. At lower temperatures, the superforce splits into three, revealing its separate electromagnetic, strong, and weak characters.

In the 1980s, cosmologists discovered that Grand Unified Theories had some remarkable implications for the very early universe. Less than 10^{-34} s after the Big Bang, parts of the cosmos could have temporarily found themselves in a very odd and unstable condition in which empty space—the very fabric of the universe—acquired **vacuum energy,** in essence becoming *excited* above its normal equilibrium state. Esoteric though this sounds, these regions are of direct interest to us—if theorists are correct, we live in one!

The appearance of vacuum energy had dramatic consequences. For a short while, as illustrated in Figure 17.21, the extra energy caused the affected region to expand at an enormously accelerated rate. (The effect is similar in concept but vastly greater in magnitude than the cosmic acceleration associated with the cosmological constant, discussed in *Discovery 17-1.*) The vacuum energy density remained almost constant as the region grew, and the expansion *accelerated* with time while this out-of-equilibrium condition persisted. In fact, the size of the region doubled many times over. This period of unchecked cosmic expansion is known as the **epoch of inflation.**

Eventually, the vacuum returned to its normal "true" state and inflation stopped. Regions of normal space began to appear within the false vacuum and rapidly spread to include the entire expanding region. The whole episode lasted a mere 10^{-32} s, but during that time the patch of the universe that had become unstable swelled in size by the incredible factor of about 10^{50}. After the inflationary phase, the universe once again resumed its (relatively) leisurely expansion, but a number of important changes had occurred that would have far-reaching ramifications for the evolution of the cosmos.

The original theory of inflation was developed in the early 1980s and associated the inflationary period (as in Figure 17.21) with the end of the so-called, *GUT epoch*, as temperatures fell below 10^{28} K and the basic forces of nature became reorganized—a little like a gas liquefying or water freezing as the temperature drops. However, researchers have since realized that conditions suitable for inflation could have occurred under many different circumstances—and possibly *many times*—during the evolution of the early universe, making inflation a rather generic prediction of current theoretical models.

Implications for the Universe

Inflation provides a natural solution for the horizon and flatness problems. The horizon problem is solved because inflation took regions of the universe that had already had time to communicate with one another—and so had established similar physical properties—and then dragged them far apart, well out of communication range of one another. In effect, the universe expanded much faster than the speed of light during the inflationary epoch, so what was once well within the horizon now lies far beyond it. (Relativity restricts matter and energy to speeds less than the speed of light, but imposes no such limit on the universe as a whole.) Regions A and B in Figure 17.19 have been out of contact with one another since 10^{-32} s after the Big Bang, but they *were* in contact before then. Their properties are the same today because they were the same long ago, before inflation separated them.

Figure 17.22 illustrates how inflation solves the flatness problem. Imagine that you are a 1-mm-long ant sitting on the surface of an expanding balloon. When the balloon is just a few centimeters across, you can easily see that its surface is curved because its circumference is comparable to your own size. However, as the balloon expands, the curvature becomes less pronounced. When the balloon is a few kilometers across, your "ant-sized" patch of the surface looks quite flat, just as Earth's surface looks flat to us. Now imagine that the balloon expands 100 trillion trillion trillion trillion times. Your local patch of the surface would be completely indistinguishable from a perfectly flat plane. Thus, any curvature the universe may have had before inflation has been expanded so much that space is now perfectly flat on all scales we can ever hope to observe.

This resolution to the flatness problem—the universe appears close to being flat because the universe *is* in fact precisely flat, to very high accuracy—has another very important consequence. Because the universe is geometrically flat, relativity tells us the total density must be exactly equal to the critical value: $\Omega_0 = 1$ (Section 17.3). This is the key result that led us to conclude in Section 17.4 that dark energy—whatever it is—must dominate total density of the universe. Not only is most matter dark (Section 17.6), but *most of the cosmic density isn't made up of matter at all.*

Even though inflation solves the horizon and flatness problems in a quite convincing way, for nearly two decades after it was first proposed the theory was resisted by many astronomers. The main reason was that its prediction of $\Omega_0 = 1$

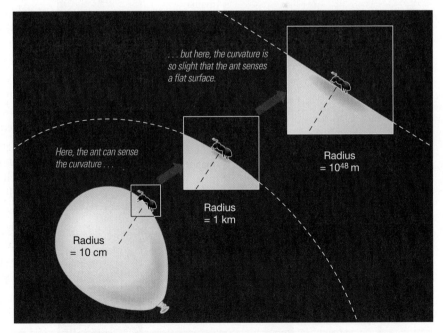

. . . *but here, the curvature is so slight that the ant senses a flat surface.*

Here, the ant can sense the curvature . . .

Radius $= 10^{48}$ m

Radius $= 1$ km

Radius $= 10$ cm

▲ FIGURE 17.22 **Inflation and the Flatness Problem** Inflation solves the flatness problem by enormously expanding a curved surface, here represented by the surface of a balloon. To an ant on the surface, the balloon looks virtually flat when the expansion is complete.

was at odds with the growing evidence that the density of matter in the universe was no more than 30 percent or so of the critical value. Many cosmologists had in fact considered the possibility that a cosmological constant might provide a way to account for the remaining 70 percent of the cosmic density, but without independent corroboration, a conclusive case could not be made. That is why the supernova observations were so important—by demonstrating acceleration in the cosmic expansion rate, they established independent evidence for the effects of dark energy and in doing so reconciled inflation with the discrepant observations.

Physicists will probably never create in terrestrial laboratories conditions even remotely similar to those that existed in the universe during the inflationary epoch. Creation of a false vacuum is (safely) beyond our reach. Nevertheless, cosmic inflation seems to be a natural consequence of many Grand Unified Theories, and it explains two otherwise intractable problems within the Big Bang theory. For these reasons, despite the lack of direct evidence for the process, inflation theory has become an integral part of modern cosmology. It makes definite, testable predictions about the large-scale geometry and structure of the universe that are critically important to current theories of galaxy formation. As we will see in the next section, astronomers are now subjecting these predictions to rigorous scrutiny.

CONCEPT CHECK ▶ Does the theory of inflation itself say anything about the composition of the universe?

17.8 Formation of Large-Scale Structure in the Universe

 ANIMATION/VIDEO Cosmic Structure

Most cosmologists think that the present large-scale structure in the universe grew from small *inhomogeneities*—slight deviations from perfectly uniform density—that existed in the early universe. Denser than average clumps of matter contracted under the influence of gravity and merged with other clumps, eventually reaching the point where stars began to form and luminous galaxies appeared. ∞ *(Sec. 16.3)*

The intense background radiation in the early universe effectively prevented clumps of normal matter from forming or growing. Just as in a star, the pressure of the radiation stabilized the clumps against the inward pull of their own gravity. ∞ *(Sec. 12.1)* Instead, the first structures to appear did so in the *dark* matter that makes up the bulk of matter in the universe. While the true nature of dark matter remains uncertain, its defining property is the fact that it interacts only very weakly with both normal matter and radiation. Consequently, clumps of dark matter were unaffected by the radiation background and began to grow as soon as matter first began to dominate the cosmic density.

Thus, as illustrated in Figure 17.23, dark matter determined the overall distribution of mass in the universe and clumped to form the observed large-scale structure. Then, at later times, normal matter was drawn by gravity into the regions of highest density, eventually forming galaxies and galaxy clusters. This picture explains why so much dark matter is found outside the visible galaxies. The luminous material is strongly concentrated near the density peaks and dominates the

The plots below are graphical representations of structure growth.

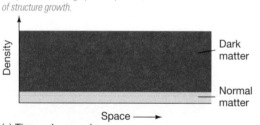

(a) Time = 1 second

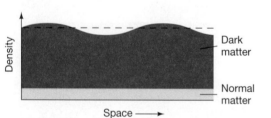

(b) Time = 1000 years

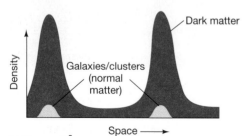

(c) Time = 10⁸ years

The maps above show what those structures might look like on the sky.

◀ **FIGURE 17.23 Structure Formation** (a) The very early universe was a mixture of (mostly) dark and normal matter. (b) A few thousand years after the Big Bang, the dark matter began to clump. (c) Eventually, the dark matter formed large structures (represented here by the two high-density peaks) into which normal matter flowed, ultimately to form the galaxies seen today.

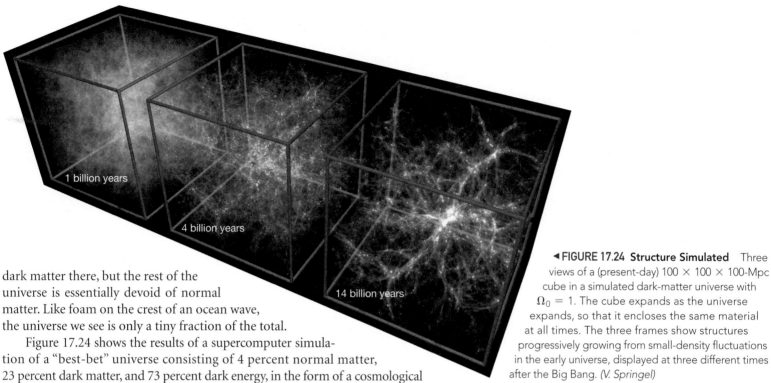

◀ FIGURE 17.24 Structure Simulated Three views of a (present-day) 100 × 100 × 100-Mpc cube in a simulated dark-matter universe with $\Omega_0 = 1$. The cube expands as the universe expands, so that it encloses the same material at all times. The three frames show structures progressively growing from small-density fluctuations in the early universe, displayed at three different times after the Big Bang. (V. Springel)

dark matter there, but the rest of the universe is essentially devoid of normal matter. Like foam on the crest of an ocean wave, the universe we see is only a tiny fraction of the total.

Figure 17.24 shows the results of a supercomputer simulation of a "best-bet" universe consisting of 4 percent normal matter, 23 percent dark matter, and 73 percent dark energy, in the form of a cosmological constant (see Figure 17.10). ⚭ *(Sec. 17.3)* The similarities with real observations of cosmic structure, shown in Figures 16.21, 16.22, and 17.1, are striking, and more detailed statistical analysis shows that these models in fact agree extremely well with reality. Although calculations like this cannot prove that these models are the correct description of the universe, the agreement in detail between simulations and reality strongly favor the cosmological constant/dark-matter model of the cosmos.

Although dark matter does not interact directly with photons, the background radiation is influenced slightly by the *gravity* of the growing dark clumps. This causes a slight gravitational redshift in the background radiation that varies from place to place depending on the dark-matter density. As a result, a key prediction of dark-matter models is that there should be tiny "ripples" in the microwave background—temperature variations of only a few parts per million from place to place on the sky. In 1992, after almost 2 years of careful observation, the *COBE* team announced that the expected ripples had been detected. The temperature variations are tiny—only 30–40 millionths of a kelvin—but they are there. The *COBE* results are displayed as a temperature map of the microwave sky in Figure 17.25.

The ripples seen by *COBE*, combined with computer simulations such as that shown in Figure 17.24, predict a present-day structure that is consistent with the superclusters, voids, filaments and Great Walls we see around us. While the *COBE* data were limited by relatively low (roughly 7°) resolution, detailed analysis of the ripples also supports the key prediction of inflation theory—that the universe is of exactly critical density and hence spatially flat. For these reasons, the *COBE* observations rank alongside the discovery of the microwave background itself in terms of their importance to the field of cosmology.

Figure 17.26 shows a more recent all-sky map made by the *Wilkinson Microwave Anisotropy Probe (WMAP)*, operated by NASA from 2001 until 2009. Permanently stationed some 1.5 million km outside Earth's orbit along the Sun–Earth line and always pointing away from the Sun to keep its delicate heat-sensitive detectors in

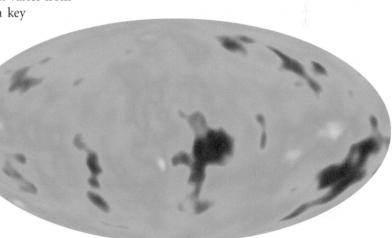

R I V U X G

▲ FIGURE 17.25 Cosmic Microwave Background Map This *COBE* map of temperature fluctuations in the cosmic microwave background over the entire sky shows hotter-than-average regions in yellow and cooler-than-average regions in blue. The temperature variation due to Earth's motion has been subtracted, as has the radio emission from the Milky Way Galaxy. (NASA)

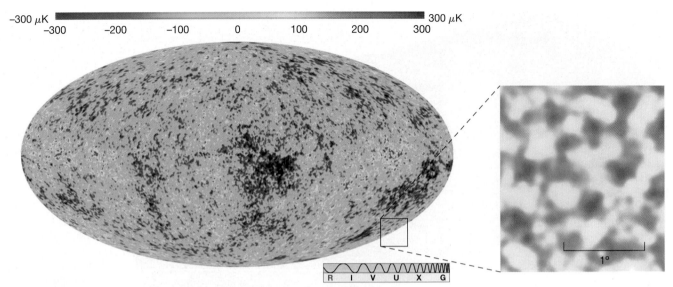

−300 μK   −300   −200   −100   0   100   200   300   300 μK

R I V U X G

INTERACTIVE ▲ FIGURE 17.26 **Early Structure** The entire microwave sky, as seen by the *WMAP* spacecraft at frequencies as high as 90 GHz (3-mm wavelength). Compare this high-resolution map with the *COBE* map shown in Figure 17.25. The inset at right shows an even higher resolution map of a small patch of sky. The bright blobs are slightly denser-than-average regions of the universe at an age of roughly 400,000 years; they will eventually contract to form clusters of galaxies. *(NASA; CBI)*

ANIMATION/VIDEO Ripples in the Early Universe

ANIMATION/VIDEO *COBE* and *WMAP* View of Cosmic Microwave Background

CONCEPT CHECK ► Why is dark matter important to galaxy formation? How do observations of the microwave background support this view?

shadow, *WMAP* completed a scan of the entire sky every 6 months. The instrument's angular resolution was roughly 20 times finer than that of *COBE* (compare Figures 17.25 and 17.26), allowing extraordinarily detailed measurements of many cosmological parameters to be made. The inset in Figure 17.26 shows a smaller-scale (just 2° wide), but even higher resolution image returned by Cosmic Background Imager, a ground-based microwave telescope located high in the Chilean Andes. (Recall from Chapter 2 that the microwave part of the spectrum is only partly transparent to electromagnetic radiation, so microwave detectors must be placed above as much of Earth's absorbing atmosphere as possible.) ∞ *(Sec. 2.3)* Both maps show temperature fluctuations of a few hundred microkelvins, with a characteristic angular scale of about 1°. This temperature range is larger than the fluctuations seen by *COBE* because *COBE*'s low resolution effectively averaged the data over a large area of the sky.

The angular scale of the fluctuations is important because detailed calculations reveal that this is a *direct* measure of Ω_0, largely independent of other cosmological "details" (such as the fraction of dark matter or the nature of the dark energy). The slightly less than 1° "blob-to-blob" angular scale, evident to the eye in the inset of Figure 17.26 and confirmed by more quantitative means, provides the strongest evidence yet that Ω_0 is very close to 1, with an extremely small margin of error, in full agreement with the inflation prediction. Each of the irregular high-temperature regions in these images represents a slightly denser than average clump of dark matter that will one day collapse to form a supercluster-sized clump of galaxies. According to the *WMAP* data, we are seeing the universe here when it was just 380,000 years old, yet the future development of the cosmos is already imprinted in these tiny ripples.

Ongoing and upcoming space-based observations will further refine these measurements, improving their errors and expanding the coverage to the entire sky. The first decade of the 21st century may well see the basic parameters of the universe measured (even if not fully understood) to an accuracy only dreamed of just a few years ago.

CHAPTER REVIEW

SUMMARY

LO1 On scales larger than a few hundred megaparsecs, the universe appears roughly **homogeneous** (p. 460; the same everywhere) and **isotropic** (p. 460; the same in all directions). In **cosmology** (p. 461)—the study of the universe as a whole—researchers usually assume that the universe is homogeneous and isotropic on large scales. This assumption is known as the **cosmological principle** (p. 461).

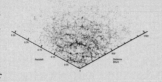

LO2 If the universe were homogeneous, isotropic, infinite, and unchanging, then the night sky would be bright because any line of sight would eventually intercept a star. The fact that the night sky is instead dark is called **Olbers's paradox** (p. 461). Its resolution is that we see only a finite part of the universe—the region within which light has had time to reach us since the universe formed. Tracing the observed motions of galaxies back in time implies that, about 14 billion years ago, the universe consisted of a single point that expanded rapidly in the **Big Bang** (p. 462). Space itself was compressed to a point at that instant—the Big Bang happened everywhere at once. The cosmological redshift occurs as a photon's wavelength is "stretched" by cosmic expansion. The extent of the observed redshift is a direct measure of the expansion of the universe since the photon was emitted.

LO3 There are only two possible outcomes to the current expansion: Either the universe will expand forever, or it will recollapse to a point. The **critical density** (p. 464) is the density of matter needed for gravity to overcome the present expansion and cause the universe to collapse. The total density of the universe (including matter, radiation, and dark energy) determines the geometry of the universe on the largest scales, as described by general relativity. In a high (greater-than-critical) density universe, the curvature of space is sufficiently large that the universe "bends back" on itself and is finite in extent, somewhat like the surface of a sphere. Such a universe is said to be a **closed universe** (p. 465). A low-density **open universe** (p. 465) is infinite in extent and has a "saddle-shaped" geometry. The **critical universe** (p. 466) has a density precisely equal to the critical value and is spatially flat.

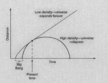

LO4 Luminous and dark matter account for only 25–30 percent of the critical cosmic density. Observations of distant supernovae suggest that the expansion of the universe may be accelerating, possibly driven by the effects of a force commonly called **dark energy** (p. 467). One candidate for this dark energy is the **cosmological constant** (p. 468), a repulsive force that may exist throughout all space. Its physical nature is unknown. Other, independent observations are consistent with the idea that the universe is flat—that is, of exactly critical density—with (mostly dark) matter making up 27 percent of the total and dark energy making up the rest. Such a universe will expand forever.

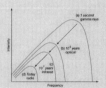

LO5 The **cosmic microwave background** (p. 471) is isotropic blackbody radiation that fills the entire universe. Its present temperature is about 3 K. Its existence is evidence that the universe expanded from a hot, dense state. As the universe has expanded, the initially high-energy radiation has been redshifted to lower and lower temperatures. At the present time, the densities of dark energy and matter in the universe both greatly exceed the equivalent mass density of radiation. The density of dark energy remains constant as the universe expands, but the matter density was much greater in the past, when the universe was smaller. However, because radiation is redshifted as the universe expands, the density of radiation was greater still. The early universe was **radiation-dominated** (p. 473).

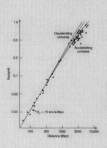

LO6 All of the hydrogen in the universe is primordial, formed from radiation as the hot early universe expanded and cooled. Most of the helium observed in the universe today is also primordial, created by fusion between protons and neutrons a few minutes after the Big Bang. This is known as **primordial nucleosynthesis** (p. 473). Other, heavier elements were formed much later, in the cores of stars. Detailed studies of this process indicate that "normal" matter can account for at most 3 percent of the critical density. By the time the universe was about 1100 times smaller than it is today, the temperature had become low enough for the first atoms to form. At that time, the (then visible) radiation background **decoupled** (p. 475) from the matter. The photons that now make up the microwave background have been traveling freely through space ever since.

LO7 According to modern **Grand Unified Theories** (p. 476), the three nongravitational forces of nature first began to display their separate characters about 10^{-34} s after the Big Bang. A brief period of rapid cosmic expansion called the **epoch of inflation**

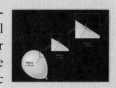

(p. 476) ensued, during which the size of the universe increased by a factor of about 10^{50}. The **horizon problem** (p. 475) is the fact that, according to the standard (that is, noninflationary) Big Bang model, there is no good reason for widely separated parts of the universe to be as similar as they are. Inflation solves the horizon problem by taking a small homogeneous patch of the early universe and expanding it enormously. Inflation also solves the **flatness problem** (p. 476), which is the fact that there is no obvious reason why the density of the universe is so close to critical. Inflation implies that the universe is spatially flat and hence that the cosmic density is exactly critical.

LO8 The large-scale structure observed in the universe today formed when density inhomogeneities in the dark matter clumped and grew to create the "skeleton" of the structure now observed.  Normal matter then flowed into the densest regions of space, eventually forming the galaxies we now see. "Ripples" in the microwave background, discovered by the *COBE* satellite, are the imprint of these early inhomogeneities on the radiation field and lend strong support to the inflationary prediction that we live in a flat, critical-density universe. Detailed observations of the microwave background, combined with studies of large-scale structure in the cosmos, provide detailed information on the basic cosmological parameters of the universe.

MasteringAstronomy®   For instructor-assigned homework go to www.masteringastronomy.com

Problems labeled POS explore the process of science. VIS problems focus on reading and interpreting visual information. LO connects to the introduction's numbered Learning Outcomes.

REVIEW AND DISCUSSION

1. **LO1** What evidence do we have that there is no structure in the universe on very large scales? How large is "very large"?

2. What is the cosmological principle?

3. What is Olbers's paradox? How is it resolved?

4. **LO2** Explain how an accurate measure of Hubble's constant can lead to an estimate of the age of the universe.

5. Why isn't it correct to say that the expansion of the universe involves galaxies flying outward into empty space?

6. Where did the Big Bang occur?

7. How does the cosmological redshift relate to the expansion of the universe?

8. What is Ω_0? What is the current best estimate of its value, and what does this imply about the geometry of the universe?

9. **LO4 POS** Why are observations of distant supernovae so important to cosmology?

10. **LO5** What is the significance of the cosmic microwave background?

11. **LO6** Why do all stars, regardless of their supply of heavy elements, seem to contain at least one-quarter helium by mass?

12. When did the universe become transparent to radiation?

13. **LO7 POS** What is cosmic inflation? How does inflation solve the horizon problem? The flatness problem?

14. **POS** What is the connection between dark matter and the formation of large-scale structure in the universe?

15. **LO8 POS** What prediction of inflation theory has since been verified by observations of the microwave background?

CONCEPTUAL SELF-TEST: TRUE OR FALSE/MULTIPLE CHOICE

1. Deep surveys of the universe indicate that the largest structures in space are about 50 Mpc in size. (T/F)

2. If the universe had an edge, that fact would violate the assumption of isotropy in the cosmological principle. (T/F)

3. Hubble's law implies that the universe will expand forever. (T/F)

4. The cosmological redshift is a direct measure of cosmic expansion. (T/F)

5. The cosmic microwave background is the highly redshifted radiation of the early Big Bang. (T/F)

6. Observations suggest that the density of the universe is made up mostly of dark matter. (T/F)

7. The microwave background radiation last interacted with matter around the time of decoupling. (T/F)

8. The theory of inflation predicts that the density of the universe is exactly equal to the critical density. (T/F)

9. Olbers's paradox is resolved by (a) the finite size of the universe; (b) the finite age of the universe; (c) light from distant galaxies being redshifted so we can't see them; (d) the fact that there is an edge to the universe.

10. The galactic distances used to measure the acceleration of the universe are determined by observations of (a) trigonometric parallax; (b) line broadening; (c) Cepheid variable stars; (d) exploding white dwarfs.

11. On the basis of our current best estimate of the present mass density of the universe, astronomers think that (a) the universe is finite in extent and will expand forever; (b) the universe is finite in extent and will eventually collapse; (c) the universe is infinite in

extent and will expand forever; (d) the universe is infinite in extent and will eventually collapse.

12. The age of the universe is estimated to be (a) less than Earth's age; (b) the same as the age of the Sun; (c) the same as the age of the Milky Way Galaxy; (d) greater than the age of the Milky Way Galaxy.

13. The horizon problem in the standard Big Bang model is solved by having the universe (a) accelerate; (b) inflate rapidly early in its existence; (c) have tiny, but significant fluctuations in temperature; (d) be geometrically flat.

14. **VIS** The data points in Figure 17.8 (Accelerating Universe) (a) prove that the universe is not expanding; (b) imply that the expansion is decelerating faster than expected; (c) allow a measurement of Hubble's constant; (d) indicate that the redshifts of distant galaxies are greater than would be expected if gravity alone were acting.

15. **VIS** According to Figure 17.16 (Changing Dominance), when the universe was 10 years old, the density of radiation was (a) much greater than that of matter; (b) much less than that of matter; (c) comparable to that of matter; (d) unknown.

PROBLEMS

The number of squares preceding each problem indicates its approximate level of difficulty.

1. ■ What is the greatest distance at which a galaxy survey sensitive to objects as faint as 20th magnitude could detect a galaxy as bright as the Milky Way (absolute magnitude –20)? ∞ *(More Precisely 10-1)*

2. ■■ If the entire universe were filled with Milky Way-like galaxies, with an average density of 0.1 per cubic megaparsec, calculate the total number of galaxies observable by the survey in the previous question if it covered the entire sky.

3. ■■ Eight galaxies are located at the corners of a cube. The present distance from each galaxy to its nearest neighbor is 10 Mpc, and the entire cube is expanding according to Hubble's law, with $H_0 = 70$ km/s/Mpc. Calculate the recession velocity (magnitude *and* direction) of one corner of the cube relative to the opposite corner.

4. ■ According to the standard Big Bang theory (neglecting any effects of cosmic acceleration), what is the *maximum* possible age of the universe if $H_0 = 50$ km/s/Mpc? 70 km/s/Mpc? 80 km/s/Mpc?

5. ■ For a Hubble constant of 70 km/s/Mpc, the critical density is 9×10^{-27} kg/m^3. (a) How much mass does this correspond to within a volume of 1 cubic astronomical unit? (b) How large a cube would be required to enclose 1 Earth mass of material?

6. ■■ The Virgo Cluster is observed to have a recessional velocity of 1200 km/s. Assuming $H_0 = 70$ km/s/Mpc and a critical-density universe, calculate the total mass contained within a sphere centered on Virgo and just enclosing the Milky Way Galaxy. What is the escape speed from the surface of this sphere? ∞ *(More Precisely 5-1)*

7. ■■ (a) What is the present peak wavelength of the cosmic microwave background? Calculate the size of the universe relative to the present size when the radiation background peaked in (b) the infrared, at 10 μm, (c) in the ultraviolet, at 100 nm, and (d) in the gamma ray, at 1 nm.

8. ■■ What were the matter density and the equivalent mass density of the cosmic radiation field when the universe was one-thousandth its present size? Assume critical density, and don't forget the cosmological redshift.

9. ■ What was the distance between the points that would someday become the center of the Milky Way and the center of the Coma Cluster at the time of decoupling? (The present distance to Coma is 100 Mpc.)

10. ■■■ The "blobs" evident in the inset to Figure 17.26 are about 20′ (arcminutes). If those blobs represent clumps of matter around the time of decoupling (redshift = 1100), and assuming Euclidean geometry, estimate the size of these clumps at the time of decoupling. ∞ *(More Precisely 15-1)*

ACTIVITIES

Collaborative

1. Make a model of a two-dimensional universe and use it to examine Hubble's law. Find a balloon that will expand into a nice, large sphere. Blow it up about halfway, and mark dots all over its surface; the dots will represent galaxies. Each group member should choose one dot as his/her home galaxy. Using a measuring tape, measure the distances to various other galaxies, numbering the dots so you will not confuse them later. Now blow the balloon up to full size and measure the distances again. Calculate the change in the distances for each galaxy; this is a measure of its velocity (change in position/change in time; the time is the same for all and is arbitrary). Plot their velocities versus their new distances as in Figure 15.16 or 17.8. Do you get a straight-line correlation, that is, a "Hubble" law? Does it matter which dot you choose as home? Demonstrate this to your class.

Individual

1. Isotropy is the extent to which things look the same in every direction. Considering buildings, geographic features, and similar objects within a few miles of your current location, is your local universe isotropic? If not, is there any scale on which isotropy applies, even approximately?

2. Write a paper on the philosophical differences between living in an open, closed, or flat universe. Are there aspects of any of these three possibilities that are hard to accept? Do you have a preference?

3. Go to your library or online and read about the *steady-state universe*, which enjoyed some measure of popularity in the 1950s and 1960s. How did it differ from the standard Big Bang model? What were its main assumptions and predictions, and why did its proponents regard it as superior to the Big Bang? Why is the steady-state model not widely accepted today—what key predictions did it make that were inconsistent with observations?

18

Life in the Universe

Are We Alone?

Are we unique? Is life on our planet the only example of life in the universe? If so, then what would be the implications of that momentous discovery? If not, then how and where should we look for other intelligent beings? These are difficult questions, for the subject of extraterrestrial life is one on which we have no data, but they are important questions, with profound implications for the human species. Earth is the only place in the universe where we know for certain that life exists. In this final chapter we take a look at how humans evolved on Earth and then consider whether those evolutionary steps might have happened elsewhere. Having done that, we assess the likelihood of our having Galactic neighbors and consider how we might learn about them if they exist.

MasteringAstronomy®   Visit www.masteringastronomy.com for quizzes, animations, videos, interactive figures, and self-guided tutorials.

◄ Is there intelligent life elsewhere in the universe? The simple answer is that we don't know. Yet astronomers keep watchful eyes on the sky, constantly aware that evidence for extraterrestrial intelligence (ET) might emerge at any moment. This fanciful painting, entitled *Galaxyrise Over Alien Planet*, suggests a plurality of life—some perhaps extinct, some perhaps exotic—on alien worlds well beyond Earth. Despite blockbuster movies, science fiction novels, and a host of claims for extraterrestrial contact, astronomers have so far found no unambiguous evidence for life, intelligent or otherwise, anywhere else in the universe. (© Dana Berry)

18.1 Cosmic Evolution

In our study of the universe, we have been careful to avoid any inference or conclusion that places Earth in a special place in the cosmos. This Copernican principle has been our essential guide in helping define our place in the "big picture." ∞ *(Sec. 1.1)* But when discussing the topic of life in the universe we face a problem: Ours is the only planet we know of on which life and intelligence have evolved, making it hard for any discussion of intelligent life *not* to treat humans as special cases.

Accordingly, in this final chapter we must adopt a decidedly different approach. We first describe the chain of events leading to the only technologically proficient, intelligent civilization we know—us. Then, we try to assess the likelihood of finding and communicating with intelligent life elsewhere in the cosmos.

Life in the Universe

With this human-centered viewpoint clearly evident, Figure 18.1 identifies seven major phases in the history of the universe: *particulate, galactic, stellar, planetary, chemical, biological,* and *cultural* evolution. Taken together, these evolutionary stages make up the grand sweep of **cosmic evolution**—the continuous transformation of matter and energy that has led to the appearance of life and civilization on Earth. The first four phases represent, in reverse order, the contents of this book up to this point. We now expand our field of view beyond astronomy to include the other three.

From the Big Bang to the evolution of intelligence and culture, the universe has evolved from simplicity to complexity. We are the result of an incredibly complex chain of events that spanned billions of years. Were those events random, making us unique, or are they in some sense *natural*, so that a technological civilization like ours is inevitable? Put another way, are we alone in the universe, or are we just one among countless other intelligent life-forms?

Before trying to answer these important questions, we need a working definition of *life*. The distinction between the living and the nonliving is not as obvious as we might at first think. Generally speaking, scientists regard the following as characteristics of living organisms: (1) they can *react* to their environment and can often heal themselves when damaged; (2) they can *grow* by taking in nourishment

▶ **FIGURE 18.1 Arrow of Time** Some highlights of cosmic history, as it relates to the emergence of life on Earth, are noted along this arrow of time, from the beginning of the universe to the present. At the bottom of the arrow are seven "windows" outlining the major phases of cosmic evolution: evolution of primal energy into elementary particles; of atoms into galaxies and stars; of stars into heavy elements; of elements into solid rocky planets; of those elements into the molecular building blocks of life; of those molecules into life itself; and of advanced life-forms into intelligence, culture, and technological civilization. *(D. Berry)*

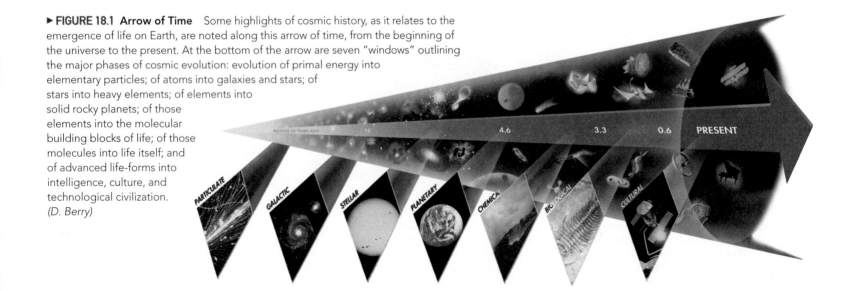

from their surroundings; (3) they can *reproduce*, passing along some of their own characteristics to their offspring; and (4) they have the capacity for genetic change and can therefore *evolve* from generation to generation to adapt to a changing environment.

These rules are not hard and fast, and a good working definition of life is elusive. Stars, for example, react to the gravity of their neighbors, grow by accretion, generate energy, and "reproduce" by triggering the formation of new stars, but no one would suggest that they are alive. A virus is inert, often crystalline in form, when isolated from living organisms, but once inside a living system, it exhibits all the properties of life, seizing control of the cell and using the cell's own genetic machinery to grow and reproduce. Most researchers think that the distinction between living and nonliving is more one of structure and complexity than a simple checklist of rules.

The general case in favor of extraterrestrial life is summed up in what are sometimes called the *assumptions of mediocrity*: (1) because life on Earth depends on just a few basic molecules, (2) because the elements that make up these molecules are (to a greater or lesser extent) common to all stars, and (3) if the laws of science we know apply to the entire universe (as we have supposed throughout this book), then—given sufficient time—life must have originated elsewhere in the cosmos. The opposing view maintains that intelligent life on Earth is the product of a series of extremely fortunate accidents—astronomical, geological, chemical, and biological events unlikely to have occurred anywhere else in the universe. The purpose of this chapter is to examine some of the scientific arguments for and against these viewpoints.

Life on Earth

What information do we have about the earliest stages of planet Earth? Unfortunately, not very much. Geological hints about the first billion years or so were largely erased by violent surface activity as volcanoes erupted and meteorites bombarded our planet; subsequent erosion by wind and water has seen to it that little evidence has survived to the present day. However, the early Earth was probably barren, with shallow lifeless seas washing upon grassless, treeless continents. Outgassing from our planet's interior through volcanoes, fissures, and geysers produced an atmosphere rich in hydrogen, nitrogen, and carbon compounds and poor in free oxygen. As Earth cooled, ammonia, methane, carbon dioxide, and water formed. The stage was set for the appearance of life.

The surface of the young Earth was a very violent place. Natural radioactivity, lightning, volcanism, solar ultraviolet radiation, and meteoritic impacts all provided large amounts of energy that eventually shaped the ammonia, methane, carbon dioxide, and water into more complex molecules known as **amino acids** and **nucleotide bases**—organic (carbon-based) molecules that are the building blocks of life as we know it. Amino acids build *proteins*, and proteins control *metabolism*, the daily utilization of food and energy by means of which organisms stay alive and carry out their vital activities. Sequences of nucleotide bases form *genes*—parts of the DNA molecule—that direct the synthesis of proteins and hence determine the characteristics of organisms (Figure 18.2). These same genes also carry an

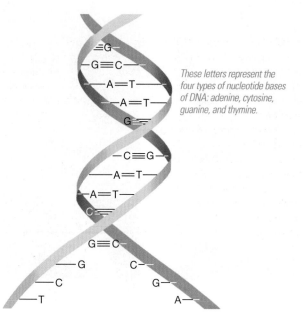

These letters represent the four types of nucleotide bases of DNA: adenine, cytosine, guanine, and thymine.

▶ **FIGURE 18.2 DNA Molecule** The DNA (deoxyribonucleic acid) molecule contains all the genetic information needed for a living organism to reproduce and survive. Often consisting of literally tens of billions of individual atoms, its double-helix structure allows it to "unzip," exposing its internal structure to control the creation of proteins needed for a cell to function. The ordering of its constituent parts is unique to each individual organism.

▶ **FIGURE 18.3 Urey-Miller Experiment**   This chemical apparatus is designed to make complex biochemical molecules by energizing a mixture of simple chemicals. Gases (ammonia, methane, carbon dioxide, water vapor) are placed in the upper bulb to simulate Earth's primordial atmosphere and then zapped by spark-discharge electrodes akin to lightening. After about a week, amino acids and other complex molecules emerge in the trap at the bottom, which simulates the primordial oceans into which heavy molecules produced in the overlying atmosphere would have fallen.

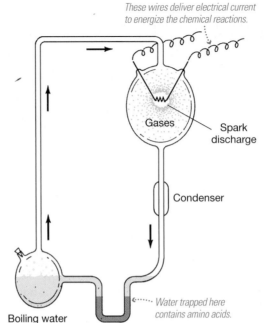

These wires deliver electrical current to energize the chemical reactions.

Gases

Spark discharge

Condenser

Water trapped here contains amino acids.

Boiling water

These three photographs, taken through a microscope, show structures on the scale of 1 micrometer, which equals 1/10,000 of a centimeter.

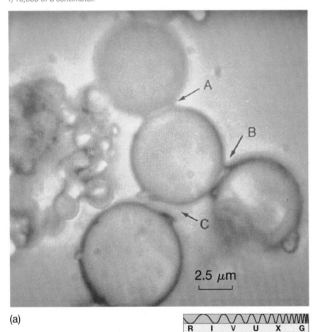

A

B

C

2.5 μm

(a)

R I V U X G

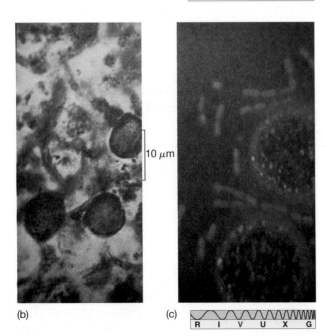

10 μm

(b)   (c)

R I V U X G

organism's hereditary characteristics from one generation to the next. In all living creatures on Earth—from bacteria to amoebas to humans—genes mastermind life, while proteins maintain it.

The first experimental demonstration that complex molecules could have evolved naturally from simpler ingredients found on the primitive Earth came in 1953. Scientists Harold Urey and Stanley Miller, using laboratory equipment somewhat similar to that shown in Figure 18.3, took a mixture of the materials thought to be present on Earth long ago—a "primordial soup" of water, methane, carbon dioxide, and ammonia—and energized it by passing an electrical discharge ("lightning") through it. After a few days, they analyzed their mixture and found that it contained many of the same amino acids found today in all living things on Earth. About a decade later, scientists succeeded in constructing nucleotide bases in a similar manner. These experiments have been repeated in many different forms, with more realistic mixtures of gases and a variety of energy sources, but always with the same basic outcomes.

None of these experiments has ever produced a living organism, and the molecules created are far less complex than the strand of DNA illustrated in Figure 18.2. Nevertheless, they demonstrate conclusively that "biological" molecules can be synthesized by strictly nonbiological means, using raw materials available on the early Earth. More advanced experiments have fashioned proteinlike blobs (Figure 18.4a) that behave to some extent like true biological cells. They resist dissolution in water and tend to cluster into small droplets called *microspheres*—a little like oil globules floating on the surface of water. The walls of these laboratory-made droplets permit the inward passage of small molecules, which then combine within the droplet to produce more complex molecules too large to pass back out through the walls. As the droplets "grow," they tend to "reproduce," forming smaller droplets.

Can we consider these proteinlike microspheres to be alive? Almost certainly not. The microspheres contain many of the basic ingredients needed to form life but are not life itself. They lack the hereditary molecule DNA. However, they do have similarities to ancient cells found in the fossil record (Figure 18.4b). Thus, while no actual living cells have yet been created "from scratch," many biochemists feel that many key steps in *chemical evolution*—the chain of events leading from simple nonbiological molecules to the building blocks of life itself—have now been demonstrated.

◀ **FIGURE 18.4 Chemical Evolution**   (a) These carbon-rich, protein-like droplets contain as many as a billion amino acid molecules in a liquid sphere. The droplets can "grow," and parts of them can separate from their "parent" droplet to become new individual droplets (as at A, B, C). (b) This photograph shows primitive fossils that display concentric spheres or walls connected by smaller spheroids; they were found in sediments radioactively dated to be about 2 billion years old. (c) For comparison, modern blue-green algae are shown, on approximately the same scale. *(S. Fox/E. Barghoorn)*

An Interstellar Origin?

Recently, a dissenting view has emerged. Some scientists argue that there may not have been sufficient energy available to power the chemical reactions in Earth's primitive atmosphere, and there may not have been enough raw material for the reactions to occur at a significant rate in any case. Instead, they suggest, most of the organic material that combined to form the first living cells was produced in *interstellar space* and subsequently arrived on Earth in the form of interplanetary dust and meteors that did not burn up during their descent through the atmosphere.

To test this hypothesis, NASA researchers carried out their own version of the Urey-Miller experiment, exposing an icy mixture of water, methanol, ammonia, and carbon monoxide—representative of many interstellar grains—to ultraviolet radiation to simulate the energy from a nearby newborn star. As shown in Figure 18.5, when they later placed the irradiated ice in water and examined the results, they found that it had formed droplets, surrounded by membranes, containing complex organic molecules. No amino acids, proteins, or DNA were found in the mix, but the results, repeated numerous times, clearly show that even the harsh, cold vacuum of interstellar space can be a suitable medium for the formation of complex molecules and primitive cellular structures.

Interstellar molecular clouds are known to contain many complex molecules, and large amounts of organic material were detected on comets Halley and Hale-Bopp when they last visited the inner solar system. ∞ *(Sec. 4.2)* We saw earlier that many scientists think cometary impacts early in our planet's history were responsible for most of Earth's water. ∞ *(Sec. 4.3)* Perhaps those comets carried complex molecules, too.

In addition, a small fraction of meteorites that survive the plunge to Earth's surface (including perhaps the controversial "Martian meteorite" discussed in Chapter 6) have been shown to contain organic compounds. ∞ *(Discovery 6-2)* Figure 18.6 shows a particularly well-studied example, which fell near Murchison, Australia, in 1969. Located soon after crashing to the ground, this meteorite contained 12 of the amino acids normally found in living cells. Such discoveries strongly suggest that complex molecules can form in an interplanetary or interstellar environment and could have reached Earth's surface unscathed after their fiery descent.

The moderately large molecules found in meteorites and in interstellar clouds are our only evidence that chemical evolution has occurred elsewhere in the universe. However, most researchers regard this organic matter as prebiotic—matter that could eventually lead to life, but that has not yet done so. Nevertheless, the hypothesis that organic matter is constantly raining down on Earth from space in the form of interplanetary debris is quite plausible. Whether this was the *primary* means by which those molecules appeared on Earth remains unclear.

Diversity and Culture

However the basic materials appeared on Earth, we know that life *did* appear. The fossil record chronicles how life on Earth became widespread and diversified over time. The study of fossil remains shows the initial appearance about 3.5 billion years ago of simple one-celled organisms, such as blue-green algae. These were followed about 2 billion years ago by more complex one-celled creatures like the amoeba.

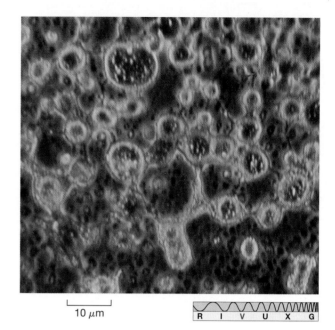

10 μm
R I V U X G

▲ FIGURE 18.5 **Interstellar Globules** These oily, hollow droplets rich in organic molecules were made by exposing a freezing mixture of primordial matter to harsh ultraviolet radiation. When immersed in water, the larger ones display cell-like membrane structure. Although they are not alive, they bolster the idea that life on Earth could have come from space. *(NASA)*

ANIMATION/VIDEO Icy Organics in Planet-Forming Disc

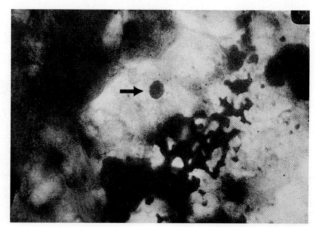

▶ FIGURE 18.6 **Murchison Meteorite** The Murchison meteorite contains relatively large amounts of amino acids and other organic material, indicating that chemical evolution of some sort has occurred beyond our own planet. In this magnified view of a fragment from the meteorite, an arrow points to a microscopic sphere of organic matter. *(Harvard-Smithsonian Center for Astrophysics)*

10 μm
R I V U X G

Multicellular organisms did not appear until about 1 billion years ago, after which a wide variety of increasingly complex organisms flourished—among them insects, reptiles, mammals, and humans. Figure 18.7 illustrates some of the key developments in the evolution of life on our planet.

The fossil record leaves no doubt that biological organisms have changed over time—all scientists accept the reality of *biological evolution*. As conditions on Earth have shifted and Earth's surface has evolved, those organisms that could best take advantage of their new surroundings succeeded and thrived—often at the expense of those organisms that could not make the necessary adjustments and consequently became extinct.

In the opinion of many anthropologists, intelligence, like any other highly advantageous trait, is strongly favored by natural selection. The social cooperation that went with coordinated hunting efforts was an important competitive advantage that developed as brain size increased. Perhaps most important of all was the development of language. Experience and ideas, stored in the brain as memories, could be passed down from one generation to the next. A new kind of evolution had begun, namely, *cultural evolution*—the changes in the ideas and behavior of society. Our more recent ancestors created, within about 10,000 years, the entirety of human civilization.

To put all this into historical perspective, let's imagine the entire lifetime of Earth to be 46 years rather than 4.6 billion years. On this scale, we have no reliable record of the first decade of our planet's existence. Life originated at least 35 years ago, when Earth was about 10 years old. Our planet's middle age is still largely a mystery. Not until about 6 years ago did abundant life flourish throughout Earth's oceans. Life came ashore about 4 years ago, and plants and animals mastered the land only about 2 years ago. Dinosaurs reached their peak about 1 year ago, only to die suddenly about 4 months later. ∞ *(Discovery 4-1)* Humanlike apes changed into apelike humans only last week, and the latest ice ages occurred only a few days ago. *Homo sapiens*—our species—did not emerge until about 4 hours ago. Agriculture was invented within the last hour, and the Renaissance—along with all of modern science—is just 3 minutes old!

CONCEPT CHECK ▶ Has chemical evolution been verified in the laboratory?

18.2 Life in the Solar System

Simple one-celled life-forms reigned supreme on Earth for most of our planet's history. It took time—a great deal of time—for life to emerge from the oceans, to evolve into simple plants, to continue to evolve into complex animals, and to develop intelligence, culture, and technology. Have those (or similar) events occurred elsewhere in the solar system? Let's try to assess what little evidence we have on the subject.

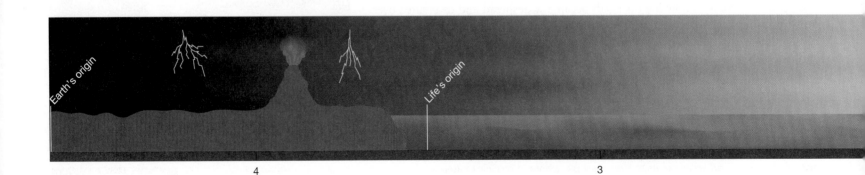

Life as We Know It

"Life as we know it" is generally taken to mean carbon-based life that originated in a liquid-water environment—in other words, life on Earth. We must always be careful in our studies not to be misled by this "Earth-centric" viewpoint—it is all too easy to attribute human or Earth-based characteristics to other life-forms—but since we have no extraterrestrial examples to guide us, at the very least this is a reasonable starting point. Thus we begin by asking, Might Earth-like life exist (or have existed) on some other body in our planetary system?

The Moon and Mercury lack liquid water, protective atmospheres, and magnetic fields, and so are subjected to fierce bombardment by solar ultraviolet radiation, the solar wind, meteoroids, and cosmic rays. Simple molecules could not survive in such hostile environments. On the other hand, Venus has far too much protective atmosphere. Its dense, dry, scorchingly hot atmospheric blanket effectively rules it out as a possible abode for life.

The jovian planets have no solid surfaces, and most of their moons (except for volcanic Io) have frozen surfaces far too cold to support Earth-like life. One possible exception is Saturn's moon Titan. With its thick atmosphere of methane, ammonia, and nitrogen gases, liquid hydrocarbon lakes, and apparent geological activity, Titan might conceivably be a place where surface life could have arisen, although the most recent results from the *Cassini-Huygens* mission suggest that the environment there would be hostile to anything remotely familiar to us. ∞ *(Sec. 8.2)*

A more promising scenario comes from the discovery that four jovian moons—Jupiter's Europa and Ganymede, and Saturn's Titan and Enceladus—may contain significant amounts of liquid water in their interiors. ∞ *(Sec. 8.1, 8.2)* This possibility has fueled speculation about the development of life within these bodies, making them prime candidates for future exploration. Europa in particular is high on the priority lists of both NASA and the European Space Agency. Again, conditions in or on these moons are far from ideal by Earthly standards, but, as we discuss below, scientists are finding more and more examples of terrestrial organisms that thrive in extreme environments once regarded as uninhabitable.

The planet most likely to harbor life (or, more likely, to have harbored it in the past) still seems to be Mars. The red planet is harsh by Earth standards—liquid water is scarce, the atmosphere is thin, and the lack of magnetism and an ozone layer allows high-energy solar particles and radiation to reach the surface unabated. But the Martian atmosphere was once thicker, and the surface warmer and much wetter. ∞ *(Sec. 6.8)* There is strong photographic evidence from orbiters such as *Viking* and *Mars Global Surveyor* for flowing and standing water on Mars in the

▼ **FIGURE 18.7 Life On Earth**   This simplified timeline of the origin and evolution of life on our planet begins at the far left with the origin of Earth about 4.6 billion years ago and extends linearly to the present at right. Notice how life-forms most familiar to us—fish, reptiles, mammals—emerged relatively recently in the history of Earth.

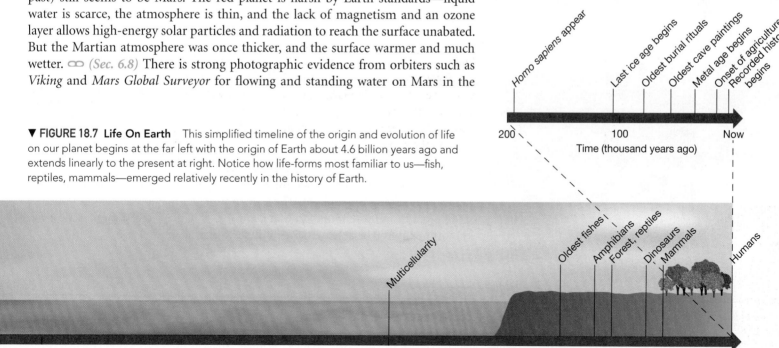

distant (and perhaps even relatively recent) past, and in 2004, the European *Mars Express* orbiter confirmed the long-hypothesized presence of water ice at the Martian poles, while NASA's *Opportunity* rover reported strong geological evidence that the region around its landing site was once "drenched" with water for an extended period. The *Viking* landers found no evidence of life, although they landed on the safest Martian terrain, not in the most interesting regions, such as near the moist polar caps, where scientists consider life to be more likely. ∞ *(Discovery 6-2) Mars Pathfinder* also surveyed part of the Martian surface, again without finding any evidence of present or past life. The recent *Mars Phoenix* mission was not equipped to test for life directly, although it did confirm the presence of water in the soil near the Martian north pole. ∞ *(Sec. 6.6)* The chemical composition of the polar soil seems less Earth-like than was at first thought, but it remains to be seen how big an obstacle this poses to the chances of life on the planet.

Some scientists think that a different type of biology might be operating, or might have operated, on the Martian surface. They suggest that Martian microbes capable of digesting oxygen-rich compounds in the Martian soil could explain the *Viking* results. This speculation will be greatly strengthened if recent announcements of fossilized bacteria in meteorites originating on Mars are confirmed. ∞ *(Discovery 6-2)* The consensus among biologists and chemists today is that Mars does not house any life similar to that on Earth, but a solid verdict regarding present or past life on Mars will likely not be reached until we have thoroughly explored our intriguing neighbor.

CONCEPT CHECK ▶ Which solar system bodies (other than Earth) are the leading candidates in the search for extraterrestrial life, and why?

Life in the Extreme

In considering the emergence of life under adversity, we should perhaps not be too quick to rule out an environment based solely on its extreme properties. Figure 18.8 shows a very hostile environment on a deep-ocean floor, where hydrothermal vents spew forth boiling hot water from vertical tubes a few meters tall. The conditions are quite unlike anything on our planet's surface, yet life thrives in an environment rich in sulfur, poor in oxygen, and completely dark. Such underground hot springs might conceivably exist on alien worlds, raising the possibility of life-forms with much greater diversity over a much wider range of conditions than those known to us on Earth.

In recent years scientists have discovered many instances of so-called **extremophiles**—life-forms that have adapted to live in extreme environments. The superheated hydrothermal vents in Figure 18.8 are one example, but extremophiles have also been found in recently confirmed frigid lakes buried tens of millions of years ago deep beneath the Antarctic glaciers; in the dark, oxygen-poor and salt-rich floor of the Mediterranean Sea; in the mineral-rich superalkaline environment of California's Mono Lake; and even in the hydrogen-rich volcanic darkness far below Earth's crust. In many cases these organisms have evolved to create the energy they need by purely chemical means, rather than relying on sunlight. These environments may present conditions not so different from those found on Mars, Europa, or Titan, suggesting that life as we know it might well be able to thrive in these hostile, alien worlds.

◀ **FIGURE 18.8 Hydrothermal Vents** A two-person submarine (the *Alvin*, partly seen at bottom) took this picture of a hot spring, or "black smoker"—one of many along the mid-ocean ridge in the eastern Pacific Ocean. As hot water rich in sulfur pours out of the top of the vent's tube (near center), black clouds billow forth, providing a strange environment for many life-forms thriving near the vent. The inset shows a close-up of the vent base, where extremophilic life thrives, including as seen here giant red tube worms and huge crabs. *(WHOI)*

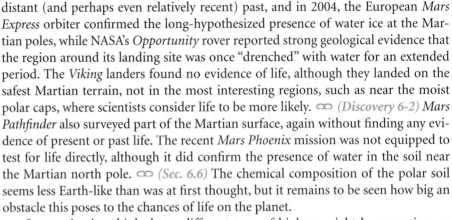

18.3 Intelligent Life in the Galaxy

With humans apparently the only intelligent life in the solar system, we must broaden our search for extraterrestrial intelligence to other stars, perhaps even other galaxies. At such distances, though, we have little hope of detecting life with current equipment. Instead, we must ask, How likely is it that life in any form—carbon-based, silicon-based, water-based, ammonia-based, or something we cannot even dream of—exists? Let's look at some numbers to develop statistical estimates of the probability of life elsewhere in the universe.

The Drake Equation

An early approach to this statistical problem is known as the **Drake equation,** after Frank Drake, the U.S. astronomer who pioneered this analysis:

| number of technological, intelligent civilizations now present in the Galaxy | = | rate of star formation, averaged over the lifetime of the Galaxy | × | fraction of stars having planetary systems | × | average number of habitable planets within those planetary systems | × | fraction of those habitable planets on which life arises | × | fraction of those life-bearing planets on which intelligence evolves | × | fraction of those intelligent-life planets that develop technological society | × | average lifetime of a technologically competent civilization. |

Of course, several of the factors in this equation are largely a matter of opinion. We do not have nearly enough information to determine—even approximately—all of them, so the Drake equation cannot give us a hard-and-fast answer. Its real value is that it subdivides a large and very difficult question into smaller pieces that we can attempt to answer separately. Figure 18.9 illustrates how, as our requirements become more and more stringent, only a small fraction of star systems in the Milky Way Galaxy have the advanced qualities specified by the combination of factors on the right-hand side of the equation.

Let's examine the factors in the equation one by one and make some educated guesses about their values. Bear in mind, though, that if you ask two scientists for their best estimates of any given one, you will likely get two very different answers!

Rate of Star Formation

We can estimate the average number of stars forming each year in the Galaxy simply by noting that at least 100 billion stars now shine in it. Dividing this number by the roughly 10-billion-year lifetime of the Galaxy, we obtain a formation rate of 10 stars per year. This may be an overestimate because astronomers think that fewer stars are forming now than formed during earlier epochs, when more interstellar gas was available. ∞ *(Sec. 16.3)* However, we do know that stars are forming today, and our estimate does not include stars that formed in the past and have since died, so our value of 10 stars per year is probably reasonable when averaged over the lifetime of the Milky Way Galaxy.

Fraction of Stars with Planetary Systems

Many astronomers regard planet formation as a natural result of the star-formation process. If the condensation theory is correct, and if there is nothing special about our Sun, as we have argued throughout this book, we would expect many stars to have a planetary system of some sort. ∞ *(Sec. 4.3)*

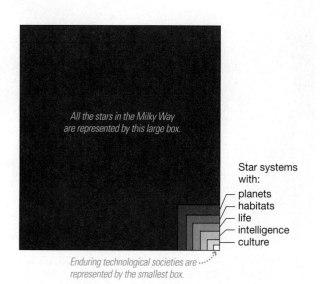

All the stars in the Milky Way are represented by this large box.

Star systems with:
- planets
- habitats
- life
- intelligence
- culture

Enduring technological societies are represented by the smallest box.

▲ FIGURE 18.9 **Drake Equation** Of all the star systems in our Milky Way Galaxy, progressively fewer and fewer have each of the qualities typical of a long-lasting technological society.

As observational techniques have improved over the past two decades, these expectations have been borne out, and there is now overwhelming evidence for planets orbiting hundreds of other stars. ∞ *(Sec. 4.4)* The first planets discovered were much larger than Earth and mostly moved on eccentric or "hot" orbits, but as we saw in Chapter 4, these were the only planets that could have been detected with the instruments available at the time. As detection technology has advanced, more and more planets with masses comparable to Earth have been discovered, to the point that, since 2010, several Earth-sized planets, some of them on roughly Earth-like orbits, have been confirmed. These observations are at the very edge of current capabilities, and many astronomers expect the numbers of "Earth-like" planets to grow rapidly as new detectors come on-line.

Only about 10 percent of the nearby stars surveyed to date have been found to have planets. However, most researchers think that this is a significant underestimate of the true fraction due to observational limitations and selection effects. ∞ *(Sec. 4.4)* Thus, accepting the condensation theory and its consequences, and without being either too conservative or naively optimistic, we assign a value near 1 to this factor—that is, we think that essentially all stars have planetary systems of some sort.

Number of Habitable Planets per Planetary System

What determines the feasibility of life on a given planet? Temperature is perhaps the single most important factor. However, the possibility of catastrophic external events, such as cometary impacts or even distant supernovae, must also be considered. ∞ *(Discovery 4-1, Sec. 12.4)*

As discussed previously (Chapter 4), the surface temperature of a planet depends on two things: the planet's distance from its parent star and the thickness of its atmosphere. Planets with a nearby (but not too close) parent star and some atmosphere (though not too thick) should be reasonably warm, like Earth or Mars. Objects far from the star and with no atmosphere, like Pluto, will surely be cold by our standards. And planets too close to the star and with a thick atmosphere, like Venus, will be very hot indeed.

Figure 18.10 illustrates how a three-dimensional *stellar habitable zone* of "comfortable" temperatures surrounds every star. ∞ *(Sec. 4.4)* (The zones are indicated as rings in this two-dimensional figure.) The habitable zone represents the range of distances within which a planet of mass and composition similar to those of Earth would have a surface temperature between the freezing and boiling points of water. (Our Earth-based bias is plainly evident here!) The hotter the star, the larger this zone. For example, an A- or an F-type star has a rather large habitable zone. G-, K-, and M-type stars have successively smaller zones. O- and B-type stars are not considered here because they are not expected to last long enough for life to develop, even if they do have planets.

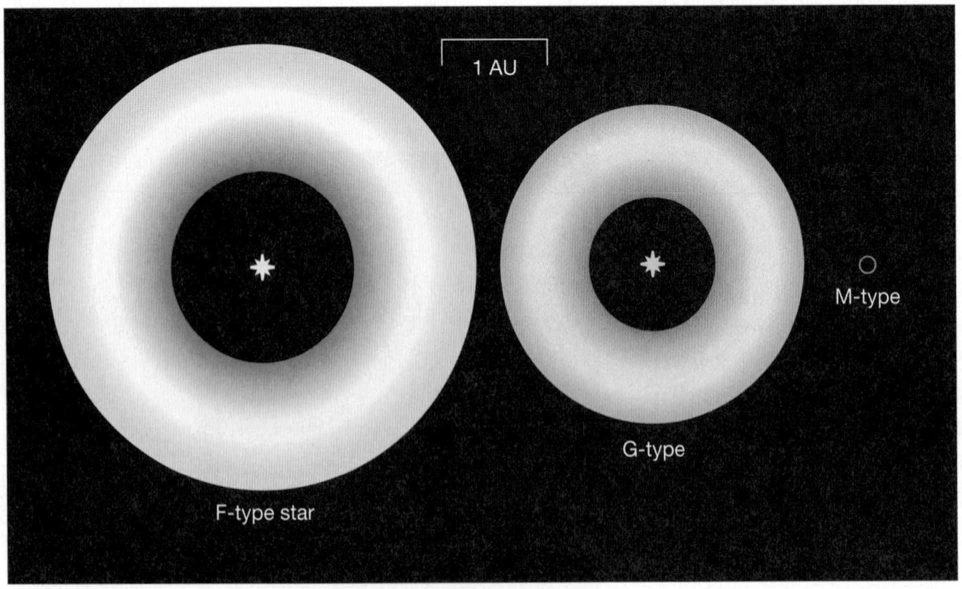

Three planets—Venus, Earth, and Mars—reside in or near the habitable zone surrounding our Sun. Venus is too hot because of its thick atmosphere and proximity to the Sun. Mars is a little too cold because its atmosphere is too thin and it is too far from the Sun. But if the orbits of Venus and Mars were swapped—not inconceivable, since chance played such a large role in the formation of the terrestrial planets—then both of these nearby planets might conceivably have evolved surface conditions

▲ **FIGURE 18.10 Stellar Habitable Zones**   Hot stars have bigger habitable zones than cool ones. For a G-type star like the Sun, the zone extends from about 0.8 AU to 2 AU. For a hotter F-type star, the range is 1.2 to 2.8 AU. For a cool M-type star, only Earth-like planets orbiting between about 0.02 and 0.06 AU would be habitable.

▶ FIGURE 18.11 **Galactic Habitable Zone** Some regions of the Galaxy are more conducive to life than others. Too far from the Galactic center, there may not be enough heavy elements for terrestrial planets to form or technological society to evolve. Too close, the radiative or gravitational effects of nearby stars may render life impossible. The result is a ring-shaped habitable zone, colored here in green, although its full extent is uncertain.

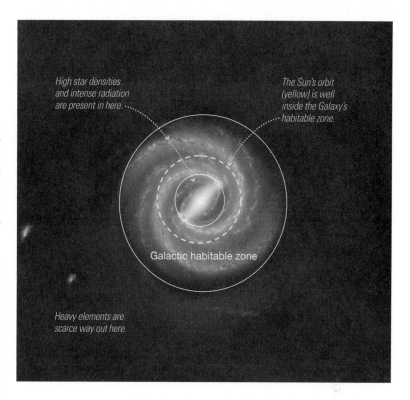

High star densities and intense radiation are present in here.

The Sun's orbit (yellow) is well inside the Galaxy's habitable zone.

Galactic habitable zone

Heavy elements are scarce way out here.

resembling those on Earth. ∞ *(Secs. 4.3, 6.8)* In that case our solar system would have had three habitable planets instead of one. Perhaps just as important, proximity to a giant planet may also render the interior of a moon (such as Europa or Titan) habitable, the planet's tidal heating making up for the lack of sunlight. ∞ *(Sec. 8.1)* Sheltered by its parent planet's gravity, such a moon might be largely immune to the habitable limitations just described for planets.

A planet moving on a "habitable" orbit may still be rendered uninhabitable by external events. Many scientists think that the outer planets in our own solar system are critical to the habitability of the inner worlds, both by stabilizing their orbits and by protecting them from cometary impacts, deflecting would-be impactors away from the inner part of the solar system. The theory presented in Chapter 4 suggests that a star with inner terrestrial planets on stable orbits would probably also need the jovian worlds to safeguard their survival. ∞ *(Sec. 4.3)* However, observations of extrasolar planets are not yet sufficiently refined to determine the fraction of stars having "outer planet" systems like our own. ∞ *(Sec. 4.4)*

Other external forces may also influence a planet's survival. Some researchers have suggested that there is a *Galactic habitable zone* for stars in general, outside of which conditions are unfavorable for life (see Figure 18.11). Far from the Galactic center, the star formation rate is low and few cycles of star formation have occurred, so there are insufficient heavy elements to form terrestrial planets or populate them with technological civilizations. ∞ *(Sec. 12.7)* Too close, and the radiation from bright stars and supernovae in the crowded inner part of the Galaxy might be detrimental to life. More important, the gravitational effects of nearby stars may send frequent showers of comets from the Oort cloud into the inner regions of a planetary system, impacting the terrestrial planets and terminating any chain of evolution that might lead to intelligent life.

Thus, to estimate the number of habitable planets per planetary system, we must first take inventory of how many stars of each type shine in the Galactic habitable zone, then calculate the sizes of their stellar habitable zones and estimate the number of planets likely to be found there. In doing so, we eliminate almost all of the stars around which planets have so far been observed, and presumably a similar fraction of stars in general. We also exclude the majority of binary star systems: Given the observed properties of binaries in our Galaxy, "habitable" planetary orbits in binary systems would be unstable in many cases, as illustrated in Figure 18.12, so there would not be time for life to develop.

The scant observational evidence currently available on Earth-like planets in habitable orbits suggests that only a few percent of the known planetary systems contain a habitable planet. ∞ *(Sec. 4.4)* However, because these planets are so close to the limits of detectability with current equipment, many astronomers think the true fraction will turn out to be significantly higher. Potentially habitable jovian moons may increase the fraction still further. However, many uncertainties remain. The inner and outer radii of the Galactic habitable zone are not known with any certainty, and the simple fact is that we still have insufficient data about most stars to make any strong statement about habitable worlds in their planetary systems.

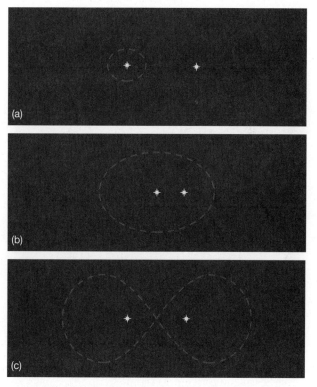

(a)

(b)

(c)

▲ FIGURE 18.12 **Binary-Star Planets** In binary-star systems, planets are restricted to only a few kinds of orbits that are gravitationally stable. (a) This orbit is stable only if the planet lies very close to its parent star, so the gravity of the other star is negligible. (b) A planet circulating at a great distance about both stars in an elliptical orbit is stable only if it lies far from both stars. (c) Another possible path interweaves between the two stars in a figure-eight pattern.

Taking the many uncertainties into account as best we can, we assign a value of 1/10 to this factor in our equation. In other words, we think that, on average, there is 1 potentially habitable planet for every 10 planetary systems that might exist in our Galaxy. Single F-, G-, and K-type stars are the best candidates.

Fraction of Habitable Planets on Which Life Arises

The number of possible combinations of atoms is incredibly large. If the chemical reactions that led to the complex molecules that make up living organisms occurred completely at random, then it is extremely unlikely that those molecules could have formed at all. In that case, life is extraordinarily rare, this factor is close to 0, and we are probably alone in the Galaxy, perhaps even in the entire universe.

However, laboratory experiments (like the Urey-Miller experiment described earlier) seem to suggest that certain chemical combinations are strongly favored over others—that is, the reactions are not random. Of the billions upon billions of basic organic groupings that could possibly occur on Earth from the random combination of all sorts of simple atoms and molecules, only about 1500 actually do occur. Furthermore, these 1500 organic groups of terrestrial biology are made from only about 50 simple "building blocks" (including the amino acids and nucleotide bases mentioned earlier). This suggests that molecules critical to life may not be assembled by pure chance. Apparently, additional factors are at work at the microscopic level. If a relatively small number of chemical "evolutionary tracks" is likely to exist, then the formation of complex molecules—and hence, we assume, life—becomes much more likely, given sufficient time.

To assign a very low value to this factor in the equation is to say that life arises randomly and rarely. Assigning a value close to 1 implies that we think life is inevitable given the proper ingredients, a suitable environment, and a long enough period of time. No easy experiment can distinguish between these extreme alternatives, and there is little or no middle ground. To many researchers, the discovery of life—past or present—on Mars, Europa, Titan, or any other object in our solar system would convert the appearance of life from an unlikely miracle to a virtual certainty throughout the Galaxy. In the absence of any objective evidence, we will simply take the optimistic view and adopt a value of 1.

Fraction of Life-Bearing Planets on Which Intelligence Arises

As with the evolution of life, the appearance of a well-developed brain is a very unlikely event if only random chance is involved. However, biological evolution through natural selection is a mechanism that generates apparently highly improbable results by singling out and refining useful characteristics. Organisms that profitably use adaptations can develop more complex behavior, and complex behavior provides organisms with the *variety* of choices needed for more advanced development.

One school of thought maintains that given enough time, intelligence is inevitable. In this view, assuming that natural selection is a universal phenomenon, at least one organism on a planet will always rise to the level of "intelligent life." If this is correct, then the fifth factor in the Drake equation equals or nearly equals 1.

Others argue that there is only one known case of intelligence, and that case is life on Earth. For 2.5 billion years—from the start of life about 3.5 billion years ago to the first appearance of multicellular organisms about 1 billion years ago—life did not advance beyond the one-celled stage. Life remained simple and dumb, but it survived. If this latter view is correct, then the fifth factor in our equation is very small, and we are faced with the depressing prospect that humans may be the smartest form of life anywhere in the Galaxy. As with the previous factor, we will be optimistic and adopt a value of 1 here.

Fraction of Planets on Which Intelligent Life Develops and Uses Technology

To evaluate the sixth factor of our equation, we need to estimate the probability that intelligent life eventually develops technological competence. Should the rise of technology be inevitable, this factor is close to 1, given long enough periods of time. If it is not inevitable—if intelligent life can somehow "avoid" developing technology (as dolphins on Earth seem to have done)—then this factor could be much less than 1. The latter possibility envisions a universe possibly teeming with civilizations, but very few among them ever becoming technologically competent. Perhaps only one managed it—ours.

Again, it is difficult to decide conclusively between these two views. We don't know how many (if any) prehistoric Earth cultures either failed to develop technology or rejected its use. We do know that tool-using societies arose independently at several places on Earth, including Mesopotamia, India, China, Egypt, Mexico, and Peru. Because so many of these ancient "technological" cultures originated *independently* at about the same time, we conclude that the chances are good that some sort of technological society will inevitably develop, given some basic intelligence and enough time. Accordingly, we take this factor to be close to 1.

Average Lifetime of a Technological Civilization

The reliability of these estimates declines markedly from the first to the last factor of the Drake equation. Our knowledge of astronomy allows us to make a reasonably good stab at the first two, but even the third is hard to evaluate, and the values adopted for the last few are more wishful thinking than science. The final factor is totally unknown. There is only one example of such a civilization that we are aware of—humans on planet Earth. Our own civilization has presently survived in its technological state for only about 100 years, and how long we will be around before a human-made catastrophe or a planetwide natural disaster ends it all is impossible to tell. ∞ *(Discovery 4-1)*

One thing is certain: If the correct value for *any one factor* in the equation is very small—and we have just seen at least two for which this could well be the case, optimistic choices notwithstanding—then few technological civilizations now exist in the Galaxy. If the pessimistic view of the development of life or of intelligence is correct, then we are unique, and that is the end of our story. However, if both life and intelligence are inevitable consequences of chemical and biological evolution, as many scientists think, and if intelligent life always becomes technological, then we can plug the higher, more optimistic values into the Drake equation. In that case, combining our estimates for the other six factors (and noting that, conveniently, $10 \times 1 \times 1/10 \times 1 \times 1 \times 1 = 1$), we can say

$$
\begin{array}{c}
\text{number of technological} \\
\text{intelligent civilizations} \\
\text{now present in the Milky} \\
\text{Way Galaxy}
\end{array}
=
\begin{array}{c}
\text{average lifetime of a} \\
\text{technologically} \\
\text{competent civilization,} \\
\text{in years.}
\end{array}
$$

Thus, if advanced civilizations typically survive for 1000 years, there should be 1000 of them currently in existence scattered throughout the Galaxy. If they survive for a million years, on average, we would expect there to be a million of them in the Galaxy, and so on.

Note that even setting aside language and cultural issues, the sheer size of the Galaxy presents a significant hurdle to communication between technological civilizations. The minimum requirement for a two-way conversation is that we can send a signal and receive a reply in a time shorter than our own lifetime. If the lifetime is short, then civilizations are literally few and far between—small in number, according

PROCESS OF ►
SCIENCE CHECK If most of the factors are largely a matter of opinion, how does the Drake equation assist astronomers in refining their search for extraterrestrial life?

to the Drake equation, and scattered over the vastness of the Milky Way—and the distances between them (in light-years) are much greater than their lifetimes (in years). In that case, two-way communication, even at the speed of light, will be impossible. However, as the lifetime increases, the distances get smaller as the Galaxy becomes more crowded, and the prospects improve.

Taking into account the size, shape, and distribution of stars in the Galactic disk (why do we exclude the halo?) and under the assumptions made above, we find that unless the life expectancy of a civilization is *at least* a few thousand years, it is unlikely to have time to communicate with even its nearest neighbor.

18.4 The Search for Extraterrestrial Intelligence

Let's continue our optimistic assessment of the prospects for life and assume that civilizations enjoy a long stay on their parent planet once their initial technological "teething problems" are past. In that case, they may be plentiful in the Galaxy. How might we become aware of their existence?

Meeting Our Neighbors

Let's assume that the average lifetime of a technological civilization is 1 million years—only 1 percent of the reign of the dinosaurs, but 100 times longer than our civilization has survived thus far and 10,000 times longer than human society has so far been in the "technological" state. The Drake equation then tells us that there are 1 million such civilizations in the Galaxy, and we can then estimate the average distance between these civilizations to be some 30 pc, or about 100 light-years. Thus, any two-way communication with our neighbors will take at least 200 years (100 years for the message to reach the planet and another 100 years for the reply to travel back to us)—a long time by human standards, but comfortably less than the lifetime of the civilization.

Might we hope to visit our neighbors by developing the capability of traveling far outside our solar system? This may never be a practical possibility. At a speed of 50 km/s, the speed of the current fastest space probes, the round-trip to even the nearest Sun-like star, Alpha Centauri, would take about 50,000 years. The journey to the nearest technological neighbor (assuming a distance of 100 pc) and back would take 1 million years—the entire lifetime of our civilization! Interstellar travel at these speeds is clearly not feasible. Speeding up our ships to near the speed of light would reduce the travel time, but this is far beyond our present technology.

Actually, our civilization has already launched some interstellar probes, although they have no specific stellar destination. Figure 18.13 is a reproduction of a plaque mounted on board the *Pioneer 10* spacecraft launched in the mid-1970s and now well beyond the orbit of Pluto, on its way out of the solar system. Similar information was also included aboard the *Voyager* probes launched in 1978. While these spacecraft would be incapable of reporting back to Earth the news that they had encountered an alien culture, scientists hope that the civilization on the other end would be able to unravel most of its contents using the universal language of mathematics. The caption to Figure 18.13 notes how the aliens might discover from where and when the *Pioneer* and *Voyager* probes were launched.

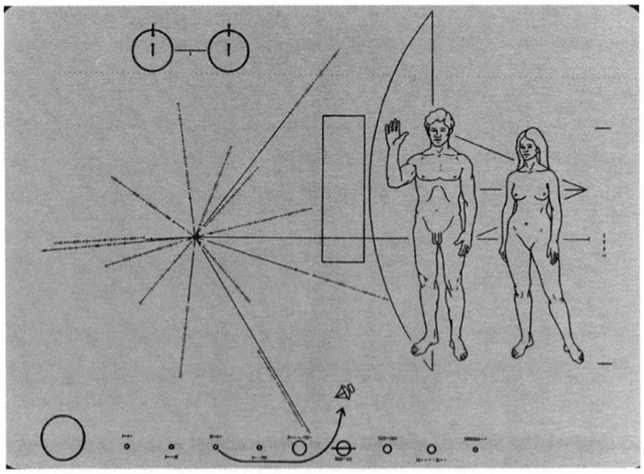

▲ FIGURE 18.13 *Pioneer 10* Plaque   This replica of a plaque mounted on board the *Pioneer 10* spacecraft shows a scale drawing of the craft, a man, and a woman; a diagram of the hydrogen atom undergoing a change in energy (top left); a starburst pattern representing various pulsars and the frequencies of their radio waves that can be used to estimate when the craft was launched (middle left); and a depiction of the solar system, showing that the spacecraft departed the third planet from the Sun and passed the fifth planet on its way into outer space (bottom). *(NASA)*

Radio Searches

A cheaper and much more practical alternative is to try to make contact with extraterrestrials using electromagnetic radiation, the fastest known means of transferring information from one place to another. Because light and other short-wavelength

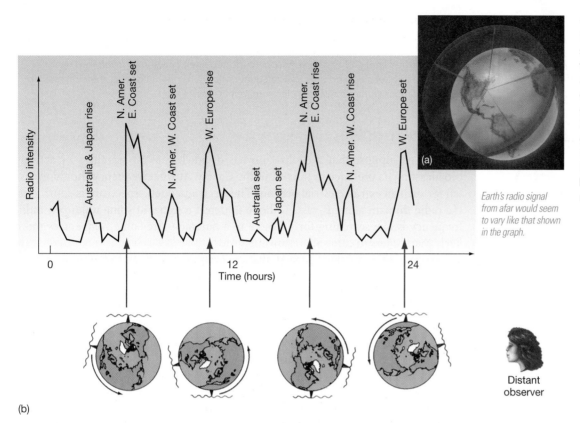

Earth's radio signal from afar would seem to vary like that shown in the graph.

(b)

◄ FIGURE 18.14 **Earth's Radio Leakage** Radio radiation leaks from Earth into space because of the daily activities of our technological civilization. (a) Most radio and television transmitters broadcast their energy parallel to Earth's surface, sending a great "sheet" of electromagnetic radiation into interstellar space. (b) Because the great majority of transmitters are clustered in the eastern United States and western Europe, a distant observer would detect blasts of radiation from Earth as our planet rotates each day.

radiation are heavily scattered while moving through dusty interstellar space, long-wavelength radio radiation seems to be the best choice. ∞ *(Sec. 2.3)* We do not attempt to broadcast to all nearby candidate stars, however—that would be far too expensive and extremely inefficient. Instead, radio telescopes on Earth listen *passively* for radio signals emitted by other civilizations.

In what direction should we aim our radio telescopes? The answer to this question at least is fairly easy. On the basis of our earlier reasoning, we should target all F-, G-, and K-type stars in our vicinity. But are extraterrestrials broadcasting radio signals? If they are not, this search technique will obviously fail. If they are, how do we distinguish their artificially generated radio signals from signals emitted naturally by interstellar gas clouds? This depends on whether the signals are produced deliberately or are simply "waste radiation" escaping from a planet.

Consider first how "waste" radio wavelengths originating on Earth would look to extraterrestrials. Figure 18.14 shows the pattern of radio signals we broadcast into space. From the viewpoint of a distant observer, the spinning Earth emits a bright flash of radio radiation every few hours as the most technological regions of our planet rise or set. In fact, Earth is now a more intense radio emitter than the Sun. Our radio and television transmissions race out into space and have been doing so since the invention of these technologies nearly seven decades ago. Another civilization as advanced as ours might have constructed devices capable of detecting this radiation. If any sufficiently advanced (and sufficiently interested) civilization resides within about 70 light-years (20 pc) of Earth, then we have already announced our presence to them.

Of course, it may very well be that, having discovered cable and fiber-optics technology, most civilizations' indiscriminate transmissions cease after a few decades. In that case, radio silence becomes the hallmark of intelligence, and we must find an alternative means of locating our neighbors.

CONCEPT CHECK ► What assumptions go into the identification of the water hole as a likely place to search for extraterrestrial signals?

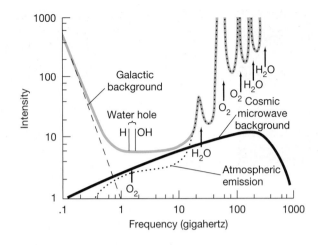

◀ FIGURE 18.15 **Water Hole** The "water hole" is bounded by the natural emission of the hydrogen (H) atom at 21-cm wavelength and the hydroxyl (OH) molecule at 18-cm wavelength. ⊂⊃ *Sec. 11.3* The topmost solid (blue) curve sums the natural emissions of our Galaxy (dashed line) and Earth's atmosphere (dotted line). This sum is minimized near the water hole frequencies, and thus all intelligent civilizations might conduct their interstellar communications within this quiet "electromagnetic oasis."

The Water Hole

Now let us suppose that a civilization has decided to assist searchers by actively broadcasting its presence to the rest of the Galaxy. At what frequency should we listen for such an extraterrestrial beacon? The electromagnetic spectrum is enormous; the radio domain alone is vast. To hope to detect a signal at some unknown radio frequency is like searching for a needle in a haystack. Are some frequencies more likely than others to carry alien transmissions?

Some basic arguments suggest that civilizations might well communicate at a wavelength near 20 cm. The basic building blocks of the universe, namely, hydrogen atoms, radiate naturally at a wavelength of 21 cm. ⊂⊃ *(Sec. 11.2)* Also, one of the simplest molecules, hydroxyl (OH), radiates near 18 cm. Together, these two substances form water (H_2O). Arguing that water is likely to be the evolution medium for life anywhere and that radio radiation travels through the disk of our Galaxy with the least absorption by interstellar gas and dust, some researchers have proposed that the interval between 18 cm and 21 cm is the best wavelength range for civilizations to transmit or listen. Called the **water hole,** this radio interval might serve as an "oasis" where all advanced Galactic civilizations would gather to conduct their electromagnetic business.

This water-hole frequency interval is only a guess, of course, but its use is supported by other arguments. Figure 18.15 shows the water hole's location in the electromagnetic spectrum and plots the amount of natural emission from our Galaxy and from Earth's atmosphere. The 18- to 21-cm range lies within the quietest part of the spectrum, where the Galactic "static" from stars and interstellar clouds happens to be minimized. Furthermore, the atmospheres of typical planets are also expected to interfere least at these wavelengths. Thus, the water hole seems like a good choice for the frequency of an interstellar beacon, although we cannot be sure of this reasoning until contact is actually achieved.

A few radio searches are now in progress at frequencies in and around the water hole. One of the most sensitive and comprehensive projects in the ongoing search for extraterrestrial intelligence (known to many by its acronym SETI) is now underway with the Allen Telescope Array (Figure 18.16a). This collection of many small dishes is currently searching millions of channels simultaneously in the 1- to 3-GHz range. Figure 18.16(b) shows what a typical narrow-band, 1-Hz signal—a potential "signature" of an intelligent transmission—would look like on a computer monitor. However, this observation was merely a test to detect the weak, redshifted radio signal emitted by the *Pioneer 10* robot, now receding into the outer realm of our solar system—a sign of intelligence, but one that we put there. Nothing resembling an extraterrestrial signal has yet been detected.

The space surrounding all of us could be, right now, flooded with radio signals from extraterrestrial civilizations. If only we knew the proper direction and frequency, we might be able to make one of the most startling discoveries of all time. The result would provide whole new opportunities to study the cosmic evolution of energy, matter, and life throughout the universe.

(a)

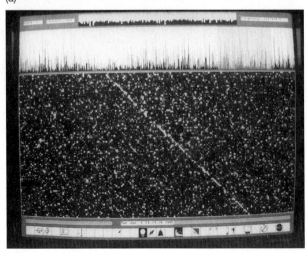

(b)

◀ FIGURE 18.16 **Project Phoenix** (a) This array of small radio dishes at the SETI Institute in California is designed to search for extraterrestrial intelligent signals. (b) A typical recording of an alien signal—here, as a test, the Doppler-shifted broadcast from the *Pioneer 10* spacecraft, now well beyond the Kuiper belt—shows a diagonal line across the computer monitor, in contrast to the random noise in the background. *(S. Shostak/SETI Institute)*

CHAPTER REVIEW

SUMMARY

L01 **Cosmic evolution** (p. 486) is the continuous process that has led to the appearance of galaxies, stars, planets, and life on Earth.  Living organisms may be characterized by their ability to react to their environment; to grow by taking in nutrition from their surroundings; to reproduce, passing along some of their own characteristics to their offspring; and to evolve in response to a changing environment. Organisms that can best take advantage of their new surroundings succeed at the expense of those organisms that cannot make the necessary adjustments. Intelligence is strongly favored by natural selection.

L02 Powered by natural energy sources, reactions between simple molecules in the oceans of the primitive Earth may have led to the formation of **amino acids** (p. 487) and **nucleotide bases** (p. 487), the basic molecules of life Amino acids build proteins, which control metabolism, while sequences of nucleotide bases make up DNA, the genetic blueprint of a living organism. Alternatively, some of these complex molecules may have been formed in interstellar space and then delivered to Earth by meteors or comets.

L03 The best hope for life beyond Earth in the solar system is the planet Mars, although no direct evidence for life—current or extinct—has yet been found. Some of the icy moons of the outer planets—Jupiter's Europa and Ganymede and Saturn's Titan and Enceladus—may also be possibilities for life of some sort. Conditions on those frozen bodies are harsh by terrestrial standards, although **extremophiles** (p. 492) on Earth have been found to thrive in hostile environments in which life had previously been thought impossible.

L04 The **Drake equation** (p. 493) provides a means of estimating the probability of other intelligent life in the Galaxy. The astronomical terms in the equation are the Galactic star-formation rate, the likelihood of planets, and the number of habitable planets. Chemical and biological terms are the probability of life appearing and the probability that it subsequently develops intelligence. Cultural and political terms are the probability that intelligence leads to technology and the lifetime of a technological civilization. Taking an optimistic view of the development of life and intelligence leads to the conclusion that the total number of technologically competent civilizations in the Galaxy is approximately equal to the lifetime of a typical civilization, expressed in years.

L05 A technological civilization would probably "announce" itself to the universe by the radio and television signals it emits into space. Observed from afar, our planet would appear as a radio source with a 24-hour period, as different regions of the planet rise and set. The **water hole** (p. 500) is a region in the radio range of the electromagnetic spectrum, near the 21-cm line of hydrogen and the 18-cm line of hydroxyl, where natural emissions from the Galaxy happen to be minimal. Many researchers regard this as the best part of the spectrum for communications purposes.

MasteringAstronomy®   For instructor-assigned homework go to www.masteringastronomy.com

Problems labeled **POS** explore the process of science. **VIS** problems focus on reading and interpreting visual information. **LO** connects to the introduction's numbered Learning Outcomes.

REVIEW AND DISCUSSION

1. **L01** Why is life difficult to define?

2. What is chemical evolution?

3. What is the Urey-Miller experiment? What important organic molecules were produced in this experiment?

4. **L02** What are the basic ingredients from which biological molecules formed on Earth?

5. **POS** How do we know anything at all about the early episodes of life on Earth?

6. What is the role of language in cultural evolution?

7. Where else, besides Earth, have organic molecules been found?

8. **L03** Where—besides the planet Mars—might we find signs of life in our solar system?

9. **POS** Do we know whether Mars ever had life at any time during its past? What argues in favor of the position that it may once have harbored life?

10. POS What is generally meant by "life as we know it"? What other forms of life might be possible?

11. LO4 How many of the terms in the Drake equation are known with any-degree of certainty? Which factor is least well-known?

12. What is the relationship between the average lifetime of Galactic civilizations and the possibility of our someday communicating with them?

13. How would Earth appear, at radio wavelengths, to extraterrestrial astronomers?

14. What are the advantages of using radio waves for communication over interstellar distances?

15. LO5 What is the water hole? What advantage does it have over other parts of the radio spectrum?

CONCEPTUAL SELF-TEST: TRUE OR FALSE?/MULTIPLE CHOICE

1. The definition of life requires only that, to be considered "alive," you must be able to reproduce. (T/F)

2. Laboratory experiments have created living cells from nonbiological molecules. (T/F)

3. Dinosaurs lived on Earth for more than a thousand times longer than human civilization has existed to date. (T/F)

4. The *Viking* landers on Mars discovered microscopic evidence of life but found no fossil evidence. (T/F)

5. The development of life and intelligence on Earth are extremely unlikely if chance is the only evolutionary factor involved. (T/F)

6. We have no direct evidence for Earth-like planets orbiting other stars. (T/F)

7. Our civilization has already launched probes into interstellar space and broadcast our presence to our neighbors. (T/F)

8. The chemical elements that form the basic molecules needed for life are found (a) in the cores of Sun-like stars; (b) commonly throughout the cosmos; (c) only on planets that have liquid water; (d) only on Earth.

9. Fossil records of early life-forms on Earth suggest that life began about (a) 6000 years ago; (b) 65 million years ago; (c) 3.5 billion years ago; (d) 14 billion years ago.

10. The least well-known factor in the Drake equation is (a) the rate of star formation; (b) the fraction of stars having planetary systems; (c) the average lifetime of a technologically competent civilization; (d) the diameter of the Milky Way Galaxy.

11. We don't expect life on planets orbiting B-type stars because the star (a) has too much gravity; (b) is too short-lived for life to evolve; (c) is at too low a temperature to sustain life; (d) would have only gas giant planets.

12. NASA's Space Shuttle orbits Earth at about 17,500 mph. If it traveled to the next Sun-like star at that speed, the trip would take at least (a) 1 week; (b) 1 decade; (c) 1 century; (d) 100 millennia.

13. The strongest radio-wavelength emission in the solar system comes from (a) human-made signals from Earth; (b) the Sun; (c) the Moon; (d) Jupiter.

14. VIS From the data shown in Figure 18.10 (Stellar Habitable Zones), the habitable zone surrounding a main sequence K-type star (a) cannot be determined; (b) extends from roughly 1 to 2 AU from the star; (c) is larger than that of an F-type star; (d) is larger than that of an M-type star.

15. VIS If Figure 18.14 (Earth's Radio Leakage) were to be redrawn for a planet spinning twice as fast, the new jagged line would be (a) unchanged; (b) taller; (c) stretched out horizontally; (d) compressed horizontally.

PROBLEMS

The number of squares preceding each problem indicates its approximate level of difficulty.

1. ■ As described in the text, imagine that Earth's 4.6-billion-year age is compressed to just 46 years. What would be your age, in seconds? How long ago (in seconds) was the end of World War II, the Declaration of Independence, Columbus's discovery of the New World? How long ago did the extinction of the dinosaurs occur (in days)?

2. ■■■ According to the inverse-square law, a planet receives energy from its parent star at a rate proportional to the star's luminosity and inversely proportional to the square of the planet's distance from the star. ⚬ *(Sec. 10.2)* According to Stefan's law, the rate at which the planet radiates energy into space is proportional to the fourth power of its surface temperature. ⚬ *(Sec. 2.4)* In equilibrium, the two rates are equal. Based on this information and given the fact that (taking into account the greenhouse effect) the Sun's habitable zone extends from 0.6 AU to 1.5 AU, estimate the extent of the habitable zone surrounding a K-type main-sequence star of luminosity one-tenth the luminosity of the Sun.

3. ■ Based on the numbers presented in the text, and assuming an average lifetime of 5 billion years for suitable stars, estimate the total number of habitable planets in the Galaxy.

4. ■■ A planet orbits one component of a binary-star system at a distance of 1 AU (see Figure 18.12a). If both stars have the same mass and their orbit is circular, estimate the minimum distance between the stars for the tidal force due to the companion not to exceed a "safe" 0.01 percent of the gravitational force between the planet and its parent star.

5. ■ Suppose that each of the "fraction" terms in the Drake equation turns out to have a value of 1/10, that stars form at an average rate of 20 per year, and that each star with a planetary system has exactly one habitable planet orbiting it. Estimate the present number of technological civilizations in the Milky Way Galaxy if the average lifetime of a civilization is (a) 100 years, (b) 10,000 years, (c) 1 million years.

6. ■■■ Adopting the estimate from the text that the number of technological civilizations in the Milky Way Galaxy is equal to the average lifetime of a civilization, in years, it follows that the distance to our nearest neighbor decreases as the average lifetime increases. Assuming that civilizations are uniformly spread over a two-dimensional Galactic disk of radius 15 kpc, and all have the same lifetime, calculate the *minimum* lifetime for which a two-way radio communication with our nearest neighbor would be possible before our civilization ends. Repeat the calculation for a round-trip personal visit, using current-technology spacecraft that travel at 50 km/s.

7. ■ How fast would a spacecraft have to travel in order to complete the trip from Earth to Alpha Centauri (a distance of 1.3 pc) and back in less than an average human lifetime (80 years, say)?

8. ■ Assuming that there are 10,000 FM radio stations on Earth, each transmitting at a power level of 50 kW, calculate the total radio luminosity of Earth in the FM band. Compare this value with the roughly 10^6 W radiated by the Sun in the same frequency range.

9. ■ Convert the water hole's wavelengths to frequencies. For practical reasons, any search of the water hole must be broken up into channels, much as you find on a television, except these channels are very narrow in radio frequency, about 100 Hz wide. How many channels must astronomers search in the water hole?

10. ■ There are 20,000 stars within 100 light-years that are to be searched for radio communications. How long will the search take if 1 hour is spent looking at each star? What if 1 day is spent per star?

ACTIVITIES

Collaborative

1. As a group, compose a paragraph everyone agrees with that defines life. It should clearly show that rocks are not alive and that plants are alive. According to your definition, are stars alive? What about viruses? Compare and contrast your group's definition with that from another group.

2. Independently, each person in your group should estimate the average lifetime of a technologically competent civilization, as used in the Drake equation. Explain the variation in your estimates.

3. If your group was appointed to "speak for Earth" upon establishing communication with an extraterrestrial world, what would you say? What questions would you ask, and what aspects of our planet would you choose to present? Write your group's speech and annotate it with explanations of why you chose to say this.

Individual

1. It has been suggested that if extraterrestrial life is discovered, it will have a profound effect on human culture. Interview as many people as you can and ask the following two questions: (1) Do you think that extraterrestrial life exists? (2) Why? From your results, try to decide whether the discovery of extraterrestrial life would indeed profoundly affect life on Earth.

2. Conduct another poll separately, or do it at the same time as the first one. Ask the following question: What one question would you like to ask an extraterrestrial life-form in a radio communication? How many responses do you receive that indicate "Earth-centered" thinking? How many responses suggest a lack of understanding of how alien an extraterrestrial life-form might be? Does this change your conclusion from the first project in any way?

3. The Drake equation should be able to "predict" at least one civilization in our Galaxy: us. Try changing the values of various factors so that you end up with at least one. What do these various combinations of factors imply about how life arises and develops? Are there some combinations that just don't make any sense?

Appendix 1
Scientific Notation

The objects studied by astronomers range in size from the smallest particles to the largest expanse of matter we know—the entire universe. Subatomic particles have sizes of about 0.000000000000001 meter, while galaxies typically measure some 1,000,000,000,000,000,000,000,000 meters across. The most distant known objects in the universe lie on the order of 100,000,000,000,000,000,000,000,000,000 meters from Earth.

Obviously, writing all those zeros is both cumbersome and inconvenient. More important, it is also very easy to make an error—write down one zero too many or too few and your calculations become hopelessly wrong! To avoid this, scientists always write large numbers using a shorthand notation in which the number of zeros following or preceding the decimal point is denoted by a superscript power, or *exponent*, of 10. The exponent is simply the number of places between the first significant (nonzero) digit in the number (reading from left to right) and the decimal point. Thus 1 is 10^0, 10 is 10^1, 100 is 10^2, 1000 is 10^3, and so on. For numbers less than 1, with zeros between the decimal point and the first significant digit, the exponent is negative: 0.1 is 10^{-1}, 0.01 is 10^{-2}, 0.001 is 10^{-3}, and so on. Using this notation we can shorten the number describing subatomic particles to 10^{-15} meter and write the number describing the size of a galaxy as 10^{21} meters.

More complicated numbers are expressed as a combination of a power of 10 and a multiplying factor. This factor is conventionally chosen to be a number between 1 and 10, starting with the first significant digit in the original number. For example, 150,000,000,000 meters (the distance from Earth to the Sun, in round numbers) can be more concisely written at 1.5×10^{11} meters, 0.000000025 meters as 2.5×10^{-8} meter, and so on. The exponent is simply the number of places the decimal point must be moved *to the left* to obtain the multiplying factor.

Some other examples of scientific notation are:

- the approximate distance to the Andromeda Galaxy
  = 2,500,000 light-years = 2.5×10^6 light-years

- the size of a hydrogen atom = 0.00000000005 meter
  = 5×10^{-11} meter

- the diameter of the Sun = 1,392,000 kilometers
  = 1.392×10^6 kilometers

- the U.S. national debt (as of June 1, 2012)
  = 15,724,907,000,000.00 = 15.724907 trillion
  = 1.5724907×10^{13} dollars.

In addition to providing a simpler way of expressing very large or very small numbers, this notation also makes it easier to do basic arithmetic. The rule for multiplication of numbers expressed in this way is simple: Just multiply the factors and add the exponents. Similarly for division, divide the factors and subtract the exponents. Thus, 3.5×10^{-2} multiplied by 2.0×10^3 is simply $(3.5 \times 2.0) \times 10^{-2+3} = 7.0 \times 10^1$—that is, 70. Again, 5×10^6 divided by 2×10^4 is just $(5/2) \times 10^{6-4}$, or 2.5×10^2 ($= 250$). Applying these rules to unit conversions, 200,000 nanometers is $200,000 \times 10^{-9}$ meter (since 1 nanometer $= 10^{-9}$ meter; see Appendix 2), or $(2 \times 10^5) \times 10^{-9}$ meter, or $2 \times 10^{5-9} = 2 \times 10^{-4}$ meter $= 0.2$ mm. Verify these rules yourself with a few examples of your own. The advantages of this notation when considering astronomical objects will soon become obvious.

Scientists often use "rounded-off" versions of numbers, both for simplicity and for ease of calculation. For example, we will usually write the diameter of the Sun as 1.4×10^6 kilometers, instead of the more precise number given earlier. Similarly, the diameter of Earth is 12,756 kilometers, or 1.2756×10^4 kilometers, but for "ballpark" estimates, we really don't need so many digits, and the more approximate number 1.3×10^4 kilometers will suffice. Very often, we perform rough calculations using only the first one or two significant digits in a number, and that may be all that is necessary to make a particular point. For example, to support the statement, "The Sun is much larger than Earth," we need only say that the ratio of the two diameters is roughly 1.4×10^6 divided by 1.3×10^4. Since 1.4/1.3 is close to 1, the ratio is approximately $10^6/10^4 = 10^2$, or 100. The essential fact here is that the ratio is much larger than 1; calculating it to greater accuracy (to get 109.13) would give us no additional *useful* information. This technique of stripping away the arithmetic details to get to the essence of a calculation is very common in astronomy, and we use it frequently throughout this text.

Appendix 2
Astronomical Measurement

Astronomers use many different kinds of units in their work, simply because no single system of units will do. Rather than the *Système Internationale* (SI), or meter-kilogram-second (MKS), metric system used in most high school and college science classes, many professional astronomers still prefer the older centimeter-gram-second (CGS) system. However, astronomers also commonly introduce new units when convenient. For example, when discussing stars, the mass and radius of the Sun are often used as reference points. The solar mass, written as $M_\odot$, is equal to 2.0×10^{33} g, or 2.0×10^{30} kg (since 1 kg = 1000 g). The solar radius, $R_\odot$, is equal to 700,000 km, or 7.0×10^8 m (1 km = 1000 m). The subscript $\odot$ always stands for Sun. Similarly, the subscript $\oplus$ always stands for Earth. In this book, we try to use the units that astronomers commonly use in any given context, but we also give the "standard" SI equivalents where appropriate.

Of particular importance are the units of length astronomers use. On small scales, the *angstrom* ($1\text{Å} = 10^{-10}$ m $= 10^{-8}$ cm), the *nanometer* (1 nm $= 10^{-9}$ m $= 10^{-7}$ cm), and the *micron* (1 μm $= 10^{-6}$ m $= 10^{-4}$ cm) are used. Distances within the solar system are usually expressed in terms of the *astronomical unit*

(AU), the mean distance between Earth and the Sun. One AU is approximately equal to 150,000,000 km, or 1.5×10^{11} m. On larger scales, the *light-year* (1 ly $= 9.5 \times 10^{15}$ m $= 9.5 \times 10^{12}$ km) and the *parsec* (1 pc $= 3.1 \times 10^{16}$ m $= 3.1 \times 10^{13}$ km $= 3.3$ ly) are commonly used. Still larger distances use the regular prefixes of the metric system: *kilo* for one thousand and *mega* for one million. Thus 1 kiloparsec (kpc) $= 10^3$ pc $= 3.1 \times 10^{19}$ m, 10 megaparsecs (Mpc) $= 10^7$ pc $= 3.1 \times 10^{23}$ m, and so on.

Astronomers use units that make sense within a context, and as contexts change, so do the units. For example, we might measure densities in grams per cubic centimeter (g/cm^3), in atoms per cubic meter (atoms/m^3), or even in solar masses per cubic megaparsec ($M_\odot$/Mpc^3), depending on the circumstances. The important thing to know is that once you understand the units, you can convert freely from one set to another. For example, the radius of the Sun could equally well be written as $R_\odot = 6.96 \times 10^8$ m, or 6.96×10^{10} cm, or 109 $R_\oplus$, or 4.65×10^{-3} AU, or even 7.36×10^{-8} ly—whichever happens to be most useful. Some of the more common units used in astronomy, and the contexts in which they are most likely to be encountered, are listed below.

Length:

1 angstrom (Å)	$= 10^{-10}$ m	atomic physics,
1 nanometer (nm)	$= 10^{-9}$ m	spectroscopy
1 micron (μm)	$= 10^{-6}$ m	interstellar dust and gas
1 centimeter (cm)	$= 0.01$ m	in widespread use
1 meter (m)	$= 100$ cm	throughout all
1 kilometer (km)	$= 1000$ m $= 10^5$ cm	astronomy
Earth radius ($R_\oplus$)	$= 6378$ km	planetary astronomy
Solar radius ($R_\odot$)	$= 6.96 \times 10^8$ m	solar system,
1 astronomical unit (AU)	$= 1.496 \times 10^{11}$ m	stellar evolution
1 light-year (ly)	$= 9.46 \times 10^{15}$ m $= 63,200$ AU	galactic astronomy,
1 parsec (pc)	$= 3.09 \times 10^{16}$ m $= 206,000$ AU $= 3.26$ ly	stars and star clusters
1 kiloparsec (kpc)	$= 1000$ pc	galaxies, galaxy
1 megaparsec (Mpc)	$= 1000$ kpc	clusters, cosmology

Mass:

1 gram (g)		in widespread use in
1 kilogram (kg)	$=1000$ g	many different areas
Earth mass ($M_\oplus$)	$=5.98 \times 10^{24}$ kg	planetary astronomy
Solar mass ($M_\odot$)	$=1.99 \times 10^{30}$ kg	"standard" unit for all mass scales larger than Earth

Time:

1 second (s)		in widespread use throughout astronomy
1 hour (h)	$=3600$ s	planetary and stellar
1 day (d)	$=86,400$ s	scales
1 year (yr)	$=3.16 \times 10^7$ s	virtually all processes occurring on scales larger than a star

Appendix 3
Tables

TABLE 1 Some Useful Constants and Physical Measurements[1]

astronomical unit	$1 \text{ AU} = 1.496 \times 10^8$ km (1.5×10^8 km)
light-year	$1 \text{ ly} = 9.46 \times 10^{12}$ km (10^{13} km; about 6 trillion miles)
parsec	$1 \text{ pc} = 3.09 \times 10^{13}$ km $= 3.3$ ly
speed of light	$c = 299{,}792.458$ km/s (3×10^5 km/s)
Stefan-Boltzmann constant	$\sigma = 5.67 \times 10^{-8}$ W/m$^2 \cdot$ K^4
gravitational constant	$G = 6.67 \times 10^{-11}$ N m^2/kg^2
mass of Earth	$M_\oplus = 5.97 \times 10^{24}$ kg (6×10^{24} kg; 6000 billion tons)
radius of Earth	$R_\oplus = 6378$ km (6500 km)
mass of the Sun	$M_\odot = 1.99 \times 10^{30}$ kg (2×10^{30} kg)
radius of the Sun	$R_\odot = 6.96 \times 10^5$ km (7×10^5 km)
luminosity of the Sun	$L_\odot = 3.90 \times 10^{26}$ W (4×10^{26} W)
effective temperature of the Sun	$T_\odot = 5778$ K (5800 K)
Hubble constant	$H_0 \approx 70$ km/s/Mpc
mass of an electron	$m_e = 9.11 \times 10^{-31}$ kg
mass of a proton	$m_p = 1.67 \times 10^{-27}$ kg

[1]The rounded-off values used in the text are shown above in parentheses.

Conversions Between Common English and Metric Units

English	Metric
1 inch	= 2.54 centimeters (cm)
1 foot (ft)	= 0.3048 meters (m)
1 mile	= 1.609 kilometers (km)
1 pound (lb)	= 453.6 grams (g) or 0.4536 kilograms (kg) (on Earth)

TABLE 2 Main-Sequence Stellar Properties by Spectral Class

Spectral Class	Typical Surface Temperature	Color	Mass*	Luminosity*	Lifetime*	Familiar Examples
	(K)		($M_\odot$)	($L_\odot$)	(10^6 yr)	
O	>30,000	Electric Blue	>20	>100,000	<2	Mintaka (O9)
B	20,000	Blue	7	500	140	Spica (B2)
A	10,000	White	3	60	500	Vega (A0) Sirius (A1)
F	7000	Yellow-white	1.5	7	2000	Procyon (F5)
G	6000	Yellow	1.0	1.0	10,000	Sun (G2) Alpha Centauri (G2)
K	4000	Orange	0.8	0.3	30,000	Epsilon Eridani (K2)
M	3000	Red	0.1	0.00006	16,000,000	Proxima Centauri (M5) Barnard's Star (M5)

*Approximate values for stars of solar composition

TABLE 3A **Planetary Data: Orbital Properties**

Planet	Semimajor Axis (AU)	Semimajor Axis (10^6 km)	Sidereal Period (tropical years)	Mean Orbital Speed (km/s)	Orbital Eccentricity	Inclination to the Ecliptic (degrees)
Mercury	0.39	57.9	0.241	47.9	0.206	7.00
Venus	0.72	108.2	0.612	35.0	0.007	3.39
Earth	1.00	149.6	1.00	29.8	0.017	0.01
Mars	1.52	227.9	1.88	24.1	0.093	1.85
Jupiter	5.20	778.4	11.86	13.1	0.048	1.31
Saturn	9.54	1427	29.42	9.65	0.054	2.49
Uranus	19.19	2871	83.75	6.80	0.047	0.77
Neptune	30.07	4498	163.7	5.43	0.009	1.77

TABLE 3B **Planetary Data: Physical Properties**

Planet	Equatorial Radius (km)	Equatorial Radius (Earth = 1)	Mass (kg)	Mass (Earth = 1)	Mean Density (kg/m^3)	Sidereal Rotation Period (solar days)[1]
Mercury	2440	0.38	3.30×10^{23}	0.055	5430	58.6
Venus	6052	0.95	4.87×10^{24}	0.82	5240	−243.0
Earth	6378	1.00	5.97×10^{24}	1.00	5520	0.9973
Mars	3394	0.53	6.42×10^{23}	0.11	3930	1.026
Jupiter	71,492	11.21	1.90×10^{27}	317.8	1330	0.41
Saturn	60,268	9.45	5.68×10^{26}	95.2	690	0.44
Uranus	25,559	4.01	8.68×10^{25}	14.5	1270	−0.72
Neptune	24,766	3.88	1.02×10^{26}	17.1	1640	0.67

Planet	Axial Tilt (degrees)	Surface Gravity (Earth = 1)	Escape Speed (km/s)	Surface Temperature (K)[2]	Number of Moons[3]
Mercury	0.0	0.38	4.2	100 to 700	0
Venus	177.4	0.91	10.4	730	0
Earth	23.5	1.00	11.2	290	1
Mars	24.0	0.38	5.0	180 to 270	2
Jupiter	3.1	2.53	60	124	16
Saturn	26.7	1.07	36	97	18
Uranus	97.9	0.91	21	58	27
Neptune	29.6	1.14	24	59	13

[1]A negative sign indicates retrograde rotation.

[2]Temperature is effective temperature for jovian planets.

[3]Moons more than 10 km in diameter

TABLE 4 The 20 Brightest Stars in Earth's Night Sky

Name	Star	Spectral Type* A	Spectral Type* B	Parallax (arc seconds)	Distance (pc)	Apparent Visual Magnitude* A	Apparent Visual Magnitude* B
Sirius	α CMa	A1V	wd[†]	0.379	2.6	−1.44	+8.4
Canopus	α Car	F0Ib–II		0.010	96	−0.62	
Arcturus	α Boo	K2III		0.089	11	−0.05	
Rigel Kentaurus	α Cen	G2V	K0V	0.742	1.3	−0.01	+1.4
(Alpha Centauri)							
Vega	α Lyr	A0V		0.129	7.8	+0.03	
Capella	α Aur	GIII	M1V	0.077	13	+0.08	+10.2
Rigel	β Ori	B8Ia	B9	0.0042	240	+0.18	+6.6
Procyon	α CMi	F5IV–V	Wd[†]	0.286	3.5	+0.40	+10.7
Betelgeuse	α Ori	M2Iab		0.0076	130	+0.45	
Achernar	α Eri	B5V		0.023	44	+0.45	
Hadar	α Cen	B1III	?	0.0062	160	+0.61	+4
Altair	α Aq1	A7IV–V		0.194	5.1	+0.76	
Acrux	α Cru	B1IV	B3	0.010	98	+0.77	+1.9
Aldebaran	α Tau	K5III	M2V	0.050	20	+0.87	+13
Spica	α Vir	B1V	B2V	0.012	80	+0.98	2.1
Antares	α Sco	M1Ib	B4V	0.005	190	+1.06	+5.1
Pollux	β Gem	K0III		0.097	10	+1.16	
Formalhaut	α PsA	A3V	?	0.130	7.7	+1.17	+6.5
Deneb	α Cyg	A2Ia		0.0010	990	+1.25	
Mimosa	β Cru	B1IV		0.0093	110	+1.25	

Name	Visual Luminosity* (Sun = 1) A	Visual Luminosity* (Sun = 1) B	Absolute Visual Magnitude[1] A	Absolute Visual Magnitude[1] B	Proper Motion (arc seconds/yr)	Transverse Velocity (km/s)	Radial Velocity (km/s)
Sirius	22	0.0025	+1.5	+11.3	1.33	16.7	−7.6[‡]
Canopus	1.4×10^4		−5.5		0.02	9.1	20.5
Arcturus	110		−0.3		2.28	119	−5.2
Rigel Kentaurus	1.6	0.45	+4.3	+5.7	3.68	22.7	−24.6
Vega	50		+0.6		0.34	12.6	−13.9
Capella	130	0.01	−0.5	+9.6	0.44	27.1	30.2[‡]
Rigel	4.1×10^4	110	−6.7	−0.3	0.00	1.2	20.7
Procyon	7.2	0.0006	+2.7	+13.0	1.25	20.7	−3.2
Betelgeuse	9700		−5.1		0.03	18.5	21.0
Achernar	1100		−2.8		0.10	20.9	19
Hadar	1.3×10^4	560	−5.4	−2.0	0.04	30.3	−12
Altair	11		+2.2		0.66	16.3	−26.3
Acrux	4100	2200	−4.2	−3.5	0.04	22.8	−11.2
Aldebaran	150	0.002	−0.6	+11.5	0.20	19.0	54.1
Spica	2200	780	−3.5	−2.4	0.05	19.0	1.0
Antares	1.1×10^4	290	−5.3	−1.3	0.03	27.0	−3.2
Pollux	31		+1.1		0.62	29.4	3.3
Formalhaut	17	0.13	+1.7	+7.1	0.37	13.5	6.5
Deneb	2.6×10^5		−8.7		0.003	14.1	−4.6
Mimosa	3200		−3.9		0.05	26.1	—

* Energy output in the visible part of the spectrum; A and B columns identify individual components of binary star systems.

[†] "wd" stands for "white dwarf."

[‡] Average value of variable velocity

TABLE 5 The 20 Nearest Stars

Name	Spectral Type* A	Spectral Type* B	Parallax (arc seconds)	Distance (pc)	Apparent Visual Magnitude* A	Apparent Visual Magnitude* B
Sun	G2V				−26.74	
Proxima Centauri	M5		0.772	1.30	+11.01	
Alpha Centauri	G2V	K1V	0.742	1.35	−0.01	+1.35
Barnard's Star	M5V		0.549	1.82	+9.54	
Wolf 359	M8V		0.421	2.38	+13.53	
Lalande 21185	M2V		0.397	2.52	+7.50	
UV Ceti	M6V	M6V	0.387	2.58	+12.52	+13.02
Sirius	A1V	wd†	0.379	2.64	−1.44	+8.4
Ross 154	M5V		0.345	2.90	+10.45	
Ross 248	M6V		0.314	3.18	+12.29	
ε Eridani	K2V		0.311	3.22	+3.72	
Ross 128	M5V		0.298	3.36	+11.10	
61 Cygni	K5V	K7V	0.294	3.40	+5.22	+6.03
ε Indi	K5V		0.291	3.44	+4.68	
Grm 34	M1V	M6V	0.290	3.45	+8.08	+11.06
Luyten 789-6	M6V		0.290	3.45	+12.18	
Procyon	F5IV–V	wd†	0.286	3.50	+0.40	+10.7
Σ 2398	M4V	M5V	0.285	3.55	+8.90	+9.69
Lacaille 9352	M2V		0.279	3.58	+7.35	
G51-15	MV		0.278	3.60	+14.81	

Name	Visual Luminosity* (Sun = 1) A	Visual Luminosity* (Sun = 1) B	Absolute Visual Magnitude* A	Absolute Visual Magnitude* B	Proper Motion (arc seconds/yr)	Transverse Velocity (km/s)	Radial Velocity (km/s)
Sun	1.0		+4.83				
Proxima Centauri	5.6×10^{-5}		+15.4		3.86	23.8	−16
Alpha Centauri	1.6	0.45	+4.3	+5.7	3.68	23.2	−22
Barnard's Star	4.3×10^{-4}		+13.2		10.34	89.7	−108
Wolf 359	1.8×10^{-5}		+16.7		4.70	53.0	+13
Lalande 21185	0.0055		+10.5		4.78	57.1	−84
UV Ceti	5.4×10^{-4}	0.00004	+15.5	+16.0	3.36	41.1	+30
Sirius	22	0.0025	+1.5	+11.3	1.33	16.7	−8
Ross 154	4.8×10^{-4}		+13.3		0.72	9.9	−4
Ross 248	1.1×10^{-4}		+14.8		1.58	23.8	−81
ε Eridani	0.29		+6.2		0.98	15.3	+16
Ross 128	3.6×10^{-4}		+13.5		1.37	21.8	−13
61 Cygni	0.082	0.039	+7.6	+8.4	5.22	84.1	−64
ε Indi	0.14		+7.0		4.69	76.5	−40
Grm 34	0.0061	0.00039	+10.4	+13.4	2.89	47.3	+17
Luyten 789-6	1.4×10^{-4}		+14.6		3.26	53.3	−60
Procyon	7.2	0.00055	+2.7	+13.0	1.25	2.8	−3
Σ 2398	0.0030	0.0015	+11.2	+11.9	2.28	38.4	+5
Lacaille 9352	0.013		+9.6		6.90	117	+10
G51-15	1.4×10^{-5}		+17.0		1.26	21.5	—

* A and B columns identify individual components of binary star systems.

† "wd" stands for "white dwarf."

Glossary

Key terms that are boldface in the text are followed by a page reference in the Glossary.

A

A ring One of three Saturnian rings visible from Earth. The A ring is farthest from the planet and is separated from the B ring by the Cassini Division. (p. 230)

aberration of starlight Small shift in the observed direction to a star, caused by Earth's motion perpendicular to the line of sight.

absolute brightness The apparent brightness a star would have if it were placed at a standard distance of 10 parsecs from Earth.

absolute magnitude The apparent magnitude a star would have if it were placed at a standard distance of 10 parsecs from Earth. (p. 274)

absolute zero The lowest possible temperature that can be obtained; all thermal motion ceases at this temperature.

absorption line Dark line in an otherwise continuous bright spectrum, where light within one narrow frequency range has been removed. (p. 56)

abundance Relative amounts of different elements in a gas.

acceleration The rate of change of velocity of a moving object. (p. 36)

accretion Gradual growth of bodies, such as planets, by the accumulation of other, smaller bodies. (p. 121)

accretion disk Flat disk of matter spiraling down onto the surface of a neutron star or black hole. Often, the matter originated on the surface of a companion star in a binary star system. (p. 331)

active galactic nucleus Region of intense emission at the center of an active galaxy responsible for virtually all of the galaxy's nonstellar luminosity. (p. 419)

active galaxies The most energetic galaxies, which can emit hundreds or thousands of times more energy per second than the Milky Way, mostly in the form of long-wavelength nonthermal radiation. (p. 418)

active optics Collection of techniques used to increase the resolution of ground-based telescopes. Minute modifications are made to the overall configuration of an instrument as its temperature and orientation change; used to maintain the best possible focus at all times. (p. 82)

active region Region of the photosphere of the Sun surrounding a sunspot group, which can erupt violently and unpredictably. During sunspot maximum, the number of active regions is also a maximum. (p. 258)

active Sun The unpredictable aspects of the Sun's behavior, such as sudden explosive outbursts of radiation in the form of prominences and flares.

adaptive optics Technique used to increase the resolution of a telescope by deforming the shape of the mirror's surface under computer control while a measurement is being taken; used to undo the effects of atmospheric turbulence. (p. 82)

aerosol Suspension of liquid or solid particles in air.

alpha particle A helium-4 nucleus.

alpha process Process occurring at high temperatures, in which high-energy photons split heavy nuclei to form helium nuclei.

ALSEP Acronym for Apollo Lunar Surface Experiments Package.

altimeter Instrument used to determine altitude.

amino acids Organic molecules that form the basis for building the proteins that direct metabolism in living creatures. (p. 487)

Amor asteroid Asteroid that crosses only the orbit of Mars.

amplitude The maximum deviation of a wave above or below zero point. (p. 45)

angstrom Distance unit equal to 0.1 nanometers, or one 10-billionth of a meter.

angular diameter Angle made between the top (or one edge) of an object, the observer and the bottom (or opposite edge) of the object.

angular distance Angular separation between two objects as seen by some observer.

angular momentum Tendency of an object to keep rotating; proportional to the mass, radius, and rotation speed of the body.

angular resolution The ability of a telescope to distinguish between adjacent objects in the sky. (p. 79)

annular eclipse Solar eclipse occurring at a time when the Moon is far enough away from Earth that it fails to cover the disk of the Sun completely, leaving a ring of sunlight visible around its edge. (p. 16)

antiparallel Configuration of the electron and proton in a hydrogen (or other) atom when their spin axes are parallel but the two rotate in opposite directions.

antiparticle A particle of the same mass but opposite in all other respects (e.g., charge) to a given particle; when a particle and its antiparticle come into contact, they annihilate and release energy in the form of gamma rays.

aphelion The point on the elliptical path of an object in orbit about the Sun that is most distant from the Sun. (p. 33)

Apollo asteroid *See* Earth-crossing asteroid.

apparent brightness The brightness that a star appears to have, as measured by an observer on Earth. (p. 272)

apparent magnitude The apparent brightness of a star, expressed using the magnitude scale. (p. 274)

association Small grouping of (typically 100 or less) bright stars, spanning up to a few tens of parsecs across, usually rich in very young stars. (p. 314)

Assumption of Mediocrity Statements suggesting that the development of life on Earth did not require any unusual circumstances, suggesting that extraterrestrial life may be common.

asteroid One of thousands of very small members of the solar system orbiting the Sun between the orbits of Mars and Jupiter. Often referred to as "minor planets." (p. 105)

asteroid belt Region of the solar system, between the orbits of Mars and Jupiter, in which most asteroids are found. (p. 106)

asthenosphere Layer of Earth's interior, just below the lithosphere, over which the surface plates slide.

astrology Pseudoscience that purports to use the positions of the planets, sun, and moon to predict daily events and human destiny.

astronomical unit (AU) The average distance of Earth from the Sun. Precise radar measurements yield a value for the AU of 149,603,500 km. (p. 34)

astronomy Branch of science dedicated to the study of everything in the universe that lies above Earth's atmosphere. (p. 4)

asymptotic giant branch Path on the Hertzsprung–Russell diagram corresponding to the changes that a star undergoes after helium burning ceases in the core. At this stage, the carbon core shrinks and drives the expansion of the envelope, and the star becomes a swollen red giant for a second time.

aten asteroid Earth-crossing asteroid with semimajor axis less than 1 AU.

atmosphere Layer of gas confined close to a planet's surface by the force of gravity. (p. 136)

atom Building block of matter, composed of positively charged protons and neutral neutrons in the nucleus surrounded by negatively charged electrons. (p. 58)

atomic epoch Period after decoupling when the first simple atoms and molecules formed.

aurora Event that occurs when atmospheric molecules are excited by incoming charged particles from the solar wind, then emit energy as they fall back to their ground states. Aurorae generally occur at high latitudes, near the north and south magnetic poles. (p. 156)

autumnal equinox Date on which the Sun crosses the celestial equator moving southward, occurring on or near September 22. (p. 11)

B

B ring One of three Saturnian rings visible from Earth. The B ring is the brightest of the three and lies just past the Cassini Division, closer to the planet than the A ring. (p. 230)

background noise Unwanted light in an image, from unresolved sources in the telescope's field of view, scattered light from the atmosphere, or instrumental "hiss" in the detector itself.

barred-spiral galaxy Spiral galaxy in which a bar of material passes through the center of the galaxy, with the spiral arms beginning near the ends of the bar. (p. 408)

basalt Solidified lava; an iron-magnesium-silicate mixture.

baseline The distance between two observing locations used for the purposes of triangulation measurements. The larger the baseline, the better the resolution attainable. (p. 17)

belt Dark, low-pressure region in the atmosphere of a jovian planet, where gas flows downward. (p. 201)

Big Bang Event that cosmologists consider the beginning of the universe, in which all matter and radiation in the entire universe came into being. (p. 462)

Big Crunch Point of final collapse of a bound universe.

binary asteroid Asteroids that have a partner in orbit around it.

binary pulsar Binary system in which both components are pulsars.

binary-star system A system that consists of two stars in orbit about their common center of mass, held together by their mutual gravitational attraction. Most stars are found in binary-star systems. (p. 285)

biological evolution Change in a population of biological organisms over time.

bipolar flow Jets of material expelled from a protostar perpendicular to the surrounding protostellar disk.

blackbody curve The characteristic way in which the intensity of radiation emitted by a hot object depends on frequency. The frequency at which the emitted intensity is highest is an indication of the temperature of the radiating object. Also referred to as the Planck curve. (p. 52)

black dwarf The endpoint of the evolution of an isolated, low-mass star. After the white dwarf stage, the star cools to the point where it is a dark "clinker" in interstellar space. (p. 329)

black hole A region of space where the pull of gravity is so great that nothing—not even light—can escape. A possible outcome of the evolution of a very massive star. (p. 361)

blazar Particularly intense active galactic nucleus in which the observer's line of sight happens to lie directly along the axis of a high-speed jet of particles emitted from the active region. (p. 423)

blue giant Large, hot, bright star at the upper-left end of the main sequence on the Hertzsprung–Russell diagram. Its name comes from its color and size. (p. 280)

blueshift Motion-induced changes in the observed wavelength from a source that is moving toward us. Relative approaching motion between the object and the observer causes the wavelength to appear shorter (and hence bluer) than if there were no motion at all.

blue straggler Star found on the main sequence of the Hertzsprung–Russell diagram, but which should already have evolved off the main sequence, given its location on the diagram; thought to have formed from mergers of lower mass stars.

blue supergiant The very largest of the large, hot, bright stars at the uppermost-left end of the main sequence on the Hertzsprung–Russell diagram. (p. 280)

Bohr model First theory of the hydrogen atom to explain the observed spectral lines. This model rests on three ideas: that there is a state of lowest energy for the electron, that there is a maximum energy beyond which the electron is no longer bound to the nucleus, and that within these two energies the electron can only exist in certain energy levels. (p. 58)

bok globule Dense, compact cloud of interstellar dust and gas on its way to forming one or more stars.

boson Particle that exerts or mediates forces between elementary particles in quantum physics.

bound trajectory Path of an object with launch speed low enough that it cannot escape the gravitational pull of a planet.

brown dwarf Remnants of fragments of collapsing gas and dust that did not contain enough mass to initiate core nuclear fusion. Such objects are then frozen somewhere along their pre-main-sequence contraction phase, continually cooling into compact dark objects. Because of their small size and low temperature, they are extremely difficult to detect observationally. (p. 312)

brown oval Feature of Jupiter's atmosphere that appears only at latitudes near 20 degrees N, this structure is a long-lived hole in the clouds that allows us to look down into Jupiter's lower atmosphere. (p. 203)

C

C ring One of three Saturnian rings visible from Earth. The C ring lies closest to the planet and is relatively thin compared to the A and B rings. (p. 230)

caldera Crater that forms at the summit of a volcano.

capture theory (Moon) Theory suggesting that the Moon formed far from Earth but was later captured by it.

carbonaceous asteroid The darkest, or least reflective, type of asteroid, containing large amounts of carbon.

carbon-based molecule Molecule containing atoms of carbon.

carbon-detonation supernova See Type I supernova. (p. 336)

cascade Process of deexcitation in which an excited electron moves down through energy states one at a time.

Cassegrain telescope A type of reflecting telescope in which incoming light hits the primary mirror and is then reflected upward toward the prime focus, where a secondary mirror reflects the light back down through a small hole in the main mirror into a detector or eyepiece. (p. 72)

Cassini Division A relatively empty gap in Saturn's ring system, discovered in 1675 by Giovanni Cassini. It is now known to contain a number of thin ringlets. (p. 230)

cataclysmic variable Collective name for novae and supernovae.

catalyst Something that causes or helps a reaction to occur, but is not itself consumed as part of the reaction.

catastrophic theory A theory that invokes statistically unlikely accidental events to account for observations.

celestial coordinates Pair of quantities—right ascension and declination—similar to longitude and latitude on Earth, used to pinpoint locations of objects on the celestial sphere. (p. 6)

celestial equator The projection of Earth's equator onto the celestial sphere. (p. 6)

celestial mechanics Study of the motions of bodies, such as planets and stars, that interact via gravity.

celestial pole Projection of Earth's North or South pole onto the celestial sphere.

celestial sphere Imaginary sphere surrounding Earth to which all objects in the sky were once considered to be attached. (p. 5)

Celsius Temperature scale in which the freezing point of water is 0 degrees and the boiling point of water is 100 degrees.

center of mass The "average" position in space of a collection of massive bodies, weighted by their masses. For an isolated system this point moves with constant velocity, according to Newtonian mechanics. (p. 38)

centigrade See Celsius.

centripetal force (literally "center seeking") Force directed toward the center of a body's orbit.

centroid Average position of the material in an object; in spectroscopy, the center of a spectral line.

Cepheid variable Star whose luminosity varies in a characteristic way, with a rapid rise in brightness followed by a slower decline. The period of a Cepheid variable star is related to its luminosity, so a determination of this period can be used to obtain an estimate of the star's distance. (p. 382)

Chandrasekhar mass Maximum possible mass of a white dwarf.

chaotic rotation Unpredictable tumbling motion that nonspherical bodies in eccentric orbits, such as Saturn's satellite Hyperion, can exhibit. No amount of observation of an object rotating chaotically will ever show a well-defined period.

charge-coupled device (CCD) An electronic device used for data acquisition; composed of many tiny pixels, each of which records a buildup of charge to measure the amount of light striking it. (p. 76)

chemical bond Force holding atoms together to form a molecule.

chemosynthesis Analog of photosynthesis that operates in total darkness.

chromatic aberration The tendency for a lens to focus red and blue light differently, causing images to become blurred.

chromosphere The Sun's lower atmosphere, lying just above the visible atmosphere. (p. 246)

circumnavigation Traveling all the way around an object.

cirrus High-level clouds composed of ice or methane crystals.

closed universe Geometry that the universe as a whole would have if the density of matter is above the critical value. A closed universe is finite in extent and has no edge, like the surface of a sphere. It has enough mass to stop the present expansion and will eventually collapse. (p. 465)

CNO cycle Chain of reactions that converts hydrogen into helium using carbon, nitrogen, and oxygen as catalysts.

cocoon nebula Bright infrared source in which a surrounding cloud of gas and dust absorb ultraviolet radiation from a hot star and reemits it in the infrared.

cold dark matter Class of dark-matter candidates made up of very heavy particles, possibly formed in the very early universe.

collecting area The total area of a telescope capable of capturing incoming radiation. The larger the telescope, the greater its collecting area and the fainter the objects it can detect. (p. 77)

collisional broadening Broadening of spectral lines due to collisions between atoms, most often seen in dense gases.

color index A convenient method of quantifying a star's color by comparing its apparent brightness as measured through different filters. If the star's radiation is well described by a blackbody spectrum, the ratio of its blue intensity (B) to its visual intensity (V) is a measure of the object's surface temperature.

color–magnitude diagram A way of plotting stellar properties, in which absolute magnitude is plotted against color index.

coma An effect occurring during the formation of an off-axis image in a telescope. Stars whose light enters the telescope at a large angle acquire comet-like tails on their images. The brightest part of a comet, often referred to as the "head." (p. 110)

comet A small body, composed mainly of ice and dust, in an elliptical orbit about the Sun. As it comes close to the Sun, some of its material is vaporized to form a gaseous head and extended tail. (p. 105)

common envelope Outer layer of gas in a contact binary.

comparative planetology The systematic study of the similarities and differences among the planets, with the goal of obtaining deeper insight into how the solar system formed and has evolved in time.

composition The mixture of atoms and molecules that make up an object.

condensation nuclei Dust grains in the interstellar medium that act as seeds around which other material can cluster. The presence of dust was very important in causing matter to clump during the formation of the solar system. (p. 120)

condensation theory Currently favored model of solar system formation that combines features of the old nebular theory with new information about interstellar dust grains, which acted as condensation nuclei. (p. 119)

conjunction Orbital configuration in which a planet lies in the same direction as the Sun, as seen from Earth.

conservation of mass and energy *See* law of conservation of mass and energy.

constellation A human grouping of stars in the night sky into a recognizable pattern. (p. 4)

constituents *See* composition.

contact binary A binary star system in which both stars have expanded to fill their Roche lobes, and the surfaces of the two stars merge. The binary system now consists of two nuclear burning stellar cores surrounded by a continuous common envelope.

continental drift The movement of the continents around Earth's surface.

continuous spectra Spectra in which the radiation is distributed over all frequencies, not just a few specific frequency ranges. A prime example is the blackbody radiation emitted by a hot, dense body. (p. 55)

convection Churning motion resulting from the constant upwelling of warm fluid and the concurrent downward flow of cooler material to take its place. (p. 141)

convection cell Circulating region of upwelling hot fluid and sinking cooler fluid in convective motion. (p. 141)

convection zone Region of the Sun's interior, lying just below the surface, where the material of the Sun is in constant convection motion. This region extends into the solar interior to a depth of about 20,000 km. (p. 246)

co-orbital satellites Satellites sharing the same orbit around a planet.

Copernican principle The removal of Earth from any position of cosmic significance.

Copernican revolution The realization, toward the end of the 16th century, that Earth is not at the center of the universe. (p. 29)

core The central region of Earth, surrounded by the mantle. (p. 136)

The central region of any planet or star. (p. 246)

core-accretion theory Theory that the jovian planets formed when icy protoplanetary cores became massive enough to capture gas directly from the solar nebula. *See* gravitational instability theory.

core-collapse supernova *See* Type II supernova.

core hydrogen burning The energy-burning stage for main-sequence stars, in which the helium is produced by hydrogen fusion in the central region of the star. A typical star spends up to 90 percent of its lifetime in hydrostatic equilibrium brought about by the balance between gravity and the energy generated by core hydrogen burning. (p. 322)

cornea (eye) The curved transparent layer covering the front part of the eye. (p. 171)

corona One of numerous large, roughly circular regions on the surface of Venus, thought to have been caused by upwelling mantle material causing the planet's crust to bulge outward (plural, *coronae*). (p. 171)

The tenuous outer atmosphere of the Sun, which lies just above the chromosphere and, at great distances, turns into the solar wind. (p. 246)

coronal hole Vast regions of the Sun's atmosphere where the density of matter is about 10 times lower than average. The gas there streams freely into space at high speeds, escaping the Sun completely. (p. 260)

coronal mass ejection Giant magnetic "bubble" of ionized gas that separates from the rest of the solar atmosphere and escapes into interplanetary space. (p. 259)

corpuscular theory Early particle theory of light.

cosmic density parameter Ratio of the universe's actual density to the critical value corresponding to zero curvature.

cosmic distance scale Collection of indirect distance-measurement techniques that astronomers use to measure distances in the universe. (p. 17)

cosmic evolution The collection of the seven major phases of the history of the universe, namely particulate, galactic, stellar, planetary, chemical, biological, and cultural evolution. (p. 486)

cosmic microwave background The almost perfectly isotropic radio signal that is the electromagnetic remnant of the Big Bang. (p. 471)

cosmic ray Very energetic subatomic particle arriving at Earth from elsewhere in the Galaxy.

cosmological constant Quantity originally introduced by Einstein into general relativity to make his equations describe a static universe. Now one of several candidates for the repulsive "dark energy" force responsible for the observed cosmic acceleration. (p. 468)

cosmological distance Distance comparable to the scale of the universe.

cosmological principle Two assumptions that make up the basis of cosmology, namely that the universe is homogeneous and isotropic on sufficiently large scales. (p. 461)

cosmological redshift The component of the redshift of an object that is due only to the Hubble flow of the universe. (p. 417)

cosmology The study of the structure and evolution of the entire universe. (p. 461)

cosmos The universe.

coudé focus Focus produced far from the telescope using a series of mirrors. Allows the use of heavy and/or finely tuned equipment to analyze the image.

crater Bowl-shaped depression on the surface of a planet or moon, resulting from a collision with interplanetary debris. (p. 152)

crescent Appearance of the Moon (or a planet) when less than half of the body's hemisphere is visible from Earth.

crest Maximum departure of a wave above its undisturbed state.

critical density The cosmic density corresponding to the dividing line between a universe that recollapses and one that expands forever. (p. 464)

critical universe Universe in which the density of matter is exactly equal to the critical density. The universe is infinite in extent and has zero curvature. The expansion will continue forever, but will approach an expansion speed of zero. (p. 466)

crust Layer of Earth that contains the solid continents and the seafloor. (p. 136)

C-type asteroid See carbonaceous asteroid.

cultural evolution Change in the ideas and behavior of a society over time.

current sheet Flat sheet on Jupiter's magnetic equator where most of the charged particles in the magnetosphere lie due to the planet's rapid rotation.

D

D ring Collection of very faint, thin rings, extending from the inner edge of the C ring down nearly to the cloud tops of Saturn. This region contains so few particles that it is completely invisible from Earth. (p. 233)

dark dust cloud A large cloud, often many parsecs across, which contains gas and dust in a ratio of about 1012 gas atoms for every dust particle. Typical densities are a few tens or hundreds of millions of particles per cubic meter. (p. 301)

dark energy Generic name given to the unknown cosmic force field thought to be responsible for the observed acceleration of the Hubble expansion. (p. 467)

dark halo Region of a galaxy beyond the visible halo where dark matter is believed to reside. (p. 394)

dark matter Term used to describe the mass in galaxies and clusters whose existence we infer from rotation curves and other techniques, but that has not been confirmed by observations at any electromagnetic wavelength. (p. 394)

dark matter particle Particle undetectable at any electromagnetic wavelength, but can be inferred from its gravitational influence.

daughter/fission theory Theory suggesting that the Moon originated out of Earth.

declination Celestial coordinate used to measure latitude above or below the celestial equator on the celestial sphere. (p. 6)

decoupling Event in the early universe when atoms first formed, after which photons could propagate freely through space. (p. 475)

deferent A construct of the geocentric model of the solar system that was needed to explain observed planetary motions. A deferent is a large circle encircling Earth, on which an epicycle moves. (p. 27)

degree Unit of angular measure. There are 360 degrees in one complete circle.

density A measure of the compactness of the matter within an object, computed by dividing the mass of the object by its volume. Units are kilograms per cubic meter (kg/m^3), or grams per cubic centimeter (g/cm^3). (p. 103)

detached binary Binary system where each star lies within its respective Roche lobe.

detector noise Readings produced by an instrument even when it is not observing anything; produced by the electronic components within the detector itself.

deuterium A form of hydrogen with an extra neutron in its nucleus.

deuterium bottleneck Period in the early universe between the start of deuterium production and the time when the universe was cool enough for deuterium to survive.

deuteron An isotope of hydrogen in which a neutron is bound to the proton in the nucleus. Often called "heavy hydrogen" because of the extra mass of the neutron. (p. 262)

differential rotation The tendency for a gaseous sphere, such as a jovian planet or the Sun, to rotate at a different rate at the equator than at the poles. More generally, a condition where the angular speed varies with location within an object. (p. 199)

differentiation Variation in the density and composition of a body, such as Earth, with low-density material on the surface and higher-density material in the core. (p. 147)

diffraction The ability of waves to bend around corners. The diffraction of light establishes its wave nature. (p. 46)

diffraction grating Sheet of transparent material with many closely spaced parallel lines ruled on it, designed to separate white light into a spectrum.

diffraction-limited resolution Theoretical resolution that a telescope can have due to diffraction of light at the telescope's aperture. Depends on the wavelength of radiation and the diameter of the telescope's mirror. (p. 79)

direct motion See prograde motion.

distance modulus Difference between the apparent and absolute magnitude of an object; equivalent to distance, by the inverse-square law.

diurnal motion Apparent daily motion of the stars caused by Earth's rotation. (p. 7)

DNA Deoxyribonucleic acid, the molecule that carries genetic information and determines the characteristics of a living organism.

Doppler effect Any motion-induced change in the observed wavelength (or frequency) of a wave. (p. 63)

double-line spectroscopic binary Binary system in which spectral lines of both stars can be distinguished and seen to shift back and forth as the stars orbit one another.

double-star system System containing two stars in orbit around one another.

Drake equation Expression that gives an estimate of the probability that intelligence exists elsewhere in the Galaxy, based on a number of supposedly necessary conditions for intelligent life to develop. (p. 493)

dust grain An interstellar dust particle, roughly 10^{-7} m in size, comparable to the wavelength of visible light. (p. 294)

dust lane A lane of dark, obscuring interstellar dust in an emission nebula or galaxy. (p. 299)

dust tail The component of a comet's tail that is composed of dust particles. (p. 110)

dwarf Any star with radius comparable to, or smaller than, that of the Sun (including the Sun itself). (p. 279)

dwarf elliptical Elliptical galaxy as small as 1 kiloparsec across, containing only a few million stars.

dwarf galaxy Small galaxy containing a few million stars.

dwarf irregular Small irregular galaxy containing only a few million stars.

dwarf planet A body that orbits the Sun, is massive enough that its own gravity has caused its shape to be approximately spherical, but is insufficiently massive to have cleared other bodies from "the neighborhood" of its orbit. (p. 105)

dynamo theory Theory that explains planetary and stellar magnetic fields in terms of rotating, conducting material flowing in an object's interior. (p. 156)

E

E ring A faint ring, well outside the main ring system of Saturn, which was discovered by *Voyager* and is believed to be associated with volcanism on the moon Enceladus. (p. 233)

Earth-crossing asteroid An asteroid whose orbit crosses that of Earth. Earth-crossing asteroids are also called Apollo asteroids, after the first asteroid of this type discovered. (p. 106)

earthquake A sudden dislocation of rocky material near Earth's surface. (p. 145)

eccentricity A measure of the flatness of an ellipse, equal to the distance between the two foci divided by the length of the major axis. (p. 33)

eclipse Event during which one body passes in front of another, so that the light from the occulted body is blocked. (p. 14)

eclipse season Times of the year when the Moon lies in the same plane as Earth and Sun, so that eclipses are possible.

eclipse year Time interval between successive orbital configurations in which the line of nodes of the Moon's orbit points toward the Sun.

eclipsing binary Rare binary star system that is aligned in such a way that from Earth we observe one star pass in front of the other, eclipsing the other star. (p. 286)

ecliptic The apparent path of the Sun, relative to the stars on the celestial sphere, over the course of a year. (p. 8)

effective temperature Temperature of a blackbody of the same radius and luminosity as a given star or planet.

ejecta (planetary) Material thrown outward by a meteoroid impact.

ejecta (stellar) Material thrown into space by a nova or supernova.

electric field A field extending outward in all directions from a charged particle, such as a proton or an electron. The electric field determines the electric force exerted by the particle on all other charged particles in the universe; the strength of the electric field decreases with increasing distance from the charge according to an inverse-square law. (p. 46)

electromagnetic energy Energy carried in the form of rapidly fluctuating electric and magnetic fields.

electromagnetic radiation Another term for light, electromagnetic radiation transfers energy and information from one place to another. (p. 44)

electromagnetic spectrum The complete range of electromagnetic radiation, from radio waves to gamma rays, including the visible spectrum. All types of electromagnetic radiation are basically the same phenomenon, differing only by wavelength, and all move at the speed of light. (p. 50)

electromagnetism The union of electricity and magnetism, which do not exist as independent quantities but are in reality two aspects of a single physical phenomenon. (p. 47)

electron An elementary particle with a negative electric charge; one of the components of the atom. (p. 46)

electron degeneracy pressure The pressure produced by the resistance of electrons to further compression once they are squeezed to the point of contact.

electrostatic force Force between electrically charged objects.

electroweak force Unification of the weak electromagnetic forces.

element Matter made up of one particular atom. The number of protons in the nucleus of the atom determines which element it represents. (p. 62)

elementary particle Technically, a particle that cannot be subdivided into component parts; however, the term is also often used to refer to particles such as protons and neutrons, which are themselves made up of quarks.

ellipse Geometric figure resembling an elongated circle. An ellipse is characterized by its degree of flatness, or eccentricity, and the length of its long axis. In general, bound orbits of objects moving under gravity are elliptical. (p. 33)

elliptical galaxy Category of galaxy in which the stars are distributed in an elliptical shape on the sky, ranging from highly elongated to nearly circular in appearance. (p. 408)

elongation Angular distance between a planet and the Sun.

emission line Bright line in a specific location of the spectrum of radiating material, corresponding to emission of light at a certain frequency. A heated gas in a glass container produces emission lines in its spectrum. (p. 55)

emission nebula A glowing cloud of hot interstellar gas. The gas glows as a result of one or more nearby young stars that ionize the gas. Since the gas is mostly hydrogen, the emitted radiation falls predominantly in the red region of the spectrum, because of the hydrogen-alpha emission line. (p. 297)

emission spectrum The pattern of spectral emission lines produced by an element. Each element has its own unique emission spectrum. (p. 55)

empirical Discovery based on observational evidence (rather than from theory).

Encke Gap A small gap in Saturn's A ring. (p. 231)

energy flux Energy per unit area per unit time radiated by a star (or recorded by a detector).

epicycle A construct of the geocentric model of the solar system that was necessary to explain observed planetary motions. Each planet rides on a small epicycle whose center in turn rides on a larger circle (the deferent). (p. 27)

epoch of inflation Short period of unchecked cosmic expansion early in the history of the universe. During inflation, the universe swelled in size by a factor of about 10^{50}. (p. 476)

equinox *See* autumnal equinox, vernal equinox. (p. 11)

escape speed The speed necessary for one object to escape the gravitational pull of another. Anything that moves away from a gravitating body with more than the escape speed will never return. (p. 136)

euclidean geometry Geometry of flat space.

event horizon Imaginary spherical surface surrounding a collapsing star with radius equal to the Schwarzschild radius, within which no event can be seen, heard, or known about by an outside observer. (p. 362)

evolutionary theory A theory that explains observations in a series of gradual steps, explainable in terms of well-established physical principles.

evolutionary track A graphical representation of a star's life as a path on the Hertzsprung–Russell diagram. (p. 309)

excited state State of an atom when one of its electrons is in a higher energy orbital than the ground state. Atoms can become excited by absorbing a photon of a specific energy or by colliding with a nearby atom. (p. 59)

exposure time Time spent gathering light from a source. (p. 77)

extinction The dimming of starlight as it passes through the interstellar medium.

extrasolar planet Planet orbiting a star other than the Sun. (p. 117)

extremophilic Adjective describing organisms that can survive in very harsh environments. (p. 492)

eyepiece Secondary lens through which an observer views an image. This lens is often chosen to magnify the image.

F

F ring Faint narrow outer ring of Saturn, discovered by *Pioneer 11* in 1979. The F ring lies just inside the Roche limit of Saturn and was found by *Voyager 1* to be made up of several ring strands apparently braided together. (p. 233)

Fahrenheit Temperature scale in which the freezing point of water is 32 degrees and the boiling point of water is 212 degrees.

false color Representation of an image in which color does not represent true visual color, but rather an invisible wavelength of radiation, or some other property, such as temperature. (p. 85)

false vacuum Region of the universe that remained in the "unified" state after the strong and electroweak forces separated; one possible cause of cosmic inflation at very early times.

fault line Dislocation on a planet's surface, often indicating the boundary between two plates.

field line Imaginary line indicating the direction of an electric or magnetic field.

fireball Large meteor that burns up brightly and sometimes explosively in Earth's atmosphere.

firmament Old-fashioned term for the heavens (i.e., the sky).

flare Explosive event occurring in or near an active region on the Sun. (p. 258)

flatness problem One of two conceptual problems with the standard Big Bang model, which is that there is no natural way to explain why the density of the universe is so close to the critical density. (p. 476)

fluidized ejecta The ejecta blankets around some Martian craters, which apparently indicate that the ejected material was liquid at the time the crater formed.

fluorescence Phenomenon in which an atom absorbs energy, then radiates photons of lower energy as it cascades back to the ground state; in astronomy, often produced as ultraviolet photons from a hot young star that are absorbed by a neutral gas, causing some of the gas atoms to become excited and give off an optical (red) glow.

flyby Unbound trajectory of a spacecraft around a planet or other body.

focal length Distance from a mirror or the center of a lens to the focus.

focus One of two special points within an ellipse, whose separation from each other indicates the eccentricity. (p. 33)

In a bound orbit, planets orbit in ellipses with the Sun at one focus. (p. 33)

forbidden line A spectral line seen in emission nebulae, but not seen in laboratory experiments because, under laboratory conditions, collisions kick the electron in question into some other state before emission can occur.

force Action on an object that causes its momentum to change. The rate at which the momentum changes is numerically equal to the force. (p. 36)

fragmentation The breaking up of a large object into many smaller pieces (for example, as the result of high-speed collisions between planetesimals and protoplanets in the early solar system). (p. 121)

Fraunhofer lines The collection of over 600 absorption lines in the spectrum of the Sun, first categorized by Joseph Fraunhofer in 1812.

frequency The number of wave crests passing any given point per unit time. (p. 45)

full When the full hemisphere of the Moon or a planet can be seen from Earth.

full Moon Phase of the Moon in which it appears as a complete circular disk in the sky. (p. 12)

fusion *See* nuclear fusion.

G

G ring Faint, narrow ring of Saturn, discovered by *Pioneer 11* and lying just outside the F ring.

galactic bulge Thick distribution of warm gas and stars around the galactic center. (p. 380)

galactic cannibalism A galaxy merger in which a larger galaxy consumes a smaller one.

galactic center The center of the Milky Way, or any other galaxy. The point about which the disk of a spiral galaxy rotates. (p. 384)

galactic disk An immense, circular, flattened region containing most of a galaxy's luminous stars and interstellar matter. (p. 380)

galactic epoch Period from 100 million to 3 billion years after the Big Bang when large agglomerations of matter (galaxies and galaxy clusters) formed and grew.

galactic habitable zone Region of a galaxy in which conditions are conducive to the development of life.

galactic halo Region of a galaxy extending far above and below the galactic disk, where globular clusters and other old stars reside. (p. 380)

galactic nucleus Small, central, high-density region of a galaxy. Many galactic nuclei are thought to contain supermassive black holes, and almost all the radiation from active galaxies is generated within the nucleus. (pp. 399, 419)

galactic rotation curve Plot of rotation speed versus distance from the center of a galaxy.

galactic year Time taken for objects at the distance of the Sun (about 8 kpc) to orbit the center of the Galaxy, roughly 225 million years.

galaxy Gravitationally bound collection of a large number of stars. The Sun is a star in the Milky Way Galaxy. (pp. 4, 380)

galaxy cluster A collection of galaxies held together by their mutual gravitational attraction. (p. 415)

Galilean moons The four large moons of Jupiter discovered by Galileo Galilei.

Galilean satellites The four brightest and largest moons of Jupiter (Io, Europa, Ganymede, Callisto), named after Galileo Galilei, the 17th-century astronomer who first observed them.

gamma ray Region of the electromagnetic spectrum, far beyond the visible spectrum, corresponding to radiation of very high frequency and very short wavelength. (p. 44)

gamma-ray burst Object that radiates tremendous amounts of energy in the form of gamma rays, possibly due to the collision and merger of two neutron stars initially in orbit around one another. (p. 356)

gamma-ray spectrograph Spectrograph designed to work at gamma-ray wavelengths. Used to map the abundances of certain elements on the Moon and Mars.

gaseous Composed of gas.

gas-exchange experiment Experiment to look for life on Mars. A nutrient broth was offered to Martian soil specimens. If there were life in the soil, gases would be created as the broth was digested.

gene Sequence of nucleotide bases in the DNA molecule that determines the characteristics of a living organism.

general theory of relativity Theory proposed by Einstein to incorporate gravity into the framework of special relativity (p. 361).

geocentric model A model of the solar system that holds that Earth is at the center of the universe and all other bodies are in orbit around it. The earliest theories of the solar system were geocentric. (p. 27)

giant A star with a radius between 10 and 100 times that of the Sun. (p. 279)

giant elliptical Elliptical galaxy up to a few megaparsecs across, containing trillions of stars.

gibbous Appearance of the Moon (or a planet) when more than half (but not all) of the body's hemisphere is visible from Earth.

globular cluster Tightly bound, roughly spherical collection of hundreds of thousands, and sometimes millions, of stars spanning about 50 parsecs. Globular clusters are distributed in the halos around the Milky Way and other galaxies. (p. 314)

gluon Particle that exerts or mediates the strong force in quantum physics.

gradient Rate of change of some quantity (e.g., temperature or composition) with respect to location in space.

Grand Unified Theories Class of theories describing the behavior of the single force that results from unification of the strong, weak, and electromagnetic forces in the early universe. (p. 476)

granite Igneous rock, containing silicon and aluminum, that makes up most of Earth's crust.

granulation Mottled appearance of the solar surface, caused by rising (hot) and falling (cool) material in convective cells just below the photosphere. (p. 251)

gravitational force Force exerted on one body by another due to the effect of gravity. The force is directly proportional to the masses of both bodies involved and inversely proportional to the square of the distance between them. (p. 37)

gravitational instability theory Theory that the jovian planets formed directly from the solar nebula via instabilities in the gas leading to gravitational contraction. *See* core-accretion theory.

gravitational lensing The effect induced on the image of a distant object by a massive foreground object. Light from the distant object is bent into two or more separate images. (p. 395)

gravitational radiation Radiation resulting from rapid changes in a body's gravitational field.

gravitational redshift A prediction of Einstein's general theory of relativity. Photons lose energy as they escape the gravitational field of a massive object. Because a photon's energy is proportional to its frequency, a photon that loses energy suffers a decrease in frequency, which corresponds to an increase, or redshift, in wavelength. (p. 369)

graviton Particle carrying the gravitational field in theories attempting to unify gravity and quantum mechanics.

gravity The attractive effect that any massive object has on all other massive objects. The greater the mass of the object, the stronger its gravitational pull. (p. 37)

gravity assist Using gravity to change the flight path of a satellite or spacecraft.

gravity wave Gravitational counterpart of an electromagnetic wave.

great attractor A huge accumulation of mass in the relatively nearby universe (within about 200 Mpc of the Milky Way).

Great Dark Spot Prominent storm system in the atmosphere of Neptune observed by *Voyager 2*, near the equator of the planet. The system was comparable in size to Earth. (p. 207)

Great Red Spot A large, high-pressure, long-lived storm system visible in the atmosphere of Jupiter. The Red Spot is roughly twice the size of Earth. (p. 201)

great wall Extended sheet of galaxies measuring at least 200 megaparsecs across; one of the largest known structures in the universe.

greenhouse effect The partial trapping of solar radiation by a planetary atmosphere, similar to the trapping of heat in a greenhouse. (p. 143)

greenhouse gas Gas (such as carbon dioxide or water vapor) that efficiently absorbs infrared radiation.

ground state The lowest energy state that an electron can have within an atom. (p. 58)

GUT epoch Period when gravity separated from the other three forces of nature.

gyroscope System of rotating wheels that allows a spacecraft to maintain a fixed orientation in space.

H

habitable zone Three-dimensional zone of comfortable temperature (corresponding to liquid water) that surrounds every star. (p. 130)

half-life The amount of time it takes for half of the initial amount of a radioactive substance to decay into something else.

Hayashi track Evolutionary track followed by a protostar during the final pre-main-sequence phase before nuclear fusion begins.

heat Thermal energy, the energy of an object due to the random motion of its component atoms or molecules.

heat death End point of a bound universe, in which all matter and life are destined to be incinerated.

heavy element In astronomical terms, any element heavier than hydrogen and helium.

heliocentric model A model of the solar system that is centered on the Sun, with Earth in motion about the Sun. (p. 29)

helioseismology The study of conditions far below the Sun's surface through the analysis of internal "sound" waves that repeatedly cross the solar interior. (p. 248)

helium-burning shell Shell of burning helium gas surrounding a nonburning stellar core of carbon ash.

helium capture The formation of heavy elements by the capture of a helium nucleus. For example, carbon can form heavier elements by fusion with other carbon nuclei, but it is much more likely to occur by helium capture, which requires less energy.

helium flash An explosive event in the post-main-sequence evolution of a low-mass star. When helium fusion begins in a dense stellar core, the burning is explosive in nature. It continues until the energy released is enough to expand the core, at which point the star achieves stable equilibrium again. (p. 325)

helium precipitation Mechanism responsible for the low abundance of helium of Saturn's atmosphere. Helium condenses in the upper layers to form a mist that rains down toward Saturn's interior, just as water vapor forms into rain in the atmosphere of Earth.

helium shell flash Condition in which the helium-burning shell in the core of a star cannot respond to rapidly changing conditions within it, leading to a sudden temperature rise and a dramatic increase in nuclear reaction rates.

Hertzsprung–Russell (H–R) diagram A plot of luminosity against temperature (or spectral class) for a group of stars. (p. 279)

hierarchical merging Widely accepted galaxy-formation scenario in which galaxies formed as relatively small objects in the early universe and subsequently collided and merged to form the large galaxies observed today. The present Hubble type of a galaxy depends on the sequence of merger events in the galaxy's past. (p. 440)

high-energy astronomy Astronomy using X- or gamma-ray radiation rather than optical radiation.

high-energy telescope Telescope designed to detect X- and gamma-ray radiation. (p. 92)

highlands Relatively light-colored regions on the surface of the Moon that are elevated several kilometers above the maria. Also called terrae. (p. 151)

high-mass star Star with a mass more than 8 times that of the Sun; progenitor of a neutron star or black hole.

HI region Region of space containing primarily neutral hydrogen.

HII region Region of space containing primarily ionized hydrogen.

homogeneity Assumed property of the universe such that the number of galaxies in an imaginary large cube of the universe is the same no matter where in the universe the cube is placed. (p. 460)

horizon problem One of two conceptual problems with the standard Big Bang model, which is that some regions of the universe that have very similar properties are too far apart to have exchanged information within the age of the universe. (p. 475)

horizontal branch Region of the Hertzsprung–Russell diagram where post-main-sequence stars again reach hydrostatic equilibrium. At this point, the star is burning helium in its core and fusing hydrogen in a shell surrounding the core. (p. 325)

hot dark matter A class of candidates for the dark matter in the universe, composed of lightweight particles such as neutrinos, much less massive than the electron.

hot Jupiter A massive, gaseous planet orbiting very close to its parent star. (p. 127)

hot longitudes (Mercury) Two opposite points on Mercury's equator where the Sun is directly overhead at perihelion.

Hubble classification scheme Method of classifying galaxies according to their appearance, developed by Edwin Hubble. (p. 429)

Hubble diagram Plot of galactic recession velocity versus distance; evidence for an expanding universe.

Hubble flow Universal recession described by the Hubble diagram and quantified by Hubble's law.

Hubble sequence Hubble's arrangement of elliptical, S0, and spiral galaxies, often used to classify galaxies but not thought to represent a true evolutionary sequence in any sense. (p. 410)

Hubble's constant The constant of proportionality that gives the relation between recessional velocity and distance in Hubble's law. (p. 417)

Hubble's law Law that relates the observed velocity of recession of a galaxy to its distance from us. The velocity of recession of a galaxy is directly proportional to its distance away. (p. 416)

hydrocarbon Molecules consisting solely of hydrogen and carbon.

hydrogen envelope An invisible sheath of gas engulfing the coma of a comet, usually distorted by the solar wind and extending across millions of kilometers of space. (p. 110)

hydrogen shell burning Fusion of hydrogen in a shell that is driven by contraction and heating of the helium core. Once hydrogen is depleted in the core of a star, hydrogen burning stops and the core contracts due to gravity, causing the temperature to rise, heating the surrounding layers of hydrogen in the star, and increasing the burning rate there. (p. 324)

hydrosphere Layer of Earth that contains the liquid oceans and accounts for roughly 70 percent of Earth's total surface area. (p. 136)

hydrostatic equilibrium Condition in a star or other fluid body in which gravity's inward pull is exactly balanced by internal forces due to pressure. (p. 248)

hyperbola Curve formed when a plane intersects a cone at a small angle to the axis of the cone.

hypernova Explosion where a massive star undergoes core collapse and forms a black hole and a gamma-ray burst. *See* supernova.

I

igneous Rocks formed from molten material.

image The optical representation of an object produced when the object is reflected or refracted by a mirror or lens. (p. 70)

impact theory (Moon) Combination of the capture and daughter theories, suggesting that the Moon formed after an impact that dislodged some of Earth's mantle and placed it in orbit.

inertia The tendency of an object to continue moving at the same speed and in the same direction, unless acted upon by a force. (p. 36)

inferior conjunction Orbital configuration in which an inferior planet (Mercury or Venus) lies closest to Earth.

inflation *See* epoch of inflation.

infrared Region of the electromagnetic spectrum just outside the visible range, corresponding to light of a slightly longer wavelength than red light. (p. 44)

infrared telescope Telescope designed to detect infrared radiation. Many such telescopes are designed to be lightweight so that they can be carried above (most of) Earth's atmosphere by balloons, airplanes, or satellites. (p. 89)

infrared waves Electromagnetic radiation with wavelength in the infrared part of the spectrum. (p. 44)

inhomogeneity Deviation from perfectly uniform density; in cosmology, inhomogeneities in the universe are ultimately due to quantum fluctuations before inflation.

inner core The central part of Earth's core, believed to be solid and composed mainly of nickel and iron. (p. 146)

instability strip Part of the Hertzsprung–Russell diagram where pulsating post-main-sequence stars are found.

intensity A basic property of electromagnetic radiation that specifies the amount or strength of the radiation. (p. 51)

intercloud medium Superheated bubbles of hot gas extending far into interstellar space.

intercrater plains Regions on the surface of Mercury that do not show extensive cratering but are relatively smooth. (p. 169)

interference The ability of two or more waves to interact in such a way that they either reinforce or cancel each other. (p. 46)

interferometer Collection of two or more telescopes working together as a team, observing the same object at the same time and at the same wavelength. The effective diameter of an interferometer is equal to the distance between its outermost telescopes. (p. 86)

interferometry Technique in widespread use to dramatically improve the resolution of radio and infrared maps. Several telescopes observe the object simultaneously, and a computer analyzes how the signals interfere with each other. (p. 86)

intermediate-mass black hole Black hole having a mass 100–1000 times greater than the mass of the Sun.

interplanetary matter Matter in the solar system that is not part of a planet or moon—cosmic "debris."

interstellar dust Microscopic dust grains that populate space between the stars, having their origins in the ejected matter of long-dead stars.

interstellar gas cloud A large cloud of gas found in the space among the stars.

interstellar medium The matter between stars, composed of two components, gas and dust, intermixed throughout all of space. (p. 294)

intrinsic variable Star that varies in appearance due to internal processes (rather than, say, interaction with another star).

inverse-square law The law that a field follows if its strength decreases with the square of the distance. Fields that follow the inverse-square law decrease rapidly in strength as the distance increases, but never quite reach zero. (p. 37)

Io plasma torus Doughnut-shaped region of energetic ionized particles emitted by the volcanoes on Jupiter's moon Io and swept up by Jupiter's magnetic field.

ion An atom that has lost one or more of its electrons. (p. 59)

ionization state Term describing the number of electrons missing from an atom: I refers to a neutral atom, II refers to an atom missing one electron, and so on.

ionosphere Layer in Earth's atmosphere above about 100 km where the atmosphere is significantly ionized and conducts electricity. (p. 141)

ion tail Thin stream of ionized gas that is pushed away from the head of a comet by the solar wind. It extends directly away from the Sun. Often referred to as a plasma tail. (p. 110)

irregular galaxy A galaxy that does not fit into any of the other major categories in the Hubble classification scheme. (p. 410)

isotopes Nuclei containing the same number of protons but different numbers of neutrons. Most elements can exist in several isotopic forms. A common example of an isotope is deuterium, which differs from normal hydrogen by the presence of an extra neutron in the nucleus.

isotropic Assumed property of the universe such that the universe looks the same in every direction. (p. 460)

J

jet stream Relatively strong winds in the upper atmosphere channeled into a narrow stream by a planet's rotation. Normally refers to a horizontal, high-altitude wind.

joule The SI unit of energy.

jovian planet One of the four giant outer planets of the solar system, resembling Jupiter in physical and chemical composition. (p. 104)

K

Kelvin scale Temperature scale in which absolute zero is at 0 K; a change of 1 kelvin is the same as a change of 1 degree Celsius.

Kelvin-Helmholtz contraction phase Evolutionary track followed by a star during the protostar phase.

Kepler's laws of planetary motion Three laws, based on precise observations of the motions of the planets by Tycho Brahe that summarize the motions of the planets about the Sun.

kinetic energy Energy of an object due to its motion.

Kirchhoff's laws Three rules governing the formation of different types of spectra. (p. 57)

Kirkwood gaps Gaps in the spacings of orbital semimajor axes of asteroids in the asteroid belt, produced by dynamical resonances with nearby planets, especially Jupiter.

Kuiper belt A region in the plane of the solar system outside the orbit of Neptune where most short-period comets are thought to originate. (p. 105)

Kuiper belt object Small icy body orbiting in the Kuiper belt. (p. 112)

L

labeled-release experiment Experiment to look for life on Mars. Radioactive carbon compounds were added to Martian soil specimens. Scientists looked for indications that the carbon had been eaten or inhaled.

Lagrangian point One of five special points in the plane of two massive bodies orbiting one another, where a third body of negligible mass can remain in equilibrium.

lander Spacecraft that lands on the object it is studying.

laser ranging Method of determining the distance to an object by firing a laser beam at it and measuring the time taken for the light to return.

lava dome Volcanic formation formed when lava oozes out of fissures in a planet's surface, creating the dome, and then withdraws, causing the crust to crack and subside.

law of conservation of mass and energy A fundamental law of modern physics that states that the sum of mass and energy must always remain constant in any physical process. In fusion reactions, the lost mass is converted into energy, primarily in the form of electromagnetic radiation. (p. 261)

laws of planetary motion Three laws derived by Kepler describing the motion of the planets around the Sun. (p. 33)

leap year Year in which an additional day is inserted into the calendar in order to keep the calendar year synchronized with Earth's orbit around the Sun.

lens Optical instrument made of glass or some other transparent material, shaped so that, as a parallel beam of light passes through it, the rays are refracted so as to pass through a single focal point.

lens (eye) The part of the eye that refracts light onto the retina.

lepton Greek word for light, referring in particle physics to low-mass particles such as electrons, muons, and neutrinos that interact via the weak force.

lepton epoch Period when the light elementary particles (leptons) were in thermal equilibrium with the cosmic radiation field.

lidar Light Detection and Ranging—a device that uses laser-ranging to measure distance.

light *See* electromagnetic radiation.

light curve The variation in brightness of a star with time. (p. 286)

light element In astronomical terms, hydrogen and helium.

light-gathering power Amount of light a telescope can view and focus, proportional to the area of the primary mirror.

lighthouse model The leading explanation for pulsars. A small region of the neutron star, near one of the magnetic poles, emits a steady stream of radiation that sweeps past Earth each time the star rotates. The period of the pulses is the star's rotation period. (p. 352)

light-year The distance that light, moving at a constant speed of 300,000 km/s, travels in 1 year. One light-year is about 10 trillion kilometers.

limb Edge of a lunar, planetary, or solar disk.

linear momentum Tendency of an object to keep moving in a straight line with constant velocity; the product of the object's mass and velocity.

line of nodes Intersection of the plane of the Moon's orbit with Earth's orbital plane.

line of sight technique Method of probing interstellar clouds by observing their effects on the spectra of background stars.

lithosphere Earth's crust and a small portion of the upper mantle that make up Earth's plates. This layer of Earth undergoes tectonic activity.

Local Bubble The particular low-density intercloud region surrounding the Sun.

Local Group The small galaxy cluster that includes the Milky Way Galaxy. (p. 414)

local supercluster Collection of galaxies and clusters centered on the Virgo cluster. *See also* supercluster.

logarithm The power to which 10 must be raised to produce a given number.

logarithmic scale Scale using the logarithm of a number rather than the number itself; commonly used to compress a large range of data into more manageable form.

look-back time Time in the past when an object emitted the radiation we see today.

low-mass star Star with a mass less than 8 times that of the Sun; progenitor of a white dwarf.

luminosity One of the basic properties used to characterize stars, luminosity is defined as the total energy radiated by a star each second, at all wavelengths. (p. 247)

luminosity class A classification scheme that groups stars according to the width of their spectral lines. For a group of stars with the same temperature, luminosity class differentiates between supergiants, giants, main-sequence stars, and subdwarfs. (p. 284)

lunar dust *See* regolith.

lunar eclipse Celestial event during which the moon passes through the shadow of Earth, temporarily darkening its surface. (p. 14)

lunar phase The appearance of the Moon at different points along its orbit.

Lyman-alpha forest Collection of lines in an object's spectra starting at the redshifted wavelength of the object's own Lyman-alpha emission line and extending to shorter wavelengths; produced by gas in galaxies along the line of sight.

M

macroscopic Large enough to be visible by the unaided eye.

Magellanic Clouds Two small irregular galaxies that are gravitationally bound to the Milky Way Galaxy. (p. 410)

magnetic field Field that accompanies any changing electric field and governs the influence of magnetized objects on one another. (p. 47)

magnetic poles Points on a planet where the planetary magnetic field lines intersect the planet's surface vertically.

magnetism The presence of a magnetic field.

magnetometer Instrument that measures magnetic field strength.

magnetopause Boundary between a planet's magnetosphere and the solar wind.

magnetosphere A zone of charged particles trapped by a planet's magnetic field, lying above the atmosphere. (p. 136)

magnitude scale A system of ranking stars by apparent brightness, developed by the Greek astronomer Hipparchus. Originally, the brightest stars in the sky were categorized as being of first magnitude, while the faintest stars visible to the naked eye were classified as sixth magnitude. The scheme has since been extended to cover stars and galaxies too faint to be seen by the unaided eye. Increasing magnitude means fainter stars, and a difference of five magnitudes corresponds to a factor of 100 in apparent brightness. (p. 274)

main sequence Well-defined band on the Hertzsprung–Russell diagram on which most stars are found, running from the top left of the diagram to the bottom right. (p. 280)

main-sequence turnoff Special point on the Hertzsprung–Russell diagram for a cluster, indicative of the cluster's age. If all the stars in the cluster are plotted, the lower-mass stars will trace out the main sequence up to the point where stars begin to evolve off the main sequence toward the red giant branch. The point where stars are just beginning to evolve off is the main-sequence turnoff. (p. 340)

major axis The long axis of an ellipse.

mantle Layer of Earth just interior to the crust. (p. 136)

marginally bound universe Universe that will expand forever but at an increasingly slow rate.

maria Relatively dark-colored and smooth regions on the surface of the Moon (singular: *mare*). (p. 151)

mass A measure of the total amount of matter contained within an object. (p. 36)

mass function Relation between the component masses of a single-line spectroscopic binary.

massive compact halo object (MACHO) Collective name for "stellar" candidates for dark matter, including brown dwarfs, white dwarfs, and low-mass red dwarfs.

mass–luminosity relation The dependence of the luminosity of a main-sequence star on its mass. The luminosity increases roughly as the mass raised to the third power.

mass–radius relation The dependence of the radius of a main-sequence star on its mass. The radius rises roughly in proportion to the mass.

mass transfer Process by which one star in a binary system transfers matter onto the other.

mass-transfer binary *See* semidetached binary.

matter Anything having mass.

matter–antimatter annihilation Reaction in which matter and antimatter annihilate to produce high energy gamma rays. *See also* antiparticle.

matter-dominated universe A universe in which the density of matter exceeds the density of radiation. The present-day universe is matter dominated.

matter era (Current) era following the radiation era when the universe is larger and cooler and matter is the dominant constituent of the universe.

Maunder minimum Lengthy period of solar inactivity that extended from 1645 to 1715.

mean solar day Average length of time from one noon to the next, taken over the course of a year—24 hours.

medium Material through which a wave, such as sound, travels.

meridian An imaginary line on the celestial sphere through the north and south celestial poles, passing directly overhead at a given location.

mesosphere Region of Earth's atmosphere lying between the stratosphere and the ionosphere, 50–80 km above Earth's surface. (p. 141)

Messier object Member of a list of "fuzzy" objects compiled by astronomer Charles Messier in the 18th century.

metabolism The daily utilization of food and energy by which organisms stay alive.

metallic Composed of metal or metal compounds.

metamorphic Rocks created from existing rocks exposed to extremes of temperature or pressure.

meteor Bright streak in the sky, often referred to as a "shooting star," resulting from a small piece of interplanetary debris entering Earth's atmosphere and heating air molecules, which emit light as they return to their ground states. (p. 114)

meteorite Any part of a meteoroid that survives passage through the atmosphere and lands on the surface of Earth. (p. 114)

meteoroid Chunk of interplanetary debris prior to encountering Earth's atmosphere. (p. 105)

meteoroid swarm Pebble-sized cometary fragments dislodged from the main body, moving in nearly the same orbit as the parent comet. (p. 114)

meteor shower Event during which many meteors can be seen each hour, caused by the yearly passage of Earth through the debris spread along the orbit of a comet.

microlensing Gravitational lensing by individual stars in a galaxy.

micrometeoroid Relatively small chunks of interplanetary debris ranging from dust-particle size to pebble-sized fragments. (p. 114)

microsphere Small droplet of protein-like material that resists dissolution in water.

midocean ridge Place where two plates are moving apart, allowing fresh magma to well up.

Milky Way Galaxy The spiral galaxy in which the Sun resides. The disk of our Galaxy is visible in the night sky as the faint band of light known as the Milky Way. (p. 380)

millisecond pulsar A pulsar whose period indicates that the neutron star is rotating nearly 1000 times each second. The most likely explanation for these rapid rotators is that the neutron star has been spun up by drawing in matter from a companion star. (p. 354)

molecular cloud A cold, dense interstellar cloud that contains a high fraction of molecules. It is widely believed that the relatively high density of dust particles in these clouds plays an important role in the formation and preservation of the molecules. (p. 304)

molecular cloud complex Collection of molecular clouds that spans as much as 50 parsecs and may contain enough material to make millions of Sun-size stars. (p. 303)

molecule A tightly bound collection of atoms held together by the atoms' electromagnetic fields. Molecules, like atoms, emit and absorb photons at specific wavelengths. (p. 62)

molten In liquid form due to high temperatures.

moon A small body in orbit around a planet.

mosaic (photograph) Composite photograph made up of many smaller images.

M-type asteroid Asteroid containing large fractions of nickel and iron.

multiple star system Group of two or more stars in orbit around one another.

muon A type of lepton (along with the electron and tau).

N

naked singularity A singularity that is not hidden behind an event horizon.

nanobacterium Very small bacterium with diameter in the nanometer range.

nanometer One billionth of a meter.

neap tide The smallest tides, occurring when the Earth–Moon line is perpendicular to the Earth–Sun line at the first and third quarters.

nebula General term used for any "fuzzy" patch on the sky, either light or dark. (pp. 119, 297)

nebular theory One of the earliest models of solar system formation, dating back to Descartes, in which a large cloud of gas began to collapse under its own gravity to form the Sun and planets. (p. 119)

nebulosity "Fuzziness," usually in the context of an extended or gaseous astronomical object.

neon–oxygen white dwarf White dwarf formed from a low-mass star with a mass close to the "high-mass" limit, in which neon and oxygen form in the core.

neutrino Virtually massless and chargeless particle that is one of the products of fusion reactions in the Sun. Neutrinos move at close to the speed of light and interact with matter hardly at all. (p. 262)

neutrino oscillations Possible solution to the solar neutrino problem, in which the neutrino has a very tiny mass. In this case, the correct number of neutrinos can be produced in the solar core, but on their way to Earth some can "oscillate," or become transformed into other particles, and thus go undetected. (p. 264)

neutron An elementary particle with roughly the same mass as a proton, but which is electrically neutral. Along with protons, neutrons form the nuclei of atoms. (p. 62)

neutron capture The primary mechanism by which very massive nuclei are formed in the violent aftermath of a supernova. Instead of fusion of like nuclei, heavy elements are created by the addition of more and more neutrons to existing nuclei.

neutron degeneracy pressure Pressure due to the Pauli exclusion principle, arising when neutrons are forced to come into close contact.

neutronization Process occurring at high densities, in which protons and electrons are crushed together to form neutrons and neutrinos.

neutron spectrometer Instrument designed to search for water ice by looking for hydrogen.

neutron star A dense ball of neutrons that remains at the core of a star after a supernova explosion has destroyed the rest of the star. Typical neutron stars are about 20 km across and contain more mass than the Sun. (p. 350)

new Moon Phase of the moon during which none of the lunar disk is visible. (p. 12)

Newtonian mechanics The basic laws of motion postulated by Newton, which are sufficient to explain and quantify virtually all of the complex dynamical behavior found on Earth and elsewhere in the universe. (p. 36)

Newtonian telescope A reflecting telescope in which incoming light is intercepted before it reaches the prime focus and is deflected into an eyepiece at the side of the instrument. (p. 72)

nodes Two points on the Moon's orbit when it crosses the ecliptic.

nonrelativistic Speeds that are much less than the speed of light.

nonthermal spectrum Continuous spectrum not well described by a blackbody.

north celestial pole Point on the celestial sphere directly above Earth's North Pole. (p. 6)

Northern and Southern Lights Colorful displays produced when atmospheric molecules, excited by charged particles from the Van Allen belts, fall back to their ground state.

nova A star that suddenly increases in brightness, often by a factor of as much as 10,000, then slowly fades back to its original luminosity. A nova is the result of an explosion on the surface of a white-dwarf star, caused by matter falling onto its surface from the atmosphere of a binary companion. (p. 330)

nuclear binding energy Energy that must be supplied to split an atomic nucleus into neutrons and protons.

nuclear epoch Period when protons and neutrons fused to form heavier nuclei.

nuclear fusion Mechanism of energy generation in the core of the Sun, in which light nuclei are combined, or fused, into heavier ones, releasing energy in the process. (p. 261)

nuclear reaction Reaction in which two nuclei combine to form other nuclei, often releasing energy in the process. *See also* fusion.

nucleotide base An organic molecule, the building block of genes that pass on hereditary characteristics from one generation of living creatures to the next. (p. 487)

nucleus Dense, central region of an atom containing both protons and neutrons and orbited by one or more electrons. (p. 58)

The solid region of ice and dust that composes the central region of the head of a comet. (p. 110)

The dense central core of a galaxy. (p. 399)

O

obscuration Blockage of light by pockets of interstellar dust and gas.

Olbers's paradox A thought experiment suggesting that if the universe were homogeneous, infinite, and unchanging, the entire night sky would be as bright as the surface of the Sun. (p. 461)

Oort cloud Spherical halo of material surrounding the solar system out to a distance of about 50,000 AU; where most comets reside. (p. 113)

opacity A quantity that measures a material's ability to block electromagnetic radiation. Opacity is the opposite of transparency. (p. 50)

open cluster Loosely bound collection of tens to hundreds of stars, a few parsecs across, generally found in the plane of the Milky Way. (p. 314)

open universe Geometry that the universe would have if the density of matter were less than the critical value. In an open universe there is not enough matter to halt the expansion of the universe. An open universe is infinite in extent. (p. 465)

opposition Orbital configuration in which a planet lies in the opposite direction from the Sun, as seen from Earth.

optical double Chance superposition in which two stars appear to lie close together but are actually widely separated.

optical telescope Telescope designed to observe electromagnetic radiation at optical wavelengths.

orbital One of several energy states in which an electron can exist in an atom.

orbital period Time taken for a body to complete one full orbit around another.

orbiter Spacecraft that orbits an object to make observations.

organic compound Chemical compound (molecule) containing a significant fraction of carbon atoms; the basis of living organisms.

outer core The outermost part of Earth's core, believed to be liquid and composed mainly of nickel and iron. (p. 146)

outflow channel Surface feature on Mars, evidence that liquid water once existed there in great quantity; believed to be the relics of catastrophic flooding about 3 billion years ago. Found only in the equatorial regions of the planet. (p. 177)

outgassing Production of atmospheric gases (carbon dioxide, water vapor, methane, and sulphur dioxide) by volcanic activity.

ozone layer Layer of Earth's atmosphere at an altitude of 20–50 km where incoming ultraviolet solar radiation is absorbed by oxygen, ozone, and nitrogen in the atmosphere. (p. 141)

P

parallax The apparent motion of a relatively close object with respect to a more distant background as the location of the observer changes. (p. 18)

parsec The distance at which a star must lie in order for its measured parallax to be exactly 1 arc second; 1 parsec equals 206,000 AU. (p. 270)

partial eclipse Celestial event during which only a part of the occulted body is blocked from view. (p. 14)

particle A body with mass but of negligible dimension.

particle accelerator Device used to accelerate subatomic particles to relativistic speeds.

particle–antiparticle pair Pair of particles (e.g., an electron and a positron) produced by two photons of sufficiently high energy.

particle detector Experimental equipment that allows particles and antiparticles to be detected and identified.

Pauli exclusion principle A rule of quantum mechanics that prohibits electrons in dense gas from being squeezed too closely together.

penumbra Portion of the shadow cast by an eclipsing object in which the eclipse is seen as partial. (p. 15)

The outer region of a sunspot, surrounding the umbra, which is not as dark and not as cool as the central region. (p. 254)

perihelion The closest approach to the Sun of any object in orbit about it. (p. 33)

period The time needed for an orbiting body to complete one revolution about another body. (p. 34)

period–luminosity relation A relation between the pulsation period of a Cepheid variable and its absolute brightness. Measurement of the pulsation period allows the distance of the star to be determined. (p. 383)

permafrost Layer of permanently frozen water ice believed to lie just under the surface of Mars. (p. 179)

phase Appearance of the sunlit face of the Moon at different points along its orbit, as seen from Earth. (p. 12)

photodisintegration Process occurring at high temperatures, in which individual photons have enough energy to split a heavy nucleus (e.g., iron) into lighter nuclei.

photoelectric effect Emission of an electron from a surface when a photon of electromagnetic radiation is absorbed.

photoevaporation Process in which a cloud in the vicinity of a newborn hot star is dispersed by the star's radiation.

photometer A device that measures the total amount of light received in all or part of the image.

photometry Branch of observational astronomy in which the brightness of a source is measured through each of a set of standard filters.

photomicrograph Photograph taken through a microscope.

photon Individual packet of electromagnetic energy that makes up electromagnetic radiation. (p. 59)

photosphere The visible surface of the Sun, lying just above the uppermost layer of the Sun's interior and just below the chromosphere. (p. 246)

photosynthesis Process by which plants manufacture carbohydrates and oxygen from carbon dioxide and water, using chlorophyll and sunlight as the energy source.

pixel One of many tiny picture elements, organized into an array, making up a digital image.

Planck curve *See* blackbody curve.

Planck epoch Period from the beginning of the universe to roughly 10^{-43} second, when the laws of physics are not understood.

Planck's constant Fundamental physical constant relating the energy of a photon to its radiation frequency (color).

planet One of eight major bodies that orbit the Sun, visible to us by reflected sunlight. (p. 4)

planetary nebula The ejected envelope of a red-giant star, spread over a volume roughly the size of our solar system. (p. 327)

planetary ring system Material organized into thin, flat rings encircling a giant planet, such as Saturn.

planetesimal Term given to objects in the early solar system that had reached the size of small moons, at which point their gravitational fields were strong enough to begin influencing their neighbors. (p. 121)

plasma A gas in which the constituent atoms are completely ionized.

plate tectonics The motions of regions of Earth's lithosphere, which drift with respect to one another. Also known as continental drift. (p. 149)

plutino Kuiper belt object whose orbital period (like that of Pluto) is in a 3:2 resonance with the orbit of Neptune.

plutoid A dwarf planet orbiting beyond Neptune. (p. 239)

polarity A measure of the direction of the solar magnetic field in a sunspot. Conventionally, lines coming out of the surface are labeled "S," while those going into the surface are labeled "N." (p. 255)

polarization The alignment of the electric fields of emitted photons, which are generally emitted with random orientations.

Population I and II stars Classification scheme for stars based on the abundance of heavy elements. Within the Milky Way, Population I refers to young disk stars, and Population II refers to old halo stars.

positron Atomic particle with properties identical to those of a negatively charged electron, except for its positive charge. This positron is the antiparticle of the electron. Positrons and electrons annihilate one another when they meet, producing pure energy in the form of gamma rays. (p. 262)

prebiotic compound Molecule that can combine with others to form the building blocks of life.

precession The slow change in the direction of the rotation axis of a spinning object, caused by some external gravitational influence. (p. 11)

primary mirror Mirror placed at the prime focus of a telescope. *See* prime focus.

prime focus The point in a reflecting telescope where the mirror focuses incoming light to a point. (p. 70)

prime-focus image Image formed at the prime focus of a telescope.

primeval fireball Hot, dense state of the universe at very early times, just after the Big Bang. (p. 462)

primordial matter Matter created during the early, hot epochs of the universe.

primordial nucleosynthesis The production of elements heavier than hydrogen by nuclear fusion in the high temperatures and densities that existed in the early universe. (p. 473)

principle of cosmic censorship A proposition to separate the unexplained physics near a singularity from the rest of the well-behaved universe. The principle states that nature always hides any singularity, such as a black hole, inside an event horizon, which insulates the rest of the universe from seeing it.

progenitor "Ancestor" star of a given object; for example, the star that existed before a supernova explosion is the supernova's progenitor.

prograde motion Motion across the sky in the eastward direction.

prominence Loop or sheet of glowing gas ejected from an active region on the solar surface that then moves through the inner parts of the corona under the influence of the Sun's magnetic field. (p. 258)

proper motion The angular movement of a star across the sky, as seen from Earth, measured in seconds of arc per year. This movement is a result of the star's actual motion through space. (p. 272)

protein Molecule made up of amino acids that controls metabolism.

proton An elementary particle carrying a positive electric charge, a component of all atomic nuclei. The number of protons in the nucleus of an atom dictates what type of atom it is. (p. 46)

proton–proton chain The chain of fusion reactions, leading from hydrogen to helium, that powers main-sequence stars. (p. 262)

protoplanet Clump of material, formed in the early stages of solar system formation, that was the forerunner of the planets we see today. (p. 121)

protostar Stage in star formation when the interior of a collapsing fragment of gas is sufficiently hot and dense that it becomes opaque to its own radiation. The protostar is the dense region at the center of the fragment. (p. 307)

protosun The central accumulation of material in the early stages of solar system formations, the forerunner of the present-day Sun. (p. 121)

Ptolemaic model Geocentric solar system model, developed by the second century astronomer Claudius Ptolemy. It predicted with great accuracy the positions of the then known planets. (p. 27)

pulsar Object that emits radiation in the form of rapid pulses with a characteristic pulse period and duration. Charged particles, accelerated by the magnetic field of a rapidly rotating neutron star, flow along the magnetic field lines, producing radiation that beams outward as the star spins on its axis. (p. 350)

pulsating variable star A star whose luminosity varies in a predictable, periodic way. (p. 382)

P-waves Pressure waves from an earthquake that travel rapidly through liquids and solids.

pyrolific-release experiment Experiment to look for life on Mars. Radioactively tagged carbon dioxide was added to Martian soil specimens. Scientists looked for indications that the radioactive material had been absorbed.

Q

quantization The fact that light and matter on small scales behave in a discontinuous manner and manifest themselves in the form of tiny "packets" of energy, called quanta.

quantum fluctuation Temporary random change in the amount of energy at a point in space.

quantum gravity Theory combining general relativity with quantum mechanics.

quantum mechanics The laws of physics as they apply on atomic scales.

quark A fundamental matter particle that interacts via the strong force; basic constituent of protons and neutrons.

quark epoch Period when all heavy elementary particles (composed of quarks) were in thermal equilibrium with the cosmic radiation field.

quarter Moon Lunar phase in which the Moon appears as a half disk. (p. 12)

quasar Starlike radio source with an observed redshift that indicates an extremely large distance from Earth. The brightest nucleus of a distant active galaxy. (p. 424)

quasar feedback The idea that some of the energy released by a quasar in an active galactic nucleus heats and ejects the gas in the surrounding galaxy, shutting off the quasar fuel supply and suppressing star formation in the galaxy, connecting the growth of the central black hole to the properties of the host galaxy. (p. 445)

quasi-stellar object (QSO) *See* quasar.

quiescent prominence Prominence that persists for days or weeks, hovering high above the solar photosphere.

quiet Sun The underlying predictable elements of the Sun's behavior, such as its average photospheric temperature, which do not change in time.

R

radar Acronym for RAdio Detection And Ranging. Radio waves are bounced off an object, and the time taken for the echo to return indicates its distance. (p. 35)

radial motion Motion along a particular line of sight, which induces apparent changes in the wavelength (or frequency) of radiation received.

radial velocity Component of a star's velocity along the line of sight.

radian Angular measure equivalent to $180/\pi = 57.3$ degrees.

radiant Constellation from which a meteor shower appears to come.

radiation A way in which energy is transferred from place to place in the form of a wave. Light is a form of electromagnetic radiation.

radiation darkening The effect of chemical reactions that result when high-energy particles strike the icy surfaces of objects in the outer solar system. The reactions lead to a buildup of a dark layer of material.

radiation-dominated universe Early epoch in the universe, when the equivalent density of radiation in the cosmos exceeded the density of matter. (p. 473)

radiation era The first few thousand years after the Big Bang when the universe was small, dense, and dominated by radiation.

radiation zone Region of the Sun's interior where extremely high temperatures guarantee that the gas is completely ionized. Photons only occasionally interact with electrons and travel through this region with relative ease. (p. 246)

radio Region of the electromagnetic spectrum corresponding to radiation of the longest wavelengths. (p. 44)

radioactivity The release of energy by rare, heavy elements when their nuclei decay into lighter nuclei. (p. 147)

radio galaxy Type of active galaxy that emits most of its energy in the form of long-wavelength radiation. (p. 420)

radiograph Image made from observations at radio wavelengths.

radio lobe Roundish extended region of radio-emitting gas, lying well beyond the center of a radio galaxy. (p. 420)

radio telescope Large instrument designed to detect radiation from space at radio wavelengths. (p. 83)

radio wave Electromagnetic radiation with wavelength in the radio part of the spectrum (p. 44)

radius–luminosity–temperature relationship A mathematical proportionality, arising from Stefan's law, that allows astronomers to indirectly determine the radius of a star once its luminosity and temperature are known. (p. 279)

rapid mass transfer Mass transfer in a binary system that proceeds at a rapid and unstable rate, transferring most of the mass of one star onto the other.

ray The path taken by a beam of radiation.

Rayleigh scattering Scattering of light by particles in the atmosphere.

recession velocity Rate at which two objects are separating from one another.

recurrent nova Star that "goes nova" several times over the course of a few decades.

reddening Dimming of starlight by interstellar matter, which tends to scatter higher-frequency (blue) components of the radiation more efficiently than the lower-frequency (red) components. (p. 295)

red dwarf Small, cool, faint star at the lower-right end of the main sequence on the Hertzsprung–Russell diagram. (p. 280)

red giant A giant star whose surface temperature is relatively low so that it glows red. (p. 279)

red giant branch The section of the evolutionary track of a star corresponding to intense hydrogen shell burning, which drives a steady expansion and cooling of the outer envelope of the star. As the star gets larger in radius and its surface temperature cools, it becomes a red giant. (p. 324)

red giant region The upper-right corner of the Hertzsprung–Russell diagram, where red giant stars are found. (p. 281)

redshift Motion-induced change in the wavelength of light emitted from a source moving away from us. The relative recessional motion causes the wave to have an observed wavelength longer (and hence redder) than it would if it were not moving.

redshift survey Three-dimensional survey of galaxies, using redshift to determine distance.

red supergiant An extremely luminous red star. Often found on the asymptotic giant branch of the Hertzsprung–Russell diagram. (pp. 279, 332)

reflecting telescope A telescope that uses a mirror to gather and focus light from a distant object. (p. 70)

reflection nebula Bluish nebula caused by starlight scattering from dust particles in an interstellar cloud located just off the line of sight between Earth and a bright star.

refracting telescope A telescope that uses a lens to gather and focus light from a distant object. (p. 70)

refraction The tendency of a wave to bend as it passes from one transparent medium to another. (p. 70)

regolith Surface dust on the moon, several tens of meters thick in places, caused by billions of years of meteoritic bombardment.

relativistic Speeds comparable to the speed of light.

relativistic fireball Leading explanation of a gamma-ray burst in which an expanding region of superhot gas radiates in the gamma-ray part of the spectrum.

residual cap Portion of Martian polar ice caps that remains permanently frozen, undergoing no seasonal variations.

resonance Circumstance in which two characteristic times are related in some simple way, e.g., an asteroid with an orbital period exactly half that of Jupiter.

retina (eye) The back part of the eye onto which light is focused by the lens.

retrograde motion Backward, westward loop traced out by a planet with respect to the fixed stars. (p. 29)

revolution Orbital motion of one body about another, such as Earth about the Sun. (p. 7)

revolving *See* revolution.

Riemannian geometry Geometry of positively curved space (such as the surface of a sphere).

right ascension Celestial coordinate used to measure longitude on the celestial sphere. The zero point is the position of the Sun at the vernal equinox. (p. 6)

rille A ditch on the surface of the Moon where molten lava flowed in the past.

ring *See* planetary ring system.

ringlet Narrow region in Saturn's planetary ring system where the density of ring particles is high. *Voyager* discovered that the rings visible from Earth are actually composed of tens of thousands of ringlets. (p. 232)

Roche limit Often called the tidal stability limit, the Roche limit gives the distance from a planet at which the tidal force (due to the planet) between adjacent objects exceeds their mutual attraction. Objects within this limit are unlikely to accumulate into larger objects. The rings of Saturn occupy the region within Saturn's Roche limit. (p. 231)

Roche lobe An imaginary surface around a star. Each star in a binary system can be pictured as being surrounded by a teardrop-shaped zone of gravitational influence, the Roche lobe. Any material within the Roche lobe of a star can be considered to be part of that star. During evolution, one member of the binary system can expand so that it overflows its own Roche lobe and begins to transfer matter onto the other star.

rock Material made predominantly from compounds of silicon and oxygen.

rock cycle Process by which surface rock on Earth is continuously redistributed and transformed from one type into another.

rotation Spinning motion of a body about an axis. (p. 5)

rotation curve Plot of the orbital speed of disk material in a galaxy against its distance from the galactic center. Analysis of rotation curves of spiral galaxies indicates the existence of dark matter. (p. 394)

R-process "Rapid" process in which many neutrons are captured by a nucleus during a supernova explosion.

RR Lyrae variable Variable star whose luminosity changes in a characteristic way. All RR Lyrae stars have more or less the same average luminosity. (p. 382)

runaway greenhouse effect A process in which the heating of a planet leads to an increase in its atmosphere's ability to retain heat and thus to further heating, causing extreme changes in the temperature of the surface and the composition of the atmosphere. (p. 188)

runoff channel Riverlike surface feature on Mars, evidence that liquid water once existed there in great quantities. They are found in the southern highlands and are thought to have been formed by water that flowed nearly 4 billion years ago. (p. 177)

S

S0 galaxy Galaxy that shows evidence of a thin disk and a bulge, but that has no spiral arms and contains little or no gas. (p. 409)

Sagittarius A/Sgr A Strong radio source corresponding to the supermassive black hole at the center of the Milky Way.

Saros cycle Time interval between successive occurrences of the "same" solar eclipse, equal to 18 years, 11.3 days.

satellite A small body orbiting another larger body.

SB0 galaxy An S0-type galaxy whose disk shows evidence of a bar. (p. 409)

scarp Surface feature on Mercury believed to be the result of cooling and shrinking of the crust forming a wrinkle on the face of the planet. (p. 169)

Schmidt telescope Type of telescope having a very wide field of view, allowing large areas of the sky to be observed at once.

Schwarzschild radius The distance from the center of an object such that, if all the mass were compressed within that region, the escape speed would equal the speed of light. Once a stellar remnant collapses within this radius, light cannot escape and the object is no longer visible. (p. 362)

science A step-by-step process for investigating the physical world based on natural laws and observed phenomena. (p. 18)

scientific method The set of rules used to guide science, based on the idea that scientific "laws" be continually tested and modified or replaced if found inadequate. (p. 18)

scientific notation Expressing large and small numbers using power-of-10 notation.

seasonal cap Portion of Martian polar ice caps that is subject to seasonal variations, growing and shrinking once each Martian year.

seasons Changes in average temperature and length of day that result from the tilt of Earth's (or any planet's) axis with respect to the plane of its orbit. (p. 9)

secondary atmosphere The chemicals that composed Earth's atmosphere after the planet's formation, once volcanic activity outgassed chemicals from the interior. (p. 187)

sedimentary Rocks formed from the buildup of sediment.

seeing A term used to describe the ease with which good telescopic observations can be made from Earth's surface, given the blurring effects of atmospheric turbulence. (p. 81)

seeing disk Roughly circular region on a detector over which a star's pointlike images is spread, due to atmospheric turbulence. (p. 81)

seismic wave A wave that travels outward from the site of an earthquake through Earth. (p. 145)

seismology The study of earthquakes and the waves they produce in Earth's interior.

seismometer Equipment designed to detect and measure the strength of earthquakes (or quakes on any other planet).

selection effect Observational bias in which a measured property of a collection of objects is due to the way in which the measurement was made, rather than being intrinsic to the objects themselves. (p. 128)

self-propagating star formation Mode of star formation in which shock waves produced by the formation and evolution of one generation of stars triggers the formation of the next. (p. 393)

semidetached binary Binary system where one star lies within its Roche lobe but the other fills its Roche lobe and is transferring matter onto the first star.

semimajor axis One-half of the major axis of an ellipse. The semimajor axis is the way in which the size of an ellipse is usually quantified. (p. 33)

SETI Acronym for Search for Extraterrestrial Intelligence.

Seyfert galaxy Type of active galaxy whose emission comes from a very small region within the nucleus of an otherwise normal-looking spiral system. (p. 419)

shepherd satellite Satellite whose gravitational effect on a ring helps preserve the ring's shape. Examples are two satellites of Saturn, Prometheus and Pandora, whose orbits lie on either side of the F ring. (p. 233)

shield volcano A volcano produced by repeated nonexplosive eruptions of lava, creating a gradually sloping, shield-shaped low dome. Often contains a caldera at its summit. (p. 171)

shock wave Wave of matter, which may be generated by a newborn star or supernova, that pushes material outward into the surrounding molecular cloud. The material tends to pile up, forming a rapidly moving shell of dense gas.

short-period comet Comet with orbital period less than 200 years.

SI Système International d'unitès, the international system of metric units used to define mass, length, time, etc.

sidereal day The time needed between successive risings of a given star. (p. 7)

sidereal month Time required for the Moon to complete one trip around the celestial sphere. (p. 14)

sidereal year The time required for the constellations to complete one cycle around the sky and return to their starting points, as seen from a given point on Earth. Earth's orbital period around the Sun is 1 sidereal year. (p. 11)

single-line spectroscopic binary Binary system in which one star is too faint for its spectrum to be distinguished, so only the spectrum of the brighter star can be seen to shift back and forth as the stars orbit one another.

singularity A point in the universe where the density of matter and the gravitational field are infinite, such as at the center of a black hole. (p. 370)

sister/coformation theory (Moon) Theory suggesting that the Moon formed as a separate object close to Earth.

soft landing Use of rockets, parachutes, or packaging to break the fall of a space probe as it lands on a planet.

solar activity Unpredictable, often violent events on or near the solar surface, associated with magnetic phenomena on the Sun. (p. 254)

solar constant The amount of solar energy reaching Earth per unit area per unit time, approximately 1400 W/m^2. (p. 247)

solar core The region at the center of the Sun, with a radius of nearly 200,000 km, where powerful nuclear reactions generate the Sun's energy output.

solar cycle The 22-year period that is needed for both the average number of spots and the Sun's magnetic polarity to repeat themselves. The Sun's polarity reverses on each new 11-year sunspot cycle. (p. 256)

solar day The period of time between the instant when the Sun is directly overhead (i.e., noon) to the next time it is directly overhead. (p. 7)

solar eclipse Celestial event during which the new Moon passes directly between Earth and the Sun, temporarily blocking the Sun's light. (p. 14)

solar interior The region of the Sun between the solar core and the photosphere.

solar maximum Point of the sunspot cycle during which many spots are seen. They are generally confined to regions in each hemisphere, between about 15 and 20 degrees latitude.

solar minimum Point of the sunspot cycle during which only a few spots are seen. They are generally confined to narrow regions in each hemisphere at about 25–30 degrees latitude.

solar nebula The swirling gas surrounding the early Sun during the epoch of solar system formation, also referred to as the primitive solar system. (p. 119)

solar neutrino problem The discrepancy between the theoretically predicted flux of neutrinos streaming from the Sun as a result of fusion reactions in the core and the flux that is actually observed. The observed number of neutrinos is only about half the predicted number. (p. 264)

solar system The Sun and all the bodies that orbit it—Mercury, Venus, Earth, Mars, Jupiter, Saturn, Uranus, Neptune, their moons, the asteroids, the Kuiper belt, and the comets. (p. 102)

solar wind An outward flow of fast-moving charged particles from the Sun. (pp. 110, 246)

solstice *See* summer solstice; winter solstice.

south celestial pole Point on the celestial sphere directly above Earth's South Pole. (p. 6)

spacetime Single entity combining space and time in special and general relativity.

spatial resolution The dimension of the smallest detail that can be seen in an image.

special relativity Theory proposed by Einstein to deal with the preferred status of the speed of light.

speckle interferometry Technique whereby many short-exposure images of a star are combined to make a high-resolution map of the star's surface.

spectral class Classification scheme, based on the strength of stellar spectral lines, which is an indication of the temperature of a star. (p. 278)

spectral window Wavelength range in which Earth's atmosphere is transparent.

spectrograph Instrument used to produce detailed spectra of stars. Usually, a spectrograph records a spectrum on a CCD detector, for computer analysis.

spectrometer Instrument used to produce detailed spectra of stars. Usually, a spectrograph records a spectrum on a photographic plate, or more recently, in electronic form on a computer.

spectroscope Instrument used to view a light source so that it is split into its component colors. (p. 55)

spectroscopic binary A binary-star system that appears as a single star from Earth, but whose spectral lines show back-and-forth Doppler shifts as two stars orbit one another. (p. 285)

spectroscopic parallax Method of determining the distance to a star by measuring its temperature and then determining its absolute brightness by comparing with a standard Hertzsprung–Russell diagram. The absolute and apparent brightness of the star give the star's distance from Earth. (p. 283)

spectroscopy The study of the way in which atoms absorb and emit electromagnetic radiation. Spectroscopy allows astronomers to determine the chemical composition of stars. (p. 57)

spectrum The separation of light into its component colors.

speed Distance moved per unit time, independent of direction. *See also* velocity.

speed of light The fastest possible speed, according to the currently known laws of physics. Electromagnetic radiation exists in the form of waves or photons moving at the speed of light. (p. 47)

spicule Small solar storm that expels jets of hot matter into the Sun's lower atmosphere.

spin–orbit resonance State that a body is said to be in if its rotation period and its orbital period are related in some simple way.

spiral arm Distribution of material in a galaxy forming a pinwheel-shaped design, beginning near the galactic center. (p. 390)

spiral density wave A wave of matter formed in the plane of planetary rings, similar to ripples on the surface of a pond, that wraps around the rings, forming spiral patterns similar to grooves in a record disk. Spiral density waves can lead to the appearance of ringlets. (p. 392)

A proposed explanation for the existence of galactic spiral arms, in which coiled waves of gas compression move through the galactic disk, triggering star formation.

spiral galaxy Galaxy composed of a flattened, star-forming disk component that may have spiral arms and a large central galactic bulge. (p. 406)

spiral nebula Historical name for spiral galaxies, describing their appearance.

spring tide The largest tides, occurring when the Sun, Moon, and Earth are aligned at new and full Moon.

S-process "Slow" process in which neutrons are captured by a nucleus; the rate is typically one neutron capture per year.

standard candle Any object with an easily recognizable appearance and known luminosity, which can be used in estimating distances. Supernovae, which all have the same peak luminosity (depending on type), are good examples of standard candles and are used to determine distances to other galaxies. (p. 412)

standard solar model A self-consistent picture of the Sun, developed by incorporating the important physical processes that are believed to be important in determining the Sun's internal structure into a computer program. The results of the program are then compared with observations of the Sun, and modifications are made to the model. The standard solar model, which enjoys widespread acceptance, is the result of this process. (p. 247)

standard time System of dividing Earth's surface into 24 time zones, with all clocks in each zone keeping the same time.

star A glowing ball of gas held together by its own gravity and powered by nuclear fusion in its core. (pp. 4, 246)

starburst galaxy Galaxy in which a violent event, such as near collision, has caused an intense episode of star formation in the recent past. (p. 419)

star cluster A grouping of anywhere from a dozen to a million stars that formed at the same time from the same cloud of interstellar gas. Stars in clusters are useful to aid our understanding of stellar evolution because, within a given cluster, stars are all roughly the same age and chemical composition and lie at roughly the same distance from Earth. (p. 313)

Stefan's law Relation that gives the total energy emitted per square centimeter of its surface per second by an object of a given temperature. Stefan's law shows that the energy emitted increases rapidly with an increase in temperature, proportional to the temperature raised to the fourth power. (p. 52)

stellar epoch Most recent period, when stars, planets, and life have appeared in the universe.

stellar nucleosynthesis The formation of heavy elements by the fusion of lighter nuclei in the hearts of stars. Except for hydrogen and helium, all other elements in our universe resulted from stellar nucleosynthesis.

stellar occultation The dimming of starlight produced when a solar system object such as a planet, moon, or ring passes directly in front of a star. (p. 234)

stratosphere The portion of Earth's atmosphere lying above the troposphere, extending up to an altitude of 40–50 km. (p. 141)

string theory Theory that interprets all particles and forces in terms of particular modes of vibration of submicroscopic strings.

strong nuclear force Short-range force responsible for binding atomic nuclei together. The strongest of the four fundamental forces of nature. (p. 261)

S-type asteroid Asteroid made up mainly of silicate or rocky material.

subatomic particle Particle smaller than the size of an atomic nucleus.

subduction zone Place where two plates meet and one slides under the other.

subgiant branch The section of the evolutionary track of a star that corresponds to changes that occur just after hydrogen is depleted in the core, and core hydrogen burning ceases. Shell hydrogen burning heats the outer layers of the star, which causes a general expansion of the stellar envelope. (p. 324)

sublimation Process by which element changes from the solid to the gaseous state, without becoming liquid.

summer solstice Point on the ecliptic where the Sun is at its northernmost point above the celestial equator, occurring on or near June 21. (p. 8)

sunspot An Earth-sized dark blemish found on the surface of the Sun. The dark color of the sunspot indicates that it is a region of lower temperature than its surroundings. (p. 254)

sunspot cycle The fairly regular pattern that the number and distribution of sunspots follows, in which the average number of spots reaches a maximum every 11 or so years, then falls off to almost zero. (p. 256)

supercluster Grouping of several clusters of galaxies into a larger, but not necessarily gravitationally bound unit. (p. 448)

super-Earth An extrasolar planet with a mass between 2 and 10 times that of Earth. (p. 127)

superforce An attempt to combine the strong and electroweak forces into one single force.

supergiant A star with a radius between 100 and 1000 times that of the Sun. (p. 279)

supergranulation Large-scale flow pattern on the surface of the Sun, consisting of cells measuring up to 30,000 km across, believed to be the imprint of large convective cells deep in the solar interior. (p. 251)

superior conjunction Orbital configuration in which an inferior planet (Mercury or Venus) lies farthest from Earth (on the opposite side of the Sun).

supermassive black hole Black hole having a mass a million to a billion times greater than the mass of the Sun; usually found in the central nucleus of a galaxy.

supernova Explosive death of a star, caused by the sudden onset of nuclear burning (Type I), or an enormously energetic shock wave (Type II). One of the most energetic events of the universe, a supernova may temporarily outshine the rest of the galaxy in which it resides. (p. 335)

supernova remnant The scattered glowing remains from a supernova that occurred in the past. The Crab Nebula is one of the best-studied supernova remnants. (p. 337)

supersymmetric relic Massive particles that should have been created in the Big Bang if supersymmetry is correct.

surface gravity The acceleration due to gravity at the surface of a star or planet. (p. 136)

S-waves Shear waves from an earthquake, which can travel only through solid material and move more slowly than p-waves.

synchronous orbit State of an object when its period of rotations is exactly equal to its average orbital period. The Moon is in a synchronous orbit and so presents the same face toward Earth at all times. (p. 138)

synchrotron radiation Type of nonthermal radiation produced by high-speed charged particles, such as electrons, as they are accelerated in a strong magnetic field. (p. 427)

synodic month Time required for the Moon to complete a full cycle of phases. (p. 14)

synodic period Time required for a body to return to the same apparent position relative to the Sun, taking Earth's own motion into account; the time between one closest approach to Earth and the next (for a planet).

T

T Tauri phase star Protostar in the late stages of formation, often exhibiting violent surface activity. T Tauri phase stars have been observed to brighten noticeably in a short period of time, consistent with the idea of rapid evolution during this final phase of stellar formation. (p. 309)

tail Component of a comet that consists of material streaming away from the main body, sometimes spanning hundreds of millions of kilometers. May be composed of dust or ionized gases. (p. 110)

tau A type of lepton (along with the electron and muon).

tectonic fracture Cracks on a planet's surface, in particular on the surface of Mars, caused by internal geological activity.

telescope Instrument used to capture as many photons as possible from a given region of the sky and concentrate them into a focused beam for analysis. (p. 70)

temperature A measure of the amount of heat in an object, and an indication of the speed of the particles that comprise it. (p. 51)

tenuous Thin, having low density.

terminator The line separating night from day on the surface of the Moon or a planet.

terrae *See* highlands.

terrestrial planet One of the four innermost planets of the solar system, resembling Earth in general physical and chemical properties. (p. 104)

theoretical model An attempt to construct a mathematical explanation of a physical process or phenomenon within the assumptions and confines of a given theory. In addition to providing an explanation of the observed facts, the model generally also makes new predictions that can be tested by further observation or experimentation. (p. 18)

theories of relativity Einstein's theories, on which much of modern physics rests. Two essential facts of the theory are that nothing can travel faster than the speed of light and that everything, including light, is affected by gravity.

theory A framework of ideas and assumptions used to explain some set of observations and make predictions about the real world. (p. 18)

thermal equilibrium Condition in which new particle–antiparticle pairs are created from photons at the same rate as pairs annihilate one another to produce new photons.

thick disk Region of a spiral galaxy where an intermediate population of stars resides, younger than the halo stars but older than stars in the disk.

threshold temperature Critical temperature above which pair production is possible and below which pair production cannot occur.

tidal bulge Elongation of Earth caused by the difference between the gravitational force on the side nearest the Moon and the force on the side farthest from the Moon. The long axis of the tidal bulge points toward the Moon. More generally, the deformation of any body produced by the tidal effect of a nearby gravitating object. (p. 137)

tidal force The variation in one body's gravitational force from place to place across another body—for example, the variation of the Moon's gravity across Earth. (p. 137)

tidal locking Circumstance in which tidal forces have caused a moon to rotate at exactly the same rate at which it revolves around its parent planet, so that the moon always keeps the same face turned toward the planet. (p. 138)

tidal stability limit The minimum distance within which a moon can approach a planet before being torn apart by the planet's tidal force.

tidally locked Condition in which tidal forces have caused a body (such as a moon) to rotate at exactly the same rate at which it orbits another body so that it always keeps the same face toward the other body. (p. 139)

tides Rising and falling motion of terrestrial bodies of water, exhibiting daily, monthly, and yearly cycles. Ocean tides on Earth are caused by the competing gravitational pull of the Moon and Sun on different parts of Earth. (p. 137)

time dilation A prediction of the theory of relativity, closely related to the gravitational reshift. To an outside observer, a clock lowered into a strong gravitational field will appear to run slow. (p. 369)

time zone Region on Earth in which all clocks keep the same time, regardless of the precise position of the Sun in the sky, for consistency in travel and communications.

total eclipse Celestial event during which one body is completely blocked from view by another. (p. 14)

transit Orbital configuration where a planet (i.e., Mercury or Venus) passes between Earth and the Sun.

transition zone The region of rapid temperature increases that separates the Sun's chromosphere from the corona. (p. 246)

transverse motion Motion perpendicular to a particular line of sight, which does not result in Doppler shift in radiation received.

transverse velocity Component of star's velocity perpendicular to the line of sight.

triangulation Method of determining distance based on the principles of geometry. A distant object is sighted from two well-separated locations. The distance between the two locations and the angle between the line joining them and the line to the distant object are all that are necessary to ascertain the object's distance. (p. 17)

triple-alpha process The creation of carbon-12 by the fusion of three helium-4 nuclei (alpha particles). Helium-burning stars occupy a region of the Hertzsprung–Russell diagram known as the horizontal branch.

triple star system Three stars that orbit one another, bound together by gravity.

Trojan asteroid One of two groups of asteroids that orbit at the same distance from the Sun as Jupiter, 60 degrees ahead of and behind the planet. (p. 106)

tropical year The time interval between one vernal equinox and the next. (p. 11)

troposphere The portion of Earth's atmosphere from the surface to about 15 km. (p. 141)

trough Maximum departure of a wave below its undisturbed state.

true space motion True motion of a star, taking into account both its transverse and radial motion according to the Pythagorean theorem.

Tully-Fisher relation A relation used to determine the absolute luminosity of a spiral galaxy. The rotational velocity, measured from the broadening of spectral lines, is related to the total mass, and hence the total luminosity. (p. 413)

turnoff mass The mass of a star that is just now evolving off the main sequence in a star cluster.

21-centimeter line Spectral line in the radio region of the electromagnetic spectrum associated with a change of spin of the electron in a hydrogen atom. (p. 303)

21-centimeter radiation Radio radiation emitted when an electron in the ground state of a hydrogen atom flips its spin to become anti-parallel to the spin of the proton in the nucleus.

twin quasar Quasar that is seen twice at different locations in the sky due to gravitational lensing.

Type I supernova One possible explosive death of a star. A white dwarf in a binary star system can accrete enough mass that it cannot support its own weight. The star collapses and temperatures become high enough for carbon fusion to occur. Fusion begins throughout the white dwarf almost simultaneously, and an explosion results. (p. 335)

Type II supernova One possible explosive death of a star, in which the highly evolved stellar core rapidly implodes and then explodes, destroying the surrounding star. (p. 335)

U

ultraviolet Region of the electromagnetic spectrum, just beyond the visible range, corresponding to wavelengths slightly shorter than blue light. (p. 44)

ultraviolet telescope A telescope that is designed to collect radiation in the ultraviolet part of the spectrum. Earth's atmosphere is partially opaque to these wavelengths, so ultraviolet telescopes are put on rockets, balloons, and satellites to get high above most or all of the atmosphere. (p. 91)

umbra Central region of the shadow cast by an eclipsing body.

The central region of a sunspot, which is its darkest and coolest part. (p. 15)

unbound An orbit that does not stay in a specific region of space, but where an object escapes the gravitational field of another. Typical unbound orbits are hyperbolic in shape.

unbound trajectory Path of an object with launch speed high enough that it can escape the gravitational pull of a planet.

uncompressed density The density a body would have in the absence of any compression due to its own gravity.

universal time Mean solar time at the Greenwich meridian.

universe The totality of all space, time, matter, and energy. (p. 4)

unstable nucleus Nucleus that cannot exist indefinitely, but rather must eventually decay into other particles or nuclei.

upwelling Upward motion of material having temperature higher than the surrounding medium.

V

vacuum energy Property of empty space when it is excited above its normal zero-energy state. Astronomers think that the temporary appearance of vacuum energy was responsible for one or more periods of inflation during the early universe. (p. 476)

Van Allen belts At least two doughnut-shaped regions of magnetically trapped, charged particles high above Earth's atmosphere. (p. 154)

variable star A star whose luminosity changes with time. (p. 382)

velocity Displacement (distance plus direction) per unit time. *See also* speed.

vernal equinox Date on which the Sun crosses the celestial equator moving northward, occurring on or near March 21. (p. 11)

visible light The small range of the electromagnetic spectrum that human eyes perceive as light. The visible spectrum ranges from about 400 to 700 nm, corresponding to blue through red light. (p. 44)

visible spectrum The small range of the electromagnetic spectrum that human eyes perceive as light. The visible spectrum ranges from about 4000 to 7000 angstroms, corresponding to blue through red light.

visual binary A binary star system in which both members are resolvable from Earth. (p. 285)

void Large, relatively empty region of the universe around which superclusters and "walls" of galaxies are organized. (p. 449)

volcano Upwelling of hot lava from below Earth's crust to the planet's surface. (p. 146)

W

wane (referring to the Moon or a planet) To shrink. The Moon appears to wane, or shrink in size, for 2 weeks after full Moon.

warm longitudes (Mercury) Two opposite points on Mercury's equator where the Sun is directly overhead at aphelion. Cooler than the hot longitudes by 150 degrees.

water hole The radio interval between 18 cm and 21 cm, the respective wavelengths at which hydroxyl (OH) and hydrogen (H) radiate, in which intelligent civilizations might conceivably send their communication signals. (p. 500)

water volcano Volcano that ejects water (molten ice) rather than lava (molten rock) under cold conditions.

watt/kilowatt Unit of power: 1 watt (W) is the emission of 1 joule (J) per second; 1 kilowatt (kW) is 1000 watts.

wave A pattern that repeats itself cyclically in both time and space. Waves are characterized by the speed at which they move, their frequency, and their wavelength. (p. 44)

wavelength The distance from one wave crest to the next, at a given instant in time. (p. 45)

wave period The amount of time required for a wave to repeat itself at a specific point in space. (p. 45)

wave theory of radiation Description of light as a continuous wave phenomenon, rather than as a stream of individual particles.

wax (referring to the Moon or a planet) To grow. The Moon appears to wax, or grow in size, for 2 weeks after new Moon.

weakly interacting massive particle (WIMP) Class of subatomic particles that might have been produced early in the history of the universe; dark-matter candidates.

weak nuclear force Short-range force, weaker than both electromagnetism and the strong force, but much stronger than gravity; responsible for certain nuclear reactions and radioactive decays. (p. 262)

weight The gravitational force exerted on you by Earth (or the planet on which you happen to be standing).

weird terrain A region on the surface of Mercury with oddly rippled features. This feature is thought to be the result of a strong impact that occurred on the other side of the planet and sent seismic waves traveling around the planet, converging in the weird region.

white dwarf A dwarf star with sufficiently high surface temperature that it glows white. (p. 279)

white dwarf region The bottom-left corner of the Hertzsprung–Russell diagram, where white dwarf stars are found. (p. 281)

white oval Light-colored region near the Great Red Spot in Jupiter's atmosphere. Like the red spot, such regions are apparently rotating storm systems. (p. 203)

Wien's law Relation between the wavelength at which a blackbody curve peaks and the temperature of the emitter. The peak wavelength is inversely proportional to the temperature, so the hotter the object, the bluer its radiation. (p. 52)

winter solstice Point on the ecliptic where the Sun is at its southernmost point below the celestial equator, occurring on or near December 21. (p. 9)

wispy terrain Prominent light-colored streaks on Rhea, one of Saturn's moons.

X

X-ray Region of the electromagnetic spectrum corresponding to radiation of high frequency and short wavelength, far beyond the visible spectrum. (p. 44)

X-ray burster X-ray source that radiates thousands of times more energy than our Sun in short bursts lasting only a few seconds. A neutron star in a binary system accretes matter onto its surface until temperatures reach the level needed for hydrogen fusion to occur. The result is a sudden period of rapid nuclear burning and release of energy. (p. 354)

X-ray nova Nova explosion detected at X-ray wavelengths.

Z

Zeeman effect Broadening or splitting of spectral lines due to the presence of a magnetic field.

zero-age main sequence The region on the Hertzsprung–Russell diagram, as predicted by theoretical models, where stars are located at the onset of nuclear burning in their cores. (p. 312)

zodiac The 12 constellations on the celestial sphere through which the Sun appears to pass during the course of a year. (p. 8)

zonal flow Alternating regions of westward and eastward flow, roughly symmetrical about the equator of Jupiter, associated with the belts and zones in the planet's atmosphere. (p. 202)

zone Bright, high-pressure region in the atmosphere of a jovian planet, where gas flows upward. (p. 201)

Answers to Check Questions
Concept Checks and Process of Science Checks

Chapter 0

1. (p. 6) (1) Because the celestial sphere provides a natural means of specifying the locations of stars on the sky. Celestial coordinates are directly related to Earth's orientation in space but are independent of Earth's rotation. (2) Distance information is lost. 2. (p. 11) Because of the tilt of Earth's rotation axis, the Sun is highest in the northern sky during northern summer, and the days are longest. 3. (p. 15) (1) The angular size of the Moon would remain the same, but that of the Sun would be halved, making it easier for the Moon to eclipse the Sun. We would expect to see total or partial eclipses, but no annular ones. (2) If the distance were halved, the angular size of the Sun would double, and total eclipses would never be seen—only partial or annular ones. 4. (p. 18) Because astronomical objects are too distant for direct ("measuring tape") measurements, we must rely on indirect means and mathematical reasoning. 5. (p. 21) A theory can never become proven "fact," because it can always be invalidated, or forced to change by a single contradictory observation. However, once a theory's predictions have been repeatedly verified by experiments over many years, it is often widely regarded as "true."

Chapter 1

1. (p. 29) In the geocentric view, retrograde motion is a real backward motion of a planet as it moves on its epicycle. In the heliocentric view, the backward motion is only apparent, caused by Earth "overtaking" the planet in its orbit. 2. (p. 30) Mainly simplicity and elegance. Both theories made testable predictions, and until Newton developed his laws, neither could explain *why* the planets move as they do. However, Copernicus's model was much simpler than the Ptolemaic version, which became more and more convoluted as observations improved. 3. (p. 32) The discovery of the phases of Venus could not be reconciled with the geocentric model. Observations of Jupiter's moons proved that some objects in the universe did not orbit Earth. 4. (p. 33) Galileo was an experimentalist; Kepler was a theorist. Galileo found observational evidence supporting the Copernican theory; Kepler found an empirical description of the motions of the planets within the Copernican picture that greatly simplified the theoretical view of the solar system. 5. (p. 35) Because those planets were unknown to Kepler. The fact that the laws apply to the outer planets then constitutes a prediction subsequently verified by observation. 6. (p. 35) Because Kepler determined the overall geometry of the solar system by triangulation using Earth's orbit as a baseline, all distances were known only relative to the scale of Earth's orbit—the astronomical unit (AU). 7. (p. 38) In the absence of any force, a planet would move in a straight line with constant velocity (Newton's first law) and therefore tends to move along the tangent to its orbital path. The Sun's gravity causes the planet to accelerate toward the Sun (Newton's second law), bending its trajectory into the orbit we observe. 8. (p. 39) Kepler's laws were accurate descriptions of planetary motion based on observations, but they contained no insight into *why* the planets orbited the Sun or why the orbits are as they are. Newtonian mechanics explained the orbits in terms of universal laws, and in addition made detailed predictions about the motion of other bodies in the cosmos— moons, comets, other stars—that were hitherto impossible to make.

Chapter 2

1. (p. 46) A wave is a way in which energy or information is transferred from place to place via a repetitive, regularly varying disturbance of some sort. A wave is characterized by its wavelength, frequency, velocity, and amplitude: the product of the first two always equals the third. 2. (p. 48) Light is an electromagnetic wave produced by accelerating charged particles. All waves on the list are characterized by their wave periods, frequencies, and wavelengths, and all carry energy and information from one location to another. Unlike waves in water or air, however, light waves require no physical medium through which to propagate. 3. (p. 50) All are electromagnetic radiation and travel at the speed of light. In physical terms, they differ only in frequency (or wavelength), although their effects on our bodies (or our detectors) differ greatly. 4. (p. 54) As the switch is turned and the temperature of the filament rises, the bulb's brightness increases rapidly, by Stefan's law, and its color shifts, by Wien's law, from invisible infrared to red to yellow to white. 5. (p. 56) They are characteristic frequencies (wavelengths) at which matter absorbs or emits photons of electromagnetic radiation. They are unique to each atom or molecule and thus provide a means of identifying the gas producing them. 6. (p. 59) Electron orbits can occur only at certain specific energies, and there is a ground state that has the lowest possible energy. Planetary orbits can have any energy. Planets can stay in any orbit indefinitely; in an atom the electron must eventually fall to the ground state, emitting electromagnetic radiation in the process. Planets may reasonably be thought of as having specific trajectories around the Sun—there is no ambiguity as to "where" a planet is. Electrons in an atom are smeared out into an electron cloud, and we can only talk of the electron's location in probabilistic terms. 7. (p. 61) The wave theory could not explain why atoms emit and absorb radiation at specific, unique wavelengths, rather than as a continuous beam. The key new insight of quantum theory was that, on microscopic scales, the properties of both matter and radiation are *quantized*. Specifically, the energy

of an electron in an atom can only take on certain definite values, unique to the type of atom, and radiation exists in the form of *photons* of specific energies. When an electron changes energy within an atom, it does so by emitting or absorbing a photon of a definite energy, and hence *color*, directly connecting observations of atomic spectra to the particular atoms producing them. 8. (p. 62) Spectral lines correspond to transitions between specific orbitals within an atom. The structure of an atom determines the energies of these orbitals, and hence the possible transitions, and thus the energies (colors) of the photons involved. 9. (p. 64) Measuring masses in astronomy usually entails measuring the orbital speed of one object—a companion star or a planet, perhaps—around another. In most cases, the Doppler effect is an astronomer's only way of making such measurements. 10. (p. 65) Because, with few exceptions, spectral analysis is the only way we have of determining the physical conditions—composition, temperature, density, velocity, etc.—in a distant object. Without spectral analysis, astronomers would know next to nothing about the properties of stars and galaxies.

Chapter 3

1. (p. 78) Because large reflecting telescopes are easier to design, build, and maintain than refracting instruments. 2. (p. 81) The need to gather as much light as possible, and the need to achieve the highest possible angular resolution. 3. (p. 82) The atmosphere absorbs some radiation arriving from space, and turbulent motion blurs the incoming rays. To reduce or overcome the effects of atmospheric absorption, instruments are placed on high mountains or in space. To compensate for atmospheric turbulence, adaptive optics techniques probe the air above the observing site and adjust the mirror surface accordingly to try to recover the undistorted image. 4. (p. 85) Radio astronomy opens up a new window on the universe, allowing astronomers to study known objects in new ways and to discover new objects that would otherwise be completely unobservable. The faintness of the sources is addressed by combining the largest available telescopes with sensitive detectors. Resolution can be greatly improved by the use of interferometry, which combines the signals from two or more separate telescopes to create the effect of a single instrument of much larger diameter. 5. (p. 92) The text presents multiwavelength observations of stars, star-forming regions, and galaxies. We must study astronomical objects at many wavelengths because most emit radiation across the entire spectrum, and much of what we know is learned from studies of invisible radiation. 6. (p. 94) Benefits: They are above the atmosphere, so they are unaffected by seeing or absorption; they can also make round-the-clock observations of the entire sky. Drawbacks: Cost/smaller size, inaccessibility, vulnerability to damage by radiation and cosmic rays.

Chapter 4

1. (p. 105) Because the two classes of planets differ in almost every physical property— orbit, mass, radius, composition, existence of rings, number of moons. 2. (p. 109) Similarities: All orbit in the inner solar system, roughly in the ecliptic plane, and are solid bodies of generally "terrestrial" composition. Differences: Asteroids are much smaller than the terrestrial planets, are generally less evolved than the planets, and move on much less regular orbits. 3. (p. 113) Most comets never come close enough to the Sun for us to see them. 4. (p. 116) Because it is thought to be much less evolved than material now found in planets, and hence it is a better indicator of conditions in the early solar system. 5. (p. 119) Because it must explain certain general features of solar system architecture, while accommodating the fact that there are exceptions to many of them. 6. (p. 124) Yes. If a star formed with a disk of matter around it, then even if it doesn't have planets like Earth, the basic processes of condensation and accretion would probably have occurred there, too. 7. (p. 131) Because the search techniques are most sensitive to massive planets orbiting close to their parent stars, which are exactly what have been observed.

Chapter 5

1. (p. 140) A tidal force is the *variation* in one body's gravitational force from place to place across another. Tidal forces tend to deform a body rather than causing an overall acceleration, and they decrease rapidly with distance. 2. (p. 145) It raises Earth's average surface temperature above the freezing point of water, which was critical for the development of life on our planet. Should the greenhouse effect continue to strengthen, however, it may conceivably cause catastrophic climate changes on Earth. 3. (p. 146) Earthquakes and seismic waves are observed on Earth's surface. These data are combined with models of how waves move through Earth's interior to construct a model of our planet's composition and physical state. We measure Earth's mass, radius, and surface composition directly. However, the composition and temperature of our planet's interior are inferred from models. 4. (p. 148) We would have far less detailed direct (from volcanoes) or indirect (from seismic studies following earthquakes) information on our planet's interior. 5. (p. 150) Convection currents in the upper mantle cause portions of Earth's crust—plates—to slide around on the surface. As the plates move and interact, they are responsible for volcanism, earthquakes, the formation of mountain ranges and ocean

trenches, and the creation and destruction of oceans and continents. 6. (p. 151) The maria are younger, denser, and are much less heavily cratered than the highlands. 7. (p. 156) It tells us that the planet has a conducting, liquid core in which the magnetic field is continuously generated. 8. (p. 158) The Moon's small size, which is the main reason that the Moon (1) has lost its initial heat, making it geologically dead today, and (2) has insufficient gravity to retain an atmosphere.

Chapter 6

1. (p. 166) The Sun's tidal force has caused Mercury's rotation period to become exactly 2/3 of the orbital period and the rotation axis to be oriented exactly perpendicular to the planet's orbit plane. A synchronous orbit is not possible because of Mercury's eccentric orbit around the Sun. 2. (p. 168) Venus's atmosphere is much more massive, hotter, and denser than that of Earth, and is composed almost entirely of carbon dioxide. 3. (p. 169) Scarps are thought to have been formed when the planet's iron core cooled and contracted, causing the crust to crack. Faults on Earth are the result of tectonic activity. 4. (p. 173) No. They are mainly shield volcanoes formed by lava upwelling through "hot spots" in the crust and are not associated with plate tectonics (which is not observed on Venus). 5. (p. 177) Mercury: improved observations of the planet's surface led to a major revision of Mercury's rotation rate. Venus: observations of the atmosphere initially suggested a relatively mild climate; however, later radio studies of the surface revealed a hot, dry, inhospitable world. Mars: with improving observations, astronomers' ideas of the surface have evolved from an arid desert irrigated by canals built by intelligent beings, to a world now dry and geologically dead, but that once had liquid water and possibly even life on its surface. 6. (p. 184) Observations reveal evidence for dried-up rivers, outflow channels, and what may be ancient oceans, strongly suggesting that the surface was once warm enough and the atmosphere thick enough for large bodies of liquid water to exist on the planet's surface. Today most of the remaining water on Mars is frozen, underground or at the poles. 7. (p. 186) Venus probably has a large liquid metal core, but it rotates too slowly for a planetary dynamo to operate. Mars rotates rapidly but lacks a liquid metal core. 8. (p. 189) Both might have climates quite similar to that on Earth, as the greenhouse effect would probably not have run away on Venus, and the water in the Martian atmosphere might not have frozen and become permafrost.

Chapter 7

1. (p. 198) Uranus's orbit was observed to deviate from a perfect ellipse, leading astronomers to try to compute the mass and location of the body responsible for those discrepancies. That body was Neptune. 2. (p. 200) The magnetic field is generated by the motion of electrically conducting liquid in the deep interior and therefore presumably shares the rotation of that region of the planet. 3. (p. 207) Like weather systems on Earth, the belts and zones are regions of high and low pressure and are associated with convective motion. However, unlike storms on Earth, they wrap all the way around the planet because of Jupiter's rapid rotation. In addition, the clouds are arranged in three distinct layers, and the bright colors are the result of cloud chemistry unlike anything operating in Earth's atmosphere. Jupiter's spots are somewhat similar to hurricanes on Earth, but they are far larger and longer-lived. 4. (p. 207) Uranus's low temperature means that the planet's clouds lie deeper in the atmosphere, making them harder to distinguish from Earth. 5. (p. 211) We have no direct observations of the interiors of these worlds. Everything we know is derived from theoretical models of their interiors, coupled with observations of their masses, radii, and surface properties. 6. (p. 213) They all combine the two essential ingredients needed to operate a planetary dynamo: (1) rapid rotation, in all cases faster than Earth's rotation, and (2) conducting liquid interiors: metallic hydrogen in the cases of Jupiter and Saturn and "slushy" ice in the cases of Uranus and Neptune.

Chapter 8

1. (p. 223) Jupiter's gravitational field, via its tidal effect on the moons. 2. (p. 227) The haze layers high in the moon's dense atmosphere are opaque to visible light. The *Huygens* probe landed on the surface, and *Cassini*'s radar and infrared sensors can penetrate the haze. 3. (p. 230) They are all very heavily cratered—a result of the heavy bombardment that marked the clearing of the comets from the outer solar system. 4. (p. 235) The Roche limit is the radius inside of which a moon will be torn apart by tidal forces. Planetary rings are found inside the Roche limit, (most) moons outside. 5. (p. 237) They have similar masses, radii, composition, and perhaps also similar origins in the Kuiper belt. 6. (p. 239) Pluto, like Ceres, Eris, and a few other bodies in the outer solar system, orbits the Sun and is large enough for its gravity to have caused it to assume an approximately spherical shape, but it is not massive enough to have cleared its orbital path of other solar system objects. In current terminology, it is called a dwarf planet, or a plutoid.

Chapter 9

1. (p. 247) When we multiply the solar constant by the total area to obtain the solar luminosity, we are implicitly assuming that the same amount of energy reaches every square meter of the large sphere in Figure 9.3. 2. (p. 250) The energy may be carried in the form of (1) radiation, where energy travels in the form of light, and (2) convection, where energy is carried by physical motion of upwelling solar gas. 3. (p. 253) The coronal spectrum shows (1) emission lines of (2) highly ionized elements, implying high temperature. 4. (p. 261) There is a strong field, with a well-defined east-west organization, just below the surface. The field direction in the southern hemisphere is opposite

that in the north. However, the details of the fields are very complex. 5. (p. 263) Because the Sun shines by nuclear fusion, which converts mass into energy as hydrogen turns into helium. 6. (p. 264) The solar neutrino problem was the fact that theory predicted that significantly more neutrinos should be detected coming from the Sun than were actually observed experimentally. Both the theory of solar nuclear fusion and the experimental apparatus on Earth were carefully checked and tested, and neither was found to be at fault. In the end, a new piece of the theory, explaining how neutrinos could change from one type to another en route from the core of the Sun to the detectors on Earth, accounted for the discrepancy and solved the problem.

Chapter 10

1. (p. 271) Because stars are so far away that their parallaxes relative to any baseline on Earth are too small to measure accurately. 2. (p. 274) Nothing—We need to know their distances before the luminosities (or absolute magnitudes) can be determined. 3. (p. 278) Even though the theory behind the classification was incorrect, it provided a very useful way to organize and categorize the large database of stellar observations, making it easy for astronomers to adapt and interpret the data when an improved understanding of atomic physics emerged some three decades later. 4. (p. 278) Because temperature controls which excited states the star's atoms and ions are in, and hence which atomic transitions are possible. 5. (p. 279) Yes, by using the radius–luminosity–temperature relationship, but only if we can find a method of determining the luminosity that doesn't depend on the inverse-square law (Sec. 10.2). 6. (p. 282) Because giants are intrinsically very luminous and can be seen to much greater distances than the more common main-sequence or white dwarfs. 7. (p. 283) All stars would be further away, but their measured spectral types and apparent brightnesses would presumably be unchanged, so their luminosities would be greater than previously thought. The main sequence would therefore move vertically upward in the H–R diagram. We would then use larger luminosities in the method of spectroscopic parallax, so distances inferred by that method would also increase. 8. (p. 284) We are assuming (1) that we can tell with some confidence whether or not an observed star lies on the main sequence, and (2) that the radius of a main-sequence star of a given spectral type is fairly well determined. Both assumptions are reasonable, although significant uncertainties exist, especially with assumption (2). 9. (p. 289) We don't. We assume that their masses are the same as similar stars found in binaries.

Chapter 11

1. (p. 297) Because the scale of interstellar space is so large that even very low densities can accumulate to a large amount of obscuring matter along the line of sight to a distant star. 2. (p. 300) Because the UV light is absorbed by hydrogen gas in the surrounding cloud, ionizing it to form the emission nebulae. The red light is H radiation, part of the visible hydrogen spectrum emitted as electrons, and protons recombine to form hydrogen atoms. 3. (p. 304) Because the cloud's main constituent, hydrogen, is very hard to observe, astronomers must use other molecules as tracers of the cloud's properties. 4. (p. 310) (1) The existence of a photosphere, meaning that the inner part of the cloud becomes opaque to its own radiation, signaling the slowing of the collapse phase. (2) Nuclear fusion in the core, and equilibrium between pressure and gravity. 5. (p. 312) No. Different parts of the main sequence correspond to stars of different masses. A typical star stays at roughly the same location on the main sequence most of its lifetime. 6. (p. 316) Because stars form at different rates—high-mass stars reach the main sequence and start disrupting the parent cloud long before lower-mass stars have finished forming.

Chapter 12

1. (p. 323) Hydrostatic equilibrium means that, if some property of the Sun changes a little, the star's structure adjusts to compensate. Small changes in the Sun's internal temperature or pressure will not lead to large changes in its radius or luminosity. 2. (p. 325) Because the nonburning inner core, unsupported by pressure, begins to shrink, releasing gravitational energy, heating the overlying layers, causing them to burn more vigorously, and increasing the luminosity. 3. (p. 331) No. It does not have a binary companion to provide mass after the Sun becomes a white dwarf, so our star will never become a nova or Type I supernova; it is too low in mass to become a Type II supernova. 4. (p. 334) Because iron cannot fuse to produce energy. As a result, no further nuclear reactions are possible, and the core's equilibrium cannot be restored. 4. (p. 338) Because the two types of supernova differ in their spectra and their light curves, making it impossible to explain them in terms of a single phenomenon. 5. (p. 342) Because a star cluster gives us a "snapshot" of stars of many different masses—but of the same age and initial composition—allowing us to directly test the predictions of the theory. 6. (p. 343) Because it is responsible for creating and dispersing all the heavy elements out of which we are made. In addition, it may also have played an important role in triggering the collapse of the interstellar cloud from which our solar system formed.

Chapter 13

1. (p. 350) No. Only Type II supernovae. According to theory, the rebounding central core of the original star becomes a neutron star. 2. (p. 352) Because (1) not all supernovae form neutron stars, (2) the emission is beamed and may not be observable from Earth, and (3) pulsars spin down and become too faint to observe after a few tens of millions of years. 3. (p. 356) Some X-ray sources are binaries containing accreting neutron stars, which may be in the process of being spun up to form millisecond pulsars. 4. (p. 360)

They are energetic bursts of gamma rays, roughly uniformly distributed on the sky, occurring roughly once per day. They pose a challenge because they are very distant, and hence extremely luminous, but their energy comes from a region less than a few hundred kilometers across. 5. (p. 362) Its gravity becomes so strong that even light cannot escape, and it becomes a black hole. 6. (p. 368) In Newton's theory, gravity is a force that exists between all massive bodies. In relativity, the acceleration we call gravity is actually our perception of motion through a curved spacetime, whose curvature depends on the local density of material. Special relativity makes some quite counterintuitive predictions about the universe—for example, time dilation, length contraction, and the loss of simultaneity—which scientists initially found hard to accept. However, all of its predictions have been repeatedly verified in laboratory experiments, and many predictions of general relativity have been verified by astronomical observations. 7. (p. 370) Because the object would appear to take infinitely long to reach the event horizon, and its light would be infinitely redshifted by the time it got there. 8. (p. 372) By observing their gravitational effects on other objects, and from the X-rays emitted when matter falls into them.

Chapter 14

1. (p. 380) The Milky Way is the thin plane of the Galactic disk, seen from within. When our line of sight lies in the plane of the Galaxy, we see many stars blurring into a continuous band. In other directions we see darkness. 2. (p. 384) No, because even the brightest Cepheids are unobservable at distances of more than a kiloparsec or so through the obscuration of interstellar dust. 3. (p. 384) The debate involved the size and location of the "spiral nebulae," such as M31 in Andromeda. Today we know them as spiral galaxies, but at the time astronomers had insufficient data to determine their distances, and hence their sizes. Shapley thought that the spiral nebulae were small, relatively nearby objects contained within our Galaxy. Curtis argued that the nebulae were distant galaxies comparable in size to our own. Shapley used observations of variable stars to argue that our Galaxy is large; Curtis thought it was much smaller. In the end, neither scientist was completely right. Shapley was correct about the size of the Galaxy, but Curtis was right in his assertion that the spiral nebulae are distant galaxies like the Milky Way. The matter was resolved a few years later when Edwin Hubble used observations of variable stars to demonstrate that many spiral nebulae did indeed reside far outside our Galaxy. 4. (p. 387) They differ greatly in their spatial distributions of stars, the types and ages of stars found in each, the amount of interstellar gas they contain, and in the orbital motions of their component stars. 5. (p. 389) Because all the gas and dust in the halo fell to the Galactic disk billions of years ago, halo star formation has long since ceased. 6. (p. 391) Because differential rotation would destroy the spiral structure within a few hundred million years. 7. (p. 395) Scientists are unwilling to suggest new physics to explain observations, and competing theories of galactic rotation curves and "missing mass" on large scales do exist. However, despite their reservations, most astronomers have concluded that dark matter, introduced as material with gravity but no interaction with electromagnetic radiation, best fits the observed facts. Several theories exist to explain how dark matter might have been formed in the early universe, and they lead to experimental tests that should some day detect dark matter if it exists. 8. (p. 396) Its emission has not been observed at any electromagnetic wavelength. Its presence is inferred from its gravitational effect on stars and gas orbiting the Galactic center. 9. (p. 400) Observations of rapidly moving stars, gas, and the variability of the radiation emitted suggest the presence of a roughly 3 million-solar-mass black hole.

Chapter 15

1. (p. 411) Most galaxies are not large spirals—the most common galaxy types are dwarf ellipticals and dwarf irregulars. 2. (p. 412) Distance-measurement techniques ultimately rely upon the existence of very bright objects whose luminosities can be inferred by other means. Such objects become increasingly hard to find and calibrate the farther we look out into intergalactic space. 3. (p. 417) It doesn't use the inverse-square law. The other methods all provide a way of determining the luminosity of a distant object, which then is converted to a distance using the inverse-square law. Hubble's law gives a direct connection between redshift and distance. 4. (p. 419) Because it means that the energy source cannot simply be the summed energy of a huge number of stars—some other mechanism must be at work. 5. (p. 425) When quasars were discovered, they were thought to be faint, relative nearby stars, or starlike objects, although their unusual spectra posed problems for astronomers. Once astronomers realized that their odd spectra actually meant that they had very large redshifts, it became clear that quasars were actually among the most distant—and hence most luminous—objects in the entire universe. 6. (p. 427) The energy is generated in an accretion disk in the central nucleus of the visible galaxy, then transported by jets out of the galaxy and into the lobes, where it is eventually emitted by the synchrotron process in the form of radio waves.

Chapter 16

1. (p. 436) First, that the galaxies are gravitationally bound to the cluster. Second, and more fundamentally, that the laws of physics as we know them in the solar system—gravity, atomic structure, the Doppler effect—all apply on very large scales, and to systems possibly containing a lot of dark matter. 2. (p. 442) Galaxies in voids are much less likely to encounter other galaxies, so their evolution will be dictated more by "internal" processes—star formation and stellar evolution—than galaxies in clusters, which will be more affected by interactions with other galaxies. 3. (p. 443) The scenario makes predictions about how

different types of galaxies were assembled over the history of the universe, and this has consequences for the types, compositions, and ages of the stars they contain. Astronomers can gauge the accuracy of the model by comparing its predictions with the properties of spiral and elliptical galaxies at different redshifts (that is, at different times in the past). 4. (p. 446) Astronomers think that the first black holes formed in young galaxies early in the history of the universe. The black holes grew by accretion of stars and gas from the surrounding galaxies, as well as mergers with other black holes when their parent galaxies merged. Eventually, they form the supermassive black holes observed in normal galaxies today. The galaxy mergers were often responsible for directing additional galactic gas onto the central black holes. The connection with active galaxies is that, during the periods of accretion, the black holes emitted large amounts of energy, becoming the active galactic nuclei we see. 5. (p. 447) Probably not. There may well be galaxies that do not harbor central black holes, and in any case, only galaxies in clusters are likely to experience the encounters that trigger activity. 6. (p. 449) Light traveling from these objects to Earth is influenced by cosmic structure all along the line of sight. Light rays are deflected by the gravitational field of intervening concentrations of mass, and gas along the line of sight produces absorption lines whose redshifts tell us the distance at which each feature formed. The light received on Earth thus gives astronomers a "core sample" through the universe, from which detailed information can be extracted.

Chapter 17

1. (p. 460) When viewed on very large scales—more than 300 Mpc—the distribution of galaxies seems to be roughly the same everywhere and in all directions. (In addition, the microwave background radiation field is observed to be isotropic to very high precision.) 2. (p. 463) Because, tracing the motion backward in time, it implies that all galaxies, and in fact everything in the entire universe, were located at a single point at the same instant in the past. The cosmological principle does not allow that point to be special, leaving a Big Bang as the only possibility. 3. (p. 464) If there is enough matter in the universe for gravity to halt and reverse the current expansion, the universe will eventually recollapse in a "Big Crunch." Otherwise, the universe will expand forever. 4. (p. 465) A low-density universe has negative curvature and is infinite in extent; a critical-density universe is spatially flat (Euclidean) and is also infinite in extent; a high-density universe has positive curvature and is finite in extent. 5. (p. 469) The observed acceleration of the expansion of the universe implies that some nongravitational force must be at work. Dark energy is our current best theory of the cause of that force. In addition, galactic and cosmological observations indicate that the universe is spatially flat, and hence has critical density, but that the density of matter (mostly dark) cannot account for the total. The amount of dark energy needed to account for the cosmic acceleration is in good agreement with the required extra density—about 70 percent of the critical value. 6. (p. 471) There doesn't seem to be enough matter to halt the collapse, and in addition, the observed cosmic acceleration suggests the existence of a large-scale repulsive force in the cosmos that also opposes recollapse. 7. (p. 472) It was formed at the time of the Big Bang. It is the electromagnetic remnant of the primeval fireball. It started off as a blackbody curve because it was the thermal emission of the hot, young cosmos. It has retained this form because the cosmological redshift has stretched all photons by the same factor, preserving the shape of the blackbody curve. 8. (p. 474) The amount of deuterium observed in the universe today implies that the present density of normal matter is at most a few percent of the critical value—much less than the density of dark matter inferred from dynamical studies. 9. (p. 478) Not directly. Inflation implies that the universe is flat and hence that the total cosmic density equals the critical value. However, the matter density seems to be only about one-third of the critical value, and the density of electromagnetic radiation (the microwave background) is a tiny fraction of the critical density. Together, these statements suggest that the remaining density is in the form of "dark energy," as described in Section 17.3. 10. (p. 480) Because dark matter is largely unaffected by the background radiation field, it was able to clump earlier than normal matter, forming the structures within which luminous matter subsequently formed the galaxies we see today. The gravity of the early dark-matter clumps caused tiny fluctuations in the temperature of the cosmic microwave background. The ripples we see today in the microwave background are the beginning of galaxy formation long ago.

Chapter 18

1. (p. 490) Not yet. The formation of complex molecules from simple ingredients by nonbiological processes has been repeatedly demonstrated, but no living cell or self-replicating molecule has ever been created. 2. (p. 492) Mars remains the most likely site because of its generally Earthlike makeup and because several lines of evidence suggest that the planet was warmer in the past and may have had liquid water on its surface. Europa and Titan also have properties that might have been conducive to the emergence of living organisms—liquid water under Europa's icy crust and an atmosphere and many complex molecules on Titan. 3. (p. 498) It breaks a complex problem up into simpler "astronomical," "biochemical," "anthropological," and "cultural" pieces, which can be analyzed separately. The analysis itself helps identify the types of stars where a search might be most fruitful. 4. (p. 499) We assume that a technological civilization will opt to work in the radio part of the spectrum, where Galactic absorption is least, and will choose a region where natural Galactic background "static" is minimized. We also assume that the 21-cm line of hydrogen will have significance to them, since hydrogen is the most common element in the universe, and that hydroxyl will also have special meaning if they evolved in a water environment as we did.

Answers to Self-Test Questions

Chapter 0
True or False?/Multiple Choice 1) F 2) F 3) T 4) F 5) T 6) T 7) F 8) b 9) b 10) c 11) a 12) c 13) d 14) b 15) a

Problems 1) Scorpius 2) 29.9 km, 1.08×10^5 km, 2.58×10^6 km 3) It would decrease by 8 minutes 4) (a) 33 arc minutes, (b) 33 arc seconds, (c) 0.55 arc seconds; 56.4 minutes 5) 1.0 km/s 6) It is! 7) 173 m 8) (a) 57,300 km, (b) 3.44×10^6 km, (c) 2.06×10^8 km 9) 2.9°

Chapter 1
True or False?/Multiple Choice 1) F 2) F 3) F 4) F 5) F 6) F 7) F 8) T 9) d 10) c 11) c 12) a 13) c 14) b 15) b

Problems 1) (a) 110 km, (b) 44,000 km, (c) 370,000 km 2) 700 s 3) 0.4 degrees retrograde 4) 3 AU, 0.333, 5.2 yr 5) 35 AU 6) Mercury's perihelion and aphelion distances are 0.307 AU and 0.467 AU, respectively. The difference is 0.159 AU. Alternatively, Mercury's aphelion distance is 52 percent larger than the perihelion distance. 7) 9.42×10^{-4} solar $= 1.88 \times 10^{27}$ kg 8) 9.50, 7.33, 1.49 m/s² 9) For a 70-kg person, force $= 684$ N $= 154$ pounds; we call it the person's weight 10) 1.7 km/s

Chapter 2
True or False?/Multiple Choice 1) F 2) T 3) T 4) F 5) T 6) T 7) T 8) a 9) d 10) a 11) b 12) c 13) d 14) a 15) d

Problems 1) 1480 m/s 2) 3 m 3) 23 Hz; radio 4) A; 3.25; 112 5) 310; 9.4 microns; infrared 6) 6.4×10^7 W/m²; 3.9×10^{26} W 7) 3×10^{10} 8) Second excited state: 3 possibilities: $2 \rightarrow 1$ (Hα, 656 nm), $2 \rightarrow$ ground (Lyβ, 103 nm), $1 \rightarrow$ ground (Lyα, 122 nm); Third excited state: 6 possibilities: those already listed, plus $3 \rightarrow 2$ (called Paschen α, 1876 nm), $3 \rightarrow 1$ (Hβ, 486 nm), $3 \rightarrow$ ground (Lyγ, 97.3 nm) 9) 137 km/s approaching 10) 1.94×10^{27} kg

Chapter 3
True or False?/Multiple Choice 1) F 2) F 3) T 4) T 5) F 6) F 7) F 8) d 9) d 10) c 11) a 12) d 13) b 14) b 15) c

Problems 1) 0.3 arc seconds; 6.8 pixels 2) 580 μm; longer than the 3 – 200 μm operating range 3) 6.7 minutes; 1.7 minutes 4) (a) 0.25″, (b) 0.01″ 5) 15 m 6) 160 light-years, using the formula in the text to compute the resolution 7) 3.3 minutes 8) (a) 36, (b) 0.61, (c) 0.012 light-years 9) 14.1 m, 16 m 10) (a) 0.003 arc seconds, (b) 0.005 arc seconds

Chapter 4
True or False?/Multiple Choice 1) F 2) T 3) F 4) F 5) F 6) T 7) F 8) d 9) b 10) d 11) a 12) c 13) b 14) a 15) c

Problems 1) 2×10^{21} kg, 0.03 percent of Earth's mass; 40 km 2) 1.1 AU, 2.0 AU 3) 3 kg 4) 1, 0.1, 0.01, 0.001 times Ceres's mass, or 1.6×10^{21} kg, 1.6×10^{20} kg, 1.6×10^{19} kg, 1.6×10^{18} kg. 5) (a) 11 million years, (b) 50 AU 6) The orbital period is 0.00836 years $=$ 3.05 days, so (for example) January 1, 2013, would be $14 \times 365 + 4$ (leap days) $+ 31 = $ 5145 days $= 1685$ periods after discovery. 7) 1.7, 0.8 Earth gravity 8) 890 AU 9) 2×10^8; once every 2.5 years 10) 0.25 AU

Chapter 5
True or False?/Multiple Choice 1) T 2) T 3) T 4) F 5) T 6) T 7) F 8) b 9) d 10) b 11) b 12) d 13) d 14) a 15) c

Problems 1) 5.3 m/s² $= 0.55$ times the actual value, 8.3 km/s 2) 30 kg 3) 21 m = 70 feet 4) 1.6×10^{-7} 5) (a) 8.4×10^{-3}, (b) 0.16, (c) 0.84, (d) 7.0×10^{-3} times Earth's mass 7) 45 percent 8) 43 minutes 9) Need at least 4.8×10^5 such craters, requiring at least 4.8 trillion years; rate must have been greater by a factor of about 1000 10) 93 m

Chapter 6
True or False?/Multiple Choice 1) T 2) F 3) T 4) T 5) T 6) F 7) F 8) b 9) b 10) b 11) c 12) b 13) b 14) a 15) c

Problems 1) 8.8 minutes 2) 4.0×10^{23} kg, (for a temperature of 700 K) or roughly 22 percent greater than it is now 3) 4.9×10^{20} kg, 98 times more massive than Earth's atmosphere, 10^{-4} times the mass of Venus 4) 400 km/h, 250 mph 5) (a) 12,100 km, (b) yes—the smallest detectable feature would be about 20 km across, much smaller than the largest impact features 6) 197 minutes, 94 minutes 7) 28 kg, assuming I weigh 70 kg on Earth 8) 10 m/s $= 22$ mph 9) 2.7×10^{17} kg; 4.5×10^{-7} times the mass of Mars; 5.5×10^{-4} times the mass of Venus's atmosphere 10) 16′, 2.7′; no—the minimum angular diameter of the Sun, as seen from Mars, is 19′

Chapter 7
True or False?/Multiple Choice 1) F 2) T 3) F 4) T 5) T 6) F 7) F 8) T 9) a 10) d 11) d 12) b 13) c 14) a 15) b

Problems 1) 150 km, or 0.002 planetary radii; 1100 km, or 0.04 planetary radii 2) 2.2×10^{17} newton; 1.4×10^{21} newton, or 6400 times larger 3) 4200 s $= 1.17$ hours; 84,000 km 4) 6×10^{-5} Jupiter masses, or 0.019 Earth masses 5) — 6) 1.5 percent of the actual mass 7) 10.5 days 8) 74 K 9) 1.7×10^{25} kg (2.8 Earth masses); 81 percent 10) 49 μm; infrared

Chapter 8
True or False?/Multiple Choice 1) T 2) T 3) T 4) T 5) F 6) T 7) T 8) F 9) a 10) c 11) c 12) b 13) d 14) c 15) b

Problems 1) 20.3 km/s, 7.6 km/s; Saturn is much more massive than Earth! 2) 2.3 Jupiter radii 3) For Io: acceleration difference $= 6.3 \times 10^{-3}$ m/s², surface gravity $= 1.8$ m/s², ratio $= 0.0035 = 0.35$ percent; For Moon: acceleration difference $= 2.2 \times 10^{-5}$ m/s², surface gravity $= 1.6$ m/s², ratio $= 1.4 \times 10^{-5}$ 4) 1.35 m/s² (Earth: 9.80 m/s²); 5) 2.6 km/s Io: 36′, Europa: 18′, Ganymede: 18′, Callisto: 9′, Sun: 6′; Yes 6) 38 km radius 7) 1.1×10^{18} 8) 6.8 times farther out 9) For a weight of 70 kg, 4.7 kg on Pluto, 2.1 kg on Charon 10) 11.1 hours; 28.700 km; 2.8 orbits

Chapter 9
True or False?/Multiple Choice 1) T 2) F 3) F 4) F 5) T 6) F 7) F 8) c 9) c 10) a 11) b 12) c 13) a 14) b 15) c

Problems 1) (a) 14,600 W/m²; (b) 52 W/m² 2) (a) 3000 km, (b) 1500, (c) 1/33 of the 167-minute orbital period 3) (a) 0.29 nm (hard X-ray), (b) 29 nm (far UV), (c) 290 nm (near UV) 4) 100 s = 17 min, comparable to the granule lifetime 5) 36 percent of the average solar value (64 percent less) 6) (a) radiation mass loss $= 4.3$ million tons/s $\approx$ twice the mass loss due to the solar wind; (b) 30 trillion years 7) 83 days 8) 300,000 years 9) 74 billion years 10) 4×10^{28}

Chapter 10
True or False?/Multiple Choice 1) F 2) F 3) F 4) F 5) T 6) F 7) T 8) d 9) b 10) b 11) c 12) d 13) c 14) b 15) b

Problems 1) 77 pc; 0.39″ 2) 47 km/s; 56 km/s 3) (a) 80 times the luminosity of the Sun; (b) 2 solar radii 4) B is 3 times farther away 5) A is 10 times farther away 6) 3.3×10^{-10} W/m²; 2.3×10^{-13} times the solar constant 7) factor of 16 trillion 8) 320 pc 9) 3.5, 2.3 solar masses 10) (a) 200 billion years, (b) 1 billion years, (c) 100 million years

Chapter 11
True or False?/Multiple Choice 1) F 2) F 3) F 4) T 5) F 6) T 7) T 8) d 9) b 10) c 11) d 12) b 13) b 14) c 15) b

Problems 1) 1.9 grams 2) 1428.6 MHz for a wavelength of 21 cm (but note that the precise wavelength is 21.11 cm, corresponding to frequency of 1420.4 MHz); frequency: 1428.8–1428.2 MHz, wavelength: 20.9965–21.0053 cm 3) 0.017 pc $= 3500$ AU 4) 4100, 9.0 5) escape speeds (km/s): 1.8, 1.1, 1.1, 0.80; average molecular speeds: 13.6, 14.0, 14.6, 14.2; No 6) yes, barely: the escape speed is 0.93 km/s, molecular speed is 0.35 km/s 7) (a) 200, (b) 2.3 solar radii 8) 8 9) 10^{-6} solar luminosities 10) 37 pc

Chapter 12
True or False?/Multiple Choice 1) T 2) T 3) F 4) T 5) F 6) T 7) T 8) F 9) a 10) b 11) a 12) b 13) b 14) c 15) d

Problems 1) 3.5×10^7 kg/m³; 5.7×10^{-4} kg/m³, smaller by a factor of 1.6×10^{-11}; solar central density is 150,000 kg/m³, smaller than the white dwarf density by a factor of 4.3×10^{-3} 2) 220 pc 3) (a) 2.9 solar masses, (b) 1.7 solar masses 4) 7.1 years, 63,000 years 5) 7300 km/s, 500,000 Earth gravities 6) 20, 3.2 Mpc; 1000 Mpc 7) 0.4 pc; no—there are no O or B stars (in fact no stars at all) within that distance of the Sun 8) apparent magnitude is −14.1, 1.6 magnitudes (4 times) brighter than the full Moon, 9.7 magnitudes (7600 times) brighter than Venus; yes—according to Appendix 3, several O and B stars, as well as Betelgeuse and some other red supergiants, lie within that distance 9) roughly 1000 km/s; not too bad an assumption—the Nebula is moving too fast to be affected much by gravity, although it is probably slowing down as it runs into the interstellar medium 10) 1.2×10^{44} J; 220 Earth masses

Chapter 13

True or False?/Multiple Choice 1) T 2) T 3) T 4) F 5) T 6) F 7) F 8) b 9) a 10) b 11) c 12) b 13) c 14) c 15) b

Problems 1) 1 million revolutions per day, or 11.6 revolutions per second 2) 3×10^{44} times higher—2.1×10^{16} for a 70-kg human; Moon's mass is greater by a factor of 3.5 million; asteroid mass is roughly 1.6×10^{12} kg (assuming spherical shape and density 3000 kg/m^3), smaller by a factor of 13,000 3) 1.9×10^{12} m/s^2 or 190 billion Earth gravities; 190,000 km/s, or 64 percent of the speed of light; 300,000 km/s 4) 1.6×10^{-5}, 1.6×10^{3}, 1.6×10^{11} solar luminosities; ultraviolet, X-rays, gamma rays; they might be very bright immediately after formation, but would become very faint as they cool; the coolest could be plotted at the bottom left of the diagram, as a very faint O-type "star" 5) 2.4×10^{44} J, roughly twice the Sun's lifetime energy output; 2.4×10^{34} J; 1.4×10^{25} J 6) 140 µs 7) 3 million km $= 4.3$ solar radii; 3 billion km $= 20$ AU 8) 2×10^{10} m/s^2 $=2 \times 10^{9}$ g; 2×10^{-3} g; 2×10^{-9} g 9) 1760 km 10) 3×10^{7} km $= 0.2$ AU

Chapter 14

True or False?/Multiple Choice 1) T 2) T 3) F 4) F 5) F 6) T 7) F 8) T 9) b 10) c 11) a 12) b 13) d 14) b 15) c

Problems 1) 2.0″, much less than the angular diameter of Andromeda 2) 100 kpc 3) 10 times 4) −6.2; 17 Mpc 5) 0.014″/yr; proper motion has in fact been measured for several globular clusters 6) 2.7×10^{11} solar masses 7) 570 million years 8) 18 9) 1 kpc 10) 1600 AU, 2.6×10^{6} solar masses

Chapter 15

True or False?/Multiple Choice 1) F 2) T 3) T 4) T 5) T 6) T 7) T 8) a 9) c 10) b 11) b 12) a 13) b 14) c 15) c

Problems 1) 3 billion years 2) 17 Mpc, taking the luminosity of the Cepheid to be 10^{4} times that of the Sun 3) 320 Mpc 4) 14,000 km/s, 57 Mpc; 10,000 km/s, 80 Mpc; 16,000 km/s, 50 Mpc 5) 14 billion years; 14 billion years 6) 1.8×10^{8} solar masses 7) 1.1 million years

8) 1.3×10^{36} W $= 3.2$ billion solar luminosities 9) 7200 km/s 10) Seyfert: ; Sirius A: ; smaller by a factor of 120; yes, if much of the emission from the nucleus is in the infrared

Chapter 16

True or False?/Multiple Choice 1) F 2) F 3) F 4) T 5) T 6) F 7) T 8) T 9) b 10) c 11) d 12) a 13) d 14) b 15) a

Problems 1) 1.2×10^{12} solar masses 2) 5.6×10^{11} solar masses, assuming a velocity of 350 km/s 3) 9.3×10^{10} solar masses 4) 700 km/s, comparable to the 660 km/s speed of the galaxy in the cluster 5) 130 million years 6) −27.2; 6.4 trillion solar luminosities; apparent magnitude of the Sun $= -26.8$; so the quasar at 10 pc is 0.4 magnitudes, or a factor of 1.4 times, brighter than the Sun at 1 AU 7) −22.5; 8.3×10^{10} solar luminosities 8) 11 million years 9) 21 Mpc 10) 48 kpc

Chapter 17

True or False?/Multiple Choice 1) F 2) F 3) F 4) T 5) T 6) F 7) T 8) T 9) a 10) d 11) c 12) d 13) b 14) c 15) a

Problems 1) 1000 Mpc 2) 4×10^{8} 3) 1200 km/s, directly away from the opposite corner 4) 20, 14, 12 billion years 5) (a) 30,000 tons, (b) 2.8 pc 6) 3×10^{15} solar masses; 1200 km/s 7) (a) 1.1 mm, (b) 0.0093, (c) 9.3×10^{-5}, (d) 9.3×10^{-7} 8) 9×10^{-18} kg/m^3; 5×10^{-19} kg/m^3 9) 91 kpc 10) 74 kpc

Chapter 18

True or False?/Multiple Choice 1) F 2) F 3) T 4) F 5) T 6) T 7) T 8) b 9) c 10) c 11) b 12) d 13) a 14) d 15) d

Problems 1) 6.3 seconds (for a 20-year-old reader); 18 seconds (in 2003); 72 seconds; 161 seconds; 237 days 2) 0.19–0.47 AU 3) 5 billion 4) 27 AU 5) (a) 0.2, (b) 20, (c) 2000 6) 3100 years, 3.1 million years 7) 32,000 km/s 8) 5×10^{8} W, 500 times the emission of the Sun 9) 1.43×10^{9} to 1.67×10^{9} Hz; 2.4×10^{6} channels 10) 2.3 years; 55 years

Credits

Chapter 0
Opener Tyler Nordgren **0.1a–b** NASA **0.1d** NOAA **0.13** Jerry Lodriguss/Photo Researchers, Inc. **0.2a** Pere Sanz/Alamy **0.5** Association of Universities for Research in Astronomy **0.13a–f** Lick Observatory Publications Office **0.14b** Glen Schneider **0.15** Bencho Angelov **0.16a** NOAA **0.16b–c** Glen Schneider **0.23a–e** Glen Schneider **p. 21** NASA/NOAA

Chapter 1
Opener Sheila Terry/Photo Researchers, Inc. **1.1** Museum of Science, Boston **1.4** Erich Lessing/Art Resource, NY **1.6** Scala/Art Resource, NY **1.9** Erich Lessing/Art Resource, NY **1.10** Newberry Library/SuperStock **1.15** Sheila Terry/Photo Researchers, Inc. **p. 28** (top left) Sergio Pitamitz/SuperStock; (bottom left) Georg Gerster/Photo Researchers, Inc.; James Cornell; Harrison Lapahie, Jr.; (right) Takyuddin and other astronomers at the Galata Observatory founded in 1557 by Sultan Suleyman, from the Sehinsahname of Murad III, c. 1581 (vellum), Turkish School, (16th century)/Istanbul University Library, Istanbul, Turkey/The Bridgeman Art Library

Chapter 2
Opener NASA/CXC/CfA/S. Wolk **2.1** Robert Gendler/Photo Researchers, Inc. **2.11a** ESO **2.11b** AURA **2.11c** SST/NASA **2.11d** Galex/NASA/John F. Kennedy Space Center **2.14** Wabash Instrument Corporation **2.20** NASA **2.22** Bausch & Lomb Incorporated **p. 66** NASA

Chapter 3
Opener ESO/NAOJ/NRAO **3.6a,c** W. M. Keck Observatory **3.7a** MIT Lincoln Lab **3.7b** NOAO **3.8a–c** NASA Headquarters **3.8d** Association of Universities for Research in Astronomy **3.9a–b** Association of Universities for Research in Astronomy **3.10a** Richard J. Wainscoat **3.10b** National Astronomical Observatory of Japan **3.11** European Southern Observatory **3.12a–d** Association of Universities for Research in Astronomy **3.14a–b** ESO **3.15a** Laurie Hatch **3.15b–c** NOAO **3.16a** NRAO **3.17a–b** D. Parker/T. Acevedo/NAIC **3.17c** Cornell University **3.18** Jack Burns **3.19a–b** National Radio Astronomy Observatory **3.21a** National Radio Astronomy Observatory **3.21b** Association of Universities for Research in Astronomy **3.22** Eric Simison/Sea West Enterprises, Inc. **3.23a** Smithsonian Astrophysical Observatory **3.24a** UCO/Lick Observatory **3.24b** UCO/Lick Observatory **3.24c–d** NASA **3–25a–b** NASA/Jet Propulsion Laboratory **3.26a** NASA Headquarters **3.26b** Galex/NASA/John F. Kennedy Space Center **3.28** NASA **3.29** CXC/SAO **3.30** NASA Headquarters **3.31a** European Space Agency **3.31b** NASA Headquarters **3.31c** Lund Observatory **3.31d** Courtesy of Konrad Dennerl and Wolfgang Voges, Max Planck Institut fur Extraterrestrische Physik, Garching, Germany **3.31e** Compton Gro/NASA Headquarters **p. 74** (top) NASA; (bottom) David Malin/Anglo Australian Telescope **p. 75** NASA **p. 95** Association of Universities for Research in Astronomy; NRAO/AUI; National Radio Astronomy Observatory; European Space Agency; NASA

Chapter 4
Opener JAXA **4.2** Jerry Lodriguss/Photo Researchers, Inc. **4.5** Palomar Observatory/California Institute of Technology/Palomar/Hale Observatory **4.6a–c** NASA Headquarters **4.7** NASA Headquarters **4.7 inset** NASA Headquarters **4.8a** NASA **4.9** National Optical Astronomy Observatories **4.10** Walter Pacholka, Astropics/Photo Researchers, Inc. **4.12a** ESA/Max Planck Institute (Courtesy H. U. Keller) **4.13a–c** NASA/Jet Propulsion Laboratory **4.14** NASA/Jet Propulsion Laboratory **4.16a** Pekka Parviainen/Science Photo Library/Photo Researchers, Inc. **4.16b** Jerry Lodriguss/Photo Researchers, Inc. **4.19** U.S. Geological Survey, Denver **4.20** NASA **4.21** Sovfoto/Eastfoto **4.22a–b** Science Graphics **4.23** VLT/ESO/European Southern Observatory **4.25a** NASA Headquarters **4.25b** Dana Berry **4.26** David Malin/Anglo Australian Telescope **4.28a** NASA/Jet Propulsion Laboratory **4.28b** NASA/John F. Kennedy Space Center **p. 108** David A. Hardy/Photo Researchers, Inc. **p. 122** Mike Powell/Getty Images **p. 131** NASA Headquarters; NASA Headquarters; National Optical Astronomy Observatories; Sovfoto/Eastfoto

Chapter 5
Opener NASA Headquarters **5.12** Megan Chaisson **5.13** David Parker/Science Photo Library/Photo Researchers, Inc. **5.16** UCO/Lick Observatory **5.17** NASA/Goddard/Arizona State University **5.18a** UCO/Lick Observatory **5.18b–c** California Institute of Technology/Palomar/Hale Observatory **5.20a–b** NASA Headquarters **5.21** NASA/John F. Kennedy Space Center **5.24** National Center for Atmospheric Research/University Corporation for Atmospheric Research/National Science Foundation **5.25a–f** Professor Willie Benz **5.26a** U.S. Geological Survey **5.26b–c** U.S. Geological Survey **5.27a** NASA Headquarters **p. 159** UCO/Lick Observatory **p. 160** Professor Willie Benz

Chapter 6
Opener NASA Headquarters **6.1b** John Sanford/Science Photo Library/Photo Researchers, Inc. **6.2** California Institute of Technology/Palomar/Hale Observatory **6.5** National Optical Astronomy Observatories **6.6** CNRS & Universite Paul Sabatier **6.8a** NASA **6.9** NASA/Goddard Space Flight Center **6.10** NASA **6.11** NASA **6.12** NASA **6.13a** NASA Headquarters **6.14** NASA Headquarters **6.15a** NASA Headquarters **6.16** NASA Headquarters **6.17** Russian Space Agency **6.18a–b** NASA Headquarters **6.19** NASA Headquarters **6.20a–b** European Space Agency **6.21** NASA **6.22** NASA **6.23a** European Space Agency **6.23b** NASA **6.24** NASA **6.25** NASA **6.26a–b** NASA **6.27** NASA **6.28** European Space Agency **6.29a–b** NASA **6.30a–b** NASA **6.31a** NASA **6.32a** NASA **6.36a–b** Kees Veenenbos **p. 174** Lowell Observatory **p. 182** NASA **p. 183** NASA **p. 189** NASA Headquarters; European Space Agency; NASA

Chapter 7
Opener NASA/JPL **7.1a** Association of Universities for Research in Astronomy **7.1b** NASA **7.2** NASA **7.3** NASA **7.4** NASA **7.5** NASA **7.6a–e** NASA **7.8** NASA **7.10** NASA **7.11** NASA **7.12** NASA **7.13a–b** NASA **7.15a–b** NASA **7.16** NASA **7.17a–d** NASA **7.18a–b** NASA **7.22** NASA **p. 196** (top right) Sovfoto/Eastfoto; (right and bottom) NASA **p. 212** NASA **p. 213** NASA; Association of Universities for Research in Astronomy

Chapter 8
Opener NASA **8.1** NASA **8.2** NASA **8.4a–c** NASA **8.5a–d** NASA **8.6a–c** NASA **8.7a–b** NASA **8.8a–b** NASA **8.9a–b** NASA/ESA **8.10a–b** NASA/ESA **8.12** NASA/ESA **8.14a–b** NASA **8.15a** UCO/Lick Observatory **8.15b–h** NASA **8.16a–f** NASA **8.17a–b** NASA **8.22** NASA/ESA **8.23** NASA **8.24a–b** NASA **8.25** NASA **8.26a–b** NASA **8.27** NASA **8.28** NASA **8.29** NASA **8.30a–b** NASA **8.32** Mike Brown (Caltech)/W.M. Keck Observatory, HI **8.33a–c** NASA **p. 239** NASA; NASA/Jet Propulsion Laboratory; UCO/Lick Observatory **p. 240** NASA

Chapter 9
Opener NASA **9.1** National Optical Astronomy Observatories **9.3** NASA **9.8** Institute for Solar Physics of the Royal Swedish Academy of Sciences, Sweden **9.10** Glen Schneider **9.11b** NASA **9.12** National Center for Atmospheric Research/University Corporation for Atmospheric Research/National Science Foundation **9.14** California Institute of Technology/Palomar/Hale Observatory **9.15a** California Institute of Technology/Palomar/Hale Observatory **9.15b** Institute for Solar Physics of the Royal Swedish Academy of Sciences, Sweden **9.16b** NASA **9.20a** European Space Agency **9.20b** NASA **9.21** USAF **9.22a** European Space Agency **9.23a–d** ISAS/Lockheed Martin **9.26a–b** Institute for Cosmic Ray Research of the University of Tokyo **p. 250** NASA **p. 266** NASA; Ernest Orlando Lawrence Berkeley National Laboratory

Chapter 10
Opener ESA/NASA **10.3a–b** Harvard College Observatory **10.7a** Pere Sanz/Alamy **10.7b** NASA/Jet Propulsion Laboratory **10.10a** National Optical Astronomy Observatories **10.10b** NASA Headquarters

Chapter 11
Opener STScI **11.1** ESO **11.2b–c** ESO, European Southern Observatory **11.03** Joyce Photographics/Photo Researchers, Inc. **11.4** ESO/S. Guisard **11.5** Harvard College Observatory **11.6** Robert Gendler/Photo Researchers, Inc. **11.7a–b** AURA/NASA **11.9a–b** NASA Headquarters **11.9c** Association of Universities for Research in Astronomy **11.9d** NASA **11.10** European Southern Observatory **11.11a–b** Charles J. Lada/Harvard Smithsonian Center for Astrophysics **11.12** Robert Gendler, Jim Misti & Steve Mazlin/Photo Researchers, Inc. **11.13a** Royal Observatory of Belgium **11.13b** European Southern Observatory **11.15** AURA **11.16** Five College Radio Astronomy Observatory **11.19a** Pere Sanz/Alamy **11.19b–c** NASA/Jet Propulsion Laboratory **11.19d** James M. Moran **11.19e** NASA Headquarters **11.23a** A. Watson/NASA Goddard Laboratory for Atmospheres **11.23b** NASA Headquarters **11.24a–b** NASA Headquarters **11.26a–c** NASA Headquarters **11.27** European Southern Observatory **11.27 inset** NASA Goddard Laboratory for Atmospheres **11.28a** National Optical Astronomy Observatories **11.29a** P. Seitzer (University of Michigan) **11.30** NASA **11.31a–b** NASA **p. 317** ESO, European Southern Observatory; NOAA Central Library Photo Collection; NASA; A. Watson/NASA Goddard Laboratory for Atmospheres; NASA; P. Seitzer/University of Michigan

Chapter 12
Opener NASA/Goddard Space Flight Center **12.9a** AURA **12.9b** AURA **12.9c** NASA Headquarters **12.11** California Institute of Technology/Palomar/Hale Observatory **12.12a** AURA **12.12b** NASA/Jet Propulsion Laboratory **12.13a–b** NASA/Association of Universities for Research in Astronomy **12.15a** California Institute of Technology/Palomar/Hale Observatory **12.15b–d** ESA **12.18** Jon Morse (University of Colorado), and NASA **12.19a–b** AURA **12.21** NASA/CXC/Chinese Academy of Sciences/F. Lu et al. **12.23a** European Southern Observatory **12.23b** NASA **12.23c** David Malin/Anglo Australian Telescope **12.25a** National Optical Astronomy Observatories **12.26a** NASA **12.27a** NASA **12.28a** NASA **12.29a** ESO **12.29b** NASA **12.29c** National Optical Astronomy Observatories **12.29d** NASA/John F. Kennedy Space Center **p. 344** NASA **p. 345** Kitt Peak National Observatory; NASA/Jet

Propulsion Laboratory; NASA Headquarters; NASA Headquarters; Space Telescope Science Institute; NASA Headquarters; David Malin/Anglo Australian Telescope; ESO; NASA; National Optical Astronomy Observatories; NASA/John F. Kennedy Space Center; NASA

Chapter 13

Opener NASA **13.1** NASA **13.4a** ESO **13.4b** NASA **13.4c** Lick Observatory **13.5a–b** NASA **13.6** NASA **13.7a** SAO **13.7b–c** NASA **13.9** NASA **13.11a–b** NASA **13.11c–d** ESO **13.12** NASA **13.13a–b** ESO **13.22a** Harvard-Smithsonian Center for Astrophysics **13.22b** NASA **13.23** L. J. Chaisson **13.24a–b** NASA **13.25** NASA **p. 359** NASA/Ames Research Center/ Ligo Project **p. 373** NASA; L. J. Chaisson

Chapter 14

Opener ESO **14.1b** Axel Mellinger, Central Michigan University **14.2a** Robert Gendler/ Photo Researchers, Inc. **14.2b** NASA **14.4** Vasily Belokurov, SDSS **14.4c** Harvard College Observatory **14.10** UMass/Caltech **14.18** AURA **14.19** Robert Gendler **14.22b** AURA **14.23a** AURA **14.23b** ESO **14.24a** NASA/JPL-Caltech/S. Stolovy (SSC/Caltech)**14.24b** National Radio Astronomy Observatory **14.24c** NASA/CXC/MIT/F.K. Baganoff et al. **14.24d** National Radio Astronomy Observatory **14.25** ESO **14.26c–f** L. J. Chaisson **p. 385** Harvard College Observatory **p. 401** Robert Gendler/Photo Researchers, Inc. **p. 402** Anglo-Australian Observatory, photograph by David Malin; National Radio Astronomy Observatory

Chapter 15

Opener NRAO/STScI **15.1a** AURA **15.1b** Hubble Space Telescope WFPC Team, NASA/ ESA, STScI **15.2a** Robert Gendler **15.2b** Space Telescope Science Institute **15.2c** Anglo-Australian Observatory, photograph by David Malin **15.3a** NASA and The Hubble Heritage Team (STScI/AURA)**15.3b** NASA/JPL-Caltech/R. Kennicutt (University of Arizona) and the SINGS Team **15.4a** NASA, ESA, and the Hubble Heritage Team (STScI/AURA) **15.4b** Anglo-Australian Observatory, photograph by David Malin **15.4c** ESO **15.5a–b** Association of Universities for Research in Astronomy **15.5c** Robert Gendler **15.6a–b** California Institute of Technology/Palomar/Hale Observatory **15.7a** Mount Stromlo & Siding Spring Observatories/Photo Researchers, Inc. **15.7b–c** Royal Observatory, Edinburgh/Science Photo Library/Photo Researchers, Inc. **15.8a** Hubble Heritage Team (AURA/STScI), J. Higdon (Cornell) ESA, NASA **15.8b** NASA, ESA, Hubble Heritage (STScI/AURA); Acknowledgment: A. Aloisi (STScI/ESA) et al. **15.10** NASA **15.12a** Matt BenDaniel **15.12b** NASA **15.13a** Matt BenDaniel **15.13b** G. Jacoby/Association of Universities for Research in Astronomy **15.14** NASA, N. Benitez (JHU), T. Broadhurst (Racah Institute of Physics/The Hebrew University), H. Ford (JHU), M. Clampin (STScI), G. Hartig (STScI), G. Illingworth (UCO/ Lick Observatory), the ACS Science Team, and ESA **15.15a–e** California Institute of Technology/ Palomar/Hale Observatory **15.19** NASA **15.20** NASA **15.21** Courtesy of NRAO/ AUI/NSF **15.22a** ESO **15.22b** NRAO **15.22c** NASA/SAO/R. Kraft et al. **15.23a** National Optical Astronomy Observatories **15.23b** NRAO/AUI; Investigators: R. Perley, C. Carilli & J. Dreher **15.24** Harvard-Smithsonian Center for Astrophysics **15.26a** National Optical Astronomy Observatories **15.26b** Association of Universities for Research in Astronomy **15.26c** NASA **15.27a–b** Association of Universities for Research in Astronomy **15.28** Adapted from Palomar/Caltech **15.29** NRAO **15–31a** NRAO **15.31b** NASA **15.32a–b** NASA

NASA **p. 429** NASA and The Hubble Heritage Team (STScI/AURA); G. Jacoby/Association of Universities for Research in Astronomy; Halton Arp, Max Planck Institute for Astrophysics, Garching, Germany

Chapter 16

Opener Space Telescope Science Institute (STScI)**16.1** NASA **16.3a** NASA **16.3b** Association of Universities for Research in Astronomy **16.3c–d** NASA/ESA **16.4a** NRAO **16.4b** Palomar/Caltech **16.5** NASA/JPL-Caltech/P.N. Appleton (SSC/Caltech) **16.6** NASA **16.7a** NASA, ESA, and the Hubble Heritage Team (STScI/AURA) ESA/Hubble Collaboration **16.7b** Association of Universities for Research in Astronomy **16.8b** R. Williams and the HDF Team (STScI) and NASA **16.8c** NASA **16.9** R. Williams and the HDF Team (STScI) and NASA **16.10a–c** NASA **16.11** Smithsonian Astrophysical Observatory **16.14** J. Moran **16.16a** NASA/STScI/R. P. van der Marel & J. Gerssen **16.16b** NASA/CXC/S/Komossa et al. **16.17a–b** NASA **16.19** data courtesy of B. Tully, University of Hawaii; visualization by S. Levy, NCSA **16.21** Harvard-Smithsonian CFA **16.24** NASA **16.26a–b** NASA **16.27a–b** Space Telescope Science Institute **16.28a–b** Deep Lens Survey team (courtesy J. A. Tyson, Alcatel-Lucent) and NSF/NOAO **16.29** X-ray: NASA/CXC/CfA/M. Markevitch et al.; Optical: NASA/STScI; Magellan/U. Arizona/D. Clowe et al.; Lensing Map: NASA/STScI; ESO WFI; Magellan/U. Arizona/D. Clowe et al. **p. 451** (left) Fermilab Visual Media Services (right) Sloan Digital Sky Survey **p. 455** NASA/ESA; NASA/JPL-Caltech/P. N. Appleton (SSC/Caltech); NASA

Chapter 17

Opener NASA **17.8a–f** NASA **17.13** Alcatel-Lucent **17.24** V. Springel **17.25** NASA **17.26a** NASA **17.26b** Cosmic Background Imager **p. 482** NASA

Chapter 18

Opener Dana Berry **18.1** Dana Berry **18.4a–b** Sidney Fox/Elso Barghoorn **18.4c** Eric Chaisson **18.5** NASA **18.6** Harvard-Smithsonian Center for Astrophysics **18.8a** Dudley Foster/ Woods Hole Oceanographic Institution **18.8b** Eric Chaisson **18.13** NASA **18.16a** Seth Shostak/SETI Institute **18.16b** SETI Institute **p. 501** NASA; Dudley Foster/Woods Hole Oceanographic Institution; Eric Chaisson

Part One

IBM Research, Almaden Research Center; Science Source/Photo Researchers, Inc.; Jerry Lodriguss/Photo Researchers, Inc.; Joyce Photographics/Photo Researchers, Inc.; NOAA

Part Two

NASA; UCO/Lick Observatory; NASA/Jet Propulsion Laboratory; NASA/JPL/Space Science Institute; NASA

Part Three

Space Telescope Science Institute; NASA

Part Four

NASA/Kennedy Space Center; Space Telescope Science Institute; NASA/Jet Propulsion Laboratory; NASA; R. Williams and the HDF Team (STScI) and NASA

Index

Note: n indicates note.

0024 + 1654 galaxy cluster, 453
1E 0657-56 ("bullet cluster"), 454–455
2M1207, 117
3C 48, 424
3C 84, 420
3C 175, 425
3C 273, 424, 425
30 Doradus, 334
47 Tucanae, 355
51 Pegasi star, 126

A0620-00, 371
A 2218, 453
A ring, 230–232
Abell 39, 328
Abell 85, 436
Abell 1689, 415
Abell 2199, 441
Absolute brightness. *See* Luminosity(ies)
Absolute magnitude, 274, 275
Absolute zero, 51
Absorption lines, 56–57
    Kirchhoff's laws and, 61–62
Absorption spectrum, 57, 61
AC114, 452
Acceleration, 36–37
    cosmic, 466–467
Accretion disk, 331, 354, 355, 358, 360, 371
    surrounding supermassive black hole, 427, 428
Accretion process, 121
    onto neutron star, 355
Active galaxies (active galactic nuclei), 402–405, 418–428. *See also*
        Quasar(s); Radio galaxies
    black holes in, 444–446
    common features of, 425
    energy of, 425–428
        emission of, 427–428
        production of, 425–426
    evolution of, 445–446
    interactions with other galaxies, 425
    radiation emitted by, 418–419
    Seyfert, 419–420, 446
Active optics, 82
Active regions, 258–259
Adams, John (mathematician), 198
Adaptive optics, 82, 83
Adenine, 487
Advanced Camera for Surveys, 458–459
Aerosol haze, 225
African plate, 149
Aine corona, 171
Aldebaran, 273, 278–282
ALH84001, 182–183
Allen Telescope Array, 500
ALMA (Atacama Large Millimeter/submillimeter Array), Chile,
        68–69
Alpha Centauri, 245, 271, 272, 274, 275, 278, 280, 281, 288
Alpha Centauri star system, 272
Alpha Eridani, 280, 284
Alpha Orionis. *See* Betelgeuse
Alpha Pegasi, 284
Altair, 280, 281
Alvin, 492
AM0644-741, 411
Amazon River, 178
Amino acids, 487, 488
Ammonia, 124, 206, 209
Ammonia ice, 202, 205
Ammonia "rain," 206
Ammonium hydrosulfide ice, 202, 205
Amplitude, 45
Andromeda Galaxy (M31), 44, 77, 79, 380, 381, 414, 438
Angstrom, 48–49
Angular diameter, 103
Angular measure, 10
Angular momentum
    concept of, 122
    conservation of, 119, 122

Angular resolution, 78–80
    of radio telescopes, 84
Annular eclipse, 16
Antarctic, ozone depletion over, 142
Antares, 279–281, 302
Antenna galaxies (NGC 4038/4039), 437–438
Antiparticle, 262
Apache Point Observatory, 451
Apennine mountains, 152
Aphelion, 33, 34
Aphrodite Terra, 170, 171, 176
Apollo program, 135, 153, 296
Apophis (asteroid), 106
Apparent brightness, 272–274, 282
Apparent magnitudes, 274
Arc degree, 10
Arc minutes, 10
Arc seconds, 10
Arcturus, 278, 280, 281, 284, 324
Arecibo telescope (National Astronomy and Ionospheric Center,
        Puerto Rico), 84, 85, 165, 225, 355
Argon, 141
Argyre Planitia, 176
Ariel, 228–229
Aristarchus of Samos, 27
Aristotle, 19, 20, 27, 36
Arp 107, 441
Arp 273, 432–433
"Artificial star," 83
Associations, 314
Assumptions of mediocrity, 487
Asteroid belt, 105, 124
Asteroids, 100–102, 104–109, 118, 239
    carbonaceous, 107
    dinosaur extinction and, 108
    Earth-crossing, 106–107
    formation of, 124–125
    properties of, 107–109
    silicate, 107
    Trojan, 106, 124
Astrologers, 26
    Chinese, 28
Astrometry, 287
Astronomical unit (AU), 34, 35
Astronomy
    in ancient societies, 28
    celestial coordinates, 6
    celestial sphere, 5–6
    constellations, 4–5, 8
    distance measurement, 17–18
        with geometry, 17–18
        triangulation, 17
    foundations of, 2–21
    Islamic, 28
    scientific theory and the scientific method, 18–21
    space-based, 89–94
Asymptote, 326n
Asymptotic giant branch, 326, 326n
Atacama Large Millimeter/submillimeter Array (ALMA), Chile,
        68–69
Atmosphere, 167–169, 201–208
    of Earth, 136, 137, 140–143, 187
        greenhouse effect and, 142–144
        ozone hole in, 142
        reddening in, 295
        solar activity and, 250
    of Io, 220
    of Jupiter, 201–204, 207
    lunar, 140, 143–145
    of Mars, 169, 188–189
    of Mercury, 167
    opacity of, 50
    of red giant, 284
    of Saturn, 204–206
    solar, 251–253
        chromosphere, 246, 252–253
        solar wind, 110, 154–156, 246, 250, 253, 260
        transition zone and corona, 14–15, 246, 250, 253, 260–261
    of Titan, 223–226

    of Uranus and Neptune, 206–207
    of Venus, 168, 187–188
Atmospheric pressure, 140, 141
Atmospheric turbulence, 81–82
Atom(s)
    Bohr model of, 58–59
    classical, 59
    excitation of, 59, 60
    formation of, 474–475
    ionization state of, 301
    modern, 59
    motion in interstellar clouds, 305
    structure of, 58–59
A-type stars, 314
AU (astronomical unit), 34, 35
Aurora/aurorae, 156
    on Jupiter, 210
Aurora australis (Southern Lights), 156
Aurora borealis (Northern Lights), 114, 156
Autumnal equinox, 9, 11
Axis of telescope mirror, 70
Azimuth, 28

B filter, 275–276
B ring, 230–232
Background radiation. *See* Cosmic microwave background
Balmer lines, 61
Bar code, 55
Barnard 68, 53, 120, 295, 343
Barnard's Star, 271–272, 278–280
    apparent magnitude of, 274
Barred-spiral galaxies, 408, 412
    properties of, 412
Barringer Crater, 106, 115, 116, 152, 153
Basalt, 137, 147, 151
Baseline
    of interferometer, 87
    of triangle, 17, 18
Bayer, Johann, 273
Becklin-Neugebauer object, 309
Bell, Jocelyn, 350, 351
Bell Telephone Laboratories, 471
Belts, atmospheric, 200
Beta Orionis. *See* Rigel
Beta Pictoris, 119, 120
Beta Tuarid meteor shower, 115
Betelgeuse, 6, 273, 275, 276, 278–281, 333, 334
    apparent magnitude of, 274
Big Bang, 462–463, 467, 470–471
    cosmic microwave background and, 471–473
    epoch of inflation and, 476–477
    flatness problem, 476, 477–478
    formation of nuclei and atoms, 473–475
    horizon problem and, 475–478
    matter and radiation constituents of universe
        and, 472–473
Big Crunch, 464
Big Horn Medicine Wheel, Wyoming, 28
Binary black hole, 445
Binary galaxy systems, 434, 435
Binary pulsar, 359
Binary stars, 285–286
    black holes in, 371–372
    close, 331, 371
    double-pulsar, 359
    gravity waves, 359
    mass of, 286–289
        correlation to other properties, 286–289
        determination of, 286
    neutron star binaries, 354–356
        pulsar planets, 355–356
        X-ray sources, 354, 355
    period of, 286
    stellar evolution in, 340
Binary-star planets, 495
Binary-star systems (binaries), 285
Biological evolution, 486, 490–491
Bipolar flow, 310
Blackbody, 52

Blackbody curve(s), 51–53
    applications of, 54
    cosmic, 471
    stellar temperatures and, 275–277
Blackbody spectrum, 51–52
Black dwarf, 329, 330
Black holes, 348–349, 353, 360–372, 443–447
    in active and normal galaxies, 372, 443–447
    binary, 445
    curved space and, 366–368
    Einstein's theories of relativity and, 362–368
        general relativity, 365–368
        special relativity, 362–364
    escape speed and, 361–362
    event horizon and, 362, 366–370
    as final stage of stellar evolution, 361
    frictional heating in, 368
    galaxy size and size of, 445, 446
    intermediate-mass, 372
    masses of, 446
    measurable properties of, 362n
    observational evidence for, 371–372
    space travel near, 368–370
        approaching the event horizon, 368–370
        singularity, 370
        tidal forces, 368, 370
    stellar-mass, 372
    supermassive, 75, 372, 399–401, 419, 426, 443–445, 447
        accretion disk surrounding, 427, 428
        origins of, 444
Black smoker, 492
Blazar, 423, 447
Blue giants, 280, 281
Blueshift, 63, 64
    galactic rotation and, 387
Blue stragglers, 340
Blue supergiants, 280, 333
Blurring
    atmospheric, 81–82
    from diffraction, 79, 80
Bode, Johann, 195
Bohr model of atom, 58–59
Bohr, Niels, 58
Boltzmann, Ludwig, 54
Bonds, chemical, 62
Bonner Durchmusterung (BD) catalogue, 273
Borealis Basin, 177
Brahe, Tycho, 32–33
Braided rings, 233
Breakwater, 46
Bright giants, 284
Brightness
    absolute (See Luminosity(ies))
    apparent, 272–275, 282
    of planet, 26, 29, 30
    ranking of stars in order of, 6
    telescope size and, 77
Bright supergiants, 284
Brown dwarfs, 117, 312–313, 315, 316, 334
    dark matter and, 395, 396
    spectral classification and, 278
Brown oval, 203
B-type stars, 287, 289, 299, 305, 312, 339–340, 344
Bulge stars, 387, 387n
"Bullet cluster" (1E 0657-56), 454–455
Buoyancy force, 140
Burst counterpart, 356–357

C ring, 230–232
Caldera, 171
Callisto, 218, 219, 222–223
Caloris Basin, 170
Canada-France-Hawaii telescope, 78
Canali, 174
Cannon, Annie, 385
Canopus, 278, 281
Capella, 279–281
Capture theory, 157
Caracol temple, Mexico, 28
Carbon, 62
Carbonaceous asteroids, 107
Carbonaceous meteorites, 116
Carbon-core star, 325–326
Carbon cycle, 187, 188
Carbon detonation (Type I) supernovae, 335–337, 336n, 350

global measurements using, 467
    as standard candles, 413
Carbon dioxide
    atmospheric, 141, 187, 189
    global warming and, 142–144
Carbon-oxygen white dwarf, 332
Carina Nebula, 292–293
"Cartwheel" galaxy, 437
Cascade, 61
Cassegrain focus, 72–73
Cassegrain, Guillaume, 72
Cassegrain telescope, 72, 74
Cassini Division, 230, 232, 233
Cassini, Giovanni Domenico, 231
Cassini-Huygens mission, 491
Cassini mission, 193–195, 197, 206, 216–217, 223–227, 229, 231, 233–234
Cassini Solstice mission, 196, 223
Cassiopeia A supernova remnant, 92–93, 348–349
Catalyst, 142
Cat's Eye Nebula (NGC 6543), 320–321, 328
Celestial coordinates, 6
Celestial equator, 6, 7
Celestial mechanics, 197
Celestial sphere, 5–6
Celsius temperature scale, 51
Centaurus A, 85–86, 420, 428
Center for High Angular Resolution Astronomy (CHARA), 88
Center of mass, 38, 39
Centrifugal force, 119
Cepheid variable stars, 382–384, 412, 413
Ceres, 104, 106, 107, 238, 239
Chaco Canyon, New Mexico, 28
Chandrasekhar mass, 335, 354
Chandrasekhar, Subramanyan, 92, 335
Chandra X-Ray Observatory, 92, 320–321, 336, 348–349, 352, 355, 372, 398, 420, 437, 445
Charge-coupled devices (CCDs), 76–77
Charged particles, 46–47
Charon, 236–237
Chemical bonds, 62
Chemical evolution, 486, 488, 489
Chicxulub crater, 108
Chlorine, ozone depletion and, 142
Chlorofluorocarbons (CFCs), 142
Chromatic aberration, 71
Chromosphere, 246, 252–253
Chryse Planitia, 175
Circinus galaxy, 419
Circle, subdivision of, 10
Circular motion, uniform, 27
Circulation currents. See Convection
Clementine spacecraft, 145
Climate, solar cycle, 250
Close binary system, 331
Closed universe, 465, 465n
Cloud(s)
    dark dust, 301–304
        21-centimeter radiation, 303
        densities of, 301
        Horsehead Nebula, 302–303
        L977, 302
        molecular gas, 303–304
        obscuration of visible light by, 301–302
        Rho Ophiuchus, 302
        star-forming regions in, 303–304
    interstellar, 306–307, 489
        contraction of, 306–308
        evolution of, 309
        fragmentation of, 306–307
    on Jupiter, 202
    molecular, 303–304, 314, 489
    on Neptune, 207
    on Saturn, 204–205
    on Uranus, 207
Clusters. See Galaxy clusters; Globular clusters; Star clusters
Coformation theory, 156
Collecting area, 77–78
    of radio telescopes, 84
Color index, 276
Color(s), 48
    false, 85–86, 92–94
Coma (halo), 110
Coma Cluster, 406, 415, 448, 449

Comet(s), 100–102, 104, 105, 110–113, 489
    formation of, 124–125
    long-period, 113
    orbits of, 112–113
    parts of, 110
    short-period, 112–113
    tail of, 110, 111
    trajectory of, 111
Composition, spectral-line analysis of, 64
Compton Gamma-Ray Observatory (CGRO), 93–94, 353, 356, 399
Computers in observational astronomy, 76–77
Conamara Chaos, 221
Condensation nuclei, 120
Condensation theory, 119–125, 129–130, 157, 493, 494
    solar system irregularities and, 125
Conservation, law of
    of angular momentum, 119, 122
    of mass and energy, 261
Constellations, 4–5, 8
Continental drift, 148–150. See also Plate tectonics
Continuous spectra, 55, 57, 61–62
Convection
    on Earth, 141
    on Jupiter, 201
    plate movement and, 149–150
    solar, 248–251
        evidence for, 251
Convection cells, 141, 249–251
Convection zone, 246
Coordinates, celestial, 6
Copernican revolution, 24–39
    defined, 29
    modern astronomy, 30–32
        ascendancy of Copernican system, 32
        Galileo's historic observations, 30–32
    Newton's laws (Newtonian mechanics), 36–39
        of gravity, 37–38
        of motion, 36–37, 39
        orbital motion, 38, 39
    planetary motions, 26–30, 32–35
        Brahe and, 32–33
        dimensions of solar system, 35
        geocentric universe, 27–28
        heliocentric model of solar system, 29–30
        Kepler's Laws, 33–35, 38–39
    scientific progress, circle of, 39
Copernicus Crater, 154
Copernicus, Nicholas, 24–25, 29, 36, 384
Cordelia, 235
Cordillera Mountains, 158
Core
    of Earth, 136, 146, 147
        magnetosphere and, 154–156
    of Jupiter, 208–209
    lunar, 137, 147–148
    of Mars, 186
    of Neptune, 209
    of Saturn, 209
    stellar, 328
    of Sun, 246
    of Uranus, 209
Core-collapse (Type II) supernovae, 334, 335–337, 344, 350
Core-dominated radio galaxy, 421
Core-hydrogen burning, 322
Corona/coronae
    solar, 14–15, 246, 250, 253, 260–261
        sunspot cycle and, 260
    of Venus, 171–172
Coronal holes, 260
Coronal hotspots, 253
Coronal mass ejection, 259
Coronal storms, 250
CoRoT (exoplanet), 128, 129
CoRoT mission, 126
Cosmic Background Explorer (COBE) satellite, 472, 479
Cosmic Background Imager, 480
Cosmic density parameter (Ω_0), 464, 466, 469–470, 480
Cosmic distance scale, 17, 383–384, 418. See also Distance measurement
Cosmic evolution, 486–490
    phases in, 486
Cosmic microwave background, 471–473, 475
    dark-matter density and, 478–479

isotropy of, 472, 475
temperature fluctuations of, 479–480
Cosmic rays, 399
Cosmological constant, 468–469, 478
Cosmological distances, 417
Cosmological principle, 461
Cosmological redshift, 416–417, 463–464
Cosmology, 458–480
  age of universe, 118, 470–471
  Big Bang, 462–463, 467, 470–471
    cosmic microwave background and, 471–473
    epoch of inflation and, 476–477
    flatness problem, 476, 477–478
    formation of nuclei and atoms, 473–475
    horizon problem and, 475–478
    matter and radiation constituents of universe and, 472–475
  cosmic composition, 469–470
  cosmological redshift, 416–417, 463–464
  decelerating universe, 467
  defined, 461
  expanding universe, 461–463, 467, 475, 476
    Olbers's paradox, 461
  fate of universe, 464–471
    cosmic acceleration/expansion, 467
    density of universe and, 466
  large-scale structures in universe, 478–480
  shape of universe, 465–466
  "static" universe, 468
Coudé (Nasmyth) focus, 72, 73
Coudé room, 73
CP1919, 351
Crab Nebula (M1), 337–338, 352
Crab pulsar, 352, 353
Crater(s)
  on Earth, 115–116
  impact
    on Callisto, 222
    on Ganymede, 222
    on Venus, 172–173
  lunar, 152–154, 158
  Martian, 175, 180
  on Mercury, 169–170
  on midsized Jovian moons, 229
  "splosh," 180
  on Titan, 224
Crescent Moon, 12, 13
Crests, wave, 45
Critical density, 464, 465, 469–470
Critical universe, 466
Crust, 146, 147
  of Earth, 136, 137
  lunar, 137, 148, 158
Cultural evolution, 486, 490
Curvature, negative and positive, 465
Curved (warped) space, 365–368, 465
Cygnus A, 421
Cygnus Loop supernova remnant, 91–92
Cygnus X-1, 371
Cytosine, 487

D ring, 231–233
Dactyl (moon), 107, 109
Dark cloud, 120
Dark, defined, 395
Dark dust clouds, 301–304
  21-centimeter radiation, 303
  densities of, 301
  Horsehead Nebula, 302–303
  L977, 302
  molecular gas, 303–304
  obscuration of visible light, 301–302
  Rho Ophiuchus, 302
  star-forming regions in, 303–304
Dark energy, 467–470
  cosmic expansion and, 467–469
  density of universe and, 478–479
Dark energy-dominated universe, 472–473
Dark galaxies, 435
Dark halos, 394, 434, 435, 441–442
Dark matter, 394–396, 434–437, 470, 474–475
  clumping of, 478–479
  density of universe and, 466
  galaxy masses calculation and, 434
  intracluster gas and, 435–436
  large-structure formation and, 478–480

mapping, 453–455
search for, 395–396
Dark nebula, 298
Dark Spot (Great), 207–208
Day
  sidereal, 7
  solar, 7
Debris, solar system, 105
Declination, 6
Decoupling, epoch of, 475
Deep Impact spacecraft, 111, 113
Deferent, 27
Deforestation, 144
Degrees, 6
Delta Aquarid meteor shower, 115
Delta Cephei, 382
Delta, Martian, 178
Deneb, 273, 281
Dense interstellar clouds. *See* Dark dust clouds
Dense matter, 328
Density
  of cosmos, curvature of space and, 465
  critical, 464, 465, 469–470
  of Earth, 103, 104, 136
  of emission nebulae, 300, 301
  escape speed and, 464–465
  of Jovian planets, 199, 209
  of Moon, 136, 147
  of planets, 103, 104
  of Sun, 247, 249
  of universe, 465–466
Density waves, spiral, 391, 392–393
Deuterium, 262, 473, 473n, 474
Deuteron, 262, 263
Diameter, angular, 103
Differential force (tidal force), 137–138, 368, 370
Differential rotation, 199–200
  of Galactic disk, 387–388, 390, 392
  of Milky Way, 390
  of Sun, 246, 255, 256
Differentiation
  of Earth, 147
  of solar system, 123–124
Diffraction, 46, 79
  resolution and, 80
Diffraction-limited resolution, 79
Dinosaurs, extinction of, 108
Dione, 227, 228
Direct (prograde) motion, 26, 106
Dirty ice, 297
Disk stars, 387
Distance, Doppler shift and, 416
Distance measurement, 17–18, 412–413, 417–418
  with geometry, 17–18
  Hubble's law, 417–418
  radar ranging, 35, 383, 384, 413
  spectroscopic parallax, 282–283, 287, 298, 383, 384, 413
  standard candles, 412–413
  stellar parallax, 29, 270, 283, 287, 383, 384, 413
  triangulation, 17
  Tully-Fisher relation, 413
  variable stars, 382–384, 412, 413
Disturbance, 45, 47
Diurnal motion, 7
DNA, 487–488
Doppler, Christian, 63
Doppler effect, 63–64
Doppler shift, 64, 126, 422
  distance and, 416
Double-line spectroscopic binary, 285, 286
Double-pulsar binary system, 359
Draconid meteor shower, 115
Drake equation, 493–497
  average lifetime of technological civilization, 497–498
  fraction of habitable planets on which life arises, 496
  fraction of life-bearing planets on which intelligence arises, 496
  fraction of planets on which intelligent life develops and uses technology, 497
  fraction of stars with planetary systems, 493–494
  number of habitable planets per planetary system, 494–496
  rate of star formation, 493
Drake, Frank, 493
Dust cocoons, 310
Dust grain, 294–295, 304
Dust, interstellar, 120, 123, 294–295, 303–304, 384

composition of, 297
detector, 111, 112
Dust lanes, 298, 299
Dust tail, 110
Dusty donut, 428
Dwarf ellipticals, 408–409, 414
  masses of, 434
Dwarf galaxy, 388–390
Dwarf irregulars, 410, 414
  masses of, 434
Dwarf planets, 102, 105–106, 238
  gas dwarfs, 128
Dwarf(s), 279, 284
  black, 329, 330
  brown, 312–313, 315.316, 334, 395, 396
  dark matter and, 395, 396
  red, 280–282
  white, 279, 328–329, 331, 332, 334, 336, 340, 395, 396
Dynamo theory, 156, 257
Dysnomia, 238

E ring, 231, 233–234
Eagle Nebula (M16), 297, 298, 300, 301
Earth, 134–158, 494
  asteroids crossing, 106–107
  atmosphere of, 136, 137, 141–143, 187
    greenhouse effect and, 142–144
    ozone hole in, 142
    reddening in, 295
    solar activity and, 250
  circumference of, 19
  core of, 146, 147
  distance to Moon, 136
  exo-Earth, 130
  formation of, 156–157
  interior of, 145–147, 157, 186
    differentiation of, 147
    modeling, 146–147
    seismology, 145–146
  life on, 487–488
    diversity and culture, 489–490
    interstellar origin of, 489
  magnetic field of, 211
  magnetic "tail," 148, 155
  magnetosphere of, 136, 137, 154–156
  orbital motion of, 7–11, 34
    day-to-day changes, 7
    long-term changes, 11
    seasonal changes, 8–11
  overall structure of, 136–137
  planets similar to, searching for, 130–131
  properties of, 104
    bulk, 198–199
    orbital, 34
    physical, 136, 164
  radio radiation from, 499
  rotation of, 167
  in Sun's habitable zone, 130
  super-Earths, 127–130
  surface activity on, 148–150
    continental drift, 148–150
    convection and plate movement, 149–150
    tides, 137–140
    gravitational deformation and, 137–138
Earth-like plants, habitable zones and, 130
Earth-mass planets, searching for, 130–131
Earthquakes, 145, 146, 148, 149
Eccentricity of ellipse, 33
Eclipse(s), 14
  annular, 16
  geometry of, 16
  lunar, 14–16, 20
  solar, 14–16, 246, 252, 253
Eclipsing binaries, 286
Ecliptic, 8–11
Eddington Crater, 154
Eddington, Sir Arthur, 367
Einstein, Albert, 39, 59, 362–368, 468
Einstein Cross, 453
Einstein's curve ball, 465
Einstein's elevator, 365
Einstein X-ray Observatory, 436
Ejecta blanket, 153, 173
Electrical force, 46–47
Electric field, 46–47

Electromagnetic force, 262
Electromagnetic radiation, 44. *See also* Light; Solar wind
    Doppler effect, 63–64
    gathering (*See* Telescopes)
    spectral-line analysis, 64–65
Electromagnetic spectrum, 48–50
    components of visible light, 48–49
    defined, 50
    full range of, 49–50
Electromagnetic waves, 47–48, 359
Electromagnetism, 47
Electron, 46
Electron cloud, 59
Electron spin, 303
Elementary particles, 46
Element(s), 62
    creation of, by dying stars, 328
    emission spectra of, 55, 56
    heavy
        formation of, 338–340, 342, 343
        fusion of, 332, 338–339
Ellipse, 33
Elliptical galaxies, 408–409, 412, 414, 427, 446, 447
    dark halos around, 435
    masses of, 434
    properties of, 412
Emission lines, 55–57
    Kirchhoff's laws and, 61–62
Emission nebulae, 60, 297–301, 327, 385
Emission, nonstellar, 419
Emission spectrum, 55–57, 61
Enceladus, 216–217, 226, 228, 229, 491
Encke gap, 231
Endurance crater, 184–185
Energy
    dark, 467–470
        cosmic expansion and, 467–469
        density of universe and, 478–479
    jets from gamma-ray bursts, 357
    law of conservation of mass and, 261
    SI unit of, 247n
    temperature and, 52, 54
    thermal, 51
    wave motion and, 45
Energy flux, 54
Energy, vacuum, 476–477
Envelope, hydrogen, 110
Envelope, stellar, 324–329, 335, 336, 338, 343, 344
Epicycle, 27
Epicyclic motion, 27, 29
Epoch of inflation, 476–477
Epsilon Eridani, 273
Epsilon ring, 234, 235
Equilibrium, hydrostatic, 248, 322, 323
Equinoxes, 9, 11
Equivalence principle, 365
Eratosthenes, 19
Eris, 106, 113, 238, 238n
Eros (asteroid), 107, 109
Erosion, lunar, 153–154
Escape speed, 136, 140, 361–362, 464
Eta Aquarid meteor shower, 115
Eta Carinae, 333
Euclid, 103
Euclidean geometry, 466, 470
Eurasian Plate, 149
Europa, 195, 217, 219–222, 491
European Space Agency, 74, 250
    search for extrasolar planets and, 126
Event horizon, 362, 366–370
"Evil Eye" (M64) galaxy, 434
Evolution. *See also* Stellar evolution
    biological, 486, 490–491
    chemical, 488, 489
    cultural, 486, 490
    planetary, 486
    of quasars, 443, 444–446
    use of term, 322
Evolutionary track of star, 309, 312
Excited state, 59
Exobiology, 183
Exo-Earth, 130
Exoplanets. *See* Extrasolar planets
Exploration, 25
Exploration Rover, 176

Exposure time, 77
Express spacecraft, 196
Extinction of dinosaurs, 108
Extrasolar planets, 117, 126–131, 356
    detecting, 125–126
    number of, 126
    orbits of, 126
    properties of, 126–128
    search for, 127, 128, 130–131
Extraterrestrial intelligence, 484–485, 498–500
Extraterrestrial life, 484–485, 487, 490–492
    on Europa, 221–222
    on Mars, 182–183
Extreme Ultraviolet Explorer (EUVE) satellite, 91, 92
Extremophiles, 492
Eyepiece, 70–71

F ring, 231–234
Fahrenheit scale, 51
"Failed" stars, 312–313
False color, 85–86, 92–94
False vacuum, 476, 478
Fermi Gamma-Ray Space Telescope, 93–94
Field
    electric, 46–47
    magnetic (*See* Magnetic field)
Filamentary structure, 449
Filters, telescope, 275–276
Fireball, relativistic, 358, 360
Fission, nuclear, 261n
Flamsteed, John, 273
Flares, solar, 258–259
Flatlanders, 465n
Flatness problem, 476, 477–478
Fleming, Williamina, 385
Flyby, defined, 167
Focal length, 70
Focus, 70, 72–73
    of ellipse, 33
Fomalhaut (star), 121
Force(s), 36
    electrical, 46–47
    Grand Unified Theories (GUTs) of, 476–478
    gravitational, 37
    instantaneous *vs.* continuous, 37
    tidal (differential force), 137–138, 368, 370
        ring formation and, 231
Forests, destruction of, 144
Formaldehyde, 304
Fornax A, 402–405
Fossil fuels, 144
Fossil record, 489–490
Fragmentation, 121
Fragment G, 212
Fraunhofer, Joseph, 56
Fraunhofer lines, 56
Frequency(ies), 45–46, 48, 50
    Doppler effect and, 63–64
    peak emission, 51–52
    photon energy and, 59–60
    temperature and, 52
Frequency scale, 49, 50
Friction, 36
F-type star, 314, 494, 496
Full Moon, 12–14, 16
Fusion, nuclear, 246, 261, 310, 322
    after Big Bang, 473
    of heavy elements, 332, 338–339
    helium, 324–325
Fuzziness from diffraction, 79, 80

G ring, 234
G1 cluster, 372
Galactic bulge, 380, 381, 384–387, 396
    black hole mass and, 444
    properties of, 388
    size of, and spiral galaxies, 408
Galactic cannibalism, 441–442
Galactic center, 380, 384, 385, 396–401
    activity in, 397–399
    central black hole (Sgr A*), 399–401
    density of matter in, 397
    Sagittarius A, 397–398
    stellar orbits near, 399
    views (artist's renditions of), 400–401
Galactic disk, 380, 381, 384–389

    gas in, 388–389
    orbital motion in, 387–388, 390, 392
    properties of, 388
    star formation on, 388
Galactic evolution, 486
Galactic habitable zone, 495
Galactic halo, 380, 381, 384, 384n, 385–389
    properties of, 388
    tidal streams in, 389
Galactic nucleus, 399, 406, 419–420
    active (*See* Active galaxies (active galactic nuclei))
Galaxy clusters, 414–415, 448
    collision of, 455
    gravitational lensing of, 452
    masses of, 434
    X-ray emission by, 435–436
Galaxy Evolution Explorer (GALEX), 91, 92, 437
Galaxy(ies), 4, 378–401, 432–455. *See also* Active galaxies (active galactic nuclei); Milky Way Galaxy; Spiral galaxy(ies)
    Antennae, 437–438
    barred-spiral, 408, 412
    binary systems, 434, 435
    black holes in, 372, 443–447
    "cartwheel," 437
    dark, 435
    defined, 380
    distance measurements to, 412–413
    distribution in space, 412–415
    dwarf, 388–390
    elliptical, 408–409, 412, 414, 427, 446, 447
        dark halos around, 435
        masses of, 434
        properties of, 412
    formation and evolution of, 439–443
        active galaxies, 446–447
        evolution and interaction, 440–442
        Hubble sequence and, 442–443
        possible evolutionary sequences, 446, 447
        quasars and, 443, 444–446
        theory of, 439
        through collisions and mergers, 437–446
    Hubble's classification of, 406–412
    Hubble's law and, 416–417, 448, 458
    irregular, 410–412, 414, 447
        masses of, 434
    masses of, 434
    motion of, 416–417
    normal (Hubble class), 406–418, 443–444
        energy emitted by, 419
    quasar-hosting, 445, 446
    radiation from, 419
    radio, 86, 419, 420–421, 423, 446, 447
        core-dominated, 421
        head-tail, 435, 437
    recession of, 416–417, 462–463
    rotation of, 413
    Seyfert, 419–420, 446, 447
    spectra of, 416
    spiral, 447
    starburst, 419, 441
    views of, 380–381
    visible, 420n
Galaxyrise Over Alien Planet (painting), 484–485
Galaxy surveys, 449, 450
Galilean moons, 31, 194, 218–223
    Europa, 195, 217, 219, 221–222, 491
    Ganymede and Callisto, 30, 218, 219, 222–223
    interiors of, 219
    Io, 218, 219–221
    as miniature solar system, 218–219
    orbits of, 219
Galilei, Galileo, 30–32, 36
Galileo mission, 107, 194, 195, 197, 202, 203, 212, 218–220
Galileo Regio, 222
Galle, Johann, 198
Gamma-ray astronomy, 93–94
Gamma rays, 44, 49, 50, 92–94, 262, 263
    bursts, 356–358, 360
        causes of, 358, 360
        distances and luminosities of, 356–357
        as energy jets, 357
    detection of, 357
    in early universe, 471
    pulsars, 352, 353
Gamma-ray telescopes, 93–94

Ganymede, 218, 219, 222, 491
Gas cloud evolution, 120
Gas dwarfs, 128
Gas(es)
    dust grains and, 120
    in Galactic disk, 388–389
    greenhouse, 142–144, 187–188
    interstellar, 294–295, 384–385
        composition of, 297
        distribution of, 297
        Galactic, 390
        interaction between stars and, 299–300
        molecular, 303–304
    intracluster, 435–436, 442
    molecular, 140
    pregalactic, 388
    primordial, 436
    volcanic, 187
Gas-exchange experiment, 182
Gaspra (asteroid), 107
Geminga pulsar, 352, 353
Geminid meteor shower, 115
Gemini North telescope, 78
General relativity, 361, 365–368, 370, 465
    black holes and, 366–368, 370
    central concept of, 365–366
    cosmological constant and, 468–469
    dark matter and, 395–396
    tests of, 367
Genes, 487–488
Geocentric model, 27–28
Geometry
    eclipse, 16
    Euclidean, 465–466
    measuring sizes with, 103
    parallax, 17, 18
    of universe, 469, 470, 475–478
Geysers on Triton, 227
Giant branch, 340
    asymptotic, 326, 326n
Giant ellipticals, 408–409
Giant(s), 279, 284
    blue, 280–282
    bright, 284
    death of, 28
    red, 279, 282, 324, 326, 328
        atmosphere of, 284
Gibbous phase, 12, 13
Giotto (spacecraft), 111
Gliese 229, 313
Gliese 623, 313
Global measurements, 467
Global Oscillations Network Group (GONG), 248
Global warming, greenhouse effect and, 144
Globular clusters, 314, 315, 340–343, 384, 385, 387
    age of, 470
    galactic system of, 384, 384n
    millisecond pulsars in, 354–355
    RR Lyrae variables in, 384
Globule, 295
Gondola, 89
Grand Canyon, 176
Grand Unified Theories (GUTs), 476–478
Granite, 147
Granulation, solar, 251
Gravitational deformation, 137–138
Gravitational lens, 395
Gravitational lensing, 395–396, 453
    of galaxy cluster, 452
    of quasar, 452
Gravitational pull (force), 136, 137, 262
Gravitational redshift, 369, 479
Gravitational "slingshots," 197
Gravity
    atmosphere and, 140
    dark energy and, 467–469
    of Jupiter, 220, 221
    Newton's law of, 37–38
    quantum, 370
    star formation and, 305–306
    of Sun, 38
    surface, 136
        Martian, 177
    theory of, 370
Gravity assist, 194, 197

Gravity Recovery and Interior Laboratory (GRAIL)
        mission, 145
Gravity waves, 359
GRB 030329, 360
GRB 080319B, 357
GRB 971214, 357
Great Dark Spot, 207–208
Great Red Spot, 201, 203, 208, 218
Great Wall, 449
    Sloan, 460
Greece, ancient, 28
Greenhouse effect, 142–143, 143n
    global warming and, 144
    runaway, 188
        reverse, 188
    on Venus, 187–188
Greenhouse gases, 142–144, 187–188
Greenwich Meridian, 6
Ground state, 58–59
G-type star, 326, 494, 496
Guanine, 487
Guest stars, 28
Gula Mons, 171

H alpha line (Hα), 61, 64
H beta line (Hβ), 61
H gamma line (Hγ), 61
Habitable zone(s), 130, 188
    galactic, 495
    stellar, 494
Hale-Bopp comet, 104, 110, 111, 489
Hall, Asaph, 174
Halley's comet, 110–112, 489
Halo (coma), 110
Halo stars, 386–387, 387n, 388
Harvard College Observatory, 385
Haumea, 238
Haute-Provence Observatory, France, 126
Hayabusa (spacecraft), 101
Haze layer, 202, 205
    on Titan, 223
Haze level, 205
HD 39801, 273
HD 95735, 273
HD 209458, 126
HDE226868, 371
Head-tail radio galaxies, 435, 437
Heat
    atmosphere and, 140
    star formation and, 305–306
Heavy elements, formation of, 340, 342, 343
    fusion, 332, 338–339
    from supernovae explosions, 338–339
Heliocentric model, 29–30
    Galileo and, 30–32
Helioseismology, 248
Helium, 57, 62, 263, 265
    in core of star, 323–325
    emission spectrum of, 56
    formation of, 473–474
    hydrogen fusion into, 263, 265, 322
    on Jovian planets, 198, 206, 209
    primordial origin of, 338
    in Saturn's atmosphere, 211
Helium flash, 325
Helium white dwarf, 329
Hellas Basin, 176–177, 179
Henry Draper catalogue, 273
Herbig-Haro 46, 53
Herbig-Haro objects, 310, 311
Herschel, William, 196, 381
Hertz (Hz), 46
Hertz, Heinrich, 46
Hertzsprung, Ejnar, 279
Hertzsprung-Russell diagram. See H-R diagram
Hewish, Anthony, 351
HH1 jet, 310, 311
HH2 jet, 310, 311
HH30 star system, 311
HI regions, 301
Hierarchical merging, 339–440
High-energy telescopes, 92–94
High Energy Transient Explorer 2 (HETE-2),
    360
Highlands

    lunar, 151, 153, 157
    Martian, 175
High-mass stars, 323, 334, 336, 337
High-resolution astronomy, 80–82
    atmospheric blurring and, 81–82
    new designs, 82
HII regions, 301
Hipparchus, 274
Hipparcos satellite, 5, 271, 281, 283
Holden Crater, 179
Homogeneity of universe, 460
Homo sapiens, 490
Horizon problem, 475–478
Horizontal branch, 325
Horsehead Nebula, 302–303
Hot Jupiters, 127–130
Hourglass, 300
Hours, 6
HR4796A, 121
H-R diagram, 279–281, 385
    asymptotic giant branch on, 326
    cluster evolution on, 339, 340
    of globular cluster, 315, 342
    high-mass evolutionary tracks on, 332
    horizontal branch on, 325
    instability strip of, 383
    main sequence on, 280–281, 283, 284
    newborn star on, 310
    of open cluster, 315
    prestellar evolutionary tracks on, 312
    protostar on, 309
    red giant branches on, 324–326, 332
    red giant region on, 281
    stellar distance and, 282–285
    variable stars on, 383
    white dwarf on, 281, 329
Hubble Deep Field, 440
Hubble diagrams, 416
Hubble, Edwin, 73, 383, 406, 416
Hubble flow, 416, 417
Hubble galaxy classification, 406–412
    ellipticals, 408–409
    irregulars, 410–412, 414
    spirals, 406–408
Hubble's constant, 417, 452, 461, 463
Hubble sequence, 410–412, 442–443
    "tuning fork" diagram in, 410, 411
Hubble's law, 416–417, 448, 461, 463, 467
Hubble Space Telescope (HST), 72–75, 77, 82, 92, 121, 204,
    205, 208, 292–293, 310, 316, 320–321, 329–331,
    333, 340, 348–349, 352, 353, 402–406, 426, 437,
    445, 451, 458–459
Hubble's "tuning fork" diagram, 410–411
Hubble Ultra-Deep Field, 77
Hues, 48
Hulse, Russell, 359
Huygens probe, 224–225
Hyades cluster, 340, 341
Hydra, 236, 237
Hydrogen
    absorption lines of, 57, 450
    atomic, 303
    Bohr's model of, 58–59
    in core of star, 323–324
    in Earth's atmosphere, 140
    emission spectrum of, 55, 56
    fusion into helium, 263, 265, 322
    fusion of nuclei, 262–263
    "heavy" (deuterium), 262, 473, 473n, 474
    on Jovian planets, 198, 206, 208, 209
    metallic, 208, 209
    molecular, 303–304
    primordial origin of, 338
    spectra of, 58–62
    structure of, 58
Hydrogen 21-cm emission, 303
Hydrogen-burning core of star, 322
Hydrogen envelope, 110
Hydrogen series, 61
Hydrogen shell-burning stage, 324
Hydrosphere, 136, 137
Hydrostatic equilibrium, 248, 322
Hydrothermal vents, 492
Hypernova, 360
Hypothesis, 20–21

Iapetus, 227–229
IC 694, 441
IC 2163, 437, 438
Icarus (asteroid), 106
Ice
  dirty, 297
  lunar, 145
Ida (asteroid), 107, 109
Images, telescopic, 70, 71
Impact craters, 107. *See also* Crater(s)
  on Callisto, 222, 223
  on Ganymede, 222
  on Venus, 172–173
Impact theory, 157
Inclination, 8
Inertia, 36, 38
Inflation, epoch of, 476–477
Infrared Astronomy Satellite (IRAS), 310
Infrared astronomy/telescopes, 89–93
Infrared interferometry, 87
Infrared radiation, 49, 50
  greenhouse effect and, 142–143, 187
Infrared waves, 44, 93
Inhomogeneities, 478
Inner core, 146
Inner planets. *See* Terrestrial planets
Inquisition, 32
Instability strip, 383
Intelligence
  extraterrestrial, 484–485, 498–500
  fraction of life-bearing planets giving rise to, 496, 497
Intensity, 51
Intercloud medium, 296
Intercrater plains, 169
Interference, 46, 86–87
Interferometry, 86–88
Intermediate-mass black holes, 372
International Astronomical Union, 105, 239, 273
Interplanetary matter, 105–117. *See also* Asteroids; Comets; Kuiper belt; Meteoroids
Interstellar cloud, 54, 306–307
  contraction of, 306–308
  evolution of, 309
  fragmentation of, 306–307
Interstellar dust, 120, 123
Interstellar medium, 292–316, 343
  composition of, 296–297
  dark dust clouds, 301–304
    21-centimeter radiation, 303
    densities of, 301
    molecular gas, 303–304
    obscuration of visible light, 301–302
    star-forming regions in, 303–304
  defined, 294
  density of, 296–297, 300, 301
  "failed" stars, 312–313
  formation of stars of other masses, 312–313
  formation of Sun-like stars, 305–311
    cloud fragment contraction (stages 2 and 3), 306, 307–308
    gravity and heat in, 305–306
    interstellar cloud (stage 1), 306–307
    interstellar cloud contraction (stage 1), 306–307
    newborn star (stages 6 and 7), 306, 310–311
    protostar (stage 4), 306, 308–309
    protostellar evolution (stage 5), 306, 309–311
  gas and dust, 294–297, 303–304
  sizes (diameter) of, 300, 301
  star clusters, 313–316
    associations and, 314
    globular, 314
    nebulae and, 314–316
    open, 314, 315
  star-forming regions of, 297–301
  ultraviolet astronomy and, 296
Interstellar molecular clouds, 489
Interstellar space, 489
Intracluster gas, 435–436, 442
Inverse-square law, 37–38, 272–275
Invisible radiation, 42–43
Io, 218, 219–221
Ion, 59
Ionization, 59
Ionization state of atom, 301
Ionosphere, 141
Ion tail, 110, 111

Iridium, 108
Iron
  in Earth's core, 146
  emission spectra of, 62
Iron-core stars, 334
Iron meteorites, 117
Iron oxide, Martian, 181
Irregular galaxies, 410–412, 414, 447
  Irr I and Irr II, 410
  masses of, 434
Ishtar Terra, 170
Islamic astronomy, 28
Isotropy
  of cosmic microwave background, 472, 475
  of universe, 460–461
Itokawa (asteroid), 100–101

James Webb Space Telescope (JWST), 75
Jets
  from accretion disk, 426, 427–428
  in active galaxies, 421, 423
  gamma-ray bursts, 357
  from protostars, 310, 311
  of quasars, 424, 425
  of radio galaxies, 421, 423
Joule (J), 54, 247n
Jovian planets, 104–105, 118, 192–213. *See also* Jupiter; Neptune; Saturn; Uranus
  atmosphere of, 201–208
  bulk properties of, 198–199
    physical characteristics, 198–199
    rotation rates, 199–200
  changes in orbits of, 124
  discoveries of Uranus and Neptune, 195, 198
  formation of, 121, 122–123
  internal heating of, 210–211, 213
  internal structure of, 208–209
  magnetospheres of, 210, 211
  observations of Jupiter and Saturn, 194, 195
    spacecraft exploration, 194–195
    view from Earth, 194
  rings of, 217, 230–235
  scales relative to Earth, 199
Juno, 239
Juno probe, 209
Jupiter, 26, 103
  atmosphere of, 201–204, 212
    appearance and composition of, 201–202
    structure of, 202
    weather, 203–204
  aurorae on, 210
  change in orbit of, 124
  differential rotation of, 199
  formation of, 218–219
  Galilean moons of, 31, 194, 218–223
    Europa, 217, 219, 221–222, 491
    Ganymede and Callisto, 218, 219, 222–223
    Io, 218, 219–221
    as miniature solar system, 218–219
  gravity of, 220, 221
  interior of, 208–209
  internal heating of, 210
  magnetic field of, 211
  magnetosphere of, 210
  observations of, 194, 195
    spacecraft exploration, 194–195
    view from Earth, 194
  properties of, 34, 104, 199
  rings of, 232, 234
  Shoemaker-Levy 9 collision with, 212

Keck Observatory, Hawaii, 73, 78–81, 87, 238, 355, 399, 440
Kelvin, Lord, 51
Kelvin temperature scale, 51
Kepler-22b, 130
Kepler, Johannes, 30, 32, 36
Kepler objects, 130
Kepler's Laws, 33–35, 38–39
  Newton's laws and, 38
  second, 112, 122
  third, 393
Kepler spacecraft, 127, 128, 130
Kirchhoff, Gustav, 57
Kirchhoff's laws, 57, 58
  spectral lines and, 61–62
Kitt Peak (U.S. National Observatory), 81, 330

KOI 961d, 130
K2-type star, 284–285
K-type stars, 287, 289, 494, 496
Kuiper belt, 105, 112, 113, 118, 120
  formation of, 124
  plutoids and, 237
Kuiper belt objects (KBOs), 102, 113, 118, 235, 237–239
Kuiper, Gerard, 112

L₁ Lagrangian point, 250
L977 dark dust cloud, 302
Labeled-release experiment, 182
Lagoon Nebula (M8), 300
Lagrange, Joseph Louis, 106
Lalande 21185, 273
Lalande, Joseph de, 273
Language, development of, 490
Laplace, Pierre Simon de, 119
Large Magellanic Cloud (LMC), 344, 371, 410
Large-scale galaxy survey, 449, 450
Large-scale structure in universe, 478–480
Las Campanas Observatory, Chile, 450
Laser Interferometric Gravity-Wave Observatory (LIGO), 359
Laser ranging, 136, 136n
Lava, 147
Lava domes, Venus, 171–172
Leavitt, Henrietta, 383, 385
Lens
  gravitational, 395
  refracting, 70
Lensing, gravitational, 395–396, 453
  of galaxy cluster, 452
  of quasar, 452
Lenticular (S0) galaxies, 409
Leonid meteor shower/storm, 114, 115
Leverrier, Urbain, 198
Lick Observatory, 83
Life, 484–500
  characteristics of, 486–487
  Drake equation, 493–497
    average lifetime of technological civilization, 497–498
    fraction of habitable planets on which life arises, 496
    fraction of life-bearing planets on which intelligence arises, 496
    fraction of planets on which intelligent life develops and uses technology, 497
    fraction of stars with planetary systems, 493–494
    number of habitable planets per planetary system, 494–496
    rate of star formation, 493
  on Earth, 487–488
    diversity and culture of, 489–490
    interstellar origin of, 489
  extraterrestrial, 484–485, 487, 490–492
    on Europa, 221–222
    on Mars, 182–183
  extraterrestrial intelligence, 484–485, 498–500
Light, 42–65. *See also* Electromagnetic radiation
  electromagnetic spectrum, 48–50
    components of visible light, 48–49
    full range of, 49–50
  electromagnetic waves and, 47–48
  interactions between charged particles and, 46–47
  obscuration of visible, 301–302
  red, 48
  spectral lines and, 58–62
    atomic structure and, 58–59
    Kirchhoff's laws and, 61–62
    particle nature of radiation, 59–60
    spectrum of hydrogen, 60–61
  speed of, 47, 418
    Doppler effect and, 64
    special relativity and, 362–364
  violet, 48
  visible, 44, 48–49
  wavelength of, 48
  wave motion, 44–46
    energy and, 45
    medium for, 46
  white, 48
Lightbulb, spectrum of light from, 55
Light curve
  of eclipsing binary, 286
  of pulsating variable stars, 382, 383
  of supernovae, 335, 344
Light echo, 333

Lighthouse model, 352
Limiting magnitudes, 274
Lin, C. C., 391
Line-of-sight velocity, 65
Little Ice Age, 257
Living and nonliving, distinction between,
    486–487
LMC X-3, 371
Local Bubble, 296
Local Group, 414, 448
Local measurement, 467
Local Supercluster (Virgo Supercluster), 448
Logarithm function, 275
Logarithmic scale, 50
Loki (volcano), 220
Long-period comets, 113
Look-back time, 422
Lorentz contraction, 364
Lorentz factor, 364
Lowell, Abbott Lawrence, 174
Lowell, Amy, 174
Lowell Observatory, 235
Lowell, Percival, 174, 235, 416
Low-mass stars, 337
Luminosity(ies)
    estimating stellar radii using, 282
    of gamma-ray bursts, 356–357
    mass-luminosity relationships, 288
    of Milky Way Galaxy, 418
    of protostar, 308–309
    of quasars, 424
    rotation speed and, 413
    of Seyfert galaxy, 420
    of stars, 272–274, 278–279, 282, 284–285
        absolute magnitude and, 275
        mass and, 287, 288
    of Sun, 247, 254, 265
    temperature and, 280
    of variable star, 383
Luminosity scale, 279
Lunar CRater Observation and Sensing Satellite (LCROSS), 145
Lunar eclipse, 14–16, 20
Lunar highlands, 151, 153, 157
Lunar maria, 137, 158
Lunar phases, 12–14
Lunar Prospector, 145, 147–148
Lunar Reconnaissance Orbiter (LRO) mission, 145, 152
Lunar regolith, 153
Lyman-alpha forests, 450
Lyrid meteor shower, 115

M4 star cluster, 302, 329–330
M8 (Lagoon Nebula), 297, 298, 300, 301, 397
M16 (Eagle Nebula), 297, 298, 300, 301
M17 (Omega Nebula), 297, 298, 301
M20 emission nebula, 297, 298, 301, 304
M31 (Andromeda) galaxy, 44, 77, 79, 380, 381, 414, 438
M32 galaxy, 414
M33 galaxy, 414
M49 galaxy, 409
M51 galaxy, 88, 407, 442
M64 ("Evil Eye") galaxy, 434
M80 globular cluster, 340, 341
M81 galaxy, 91, 92, 407
M82 galaxy, 91, 92, 372
M84 galaxy, 409
M86 galaxy, 415, 421
M87 galaxy (Virgo A), 415, 423, 427, 428, 446
M100 galaxy, 75
M101 galaxy, 381
M104 (Sombrero) galaxy, 407–408
M110 galaxy, 409
Magellan, Ferdinand, 410
Magellanic Clouds, 390, 410, 414, 442
    Large, 344, 371, 410
    Small, 410
Magellan spacecraft, 170, 172, 196
Magnetic carpet, 260
Magnetic field, 47
    of Earth, 47
    of Europa, 221
    of Ganymede, 222
    of Jovian planets, 199–200
    lunar, 156

of Mars, 186
of Mercury, 185
of neutron stars, 350
produced within the accretion disk, 426
spectral-line analysis of, 65
of Sun, 252, 255–256
of sunspots, 255–256
Magnetic field lines, 154
Magnetic poles, 154
Magnetopause, 155
Magnetosphere(s)
    of Earth, 136, 137, 154–156
    of Jovian planets, 210, 211
        rotation rates and, 199–200
Magnitude scale, 274
Main sequence, 280–281, 283, 284
    core-hydrogen burning on, 322–323
    leaving, 322–323
    stellar masses and, 286–288
    zero-age, 312
Main-sequence turnoff, 340
Major axis, 33
Major mergers, 443, 447
Makemake, 238
Manicouagan Reservoir, 116
Mantle, 136, 137, 146, 147
    lunar, 147–148
Mare Imbrium, 151, 152
Maria, 151–153
    on Dione, 227
    on Ganymede, 222
    lunar, 137, 158
Mariner 10 mission, 169, 196
Mariner mission, 167, 170, 174, 186
Mariner Valley (Valles Marineris), 175, 176
Mars, 26, 103, 162–163, 494
    atmosphere of, 169, 188–189
    exploration of, 180–182
    internal structure and geological history of, 186
    life on, 163, 182–183, 491–492
    properties of, 104, 199
        orbital and physical, 34, 164, 165
    rotation of, 166–167
    spaceflights to, 196–197
    in Sun's habitable zone, 130
    surface of, 173–184
        "canals," 174
        exploration by landers, 180–182
        gullies, 180, 181
        large-scale topography, 174–177
        polar caps, 173, 178, 179–180
        Valles Marineris (Mariner Valley), 175, 176
        view from Earth, 173
        volcanism, 177, 189
        water on, 163, 177–182
Mars Exploration Rover mission, 182
Mars Express orbiter, 175, 180, 492
Mars Global Surveyor satellite, 175, 177, 181, 186, 491
Mars Odyssey orbiter, 179
Mars Pathfinder, 181, 492
Mars Phoenix, 492
Mars Reconnaissance Orbiter, 177
Mass, 36
    angular momentum and, 122
    of asteroids, 107
    center of, 38, 39
    Chandrasekhar, 335, 354
    of comets, 111
    of Earth, 136
    escape speed and, 464
    gravity and, 37
    of the Moon, 136
    of nebulae, 301
    of planets, 103
    of stars, 285–289
        evolution and, 323
    of Sun, 104, 247
    turnoff, 340
Mass and energy, law of conservation of, 261
Massive Compact Halo Objects (MACHOs),
    395n
Mass–luminosity relationships, 288
Mass–radius relationships, 288
Mathilde (asteroid), 107, 109
Matter, dark. *See* Dark matter

Matter-dominated universe, 472–473
Mauna Kea Observatory, Hawaii. *See* Keck Observatory, Hawaii
Mauna Loa volcano, Hawaii, 177
Maunder minimum, 257
Maury, Antonia, 385
Maxwell, James Clerk, 231
Mead crater, 173
Mead, Margaret, 173
Measurement(s). *See also* Distance measurement
    angular, 10
    global, 467
    local, 467
Mecca, 28
Mechanics, Newtonian, 36
Mediocrity, assumptions of, 487
Megatons, 108
Mercury, 26, 103
    atmosphere of, 167
    emission spectrum of, 56
    internal structure and geological history of, 185–186
    orbit of, 367
    properties of, 102, 104, 199
        orbital and physical, 34, 164, 165
    rotation of, 164–167
    spaceflights to, 196
    surface of, 169–170
Merging, hierarchical, 339–440
Mesosphere, 141
Messenger mission, 167, 169, 170, 185, 196
Messier 2, 53
Messier, Charles, 297
Metabolism, 487
Metallic hydrogen, 208, 209
Meteorites, 114, 116–117
    carbonaceous, 116
    iron, 117
    organic material on, 489
    stony (silicate), 116, 117
Meteoroids, 102, 105, 114–117
    lunar cratering and, 153–154
Meteoroid swarm, 114
Meteors, 114
Meteor shower, 114–115
Methane, 124, 206–207
    on Titan, 225, 226
Michelson, A. A., 362–363
Michelson-Morley experiment, 362–364
Microlensing, 452
Micrometeoroids, 114, 153, 154
Microspheres, 488
Microwave background. *See* Cosmic microwave background
Microwaves, 49
Mid-Atlantic Ridge, 149, 171
Midocean ridges, 149
Milky Way Galaxy, 2–3, 94, 294, 296–298, 304,
    378–401, 414
    diameter of, 385
    differential rotation of, 390
    formation of, 388–389
    Galactic center, 380, 384, 385, 396–401
        activity in, 397–399
        central black hole (Sgr A*), 399–401
        density of matter in, 397
        Sagittarius A, 397–398
        stellar orbits near, 399
        views (artist's renditions) of, 400–401
    Galactic disk, 380, 381, 384–389
        gas in, 390
        orbital motion in, 387–388
        properties of, 388
        star formation on, 388
    Galactic plane, 298
    infrared views of, 386, 397
    luminosity of, 418
    mapping, 384–386
    mass of, 394–396
        dark matter and, 394–396
    measuring, 381–384
        observations of variable stars, 382–383
        period-luminosity relationship, 383
        star counts, 381–382
    rotation curve of, 394
    size and shape of, 384
    spiral arms, 390–393
    structure of, 384–387

orbital motion, 387–388
   stellar populations, 386
   tidal stream in, 442
Miller, Stanley, 488
Millimeter-wavelength interferometry, 87
Millisecond pulsars, 354–356
Mimas, 228, 230, 233
Minor mergers, 443, 447
Minor planets. *See* Asteroids
Mintaka, 278
Minutes, 6
Mira, 281
Miranda, 228–230
Mirrors in telescopes, 70, 71–72
Model-building, 208
Modern astronomy, influence on, 28
Modern science, birth of, 20
Molecular cloud complexes, 303–304
Molecular clouds, 303–304, 314, 489
Molecule(s), 62
   DNA, 487–488
   gas, 140
Momentum, angular, 119
Month
   sidereal, 14
   synodic, 14
Moon, the, 134–158, 218
   angular diameter of, 103
   apparent magnitude of, 274
   atmosphere of, 140, 143–145
   crust of, 137, 148, 158
   density of, 147
   distance from Earth, 136
   evolution of, 157–158
   far side of, 152
   formation of, 156–157
   Galileo's observations of, 30
   interior of, 147–148, 157–158
   magnetism on, 156
   Mercury compared to, 169–170, 185
   motion of, 12–16
      eclipses and, 14–16
      lunar phases, 12–14
   near side of, 151
   orbit of, 138–140
   overall structure of, 137
   plate tectonics on, 150–151
   properties of, 104, 136, 164, 165
   Saturnian moons compared to, 228
   surface of, 151–154
      cratering, 152–154
      erosion on, 153–154
      large-scale features of, 151–152
   tides, 137–140
      gravitational deformation, 137–138
      tidal locking, 138–140
   trans-Neptunian objects compared to, 238
Moonlets, 232–233
Moonquakes, 147
Moons, 102, 104, 218–230
   direction of rotation, 118
   Enceladus, 491
   Galilean, 31, 194, 210, 218–223
      Europa, 217, 219, 221–222, 491
      Ganymede, 491
      Ganymede and Callisto, 218, 219, 222–223
      Io, 218, 219–221
      as miniature solar system, 218–219
   Jovian, 123, 218–230
      medium-sized, 218, 227–230
   of Pluto, 236–237
   shepherd, 233, 235
   Titan, 218, 223–226, 491
   Triton, 110, 218, 226–227
Morley, E. W., 362–363
Motion
   epicyclic, 27, 29
   Newton's laws of, 36–37, 39
      orbital motion, 38, 39
   prograde (direct), 26, 106
   proper, 272
   radial, 63–64
   retrograde, 26, 27, 29, 30, 39
   transverse, 63–64
   uniform circular, 27

wave, 44–46
   energy and, 44–45
   medium for, 46
Mount Everest, 170
Mount Hadley, 134–135
Mount Wilson Observatory, California, 88, 406, 419
M-type stars, 287, 289, 312, 339, 494, 496
Multiple-star systems, 285
Murchison meteorite, 489
Muslim world, astronomy in, 28

Nanometer (nm), 48–49
NASA, 74, 126
NASA infrared telescope, 78
Nasmyth (coudé) focus, 72, 73
National Astronomy and Ionospheric Center, Arecibo, Puerto Rico, 84–85, 165, 225, 335
National Radio Astronomy Observatory, West Virginia, 83, 84
Native Americans, 28
Natural selection, 490, 496
Neap tides, 138
Near Earth Asteroid Rendezvous (NEAR)-Shoemaker, 109
Near-infrared spectrum, 142
Nebula/nebulae, 297–301
   clusters and, 314–316
   contraction of, 119–120
   dark, 298
   defined, 119, 297
   dust lanes of, 298, 299
   emission, 60, 297–301, 327, 385
   NGC 281, 42–43
   planetary, 326–328
      as standard candles, 413
   properties of, 301
   reflection, 298, 299
   solar, 119, 309
      random encounters in, 125
      temperature in, 124
   spectrum of, 301
   structure of, 299
Nebular theory, 119–124
Negative curvature, 465
Neon, emission spectrum of, 56
Neon-oxygen white dwarf, 330
Neptune
   atmosphere of, 206–207
   discovery of, 195, 198
   distance from Sun, 102, 124
   interior of, 209
   internal heating of, 207, 213
   magnetic field of, 211
   magnetosphere of, 210
   moons of, 229
      medium-sized, 228–229
      Triton, 118, 218, 226–227, 235
   orbit of, 102, 104
   properties of, 34, 102, 104, 199
   rings of, 232, 234, 235
   rotation of, 199–200
   weather on, 207–208
Neutrinos, 261–263, 334, 344
   oscillations of, 264
   solar, 264
   from supernova explosion, 344
Neutrino telescope, 264
Neutrons, 62, 328, 334, 474
Neutron stars, 350. *See also* Black holes
   binaries, 354–356
      pulsar planets, 355–356
      X-ray sources, 354, 355
   compression of, 360
   defined, 350
   gamma-ray bursts, 356–358, 360
      causes of, 358, 360
      distances and luminosities, 356–357
   merger of, 359, 360–361
   pulsars, 350–353
      millisecond pulsars, 354–356
      model of, 351–352
      pulsating variable stars vs., 382n
New Horizons mission, 197
New Moon, 12–14, 16
Newtonian focus, 72
Newtonian telescope, 72
Newton, Isaac, 36–37, 48, 72, 394

Newton's laws (Newtonian mechanics), 36–39, 363, 464
   of gravity, 37–38, 395
   of motion, 36–37, 39
      orbital, 38, 39
   planetary orbits, 118
NGC 265, 268–269
NGC 281, 42–43
NGC 1201, 409
NGC 1265, 435, 437
NGC 1275, 441
NGC 1300, 408
NGC 1316, 402–405
NGC 1365, 408
NGC 1566, 392
NGC 1569, 411
NGC 1620, 434
NGC 2207, 437, 438
NGC 2808, 342
NGC 2859, 409
NGC 2997, 407
NGC 3145, 434
NGC 3603, 314
NGC 3690, 441
NGC 4038/4039 (Antennae galaxies), 437–438
NGC 4258, 443
NGC 4261, 427
NGC 4314, 393
NGC 4378, 434
NGC 4565, 381
NGC 4603, 413
NGC 4984, 434
NGC 6240, 445
NGC 6520, 120
NGC 6543 (Cat's Eye Nebula), 320–321, 328
NGC 6872, 408
NGC 6934 (star cluster), 83
NGC 7635, 343
NGC 7664, 434
NGC 7742, 419
Nickel in Earth's core, 146
Nitrogen geysers, 227
Nitrogen on Earth, 140, 141
Nix, 236, 237
Nonliving and living, distinction between, 486–487
Nonrelativistic speeds, 363
Nonstellar radiation, 419
Nonthermal radiation, 428
Normal galaxies (Hubble-class), 406–418
   black holes in, 443–444
   energy emitted by, 419
North American plate, 149
North celestial pole, 6, 7
Northern Lights (aurora borealis), 114, 156
Northern sky, 6
North Star, 6
Nova Cygni, 331
Novae, 330–331
   recurrent, 331
   supernovae *vs.*, 335, 335n
Nova Herculis 1934, 330
Nova Persei, 331
Nuclear fission, 261n
Nuclear force
   strong, 261, 262
   weak, 262
Nuclear fusion, 246, 261, 310, 322
   after Big Bang, 473
   of heavy elements, 332, 338–339
   helium, 324–325
Nuclear Test Ban Treaty, 356
Nucleosynthesis, primordial, 473
Nucleotide bases, 487, 488
Nucleus/nuclei
   atomic, 58
   of comet, 110
   condensation, 120, 121
   formation of, 473, 473n
   galactic, 399, 406, 419–420

Oberon, 228, 229
Objectivity, scientific method and, 32
Observation, 21
Occam's razor, 20, 27
Ockham, William of, 20
Olbers, Heinrich, 461

Olbers's paradox, 461
Olympus Mons, 177
Omega Centauri, 315
Omega Nebula (M17), 297, 298, 301
Omens, 28
On the Revolution of Celestial Spheres (Copernicus), 29
Oort cloud, 113, 118, 230
    formation of, 124
Oort, Jan, 113
Opacity, 50
Open cluster, 314, 315, 343, 385
Open universe, 465
Ophelia, 235
Opportunity rover, 182, 184, 492
Opposite-spin state, 303
Optical interferometry, 88
Optical telescopes, 70–77, 93
    detectors and image processing, 76–77
    reflecting and refracting, 70–72
    reflecting, types of, 72–73
Optics
    active, 82
    adaptive, 82, 83
Orbital eccentricity, 34
Orbital motion
    of Earth, 7–11, 34
        day-to-day changes, 7
        long-term changes, 11
        seasonal changes, 8–11
    in galactic disk, 387–388, 390, 392
Orbital period, 34
Orbitals, atomic, 59
Orbit(s)
    of asteroids, 106–107
    of comets, 112–113
    extrasolar, 127–128
    of planets, 102, 104, 118, 121, 125 (*See also individually named planets*)
        prograde, 106
        retrograde, 26, 27, 29, 30, 39
        synchronous, 138–140
    of Triton, 227
Orcus, 238
Orientale Basin, 158
Orion constellation, 5, 308
Orionid meteor shower, 115
Orion molecular cloud, 310, 311
Orion Nebula, 90, 275, 307–308, 316
O-type stars, 287, 289, 312, 314, 339
Outer core, 146, 147
Outer planets. *See* Jovian planets
Outflow channels, 177, 178, 180
Outflow, protostellar, 310, 311
Outgassing, 187
Oval storm systems, 203–204
Oxygen, atmospheric, 140–142
Ozone hole, 142
Ozone layer, 141–142

Pacific plate, 149
Pallas, 239
Pandora, 233
Pangaea, 150
Parallactic shift, 270
Parallax (parallactic angle), 17, 18, 136
    spectroscopic, 282–283, 287, 298, 413
    stellar, 29, 270, 271, 283, 287, 383, 384, 413
Paranal Observatory, Chile, 79, 81
Parsec, 270
Partial eclipse, 14, 15
Particle nature of radiation, 59–60
Particles
    charged, 46–47
    elementary, 46
Particulate evolution, 486
Pathfinder mission, 176, 181
Pauli exclusion principle, 328
Peak emission frequency, 51–52
Penumbra, 15, 254
Penzias, Arno, 471
Perihelion, 33, 34
Period–luminosity relationship, 383, 385
Period, orbital, 34
Permafrost, 179, 180
Perseid meteor shower, 115

Persei double cluster, 340
Perseus Cluster, 435
PG0052 + 251, 446
Phases, lunar, 12–14
PHL 938, 450
Phoenix spacecraft, 162–163, 176, 184, 196
Photodisintegration, 334
Photometry, 287
Photon(s), 59–61
Photosphere, 246, 250, 252, 253, 307, 308
Pioneer 10 plaque, 498
Pioneer 10 spacecraft, 196, 210, 500
Pioneer 11, 233
Pioneer Venus spacecraft, 168, 170
Pixels, 76
Plains, intercrater, 169
Planetary motions, 26–30, 32–35
    Brahe and, 32–33
    dimensions of solar system, 35
    geocentric universe, 27–28
    heliocentric model of solar system, 29–32
    Kepler's Laws, 33–35, 38–39
        Newton's laws and, 38
        second, 122
        third, 393
Planetary nebulae, 326–328
    as standard candles, 413
Planetary rings. *See* Rings, planetary
Planetary systems, 4
    extrasolar, 117, 126–131
    number of habitable planets per, 494–496
    stars with, 494
Planetesimals, 121, 124–125. *See also* Meteoroids
    distance from Sun and, 124
    ejection of, 124–125
Planets, 4, 26. *See also* Jovian planets; Terrestrial planets; *specific planets*
    alignment of, 102, 103
    binary-star, 495
    brightness of, 26, 29, 30
    density of, 104
    direction of rotation, 118
    distance from Sun, 102
    dwarf, 102, 106
    evolution of, 486
    extrasolar, 117, 126–131
        detecting, 125–126
        number of, 126
        orbits of, 128
        properties of, 126–128
        search for, 127, 128, 130–31
    formation of, 120–123
    isolation of, 118
    life on (*See* Drake equation)
    mass of, 103, 104
    orbits of, 102, 104, 118, 121, 124–125
    properties of, 34, 102–103, 118
    pulsar, 355–356
    radii of, 102, 104
    relative sizes of, 102, 103
    sinking, 130
Plants, Earth-like, habitable zones and, 130
Plates, 149
Plate tectonics, 148, 149
    carbon dioxide recycling through, 189
    convection and, 149–150
    lunar, 150–151
    on Mars, 186
    on Venus, 186
Pleiades (Seven Sisters), 5, 314, 315
Pluto, 106, 235–236, 250
    discovery of, 235–236
    mass of, 236–237
    moons of, 236–237
    Pluto–Charon system, 236–237
    properties of, 104
    status as dwarf planet, 113
    status as planet, 238–239
Plutoids, 217, 239
Polar caps, Martian, 173, 178–180
Polaris (Pole Star), 6, 274
Polarity of sunspot, 255–256
Polar vortex, 168
Population I stars, 386
Population II stars, 386

Positive curvature, 465
Positrons, 262, 263, 265
Power, 247n
Precession, 11
Pressure, 140
    atmospheric, 140, 141
    spectral-line analysis of, 65
Pressure waves, 145–146
Prestellar evolutionary tracks, 312
Primary mirror, 70
Primary waves (P-waves), 145–146
Prime focus, 70, 71
Primeval fireball, 462
Primordial gas, 436
Primordial nucleosynthesis, 473
"Primordial soup," 488
Princeton University, 471, 472
Prism, 48
    refraction by, 70
Procyon, 273
Procyon A, 280, 281
Procyon B, 280
Prograde orbits, 26, 106
Project Phoenix, 500
Prometheus (volcano), 220
Prominences, solar, 252, 256, 258
Proper motion, 272, 399
Proteins, 487
Proteus, 228, 229
Protomoons, 121
Proton–proton chain, 261–263, 308
    energy generation in, 265
Protons, 46, 474
Protoplanets, 121, 122–123
Protostars, 306–309, 311
    on H-R diagram, 309
Protostellar collisions, 316
Protostellar evolution, 306, 309–311
Protostellar winds, 309–310
Protosun, 121
Proxima Centauri, 271, 279, 280, 288
Pseudoscience, 20
Ptolemaeus, Claudius (Ptolemy), 27
Ptolemaic model, 27, 31
Pulsar planets, 355–356
Pulsars, 350–353
    binary, 359
    millisecond, 354–356
    model of, 351–352
    pulsating variable stars *vs.*, 382n
Pulsar wind, 352
Pulsating variable stars, 382–384
P-wave, 145–146
P/Wild 2 comet, 111, 112
Pyrolitic-release experiment, 182

QSO 1422 + 2309, 450
Quadrantid meteor shower, 115
Quantum gravity, 370
Quantum mechanics, 58, 370
Quaoar, 238
Quarter Moon, 12, 13
Quasar epoch, 444–446, 471
Quasar feedback, 445–446
Quasar(s), 419, 424–425, 447, 471
    absorption lines of, 449–450
    evolution of, 443, 444–446
    galactic evolution and, 443, 444–446
    galaxies hosting, 445, 446
    twin, 452

R136 star cluster, 77, 82
Radar, 35
    planetary, 165
Radar gun, 63, 64
Radar ranging, 35, 136, 136n, 283, 383, 384, 413
Radial motion, 63–64
Radial velocity, 63–64, 271, 272, 287
Radiant, 115
Radiation, 423. *See also* Waves
    defined, 44
    electromagnetic, 44
        Doppler effect and, 63–64
        spectral-line analysis, 64–65
    full range of, 49–50
    galactic, 418–419

invisible, 42–43
nonstellar, 419
particle nature of, 59–60
pulsar, 351
from quasars, 419
spectroscopy of, 55–58, 276
    absorption lines, 56–57
    astronomical applications of, 57–58
    emission lines, 55–57
synchrotron (nonthermal), 427, 428
thermal, 51–54, 428
    astronomical applications of, 54
    blackbody spectrum, 51–52
    laws of, 52–54
21-centimeter, 303
in universe, 472–473
Radiation darkening, 227, 228
Radiation-dominated universe, 472–473
Radiation-matter decoupling, 475
Radiation zone, 246
Radioactive dating, 151, 151n
Radioactivity, differentiation and, 147
Radio astronomy/telescopes, 83–88, 93
    angular resolution of, 80
    essentials of, 83–84
    interferometry, 86–88
    search for extraterrestrial intelligence with, 498–500
    value of, 85–86
Radio frequencies, 49
Radio galaxies, 86, 419, 420–421, 423, 446, 447
    core-dominated, 421
    head-tail, 435, 437
Radio lobes, 86, 420–421, 423, 425, 427
Radio-loud quasars, 425
Radio radiation from Earth, 499
Radio waves, 44, 93
Radius-luminosity-temperature relationship, 279, 280
Radius/radii
    angular momentum and, 122
    of atomic orbital, 59
    of Earth, 136
    of ellipse, 33
    of the Moon, 136
    of planets, 104
    Schwarzschild, 362, 370
    of stars, 278–279, 282, 284
        mass and luminosity and, 288
    of Sun, 246, 247
RCW 38, 343
Recession, 63
    universal, 416–417, 462–463
Recessional velocity, 63, 64, 416–418, 422
Reconnection, 259
Recurrent novae, 331
Reddening, 295
Red dwarfs, 280, 281
Red giant, 279, 324, 326, 328
    atmosphere of, 284
Red giant branch, 324–326, 332
Red giant region, 281, 282, 323–324
Red light, 48
Red River, 178
Redshift, 63, 64, 467
    astronomical definition of, 418
    cosmological, 416–417, 463–464
    galactic rotation and, 387
    gravitational, 369, 479
    relativistic, 422
Redshift surveys, 448, 451, 460
Red Spot (Great), 201, 203, 208, 218
Red supergiant, 279, 332, 333
Reflecting mirror, 70
Reflecting telescopes, 70–72
Reflection nebula, 298, 299
Refracting lens, 70
Refracting telescopes, 70–72
Refraction, defined, 70
Region of totality, 15–16
Regolith, lunar, 153
Reinhold Crater, 154
Relativistic fireball, 358, 360
Relativistic redshifts, 422
Relativity theories, 47, 359
    general, 361, 365–368, 370, 465
        black holes and, 366–368, 370

central concept of, 365–366
cosmological constant and, 468–469
dark matter and, 395–396
tests of, 367
principle of, 363
special, 362–364, 423
Religion, science distinguished from, 19–20
Remote sensing, 180
Renaissance, 20
Residual caps, 180
Resolution
    angular, 78–80, 84
    defined, 78–79
    diffraction-limited, 79
Retrograde motion, 26, 27, 29, 30, 38, 166, 167
Reverse runaway greenhouse effect, 188
Revolving around sun, 7
Rhea, 227, 228
Rho Ophiuchus dark dust cloud, 302
Ridges, midocean, 149
Rigel, 6, 273, 275, 278, 280, 281, 333, 334
Rigel Kentaurus. See Alpha Centauri
Right ascension, 6, 10
Ringlets, 232
Ring Nebula, 327, 328
Rings, planetary, 194, 230–235
    fine structure of, 232–234
    formation of, 235
    Jovian, 217, 230–235
    Roche limit, 231–232
Robot probes, 197, 369
Roche, Edouard, 231
Roche limit, 231–232
Rocky Mountains, 149
Roman Catholic Church, 29, 31–32
Rotation, 5–6, 164–167
    differential, 199
        of Galactic disk, 387–388, 390, 392
        of Milky Way, 390
        of Sun, 246, 247, 255, 256
    of a galaxy, 413
    of Jovian planets, 199–200
    of Mercury, 164–167
    rate of
        angular momentum and, 122
        spectral-line analysis of, 65
    sidereal, 136
    solar, 246, 247, 255, 256
    synchronous, 14
    of Uranus, 207
    of Venus and Mars, 166–167
Rotation curves, 394, 434
RR Lyrae variable stars, 382–384
Runaway greenhouse effect, 188
    reverse, 188
Runoff channels, 177–178
Russell, Henry Norris, 279

S0 (lenticular) galaxies, 409
S2 star, 399
S16 star, 399
Sagittarius, 297
Sagittarius A, 397–398
Sagittarius galaxy, 442
San Andreas Fault, 149
SAO 113271, 273
Saturn, 26, 103, 192–195
    atmosphere of, 204–206
    density of, 198
    interior of, 209
    internal heating of, 211
    magnetic field of, 211
    magnetosphere of, 210
    moons of, 216–217
        medium-sized, 227
        Titan, 218, 223–226, 491
    observations of, 194, 195
        spacecraft exploration, 194–195
        view from Earth, 194
    properties of, 34, 104, 199
    ring system of, 194, 230–234
        E ring, 229
        formation of, 235
    Roche limit of, 231–232
    rotation of, 199–200

SB0 (lenticular) galaxies, 409
Scarps, 169–170, 186
Schiaparelli, Giovanni, 165, 174
Schwarzschild, Karl, 362
Schwarzschild radius, 362, 370
Scientific method, 18–21, 32, 39, 174
    spectroscopy and, 57–58
    stars and, 322
Scientific theory, 18–21
Search for extraterrestrial intelligence (SETI), 500
Seasonal caps, 179–180
Seasonal changes, 8–11
Seasons, 9–11
    on Uranus, 200
Secondary waves (S-waves), 145–146
Seconds, 6
Sedna, 238
Seeing, 81
Seeing disk, 81
Seismic waves, 145–146
Seismograph, 145
Seismology, 145–146
Selection effect, 128–130
Self-propagating star formation, 393
Semimajor axis, 33, 34
Sensitivity of telescope, 77–78
Seven Sisters (Pleiades), 5, 314, 315
Seyfert, Carl, 419
Seyfert galaxies, 419–420, 446, 447
Sgr A*, 399–401
Shadow zones, 146
Shane telescope, 83
Shapley, Harlow, 384
Shear waves, 146
Shepherd satellites, 233, 235
Shield volcanoes, 171, 177
Shoemaker-Levy 9, 212
Short-period comets, 112–113
Shu, Frank, 391
Siberian meteoroid (Tunguska event), 108, 116
Sidereal day, 7
Sidereal month, 14
Sidereal rotation, 136
Sidereal year, 11
Silicate asteroids, 107
Silicate (stony) meteorite, 116, 117
Single-line binary systems, 285–286
Singularity, 370
Sinking planet, 130
Sirius, 273, 274, 278, 288
Sirius A, 279–281, 287, 329
Sirius B, 279, 280, 282, 287, 329
SI units of energy, 54
Skylab, 260
"Slingshots," gravitational, 197
Slipher, Vesto M., 416
Sloan Digital Sky Survey (SDSS), 451, 460
Sloan Great Wall, 460
Small Magellanic Cloud, 269, 410
Smithsonian Astrophysical Observatory catalogue, 273
SN1987A, 334, 344
Sodium, emission spectrum of, 56
Sojourner rover, 181
Solar activity, 254
Solar and Heliospheric Observatory (SOHO), 248, 250, 253, 258–260
Solar constant, 247, 272
Solar convection, 249–251
Solar corona, 246, 250
Solar cycle, 250, 256–257
Solar day, 7
Solar Dynamics Observatory (SDO), 245, 250
Solar eclipses, 14–16, 246, 252, 253
Solar interior, 246
Solar nebula, 119, 121, 309
    random encounters in, 125
    temperature in, 124
Solar neutrino problem, 264
Solar neutrinos, 264
Solar system, 100–131
    age of, 118
    Copernican model of, 30
    dimensions of, 35
    formation of, 117–125
        asteroids and comets, 124–125
        condensation theory of, 119–125, 129, 157, 493, 494

differentiation, 123–124
model of, 118–119
nebular contraction, 119–120
nebular theory of, 119–124
planet formation, 120–123
random encounters in solar nebula, 125
Galilean moons as miniature, 218–219
geocentric model of, 27–28
heliocentric model of, 29–30
interplanetary matter, 105–117
asteroids, 102, 106–109
asteroids and comets, 100–101
comets, 102, 104, 110–113
meteoroids, 102, 105, 114–117
inventory of, 102–105
planetary properties, 102–103
solar system debris, 105
terrestrial and Jovian planets, 104–105
life in, 490–492
planets beyond (extrasolar planets), 117, 126–131
detecting, 125–126
orbits of, 127–128
properties of, 126–128
search for, 127, 128, 130–131
spaceflight in, 196–197
ten known facts about, 118
uniqueness of, 129–130
Solar wind, 110, 154–156, 210, 246, 250, 253, 260
Sombrero (M104) galaxy, 407–408
Sound waves, 46
South American plate, 149
South celestial pole, 6
Southern Lights (aurora australis), 156
Soyuz, 296
Space, curved (warped), 365–368
Spaceflight, in solar system, 196–197
Spacetime, 363, 365–366
Space warping, 365–368
Special relativity, 362–364
Spectral classes (spectral types), 277–278, 278n
Spectral classification system, 385
Spectral lines, 58–62
analysis of, 64–65
atomic structure and, 58–59
Kirchhoff's laws and, 61–62
particle nature of radiation and, 59–60
spectrum of hydrogen, 60–61
widths of, 28
Spectroscope, 55
Spectroscopic binaries, 285, 286
Spectroscopic parallax, 282–283, 287, 298, 383, 384, 413
Spectroscopy, 55–58, 287
absorption lines, 56–57
astronomical applications of, 57–58
defined, 57
emission lines, 55–57
scientific method and, 57–58
Spectrum (spectra), 48, 77
continuous, 61–62
stellar, 276–278
Spectrum icon, 50
Speed
escape, 136, 140, 361–362, 464
of galactic rotation, 413
of gas particles, 140
of light, 47, 418
Doppler effect and, 64
special relativity and, 362–364
recessional, 422
Spica A, 279, 280
Spica B, 288
Spicules, solar, 252
Spin, electron, 303
Spiral arms, 390–393
Spiral density waves, 391, 392–393
Spiral galaxy(ies), 386, 406–408, 446, 447. See also Milky Way
Galaxy
barred-spiral galaxy, 408
classification of, 406–408
dark halos around, 435
destruction of disk, 443
distances to, 413
evolution of, 442–443
masses of, 434

properties of, 412
recession of, 416
Spirit rover, 182, 183–184, 196
Spitzer, Lyman, Jr., 90
Spitzer Space Telescope, 89–91, 121, 298, 313, 316, 348–349, 408, 437, 441
"Splosh" crater, 180
Spring tides, 138
Stadium (unit of length), 19
Standard candles, 412–413
Standard solar model, 247–249, 264
Starburst galaxies, 419, 441
Starbursts, 441, 443
Star clusters, 313–316, 378–379
associations and, 314
globular, 314
nebulae and, 314–316
NGC 6934, 83
open, 314, 315
R136, 77, 82
stellar evolution in, 339–342
Stardust mission, 111, 112
Stars, 4–6, 245, 246, 268–289. See also Binary stars; Black holes;
H-R diagram; Neutron stars; Star clusters; Stellar evolution; Sun
apparent brightness of, 272–274, 282
"artificial," 83
A-type, 314
blackbody curves of, 53, 54
bulge, 387
collisions between, 340, 342
composition of, 287
disk, 387
distances to, 270, 274, 282–285, 287
luminosity class and, 284–285
spectroscopic parallax and, 282–283, 287, 298, 383, 384, 413
in elliptical galaxies, 409
"failed," 312–313
formation of, 119–120, 305–311, 342–343, 441
in Galactic bulge, 386
in Galactic disk, 388
in Galactic halo, 388
galaxy collisions and, 437–439
gravity and heat in, 305–306
from interstellar medium, 305–311
of masses greater than or less than the Sun, 312–313
rate of, 493
regions for, 297–301
self-propagating, 393
stage 1 (interstellar cloud), 306–307
stage 4 (protostar), 306, 308–309
stage 5 (protostellar evolution), 306, 309–311
stages 2 and 3 (cloud fragment contraction), 306, 307–308
stages 6 and 7 (newborn star), 306, 310–311, 327
F-type, 494, 496
Galactic, color differences in, 386
giants and dwarfs, 279–281
G-type, 326, 494, 496
habitable zones around, 494, 496
habitable zones surrounding, 130
halo, 386–387, 387n, 388
high-mass, 323
interaction between gas and, 299–300
K-type, 494, 496
lifetime of, 288–289
low-mass, 323, 336
luminosity of, 272–274, 278–279, 282, 284–285, 287, 288
magnitude scale for, 274
mergers of, 340
motion of, 271–272, 287
M-type, 312, 494, 496
naming, 273
nearest to Earth, 271
orbits around Milky Way, 387–388
origins of, 246
O-type, 312, 339
with planetary systems, 129, 130, 493–494
Population I, 386
Population II, 386
properties of, 270
radial velocity, 63–64, 126
ranking in order of brightness, 6
reddening of, 295
scientific method and, 322

sizes (radii) of, 278–279, 284, 287, 288
estimating, 282
stellar parallax, 29, 270, 283, 287, 383, 384, 413
temperatures (colors) of, 276–278, 284, 287, 288
blackbody curve and, 275–277
spectral classification, 277–278, 278n
stellar spectra and, 276–278
twinkling of, 81
variable, 382–383, 412, 413
wobble of, and detection of planets, 126
Stefan-Boltzmann constant (σ), 54
Stefan-Boltzmann equation, 54
Stefan-Boltzmann law, 278, 282
Stefan, Josef, 52
Stefan's law, 52, 54, 143
Stellar balance, 248
Stellar envelope, 324–329, 335, 336, 338, 343, 344
Stellar evolution, 246, 320–344, 486. See also Black holes; Neutron
stars; Stars; Supernova/Supernovae
in binary systems, 340
cycle of, 342–343
death of low-mass star, 326–331
black dwarf (stage 14), 327, 329
dense matter, 328
novae, 330–331
planetary nebula (stage 12), 326–328
white dwarf (stage 13), 327, 328–329
defined, 322
end points of, 334
evolutionary track, 309, 312
final stage of, 361
leaving the main sequence, 322–323
mass and, 323
in star clusters, 339–342
of stars more massive than the Sun, 332–334
end point of, 333–334
heavy element fusion, 332
supergiants, 332–334
of Sun-like stars, 323–327
back to giant branch (stage 11), 325–327
change in composition, 323
helium fusion (stage 10), 324–325, 327
key stages, 326, 327
subgiant and red giant branches (stage 8), 323–324, 327, 340
Stellar-mass black holes, 372
Stellar occultation, 234
Stellar parallax, 29, 270, 283, 287, 383, 384, 413
Stellar spectra, 276–278
classification of, 277–278, 278n
spectroscopic parallax and, 283
Stellar winds, 333, 344
Stonehenge, England, 28
Stony (silicate) meteorite, 116, 117
Storm systems
coronal, 250
on Jupiter, 203–204
of Neptune, 207–208
oval, 203–204
on Saturn, 205–206
Stratosphere, 141
of Jupiter, 202
of Saturn, 205
Strong nuclear force, 261, 262
Subaru telescope, 78
Subatomic particles, as dark matter, 395
Subgiant branch, 323–324, 340
Subgiants, 284
Sudbury Neutrino Observatory (SNO), 264
Sulfur dioxide, 172
Summer solstice, 8–9, 28
Sun, 244–265
active regions of, 258–259
apparent magnitude of, 274
atmosphere of, 251–253
chromosphere, 246, 252–253
solar wind, 110, 154–156, 246, 250, 253, 259
transition zone and corona, 14–15, 246, 250, 253, 260–261
blackbody curves of, 53, 54
composition of, 252
coronal holes, 260
differential rotation of, 246, 255, 256
distance to, 35
Galileo's observations of, 30
granulation of surface of, 251
gravity of, 38

habitable zone of, 130
heart of, 261–264
    nuclear fusion, 261
    proton-proton chain, 261–263, 265, 308
    solar neutrinos, 264
H-R diagram of, 280, 281
interior of, 247–251
    circulation patterns in, 248–249
    density and temperature distributions, 247, 249, 253
    energy transport, 249–250
    standard solar model of, 247–249, 264
in the Local Bubble, 296
luminosity of, 247, 254, 265
magnetic field of, 252, 255–256
mass of, 104, 247
Mercury's rotation and, 165–166
oscillations of, 248
overall structure of, 246
photosphere, 246, 252, 253, 307, 308
properties of, 247, 288
radius of, 246, 247
size of, 279
solar cycle, 250, 256–257
spectral analysis of, 251
spectral classification of, 278, 280, 281
supergranulation of, 251
temperature of, 247, 249, 253
tidal forces exerted by, 138, 139
visible spectrum of, 56
in X-rays, 260
Sun Dagger, Chaco Canyon, 28
Sunspots, 30, 246, 254–255
    magnetic field of, 255–256
    polarity of, 255–256
    sunspot cycle, 249, 250, 256–257, 260
Superclusters, 448
Super-Earths, 127–130
Superforce, 476
Supergiants, 279–281, 284, 333
    blue, 333
    evolution of, 332–334
    red, 332, 333
Supergranulation, 251
Super Kamiokande, 264
Supermassive black holes, 75, 372, 398–401, 419, 426, 443–445, 447
    accretion disk surrounding, 427, 428
    origins of, 444
Supernova/supernovae, 28, 335–339, 343, 344
    Cassiopeia A remnant, 92–93
    core-collapse, 334
    Cygnus Loop remnant, 91–92
    "failed," 360
    heavy elements formation and, 338–339
    light curves of, 335, 344
    remnants of, 337–338, 348–349
        pulsars associated with, 352
    Type I (carbon-detonation), 335–337, 336n, 350
        global measurements using, 467
        as standard candles, 413
    Type II (core-collapse), 334–337, 344, 350
Super star clusters, 438
Surface-burning stage, 331
Surface gravity, 136
    Martian, 177
Surface of last scattering, 475
Surface temperature, 494
    of Mercury, 167
    of Venus, 168
S-wave, 145–146
Swedish Solar Telescope, 251
Swift mission, 357, 360
Synchronous orbit, 14, 138–140
Synchrotron radiation, 427, 428
Synodic month, 14

Tail of comet, 110, 111
Taurid meteor shower, 115
Taylor, Joseph, 359
Techniques, distance-measurement, 283
Technology, ozone depletion and, 142
Tectonic activity. See Plate tectonics
Tectonic fractures, 227
Telescope filters, 275–276
Telescopes, 30, 68–94
    collecting area of, 77–78

defined, 70
exposure time, 77
full-spectrum coverage, 94
high-energy astronomy, 92–94
high-resolution astronomy, 80–82
    atmospheric blurring and, 81–82
    new designs, 82
infrared and ultraviolet astronomy, 89–93
limiting magnitudes of, 274
neutrino, 264
optical, 70–77, 93
    detectors and image processing, 76–77
    reflecting and refracting, 70–72
    reflecting, types of, 72–73
radio astronomy, 83–88, 93
    essentials of, 83–84
    interferometry, 86–88
    search for extraterrestrial intelligence with, 498–500
    value of, 85–86
sensitivity of, 77–78
size of, 77–79
    light-gathering power and, 77–78
    resolving power and, 77, 78–79
X-ray, 259–260, 354
Temperature(s), 51
energy and, 52, 54
estimating stellar radii using, 282
feasibility of life and, 494
frequency or wavelength and, 52
of gas, 140
luminosity and, 280
lunar, 144
of nebulae, 301
spectral-line analysis of, 65
of stars, 275–278, 284, 288
    blackbody curve and, 275–277
    spectral classification, 277–278, 278n
    stellar spectra and, 276–278
of Sun, 247, 249, 253
surface, 494
    of Mercury, 167
    of Venus, 168
Terminator, 152
Terrae, 151
Terrestrial planets, 104–105, 118, 162–189. See also Earth; Mars; Mercury; Venus
    atmospheres of, 167–169, 187–189
    internal structure and geological history, 185–186
    orbital and physical properties of, 164, 165
    rotation rates, 164–167
    searching for, 130–131
    surface of, 169–184
    visibility of, 164
Terzan 2, 354
Testing, 21
Tethys, 228
Tharsis bulge, 175–177, 186
Theory, 19–21
Thermal energy, 51
Thermal radiation, 51–54, 428
    astronomical applications of, 54
    blackbody spectrum, 51–52
    radiation laws, 52–54
Third Cambridge Catalog, 424
Thorium, 147
Thymine, 487
Tidal bulge, 137–140
    on Io, 220–221
Tidal force (differential force), 137–138, 368, 370
    on Enceladus, 229
    ring formation and, 231
Tidal streams, 389, 442
Tides, 137–140
    gravitational deformation and, 137–138
    neap, 138
    spring, 138
    tidal locking, 138–140
Time
    arrow of, 486
    look-back, 422
Time dilation, 364, 369
Titan, 218, 223–226, 491
Titania, 228, 229
Titanquakes, 224
Tombaugh, Clyde, 235

Total eclipse, 14–15
Transition Region and Coronal Explorer (TRACE) satellite, 255, 258
Transition zone, 246, 253
Transits, planetary, 126, 127, 130
Trans-Neptunian objects. See Kuiper belt objects (KBOs)
Transparency, 50
Transverse motion, 63–64
Transverse velocity, 271, 272, 287
Trapezium, 316
Triangulation, 17
Trifid Nebula, 298
Triple-star system, 271
Triton, 118, 218, 226–227
    orbit of, 227
Trojan asteroids, 106, 124
Tropical year, 11
Troposphere
    of Earth, 141
    of Jupiter, 202
    of Saturn, 205
Troughs, wave, 45
T-Tauri phase, 123, 309–310
Tully-Fisher relation, 413
Tunguska event (1908), 108, 116
"Tuning fork" diagram, in Hubble sequence, 410, 411
Turbulence, atmospheric, 81–82
Turnoff mass, 340
21-centimeter line, 303
Twinkling of stars, 81
Twin quasars, 452
Two Micron All Sky Survey, 386
Type E (elliptical) galaxies, 408–409
Type I (carbon-detonation) supernovae, 335–337, 336n, 350
    global measurements using, 467
    as standard candles, 413
Type II (core-collapse) supernovae, 334–337, 344, 350
Type SB (barred-spiral) galaxies, 408
Type S (spiral) galaxies, 407

Uhuru satellite, 371
Ultra Deep Field, 458–459
Ultraviolet astronomy, 91–92, 296
Ultraviolet radiation, 49–50, 142
Ultraviolet telescope, 91
Ultraviolet waves, 44, 93
Ulysses spacecraft, 260
Umbra, 15–16, 254
Umbriel, 228, 229
Uncompressed density, 105
Uniform circular motion, 27
United Kingdom infrared telescope, 78
Universal recession, 416–417, 462–463
Universe, 4, 448–455. See also Cosmology
    age of, 118, 470–471
    blackbody curves for, 471
    closed, 465, 465n
    composition of, 469–470
        nucleosynthesis and, 473–474
    critical, 466
    dark energy-dominated, 472–473
    density of, 465, 466
        critical, 469–470
    expansion of, 461–463, 467, 475, 476
        Olbers's paradox, 461
    fate of, 464–471
    filamentary structure of, 449
    flat geometry of, 469, 470, 475–478
    Great Wall, 449
    homogeneity of, 460
    isotropicity of, 460–461
    large-scale structure of, 478–480
    large-scale surveys of, 460–461
    life in (See Life)
    mapping dark matter, 453–455
    matter-dominated, 472–473
    modern view of, 4
    open, 465
    phases in history of, 486
    quasar absorption lines, 449–450
    radiation-dominated, 472–473
    redshift surveys of, 448–449
    shape of, 465–466
    size and scale in, 4
    superclusters, 448

twin quasars, 452
voids, 449
Upsilon Andromedae, 126
Uraniborg Observatory, Denmark, 32, 33
Uranium, 147
Uranus
  atmosphere of, 206–207
  discovery of, 195, 198
  interior of, 209
  internal heating of, 213
  magnetic field, 211
  magnetosphere of, 210
  moons of, 229, 230
  orbit of, 200
  properties of, 34, 104, 199
  rings of, 232, 234
  rotation of, 199–200, 207
  seasons on, 200
  weather on, 207–208
Urey, Harold, 488
Urey-Miller experiment, 488, 489
U.S. National Observatory (Kitt Peak), 81
U.S. Naval Observatory, 236

V filter, 276
Vacuum energy, 476–477
Vacuum, false, 476, 478
Valhalla, 223
Valles Marineris (Mariner Valley), 175, 176
Valley networks, 178
Van Allen belts, 154–156
Variable stars, 382–383, 412, 413
Varuna, 238
Vega, 278, 280, 281, 288
Vela satellites, 93
Vela supernova remnant, 337, 338
Velocity
  line-of-sight, 65
  radial, 63–64, 271, 272, 287
  recessional, 63, 64, 416–418, 422
  transverse, 271, 272, 287
  wave, 46
Venera spacecraft, 172, 186, 196
Venus, 26, 28, 35, 103, 494
  apparent magnitude of, 274
  atmosphere of, 168, 187–188
  greenhouse effect on, 187–188
  internal structure and geological history, 186
  orbit of, 34
  phases of, 31
  properties of, 104, 199
    orbital and physical, 34, 164, 165
  rotation of, 125, 166–167
  spaceflights to, 196
  in Sun's habitable zone, 130
  surface of, 170–173
    large-scale topography, 170–171
    volcanism and cratering, 171–173
  water on, 187–188
Venus Express orbiter, 168
Vernal equinox, 9, 11

Very Large Array (VLA), New Mexico, 86–88, 402–405
Very Large Telescope (VLT), Chile, 78, 303, 337, 399, 451
Very Long Baseline Array, 444
Very-long-baseline interferometry (VLBI), 87
Vesta, 239
Viking missions, 174–177, 181, 182–183, 196, 491–492
Violet light, 48
Virgo A (M87 galaxy), 423, 425, 427, 428
Virgo Cluster, 415, 416, 446, 448
Virgo Supercluster (Local Supercluster), 448, 449
Visible galaxy, 420n
Visible light
  components of, 48–49
  defined, 44
  obscuration of, 301–302
Visible spectrum, 48, 93, 142–143
Visual binaries, 286
Voids, 449
Volcanic gases, 187
Volcanism
  on Io, 220
  on Mars, 177, 189
  on Mercury, 186
  on the Moon, 158
  on Titan, 224
  on Venus, 171–173
Volcanoes, 146–148
  shield, 171, 177
  water, 229
Vortex, polar, 168
Voyager missions, 194, 195, 197, 200, 201, 203, 218, 219–220, 223, 226–227, 231–234, 498

W44 region, 94
Waning Moon, 12, 13
Warped (curved) space, 365–368
Water
  on Mars, 163, 177–180
  on Moon, 144–145
  on Titan, 224
  on Triton, 226
  on Venus, 187–188
  waves in, 45
Water hole, 500
Water ice
  on Jovian moons, 219, 221–222, 224, 226, 231
  on Jupiter, 202
  in Saturn's atmosphere, 205
Water vapor, 124, 141, 143, 144
Water volcanoes, 229
Watt (W), 247n
Wavelength (l), 45, 50, 54, 93, 94
  Doppler effect and, 63–64
  fuzzy effects of diffraction and, 79, 80
  of light, 48
  temperature and, 52, 54
Wave motion, 44–46
  energy and, 44–45
  medium for, 46
Waves. See also Radiation
  electromagnetic, 47–48, 359

gravity, 359
  period of, 45
  properties of, 45
  seismic, 145–146
  sound, 46
  velocity of, 46
  in water, 45
Waxing Moon, 12, 13
Weakly Interacting Massive Particles (WIMPs), 395n
Weak nuclear force, 262
Weather
  on Jupiter, 203–204
  on Neptune, 207–208
  on Saturn, 205–206
  on Uranus, 207–208
Wheeler, John Archibald, 365n
White-dwarf region, 281
White dwarfs, 279, 328–329, 331, 332, 334, 336, 340, 341
  dark matter and, 395, 396
White light, 48
White ovals, 203
Wien's law, 52, 54, 89–90, 143
Wien, Wilhelm, 52
Wilkinson Microwave Anisotropy Probe (WMAP), 479–480
Wilson, Robert, 471
Wind
  protostellar, 309–310
  pulsar, 352
  solar, 110, 154–156, 246, 250, 253, 260
  stellar, 333, 344
Winter solstice, 9
Wispy terrain, 227
Women, early research contributions by, 385
WW Cygni, 382

X-ray binaries, 354, 355
X-ray emissions, 44, 49, 50, 93
  bursters, 354
  from galaxy cluster, 435–436
  from neutron stars, 354, 355
  from Sun, 259–260
X-Ray Multi-Mirror satellite (XMM-Newton), 93
X-ray telescopes, 92–93, 259–260, 354

Year
  sidereal, 11
  tropical, 11
Yogi (Martian rock), 181
Yohkoh satellite, 260
Yohkoh X-ray solar observatory, 260
Yuty impact crater, 179, 180

Zeeman effect, 65
Zenith, 28
Zero-age main sequence (ZAMS), 312
Zodiac, 8
Zonal flow
  on Jupiter, 202
  on Saturn, 205
Zones, atmospheric, 201

Star Charts

Have you ever become lost in an unfamiliar city or state? Chances are you used two things to get around: a map and some signposts. In much the same way, these two items can help you find your way around the night sky in any season. Fortunately, in addition to the seasonal Star Charts on the following pages, the sky provides us with two major signposts. Each seasonal description will talk about the Big Dipper—a group of seven bright stars that dominates the constellation Ursa Major the Great Bear. Meanwhile, the constellation Orion the Hunter plays a key role in finding your way around the sky from late autumn until early spring.

Each chart portrays the sky as seen from near 35° north latitude at the times shown at the top of the page. Located just outside the chart are the four directions: north, south, east, and west. To find stars above your horizon, hold the map overhead and orient it so a direction label matches the direction you're facing. The stars above the map's horizon now match what's in the sky.

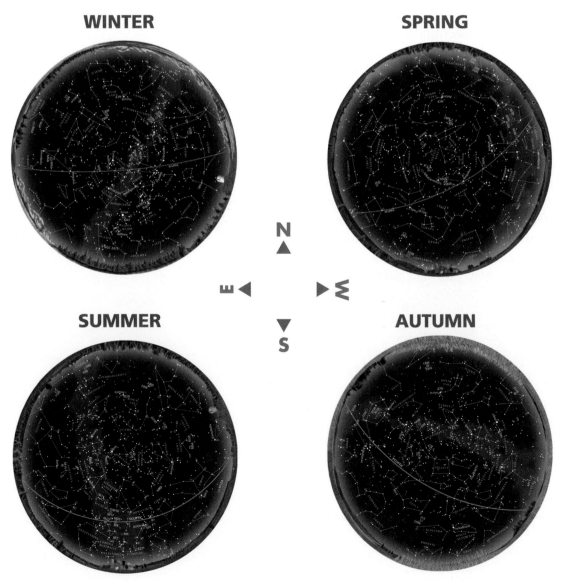

WINTER **SPRING**

N ▲

E ◄ ► W

SUMMER ▼ S **AUTUMN**

Exploring the Winter Sky

Winter finds the Big Dipper climbing the northeastern sky, with the three stars of its handle pointing toward the horizon and the four stars of its bowl standing highest. The entire sky rotates around a point near Polaris, a second-magnitude star found by extending a line from the uppermost pair of stars in the bowl across the sky to the left of the Dipper. Polaris also performs two other valuable functions: The star's altitude above the horizon equals your latitude north of the equator, and a vertical line dropped from Polaris to the horizon indicates the direction of true north.

Turn around with your back to the Dipper and you'll be facing the diamond-studded winter sky. The second great signpost in the sky, Orion the Hunter, is central to the brilliant scene. Three closely spaced, second-magnitude stars form a straight line that represents the unmistakable belt of Orion. Extending the imaginary line joining these stars to the upper right leads to Taurus the Bull and its orangish first-magnitude star, Aldebaran. Reverse the direction of your gaze to the belt's lower left and you cannot miss Sirius the Dog Star—brightest in all the heavens at magnitude −1.5.

Now move perpendicular to the belt from its westernmost star, Mintaka, and find at the upper left of Orion the red supergiant star Betelgeuse. Nearly a thousand times the Sun's diameter, Betelgeuse marks one shoulder of Orion. Continuing this line brings you to a pair of bright stars, Castor and Pollux. Two lines of fainter stars extend from this pair back toward Orion—these represent Gemini the Twins. At the northeastern corner of this constellation lies the beautiful open star cluster M35. Head south of the belt instead and your gaze will fall on the blue supergiant star Rigel, Orion's other luminary.

Above Orion and nearly overhead on winter evenings is brilliant Capella, in Auriga the Charioteer. Extending a line through the shoulders of Orion to the east leads you to Procyon in Canis Minor, the Little Dog. Once you have these principal stars mastered, using the chart to discover the fainter constellations will be a whole lot easier. Take your time, and enjoy the journey. Before leaving Orion, however, aim your binoculars at the line of stars below the belt. The fuzzy "star" in the middle is actually the glorious Orion Nebula (M42), a stellar nursery illuminated by bright, newly formed stars.

WINTER
2 A.M. on December 1; midnight on January 1; 10 P.M. on February 1

Exploring the Spring Sky

The Big Dipper, our signpost in the sky, swings high overhead during the spring and lies just north of the center of the chart. This season of rejuvenation encourages us to move outdoors with the milder temperatures, and with the new season a new set of stars beckons us.

Follow the arc of stars outlining the handle of the Dipper away from the bowl and you will land on brilliant Arcturus. This orangish star dominates the spring sky in the kite-shaped constellation Boötes the Herdsman. Well to the west of Boötes lies Leo the Lion. You can find its brightest star, Regulus, by using the pointers of the Dipper in reverse. Regulus lies at the base of a group of stars shaped like a sickle or backward question mark, which represents the head of the lion.

Midway between Regulus and Pollux in Gemini, which is now sinking in the west, is the diminutive group Cancer the Crab. Centered in this group is a hazy patch of light that binoculars reveal as the Beehive star cluster (M44).

To the southeast of Leo lies the realm of the galaxies and the constellation Virgo the Maiden. Virgo's brightest star, Spica, shines at magnitude 1.0.

During springtime, the Milky Way lies level with the horizon, and it's easy to visualize that we are looking out of the plane of our Galaxy. In the direction of Virgo, Leo, Coma Berenices, and Ursa Major lie thousands of galaxies whose light is unhindered by intervening dust in our own Galaxy. However, all these galaxies are elusive to the untrained eye and require binoculars or a telescope to be seen.

Boötes lies on the eastern border of this galaxy haven. Midway between Arcturus and Vega, the bright "summer" star rising in the northeast, is a region where no star shines brighter than second magnitude. A semicircle of stars represents Corona Borealis the Northern Crown, and adjacent to it is a large region that houses Hercules the Strongman, the fifth-biggest constellation in the sky. It is here we can find the northern sky's brightest globular star cluster, M13. A naked-eye object from a dark site, it looks spectacular when viewed through a telescope.

Returning to Ursa Major, check the second-to-last star in the Dipper's handle. Most people will see it as double, and binoculars show this easily. The pair is called Mizar and Alcor, and they lie just 0.2° apart. A telescope reveals Mizar itself to be a double. Its companion star shines at magnitude 4.0 and lies 14 arcseconds away.

SPRING
1 A.M. on March 1; 11 P.M. on April 1; 9 P.M. on May 1. Add 1 hour for daylight-saving time

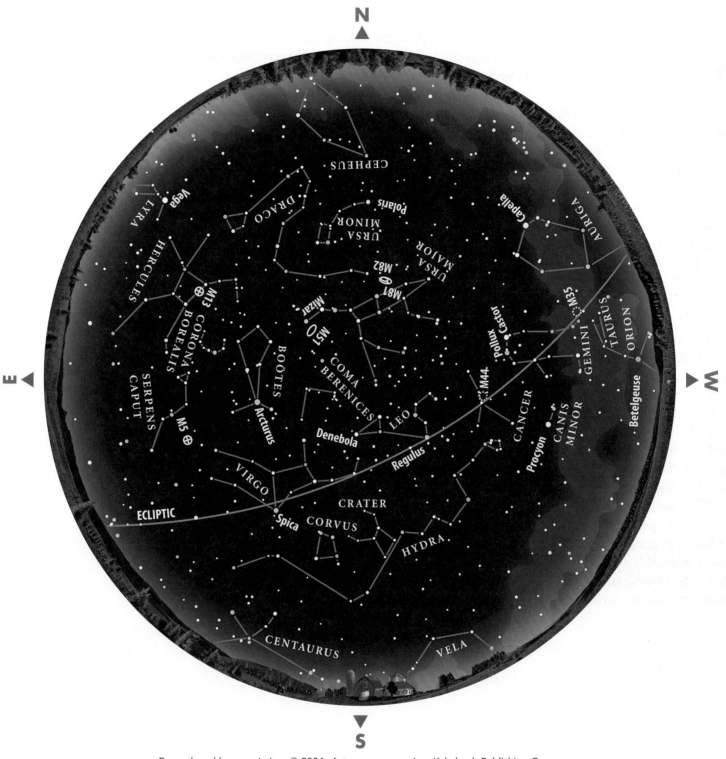

Reproduced by permission. © 2004, *Astronomy* magazine, Kalmbach Publishing Co.

Exploring the Summer Sky

The richness of the summer sky is exemplified by the splendor of the Milky Way. Stretching from the northern horizon in Perseus, through the cross-shaped constellation Cygnus overhead, and down to Sagittarius in the south, the Milky Way is packed with riches. These riches include star clusters, nebulae, double stars, and variable stars.

Let's start with the Big Dipper, our perennial signpost, which now lies in the northwest with its handle still pointing toward Arcturus. High overhead, and the first star to appear after sunset, is Vega in Lyra the Harp. Vega forms one corner of the summer triangle, a conspicuous asterism of three stars. Near Vega lies the famous double-double, Epsilon (ε) Lyrae. Two fifth-magnitude stars lie just over 3 arcminutes apart and can be split when viewed through binoculars. Each of these two stars is also double, but you need a telescope to split them.

To the east of Vega lies the triangle's second star: Deneb in Cygnus the Swan (some see a cross in this pattern). Deneb marks the tail of this graceful bird, the cross represents its outstretched wings, and the base of the cross denotes its head, which is marked by the incomparable double star Albireo. Albireo matches a third-magnitude yellow star and a fifth-magnitude blue star and offers the finest color contrast anywhere in the sky. Deneb is a supergiant star that pumps out as much light as 60,000 Suns. Also notice that the Milky Way splits into two parts in Cygnus, a giant rift caused by interstellar dust blocking starlight from beyond.

Altair, the third star of the summer triangle and the one farthest south, is the second brightest of the three. Lying 17 light-years away, it's the brightest star in the constellation Aquila the Eagle.

Frequently overlooked to the north of Deneb lies the constellation Cepheus the King. Shaped rather like a bishop's hat, the southern corner of Cepheus is marked by a compact triangle of stars that includes Delta (Δ) Cephei. This famous star is the prototype of the Cepheid variable stars used to determine the distances to some of the nearer galaxies. It varies regularly from magnitude 3.6 to 4.3 and back again with a 5.37-day period.

Hugging the southern horizon, the constellations Sagittarius the Archer and Scorpius the Scorpion lie in the thickest part of the Milky Way. Scorpius's brightest star, Antares, is a red supergiant star whose name means "rival of Mars" and derives from its similarity to the planet in both color and brightness.

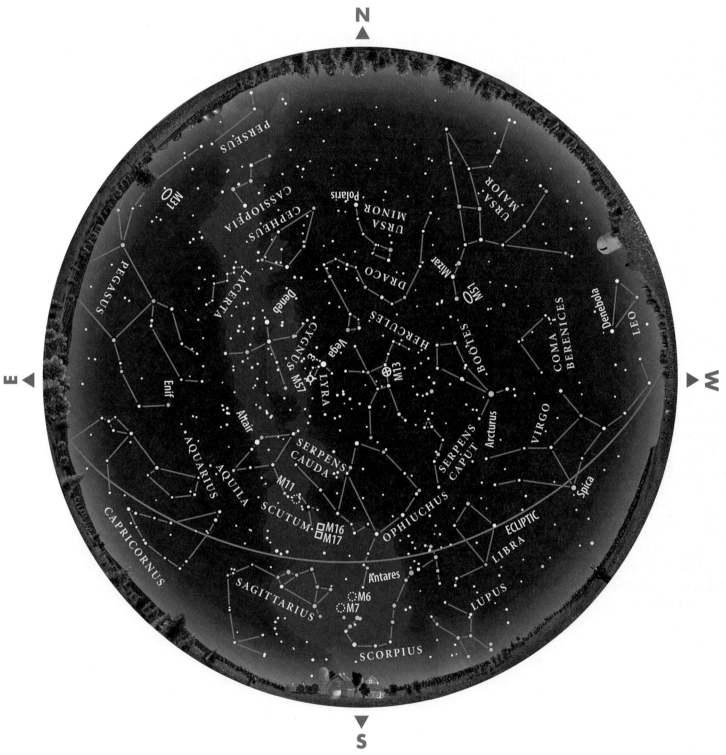

Exploring the Autumn Sky

The cool nights of autumn are here to remind us the chill of winter is not far off. Along with the cool air, the brilliant stars of the summer triangle descend in the west to be replaced with a rather bland-looking region of sky. But don't let initial appearances deceive you. Hidden in the fall sky are gems equal to summertime.

The Big Dipper swings low this season, and for parts of the Southern United States it actually sets. Cassiopeia the Queen, a group of five bright stars in the shape of a "W" or "M," reaches its highest point overhead, the same spot the Big Dipper reached 6 months ago. To the east of Cassiopeia, Perseus the Hero rises high. Nestled between these two groups is the wondrous Double Cluster—NGC 869 and NGC 884—a fantastic sight in binoculars or a low-power telescope.

Our view to the south of the Milky Way is a window out of the plane of our Galaxy in the opposite direction to that visible in spring. This allows us to look at the Local Group of galaxies. Due south of Cassiopeia is the Andromeda Galaxy (M31), a fourth-magnitude smudge of light that passes directly overhead around 9 P.M. in mid-November. Farther south, between Andromeda and Triangulum, lies M33, a sprawling face-on spiral galaxy best seen in binoculars or a rich-field telescope.

The Great Square of Pegasus passes just south of the zenith. Four second- and third-magnitude stars form the square, but few stars can be seen inside of it. If you draw a line between the two stars on the west side of the square and extend it southward, you'll find first-magniude Fomalhaut in Piscis Austrinus, the Southern Fish. Fomalhaut is the solitary bright star low in the south. Using the eastern side of the square as a pointer to the south brings you to Diphda in the large, faint constellation of Cetus, the Whale.

To the east of the Square lies the Pleiades star cluster (M45) in Taurus, which reminds us of the forthcoming winter. By late evening in October and early evening in December, Taurus and Orion have both cleared the horizon, and Gemini is rising in the northeast. In concert with the reappearance of winter constellations, the view to the northwest finds summertime's Cygnus and Lyra about to set. The autumn season is a great transition period, both on Earth and in the sky, and a fine time to experience the subtleties of these constellations.

AUTUMN

1 A.M. on September 1; 11 P.M. on October 1; 9 P.M. on November 1. Add 1 hour for daylight-saving time

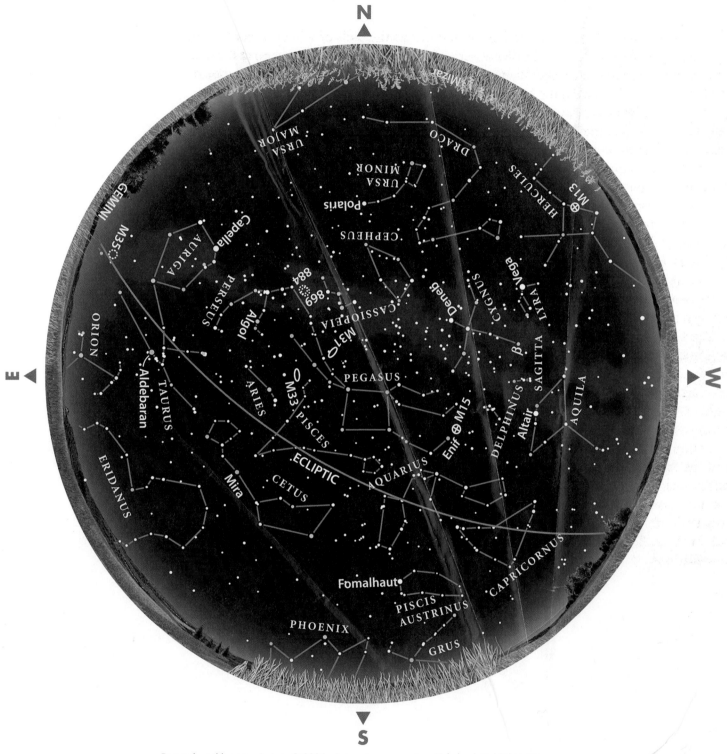